CONTROL SYSTEM APPLICATIONS

edited by

William S. Levine

CRC Press
Taylor & Francis Group
Boca Raton London New York

CRC Press is an imprint of the
Taylor & Francis Group, an **informa** business

CRC Press
Taylor & Francis Group
6000 Broken Sound Parkway NW, Suite 300
Boca Raton, FL 33487-2742

First issued in paperback 2019

ISBN-13: 978-0-8493-0054-7 (hbk)
ISBN-13: 978-0-367-39906-1 (pbk)

Visit the Taylor & Francis Web site at
http://www.taylorandfrancis.com

and the CRC Press Web site at
http://www.crcpress.com

Preface

Control technology is remarkably varied; control system implementations range from float valves to microprocessors. Control system applications include regulating the amount of water in a toilet tank, controlling the flow and generation of electrical power over huge geographic regions, regulating the behavior of gasoline engines, controlling the thickness of rolled products as varied as paper and sheet steel, and hundreds of controllers hidden in consmer products of all kinds. The different applications often require unique sensors and actuators. This book illustrates the diversity of control systems, and it provides examples of how theory can be applied to specific practical problems. In addition, it contains important information about aspects of control that are not fully captured by theory, such as techniques for protecting against controller failure and the role of cost and complexity in specifying controller designs.

Contributors

Brian Armstrong
Department of Electrical Engineering and Computer Science
University of Wisconsin
Milwaukee, Wisconsin

W.L. Bialkowski
EnTech Control Engineering Inc.

Okko H. Bosgra
Mechanical Engineering Systems and Control Group
Delft University of Technology
Delft, The Netherlands

S. Boyd
Department of Electrical Engineering
Stanford University
Stanford, California

Herman Bruyninckx
Department of Mechanical Engineering
Katholieke Universiteit Leuven
Leuven, Belgium

Y. Cho
Department of Electrical Engineering
Stanford University
Stanford, California

J.A. Cook
Scientific Research Laboratory
Control Systems Department
Ford Motor Company
Dearborn, Michigan

Vincent J. Coppola
Department of Aerospace Engineering
University of Michigan
Ann Arbor, Michigan

Bruce G. Coury
Applied Physics Laboratory
Johns Hopkins University
Laurel, Maryland

C. Davis
Semiconductor Process and Design Center
Texas Instruments
Dallas, Texas

Rik W. De Doncker
Silicon Power Corporation
Malvern, Pennsylvania

Clifford C. Federspiel
Johnson Controls, Inc.
Milwaukee, Wisconsin

G. Franklin
Department of Electrical Engineering
Stanford University
Stanford, California

J.W. Grizzle
Department of EECS
Control Systems Laboratory
University of Michigan
Ann Arbor, Michigan

P. Gyugyi
Department of Electrical Engineering
Stanford University
Stanford, California

David Haessig
GEC-Marconi Systems Corporation
Wayne, New Jersey

R.A. Hess
University of California
Davis, California

Constantine H. Houpis
Air Force Institute of Technology
Wright-Patterson Air Force Base, Ohio

Thomas M. Jahns
Corporate Research and Development
General Electric Corporation
Schenectady, New York

Hodge Jenkins
George W. Woodruff School of Mechanical Engineering
Georgia Institute of Technology
Atlanta, Georgia

S.M. Joshi
Langley Research Center
NASA
Langley, Virginia

T. Kailath
Department of Electrical Engineering
Stanford University
Stanford, California

A.G. Kelkar
Langley Research Center
NASA
Langley, Virginia

Petar V. Kokotovic
University of California
Santa Barbara

Thomas R. Kurfess
George W. Woodruff School of Mechanical Engineering
Georgia Institute of Technology
Atlanta, Georgia

Harry G. Kwatny
Drexel University
Philadelphia, Pennsylvania

Einar V. Larsen
Power Systems Engineering
General Electric Corporation
Schenectady, New York

M.K. Liubakka
Advanced Vehicle Technology
Ford Motor Company
Dearborn, Michigan

Claudio Maffezzoni
Politecnico Istituto di Milano
Milan, Italy

N. Harris McClamroch
Department of Aerospace Engineering
University of Michigan
Ann Arbor, Michigan

M. Moslehi
Semiconductor Process and Design Center
Texas Instruments
Dallas, Texas

S. Norman
Department of Electrical Engineering
Stanford University
Stanford, California

M. Pachter
Department of Electrical and Computer Engineering
Air Force Institute of Technology
Wright-Patterson Air Force Base, Ohio

P. Park
Department of Electrical Engineering
Stanford University
Stanford, California

John J. Paserba
Power Systems Engineering
General Electric Corpration
Schenectady, New York

D.S. Rhode
Advanced Vehicle Technology
Ford Motor Company
Dearborn, Michigan

Juan J. Sanchez-Gasca
Power Systems Engineering
General Electric Corpration
Schenectady, New York

K. Saraswat
Department of Electrical Engineering
Stanford University
Stanford, California

C. Schaper
Department of Electrical Engineering
Stanford University
Stanford, California

Gerrit Schootstra
Philips Research Laboratories
Eindhoven, The Netherlands

Joris De Schutter
Department of Mechanical Engineering
Katholieke Universiteit Leuven
Leuven, Belgium

John E. Seem
Johnson Controls, Inc.
Milwaukee, Wisconsin

F. Greg Shinskey
Process Control Consultant
North Sandwich, New Hampshire

Mark W. Spong
Coordinated Science Laboratory
University of Illinois
Urbana-Champaign, Illinois

Maarten Steinbuch
Philips Research Laboratories
Eindhoven, The Netherlands

J. Sun
Scientific Research Laboratory
Control Systems Department
Ford Motor Company
Dearborn, Michigan

Jacob Tal
Galil Motion Control, Inc.

David G. Taylor
School of Electrical and Computer Engineering
Georgia Institute of Technology
Atlanta, Georgia

George C. Verghese
Massachusetts Institute of Technology
Cambridge, Massachusetts

John Ting-Yung Wen
Department of Electrical, Computer, and Systems Engineering
Rensselaer Polytechnic Institute
Troy, New York

J.R. Winkelman
Advanced Vehicle Technology
Ford Motor Company
Dearborn, Michigan

Carlos Canudas de Wit
Laboratoire d'Automatique de Grenoble
ENSIEG
Grenoble, France

Contents

SECTION I Process Control

1 Water Level Control for the Toilet Tank: A Historical Perspective *Bruce G. Coury* . . . 3
- 1.1 Introduction . . . 3
- 1.2 Control Technology in the Toilet . . . 4
- 1.3 Toilet Control . . . 5
- 1.4 The Concept of Flushing . . . 7
- 1.5 The Need for a Sanitation System . . . 8
- 1.6 Concerns About Disease . . . 9
- 1.7 Changing Attitudes About Health and Hygiene . . . 10
- 1.8 The Indoor Toilet . . . 10
- 1.9 Historical Approaches to Systems . . . 11
- 1.10 Summary . . . 13

2 Temperature Control in Large Buildings *Clifford C. Federspiel and John E. Seem* . . . 15
- 2.1 Introduction . . . 15
- 2.2 System and Component Description . . . 15
- 2.3 Control Performance Specifications . . . 20
- 2.4 Local-Loop Control . . . 21
- 2.5 Logical Control . . . 23
- 2.6 Energy-Efficient Control . . . 24
- 2.7 Configuring and Programming . . . 27

3 Control of pH *F. Greg Shinskey* . . . 29
- 3.1 Introduction . . . 29
- 3.2 The pH Measuring System . . . 29
- 3.3 Process Titration Curves . . . 30
- 3.4 Designing a Controllable Process . . . 33
- 3.5 Control System Design . . . 37
- 3.6 Defining Terms . . . 40

4 Control of the Pulp and Paper Making Process *W.L. Bialkowski* . . . 43
- 4.1 Introduction . . . 43
- 4.2 Pulp and Paper Manufacturing Background . . . 44
- 4.3 Steady-State Process Design and Product Variability . . . 46
- 4.4 Control Equipment Dynamic Performance Specifications . . . 56
- 4.5 Linear Control Concepts — Lambda Tuning — Loop Performance Coordinated with Process Goals . . . 56
- 4.6 Control Strategies for Uniform Manufacturing . . . 64
- 4.7 Conclusions . . . 65
- 4.8 Defining Terms . . . 65

5 Control for Advanced Semiconductor Device Manufacturing: A Case History *T. Kailath, C. Schaper, Y. Cho, P. Gyugyi, S. Norman, P. Park, S. Boyd, G. Franklin, K. Saraswat, M. Moslehi, and C. Davis* . . . 67
- 5.1 Introduction . . . 67
- 5.2 Modeling and Simulation . . . 70
- 5.3 Performance Analysis . . . 71
- 5.4 Models for Control . . . 72
- 5.5 Control Design . . . 76
- 5.6 Proof-of-Concept Testing . . . 77
- 5.7 Technology Transfer to Industry . . . 79

5.8 Conclusions . . . 83

SECTION II Mechanical Control Systems

6 Automotive Control Systems . . . *87*
6.1 Engine Control *J. A. Cook, J. W. Grizzle and J. Sun* . . . 87
6.2 Adaptive Automotive Speed Control *M. K. Liubakka, D.S. Rhode, J. R. Winkelman and P. V. Kokotović* . . . 100
6.3 Performance in Test Vehicles . . . 105

7 Aerospace Controls . . . *113*
7.1 Flight Control of Piloted Aircraft *M. Pachter and C. H. Houpis* . . . 113
7.2 Spacecraft Attitude Control *Vincent T. Coppola and N. Harris McClamroch* . . . 129
7.3 Control of Flexible Space Structures *S. M. Joshi and A. G. Kelkar* . . . 142
7.4 Line-of-Sight Pointing and Stabilization Control System *David Haessig* . . . 152

8 Control of Robots and Manipulators . . . *165*
8.1 Motion Control of Robot Manipulators *Mark W. Spong* . . . 165
8.2 Force Control of Robot Manipulators *Joris De Schutter and Herman Bruyninckx* . . . 177
8.3 Control of Nonholonomic Systems *John Ting-Yung Wen* . . . 185

9 Miscellaneous Mechanical Control Systems . . . *195*
9.1 Friction Modeling and Compensation *Brian Armstrong and Carlos Canudas de Wit* . . . 195
9.2 Motion Control Systems *Jacob Tal* . . . 208
9.3 Ultra-High Precision Control *Thomas R. Kurfess and Hodge Jenkins* . . . 212
9.4 Robust Control of a Compact Disc Mechanism *Maarten Steinbuch, Gerrit Schootstra and Okko H. Bosgra* . . . 231

SECTION III Electrical and Electronic Control Systems

10 Power Electronic Controls . . . *241*
10.1 Dynamic Modeling and Control in Power Electronics *George C. Verghese* . . . 241
10.2 Motion Control with Electric Motors by Input-Output Linearization *David G. Taylor* . . . 252
10.3 Control of Electrical Generators *Thomas M. Jahns and Rik W. De Doncker* . . . 265

11 Control of Electrical Power . . . *281*
11.1 Control of Electric Power Generating Plants *Harry G. Kwatny and Claudio Maffezzoni* . . . 281
11.2 Control of Power Transmission *John J. Paserba, Juan J. Sanchez-Gasca, and Einar V. Larsen* . . . 311

SECTION IV Control Systems Including Humans

12 Human-in-the-Loop Control *R. A. Hess* . . . *327*
12.1 Introduction . . . 327
12.2 Frequency-Domain Modeling of the Human Operator . . . 328
12.3 Time-Domain Modeling of the Human Operator . . . 329
12.4 Alternate Modeling Approaches . . . 330
12.5 Modeling Higher Levels of Skill Development . . . 331
12.6 Applications . . . 331
12.7 Closure . . . 334
12.8 Defining Terms . . . 334

Index . . . *337*

I

Process Control

1

Water Level Control for the Toilet Tank: A Historical Perspective

Bruce G. Coury
The Johns Hopkins University, Applied Physics Laboratory, Laurel, MD

1.1 Introduction 3
1.2 Control Technology in the Toilet 4
1.3 Toilet Control 5
1.4 The Concept of Flushing 7
1.5 The Need for a Sanitation System 8
1.6 Concerns About Disease 9
1.7 Changing Attitudes About Health and Hygiene 10
1.8 The Indoor Toilet 10
1.9 Historical Approaches to Systems 11
1.10 Summary 13
References 14

1.1 Introduction

Control technologies are a ubiquitous feature of everyday life. We rely on them to perform a wide variety of tasks without giving much thought to the origins of that technology or how it became such an important part of our lives. Consider the common toilet, a device that is found in virtually every home and encountered every day. Here is a device — hidden away in its own room and rarely a major topic of conversation — that plays a significant role in our life and depends on several control systems for its effective operation. Unlike many of the typical household systems which depend on control technologies and are single purpose, self-contained devices (e.g., coffee makers), the toilet is a relatively sophisticated device that uses two types of ancient controllers. For most of us, the toilet is also the primary point of daily contact with a large distributed system for sanitation management; a system that is crucially dependent on control and has extremely important social, cultural, environmental, and political implications. In addition, we have some very clear expectations about the performance of that technology and well-defined limits for the types of tolerable errors should the technology fail. To imagine life without properly functioning modern toilets conjures up images of a lifestyle that most of us would find unacceptable and very unpleasant.

One need not look too far back in time to discover what life was *really* like without toilets linked to a major sanitation system. The toilet as we know it today is a recent technological development, coming into widespread use only at the end of the 19th century. Prior to that time, indoor toilets were relatively rare in all but the most wealthy homes, and the disposal and management of sewage was a rather haphazard affair. Only as a result of the devastating cholera epidemics of the mid-19th century in the United States and Europe were significant efforts made to control and manage waste products. Surprisingly, the technologies for sanitation had been available for quite some time. For instance, the toilet has a rich technological history with roots dating back to antiquity. The Greeks and Romans had developed the types of control technologies necessary for the operation of modern toilets, but the widespread adoption of that technology did not occur for another 1800 years.

One wonders, then, why a control technology so obviously useful took centuries to be adopted. To ponder the factors that motivate developing and adopting a technology is the essence of a historical analysis of that technology and is the first step in tracing the roots of technological development. One quickly realizes that the answers to such questions are rather complex, requiring one to explore not only the development of the technology, but also the economic, political, and social influences that shaped the development and implementation of that technology. In this respect, toilets and their controls are an ideal topic. First, it is a technology that is common, familiar, and classic in its control engineering. Although writing about the control technology of toilets is fraught with many pitfalls (one can easily fall victim to bathroom humor), toilets are a technology that people encounter every day that is part of a very large and highly distributed system.

Second, the history of the development of control technologies for toilets has many dimensions. There is the obvious chronicle of the evolution from latrine to commode to water closet to

0-8493-0054-3/96/$0.00+$.50

toilet. The history will quickly reveal that the development of toilet technology was inexplicably linked to the development of an entire water supply and sanitation system. Perhaps more important, we will see that the history of toilet technology and the entire sanitation system must be considered in the context of significant parallel social, cultural, and political trends that describe changing attitudes towards health and sanitation. Hopefully, by the end of this article, the reader will begin to grasp the complexity of technological development and the myriad forces that shape the development and use of a technology.

To consider the historical development of technology requires a specific framework of analysis. To focus only on the development of toilet technology and the adaptation of sensory and control mechanisms to the operation of toilets is to miss the larger context of the history of technological development. As Hughes [4] points out in his studies of the historical development of systems (he concentrates on electric power systems), "systems embody the physical, intellectual, and symbolic resources of the society that constructs them." As we shall see with the toilet, its development and use was driven by the threat of disease, the appalling conditions resulting from uncontrolled waste disposal, changing attitudes towards health and cleanliness, and the fashionable features of a "modern" home made possible by an abundant water supply.

The discussion begins with a description of the toilet technology relevant to this chapter, the valves for maintaining the water level in the toilet tank. What may come as a surprise to the reader is the similarity of modern toilet technology to the water closets of the late 19th century and how the same technological challenges have persisted for more than 150 years. We will then turn to the historical roots of those technologies and trace the events of the last few centuries that resulted in today's toilet and sanitation system.

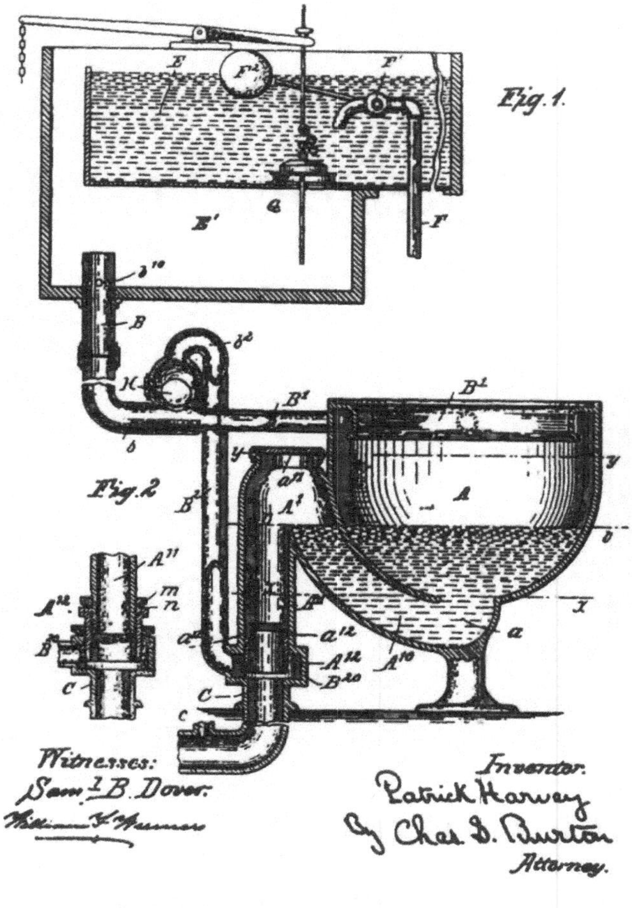

Figure 1.1 U.S. Patent drawing of the design of a water closet showing the basic components of the toilet tank, flush and float valves, and the toilet basin. *Source:* U.S. Patent No. 349,348 filed by P. Harvey, Sept. 21, 1886.

1.2 Control Technology in the Toilet

Lift the lid from the top of a toilet tank and inside is the basic control technology for the entry to a widely dispersed sanitation system. Although simple in operation (e.g., as shown in Figures 1.1, 1.2, and 1.3), the mechanisms found in the toilet tank serve one very important function: maintaining a constant water level in the tank. Modern toilets operate on the concept of flushing where a preset amount of water is used to remove waste from the toilet bowl. The tank stores water for use during flushing and controls the amount of water in storage by two valves: one controlling the amount of water entering the tank and another controlling the amount of water leaving the tank.

Push the handle on the tank to activate the flush cycle. Pushing the handle pops open the valve at the bottom of the tank (the toilet flush valve), allowing the water to rush into the toilet bowl (assuming well-maintained and nonleaking equipment). Notice that this valve (in most toilet designs) also floats, thereby keeping the valve open after the handle is released. As the water level drops, the float attached to the second valve (the float valve) also descends, opening that valve and allowing water to enter the tank. When the water level drops below a certain level, the flush valve closes, allowing the tank to refill. The water level rises, carrying the float with it until there is sufficient water in the tank to close the float valve.

Within a toilet tank we find examples of classic control devices. There are valves for controlling input and output, activation devices for initiating a control sequence, feedback mechanisms that sense water level and provide feedback to the control devices, and failure modes that minimize the cost of disruptions in control. The toilet tank is comprised of two primary, independent control mechanisms, the valve for controlling water flow into the tank and the valve for controlling water flow out of the tank, both relying on the same variable (tank water level) for feedback control. Using water level as the feedback parameter is very useful. If all works well, the water level determines when valves should be opened and closed for proper toilet operation.

The amount of water in the tank is also the measure of performance required to minimize the adverse effects of a failure. Should one of the valves fail (e.g., the float valve in the open position), the consequences of the failure (water running all over the floor) are minimized (water pours down the inside of the flush

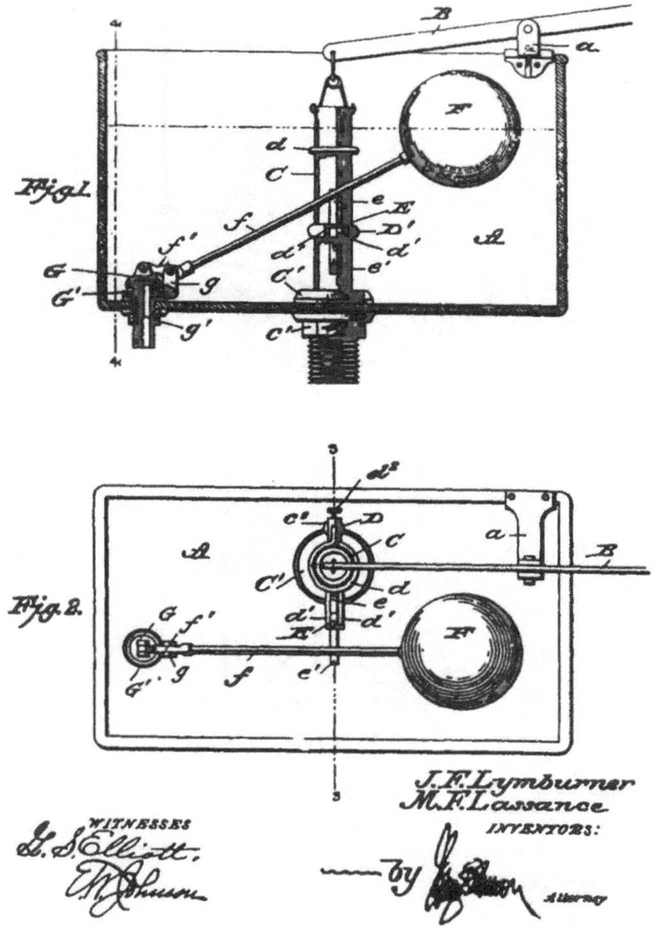

Figure 1.2 U.S. Patent drawing of the flush valve and the float valve for the toilet tank. *Source:* U.S. Patent No. 549,378 filed by J.F. Lymburner and M.F. Lassance, Nov. 5, 1895.

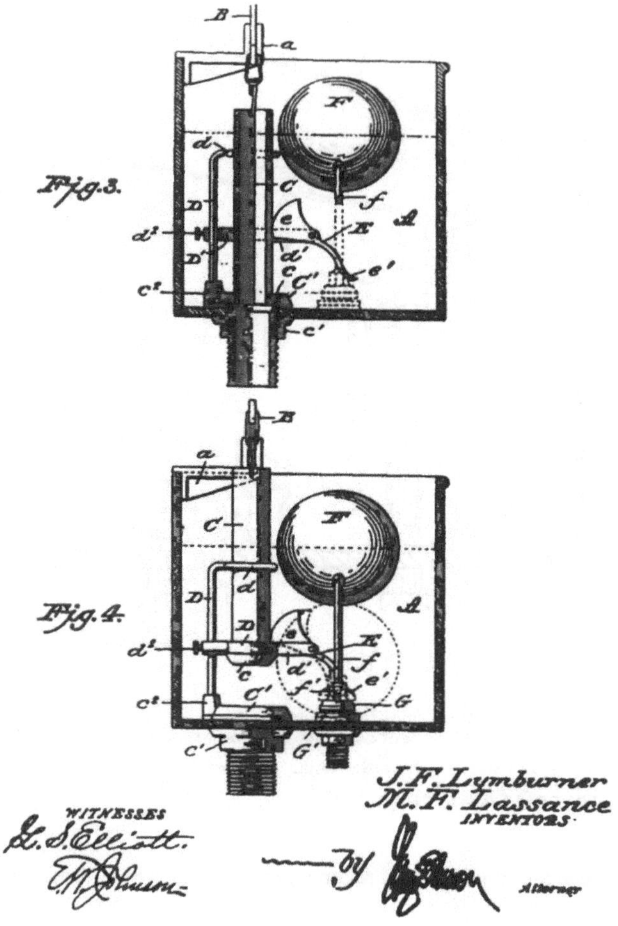

Figure 1.2 *(Continued.)* U.S. Patent drawing of the flush valve and the float valve for the toilet tank. *Source:* U.S. Patent No. 549,378 filed by J.F. Lymburner and M.F. Lassance, Nov. 5, 1895.

valve tube). Not all of the solutions to the failsafe maintenance of water level are the same. In the design shown in Patrick Harvey's U.S. Patent dated 1886 (Figure 1.1), a tank-within-a-tank approach is used so that the overflow from the cistern due to float valve failure goes directly to the waste pipe. The traditional approach to the same problem is shown in Figures 1.2 and 1.3; although the two designs are almost 100 years apart (Figures 1.2a and 1.2b are from the U.S. Patent filed by Joseph F. Lymburner and Mathias F. Lassance in 1895; Figure 1.3 is from the U.S. Patent filed in 1992 by Olof Olson), each uses a hollow tube set at a predetermined height for draining overflow should the float valve fail. In other words, the toilet is a control device characterized by relatively fail-safe operation.

1.3 Toilet Control

The toilet flush valve and the float valve operate on two different principles. The float valve has separate mechanisms for sensing the level of water in the tank (the float) and for controlling the input of water into the tank (the valve). The flush valve, on the other hand, combines both mechanisms for sensing and control into a single device. Both types of control technology have their origins in antiquity. The requirement for control of the level and flow of water was recognized by the Greeks in the design of water clocks. Water clocks were a rather intriguing device built on the principle that a constant flow of water could be used to measure the passage of time. In the design by Ktesibious (Figure 1.4), a mechanician serving under King Ptolemy II Philadelphus (285–247 B.C.), water flows into a container through an orifice of predetermined size in which the water level slowly rises as time passes [6]. Riding on the surface of the water is a float (labeled *P* in the diagram) attached to a mechanism that indicates the time of day; as the float rises, the pointer moves up the time scale. To assure an adequate supply of water to the clock, the orifice controlling the flow of water into the clock container was attached to a holding tank.

Ktesibious recognized that accurate time keeping requires precise control over water flow and maintenance of a constant water level in the container. Consequently, it was necessary to develop a method of control assuring a constant flow rate of water into the container. The Greeks clearly understood the relationship

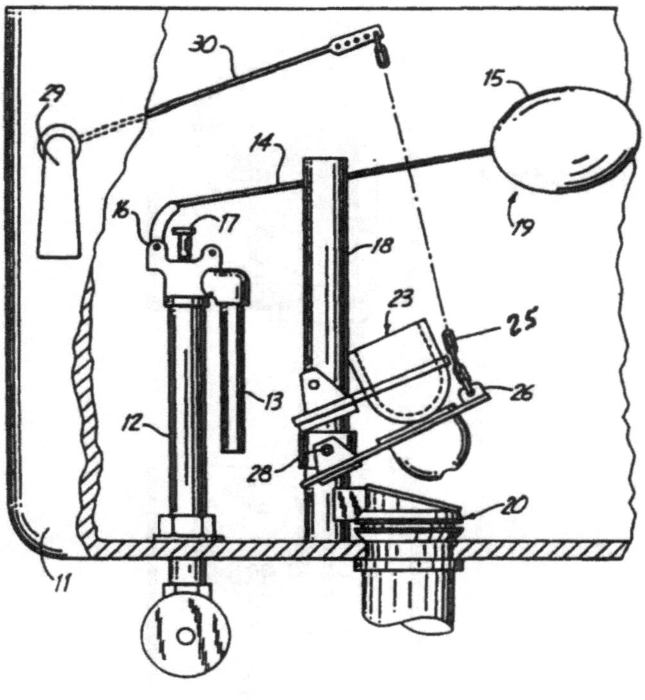

Figure 1.3 U.S. Patent drawing of a modern design for the flush valve and float valve for the toilet tank. *Source:* U.S. Patent No. 5,142,710 filed by O. Olson, Sept. 1, 1992.

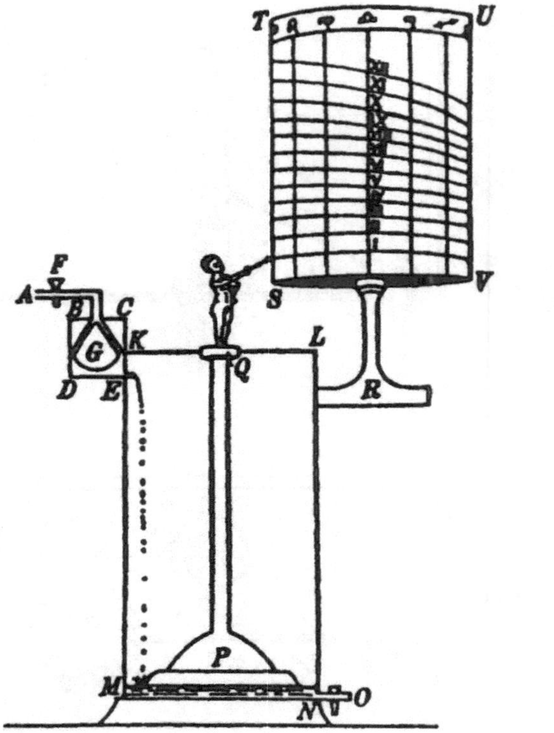

Figure 1.4 Control technology in holding vessels designed by Ktesibious (285–247 B.C.) showing the use of a float valve for regulating water flow. *Source:* Mayr, O., *The Origins of Feedback Control.* MIT Press, Cambridge, MA, 1969.

between flow rate and relative pressure, and recognized that the flow rate of water into the clock would decrease as the holding tank emptied. By maintaining a constant water level in the holding tank, the problem of variations in flow rate could be solved. There are, however, a number of possible passive or active solutions to the constant water level problem. A passive approach to the problem could use either a constantly overflowing holding tank or an extremely large reservoir of water relative to the size of the water clock container. Active control solutions, on the other hand, would use some form of water level sensing device and a valve to regulate the amount of water entering the holding tank (thereby reducing the need for a large reservoir or a messy overflow management system). Ktesibious chose the active control solution by designing a float valve for the inlet to the holding tank that assured a constant water level. From the descriptions of Ktesibious' water clock [6], the valve was a solid cone (labeled *G* in Figure 1.4) that floated on the surface of the water in the holding tank serving the orifice of the water clock. The valve stopped the flow of water into the holding tank when the level of the water forced the tip of the cone into a similarly shaped valve seat. Notice again that the functions of sensing the level of water and controlling the flow of water are both contained in the float valve.

The flush valve of a modern toilet uses a similar principle. Combining both sensing and control in the same device, the flush valve is a modern variation of Ktesibious cone-shaped float that controls the amount of water flowing out of the toilet tank. Unlike the Greek original, however, the control sequence of flush valve action is initiated by the mechanism attached to the flushing handle on the outside of the toilet tank. Thus, the toilet flush valve is a discrete control device that seeks to control the output of water after some external event has resulted in a drop in water level.

Subsequent developments of the float valve by Heron in the first century A.D. improved on the relationship between float and valve. Described in the *Pneumatica*, Heron developed float valves to maintain constant fluid levels in two separate vessels (e.g., as in Figure 1.5).

In his design, he employed a float connected to a valve by a series of jointed levers. The rise and fall of the float in one vessel would close or open a valve in another vessel. This design effectively separates the sensing of fluid level from the actions to control the flow of fluid, and most closely resembles the float valve for regulating water level in the toilet tank. Although the purpose of the toilet float valve is to maintain a constant level of water in the toilet tank, variations in that water level are usually step functions resulting from discrete, external events. The technology can be applied to continuous control, as is so audibly illustrated when the toilet flush valve leaks.

The float valve for the water clock and the holding vessels illustrates a number of important concepts critical to feedback and control. First, feedback is a fundamental component of control. Without a means for sensing the appropriate performance parameter, automatic regulation of input and output cannot be

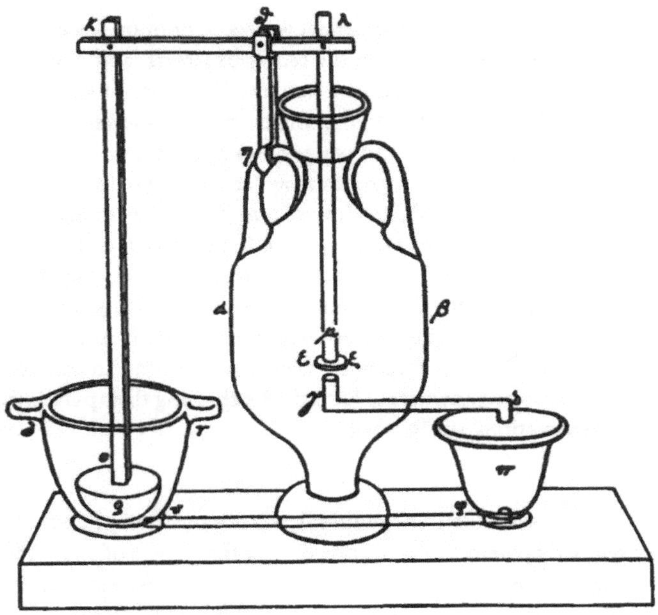

Figure 1.5 Control technology in the waterclock designed by Heron (100 A.D.) showing the use of a float connected to a valve for maintaining a constant fluid level. *Source:* Mayr, O., *The Origins of Feedback Control.* MIT Press, Cambridge, MA, 1969.

accomplished. Second, separation of the mechanisms for sensing and control provide a more sophisticated form of control technology. In the example of Heron's holding vessels, using a float to sense fluid level that was connected by adjustable levers to a valve for controlling fluid flow allowed adjustments to be made to the float level. Thus, the level of fluid required to close the valve could be varied by changing the required float level. The float valve in Ktesibious' design, on the other hand, was not adjustable; to change the water level resulting in valve closure required substituting an entirely new float valve of the proper dimensions.

1.4 The Concept of Flushing

The Greeks and Romans did not immediately recognize the relevance of this control technology to the development of toilet like devices. The lack of insight was certainly not due to the inability to understand the relevance of water to waste removal. Flowing streams have always been a source of waste removal, especially in ancient history when people tended to settle near water. By the time the Greeks and Romans built public baths, the use of water to cleanse the body and remove excrement was an integral part of the design. Even the frontier Roman forts in Britain had elaborate latrines that used surface water to flush away human wastes [8]. There was, however, no explicit mechanism for controlling the flow of water other than devices for diverting streams, pipes for routing the flow, or reservoirs to assure an ample supply of water [10].

The use of a purpose-built *flushing* mechanism (where water is stored in a holding tank until called upon for the removal of waste) was slow to develop and was relatively rare until more recent history (although Reynolds discusses the possibility for such a water closet at Knossos). For instance, 15th century evidence of such a mechanism for a latrine was found during excavation of St. Albans the Abbot in England [10]. In the most simple form of an 18th century flushing toilet, a cistern captured and stored rain water until a valve was opened (usually by pulling on a lever), allowing the water to flush away waste products. In most early applications of flushing, the amount of water used was determined by the size of the cistern, the source of water, and the patience of the user. Although such devices persisted until the latter part of the 19th century, they were considered to be a rather foul and obnoxious solution to the problem.

In general, the collection and disposal of human waste was accomplished without sophisticated technology until the mid-1800s. When a special purpose device for collecting human waste existed in a home, it was usually a commode or chamber pot (although the *close stools* built for royalty in 16th century Europe could be quite throne-like in appearance, complete with seat and arms and covered with velvet). Otherwise, residents trekked outside to a latrine or privy in the backyard or garden. The first evidence of a recognizable predecessor of the modern toilet is found in the British patents filed by Alexander Cummings in 1775 and Joseph Bramah in 1778 (although Sir John Harrington's valve closet of 1596 had some characteristics similar to 18th century water closets). Cummings proposed a valve closet that had an overhead supply cistern, a handle activated valve for the flush mechanism, and a valve that opened in the bottom of the basin to allow waste to escape. All valves and flushing mechanisms were activated and controlled by the user. Bramah's contribution was a crank activated valve for emptying the basin. No control devices that relied on water level were evident, and the valve separating the basin from the discharge pipe was not a very effective barrier against noxious odors and potentially dangerous sewer gases [10]. An early version of an American water closet of similar design is shown in Figure 1.6. Patented by Daniel Ryan and John

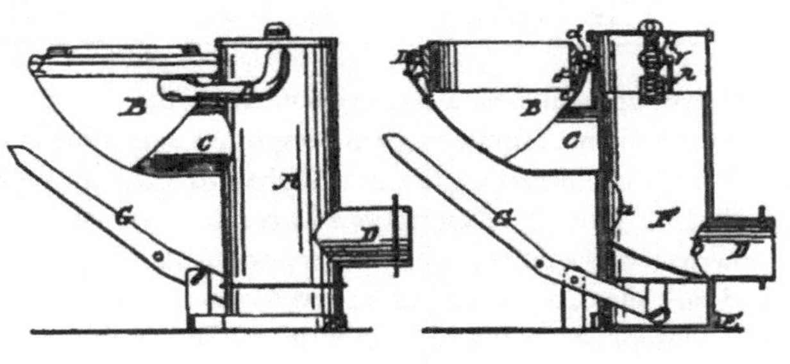

Figure 1.6 U.S. Patent drawing of an early water closet design showing manual operation of flushing cycle. *Source:* U.S. Patent No. 10,620 filed by D.Ryan and J. Flanagan, Mar. 7, 1854.

Flanagan in 1854, the basic components of a manually operated water closet are shown. The lever (labeled *G*) operates the water closet. When the lever is depressed, it pushes up the sliding tube

F until the opening *a* coincides with the pipe *C* from the toilet bowl. At the same time, the valve labeled *V* is opened to allow water to enter the basin and flush away waste.

In all situations, commode, latrine, privy, or water closet, no organized or structured system of collecting and managing the disposal of human waste products existed. Human household waste was typically dumped directly into vaults or cesspits (some of which were in the basements of homes) or heaved into ditches or drains in the street. The collection and disposal of that waste was usually handled by workers who came to the home at night with shovel and bucket to empty out the vault or cesspit and cart off the waste for disposal (usually into the most convenient ditch or source of flowing water). When a system of sewers did exist, household waste was specifically excluded. Even the Romans, who were quite advanced in building complex water and sanitation systems, used cesspits in the gardens of private homes to collect human waste products [8]. Not until the link between sanitation and health become evident and a universal concern arose in response to the health hazards of uncontrolled sewage did the need arise to develop an infrastructure to support the collection and removal of human waste.

At this point, it should be evident to the reader that the development of toilet technology did not proceed along an orderly evolutionary path from control principle to technological development. For instance, the development of critical water level control technologies by the Greeks and Romans did not immediately lead to the use of that technology in a sanitation system (despite the fact that the need for the removal of waste products had been well-established for a long time). Nor did the existence of critical technologies immediately lead to the development of more sophisticated devices for the collection and disposal of human excrement. Even where conditions surrounding the disposal of human waste products was recognized as deplorable for many centuries, little was done to develop technological solutions that remotely resembled the modern toilet until other factors came into play. Understanding those factors will lead us to a better understanding of the development of toilet control technology.

1.5 The Need for a Sanitation System

The development of sanitation systems was a direct response to health concerns. From a modern perspective, sanitation and the treatment of human waste was a haphazard affair through the early 19th century. Descriptions of conditions in British and American cities portray open sewers, waste piling up in streets and back alleys, cesspits in yards or even under living room floors, and outdoor privies shared by an entire neighborhood. Mortality rates during that period were shocking: the death rate per 1,000 children under five years of age was 240 in the English countryside and 480 in the cities [10]. In 1849, sewers delivered more than 9 million cubic feet of untreated sewage into the Thames, the same river used as a source of water for human consumption.

In his comprehensive treatment of the conditions in 19th century Newark, NJ, Galishoff [3] provides a vivid account of the situation in an American city. Streets were unpaved and lined with household and commercial wastes. Common sewers were open drains that ran down the center of streets, and many residents built drains from their home to the sewer to carry away household wastes. Prior to the construction of a sewer system in 1854, Newark's "privy and cesspool wastes not absorbed by the soil drained into the city's waterways and into ditches and other open conduits" [3]. When heavy rains came, the streets turned into filthy quagmires with the runoff filling cellars and lower levels of homes and businesses in low-lying areas. States Galishoff [3], "Despite greater and greater accumulations of street dirt (consisting mainly of garbage, rubbish, and manure), gutters and streets were cleaned only twice a year. An ordinance compelling property owners to keep streets abutting their parcels free of obstructions was ignored."

Given such horrible conditions, it is not surprising that cholera reached epidemic proportions during the period of 1832-1866; 14,000 Londoners died of cholera in 1849, another 10,000 in 1854, and more than 5,000 in the last major outbreak in 1866 [10]. Newark was struck hard in 1832, 1849, and 1866, with minor outbreaks in 1854 and 1873. Although Newark's experience was typical of many American cities (the number of deaths represented approximately 0.5 percent of the city populations), and only Boston and Charleston, SC, escaped relatively unharmed, cholera struck some cities devastating blows. Lexington, KY, for instance, lost nearly a third of its population when the disease swept through that city in 1833 [3].

Until legislative action was taken, the private sector was responsible for constructing sewage systems through most of the 19th century, especially in the United States. This arrangement resulted in highly variable service and waste treatment conditions. For example, Boston had a well developed system of sewers by the early 1700s, whereas the sewers in the city of New York were built piecemeal over a six decade period after 1700 [1]. At that time, and until the recognition of the link between disease and contaminated water in the mid-19th century, the disposal of general waste water and the disposal of human waste were treated as separate concerns. Private citizens were responsible for the disposal of the waste from their own privies, outbuildings, and cesspools. If the waste was removed (and in the more densely populated sections of American cities, such was hardly ever the case), it usually was dumped into the nearest body of water. As a result, American cities suffered the same fate as European cities during the cholera years of 1832–1873. Needless to say, lawmakers were motivated to act, and the first public health legislation was passed. London enacted its Public Health Act in 1848, and efforts were made to control drainage, close cesspits, and repair, replace, and construct sewers. Resolution of the problem was a Herculean task; not until the 1870s did the situation improve sufficiently for the death rate to decline significantly. The Newark Board of Health was created in 1857 to oversee sanitary conditions. It, too, faced an uphill battle and was still considered ineffectual in 1875.

1.6 Concerns About Disease

Fundamental to the efforts to control sewage was the realization of a direct link between contaminated water and the spread of disease. This realization did not occur until the medical profession established the basis for the transmission of disease. Prior to the mid-1800s, the most common notion of the mechanism for spreading disease was based on atmospheric poisoning (referred to as the miasmic theory of disease) caused by the release of toxic gases during the fermentation of organic matter. In this view, diseases such as cholera were spread by the toxic gases released by stagnant pools of sewage and accumulations of garbage and waste. Urban congestion and squalor appeared to confirm the theory; the highest incidence of cholera (as well as most other types of communicable diseases) occurred in the most densely populated and poorest sections of a city where sanitation was virtually nonexistent. Some believed, even in the medical community, that sewers were especially dangerous in congested, urban areas because of the large volumes of sewage that could accumulate and emit deadly concentrations of sewer gases.

The higher incidence of disease in the poorest sections of a city also contributed to the notion that certain factors predisposed specific segments of the populace to infection, a convenient way to single out immigrants and other less fortunate members of the community for discriminatory treatment in the battle against disease. Such attitudes also contributed to the slow growth in the public health movement in the United States because cleaning up the squalor in the poorest sections of a city could potentially place a significant economic burden on the city government and business community [3]. By associating the disease with a particular class of people, costly measures for improving sanitation could be ignored. Class differences in the availability of and access to water and sanitation facilities persisted in Newark from 1850–1900, with much of the resources for providing water and sanitation services directed towards the central business district and the more affluent neighborhoods [3].

The medical profession slowly realized that "atmospheric poisoning" was not the mechanism for spreading diseases. During the 19th century, medical science was evolving and notions about the spread and control of disease were being formalized into a coherent public health movement. Edwin Chadwick, as secretary of the British Poor Law Commission, reported to Parliament on the social and environmental conditions of poor health in 1848. The influence of his report was so great that Chadwick is credited with initiating the "Great Sanitary Awakening"[9]. In 1854, John Snow, a London anesthesiologist, determined that a contaminated water supply led to the deaths of 500 people when he traced the source of the infected water to a single common well. Snow had argued earlier that cholera was spread by a "poison" that attacked the intestines and was carried in human waste. His study of the 500 deaths caused by the contaminated well provided strong support for his theory. The first recorded study in the United States of the spread of disease by contaminated water was conducted by Austin Flint in 1855. In that study he established that the source of a typhoid epidemic in North Boston, NY was a contaminated well [9].

During the 1860s, the miasmic theory of disease was slowly displaced by the germ theory. Louis Pasteur established the role of microorganisms in fermentation. At about the same time, Joseph Lister introduced antiseptic methods in surgery. These changes in medicine were due to an increasing awareness of the link between microorganisms and disease and the role of sterilization and cleanliness in preventing the spread of disease. The germ theory of disease and the role of bacteria in the spread of disease were established in the 1880s by Robert Koch. His research unequivocally demonstrated that bacteria were the cause of many types of disease. By firmly establishing the link between bacteria and disease, germ theory provided the basis for understanding the link between contaminated water and health.

Once it was discovered that disease could spread as a result of ground water seepage and sewage leachate, uncontrolled dumping of waste was no longer acceptable. To eliminate water contamination, potable water had to be separated from sewage. Such an objective required that an extensive, coordinated sewer system be constructed to provide for the collection of waste products at the source. Water and drainage systems did exist in some communities and wealthy households in the mid-18th century [e.g., [10]], but, in general, widespread water and sanitation systems were slow to develop. One major force behind the development of sanitation systems was the rapid growth of cities in the years leading up to the Civil War. For example, Newark became one of the nation's largest industrial cities when its population grew from 11,000 to 246,000 during the period 1830–1900. In the three decades following 1830, Newark's population increased by more than 60,000 people. Much of the increase was due to a large influx of Irish and German immigrants. After 1900, the population of Newark almost doubled again (to 435,000 in 1918) after undergoing a new wave of immigration from eastern and southern Europe [3]. These trends paralleled urban growth in other parts of the nation; the total urban population in the United States, 322,371 in 1800, had grown to more than 6 million by 1860, and exceeded 54 million by 1920 [7].

As a consequence of such rapid growth, the demand for water for household and industrial use also increased. In the United States during the late 1700s, private companies were organized in a number of cities to provide clean water. The companies relied on a system of reservoirs, pumping stations, and aqueducts. For instance, by 1840 Philadelphia had developed one of the best water systems in the nation, delivering more than 1.6 million gallons of water per day to 4, 800 customers. The enormous population growth during and after the Civil War (and the concomitant increase in congestion) overwhelmed the capacity of these early systems and outstripped the private sector's ability to meet the demands for water and sanitation. Worsening conditions finally forced the issue, resulting in large scale public works projects to meet the demand for water. In 1842, for example, the Croton Aqueduct was completed to provide New York City with a potential capacity of 42 million gallons of water per day. That system became obsolete by the early 1880s when demand exceeded nearly 370 million gallons per day for 3.5 million people and a more extensive municipal system had to be built using public funds [1].

The abundance of water allowed for the development of indoor plumbing, thereby increasing the volumes of waste water. In the late 1800s, most of the increase in domestic water consumption was due to the installation of bathroom fixtures [3]. In Chicago, per capita water consumption increased from 33 gallons per day in 1856 to 144 in 1872. By 1880, approximately one-fourth of all Chicago households had water closets [7]. Unfortunately, early sewage systems were constructed without allowance for human and household wastes. As demand increased and both residential and industrial effluent was being diverted into storm drainage systems, it became clear that municipalities would have to build separate sewers to accommodate household wastes and adopt the "water-carriage" system of waste removal (rather than rely on cesspools and privy vaults). The first separate sewer system was built in Memphis, TN, in 1880. Because of the enormous economic requirements of such large scale sewer systems, cities assumed the responsibility for building sewer systems and embarked on some of the largest construction projects of the 19th century [1], [7]. In Newark, more than 200 miles of sewers were constructed during the period 1894–1920 at a cost exceeding several million dollars, providing service to 95 percent of the improved areas of the city [3]. The technology to treat sewage effectively, however, (other than filtering and irrigation) would not be widely available for another 10 years.

1.7 Changing Attitudes About Health and Hygiene

The major source of the increase in household water consumption was personal hygiene. Accompanying the efforts to curb the spread of disease was a renewed interest in bathing. Although a popular activity among the Greeks and Romans, the frequency of bathing and an emphasis on cleanliness has ebbed and flowed throughout history. By the Dark Ages, cleanliness had fallen out of favor, only to become acceptable once again during the time of the Crusades. Through the 16th and 17th centuries bathing was rare, except among members of the upper class and then on an infrequent basis. By the first half of the 19th century, bathing was largely a matter of appearance and daily sponge bathing was uncommon in middle-class American homes until the mid-1800s [5]. As awareness of the causes of disease increased at the end of the 19th century, cleanliness became a means to prevent the spread of disease. Bathing and personal hygiene to prevent disease, states Lupton and Miller, "was aggressively promoted by health reformers, journalists, and the manufacturers of personal care products." Thus, hygienic products and the popular press became important mechanisms for defining the importance of cleanliness for Americans in the latter half of the 19th century.

In this period of technological development, social forces significantly influenced the construction, adaptation, use, and acceptance of technologies related to hygiene. The importance of cleanliness and the appropriate solutions to the hygiene problem were defined for people by specific agents of change, namely, health professionals, journalists, and commercial interests. Thus, the meaning of and need for health care products and home sanitary systems were defined by a number of influential social groups. In the history of technology literature, this is referred to as social constructionism where the meaning of a technology (especially in terms of its utility or value) is defined through the interaction of relevant social groups [2]. In the social constructionist's view, technological development is cast in terms of the problems relevant for each social group, with progress determined by the influence exerted by a particular social group to resolve a specific problem. For instance, the health professionals were instrumental in defining the need for cleanliness and sanitation, with journalists communicating the message to the middle class in the popular press. Through marketing and advertisements, those groups concerned with producing plumbing and bathroom fixtures significantly influenced the standards (and products) for personal hygiene in the home. By increasing awareness of the need for a clean and sanitary home and defining the dominant attitudes towards health standards, these influential social forces had defined for Americans the importance of the components of a bathroom [5].

With the arrival of indoor plumbing, both hot and cold water could be piped into the home. Previous to the introduction of indoor plumbing, appliances for bathing and defecating were portable and similar in appearance to furniture (to disguise their purpose when placed in a room). The fixed nature of pipes and the use of running water required that the formerly portable equipment become stationary. To minimize the cost of installing pipes and drains, the bath, basin, and commode were placed in a single room; consequently, the "bathroom" became a central place in the home for bathing and the elimination of waste. Early designs of bathroom furniture retained the wood construction of the portable units, but the desire for easy cleaning and a nonporous surface that would not collect dirt and grime led to the use of metal and china. The bathroom became, in effect, a reflection of the modern desire for an antiseptic approach to personal hygiene. Porcelain-lined tubs, china toilets, and tiled floors and walls (typically white in color) emphasized the clinical design of the bathroom and the ease with which dirt could be found and removed. As Lupton and Miller [5] point out, the popular press and personal hygiene guides of the period compared the design of the modern bathroom to a hospital. The cleanliness of the bathroom became a measure of household standards for hygiene and sanitation. By the late 1880s, indoor plumbing and bathrooms were a standard feature in homes and a prerequisite for a middle-class lifestyle.

1.8 The Indoor Toilet

The growing market in bathroom fixtures stimulated significant technological development. Before 1860, there were very few U.S. Patents filed for water closets and related toilet components (the Ryan and Flanagan design shown in Figure 1.6 is one of only three patents for water closets filed in the period 1847–1855). Once the toilet moved into the home and became a permanent fixture in the bathroom, the technology rapidly developed. The primary concern in the development of the technology was the amount of water used by the toilet and its subsequent impact on the water supply and sewage collection and treatment systems. The fact that water could be piped directly to the cistern of the water closet (previous designs of water closets had required that they be filled by hand) necessitated some mechanism to control the flow of water to the toilet and minimize the impact of toilet use on the water supply. In addition, there was a more systemwide concern to provide some automatic way to control the amount of water used by the toilet and discharged into the sewerage system (and thereby eliminating the unacceptable continuous flow method of flushing used by the Greeks and Romans). As a consequence, much effort was devoted to developing mechanisms for controlling water flow into and out of the toilet.

During the period 1870–1920, the number of U.S. patents filed for toilet technology rapidly escalated. In the 1880s, more than 180 U.S. patents were issued for water closets. For many of the inventors concerned with the design of toilet technology, the flush and float valves became the center of attention. In 1879, William Ross of Glasgow, Scotland, filed the U.S. Patent "Improvement in Water-Closet Cisterns" shown in Figure 1.7. In the patent documents, Ross put forward an eloquent statement of the design objectives for the valves in the cistern: "The object I have in view is to provide cisterns for supplying water to water-closets, urinals, and other vessels with valves for controlling the inlet and outlet of water, the valves for admitting water to such cisterns being balanced, so that they can be operated by small floats, and also being of simple, light, and durable construction, while the valves for governing the outflow of water from such cisterns allow a certain amount of water only to escape, and thus prevent waste, and are more simple and efficient in their action than those used before for the same purpose, and are portable and self-contained, not requiring outer guiding cylinders or frames, as heretofore." It is interesting to note that the basic design put forward by Ross for the float and flush valves has changed little in the past 116 years. By 1880 the basic design of the toilet, as we know it today, was well-established (as shown in Figures 1.1 and 1.2). A cistern with flush and float valves and a basin using a water trap are clearly evident.

Some of the designs were quite complex. The patent filed in 1897 by David S. Wallace of the Denver, CO, Wallace Plumbing Improvement Co. depicts a design (shown in Figure 1.8) that would allow both manual and automatic control of the open and close cycle of the flush valve. Such an approach would allow the toilet user to adjust the amount of water to be used during the flushing cycle (a concept that would be resurrected almost 100 years later as a water saving device). Not all inventors chose to use the same approach to the flush valve. David Craig and Henry Conley proposed a quite different solution in 1895 (Figure 1.9). In operation, the valve was quite simple, using a ball to close the water outlet from the toilet tank. The flushing operation is initiated by pulling on the cord 2, which raises the ball inside the tube d', allowing the ball to roll to the f' end of the tube and the tank to empty. The ball rolls slowly back down the tube, settling into the collar c and closing the water outlet to the toilet bowl. This approach to control was not widely adopted.

Throughout the history of the development of toilet control technology, much of the technological effort was directed towards improving the mechanisms for assuring correct closure of the flush valve. More recently, however, concerns about conservation of water have motivated technologists to consider new designs that increase the effectiveness of the flush valve. The design by Olson (Figure 1.3) is a good example of recent concerns in the development of control technologies for the toilet. The design allows the user to control the amount of water used during the flushing cycle. Characterized as a water saving system that uses only several pints of water rather than the several gallons of a conventional toilet, the actual amount of water used during the flushing cycle is actively controlled by the user. Recall that this is similar in concept to the approach taken by Wallace in 1897 (Figure 1.8). Notice, too, that the approach to control remains the same; once the flushing cycle has been initiated, the level of water in the toilet tank is controlled by the float valve.

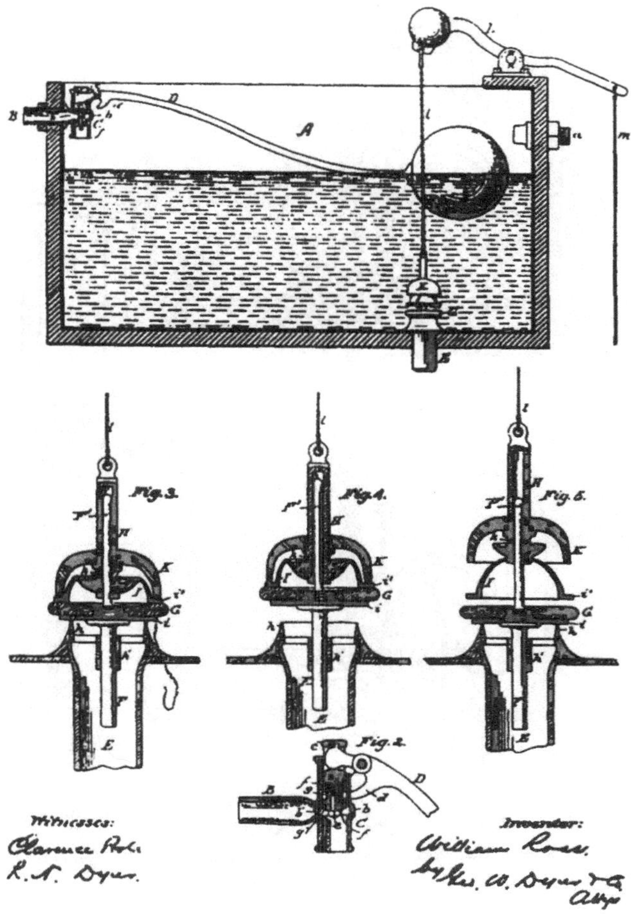

Figure 1.7 U.S. Patent drawing of the flush and float valves, including drawings of the operation of the flush valve, for a toilet tank. *Source:* U.S. Patent No. 211,260 filed by W. Ross, Jan. 7, 1879.

1.9 Historical Approaches to Systems

It should be clear that understanding the historical development of toilet technology requires understanding the historical devel-

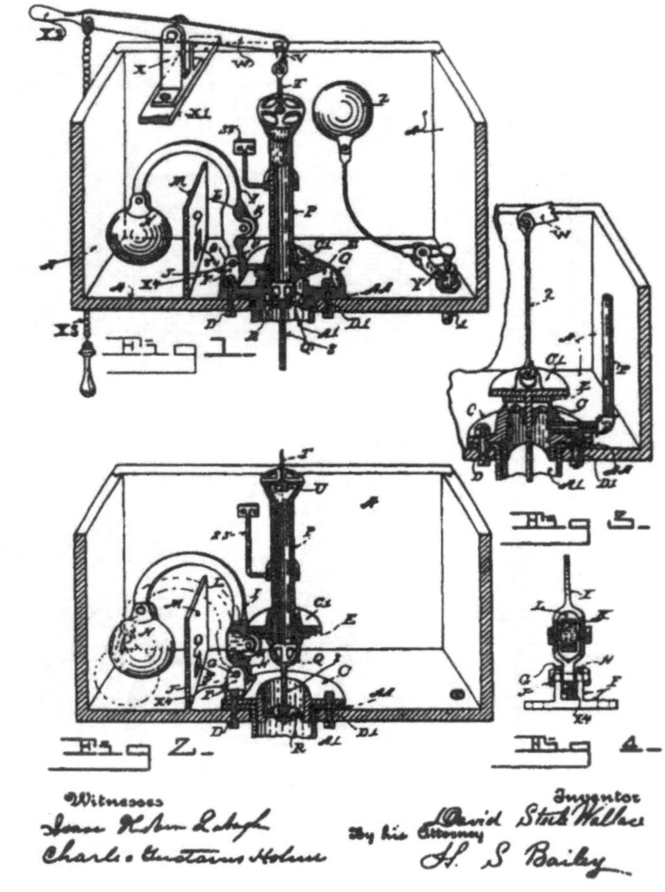

Figure 1.8 U.S. Patent drawing of the flush and float valves that allows both manual and automatic control of the open and close cycle of the flush valve. *Source:* U.S. Patent No. 577,899 filed by D.S. Wallace, Mar. 2, 1897.

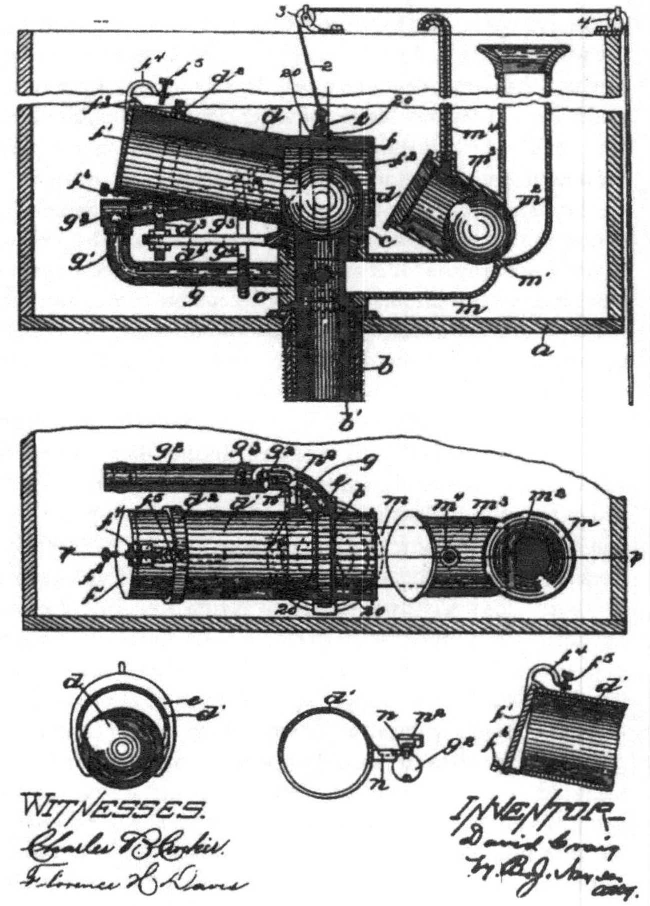

Figure 1.9 U.S. Patent drawing of a variation on the flush valve for the toilet tank. The ball labeled *d* provides the means for closing and sealing the outlet to the toilet basin during the flush cycle. *Source:* U.S. Patent No. 543,570 filed by D. Craig and H. Conley, Jul. 30, 1895.

opment of an entire system of sanitation. Historians have recognized for some time that the history of technology must be considered in terms of entire systems, and considerable recent effort has been devoted to understanding the ways in which systems and the technology required to support them develop over time.

The Thomas Hughes [4] model for the historical development of systems has been especially influential. Hughes extensively studied the development of electric power networks in the period 1880–1930 and, as a result of that research, constructed a four-phase model of the historical development of large-scale distributed systems. Each phase of development is defined by its own characteristics and a group of professionals who play a dominant role in developing the technologies employed in that phase. In the first phase, invention predominates with significant technical effort and intellectual resources devoted to developing the technologies comprising the system. The second phase is primarily concerned with the transfer of technologies from one region and society to another. Although inventors still play a predominant role in this phase, entrepreneurs and financiers become increasingly important to the growth and survival of the system.

In the third phase of the model, regional systems grow into national systems and the scalability of regional technologies becomes a major concern. In the third phase of the model, Hughes links system growth and development to reverse salients and critical problems. Borrowed from military terminology, a reverse salient describes a situation where a section of an advancing line (in this context, the development of some aspect of system technology) is slowed or halted while the remaining sections continue movement in the expected direction. As Hughes [4] states, "A reverse salient appears in an expanding system when a component of the system does not march along harmoniously with other components" and reveals "imbalances or bottlenecks within the system. The imbalances were seen as drags on the movement of the system towards its goals, especially those of lower costs or larger size."

A reverse salient has two major effects on technological development: the growth of technology slows and action must be taken for sustained growth to continue. It is the occurrence of a reverse salient and the effort expended to eliminate that bottleneck which, according to Hughes, captures the essence of technological development. Innovations in technology occur because a reverse salient and the imbalance in system development motivates inventors, industrial scientists, and engineers to concentrate their efforts in finding solutions to those problems. The imbalance defines the problem for the engineers, and the removal of that problem leads to innovations in technology, management policy, and methods of control. Hughes cites a number of examples of inventions and innovations in electrical technologies to illustrate the concept of a reverse salient. One such example is the complex information and control networks established in the 1920s to collect data, monitor plant performance, and control plant output. These networks developed in response to an increase in demand for electricity and a need for better scheduling of plant utilization.

In the fourth phase of the model, the system develops substantial momentum with "mass, velocity, and direction." Much of that momentum can be characterized by the significant investment in capital equipment and infrastructure necessary to maintain the system. At this stage of development, the system's "culture" develops, defined by the professionals, corporate entities, government agencies, investors, and workers whose existence depends upon developing and perpetuating the system.

The development of a comprehensive water and sanitation system (with the toilet as an integral part of it), fits well into Hughes model. Once the need for a system to meet the demand for potable water and provide for the collection and disposal of human waste was identified, rapid technological development occurred. The necessary components of the system were developed and employed in building regional water supply and sewerage systems during the last half of the 19th century (phases one and two of the model). During that period there was considerable transfer of knowledge about requisite sanitation methods from one community to another and among the various medical, engineering, and public health professionals concerned with sanitation policy and the construction of water supply and sewerage systems. As cities grew and demand outstripped capacity, much of the effort was directed towards transforming small-scale private systems into the large-scale municipal systems that could accommodate the requirements of rapidly expanding urban populations (phase three). Because the means for supplying water typically preceded systems of disposal, the collection and disposal of household and industrial waste products were a continual source of problems. When indoor plumbing provided the means to distribute water throughout the home, the market for sinks, baths, and toilets blossomed and the rate of domestic water consumption increased dramatically.

In general, water supply preceded waste disposal, so that an increase in the ability to supply water resulted in significantly more volumes of waste water than could be handled by existing sewer systems. The increase in waste water created a reverse salient that could not be ignored and required a massive municipal response to construct the necessary collection and disposal system. The ability to pipe water directly to bathroom fixtures also created a reverse salient that was directly related to the control problem in the toilet tank. Both the amount of water flowing into the toilet and the amount of water used by the toilet during flushing had to be controlled to minimize the impact on the water supply and sewer systems (in fact, the control technology in the toilet tank became the means for linking the two systems). As a result, significant effort was expended in the two decades following 1875 to refine the operation of the toilet flush and float valves, especially to minimize the potential for incomplete valve closure and resultant water leakage. As we have seen, many designs were proposed, but the solutions have remained fairly consistent through the years.

Once work began on the basic infrastructure for supplying water and disposing of wastes, the development of sanitation systems was well into Hughes' fourth stage. The increase in the ability to supply water created demand for water and bathroom technologies and the subsequent need for waste water disposal. A public health movement and public works facilities and services grew out of the concern for clean water and sanitary living conditions. Underlying the desire for clean water and sanitary conditions were changing attitudes towards health, cleanliness, and the causes of disease. Once urban congestion reached a point where the health and well-being of people living in cities was threatened by disease, change occurred and new norms for cleanliness and standards of living were adopted. One of the major technological developments arising from that period of change was the modern toilet.

1.10 Summary

The goal of this chapter was to place the development of a control-based technology in its historical context. In the process of attaining that goal, an attempt has been made to show how the development of such a technology, even in its simplest form, has many dimensions. The toilet and the basic control technology that resides in its water tank found its way into the home as the result of a number of important social, political, economic, and health reasons. The chronology shown in Figure 1.10 captures the multitude of events that occurred between 1830 and 1905 in the areas of disease, health, medicine, public works, and toilet technology. The chronology focuses on events in Newark, NJ during that period. Prior to 1832 and the first cholera epidemic in Newark, there was little activity in the areas of public health or public works. Until the community was motivated to do something about the appalling conditions, no significant efforts were made to supply clean water and effectively remove waste products. As far as toilet technology was concerned, there was very little development prior to 1854. The development of control technology for the toilet rapidly followed the development of water supply and sewer systems, so that, by the 1880s, considerable progress had been made in controlling water flow in the toilet tank.

There is no one single event that can be identified as the motivating factor for putting control technologies into the toilet. Clearly, the availability of an adequate water supply, indoor plumbing, and a sewer system for carrying away waste were prerequisites for developing water flow control in the toilet tank.

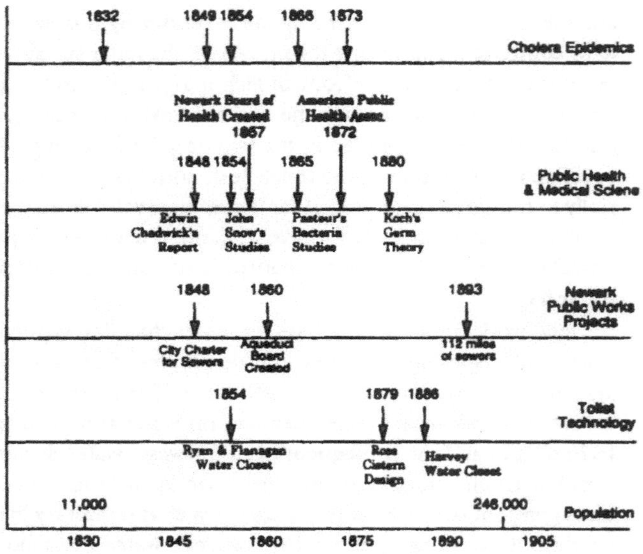

Figure 1.10 Chronology of events in Newark, NJ between 1830 and 1905 in the areas of disease, health, medicine, public works, and toilet technology.

However, those factors were not sufficient motivation for moving the toilet into the home nor adequate stimulus for the development of toilet technology. The threat of disease, changing attitudes towards health and sanitation, and a redefinition of the requirements of a modern home also contributed to the demand for bathroom technology. History has shown us how the development of a control technology, as well as the attitudes towards health, cleanliness, and standards for a middle class home, can be influenced by a number of strong, interacting forces. As Hughes' model points out, however, once development attained momentum, there was no stopping toilet technology.

References

[1] Armstrong, E.L., *History of Public Works in the United States 1776-1976*, American Public Works Association, Chicago, IL, 1976.

[2] Bijker, W.E., Hughes, T.P., and Pinch, T.J., *The Social Construction of Technological Systems*, The MIT Press, Cambridge, MA, 1987.

[3] Galishoff, S., *Newark: The Nation's Unhealthiest City, 1832-1895*, Rutgers University Press, New Brunswick, NJ, 1975.

[4] Hughes, T.P., *Networks of Power: Electrification in Western Society, 1880-1930*, The Johns Hopkins University Press, Baltimore, MD, 1983.

[5] Lupton, E. and Miller, J.A., *The Bathroom, The Kitchen, and the Aesthetics of Waste*, MIT List Visual Arts Center, Cambridge, MA, 1992.

[6] Mayr, O., *The Origins of Feedback Control*, MIT Press, Cambridge, MA, 1969.

[7] Melosi, M.V., *Pollution and Reform in American Cities, 1870-1930*, University of Texas Press, Austin, TX, 1980.

[8] Reynolds, R., *Cleanliness and Godliness*, Doubleday, New York, 1946.

[9] Winslow, C.E.A., *Man and Epidemics*, Princeton University Press, Princeton, NJ, 1952.

[10] Wright, L., *Clean and Decent*, Revised ed., Routledge & Kegan Paul, London, 1980.

2

Temperature Control in Large Buildings

2.1 Introduction 15
2.2 System and Component Description 15
A Prototypical System • Components • Controls
2.3 Control Performance Specifications 20
Time-Domain Specifications • Performance Indices
2.4 Local-Loop Control 21
Mathematical Models • Feedback Control Algorithms • Tuning
2.5 Logical Control 23
Discrete-State Devices • Modulating Devices
2.6 Energy-Efficient Control 24
Economizer Cooling • Electrical Demand Limiting • Predicting Return Time from Night or Weekend Setback • Thermal Storage • Setpoint Resetting
2.7 Configuring and Programming 27
Textual Programming • Graphical Programming • Question-and-Answer Sessions
References 27

Clifford C. Federspiel
Johnson Controls, Inc., Milwaukee, WI

John E. Seem
Johnson Controls, Inc., Milwaukee, WI

2.1 Introduction

Heating, ventilating, and air-conditioning (HVAC) systems in large buildings are large-scale processes consisting of interconnected electromechanical and thermo-fluid subsystems. The primary function of HVAC systems in large commercial buildings is to maintain a comfortable indoor environment and good air quality for occupants under all anticipated conditions with low operational costs and high reliability.

The importance of well-designed controls for HVAC systems becomes apparent when one considers the impact of HVAC systems on the economy, the environment, and the health of building occupants. One of the most significant operational costs of HVAC systems is energy. There is a direct cost associated with purchasing energy, and an indirect environmental cost of generating energy. It has been estimated that HVAC systems consume 18% of the total energy used in the U.S. each year [1]. Energy utilization in buildings is strongly influenced by the HVAC control systems, so the potential impact of HVAC controls on the economy and the environment is significant.

In addition to the impact on national energy consumption and the environment, HVAC systems have an impact on human health because people spend more than 90% of their lives indoors. Based on data from residential and commercial buildings, the Environmental Protection Agency (EPA) has placed indoor radon first and indoor air pollution fourth on a list of 31 environmental problems that pose cancer risks to humans [2]. Additionally, indoor air pollution other than radon was assessed as a high noncancer risk. Therefore, HVAC control systems can have a significant impact on human health because they can play a central role in controlling indoor air pollution.

In this chapter, building temperature controls and the associated flow and pressure controls are described. The emphasis of the chapter is on the current practice of control system design for large commercial buildings. Therefore, the controls described in this chapter may not reflect those for residential buildings or industrial buildings.

2.2 System and Component Description

Before one can design a control system for a process, it is necessary to understand how the process works. In this section, a prototypical HVAC system for a large building is described. Important features of subsystems that affect the control system design and performance are also described. Also, a description of the control system design is provided.

2.2.1 A Prototypical System

In large buildings, there is typically a central plant that supplies chilled and hot water or steam to a number of heating and cooling coils in air-handling units, terminal units, and perimeter heat-

0-8493-0054-3/00/$0.00+$.50

ing coils. The central plant uses one or more boilers, chillers, and cooling towers to perform this task. Primary and secondary pumping systems such as shown in Figure 2.1 may be used to lower the pumping power requirements. Chillers generally re-

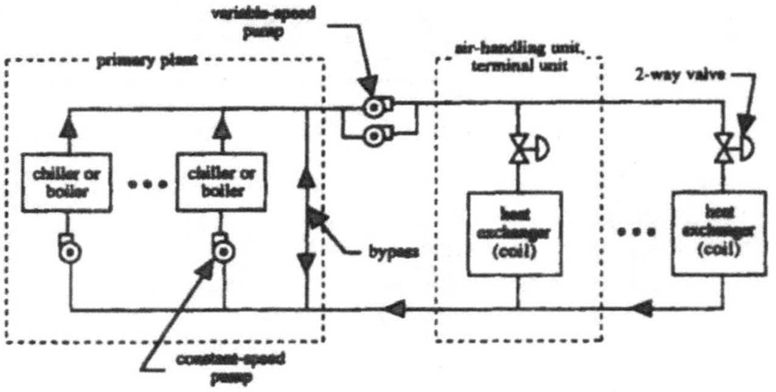

Figure 2.1 Schematic diagram of a piping distribution system. The primary system is constant-flow while the secondary systems are variable-flow.

quire a constant water flow rate, so the primary pumps are run at constant speed. The speed of the secondary pumps may be modulated to reduce pumping costs.

Air-handling units distribute conditioned air to a number of zones, which is a volume within the building that is distinct from other volumes. Often, but not always, physical partitions, walls, or floors separate zones. In large buildings, air-handling units typically supply cool air to all zones during all seasons. Core zones in buildings must be cooled even when the outdoor temperature is well below freezing because of the large number of heat sources such as lights, equipment, and people, and because core zones are insulated from weather conditions by perimeter zones. As Figure 2.2 shows, air-handling units have a number of discrete components. Not all of the components shown in the figure are

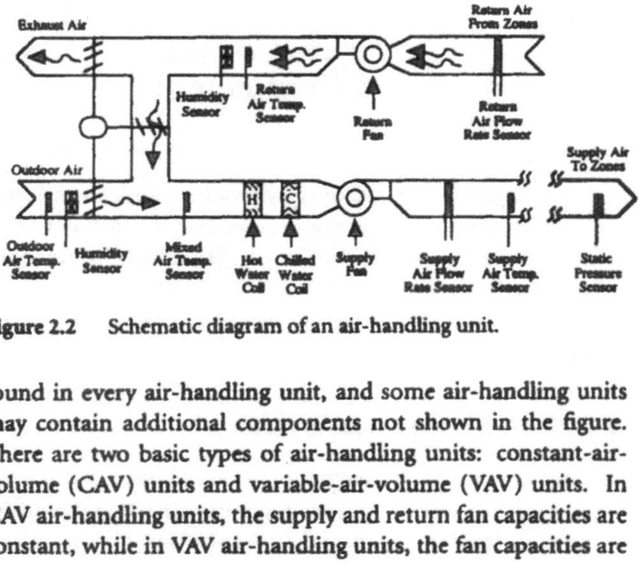

Figure 2.2 Schematic diagram of an air-handling unit.

found in every air-handling unit, and some air-handling units may contain additional components not shown in the figure. There are two basic types of air-handling units: constant-air-volume (CAV) units and variable-air-volume (VAV) units. In CAV air-handling units, the supply and return fan capacities are constant, while in VAV air-handling units, the fan capacities are modulated.

Terminal units, such as the one shown in Figure 2.3, are used to control the temperature in each zone. Terminal units are typically

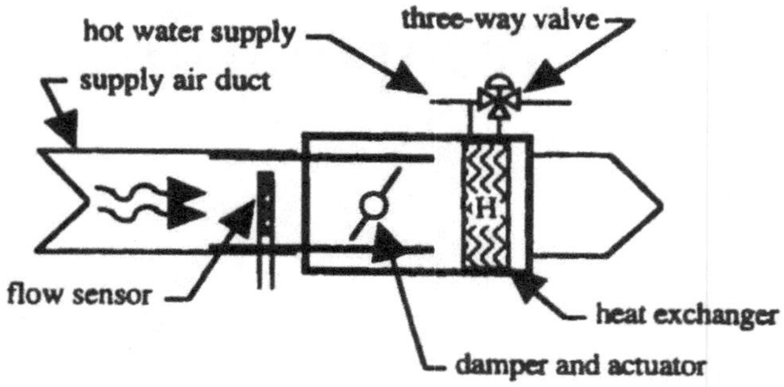

Figure 2.3 Schematic diagram of one type of VAV terminal unit. The zone is cooled by modulating the air flow rate, and the zone is heated by modulating the valve on the reheat coil.

installed in the supply air duct for each zone, where they are used to modulate the flow rate and/or the temperature of air supplied to a zone.

2.2.2 Components

An HVAC system consists of a number of interconnected components that each have an impact on the behavior of the system and on the ability of the control system to affect that behavior. Components that have a key influence on the control system design are described next.

Actuators

Actuators are used to modulate the flow through dampers and valves. The two most common types of actuators are electric motors and pneumatic actuators. Most of the electric motors used in HVAC controls are synchronous ac motors, some of which are stepper motors. To reduce motor power consumption and cost, gear ratios are used such that the load reflected onto the motor through the gear reduction is small.

Pneumatic actuators are usually linear motion devices consisting of a spring-loaded piston in a cylinder. Air pressure in the cylinder pushes against the spring and any other forces acting on the piston. The spring returns the piston to the "normal" position in case of a loss of pressure. Pressure is supplied to the piston from a main, typically at a pressure of 20 (psi). Typical hardware for throttling the main pressure is a dual set of solenoid valves, an electric-to-pneumatic transducer, or a pilot positioner. Pneumatic actuators are prone to have substantial friction and hysteresis nonlinearities. A detailed description of pneumatic actuators can be found in [3].

Heat Exchangers

It is important to understand the steady-state behavior of heat exchangers because it affects the process gain. There are different ways to mathematically model the steady-state behavior including the log mean temperature difference (LMTD) method and the effectiveness-number of heat transfer units (ϵ-NTU) method. These relations are used to design and size heat exchangers. For control system design, one needs to model and understand the relationship between the heat transfer rate and the controlled flow rate. The steady-state characteristics of heat exchangers can be derived from the ϵ-NTU relations, which can be found in most texts on heat transfer (e.g., [4]). For example, consider a cross-flow, water-to-air heat exchanger with unmixed flow on both sides in which the air flow rate is a constant, the water flow rate is modulated, and the water is heating the air. For such a heat exchanger, the ϵ-NTU relation is [4]

$$\epsilon = 1 - exp((\frac{1}{C_r})(NTU)^{0.22}(exp(-C_r(NTU)^{(0.78)}) - 1)) \tag{2.1}$$

where ϵ is the heat exchanger effectiveness, C_r is the ratio of the minimum to maximum heat transfer capacity rates, and NTU is the number of heat transfer units. All three of these quantities are nondimensional. Mathematically, they are defined as

$$\epsilon \equiv \frac{q}{q_{max}} \tag{2.2}$$

$$q_{max} = C_{min}(T_{w,i} - T_{a,i}) \tag{2.3}$$

$$C_{min} \equiv \min[C_a, C_w] \tag{2.4}$$

$$C_r \equiv \frac{C_{min}}{C_{max}} \tag{2.5}$$

$$C_{max} = \max[C_a, C_w] \tag{2.6}$$

$$C_a \equiv \dot{m}_a c_{p_a} \tag{2.7}$$

$$C_w \equiv \dot{m}_w c_{p_w} \tag{2.8}$$

$$NTU \equiv \frac{UA}{C_{min}} \tag{2.9}$$

where q denotes the heat transfer rate in units of energy per unit time, q_{max} denotes the maximum achievable heat transfer rate, C_{min} is the minimum heat transfer capacity rate of the two fluids in units of energy per unit time per unit temperature, $T_{w,i}$ is the temperature of the water entering the heat exchanger, $T_{a,i}$ is the temperature of the air upstream of the heat exchanger, C_{max} is the maximum heat transfer capacity rate of the two fluids, C_a is the heat transfer capacity rate of the air, C_w is the heat transfer capacity rate of the water, $\dot{m}_a$ is the mass flow rate of the air, c_{p_a} is the specific heat at constant pressure of the air in units of energy per unit mass per unit temperature, $\dot{m}_w$ is the mass flow rate of the water, c_{p_w} is the specific heat at constant pressure of the water, U is the heat transfer coefficient at the surface of the heat exchanger in units of energy per unit time per unit temperature per unit area, and A is the area of the heat exchanger (U and A must correspond to the same surface).

The steady-state relation between the water flow rate and the heat transfer is

$$q = q_{max}\epsilon \tag{2.10}$$

Figure 2.4 shows the characteristic of Equation 2.10 normalized by the heat transfer at the maximum water flow rate when $NTU_{max} = 2$, $\dot{m}_a = 5.8$ kg/s (10000 ft^3/min), and $\dot{m}_{w_{max}} = 3\dot{m}_a\frac{c_{p_a}}{c_{p_w}}$. The large slope at low flow and small slope at high flow

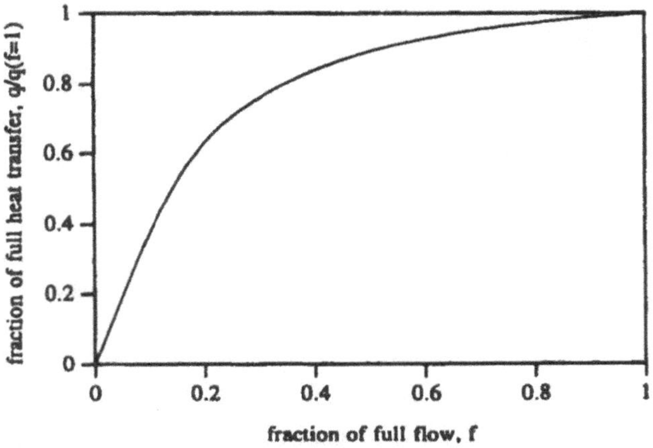

Figure 2.4 Heat exchanger characteristic for a cross-flow heat exchanger with unmixed flow in both fluids.

is typical of heat exchanger characteristics. This fact is important when control valves are coupled with heat exchangers, and it will be discussed in more detail later.

The dynamic behavior of heat exchangers is complex and may be difficult to model accurately because heat exchangers are nonlinear, distributed-parameter subsystems. Qualitatively, a heat exchanger introduces a phase lag and time delay into the overall system. This lag and delay vary as the flow rate through the heat exchanger is modulated.

Dampers and Valves

The flow through a damper or valve can be determined from

$$Q = fC_v\sqrt{\frac{\Delta p_v}{g_s}} \tag{2.11}$$

where Q is the flow rate, f is the flow characteristic of the valve, C_v is the flow coefficient, Δp_v is the pressure drop across the device, and g_s is the specific gravity of the fluid. The flow characteristic is the relation between the position of the damper blades or valve plug and the fraction of full flow. When the pressure difference across the damper or valve is held constant for all positions, the flow characteristic is referred to as the inherent characteristic and is denoted by f_i. The flow coefficient, C_v, is the ratio of the flow rate in the fully open position to the ratio of the square root of the pressure drop to the specific gravity. The C_v increases as the size of the device increases. For water valves in which the flow rate is measured in units of gallons per minute and pressure measured in units of pounds per square inch, typical values for the C_v are 1 for a $\frac{1}{2}$-inch valve, 37 for a $2\frac{1}{2}$-inch valve, and 150 for a 4-inch valve.

Figure 2.5 shows an example of the inherent characteristic of a damper or valve. The quantity f_i is the fraction of full flow, and L is the fraction of the fully open position of the damper blades or valve stem. The important features of a damper's inherent char-

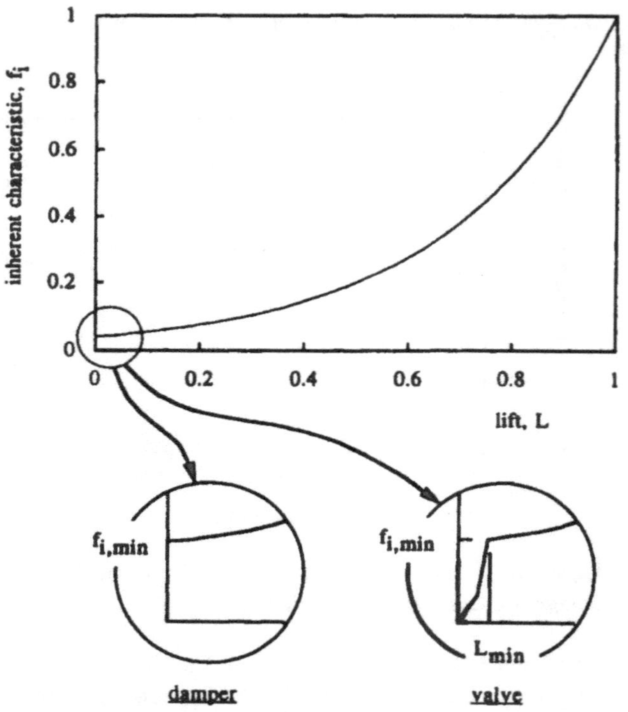

Figure 2.5 Inherent characteristic of a damper or valve.

acteristic are (1) the flow rate at the closed position (dampers leak, although seals may be used to reduce leakage) and (2) the shape of the inherent characteristic between the fully closed and fully open positions. Parallel-blade dampers have a faster-opening inherent characteristic than opposed-blade dampers.

Like dampers, one of the important features of a valve's inherent characteristic is the behavior when the lift is nearly zero. Unlike dampers, many valves do not leak when fully closed. However, when the lift is nearly zero, the characteristic changes dramatically as shown in Figure 2.5. When $L < L_{min}$, the inherent characteristic is dependent on the valve construction, seat material, manufacturing tolerances, etc. The valve may shut off quickly and completely as shown in the figure, or it may shut off quickly but not completely. If the valve seat has rubberized seals, then L_{min} will be larger than without rubberized seals.

All actuation devices have a limited resolution for positioning the valve stem. Pneumatic actuators, which are commonly used in HVAC systems, have low resolution because they are prone to stiction and hysteresis. Although the characteristic of the valve may be a continuous function of the lift for flow rates less than f_{min}, the limited resolution of the valve stem positioning implies that the characteristic may be accurately modeled as a discontinuity at zero lift. The magnitude of the discontinuity is characterized by the rangeability of the valve. The inherent rangeability of a valve is defined as the ratio of the maximum to the minimum controllable flow. Mathematically, it is

$$R_i = \frac{1}{f_{i,min}} \tag{2.12}$$

Control valves in HVAC systems typically have an inherent rangeability between 20 and 50.

The other important feature of a valve's inherent characteristic is the relationship between L and f_i for $L_{min} \leq L \leq 1$. Valves are often designed to have one of the following inherent characteristics: quick-opening, linear, or equal-percentage. Assuming the inherent characteristic is discontinuous at $L = 0$ (i.e., $L_{min} = 0$) due to limited resolution of the lift, then the inherent characteristic of a linear valve is

$$f_i = (1 - f_{i,min})L + f_{i,min} \tag{2.13}$$

and the inherent characteristic of an equal-percentage valve is

$$f_i = R_i^{L-1} \tag{2.14}$$

In most systems, the pressure drop across a damper or valve varies with the flow rate because of additional pressure losses in ducts or pipes. The authority of a damper or valve is defined to be

$$\alpha = \frac{p_v}{p_s} \tag{2.15}$$

where p_v is the pressure drop across the damper or valve at the fully open position and p_s is the total system pressure drop (e.g., the combined pressure drop across a valve, piping, and heat exchanger). The relationship between the lift and the fraction of full flow when a damper or valve is installed in a system is called the installed characteristic. The installed characteristic is dependent on both the lift and the authority. Mathematically, the installed characteristic is

$$f_s = \sqrt{\frac{1}{1 + \alpha(f_i^{-2} - 1)}} \tag{2.16}$$

Figure 2.6 shows the installed characteristics of equal-percentage valves for several different values of authority assuming that the inherent characteristic is discontinuous when the lift is zero. When the authority is unity, the installed characteristic is the same as the inherent characteristic. Note that the installed value of f_{min} is dependent on the authority. This means that the range-

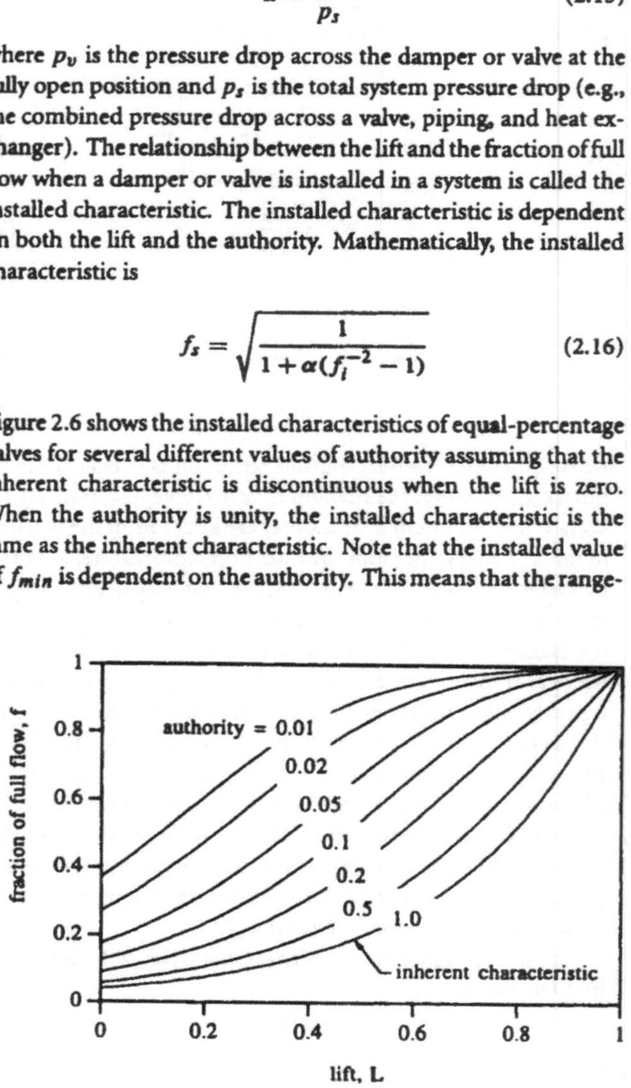

Figure 2.6 Installed characteristics of equal-percentage valves with R_i = 25.

ability of an installed valve or damper is not the same as the inherent rangeability. The installed rangeability is the ratio of the maximum installed flow to the minimum controllable installed flow and is mathematically defined as

$$R_s = \frac{1}{f_{s,min}} \tag{2.17}$$

where $f_{s,min}$ is the smallest controllable fraction of the maximum installed flow. For an equal percentage valve, the installed rangeability is

$$R_s = \sqrt{1 + \alpha(R_i^2 - 1)} \tag{2.18}$$

See [5] for complete details on valve characteristics and terminology.

Valves need to be properly sized to ensure proper operation. The size of the valve affects the authority of the valve; oversized valves have a low authority. Undersized valves cannot deliver sufficient flow rates to meet the maximum load conditions, but oversized valves require less pumping power. In [6] it is recommended that valves be sized so that the authority is between 25 and 33%. In the HVAC industry, oversized valves are more prevalent than undersized valves because large safety factors are used to ensure adequate capacity.

The inherent rangeability and the authority affect flow control loops at low flow rates because it is impossible to smoothly modulate the flow rate below the value of $f_{s,min}$. Low authority and/or low rangeability make this problem worse. Problems caused by low authority and low rangeability are worse when valves are coupled with heat exchangers. Consider the equal-percentage characteristic such as shown in Figure 2.6. When an equal-percentage valve is coupled with the cross-flow heat exchanger described in Section 2.2.2, the resulting steady-state characteristic of the valve and heat exchanger combination is as shown in Figure 2.7 for two different combinations of authority and inherent rangeability. One can define the heat transfer rangeability for a valve and heat exchanger combination as the ratio of the maximum to minimum controllable heat transfer at steady-state. Mathematically, the heat transfer rangeability is

$$R_q = \frac{1}{q_{min}} \tag{2.19}$$

Note that the value of R_q corresponding to $\alpha = 0.1$ and $R_i = 25$ is 2.2. The implication of the heat transfer rangeability is that a heating or cooling load less than q_{min} cannot be matched without cycling. Since the steady-state characteristic of heat exchangers has a large slope at low flow rates, the value of R_q will be very low and cycling will occur unless the inherent rangeability of the valve and the authority of the valve are high.

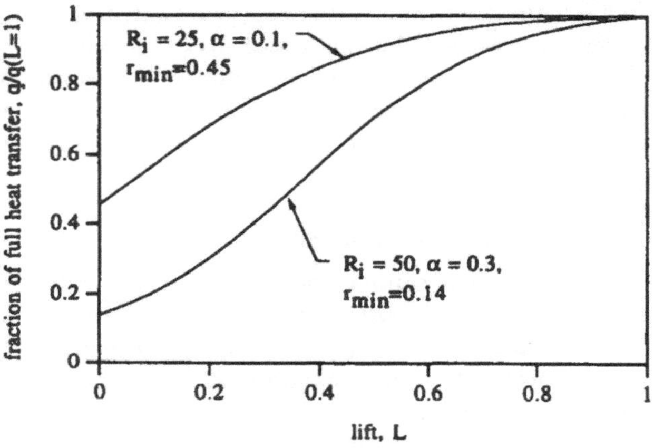

Figure 2.7 Steady-state installed heat transfer characteristic for an equal-percentage valve with $R_i = 25$ and $\alpha = 0.1$.

Fans and Pumps

Steady-state relations between head, flow rate, speed, and power consumption are nonlinear. The following are idealized similarity relations for pumps moving incompressible fluids [7]

$$\frac{Q_2}{Q_1} = \frac{\omega_1}{\omega_2}\left(\frac{D_2}{D_1}\right)^3 \tag{2.20}$$

$$\frac{H_2}{H_1} = \left(\frac{\omega_2}{\omega_1}\right)^2\left(\frac{D_2}{D_1}\right)^2 \tag{2.21}$$

$$\frac{P_2}{P_1} = \frac{\rho_2}{\rho_1}\left(\frac{\omega_2}{\omega_1}\right)^3\left(\frac{D_2}{D_1}\right)^5 \tag{2.22}$$

where Q denotes flow rate, ω denotes speed, D denotes the characteristic size, H denotes head, P denotes power consumption, and ρ denotes density.

These relations have two important consequences for control systems. The first is that the cubic relation between power and speed implies that capacity modulation by speed modulation is more efficient than by throttling; throttling results in only minor reductions in power consumption.

The second implication of the similarity relations is that the gain of pressurization loops will be nonlinear due to the quadratic relation between pressure and speed. Since the load on a modulating pump or fan varies with time due to changes in the position of control valves or dampers, the quadratic relation varies with time.

Sensors and Transmitters

One feature of sensors that affects control performance is the response time of the sensors. In [6] it is recommended that measurement and transmission time constants be less than one tenth the largest process time constant. In some cases, such as flow control loops, the process dynamics are extremely fast, so it may not be possible to meet this recommendation.

Design decisions for HVAC systems and the associated controls are strongly influenced by initial cost or installed cost. Therefore, many buildings have just the minimum number of sensors required for operation. In other words, many buildings are "sensory-starved." It is often possible to reduce operational costs when additional sensors are included in the design. Additional sensory information can be used by control systems to improve energy utilization and to increase diagnostic capabilities.

2.2.3 Controls

HVAC control systems are partially decentralized. Typically there is a separate controller for each subsystem. For example, there may be a controller for the central plant, a controller for each air-handling unit, and a controller for each terminal unit. Each subsystem may have multiple inputs and multiple outputs. The commands to the process inputs are usually generated through a logical coordination of single-input, single-output (SISO) proportional-integral-derivative (PID) algorithms. There may be multiple PID algorithms operating in a controller at one time.

Historically, pneumatic devices were used exclusively to compute and execute control commands. Today, many control systems in buildings are still partially or completely pneumatic, but the trend is toward the use of distributed digital controllers. The most modern control systems for HVAC equipment contain a network of digital controllers connected on a communication trunk or bus. There are several different methods for connecting various controllers. Figure 2.8 shows an architecture that uses peer-to-peer communications. The digital controllers and operator workstations all communicate on the same network. There may be hundreds of digital controllers on a single network. Due to the historic use of pneumatic controls, many of the digital control systems in buildings are programmed to emulate pneumatic controllers. The communications networks are used primarily for alarm reporting, monitoring, and status communications.

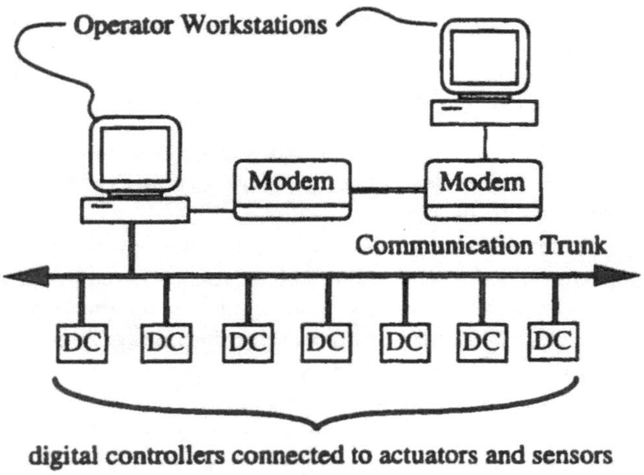

Figure 2.8 Communication architecture with peer-to-peer communications.

2.3 Control Performance Specifications

In many engineering design specifications for HVAC systems, the performance of the control system is not rigorously specified. Often there is little reference made to quantifiable measures of performance. In this section, guidelines for certain measures of performance are described. These guidelines are based on common practice and inferences from industry standards. Since most HVAC subsystems are controlled using a decentralized SISO design, this section is focused on control performance specifications for SISO systems.

2.3.1 Time-Domain Specifications

Zone Temperature

For zone temperature control, the steady-state error has a strong impact on the comfort of occupants. In [8] the ASHRAE standard for an acceptable range of zone temperatures is described. If temperatures remain within this range, then 80% of the occupants should be comfortable. The temperature range is 7°F and the center of the range is dependent on the season. The implication of this standard is that the steady-state error for zone temperature control is dependent on the setpoint. If the setpoint is chosen as the center of the ASHRAE acceptable temperature range, then the maximum allowable steady-state error is 3.5°F. It is common practice for building operators to raise the zone temperature setpoints when it is hot outside and lower the setpoints when it is cold outside to conserve energy. If the setpoint is raised or lowered to the limit of the ASHRAE standard, then the steady-state error for the zone temperature controls should be zero.

Flow Control

Flow control loops are cascaded with temperature control loops in pressure-independent VAV boxes and are used in air-handling units to regulate the outdoor air flow rate. Typically, it is expected that flow control loops will be controllable from 5 to 100% of the maximum rated flow.

Flow controllers may be operated manually during commissioning or when troubleshooting. During manual operation, the settling time of flow controllers is important because it is expensive to have an operator wait for transients to settle out. A reasonable settling time is twice the stroke time of the actuator.

For outdoor air flow control loops and cooling based on the use of cool outdoor air (economizer cooling), overshooting may be a problem when the outdoor air temperature is below freezing. If the controls overshoot, then freeze protection devices may be activated, which will shut down an air-handling unit. The maximum allowable overshoot on an outdoor air flow control or economizer control loop depends on the outdoor air temperature. When the outdoor air temperature is well below freezing, overshoot may be intolerable.

Pressurization Control

For pressure control loops, overshoot can be dangerous. Overshooting pressurization controls can trip pressure relief devices or rupture ducts or pipes. The maximum allowable percent overshoot on a pressurization control loop depends on how close the loop is being controlled to the high limit. For example, if the high limit on a duct pressurization loop is 2.5 inches of water and the setpoint is 2 inches, then the maximum allowable percent overshoot for a safety factor of 5 is 5%. However, if the setpoint is 1 inch, then the maximum allowable percent overshoot for the same safety factor is 30%.

2.3.2 Performance Indices

It is often easier to specify control performance in terms of time-domain specifications because the physical meaning of the quantities such as percent overshoot and steady-state error are clear and are particularly relevant to certain control problems. However, it is often easier to design a control system to optimize a single index of the control system performance.

Three commonly used performance indices for designing single-loop control algorithms are the integrated absolute error (IAE), the integrated squared error (ISE), and the integrated time absolute error (ITAE). These indices are defined as

$$IAE = \int_0^\infty |e(t)|dt \tag{2.23}$$

$$ISE = \int_0^\infty e^2(t)dt \tag{2.24}$$

$$ITAE = \int_0^\infty t|e(t)|dt \tag{2.25}$$

These indices are rarely used to determine parameters of control loops in the field, but are commonly used to design automated tuning algorithms or adaptive control algorithms. They are also used as benchmarks for simulation of new control algorithms. For low-order linear systems, design relations that minimize one of these indices are available for certain types of linear SISO controllers.

2.4 Local-Loop Control

This section describes algorithms for modulating valves, dampers, and other devices to control temperature, pressure, or flow. The emphasis is on SISO systems and PID control algorithms.

2.4.1 Mathematical Models

Despite the fact that many of the components comprising an HVAC system are most accurately modeled as nonlinear distributed-parameter systems, low-order linear models are used to determine parameters for the local-loop controllers. Many of the most commonly used process transfer function models for designing the local-loop controllers can be represented by the following transfer function:

$$G(s) = \frac{Ke^{-sT}}{(s+p_1)(s+p_2)} \tag{2.26}$$

For example, one can get a first-order delayed transfer function from Equation 2.26 by letting $p_2 \to \infty$, or one can get a second-order Type 1 system (an integrating system) by letting $p_2 \to 0$. The control systems are typically designed based on the worst-case model parameters.

2.4.2 Feedback Control Algorithms

Figure 2.9 shows the basic structure of a feedback controller. The objective of the feedback controller is to maintain the process output at the setpoint. The feedback controller uses the error to determine the control signal. The control signal is used to adjust the manipulated variable.

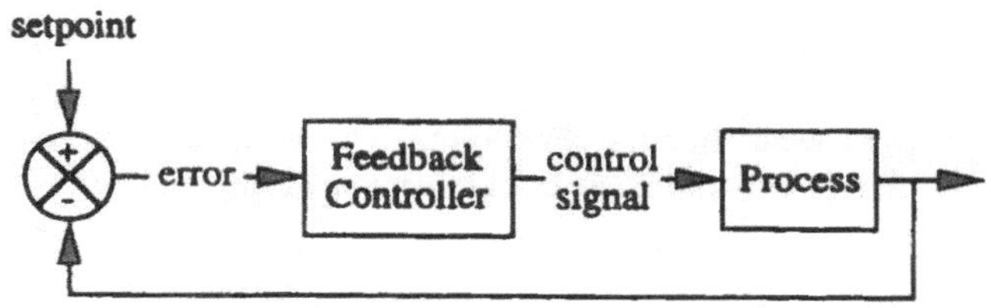

Figure 2.9 Structure of feedback controller.

According to [3], proportional-only (P) control is used in most pneumatic and older electric systems. The control signal for a P-controller is determined from

$$u(t) = \bar{u} + K_P e(t) \tag{2.27}$$

where $u(t)$ is the controller output at time t, $\bar{u}$ is a constant bias, K_P is the controller gain, and $e(t)$ is the error signal at time t.

Equation 2.27 describes the ideal behavior of a P-controller. In an actual application, the actuator usually has upper and lower limits (e.g., a valve has limits of completely open or closed). P-controllers exhibit a steady-state offset if the process is a Type 0 system (i.e., nonintegrating process) and if the controller output is not biased. The amount of steady-state offset can be reduced by setting the bias value equal to the expected nominal steady-state value of the controller output.

The steady-state offset of Type 0 systems can be eliminated by using a proportional plus integral (PI) controller. The time-domain representation of a PI controller is

$$u(t) = \bar{u} + K_P e(t) + K_I \int_0^t e(\tau)d\tau \tag{2.28}$$

where K_I is the integration gain.

There is a disadvantage of integral control action under certain conditions. If the actuator saturates and the error continues to be integrated, then the integral term will become very large and it will take a long time to bring the integral term back to a normal value. Integrating a sustained error while the actuator is saturated is commonly called integral windup or reset windup. Integral windup can occur during start-up, after a large setpoint change, or during a large disturbance in which the process output cannot be controlled to the setpoint. In [6, 9, 10], methods for reducing the effect of integral windup are described. The feature of reducing the effect of integrator windup is called anti-reset windup. An anti-reset windup strategy will significantly improve the performance of a controller with integral action.

For some systems, the performance of a PI controller can be improved by adding a derivative (D) term. The classical equation for a PID control algorithm is

$$u(t) = \bar{u} + K_P e(t) + K_I \int_0^t e(\tau)d\tau + K_D \frac{d}{dt} e(t) \quad (2.29)$$

where K_D is the derivative gain.

In commercial controllers, one does not use the classical form of the PID control algorithm because of a phenomenon known as derivative kick. If the classical PID controller is used, then the controller output goes through a large change following a step change in the controller output. This phenomenon is called derivative kick and it can be eliminated by using

$$u(t) = \bar{u} + K_P e(t) + K_I \int_0^t e(\tau)d\tau - K_D \frac{d}{dt} y(t) \quad (2.30)$$

Equation 2.30 determines the derivative contribution based on the process output rather than the error.

Derivative action generally improves the response of control loops in buildings. However, it is not commonly used in HVAC control loops because a three-parameter PID controller is more difficult to tune than a two-parameter PI controller and because derivative action makes the controller more sensitive to noise. Therefore, the additional improvement in the response time of the controller attributed to derivative action is generally outweighed by the additional cost of tuning and by the increased sensitivity to noise.

Figure 2.10 shows a block diagram for a practical feedback control algorithm in a digital controller. Next, we describe the purpose of the analog filter, noise spike filter, digital filter, and the deadzone nonlinearity.

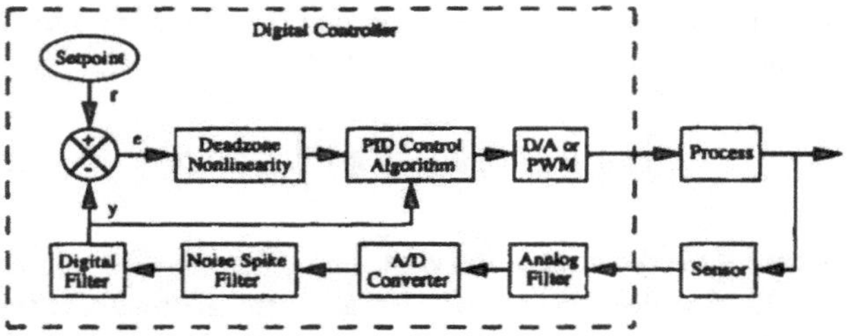

Figure 2.10 Structure of digital feedback controller.

Analog Filter

The purpose of the analog filter is to remove high frequency noise before sampling by the analog-to-digital (A/D) converter. This filter is commonly called an anti-aliasing filter or an analog prefilter. It is especially important to use an anti-aliasing filter when controlling flow or pressure in a duct or pipe because turbulence can generate a large amount of noise. The bandwidth of the analog filter should be selected based on the sampling period for the A/D converter. In the HVAC industry, a sampling period of 1 second is typical.

Noise Spike Filter

Noise spike filters improve the performance of digital control systems by removing outliers in the process output measurement. Outliers can be caused by a brief communication failure, instrument malfunction, electrical noise, or brief power failures. If noise spike filters are not used, then an outlier in the process output measurement will introduce a large disturbance into the control system. Also, if integration is used in the feedback controller, it may take a long time to recover from the disturbance. In [6, 9] simple noise spike filters are described.

Digital Filters

The sampling time for digital feedback controllers should be selected based on the dynamics of the control system. When controlling slow loops, such as the temperature in a room, a sampling period of 1 minute may be adequate for the feedback controller. However, the A/D converter may have a fixed sampling period of 1 second. For this case a combination of analog and digital filters should be used to remove aliases. The analog filter removes higher-frequency noise than the digital filter. Also, the cutoff frequency of the digital filter is adjusted in software. Thus, digital filters are especially useful for digital control algorithms that use an adjustable sampling period.

Deadzone Nonlinearity

The deadzone nonlinearity shown in Figure 2.11 is used to reduce actuator movement for small errors. The deadzone nonlinearity can be used to increase actuator life without sacrificing control performance. The input-output characteristic curve for the deadzone nonlinearity is shown in Figure 2.11. The output from the deadzone block can be determined from the following relationship:

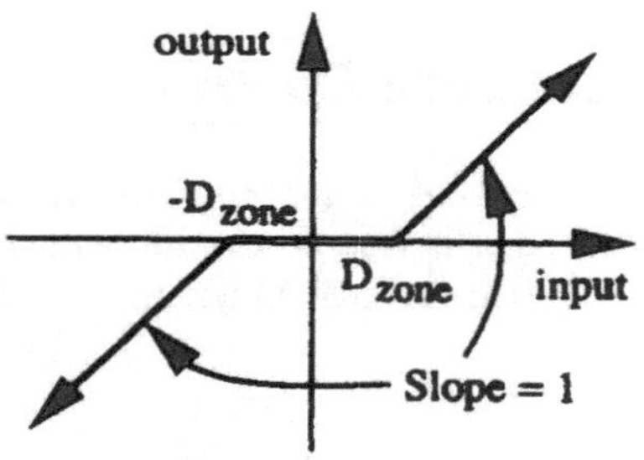

Figure 2.11 Input-output characteristic curve for a dead-zone nonlinearity.

$$\bar{e} = \begin{cases} e - D_{zone} & \text{when} \quad e > D_{zone} \\ 0 & \text{when} \quad |e| \leq D_{zone} \\ e + D_{zone} & \text{when} \quad e < -D_{zone} \end{cases} \quad (2.31)$$

where $\bar{e}$ is the modified error, D_{zone} is the size of the dead zone, and e is the error. The dead zone nonlinearity is especially useful when controlling processes with noisy signals, such as the flow through pressure-independent VAV boxes. The size of the dead zone should be selected based on the noise in the sampled signal.

2.4.3 Tuning

Tuning methods are used to determine appropriate values of controller parameters. Since the HVAC industry is cost driven, people installing and commissioning systems do not have a long time to tune systems. Consequently, a number of systems use the default parameters shipped with the controller. For some systems, the default parameters may not be appropriate. HVAC systems are nonlinear and time-varying. Thus, fixed-gain controllers may be poorly tuned some of the time even if they were properly tuned when they were installed. Control loops for HVAC systems may require retuning during the year for the processes that have nonlinear characteristics that are dependent upon the season. If a feedback control loop is not retuned, the control response may be "poor" (e.g., the closed-loop response for the controlled variable is oscillatory). An operator would have to retune the controller to maintain "good" control. In buildings, it is important to tune loops because properly tuned loops improve indoor air quality, decrease energy consumption, and increase equipment life.

Computer control systems can automate control loop tuning. One automated method for tuning is autotuning. Autotuning involves automatic tuning of a loop on command from an operator or a computer. With autotuning, the operator or control system must issue a new command to determine new control parameters. Since HVAC processes are nonlinear and time-varying, loops may require retuning after the system load changes. Another automated method for determining control parameters is adaptive control. With adaptive control, the control parameters are continually being updated or adjusted.

In the HVAC industry, people installing and commissioning control systems often have limited analytical skills. Thus, it is necessary that autotuning and adaptive control methods be user friendly. Also, the methods should be robust and have low computational and memory requirements.

Prior to tuning a system, the operator should make sure that all elements of the control system are working properly. The controller should be reading the signal for the process output, and the actuator should be working properly. If the control system uses an electric-to-pneumatic (E-P) converter to drive the actuator, then the zero and span of the E-P converter should be properly adjusted.

Several manual tuning and autotuning methods are based on the response to a step change in the controller output. Open-loop step tests should be run in the high-gain region of the process. Running the step test in the high-gain region helps ensure that the control system remains stable for other regions of operation. (The process gain is equal to the ratio of the steady-state change in the process output to the change in the controller output following a step change in the controller output). For systems with a large amount of hysteresis in the controlled device, the user should stroke the controlled device to reduce the effect of hysteresis on the step test. Figure 2.12 shows how to reduce the effect of hysteresis for a step change in the positive direction. Initially,

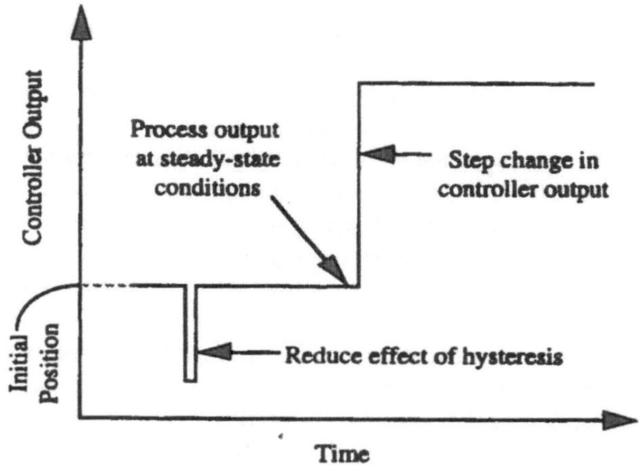

Figure 2.12 Controller output during a step test.

the controller output should move in a negative direction. Then, the controller output should move in the positive direction until it is back to the initial position. The step change in controller output should take place after the process output returns to a nearly steady-state condition. A similar procedure should be used for a step change in the negative direction.

2.5 Logical Control

Logical operations play a central role in the control of HVAC systems. Logical operations are used to ensure the safe start-up and shutdown of subsystems such as boilers, chillers, and air-handling units. They are also used to adjust the capacity of subsystems with discrete states and to sequence modulating controls.

2.5.1 Discrete-State Devices

Discrete-state devices are common in HVAC control systems. These devices may be mechanical devices, such as a pump, or electronic or software devices, such as an alarm.

It is common to use combinations of discrete-state devices to modulate the capacity of a subsystem. For example, condenser water temperature is often controlled by sequencing fans, which may have different capacities and different discrete speeds.

When sequencing discrete-state devices to modulate capacity there is a trade-off between the switching rate and the maximum deviation from the setpoint. The faster the switching rate, the smaller the deviations. This trade-off is affected by the capacitance or inertia of the system. If the system has a large capacitance, then the switching rate required to maintain the same maximum deviation from setpoint is lower.

2.5.2 Modulating Devices

Often the capacity of a single modulating device such as a control valve is insufficient to maintain control over the entire range of operating conditions. In such cases, multiple modulating devices are operated in sequence. A common example is the control of supply air temperature in an air-handling unit or similar subsystem through the sequencing of the cooling valve, outdoor air dampers, and heating valve. When it is sufficiently cool outdoors, the desired sequence involves modulating only one device at a time and placing the outdoor air damper control between the heating valve control and the cooling valve control. This sequence guarantees that the system does not heat and cool at the same time and that the relatively inexpensive cooling with outdoor air (called economizer control) is used before resorting to the relatively expensive cooling provided with the cooling valve.

A common method of sequencing modulating devices is to span the output of a single controller over the range of the modulating devices. For example, the heating valve, economizer, and cooling valve are often controlled as shown in Figure 2.13. While

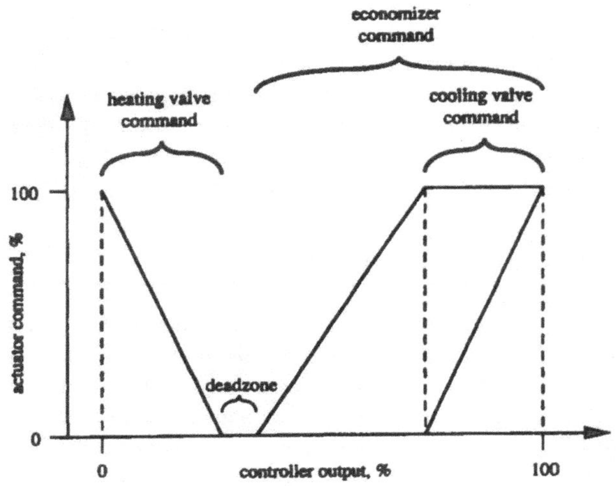

Figure 2.13 Sequencing control for heating, economizer and cooling by spanning the output of a single controller.

this method can result in a logically correct control sequence, it has a basic problem. It is difficult to tune a single fixed-gain PID controller for three different modes of operation (heating, cooling, and economizer) because the system dynamics in each mode are different. A way to get better control performance from a sequence of modulating devices is to use separate controllers for each device. While this method improves the control performance, it complicates the sequencing logic.

A problem with graphically representing sequencing logic as in Figure 2.13 is that the representation of the logic is incomplete. For example, the economizer may be disabled due to the outdoor air temperature, an override by the ventilation controls, or an override by the mixed-air temperature controls.

2.6 Energy-Efficient Control

Most of the local control loops in an HVAC system do not directly affect the zone temperatures. The purpose of these loops is to ensure the safe, reliable, and efficient operation of the subsystems. In this section, strategies for controlling such loops are described.

2.6.1 Economizer Cooling

The use of outdoor air for cooling is commonly referred to as economizer cooling or free cooling. In [11] simulations were used to estimate the energy savings with economizer cooling for two different building types in five different locations. Energy savings for cooling varied from 5 to 52%.

Figure 2.14 is a psychrometric chart that shows the different regions of operation for the air-handling unit as a function of the thermodynamic state of the outdoor air. The thermodynamic

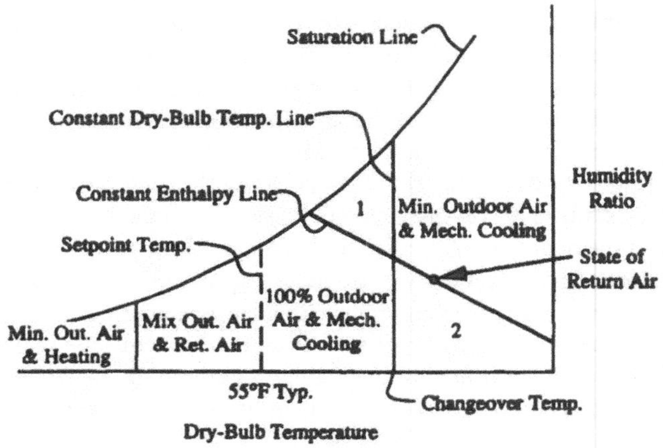

Figure 2.14 Psychrometric chart showing the decision boundaries for temperature-based and enthalpy-based economizer control.

state of the return air also is shown in Figure 2.14. There are two commonly used strategies for switching between 100% outdoor air and mechanical cooling to minimum outdoor air and mechanical cooling. One strategy compares outdoor air temperature with a changeover temperature (e.g., return air temperature), and the other strategy compares outdoor air enthalpy with a changeover enthalpy (e.g., return air enthalpy). With the temperature economizer strategy, the controls switch from minimum outdoor air to 100% outdoor air when the outdoor air temperature drops below the changeover temperature. With the enthalpy economizer cycle, the controls switch from minimum outdoor air to 100% outdoor air when the enthalpy of the outdoor air drops below the changeover enthalpy. For example, in region 1 of Figure 2.14, the temperature economizer cycle uses 100% outdoor air while the enthalpy economizer cycle uses the minimum amount of outdoor air. Also, in region 2 of Figure 2.14, the temperature economizer uses the minimum amount of outdoor air while the enthalpy economizer cycle uses 100% outdoor air.

Accurate humidity sensors should be used with the enthalpy economizer cycle because moderate sensor errors can result in significant energy penalties. Problems with low-cost humidity sensors are drift and slow recovery from saturated conditions.

It is often assumed that the energy required for cooling will always be less with the enthalpy economizer cycle. However, this is not true. In [11] it was shown that in dry climates the temperature economizer cycle may use less energy than the enthalpy economizer cycle.

2.6.2 Electrical Demand Limiting

Owners or operators of commercial buildings are commonly billed for electric power based upon energy consumed (i.e., kWh) and peak consumption over a demand interval. The rate structure of utilities varies considerably. For example, some utilities use a 30-minute demand interval and other utilities use a 5-minute demand interval.

Computer control systems can reduce the charges for peak demand by shutting off non-essential equipment during times of peak consumption. The strategy of shutting off the nonessential electric loads is commonly called demand limiting. Demand-limiting control strategies can reduce electric bills by 15 to 20% [12]. The building operator selects the desired electric demand target and enters a table of nonessential loads that can be turned off into the computer control system. Also, minimum and maximum times the load can remain off should be entered into the computer control system. A computer control system measures the electrical demand and determines the electrical loads to turn off to maintain the electrical consumption below the target.

In [13] an adaptive demand-limiting strategy with three important features is developed. First, the algorithm is easy to use because statistical methods automatically determine the characteristics of electric energy consumption for a particular building. Second, the algorithm controls the energy consumption just below the target level when there are enough nonessential loads to turn off. Third, the algorithm has low computational and memory requirements. Figure 2.15 shows the target energy consumption, actual energy consumption data with the demand-limiting algorithm, and estimated energy consumption without demand-limiting control from a manufacturing facility. Notice that the

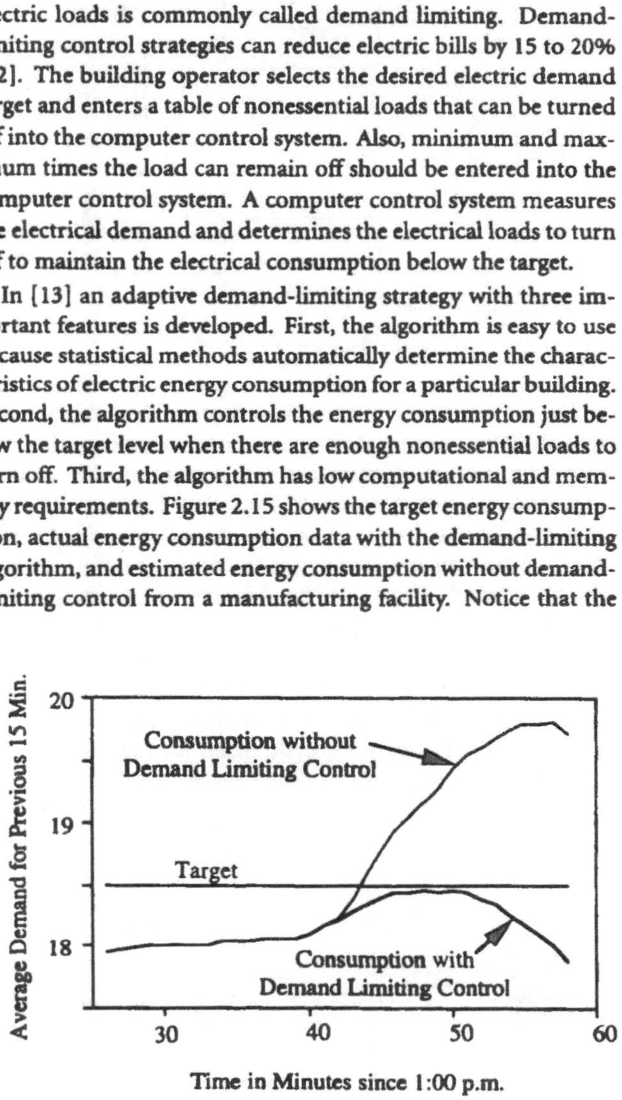

Figure 2.15 Electric consumption with and without demand limiting control.

demand stays just below the target. Figure 2.16 shows the calculated load to shed from the demand-limiting algorithm and the actual load shed. To limit the electrical demand below the

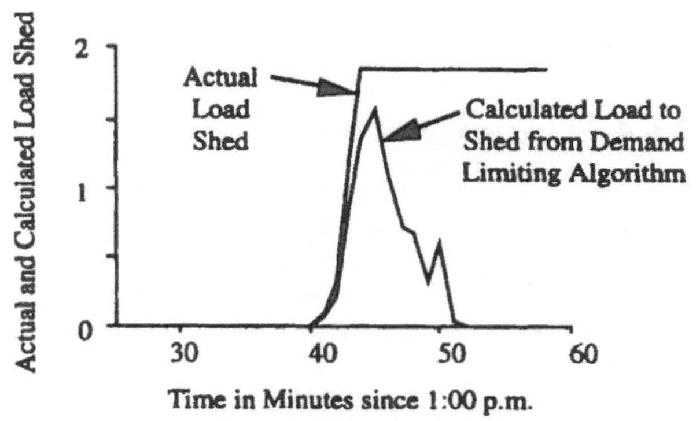

Figure 2.16 Actual amount of load shed and desired amount to shed.

target level, the actual load shed is larger than the calculated load to shed from the demand limiting algorithm. Also, the actual load shed remains larger than the calculated load to shed from the demand limiting algorithm because of a minimum off time for the sheddable loads.

2.6.3 Predicting Return Time from Night or Weekend Setback

For buildings that are not continuously occupied, energy costs can be reduced by adjusting the setpoint temperatures during unoccupied times. One common strategy for adjusting setpoint temperatures is called night setback. Night setback involves lowering the setpoint temperature for heating and raising the setpoint temperature for cooling. In [14] it is shown that night setback can result in energy savings of 12% for heavyweight buildings and 34% for lightweight buildings.

When using a night setback strategy, space conditions need to be brought to comfortable conditions prior to the time when occupants return. Ideally, the space conditions would just be in the comfortable range as the building occupants return. In [15] seven different methods for predicting return time from night or weekend setback were compared. For nights that required cooling to return, the following equation can be used to predict return time

$$\hat{\tau} = a_0 + a_1 T_{r,i} + a_2 T_{r,i}^2 \qquad (2.32)$$

where $\hat{\tau}$ is the estimate of the return time from night or weekend setback, a_i is a coefficient, and $T_{r,i}$ is the initial room temperature at the beginning of the return period.

For nights that require heating, the return time can be estimated from

$$\hat{\tau} = a_0 + (1 - w)(a_1 T_{r,i} + a_2 T_{r,i}^2) + w a_3 T_a \qquad (2.33)$$

$$w = 1000^{-\frac{T_{r,i} - R_{h,u}}{R_{h,o} - R_{h,u}}} \qquad (2.34)$$

where $R_{h,u}$ is the setpoint temperature for heating during unoccupied times, $R_{h,o}$ is the setpoint temperature for heating during occupied times, and T_a is the outdoor air temperature at the beginning of return from night setback. Recursive linear least squares with exponential forgetting is used to determine the coefficients in Equations 2.32 and 2.33.

2.6.4 Thermal Storage

Thermal storage systems use electrical energy to cool some storage media at off-peak hours, then use the cooling capacity of the media during on-peak hours. Storage systems are used for some of the same reasons as the electrical demand-limiting algorithm described previously; they allow the electrical demand at peak hours to be reduced. They also allow the chiller load to be shifted to times when electrical energy is inexpensive. Thermal storage systems may also allow one to use lower-capacity chillers.

The two most common storage media are chilled water and ice. Ice storage tanks are becoming more popular because the energy density of ice is much greater than that of water.

Water and Ice Storage Control Strategies

The simplest control strategy for water or ice storage systems is chiller priority control. With chiller priority control, the cooling load is matched by the chiller until the load exceeds the chiller capacity. Then the storage media is used to augment the chiller capacity. During the off-peak hours, the storage media is fully recharged. Chiller priority control is sometimes referred to as demand-limiting control. Chiller priority control is simple to implement, but it is substantially suboptimal except when cooling loads are high.

Storage priority control strategies make better use of the storage capacity than chiller priority control. The objective of storage priority control is to use mainly stored energy during the on-peak period when energy is expensive and to use all of the stored energy by the end of the on-peak period. An example of storage priority control is load-limiting control [16]. During the off-peak unoccupied hours, the storage media is charged as much as possible. During the on-peak unoccupied hours, the building is cooled using chiller priority control. During the on-peak hours, the chiller is run at a constant capacity such that the storage media is completely used by the end of the on-peak period. Storage priority control strategies such as load-limiting control require a forecast of the cooling load for the on-peak period.

Optimal control of thermal storage systems requires determining a sequence of control commands for charging and discharging the storage media such that the total cost of supplying chilled water is minimized subject to operating constraints. To determine the truly optimal sequence requires perfect knowledge of the behavior of the process and perfect knowledge of future cooling loads. Even with perfect knowledge of the process behavior and of future cooling loads, determining the optimal sequence is computationally intensive (e.g., requires solving a dynamic programming problem). Consequently, most of the practical work on optimal control of thermal storage systems involves the development of heuristic strategies that are nearly optimal. For example, a nearly optimal strategy for the control of ice storage systems is developed in [17].

2.6.5 Setpoint Resetting

One way to improve the operational efficiency of an HVAC system is to make adjustments to the setpoints of the local-loop controllers. This is commonly called setpoint resetting. In this section, two common setpoint resetting strategies are described.

Chiller Water Temperature

Figure 2.17 shows a schematic diagram of a chiller. The

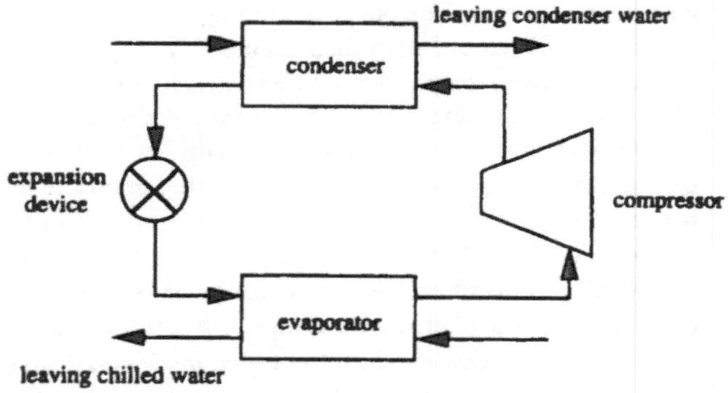

Figure 2.17 Schematic diagram of a chiller.

chiller power consumption is a function of the refrigerant flow rate and the pressure difference across the compressor. The larger the pressure difference across the compressor, the larger the chiller power consumption. The temperature difference between the water leaving the condenser and the water leaving the evaporator is strongly correlated with the pressure difference across the compressor. Therefore, the chiller power can be reduced if the leaving water temperature difference can be reduced.

One way to reduce the leaving water temperature difference is to reset the chilled water supply setpoint to a higher value whenever possible. There are numerous strategies for doing this. One is to control the position of the most-open valve in the chilled water distribution circuit to a nearly open position as shown in Figure 2.18 and described in [18].

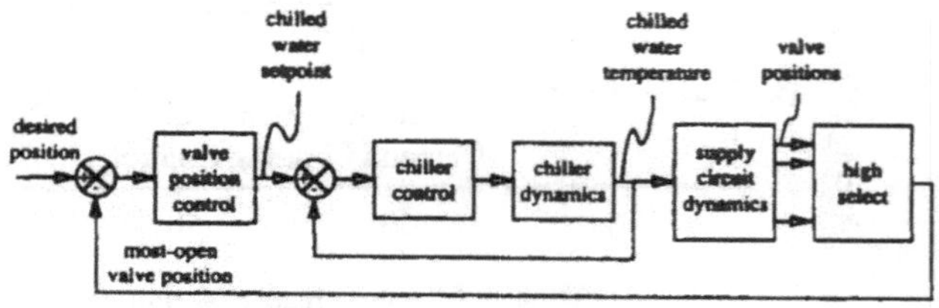

Figure 2.18 Block diagram of chilled water reset control.

An advantage of this strategy over others is that it allows the chilled water temperature to go as high as possible without forcing a cooling valve into saturation. A potential disadvantage is that, if the controller is not properly tuned, the chiller controls are sensitive to disturbances on a single cooling coil. Therefore, the controller should be tuned so that the bandwidth of the resetting loop is much smaller than the bandwidth of the slowest cooling coil control loop.

Duct Static Pressure

In VAV systems, the most common method of modulating the total supply air flow rate is to modulate the supply fan so as to regulate the static pressure at some point in the supply air duct. The operator must select a static pressure setpoint. If the setpoint is too low, one or more terminal units will open completely and still not be able to deliver the required quantity of supply air. If the setpoint is too high, the fan power consumption will be greater than necessary to supply air to the terminal units. Therefore, there is an optimal static pressure that minimizes the fan power yet supplies the necessary amount of air to all of the terminal units.

One method to control the fan efficiently is to reset the static pressure setpoint based on the position of the terminal unit that is open the most. In [19], a strategy is described in which the setpoint is increased by a fixed amount when one or more of the terminal units is saturated fully open and decreased by a fixed amount when one or none of the terminal units is saturated. It is reported that this strategy is able to reduce the required fan power by 19 to 42% over a fixed-setpoint control strategy.

2.7 Configuring and Programming

There are several different ways that digital HVAC controllers are configured or programmed. In this section, three common methods are described.

2.7.1 Textual Programming

Perhaps the oldest approach to configuring and programming controllers is textual programming. For example, BASIC, FORTRAN, or C may be used to program logical controls, local-loop controls, and supervisory controls and diagnostics. The advantage of using textual programming is that it allows the programmer great flexibility in constructing the control software. Textual programming requires substantial programming expertise. Since many controllers must be custom-programmed, the expertise required for textual programming often makes the cost of using textual programming prohibitive.

2.7.2 Graphical Programming

Control algorithms and logic may be programmed graphically through the construction of diagrams. One common type of control diagram is the ladder logic diagram. Originally, ladder logic diagrams were designed to resemble electrical relay circuits and, therefore, could perform only limited types of operations. Today, additional functions can be included in ladder logic diagrams so that the controllers can perform functions such as arithmetic and data manipulation operations.

Another common method of graphical programming involves the construction of block diagrams. Function blocks are connected by lines or wires. The blocks may perform basic logical operations or more complex functions (e.g., implement PID control algorithms). Also, the programmer may construct new blocks for a specialized purpose.

The advantage of using graphical programming is that the controller can be configured and programmed for a nonstandard system or subsystem. Specialized control functions can often be constructed quickly, and the required level of programming expertise may be lower than other programming methods.

2.7.3 Question-and-Answer Sessions

Many HVAC subsystems are designed in such a way that standardized controls can be used. In such cases, the systems are preprogrammed. To be made fully operational, only information about the size or type of subsystem and the desired control strategy need be supplied. Therefore, these types of controllers may be configured through a simple question-and-answer session. Control strategies such as those described in Section 2.6 may be preprogrammed for a variety of different equipment and can be selected to match the particular system.

The obvious disadvantage of using preprogrammed controls is that they are inflexible. The user cannot make changes to the way in which the controller operates. However, there are significant advantages to preprogrammed controls. One advantage is that it reduces the commissioning time. Furthermore, it does not require programming expertise. Another advantage is that preprogrammed controls can be tested in a laboratory so that they are guaranteed to be safe, reliable, and efficient. If the controls are custom-programmed for a specific system, rigorous testing cannot be done.

References

[1] Hirst, E., Clinton, J., Geller, H., and Kroner, W., *Energy Efficiency in Buildings: Progress and Promise*, O'Hara, F.M., Jr., Ed., American Council for an Energy Efficient Economy, 1986, 3.

[2] Environmental Protection Agency, *Unfinished Business: A Comparative Assessment of Environmental Problems*, Vol. 1, February 1987.

[3] Haines, R. W. and Hittle, D.C.,*Control Systems for Heating, Ventilating, and Air Conditioning*, Van Nostrand Reinhold, New York, 1993.

[4] Incropera, F. P. and DeWitt, David P., *Fundamentals of Heat and Mass Transfer*, Wiley, New York, 1990.

[5] Instrument Society of America, *ISA Handbook of Control Valves*, Pittsburgh, 1976.

[6] Seborg, D. E., Edgar, T.F., and Mellichamp, D.A., *Process Dynamics and Control*, Wiley, New York, 1989.

[7] White, F. M., *Fluid Mechanics*, McGraw-Hill, New York, 1979.

[8] *1993 ASHRAE Handbook: Fundamentals*, ASHRAE, Atlanta, 1993.

[9] Clarke, D. W., PID algorithms and their computer implementation, Oxford University Engineering Laboratory Report No. 1482/83, Oxford University, 1983.

[10] Shinskey, F. G., *Process Control Systems: Application, Design, and Tuning*, McGraw-Hill, New York, 1988.

[11] Spitler, J. D., Hittle, D.C., Johnson, D.L., and Pedersen, C.O., A comparative study of the performance of temperature-based and enthalpy-based economy cycles, *ASHRAE Trans.*, 93(2), 13–22, 1987.

[12] Stein, B., Reynolds, J.S., and McGuinness, W.J., *Mechanical and Electrical Equipment for Buildings*, 7th ed., Wiley, New York, 1986.

[13] Seem, J. E., Adaptive demand limiting control using load shedding, *Int. J. Heat., Vent., Air-Cond. Refrig. Res.*, 1(1), 21–34, January 1995.

[14] Bloomfield, D. P. and Fisk, D.J., The optimization of intermittent heating, *Buildings and Environment*, Vol. 12, 1997, 43–55.

[15] Seem, J. E., Armstrong, P.R., and Hancock, C.E., Algorithms for predicting recovery time from night setback, *ASHRAE Trans.*, 95(2), 1989.

[16] Braun, J. E., A comparison of chiller-priority, storage-priority, and optimal control of an ice-storage system," *ASHRAE Trans.*, 98(1), 1992.

[17] Drees, K., *Modeling and Control of Area Constrained Ice Storage Systems*, Master's thesis, Mechanical Engineering Department, Purdue University, 1994.

[18] *1991 ASHRAE Handbook: HVAC Applications*, ASHRAE, Atlanta, 1991, chap. 41.

[19] Lorenzetti, D. M., and Norford, L.K., Pressure setpoint control of adjustable speed fans, *J. Sol. Energy Eng.*, 116, 158–163, August 1994.

3

Control of pH

3.1 Introduction 29
3.2 The pH Measuring System 29
The pH Scale • Electrodes
3.3 Process Titration Curves 30
The Strong Acid-Strong Base Curve • Distance from Target • Weak Acids and Bases
3.4 Designing a Controllable Process 33
A Dynamic Model of Mixing • Vessel Design • Effective Backmixing • Reagent Delivery Systems • Protection Against Failure
3.5 Control System Design 37
Valve or Pump Selection • Linear and Nonlinear Controllers • Controller Tuning Procedures • Feedforward pH Control • Batch pH Control
3.6 Defining Terms 40
References 41

F. Greg Shinskey
Process Control Consultant, North Sandwich, NH

3.1 Introduction

The pH loop has been generally recognized as the most difficult single loop in process control, for many reasons. First, the response of pH to reagent addition tends to be nonlinear in the extreme. Second, the sensitivity of pH to reagent addition in the vicinity of the set point also tends to be extreme, in that a change of one pH unit can result from a fraction of a percent change in addition. Thirdly, the two relationships above are often subject to change, especially when treating wastewater. And finally, reagent flow requirements may vary over a range of 1000:1 or more, especially when treating wastewater.

As a result of these unusual characteristics, many, if not most pH control loops are unsatisfactory, either limit-cycling or slow to respond to upsets or both. Considerable effort and ingenuity have gone into designing advanced controls, nonlinear, feedforward, and adaptive, to solve these problems. Not enough has gone into designing the neutralization process to be controllable. Therefore, after describing the pH control problem and before developing control-system design guidelines, process design is covered in substantial detail. If the process is designed to be controllable, the control system can be more effective and simpler, and, therefore easier to operate and maintain.

3.2 The pH Measuring System

Most of the unique characteristics of the pH control loop center around its measuring system with its unusual combination of high sensitivity and wide range. The fundamentals of the measuring system are firm and not likely to change with time or technology. However, the methods and devices used to measure pH can be expected to continue to improve, as obstacles are overcome. Still, failures can occur, and recognizing the possible failure modes is important in anticipating the response of the controls to them.

3.2.1 The pH Scale

The pH measuring system is expressly designed to report the activity of hydrogen ions in an aqueous solution. (While pH measurements can be made in some nonaqueous media such as methanol, the results are not equivalent to those obtained in water and require independent study.) The true *concentration* of the hydrogen ions may differ from its measured activity at pH levels below 2, where ion mobility may be impeded. Most pH loops have control points in the pH 2–12 range, however, where activity and concentration are essentially identical. Consequently, the pH measurement will be used herein as an indication of the concentration of hydrogen ions $[H^+]$ in solution.

That concentration is expressed in terms of gram-ions of hydrogen per liter of solution. A *Normal* solution will contain one gram-ion per liter. Laboratory solutions are typically prepared using the Normal scale, e.g., 1.0 N, 0.01 N, etc., representing the number of replaceable gram-ions of hydrogen or hydroxyl ions per liter.

When a pH measuring electrode is placed in a solution with a reference electrode and a liquid junction between them, a potential difference E may be measured between the two electrodes:

0-8493-0054-3/00/$0.00+$.50

$$E = 59.16 \log[H^+] - E_{ref} - E_j \tag{3.1}$$

where E_{ref} is the potential of the reference cell and E_j that of the liquid junction. The commonly accepted method of expressing base-10 logarithms is to use p to indicate the negative power. Thus, $pH = -\log[H^+]$ and $[H^+] = 10^{-pH}$. The potential developed by the pH electrode is therefore 59.16 millivolts per pH unit.

Actually, this coefficient of 59.16 is precise only at 25°C, varying with $T/298$ where T is the absolute temperature in Kelvin. The reference cell is designed to match the internal cell of the pH electrode, buffered to pH 7. Therefore at pH 7, there is no potential difference between the measuring and reference electrodes, and temperature has no effect there, the *isopotential* point. Temperature compensation may be applied either automatically or manually, as desired.

The liquid-junction potential represents any voltage developed by diffusion of ions across the junction between the solution being measured and that of the reference electrode. Potassium chloride is selected for the reference solution, because its ions have approximately the same ionic mobility, thereby minimizing the junction potential. For most solutions, E_j is less than 1 mV, and is not considered a major source of error.

The net result of the above is a linear scale of pH vs. millivolts, with zero voltage corresponding to pH 7. The significance of pH 7 is that it represents the neutral point for water, where hydrogen and hydroxyl ions are equal. The two ions are related via the equilibrium constant, which is 10^{-14} at 25°C:

$$[H^+][OH^-] = 10^{-14}. \tag{3.2}$$

The equilibrium constant is also a function of temperature, causing a given solution to change pH with temperature [1]. This effect is only significant above pH 6, and is not normally included in the temperature compensation applied to the electrodes.

3.2.2 Electrodes

The pH measuring electrode in most common use is the glass electrode. It consists of a glass bulb containing a solution buffered to pH 7, with an internal reference cell identical to that in the reference electrode. This combination produces the null voltage at pH 7. The sensitive portion of the electrode is a thin glass membrane saturated with water. The membrane must be thin enough to keep its electrical impedance in a reasonable range and yet resist breakage under normal use. The impedance of the glass can exceed 100 megohms, requiring extremely high-impedance voltage-measuring devices to produce accurate results. In view of these special requirements, ordinary voltmeters and wiring cannot be used to measure pH, as circuits are very sensitive to electrical leakage and grounding.

The reference electrode forms the other connection between the solution to be measured and the voltage measuring device. It must have a stable voltage (where that of the measuring electrode varies with solution pH) and a low impedance.

The extremely high impedance of the glass electrode makes it susceptible to *short-circuiting*. Short-circuiting reduces the measured millivolts relative to the true pH of the solution, producing pH readings closer to 7 than actually the case. This is an unfortunate failure mode, for 7 is the most common set point, especially in wastewater treatment, and will not cause alarm. Rovira [2] describes a microprocessor-based pH transmitter with the capability of checking the impedance of the glass electrode and thereby warning of this type of failure.

There is only one adjustment available to calibrate or *standardize* a pH measuring system. If the electrodes are in good working order, calibration against one buffer should produce accurate results against other buffers without further adjustment. If it does not, then the millivoltage change per pH unit is not 59.16 at 25°C, indicating that the glass electrode is damaged or that there is an electrical leak.

Coating of the pH electrode introduces a time lag between the ions in the process solution and those at the surface of the electrode. Coating can be verified by the appearance of the electrodes, but be aware that a film as thin as a millimeter could produce a time constant of several minutes. McMillan [1] recommends increasing the velocity of the process fluid past the electrodes to 7 ft/sec. to minimize fouling, although velocities exceeding 10 ft/sec. may cause excessive noise and erosion.

3.3 Process Titration Curves

What we recognize as the nonlinearity in pH control loops is the process *titration curve*, the relationship between measured pH and the amount of acid or base reagent added to a solution. The reagents are usually strong acids or bases of known concentration added to a solution that contains variable concentrations of possibly unknown substances. Laboratory titrations are conducted batchwise by adding reagent incrementally to a measured volume of process solution. In a process plant, pH control can be applied to a batch of solution, but more often is applied to a flowing stream, matching reagent flow to it. In a continuous operation, only that portion of the titration curve near the set point may be visible, but the slope of the curve in this region determines the steady-state process gain, and the amount of reagent required to reach that point determines the process load. Therefore, titration curves are as important in continuous processing as in batch.

Laboratory titrations are usually performed using standard solutions prepared to 0.01 N or similar concentration. Process reagents are usually much more concentrated, but their concentration must be known in the same terms to estimate required delivery based on laboratory titrations of process samples. The normality of 32% HCl is 10.17 N, that of 98% H_2SO_4 is 36.0 N, that of 25% NaOH is 7.93 N, and that of 10% $Ca(OH)_2$ is 2.86 N. For other solutions, normality can be calculated by multiplying the weight percent concentration of the reagent by $10n\rho/M$, where n is the number of replaceable H^+ or OH^- ions per molecule, ρ is its density in g/ml, and M is its molecular weight.

3.3.1 The Strong Acid-Strong Base Curve

A "strong " agent is one which ionizes completely in solution. Examples of strong acids are HCl and HNO_3, and of bases are the alkalies NaOH and KOH. Surprisingly, H_2SO_4 and HF are not classified as strong acids, despite their tendency to dehydrate and etch glass, respectively. A common basic reagent, $Ca(OH)_2$, similarly is not strong, and also has limited solubility.

Consider a reaction between NaOH and HCl:

$$HCl + NaOH + H_2O \rightarrow H^+ + OH^- + Na^+ + Cl^- \quad (3.3)$$

The sum of the negative and positive charges must be equal:

$$[Na^+] + [H^+] = [Cl^-] + [OH^-] \quad (3.4)$$

With both reagents being completely ionized, the concentration of the acid x_A can be substituted for $[Cl^-]$ and the concentration of the base x_B for $[Na^+]$:

$$x_B - x_A = [OH^-] - [H^+]. \quad (3.5)$$

Due to the equilibrium of hydrogen and hydroxyl ions in water, the former may be substituted for the latter, using Equation 3.2:

$$x_B - x_A = 10^{-14}/[H^+] - [H^+]. \quad (3.6)$$

Placed in terms of pH, the relationship becomes

$$x_B - x_A = 10^{pH-14} - 10^{-pH}. \quad (3.7)$$

This is the fundamental pH relationship in the absence of buffering. Although buffers are common to a limited extent in most solutions, Equation 3.7 describes the most severe titration curve that a control system must handle. It is shown for different scales of normality in Figure 3.1. Although the curves appear to have different breakpoints and therefore to be different curves, they are, in fact, a single curve shown at two different ranges of sensitivity.

Both curves have the same characteristic in that a base-acid mismatch of one-tenth of fullscale leaves the pH only one unit from the end of the curve. Similarly, a mismatch of 1/100 of full scale leaves the pH two units from the end of the curve. This logarithmic relationship holds everywhere that one of the terms on the right of Equation 3.7 is negligible compared to the other, i.e., everywhere except in the range of pH 6–8.

3.3.2 Distance from Target

The essence of the pH-control problem is the matching of acid and base at neutrality. In the strong acid-strong base system, neutrality occurs at pH 7, and therefore a perfect match will bring this result. In some process work, pH must be controlled within ±0.1 unit, but wastewater limits are usually pH 6–9, a much broader target. Consider how difficult it can be to control wastewater pH in the absence of buffering. If the wastewater enters at pH 3, as in the inner curve of Figure 3.1, a mismatch between acid and base of 10% will leave the pH at 4 or 10; a mismatch of 1.0% will leave it at 5 or 9; a mismatch of 0.1% is required to produce a pH in the 6–8 range.

The size of the target does not change with the scale, but the size of the reagent valve does. For example, if the wastewater were to enter one pH unit farther away from neutral, ten times as much reagent would be required to neutralize it. The accuracy of delivery must then be ten times as great percentagewise to control to the same target. For example, a solution represented by the outer curve in Figure 3.1 would require a percentage accuracy in reagent delivery 100 times as great as for the inner curve.

A graphic illustration of the problem can be envisioned by comparing the size of the target to our distance from it. Think of the range of pH 6–8 as a target one foot in diameter, $x_B - x_A$ being essentially $\pm 10^{-6}$ at the edge. If the entering solution is at pH 5 or 9, the distance to the target is ten times its radius or five feet away. From this distance, the target could be easily hit by throwing darts at it. If the entering pH were 3 or 11, the same target would be 500 feet away, requiring an expert with a high-powered rifle. If the pH entering were 1 or 13, the same target would be ten miles away, requiring a guided missile.

To maximize the accuracy of reagent delivery, *dead band* in the control valves must be eliminated through the use of *valve positioners*. Dead band has been observed to cause limit-cycling in control of pH and reduction-oxidation potential, which was eliminated by installing positioners on the valves.

3.3.3 Weak Acids and Bases

Most acids and bases do not completely ionize in solution, but establish an equilibrium between the concentration of the undissociated molecule and its ions. A *monoprotic* agent is one that has a single ionizable hydrogen or hydroxyl group. An example is acetic acid, which dissociates into hydrogen and acetate ions in accordance with the following relationship:

$$[H^+][Ac^-] = K_A[HAc] \quad (3.8)$$

where K_A is the ionization constant for acetic acid at $10^{-4.75}$. A pH electrode senses only the $[H^+]$, leaving the undissociated [HAc] unmeasured. However, any adjustment to solution pH made by adding either base or another acid will shift the equilibrium. For example, increasing $[H^+]$ with a strong acid will convert some of the $[Ac^-]$ into [HAc], leaving more of it unmeasured. Conversely, reducing $[H^+]$ with a base will shift the equilibrium the other way, ionizing more of the [HAc]. This behavior tends to reduce the change in pH to additions of acids and bases, and is called buffering.

Adding acetic acid to the former solution of HCl and NaOH adds acetate ions to the charge balance:

$$[H^+] + [Na^+] = [OH^-] + [Cl^-] + [Ac^-]. \quad (3.9)$$

Let the total concentration of acetic acid be represented by

$$x_{Ac} = [HAc] + [Ac^-] \quad (3.10)$$

The [HAc] in the above expression can be replaced by its value from Equation 3.8:

$$x_{Ac} = [Ac^-](1 + [H^+]/K_A). \quad (3.11)$$

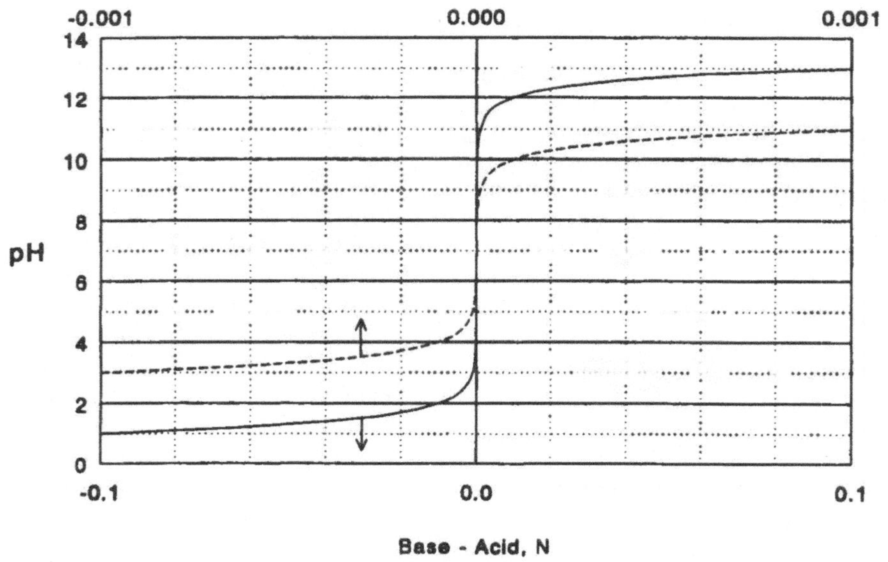

Figure 3.1 Titration curve for strong base against strong acid, two scales.

Replacing all of the ions in the charge balance with equivalent values of the three ingredients, and [H$^+$] with pH, gives

$$x_B - x_A = 10^{\text{pH}-14} - 10^{-\text{pH}} + \frac{x_{Ac}}{1 + 10^{\text{p}K_A - \text{pH}}} \tag{3.12}$$

where pK_A (like pH) is the negative base-10 logarithm of the ionization constant.

The titration curve for a monoprotic weak base is derived in a fashion similar to that of the acid. As Equation 3.12 simply had the weak-acid term added to the strong acid-strong base relationship of Equation 3.7, so the following weak-base term may be added in the same way:

$$\cdots - \frac{x_{Bw}}{1 + 10^{\text{p}K_B + \text{pH} - 14}} \tag{3.13}$$

where x_{Bw} represents the concentration of weak base whose ionization constant has the negative logarithm pK_B. Ammonia is a weak base having a pK_B of 4.75, identical to the pK_A value for acetic acid.

Most of the weak acids and bases encountered in industry have two ions, and are hence classified as *diprotic*. A very common diprotic weak acid is carbon dioxide, encountered in groundwater in the form of carbonate and bicarbonate ions. Dissociation of CO_2 and water into H^+ and HCO_3^- ions has a pK_{A1} value of 6.35, so that carbonic acid is weaker than acetic acid by a factor of about 40. A second equilibrium takes place between HCO_3^- ions and CO_3^{2-} ions with a pK_{A2} value of 10.25. Because this second value exceeds 7, the carbonate ion effectively acts as a weak base having a pK_B value of 14 – 10.25 or 3.75.

Derivation of the equations representing the titration curves for diprotic agents follows the same procedures used for monoprotic agents. The result is adding to Equation 3.7 the following for weak diprotic acids:

$$\cdots + \frac{x_{Aw}(1 + 0.5 \times 10^{\text{p}K_{A2} - \text{pH}})}{1 + 10^{\text{p}K_{A2} - \text{pH}}(1 + 10^{\text{p}K_{A1} - \text{pH}})} \tag{3.14}$$

where x_{Aw} represents the concentration of the weak acid.

Figure 3.2 is a titration curve for water containing 0.001 N CO_2, equivalent to a total alkalinity of 50 ppm as $CaCO_3$, the customary method of reporting water hardness. On the larger scale of ±0.01 N, the bicarbonate shows its presence with a slight moderation of the titration curve around pH 6.35. This makes the pH of wastewaters easier to control than it would be otherwise, and most water does contain significant amounts of bicarbonate ions. The second breakpoint associated with the carbonate ionization is not visible on this scale.

On the smaller scale of ±0.001 N, the bicarbonate buffering is even more pronounced. This characteristic makes bicarbonate useful for neutralizing stomach acids. The neutral point for this solution is pH 8, however only half as much strong base as acid is required to reach the neutral point. The reason for this is that the value of pK_{A2} exceeds 7, resulting in carbonates being weak bases. Na_2CO_3 is a useful reagent for neutralizing acids, because it produces a titration curve well-buffered in the region of pH 6–7; unfortunately, twice as much sodium is required per mol of acid than if NaOH were used, making it too costly for most waste treatment applications.

Carbonates are common in cleaning solutions, and therefore in wastewater, but their concentration may fluctuate considerably, as most cleaning is done batchwise. Other cleaning agents such as phosphates and silicates, ammonia and amines, and soaps and detergents of all kinds are weak agents. Consequently, the degree of buffering in industrial wastewater may be expected to be quite variable.

The term added to the pH relationship for weak diprotic bases is similar to that for acids:

$$\cdots - \frac{x_{Bw}(1 + 0.5 \times 10^{\text{p}K_{B2} + \text{pH} - 14})}{1 + 10^{\text{p}K_{B2} + \text{pH} - 14}(1 + 10^{\text{p}K_{B1} + \text{pH} - 14})}. \tag{3.15}$$

A list of ionization constants for weak acids and bases is given in Table 3.1. For a more complete list, consult Weast [7].

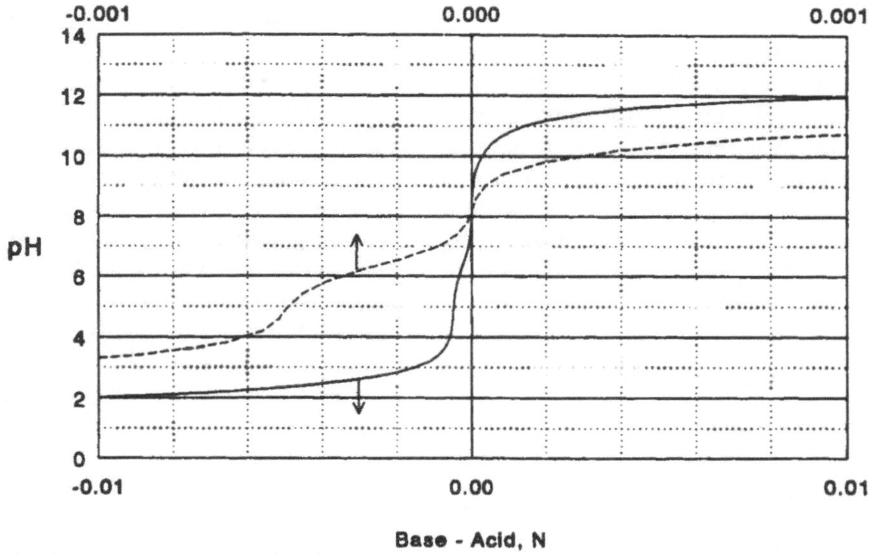

Figure 3.2 Titration of 0.001 N CO_2 by strong base on two scales.

TABLE 3.1 Ionization Constants for Common Acids and Bases.

Acid	pK_{A1}	pK_{A2}
Acetic acid	4.75	
Boric acid	4.7	9.1
Carbon dioxide	6.35	10.25
Chromic acid	0.7	6.2
Ferric ion	2.5	4.7
Ferrous ion	6.8	
Formic acid	3.65	
Hydrogen fluoride	3.17	
Hydrogen sulfide	7.0	12.9
Hypochlorous acid	7.5	
Oxalic acid	1.1	4.0
Phosphoric acid	2.23	7.21
Sulfuric acid	−3	1.99
Sulfur dioxide	1.8	6.8
Base	pK_{B1}	pK_{B2}
Aluminate ion	1.6	
Ammonia	4.75	
Barium hydroxide	0.7	
Calcium hydroxide	1.40	2.43
Cupric hydroxide	7.2	
Ethylamine	3.3	
Hydrazine	5.5	
Hydroxylamine	7.97	
Magnesium hydroxide	2.6	
Methylamine	3.28	
Phenol	4.2	
Silicon hydroxide	1.3	4.4
Urea	13.9	
Zinc hydroxide	3.02	

Any of the agents in the table could be listed either as an acid or base. For example, the ferric ion is listed as an acid with pK_A values of 2.5 and 4.7; it could have been listed as ferric hydroxide with pK_B values complementary at 9.3 and 11.5, the effect is the same. Most metal ions are weak agents, producing titration curves with multiple breakpoints like the inner curve of Figure 3.2. Therefore, effluents from metal treating operations, like pickling and plating, tend to be heavily buffered at all times, and their pH is generally easy to control.

While it is possible to create titration curves from the equations and pK values above, it is strongly recommended that several samples be taken at different times of day from the stream whose pH is to be controlled and titrated to identify how extreme the curves are and how much they may vary. This information will be useful in sizing valves and reducing the likelihood of surprises when the control system is commissioned.

3.4 Designing a Controllable Process

The extreme gain changes presented by most titration curves, and their variablilty, pose special problems for the pH control system. The typically very high gain of the titration curve in the region about set point is its most demanding feature, causing most pH loops to cycle about the set point. Uniform cycling develops when the gain of the control loop is unity. Loop gain is the product of all of the gains in the loop: process and controller, steady-state and dynamic. The high steady-state gain of the typical titration curve requires a proportionately low controller gain if cycling is to be avoided. But a low controller gain means poor response to disturbances in the flow and composition of the stream whose pH is being controlled.

To maximize the controller gain, the process gains need to be minimized. The steady-state gain terms for the process consist of the titration curve converting concentration to pH, the reagent valve converting percent controller output to reagent concentration, and the pH transmitter converting pH to percent controller input. Other than selecting a buffering reagent such as Na_2CO_3 or CO_2, which may not be feasible, we must accept the process titration curve as it exists. Similarly, the valve size is determined by process needs, as is the transmitter span.

But the process has a dynamic gain, too, and this can be minimized by proper process design. While this is always good practice, it is especially critical in pH control. Furthermore, in most processes, the design is determined by other considerations such as reaction rate, heat transfer, etc. But most acid-base reactions are zero order and generate negligible amounts of heat, so these considerations do not apply. The process designer is then free to choose vessel size, layout, mixing, piping, etc., and should choose those which favor dynamic response.

3.4.1 A Dynamic Model of Mixing

Consider an acid-base reaction performed by combining the process stream and reagent in a *static mixer* as shown in Figure 3.3 (but without the circulating pump, whose function is described later). A static mixer consists of a pipe fitted with internal baffles which alternately split and rotate the fluids, resulting in a homogeneous blend at the exit. The principal problem with the static mixer is that it presents pure deadtime to the controller. A step change in the reagent flow will produce no observable response in the pH at the exit until the new mixture emerges, at which time the full extent of the step appears at once. The dynamic gain of the mixer is 1.0 because the step at the inlet appears as a step at the exit, not diminished or spread out over time.

The time between the initiation of the step and its appearance in the response of the pH transmitter is the deadtime caused by transportation through the mixer. It is equal to the length of the mixer divided by the velocity of the mix, or the volume of the mixer divided by the flow of the mix. (The connection between the mixer exit and the pH electrodes must be included in this calculation.) In addition to the high dynamic gain of the mixer, its deadtime is also *variable*, as a function of flow. This variablilty presents an additional problem for the controller, in that its optimum integral time is contingent on the process deadtime. If tuned for the highest flow (and hence shortest deadtime), the controller will integrate too fast for the process to respond at lower flow rates, causing cycling to develop. Therefore, the controller must be tuned at the lowest expected flow (where deadtime is the longest) for stability across the flow range. However, control action will then be slower than desired at higher flow rates. And if flow should ever stop, the control loop will be open. For the above reasons, static mixers are *not* recommended for pH control, and transportation delay of effluent to the pH electrodes should be eliminated or minimized.

The undesirable properties of the static mixer can be mitigated, however, by adding the circulating pump also shown in Figure 3.3. The pump applies suction downstream of the pH electrodes, recirculating treated product at a flow rate F_a relative to the discharge rate F. This reduces the deadtime through the process from V/F to $V/(F+F_a)$, where V is the volume of the process, and also places an upper limit on the deadtime in the loop at V/F_a. But it reduces the dynamic gain of the process as well.

Consider the case where $F_a = F$, and the feed is being mixed with enough reagent introduced stepwise to change the product concentration by one unit in the steady state. As the step in reagent flow is introduced, however, the reagent is being diluted with twice as much flow as without recirculation, so that its effect on product concentration after the lapse of the deadtime is reduced by a factor of two. At this point, this half-strength product is recirculated and mixed in equal volumes with full-strength feed, causing the blend to reach 75 percent of full strength after the lapse of another deadtime. Then the 75% product is recirculated to produce 87.5% product after the next deadtime, etc. The product concentration approaches its steady-state value covering 50% of the remaining distance each deadtime. The dynamic gain of the process has effectively been reduced by 50%. Increasing recirculation F_a to $4F$ reduces deadtime to $0.2V/F$ and the dynamic gain to 0.2, as illustrated by the staircase response in Figure 3.4. The effect of recirculation is quite predictable.

Next consider a *backmixed* tank with approximately equal diameter and depth of liquid, as shown in Figure 3.5, and let the agitator be sized to recirculate liquid at a rate F_a equal to $4F$. The same step response will be obtained as with the static mixer, except that the tank itself will not produce the same plug-flow profile as the baffled static mixer. As a result, the duration of the molecule travel from the reagent valve to the pH sensor has a wider distribution. Some will take longer than the deadtime estimated as if it were a static mixer, and some will take less time, although the average will be the same. The resulting step response will be better described by the smooth curve passing through the midpoints of all of the steps in Figure 3.4. This has the advantage of reducing the deadtime by a factor of two. Therefore, for the backmixed vessel, deadtime τ_d can be estimated as

$$\tau_d = V/2(F + F_a). \tag{3.16}$$

The average time that a molecule remains in the vessel is called its *residence time*, calculated as V/F. Observe that the exponential curve of Figure 3.4 passes through 63.2% response at time V/F. This curve represents the familiar step response of a first-order lag, which is 63.2% complete when the elapsed time from the beginning of the curve is equal to the time constant of the lag. The difference between the residence time and the deadtime is therefore the time constant of the vessel's first-order lag, τ_1 :

$$\tau_1 = V/F - \tau_d. \tag{3.17}$$

A first-order lag has a phase angle ϕ_1 and a dynamic gain G_1 which are functions of its time constant and the frequency or period at which the loop is cycling. In process control, the period, the duration of a complete cycle, is easier to measure than the frequency in cycles per minute or radians per minute, which in practice is calculated from the observed period. Therefore, the phase lag and dynamic gain are given here as functions of the period of oscillation τ_o.

$$\phi_1 = -\tan^{-1}(2\pi\tau_1/\tau_o) \quad G_1 = \cos\phi_1 \tag{3.18}$$

The period of oscillation is a dependent variable, produced where the sum of the dynamic phase lags in the loop reaches 180 degrees. Process deadtime also contributes to the loop phase lag, but its dynamic gain is always unity:

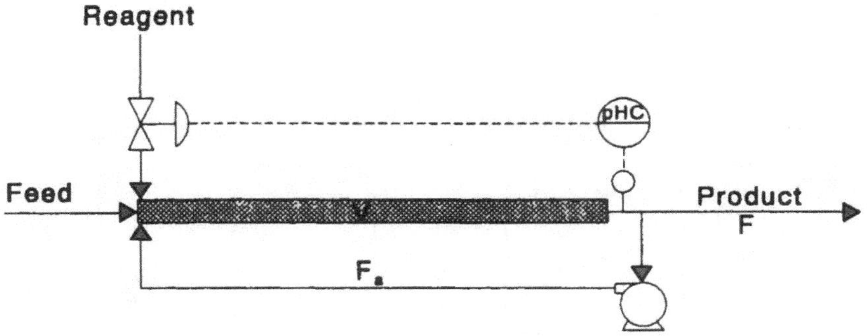

Figure 3.3 Controlling pH using a static mixer features variable deadtime and should not be used without recirculation.

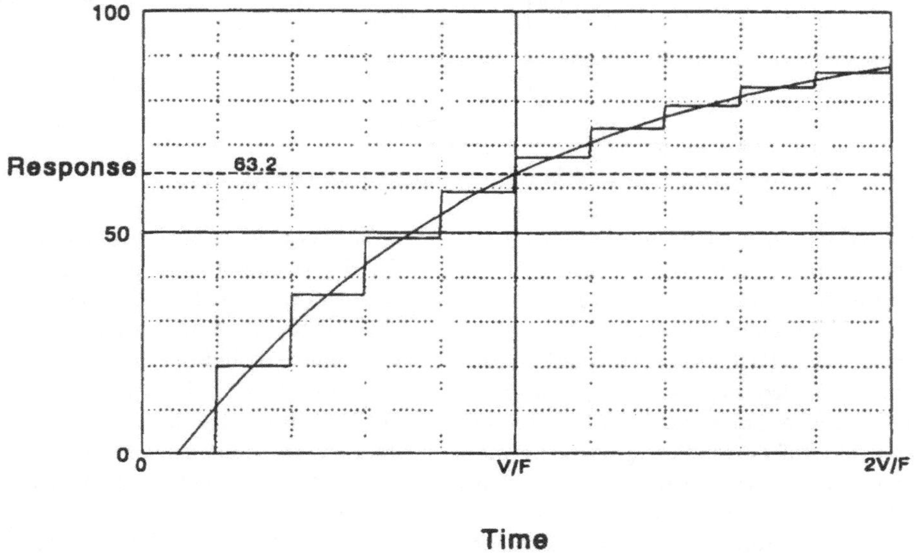

Figure 3.4 The steps are produced by recirculation through a plug-flow process; the curve is produced by a backmixed tank.

$$\phi_d = -360°\tau_d/\tau_o \quad G_d = 1.0 \tag{3.19}$$

If a proportional controller is used, which has no dynamic phase lag of its own, the period of oscillation determined by the process dynamics alone is then called the *natural* period, τ_n. Solution of $\phi_1 + \phi_d = -180°$ for τ_o by trial and error can give the natural period of the process; then its dynamic gain G_1 can be expressed as a function of the ratio of τ_d/τ_1 produced by a given degree of mixing. Table 3.2 displays the results of these estimates for a selected set of τ_d/τ_1 ratios.

TABLE 3.2 Natural Period and Dynamic Gain with Mixing.

			F_a/F	
τ_d/τ_1	$\tau_n/(V/F)$	G_1	Plug-Flow	Backmixed
∞	2.0	1.000	0	
2.0	1.83	0.657	0.5	
1.0	1.55	0.441	1.0	0
0.5	1.14	0.262	2.0	0.5
0.2	0.62	0.117	5.0	2.0
0.1	0.35	0.061	10.0	4.5

Reducing the τ_d/τ_1 ratio by recirculation has *two* advantages: it shortens the natural period relative to the residence time of the vessel, and it lowers its dynamic gain as well. This allows an increase in controller gain (commensurate with the reduction in process dynamic gain) and a reduction in its integral time (commensurate with the reduction in natural period). For a given upset then, the deviation from set point will be smaller and of shorter duration. The corresponding recirculation ratios F_a/F for both the plug-flow (static) mixer and the backmixed tank are included in the table.

3.4.2 Vessel Design

The principal elements to be determined in designing a vessel are its volume, its dimensions, and the location of inlet and outlet ports. Its volume is determined by the residence time needed for the reaction to go to completion at the maximum rate of feed. Since acid-base reactions are essentially zero order, the residence time on the basis of reaction rate is indeterminant. Then the limiting factor may be the response time of the pH electrode, which could be from a few seconds to a minute or more depending on the thickness of the glass and the presence of any coating [1]. Neutralization reactions have been successfully conducted within

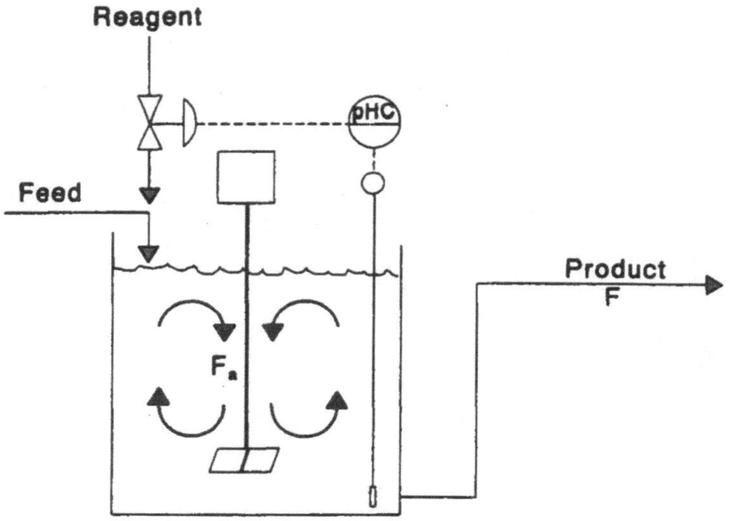

Figure 3.5 This is the preferred location of ports and of the pH measurement in a backmixed tank.

the casing of a centrifugal pump, using the impeller as a mixer.

When reacting species are not in the liquid phase, however, time must be allowed for mass transfer between phases. The most common example of this type is the use of a slurry of hydrated lime as a reagent. The $Ca(OH)_2$ already in solution reacts immediately, but must be replaced by mass transfer from the solid phase, which takes time. Neutralization of acid wastes with lime slurry has been successfully controlled in vessels having a residence time of 15 minutes. If the residence time is too short, the effluent will leave the vessel with some lime still undissolved, which will raise the pH when dissolution is complete. In the absence of mass transfer limitations such as this, a residence time of 3–5 minutes is ordinarily adequate.

If the pH of the feed stream is within the 2–12 range, neutralization can ordinarily be controlled successfully within a single vessel. Should the pH of the feed fall outside this range, it should be first pretreated in a smaller vessel, where the pH is controlled in the 2–3 range for acid feeds or 11–12 for basic feeds, and then brought to neutral in a second, larger vessel. The reason for recommending a smaller vessel for pretreatment is to separate the periods of the two pH loops. A control loop is most sensitive to cyclic disturbances of its own period, and can actually amplify those cycles rather than attenuate them. Given equal recirculation ratios in the two vessels, the periods of the two pH loops will be proportional to their residence times. The sensitivity of the second loop to cycles in the first can be reduced by a factor of ten if the ratio of their residence times is 3:1.

The vessel should be of equal dimensions in all directions, to minimize deadtime between inlet and outlet. Trying to control pH in a long narrow channel is ineffective, no matter how thoroughly mixed, because it is essentially all deadtime. If the long narrow channel is already fixed, or if pH is to be controlled in a pond (e.g., a flocculation basin or an ash pond at a power plant), then a cubic section should be walled off at the inlet as shown in Figure 3.6, with pH controlled at the exit from that section. The remainder of the channel or pond can be used to attenuate fluctuations, precipitate solids, etc.

The vessel where pH is to be controlled should be either cylindrical or cubic, with inlet and outlet ports diametrically opposed as shown in Figure 3.5. This avoids short-circuiting the vessel, allowing use of its full capacity. The pH electrodes should be located directly in front of the exit port, to give accurate representation of the product quality. While faster response may be obtained by moving them closer to the inlet, the measurement no longer represents product quality. However, avoid placing the electrodes within the exit pipe itself, as that adds deadtime to the loop, which varies inversely with feed rate; it also results in an open loop whenever feed is discontinued.

Reagent should be premixed with the feed at the feed point, rather than introduced anywhere else, because reagent introduced at other locations has produced erratic results. The feed ordinarily enters through a free fall, whose turbulence affords adequate premixing if the reagent is introduced there.

Feed *could* be introduced near the bottom of the vessel and product withdrawn from the top on the other side, but this increases the deadtime between reagent addition and electrode response. Agitators are designed to pump downward (assuming aeration is not the purpose): the general pattern of flow in Figure 3.5 carries reagent down the shaft of the mixer over to the electrodes in a short path. If feed and reagent are introduced near the bottom, however, the pattern of flow takes reagent up the wall of the vessel, then down the center, and finally up the wall on the other side where the electrodes are located; this path is twice as long, doubling the deadtime in the loop. Equation 3.16 applies to the arrangement shown in Figure 3.5.

3.4.3 Effective Backmixing

For pH control, high-velocity mixing is most effective. The vessel should be fitted with a direct-drive *axial* turbine (or marine propeller in smaller sizes) to maximize circulation rate. Low-speed radial mixers intended to keep solids in suspension increase deadtime excessively.

If a vortex forms, the solution is rotating like a wheel, without

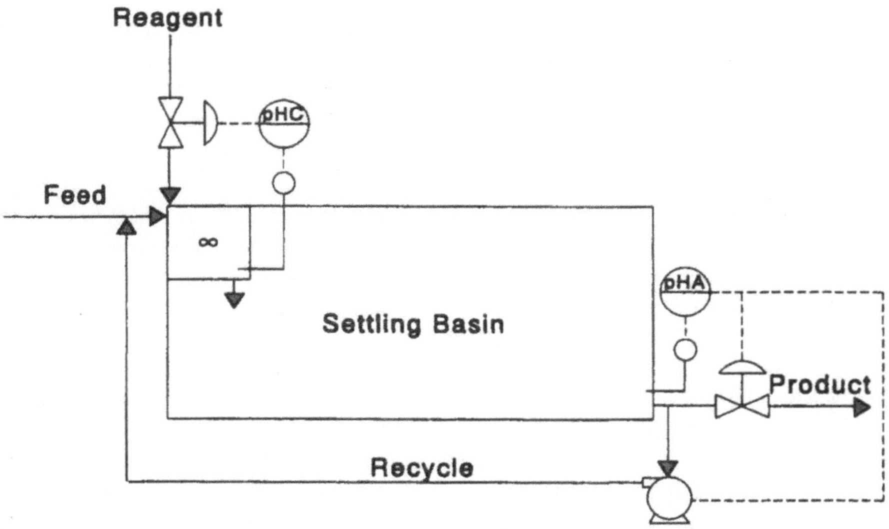

Figure 3.6 A large settling basin should be partitioned for pH control at the inlet; a pH alarm at the outlet starts recirculation of off-specification product.

distributing the reagent uniformly. In smaller cylindrical vessels, tilting the shaft of the mixer or offsetting it from center can eliminate vortex formation. In smaller rectangular vessels, the corners may be sufficient to prevent it, but in large, especially cylindrical vessels, vertical baffles are needed.

Horsepower requirements per gallon of vessel capacity decrease as the vessel capacity increases from 2.3 Hp for vessels of 1,000 gal to 0.8 Hp/1,000 gal for 100,000 gal. This level of agitation gives τ_d/τ_1 ratios of about 0.05.

3.4.4 Reagent Delivery Systems

Although Figure 3.5 shows the reagent valve located atop the vessel, it may be located in a more protected environment. However, its discharge line must not be allowed to empty when the valve is closed, as this adds deadtime to the loop. To avoid this problem, place a loop seal (pigtail) in the reagent pipe at the point of discharge, even if the discharge is below the surface of the solution.

If the reagent is a slurry such as lime, it must be kept in circulation at all times to avoid plugging. A circulation pump must convey the slurry from its agitated supply tank to the reaction vessel and back. At the reaction vessel a "Y" connection then brings slurry up to a ball control valve located at the high point in its line, from which reagent drops into the vessel [3, p. 178]. Then the ball valve will not plug on closing, as solids will settle away from it.

Often, feed is supplied by a pump driven by a level switch in a sump. In this case, feed rate is constant while it is flowing, but there are also periods of no flow. To protect the system against using reagents needlessly, the air supply to their control valves should be cut off whenever the feed pump is not running. At the same time, the pH controller should be switched to the manual mode to avoid integral *windup*, which would cause overshoot when the pump starts. When the pump is restarted, the controller is returned to the automatic mode at the same output it had when last in automatic; its set point should *not* be initialized while in manual.

3.4.5 Protection Against Failure

Failures are unavoidable. The pH electrodes could fail, reagent could plug or run out, or the feed could temporarily overload the reagent delivery system. Because of the damage which could result from off-specification product, protection is essential. A very effective strategy is shown in Figure 3.6.

A second pH transmitter is located at the outfall, downstream of the point where pH is controlled. It actuates an alarm if the product is out of specification, alerting operators to a failure. At the same time, the alarm logic closes the discharge valve and starts a recirculation pump to treat the effluent again. The process must have sufficient extra capacity to operate under these conditions for an hour or more, to allow time for diagnosis and repair of the failure.

3.5 Control System Design

If all of the above recommendations are carefully implemented, control system design is a relatively simple matter. But if they are not, even the most elaborately designed control system may not produce satisfactory results. This section considers those elements of design which will provide the simplest, most effective system.

3.5.1 Valve or Pump Selection

The first order of business is estimating the maximum reagent flow required to meet the highest demand. (This does not necessarily correspond to the highest feed rate, because in some plants, the highest feed rate is stormwater which may require no reagent at all.) Then the minimum reagent demand must also be estimated, and divided into the maximum to determine the rangeability required for both conditions. If full flow of reagent is insufficient to match the demand, the pH will violate specifications and the product will have to be impounded and retreated, as described above. If the minimum controllable flow of reagent is too high to match the lowest demand, the pH will tend to limit-cycle between that produced by the two conditions of minimum controllable flow and zero flow. The cycle will tend to be saw-toothed, with a relatively long period. Its amplitude could be within specification limits, however.

Control valves typically have a rangeability of 35–100:1, higher than metering pumps at 10–20:1. A linear valve will produce a linear relationship between controller output and reagent flow if it operates under constant pressure drop, as for example, supplied from a head tank. For higher rangeability, large and small valves must be operated in sequence, with the one not in use completely closed. Proper sequencing requires that the valves have equal-percentage (logarithmic) characteristics, yielding an overall characteristic which is also equal percentage and must be linearized by a matching characterizer. Ball valves used to manipulate slurry flow ordinarily have equal-percentage characteristics.

If the pH of the feed could be on either side of the set point, then both acid and base valves must be sequenced, with both closed at 50% controller output. Since both valves must fail closed, the base valve should be fitted with a reverse-acting positioner, opening the valve as the controller output moves from 50% to zero.

3.5.2 Linear and Nonlinear Controllers

In the case of a well-buffered titration curve, or controlling at a set point away from the region of highest gain, a linear pH controller may be used quite satisfactorily. Where neither of these mitigating factors apply, however, a linear controller will not give satisfactory results. Many pH loops limit-cycle between pH 4–10 or 5–9 using a linear PID controller having a proportional band below 400% (proportional gain > 0.25). If the proportional band is increased (gain lowered) until cycling stops, the controller will not respond satisfactorily to load upsets, allowing large deviations to last for exceptionally long durations.

The reason for this behavior is the very low gain of the titration curve beyond pH 4–10. When a disturbance forces the pH out of the 4–10 range, the loop gain is so low that very little correction is applied by the controller, amounting to integration of the large deviation over time. Eventually, reagent flow will match the load, but it is likely to cause overshoot as shown for the step load change in Figure 3.7. The long durations during which the pH is out of specification could cause damage to the environment downstream unless protection is provided, in which case the protective system may severely curtail production.

The simplest characterizer for pH control is a three-piece function which produces a low gain within an adjustable zone around the set point, with normal gain beyond. The width of the zone and the gain within it must be adjustable to match the titration curve. (Note that this is not the same as switching the controller gain when the deviation changes zones, as this would bump the output.) Although an imperfect match for the typical curve, this compensation provides the required combination of effective response to load changes with stability at the set point. The step load response in Figure 3.8 for the same process as in Figure 3.7 shows the effectiveness of the three-piece compensator. The width of the low-gain zone is ±2.2 pH around the set point, and its gain is 0.05 compared to a gain of 1.0 outside the zone. The proportional band of the controller is a factor of 14 lower (gain 14 times higher) than that of the linear controller of Figure 3.7. This combination actually gives more stability (lower loop gain) around the set point.

For very precise pH control of a process having a single well-defined titration curve, more accurate characterization is needed. In building such a characterizer, remember that input and output axes are reversed, pH being the input to the characterizer and equivalent concentration the output. The required characterizer simply has more points connected by straight lines than the three-piece characterizer described above. If the set point is always to remain in a fixed position, this characterizer can relate input and output in terms of deviation, as was done with the three-piece characterizer. However, if set-point adjustments are likely, then there must be two identical characterizers used, one applied to the pH measurement and the other to the set point. (In a digital controller, the same characterizer can be used for both.)

3.5.3 Controller Tuning Procedures

When a characterizer is not required, tuning of the PID controller should be carried out following the same rules recommended for other loops. Shinskey [6] gives tuning rules in terms of gain, deadtime, and time constant, or the natural period and gain at that period. For a PI controller, integral time should be set at $4.0\tau_d$ or τ_n; for a PID controller, integral time should be $2.0\tau_d$ or $0.5\tau_n$ and derivative time $0.5\tau_d$ or $0.12\tau_n$. Because pH loops tend to limit-cycle, the natural period is easy to obtain and should give better results.

Derivative action is strongly recommended for pH control. Derivative action amplifies noise, however, so that if measurement noise is excessive, derivative action may not be useful, at least without filtering. A viable alternative is the PIτ_d controller, a PI controller with deadtime compensation [6]. When properly tuned, it provides integration without phase lag and without amplifying noise. Optimum integral time is the same as for the PID controller, and optimum controller deadtime is 25% less.

Proportional gain should be adjusted for quick recovery from disturbances in the process load without excessive overshoot or cycling. Rules for fine-tuning based on observed overshoot and decay ratio in response to disturbances in the closed loop are also provided in Shinskey [6]. Most pH loops operate at the same set point all the time. If a loop need not contend with set point disturbances, it should not be tested by disturbing the set point. Instead, a simulated load change should be introduced by manually stepping the controller output from a steady state

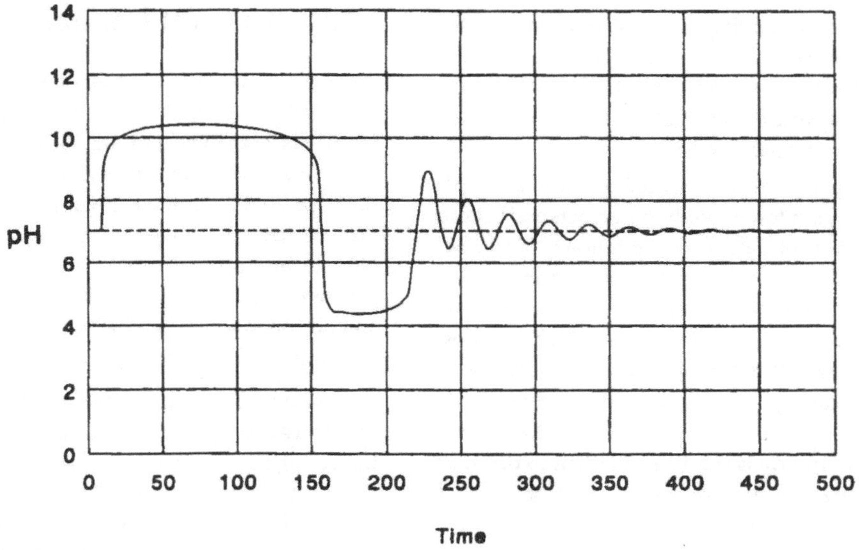

Figure 3.7 A linear PID controller tuned to avoid cycling will be slow in responding to load changes.

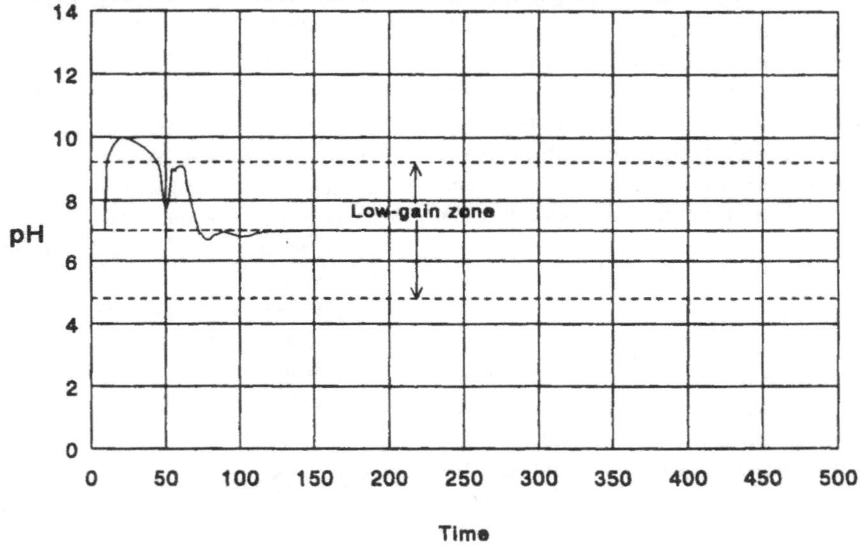

Figure 3.8 A three-piece nonlinear characterizer adds fast recovery and stability to the PID loop.

at set point, and immediately transferring the controller to the automatic mode. The effect of such a disturbance will be identical to that of the step load changes introduced in Figures 3.7 and 3.8.

If a three-piece characterizer is used to compensate the titration curve, it should be adjusted to position its breakpoints about 1 pH unit inside the knees of the curve. The slope of the characterizer between the breakpoints should be in the order of 0.05: if set $>$ 0.1, it will not provide enough gain change for effective compensation, and, if set at zero, it will cause slow cycling between the breakpoints. In the event the titration curve is unknown, the width of the low-gain zone should be initially set at zero and the proportional band decreased (gain increased) until cycling begins. The width should then be increased until the cycling abates. There is no hope of adjusting more than two breakpoints on-line.

An encounter with a particularly variable titration curve led the author to develop a self-tuning characterizer [4]. It was a three-piece characterizer with remotely adjustable width for the low-gain zone. In the presence of an oscillation, the zone would expand until the oscillation stopped; in the presence of prolonged deviation it would contract. This solved the immediate problem, but required a constant period of oscillation, because the PID settings of the controller and the dynamics of the tuner were manually set.

Now there are autotuning and self-tuning PID controllers on the market. *Autotuning* generally means that the controller tests the process on request, either with a step in the open loop or by relay-cycling in the closed loop to generate initial PID settings (which may or may not be effective). These controllers have no means for improving on the initial settings, however, or of automatically adapting them to observed variations in process parameters. A true *self-tuning* controller observes closed-loop responses to naturally occurring disturbances, adjusting the controller settings to converge to a response that is optimum or represents a desired overshoot or decay ratio. The self-tuning controller therefore is capable of automatically adapting its settings to variations in process parameters.

If a linear self-tuning controller is applied to a nonlinear pH process, it will adjust its gain low enough to dampen oscillations about set point. When a load disturbance follows, it will begin to respond like the low-gain linear controller in Figure 3.7. Observing the slow recovery of the loop, it will then increase controller gain until the set point is crossed, at which point oscillation begins, requiring that the gain be reduced to its former value. Although the resulting response is an improvement over that in Figure 3.7, it is not as effective as that in Figure 3.8. Therefore, nonlinear characterization is recommended, fit to the most severe titration curve likely to be encountered, with self-tuning used to adapt the controller gain to more moderate curves.

Self-tuning of the integral and derivative settings of the controller can adapt to a period increasing with fouling of the electrodes. While this can keep the loop stable as fouling proceeds, response to disturbances will nonetheless deteriorate due to the increasing time constants of the measurement and the controller. Much more satisfactory behavior will be achieved by keeping the electrodes clean. In the event that fouling is not continuous and cleaning is required only periodically, an alarm can be activated when the adapted integral time of the controller indicates that fouling is excessive.

3.5.4 Feedforward pH Control

Feedforward control uses a measurement of a disturbing variable to manipulate a correcting variable directly, without waiting for its effect on the controlled variable. In a pH loop, disturbing variables are the flow and composition of the feed. It is quite effective to set the flow of reagent in ratio to the feed flow, and thereby eliminate this source of upset. However, the ratio between the two flows depends on the composition of the feed, which can be quite variable. The feedback pH controller must adjust this ratio, as needed, to control pH and thereby compensate for variations in feed composition. This is done by *multiplying* the measured feed flow by the output of the pH controller, with the product being the required flow of reagent. The reagent flow must then respond linearly to this signal. This system works well when the rangeability of the reagent flow is 10:1 or less, because it depends on the rangeability of flowmeters and metering pumps.

Feedforward systems have been implemented using the pH of the feed as an indication of its composition, but with very limited success. This measurement is only representative of composition if the titration curve is fixed, which is never true for wastewater. Efforts have been made to titrate the feed to determine its composition or estimate the shape of the titration curve, but this takes time and these complications reduce the reliability of the system. Far more satisfactory performance will be realized if the process has been designed to be controllable, and a nonlinear feedback controller with self-tuning is used.

3.5.5 Batch pH Control

Some pH adjustments are carried out batchwise: this is the case for small volumes, or solutions too precious or toxic to be allowed discharge without assurance of precise control. The solution is simply impounded until its composition is satisfactory. Batch processes require a different controller than their continuous counterparts, however. In a continuous process, flow in and out can be expected to vary over time, constituting the major source of load change. In a batch process, there is no flow out, and so the load is essentially zero, while composition changes with time as the neutralization proceeds.

If a controller with integral action is used to control the endpoint on a batch process, the controller will *windup*, an expression indicating that the integral term has reached or exceeded the output limit of the controller. The cause of windup is the typically large deviation from set point experienced while reagent is being added to bring the deviation to zero. When the deviation is then finally reduced to zero, this integral term will keep the reagent valve open, causing a large overshoot. In a zero-load process, overshoot is permanent, and so must be avoided.

The recommended controller for a batch pH application is proportional-plus-derivative (PD), whose output bias is set to zero. The controller output will then be zero when the pH rests at the set point, and will be zero as the pH approaches the set point if it is moving. This controller can thereby avoid overshoot, when properly tuned. Proportional gain should be set to keep the reagent at full flow until the pH begins to approach the set point, and the derivative time set to avoid overshoot. An equal-percentage reagent valve will help to compensate for the nonlinearity of the titration curve, because its gain decreases as flow is reduced and, therefore, as the set point is approached.

3.6 Defining Terms

Activity: The effect of hydrogen ions on a hydrogen-ion electrode.

Autotuning: Tuning a controller automatically based on the application of rules to the results of a test.

Backmixing: Recirculating the contents of a vessel with its feed.

Concentration: The amount of an ingredient per unit of mixture.

Dead band: The largest change in signal which fails to cause a valve to move upon reversal of direction.

Diprotic: Denoting an acid or base which yields two hydrogen or hydroxyl ions per molecule.

Feedforward control: Conversion of a measurement of a disturbing variable into changes in the manipulated variable which will cancel its effect on the controlled variable.

Isopotential point: The pH at which changes in temperature have no effect on output voltage.

Monoprotic: Denoting an acid or base which yields one hydrogen or hydroxyl ion per molecule.

Normal: A Normal solution contains 1.0 gram-ion of hydrogen or hydroxyl ions per liter of solution.

Residence time: The average time that molecules spend in a vessel.

Self-tuning: Tuning a controller to improve the response observed to changes in set point or load.

Solution ground: A connection between the process solution and the instrument ground.

Standardize: To calibrate a pH instrument against a standard solution.

Static mixer: An in-line mixer with no moving parts.

Titration curve: A plot of pH vs. reagent added to a solution.

Valve positioner: A device which forces the position of a valve stem to match the control signal.

Windup: Saturation of the integral mode of a controller, causing overshoot of the set point.

References

[1] McMillan, G.K., Understanding Some Basic Truths of pH Measurements, *Chem.Eng.Prog.*, October, 30–37, 1991.

[2] Rovira, W.S., Microprocessors Bring New Power and Flexibility to pH Transmitters, *Contr.Eng.*, September, 24–25, 1990.

[3] Shinskey, F.G., *pH and pIon Control in Process and Waste Streams,* John Wiley & Sons, New York, 1973.

[4] Shinskey, F.G., Adaptive pH Controller Monitors Nonlinear Process, *Contr.Eng.*, February, 57–59, 1974.

[5] Shinskey, F.G., Process Control Systems, 3rd ed., McGraw-Hill, New York, 1988.

[6] Shinskey, F.G., Manual Tuning Methods, in *Feedback Controllers for the Process Industries,* McGraw-Hill, New York, 1994, pp. 143–183.

[7] Weast, R.C., *Handbook of Chemistry and Physics,* The Chemical Rubber Company, Cleveland, OH, 1970.

4

Control of the Pulp and Paper Making Process

4.1 Introduction 43

4.2 Pulp and Paper Manufacturing Background 44
Wood Species and Raw Materials • Products • Manufacturing Processes • Mechanical Pulping • Chemical Pulping • Bleaching • Paper Making • The Pulp and Paper Mill • Competitive Marketplace

4.3 Steady-State Process Design and Product Variability 46
Mill Variability Audit Findings • Variability Audit Results • Impact on Papermaking • Final Product Variability • Variability Audit Typical Findings • Control Equipment • Loop Tuning • Control Strategy Redesign, Process Redesign, and Changes in Operating Practice • Pulp and Paper Mill Culture and Awareness of Variability

4.4 Control Equipment Dynamic Performance Specifications 56

4.5 Linear Control Concepts — Lambda Tuning — Loop Performance Coordinated with Process Goals 56
Design Example — Blend Chest Dynamics using Lambda Tuning • Pulp and Paper Process Dynamics • Lambda Tuning • Choosing the Closed-loop Time Constant for Robustness and Resonance • Multiple Control Objectives • Algorithms

4.6 Control Strategies for Uniform Manufacturing 64
Minimizing Variability — Integrated Process and Control Design

4.7 Conclusions 65

4.8 Defining Terms 65

References 66

W.L. Bialkowski
EnTech Control Engineering Inc.

4.1 Introduction

Pulp and paper products are consumed by almost every sector of modern society—a society that shows no sign of becoming "paperless," especially with the advent of the photocopier and laser printer. Paper is manufactured from wood, a naturally renewable resource, by a large industry with significant economic impact on the world economy. The manufacture of pulp and paper products represents a particularly challenging environment for the control engineer, as it is large in scale, highly nonlinear, highly stochastic, and dominated by time delays. This is not a linear-time-invariant system, but rather a "near-chaotic" environment with all of the implications that this has for process and product variability. The market for paper products, however, is increasingly quality conscious and customer driven, one in which product uniformity (the opposite of product variability) is increasingly being demanded by the customer as the prime ingredient of "quality" and the "price of admission."

This chapter reflects the experience gained [3] in the pulp and paper industry over the last decade, while attempting to improve product and process variability throughout the industry. The control engineer's challenge is not an academic one, but rather it is nothing short of enhancing the ability of his or her company to continue existing in such a competitive world. This challenge extends far beyond linear control theory and involves the control engineer as a critical member of an interdisciplinary team, who brings the essential understanding of dynamics to the manufacturing problem. Without this understanding of dynamics, it is impossible to understand variability. In the pulp and paper industry the control engineer is likely to be the only person who understands dynamics. Hence the scope of interest must extend far beyond the narrow confines of linear control theory or the closed-loop dynamic performance of single loops and must encompass the characteristics of the product required by the customer, the characteristics of the raw materials, the nonlinear and multivariable nature of the process, the effectiveness of the control strategy to remove variability, and the information and training needs of the rest of the team.

0-8493-0054-3/00/$0.00+$.50

This chapter will attempt to present the essence of this challenge by developing the subject in the following sequence: pulp and paper background, the impact of steady state design, mill variability results to date, actuator nonlinearities, linear control concepts — "Lambda Tuning," algorithms in use, control strategies for uniform manufacturing, minimizing variability by integrating process and control design, and finally conclusions. The reader is assumed to have an electrical engineering background, a strong interest in automatic control, and no prior exposure to the pulp and paper industry.

4.2 Pulp and Paper Manufacturing Background

Paper products, ever present in our daily lives, are manufactured primarily from wood in pulp and paper mills which are large, industrial complexes, each containing thousands of control loops. Let us describe some of this background before developing the control engineering perspective.

4.2.1 Wood Species and Raw Materials

Wood from trees, the primary raw material for pulp and paper products, is a renewable natural resource. In North America alone, there are nearly 1000 pulp and paper mills which depend on forest harvesting activities extending from northern Canada, most of the U.S., northern Mexico, and from Newfoundland in the east, to Alaska in the west. Some mills produce and sell pulp as a product. Others are integrated mills that produce pulp and then paper. Some paper mills buy pulp as a raw material, and make paper. Recycling mills take paper waste and make paper. The pulp and paper industry exists in many parts of the world.

The prime ingredient of wood, which makes it useful as a raw material, is **cellulose**, a long chain polymer of the sugar, glucose, which constitutes much of the wall structure of the wood cell, or fiber. Trees grow as a result of the natural process of photosynthesis, in which the energy from the sun converts atmospheric CO_2 into organic compounds, such as cellulose, with the release of O_2 gas. Growth occurs in the tree underneath the bark, as layers of cells are deposited in the form of annular rings (the age of a tree can be easily determined by counting these annular rings).

The wood cell, or **fiber**, is "cigar" shaped, roughly square in cross-section and varies in length from 0.5 mm to well over 4 mm. The center of the cell is hollow. Fiber length is a key property for paper making and is primarily determined by the wood species, which divide into two main types: softwoods and hardwoods. Softwoods are mainly coniferous trees which have needles instead of leaves, such as pines and firs. These species have long fibers: typically 3 mm long (redwoods have 7 mm long fibers) and about 30 μm wide. Hardwoods are primarily the deciduous trees with leaves, such as maple, oak, and aspen. Their fibers tend to be short and "stubby" (typically 0.5 mm).

The fiber structure consists of many layers of cellulosic "fibrils," crystalline bundles of cellulose molecules. Wood also contains **hemicellulose**, polymers of other sugars with chain lengths shorter than cellulose. The cell structure is bonded together by **lignin**, the third wood ingredient, which acts very much like a glue. Typical wood composition by fraction on a dry basis is shown in Table 4.1

TABLE 4.1 Wood Composition (Dry Basis).

Wood Species	Hardwood	Softwood
Cellulose %	50	50
Hemicellulose %	30	20
Lignin %	20	30

The process of removing fibers from solid wood is called "pulping." There are two process for pulping: mechanical and chemical. Mechanical pulping involves "ripping" the fiber out of the solid wood structure by mechanical and thermal means. Mechanical pulp fibers include all of the above ingredients, hence the pulping process has a very high yield (90%+). However, due to the nature of the pulping process, there is substantial fiber damage with a resulting loss in strength. Chemical pulping involves chemically dissolving the lignin fraction (and usually the hemicellulose fraction as well). This liberates the cellulose fraction, still in the shape of the original fiber, with relatively little damage, and results in high strength and the potential for high brightness, once all traces of lignin have been removed and the pulp has been bleached. The yield of chemical pulp is low, < 50% to 60%, because the lignin and hemicellulose have been dissolved. Typically, these organic compounds are used as a fuel in a chemical pulp mill.

4.2.2 Products

Pulp and paper products cover a broad spectrum. An incomplete list includes newspaper, paper for paperback books, photocopier paper, photographic paper, paper for books, coated paper for magazines, facial tissue, toilet tissue, toweling, disposable diapers, sanitary products, corrugated boxboard, linerboard, corrugating medium, food board (e.g., cereal boxes, milk carton), Bristol board, bleached carton (cigarette carton, beer cases), molded products (paper plates), insulation board, roofing felt, fiberboard, market pulp for paper manufacture, and pulp for chemical feed stock such as the manufacture of rayon, film, or food additives.

Each of these products requires certain unique properties which must be derived from the raw material. For instance, newspaper must be inexpensive, strong enough to withstand the tension imposed by the printing press without breaking, and have good printing properties. It is often made from recycled paper with added mechanical or chemical softwood pulp for strength. Photocopier paper must have excellent brightness, a superb printing surface and must not curl or jam in the photocopier. Highly bleached chemical pulp is used. This type of paper is made by blending hardwood pulp (for a very smooth printing surface) with softwood pulp (for strength). Additives such as clay and titanium dioxide are added to enhance the printing surface. Facial tissue must be soft and absorbent. It is made from chemically bleached pulp. Market pulp is sold by pulp grade. Each grade having specifications for species, brightness, viscosity, cleanliness, etc. In most cases market pulps are made chemically and are usually bleached.

4.2.3 Manufacturing Processes

Pulp and paper mills vary greatly in design, as they reflect the product being made. The simplest division is to separate pulping from papermaking. We will discuss each of the significant areas as unit operations. The degree of control and instrumentation which exists varies widely depending on both the age and design of the plant.

4.2.4 Mechanical Pulping

Mechanical pulping can be achieved by grinding or refining. *Groundwood* mills mechanically grind whole logs against an abrasive surface. The pulping action is a combination of raising the temperature, and mechanically "ripping" the fiber from the wood surface. This is done by feeding logs to the "pockets" of grinders, which are powered by large synchronous motors (typically 5,000 to 10,000 HP). A typical groundwood mill may have 10 to 20 grinders (50 to 150 control loops). The *Thermal-Mechanical Pulping* (TMP) process (100 to 300 control loops) is the more modern way of mechanical pulping. It consists of feeding wood chips into the "gap" of rotating pressurized machines, called chip refiners, to produce pulp directly. Normally there are two refining stages. The chips disintegrate inside the refiner as they pass between the "teeth" of the refiner plates. Each refiner is powered by a synchronous motor of 10,000 to 30,000 HP. The amount of refining can be controlled by adjusting the gap between the rotating plates. Mechanical pulps can be brightened (bleached) to some degree.

4.2.5 Chemical Pulping

Chemical pulping can be divided into two main processes, sulfite and Kraft. The *sulfite process* is an acid based cooking process, whose use is in decline. The *Kraft process* is an alkali-based process in which the active chemicals are fully recycled in the *Kraft liquor cycle*. The Kraft process itself consists of eight individual unit operations. It starts with wood chips being fed to a *digester house* (unit process with 50 to 300 control loops) in which the chips are fed to a digester and impregnated with **white cooking liquor** (a solution of NaOH and Na_2S), and "cooked" at about 175°C for about an hour. In this process the lignin and hemicellulose are dissolved. The spent cooking liquor is then extracted and the pulp is "blown" into the "blow" tank. Modern digesters are continuous vertical columns, with the chips descending down the column. The impregnation, cooking, and extraction processes take about three hours or so. Some digester houses use batch digesters instead, with 6 to 20 batch digesters. From the blow tank the pulp is pumped to the *brown stock washers* (unit process with 20 to 200 control loops), in which the pulp is washed in a multistage countercurrent washing process to remove the spent cooking liquor, including the dissolved organic compounds (lignin, hemicellulose) and the spent sodium compounds which must be removed from the pulp stream and reused. The two output streams consist of: washed pulp (which goes to the bleach plant for bleaching or to the paper machines if bleaching is not required) and the spent liquor which is called *black liquor* and is returned to the Kraft liquor cycle.

The *Kraft liquor cycle* starts with pumping black liquor from the brown stock washers to the *multiple effect evaporators* (unit process with 20 to 100 control loops) where the black liquor solids are raised from about 15% to about 50% by evaporation with steam. These solids are then fed to the **recovery boiler** (unit process with 100 to 500 control loops) in which the black liquor solids are further concentrated and then the liquor is fired into the recovery boiler as a "fuel." The bottom of this boiler contains a large smoldering "char-bed" of burning black liquor solids, below which the sodium compounds form a molten pool of "smelt," containing Na_2CO_3, Na_2S and Na_2SO_4. The smelt pours from the bottom of the recovery boiler into a dissolving tank, where it is dissolved in water, and becomes *green liquor*. The upper part of the recovery boiler is conventional in design and produces about 50% of the total steam consumed by the mill. The green liquor is pumped to the **causticizing area** (unit process with 50 to 200 control loops) where "burned lime" (CaO) is reacted with the green liquor in order to re-constitute the white cooking liquor (first reaction is: $CaO+H_2O=Ca(OH)_2$; second reaction is: $Na_2CO_3+Ca(OH)_2 = 2NaOH+CaCO_3$). The resulting white liquor is sent to the digester for cooking. The calcium carbonate is precipitated and sent to the **lime kiln** (unit process with 50 to 200 control loops) where the $CaCO_3$ plus heat produces burned lime (CaO) plus carbon dioxide gas. The burned lime is used in the causticizing area to reconstitute the white cooking liquor.

4.2.6 Bleaching

After cooking, the pulp may need to be bleached. This is necessary for all products which require high brightness or the complete removal of lignin. A typical bleach plant (unit process with 50 to 300 control loops), consists of the sequential application of specific chemicals, each followed by a reaction vessel and a washing stage. A bleaching sequence which has been used often in the past involves pumping unbleached "brown stock" from the pulp mill and applying the following chemicals in turn: chlorine (to dissolve lignin), caustic (to wash out lignin by-products), chlorine dioxide (to brighten), caustic (to dissolve by-products), and finally chlorine dioxide to provide final brightening. Bleach plant technology is currently in a state of flux as a result of the environmental impact of chemicals, such as chlorine. As a result, the industry is moving towards new chemicals, chiefly oxygen and hydrogen peroxide.

4.2.7 Paper Making

Paper is made on a paper machine (unit process with 100 to 1000 control loops). Paper making involves **refining** various types of pulps (mechanical work to improve bonding tendencies) individually, and then blending them together in combination with specific additives. This "stock" is then screened and cleaned to remove any dirt, and is diluted to a very lean pulp slurry before being ejected onto a moving drainage medium, known as "the wire," Fourdrinier, or former, where a sheet of wet paper is formed. This wet paper is then pressed, dried, and calendered (pressed between smooth polished surfaces) to produce a smooth final product with a specified basis weight (weight per unit area), moisture content, caliper (thickness), smoothness, brightness, color, ash content, and many other properties. Figure 4.1 shows a diagram of a simple fine paper machine (say for photocopier paper): Hardwood and softwood stocks are stored in High Density (HD) storage tanks at about 15% consistency (mass percentage of pulp in a slurry). As the stock is pumped from the HD's it is diluted to 4% and refined. Paper machine refiners have rotating discs which fibrilate or "fluff" the fiber, thereby giving it better bonding strength. The hardwood and softwood stocks are then blended together in a desired ratio in the blend chest (subject of case study later), and sent to the machine chest, the final source of stock for the paper machine. From here the stock is diluted with "white water" (filtrate which has drained through the forming section), and is sent to the headbox (a pressurized compartment with a wide orifice or "slice lip" which allows the "stock" to be ejected as a wide jet). The stock discharges from the headbox at the required speed onto the "wire," a wide porous "conveyor belt" consisting of synthetic woven material. Most of the fiber stays on, and most of the water drains through to be recycled again as "white water."

4.2.8 The Pulp and Paper Mill

The pulp and paper mill consists of between one and some thirty unit operations, with a total of about 500 to 10,000 control loops. A typical mill produces 1000 tons/day of product (although the capacity of mills can vary from 50 to over 3000 tons). Each of the operations involving wood chips or fiber is essentially hydraulic in nature, and involves a sudden application of chemical, dilution water, mechanical energy, or thermal energy. Sometimes this is followed by a reaction vessel with long residence time. As a result, the fast hydraulic dynamics are very important in determining the final pulp or paper uniformity. A typical mill is operated some 355 days per year, 24 hours per day, by a staff of a few hundred people, often with a unionized workforce. The traditional responsibility for control loop tuning lies with the instrumentation department, which may have between 5 and 30 instrument technicians. The number of control engineers varies from a dozen in some mills to none in others.

This has been a very brief description of pulp and paper science, a well established field with a rich literature. A good introduction to this work is available in the textbook series on pulp and paper manufacture edited by Kocurek [14].

4.2.9 Competitive Marketplace

Regardless of the imperfections which might exist in the pulp and paper manufacturing environment, the marketplace now demands ever greater product uniformity. Warnings and recommendations about control performance and control engineering skill have been reported [3] in pulp and paper industry literature. Such problems are not unique to pulp and paper, and probably apply to most process industries. For instance, similar concerns have been expressed in the chemical industry [7], where the following recent prediction was made: *"In ten years the ability to produce highly consistent product will not be an issue because those companies that fail in this effort will be out of the market or possibly out of business. Among the remaining companies, the central issue ten years from now will be the efficient manufacture of products that conform to customer variability expectations"*. To a great extent the issues are not technological but human. A higher level of competency in process control is required to reduce variability [9], which requires awareness creation, training, and education. To this end, there has been extensive training of both instrument technicians and control engineers in recent years [4], with nearly 2000 industry engineers and technicians having received training in dynamics and control.

4.3 Steady-State Process Design and Product Variability

Pulp and paper mills, like most of heavy industry, have been designed only in steady state. Design starts with a steady-state flow sheet, proceeds to mass and energy balances, following which major equipment is selected and the design proceeds to the detailed stage. It is in the detailed design stage that control of the process is often considered for the first time, when sensors and valves are selected and a general purpose control system is purchased. It is generally assumed that only single PID loops will be required to control the process adequately and that these controllers can handle whatever dynamics are encountered when the process starts up. Until now, the prime focus has been on production capacity.

The last ten years have witnessed increasing concern about high product variability, as the quality "revolution" has started to engulf the pulp and paper industry. Customer loyalties can change and large long term contracts can be canceled when the product does not perform well in the customer's plant. As an example, Japanese newspaper print-rooms are now demanding paper which is guaranteed to have only one web break or less, per one thousand rolls of paper [13]. Such intense competition has not yet reached North America, where newsprint suppliers are still being rated by their customers in breaks per hundred rolls, with 3 being good — this is 30 breaks per thousand rolls! As competition intensifies, pressures such as these will create an increasing need for closer attention to variability. Such market forces are creating new demands. For instance, the consulting company, EnTech Control Engineering Inc., was formed in response to the needs for an independent variability and control auditing service with design and training expertise to help pulp

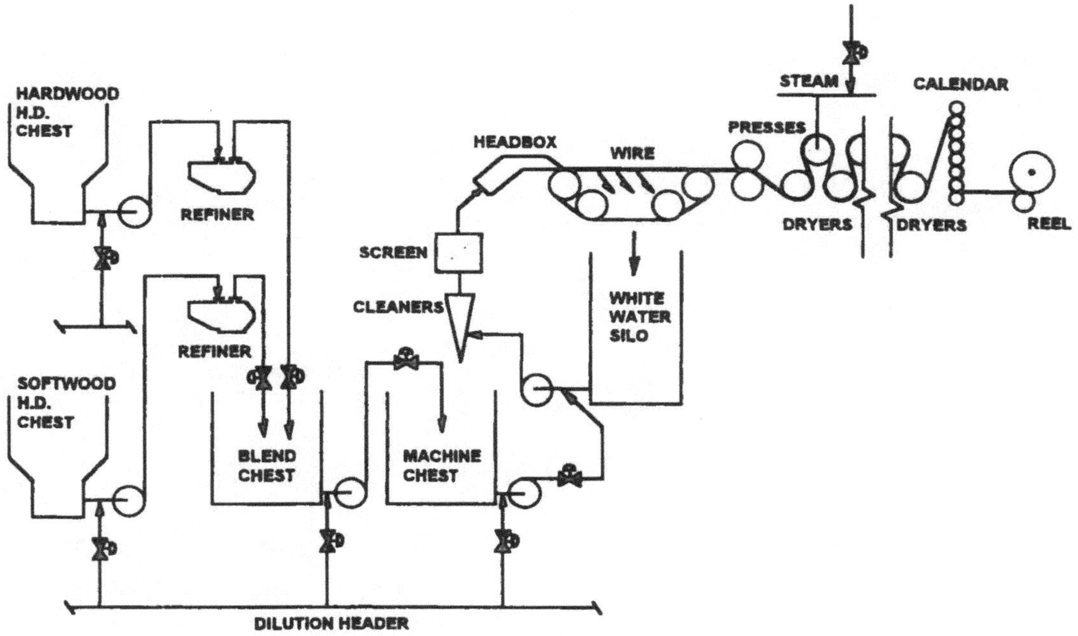

Figure 4.1 A paper machine.

and paper companies acquire the skills to become more effective competitors in the marketplace. About 200 mill variability audits have been carried out to date. The results of these process and product variability audits [3] raise serious questions about the quality of mill design, maintenance practice, control equipment (such as control valves) and control knowledge, as these issues pertain to the ability of achieving effective process dynamics and low product variability.

4.3.1 Mill Variability Audit Findings

The results shown are fairly typical of audit results which may be obtained from any part of the pulp and paper industry prior to corrective action. The blend chest area of a paper machine has been chosen as a small case study to illustrate typical mill variability problems. A design discussion of the blend chest area is continued throughout later sections of the chapter. The blend chest is included in the paper machine sketch of Figure 4.1, and is also shown in more detail in Figure 4.2 below, which illustrate a fine paper machine (fine paper refers to high quality, publication grade paper) manufacturing paper, such as photocopy paper, using 70% hardwood and 30% softwood and typically producing about 500 tons per day. The hardwood pulp is needed for surface smoothness and printability, and the softwood pulp is needed for strength. These two pulp slurries (stocks) are pumped out of their respective high density (HD) storage chests at about 5% *consistency* (mass percentage fiber slurry concentration). The hardwood stock is pumped out of the hardwood HD chest and is diluted further to 4.5% consistency at the suction of the stock pump, by the addition of dilution water. The dilution water is supplied from a common dilution header (pipe) for this part of the paper machine. In the example of Figure 4.2, the dilution water addition is modulated by the consistency control loop, NC104 (typical loop tag based on ISA terminology), regulating the consistency to a set point of about 4.5%. The consistency sensor is normally located some distance after the pump, per the manufacturer's installation instructions. A time delay of 5 seconds or longer, is typical. After consistency control, the hardwood stock is pumped into the hardwood refiner. The purpose of the refiner is to "fibrillate" the fiber, increasing its surface area and tendency to bond more effectively in the forming section. The refining process is sensitive to the mass flow of fiber and the mechanical energy being supplied to the refiner motor (specific energy). After the refiner, the stock flow is controlled by flow controller FC105. The softwood line is identical to the hardwood line. The two flow controllers FC105 and FC205 are in turn part of a cascade control structure, in which both set points are adjusted together to maintain the blend chest level at the desired set point (typically at 70%), while maintaining the desired blend ratio (70% hardwood and 30% softwood).

4.3.2 Variability Audit Results

Figure 4.3 shows the blend chest consistency NC302 on automatic control (upper plot) and on manual (lower plot) for a period of 15 minutes (900 sec). On automatic, the mean value is 3.82% consistency, and two standard deviations ("2Sig") are 0.05093% consistency. This number (which represents 95% of all readings assuming a Gaussian distribution) can also be expressed as 1.33% of mean. It is this number which is usually referred to as "variability" in the pulp and paper industry (approximately half the range of the variable expressed as a percent). For the manual data, the mean value is virtually unchanged (the loop was merely turned off control). The variability however is now 0.848%, 50% lower than on automatic. NC302 is controlled by a PI controller which has been tuned with default tuning settings consisting of a

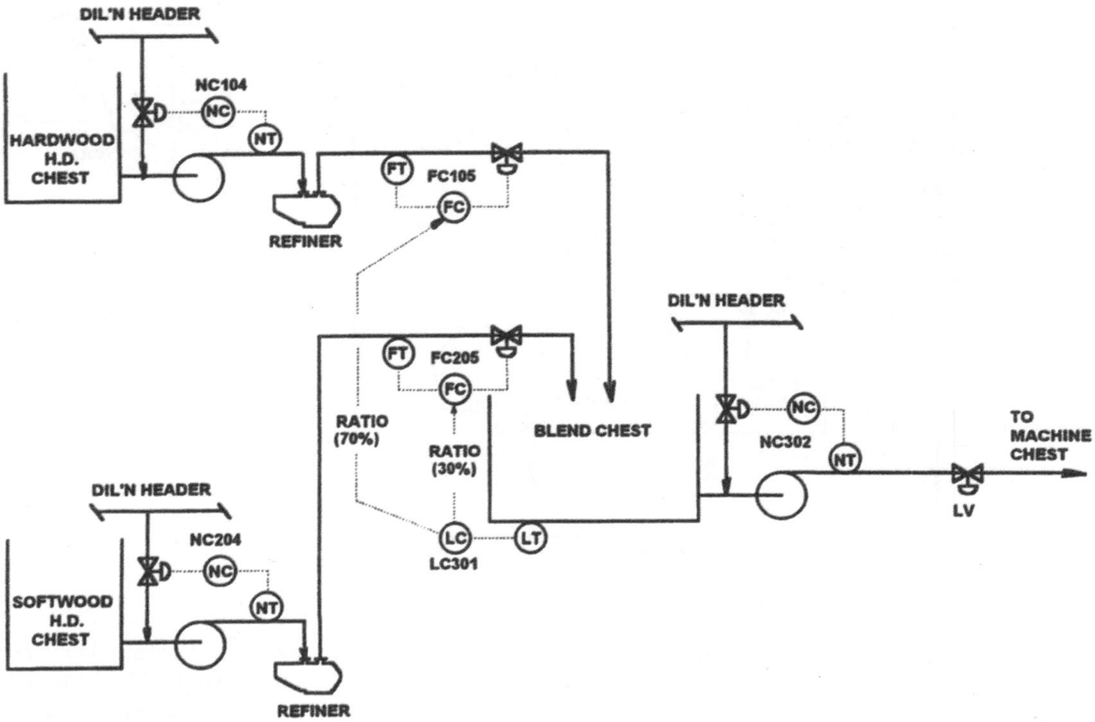

Figure 4.2 The blend chest area of a paper machine.

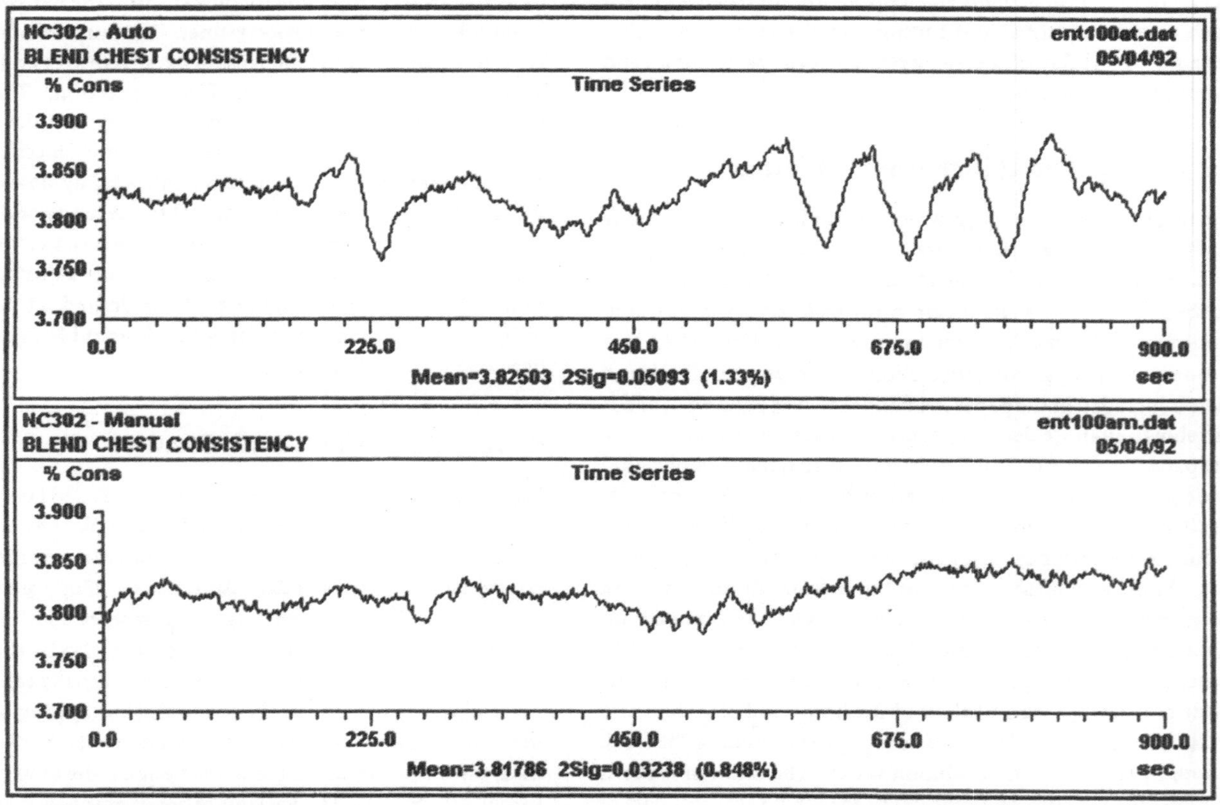

Figure 4.3 Blend chest consistency NC302 time Series — on and off control.

controller gain of 1.0 and an integral time of 0.5 minutes/repeat (standard PI controller form). These tuning parameters occur frequently, are entered as default values prior to the shipping of controller equipment, and are "favorite numbers" for people tuning by "trial and error." However, the variabilty problems of NC302 are not all tuning related; the control valve was found to have about 1% stiction. It is this tendency which induces the limit-cycling behavior seen in Figure 4.3 at second 225 and from seconds 500 to 800.

Variability of 1.33% for a blend chest consistency is not very high, but, considering that paper customers are starting to demand paper variability approaching 1%, this represents potential for improvement, especially because the variability on manual control is 50% lower than on automatic. Figure 4.4 shows two open-loop step tests *(bump tests)* for the blend chest consistency loop. Also shown is the impact of these on the dilution header pressure control loop, PC309 (not shown in Figure 4.2). Excursions in the consistency control valve of 2% cause header pressure excursions of over 1% and pressure control valve excursions of about 1%. This example illustrates the degree of coupling that exists between these two variables and poses questions about the tendency of this dilution header design to cause interaction between all of the consistency loops which feed from it. Every time that one consistency loop demands more dilution water, it will cause a negative dilution header pressure upset and disturb the other consistency loops. Not visible in Figure 4.4 is the fact that the NC302 dilution control valve also has about 1% stiction which was mentioned as the cause of the limit cycle in Figure 4.3.

Figure 4.5 shows an expanded view of the first NC302 bump test of Figure 4.4. The top trace shows the controller output, while the lower shows the process variable. A fit of a first-order plus deadtime process transfer function model is superimposed over this data. The lower left hand of Figure 4.5 shows the resulting parameters for the transfer function form,

$$G_P(s) = \frac{K_P e^{-sT_d}}{(\tau_1 s+1)} \tag{4.1}$$

The time delay of 8.3 seconds is excessively long. It results from locating the consistency sensor too far from the pump discharge (could be less than 5 sec.). The time constant of 18.8 seconds is also excessively long (could be 3 sec.). This parameter is primarily determined by the adjustment of a low-pass filter in the consistency sensor itself. Both of these issues contribute to the sluggish behavior of NC302.

Figure 4.6 shows a series of open-loop bump tests performed on the hardwood flow FC105. There are seven steps performed on the controller output. The first five have a magnitude of about 0.7% (one negative, four positive), while the last two have a magnitude of about 1.5%. It is clear that steps 3, 4, and 5 produce no noticeable flow response. The response of step number six is 1.5% producing a flow response of about 50 gallons per minute, and the last step produces no response at all. The combined backlash and stiction in this valve assembly is about 5%. Clearly, it is impossible to control this flow to within 1%. Similar bump tests on the softwood flow FC205 indicated that the combined backlash and stiction are in the order of about 1.7%. Figure 4.7 shows the softwood flow FC205 on automatic control. There is a limit cycle induced by the combined stiction and backlash of more than ±5%. The controller output appears like a triangular wave of amplitude 1.7%. This is a characteristic of stiction-induced limit cycles and is caused by the PI controller integral term attempting to induce valve motion. The valve, however, is stuck and will not move until the controller output has changed by about 1.7%. The valve will then release suddenly and move too far, thereby, inducing the next cycle. The period of the limit cycle is about 1.4 minutes. The period is a function of the stiction magnitude, the PI controller tuning, and the process gain. Figure 4.8 shows what happens when FC205 is taken off control. The upper plot of Figure 4.8 is the data from the last half of Figure 4.7. The bottom half of Figure 4.8 shows FC205 on manual control. It is clear that the limit cycling stops and the variability drops significantly. This is quite typical of many flow control loops.

Figure 4.9 shows the operation of the blend chest level controller, together with the two flows entering the chest. The upper left hand plot shows the blend chest level LC301. The upper right hand plot shows the LC301 power spectrum plotted on a log-log scale. There is a tendency for the level control to cycle with a period of about 17 minutes, as indicated by the time series plot and by the low-frequency lobe of the power spectrum which contains the fundamental frequency (period of 1024 seconds). Level controllers frequently cycle in industry because their tuning is not intuitive. There is a tendency for people to use tuning parameters such as the default settings of a gain of 1.0 and an integral time of 0.5 minutes. This represents far too aggressive an integral time and far too low a gain for an integrating process such as the blend chest, which requires quite different tuning. The two lower plots show the softwood FC205 (left) and hardwood FC105 (right) flows. The variability of both is about 5%. However, this is where the similarity ends. Both flows limit cycle in their own characteristic way with large amplitude swings. The softwood flow is limit cycling in a manner similar to that shown in Figure 4.7 (period now 2.6 minutes). The hardwood flow FC105, on the other hand appears to break into a violent limit cycle on a periodic basis. The amplitude of this limit cycle is more than 8% with a period of about one minute. The violent limit cycling stops periodically, followed by periods of little action lasting more than five minutes. It is interesting how such nonlinearities cause frequency shifts and multiple frequency behavior reminiscent of chaos theory.

4.3.3 Impact on Papermaking

The impact of the variability described above is quite damaging to papermaking. The variability in hardwood and softwood flows causes a number of problems. First of all, flow variability causes variable consistency, which the consistency controller cannot correct due to its time-delay dominant dynamics. In addition, the consistency controller tends to have a resonant mode due to its time-delay dominant dynamics. The net result is that the variability in the mass flow of fiber is amplified by the joint action of consistency control and stock flow control. This variability is driven by the unique limit-cycle behavior of each flow

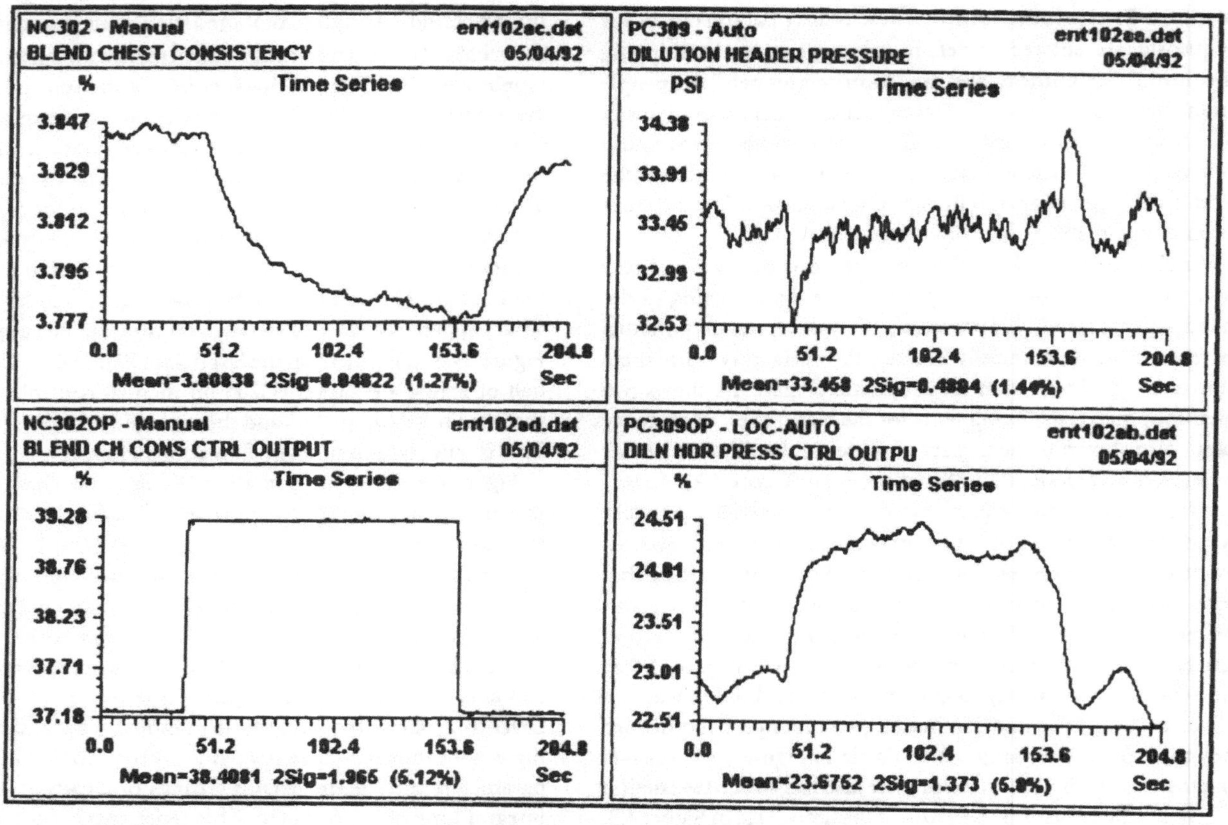

Figure 4.4 Interaction between blend chest consistency NC302 and dilution header pressure PC309.

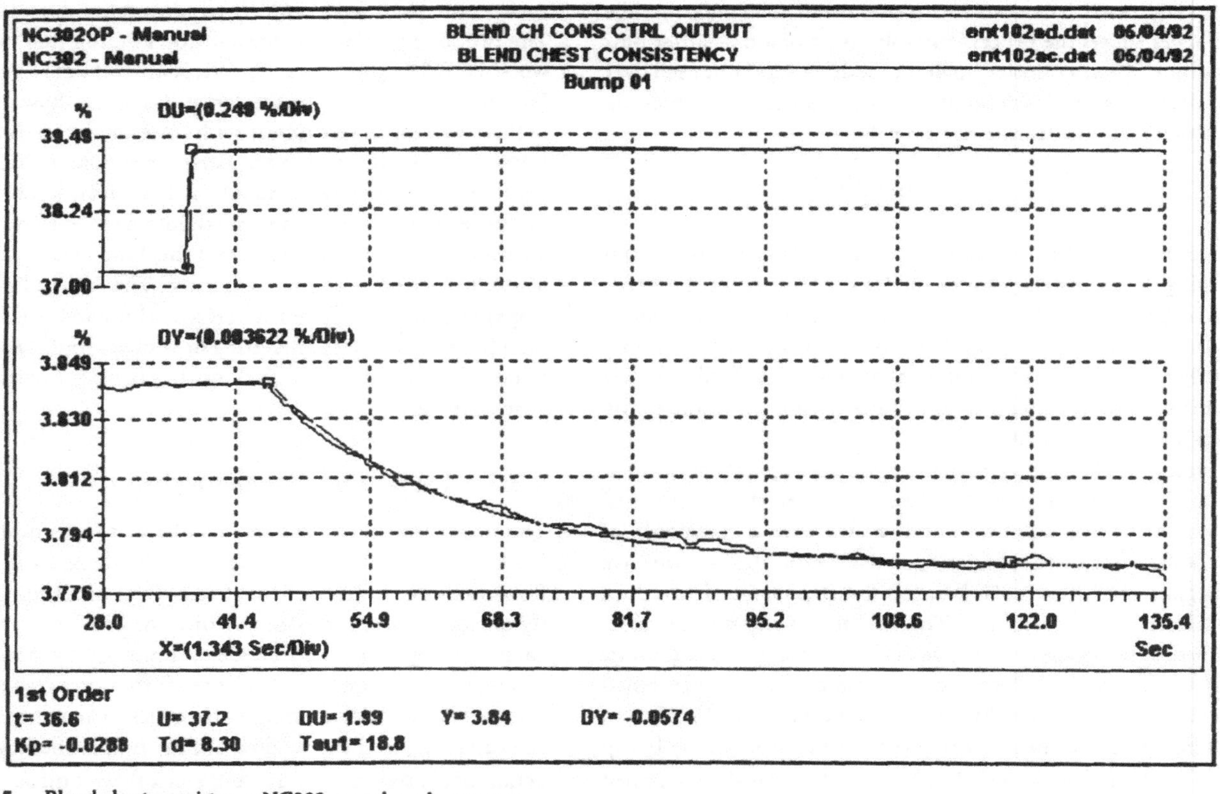

Figure 4.5 Blend chest consistency NC302 open loop bump test.

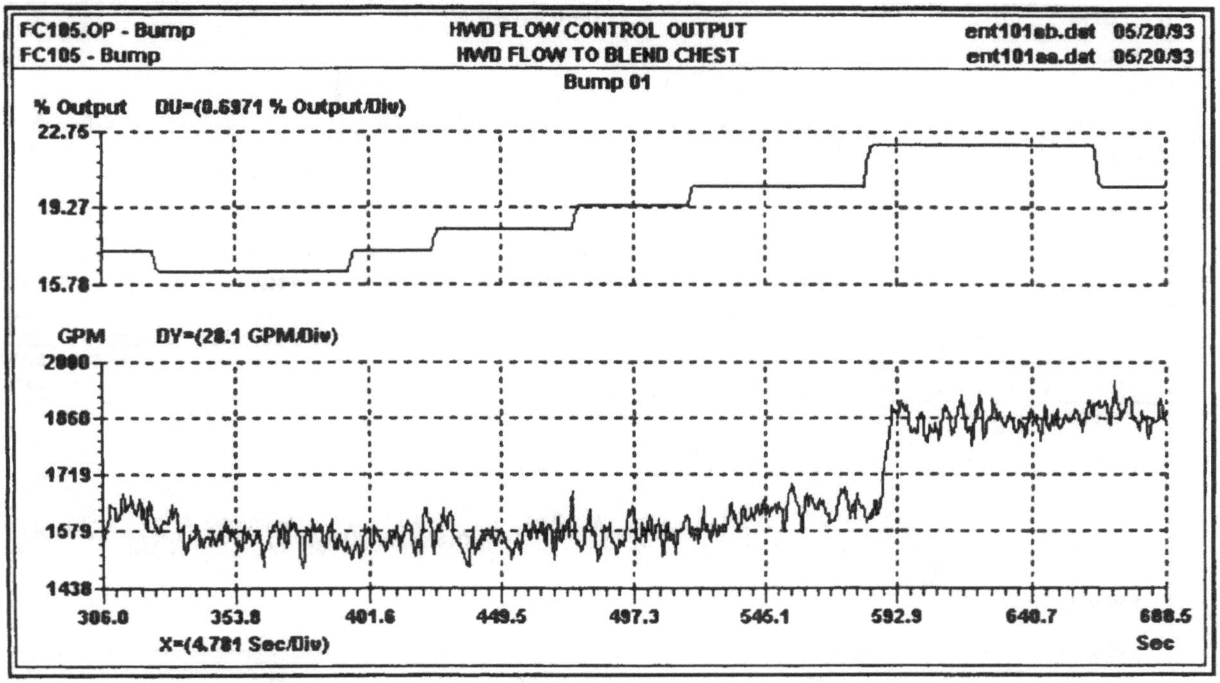

Figure 4.6 Hardwood flow FC105 bump tests showing about 5% backlash/stiction.

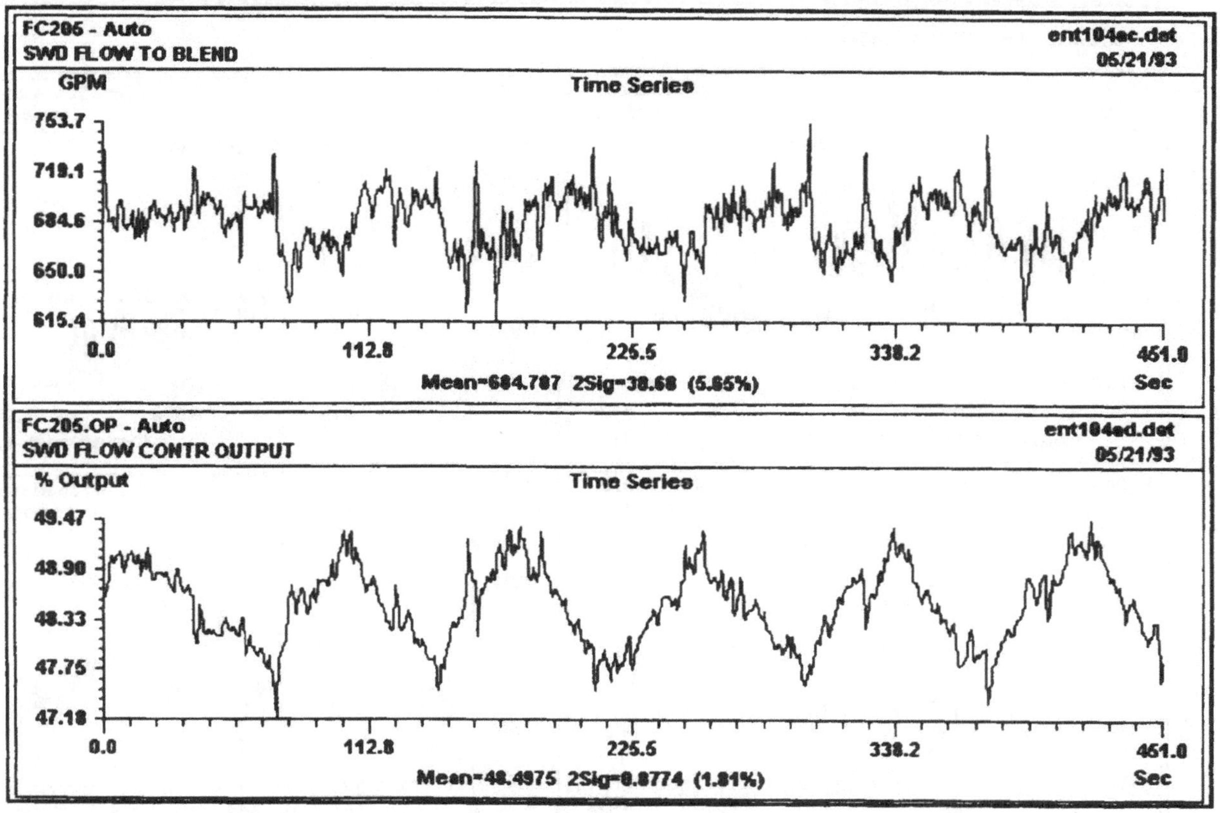

Figure 4.7 Softwood flow FC205 with 1.2% stiction showing a limit cycle.

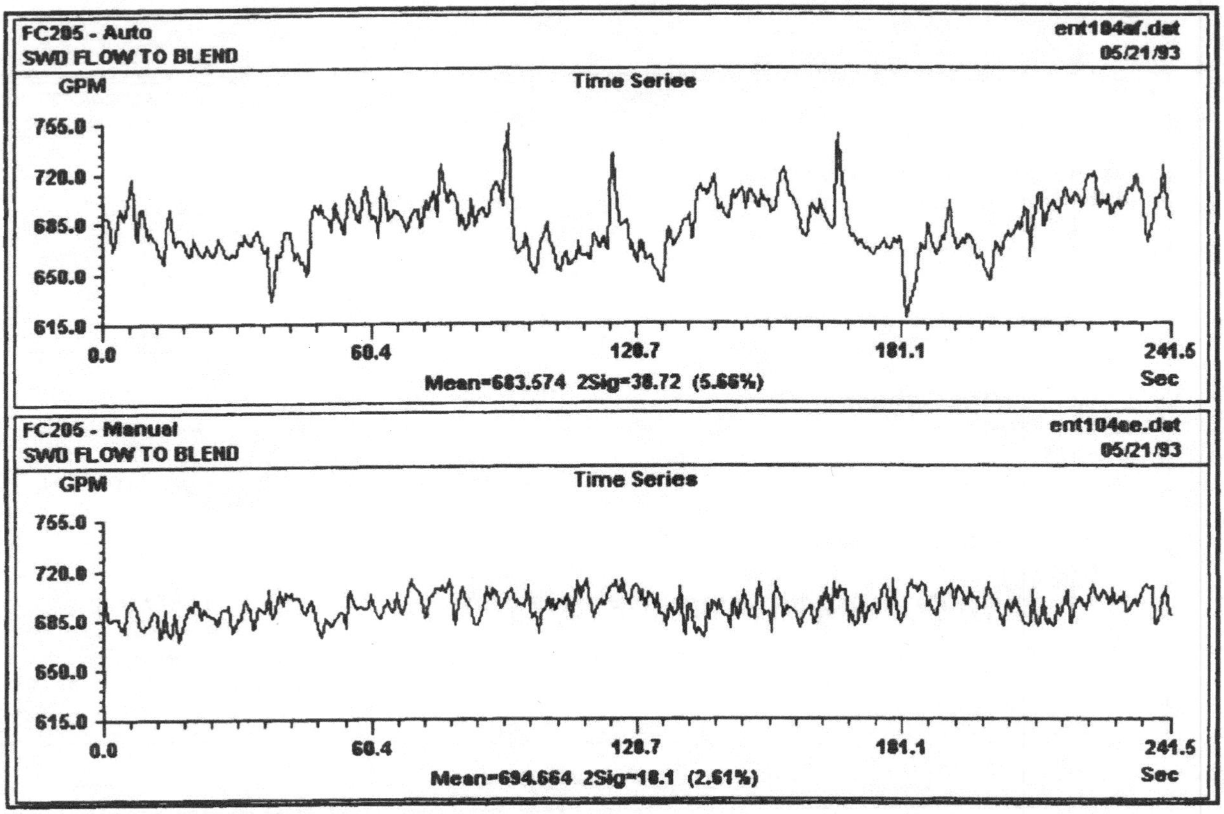

Figure 4.8 Softwood flow FC205 on and off control.

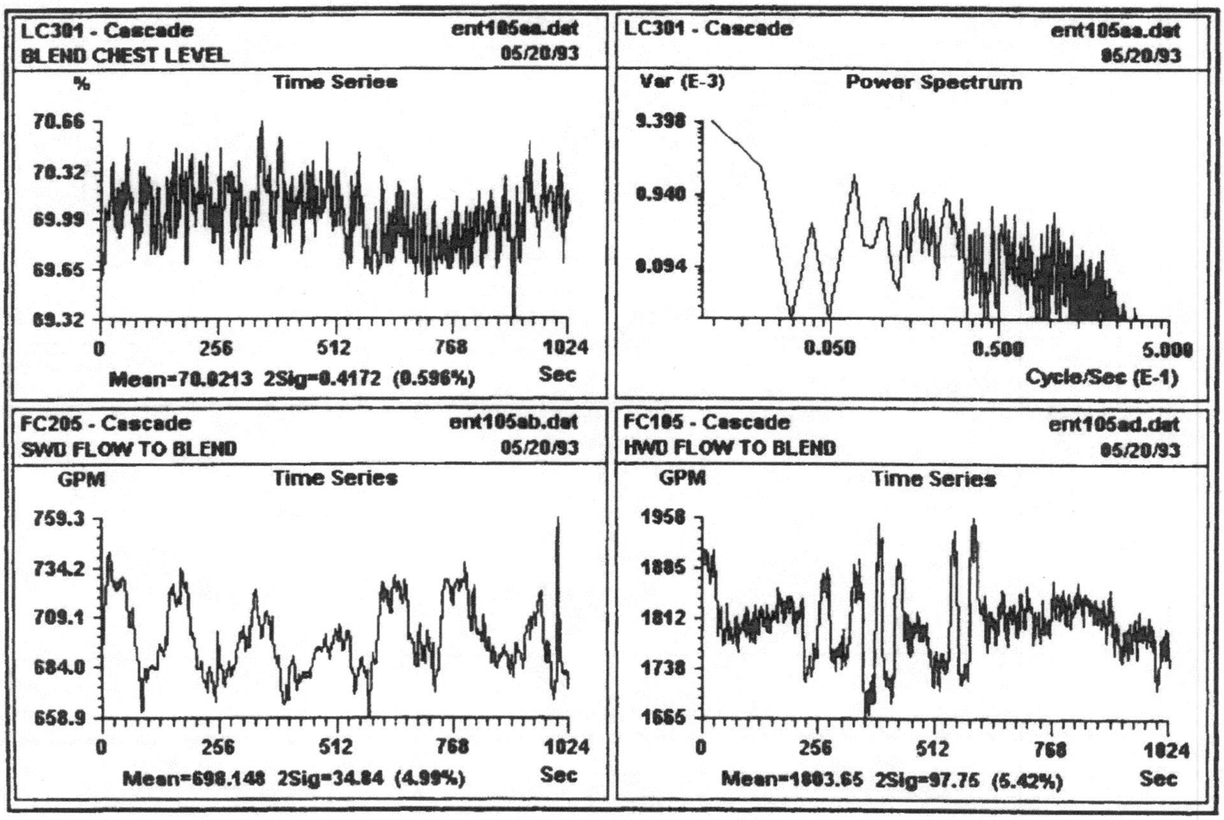

Figure 4.9 Blend chest level LC301 on automatic control.

controller. As a result the refining process will suffer, with the bonding strength developed by each fiber species varying in different ways. In addition, the stock blend (e.g., 70% hardwood, 30% softwood) uniformity is also compromised by the different limit cycles. At times, the hardwood/softwood ratio is off by 5% in Figure 4.8. Clearly this cannot help to deliver uniformly blended stock to the paper machine.

4.3.4 Final Product Variability

Final paper product measurements include: basis weight (mass per unit area), moisture content, caliper (thickness), smoothness, gloss, opacity, color, and brightness. Of these, basis weight is the most important and the easiest to measure, as the sensor is essentially an analog sensor. Almost all of the other sensors are digital with limited bandwidth. The final quality sensors of most paper machines are located on the scanning frame, which allows the sensors to traverse the sheet at the *dry end*. Typically each traverse takes 20 to 60 seconds. During each traverse, sensor averages are calculated to allow feedback control in the "machine-direction" (MD). At the same time, data vectors are built during the sheet traversing process to measure and control the "cross-direction" (CD) properties. To measure "fast" basis weight variability the traversing must be stopped by placing the scanning sensors in "fixed-point" (MD and CD controls are suspended for the duration). While in fixed point, it is possible to collect data from the basis weight sensor (beta gauge) at data rates limited only by the bandwidth of the data collection equipment. Figure 4.10 shows a fixed-point basis weight data collection run. The average basis weight is 55 g/m^2 (gsm), with variability of 3.29%. Clearly there is a strong tendency to cycle at about 0.2 Hz, or 5 seconds per cycle. This is evident in the time series plot (upper left), in the power spectrum plot (upper right), in the autocorrelation function (bottom left), and in the cumulative spectrum (bottom right), which shows that about 25% of the variance is caused by this cycle. The cause of the cycle is not related to the blend chest data but rather to variability in the headbox area. Nevertheless, the time series data indicates that it is worthwhile to identify and eliminate this cycle. Figure 4.11 shows similar data collected at a slower rate for a longer period of time. Once again the mean basis weight is 55 gsm, and the variability is 3.7%. The time series data indicates a relatively confused behavior. From the cumulative spectrum it is clear that 50% of the variance is caused by variability slower than 0.02 Hz (50 seconds/cycle). From the power spectrum it is evident that there is significant variability from 0.003 Hz (5.6 minutes/cycle) through to about 0.0125 Hz (1.3 minutes/cycle). In addition there is considerable power at the fundamental period of 17 minutes. All of these frequencies fit the general behavior of the blend chest area, including the hardwood and softwood flows, as well as the blend chest level. Proof of cause and effect can be established by preventing the control-induced cycling (placing the control loops on manual) and collecting the basis weight data again. Figure 4.12 shows the behavior of basis weight over a period of about 2.5 hours based on scan average data (average of each traverse) collected every 40 seconds. The mean is still 55 gsm, and the variability is now 1.39%. The Nyquist frequency of 0.758 cycles/minute corresponds to a scanning time of 40 seconds. There is significant power at 0.38 cycles/minute (2.6 minutes/cycle), and at 0.057 cycles/minute (17 minutes/cycle). Once again these frequencies correspond to some of the behavior in the blend chest area.

The purpose of this small case study was only to illustrate typical behavior of automatic control loops in a fairly realistic case study. In practice, an operating unit process, such as the example paper machine, has several hundred control loops of which about a hundred would be examined carefully during a variability audit to investigate the causal interrelationships between these individual variables and the final product.

4.3.5 Variability Audit Typical Findings

The case study results are typical of audit results. As of now, our audit experience extends over approximately 200 audits, with 50 to 100 control loops analyzed in each audit. Audit recommendations frequently include 20 to 50 items covering process design, control strategy design, valve replacement, and loop tuning issues. Findings by loop are categorized into several categories as listed in Table 4.2.

TABLE 4.2 Variability Audit Findings by Category.

Loop category: Loops which...	%
reduce variablility (have adequate/good design, equipment, and tuning)	20
cycle and increase variability due to control equipment (valve backlash, stiction, etc.)	30
cycle and increase variability due to poor tuning	30
require control strategy redesign to work properly	10
requires process redesign or changes in operating practice to reduce variability	10
Total	100

These findings say that only 20% of the control loops surveyed actually reduce variability over the "short term" in their "as-found" condition. "Short term" means periods of about 15 minutes for fast variables, such as flow loops, and periods of one or two hours for loops with slow dynamics, such as basis weight. Let us examine the chief reasons for inadequate performance which are control equipment, tuning, and design.

4.3.6 Control Equipment

The control equipment category accounts for about 30% of all loops and represents a major potential for improvement. Control valve nonlinear behavior is the primary problem. However, also included are other problems, such as installation (e.g., sensor location), excessive filtering, or equipment characteristics (e.g., inappropriate controller gain adjustment ranges or control mode features). The blend chest case study discussed several problems. The most serious were the control valve backlash/stiction problems exhibited by FC105 and FC205. To some extent, the blend

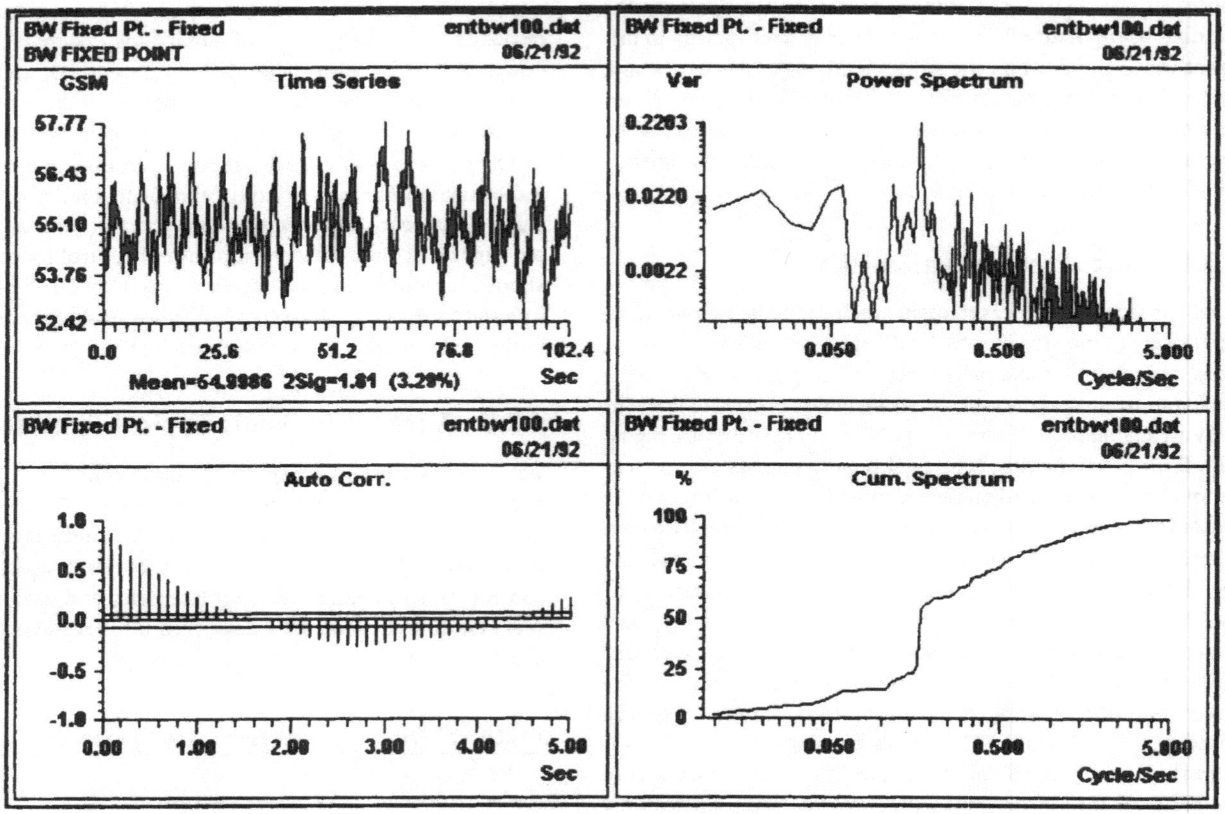

Figure 4.10 Paper final product—Basis weight (0.01 to 5Hz).

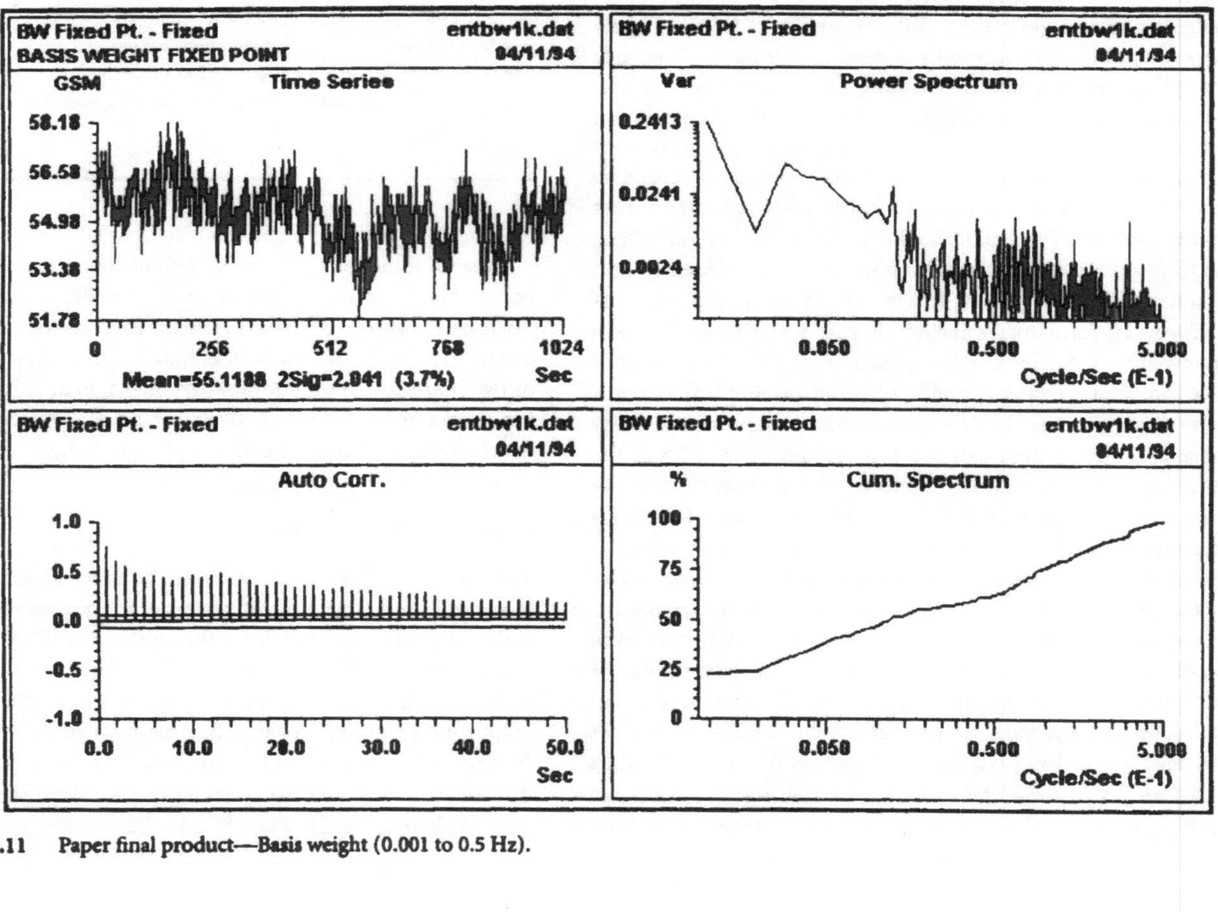

Figure 4.11 Paper final product—Basis weight (0.001 to 0.5 Hz).

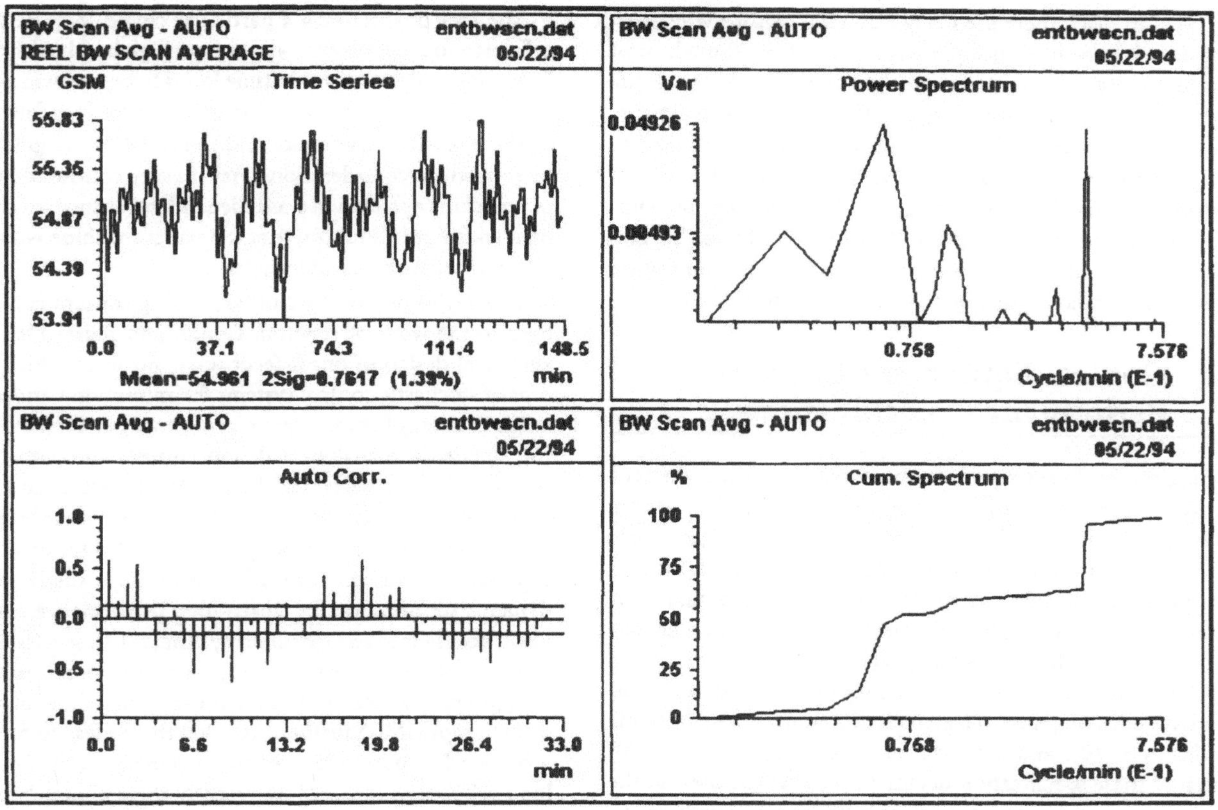

Figure 4.12 Paper final product — Basis weight (0.0001 to 0.0125 Hz).

chest consistency loop NC302 also suffered from control valve induced limit cycling. Lesser problems included the inappropriate consistency sensor location for NC302 (time delay too long) and the excessive filter time constant.

The serious problems in this category are the control valve nonlinearities such as backlash, stiction, and issues relating to control valve positioners (local pneumatic feedback servo) which react to valve stiction with time delays that are inversely proportional to the change in the valve input signal. These nonlinearites cannot be modeled through "small-signal" linearization methods and are extremely destabilizing for feedback control. The only recourse is to eliminate such equipment from the process (best way of linearizing). This is discussed further in the section on dynamic performance specifications.

4.3.7 Loop Tuning

This category accounts for 30% of all loops and represents a major potential for improvement. In the case study of the blend chest consistency, NC302 and the blend chest level, LC301 were in this category. Until recently, loop tuning has been an "art" in the process industries done by trial and error. The responsibility for loop tuning in a pulp and paper mill normally rests with the Instrumentation Shop and is carried out by Instrument Technicians. Until recently, the only training that most instrument technicians have received is the Quarter-Amplitude-Damping method of Ziegler and Nichols [21]. This method is not robust (gain margin of 2.0 and phase margin of 30°) and tends to produce excessive cycling behavior. As a result most people have taught themselves to "tune by feel." Recent training [4] throughout the industry has introduced many engineers and technicians to "Lambda Tuning," discussed in detail in a later section.

4.3.8 Control Strategy Redesign, Process Redesign, and Changes in Operating Practice

Control equipment and loop tuning problems can be corrected at relatively little cost. The remaining opportunity to gain another 20% advantage in variability reduction is spread over a number of areas including control strategy redesign, process redesign, and changes in operating practice. Issues relating to process and control design are discussed in a later section. Changing operating practice involves people and how they think about their operation. Making changes of this kind involves creating awareness and changing "culture."

4.3.9 Pulp and Paper Mill Culture and Awareness of Variability

The reader can be forgiven for the obvious question: Are people aware of the variability that they actually have? The answer is generally "No." Space does not permit a thorough discussion of this subject which determines a mill's ability to move forward (see [9]). In brief, there are many historical reasons why operating data is presented in a particular way. Even though most mills use digital "distributed control systems" (DCS) today, the psychology of the past has carried over to the present. In current DCSs, process data is manipulated in ways which tend to strip the true nature of the process variability from it. Some of these include slow input sampling rates, data aliasing, slow update rates for operator consoles and mill data archives, "report-by-exception" techniques (to minimize digital communication traffic), and display of data on a scale of 0–100% of span [11].

4.4 Control Equipment Dynamic Performance Specifications

Nothing can be done to improve control performance when the control equipment does not perform adequately. The control valve has been identified in audit findings as the biggest single cause of process variability in the pulp and paper industry. Another problem identified in audits is that some automatic controllers have inappropriate design features, such as gain adjustment ranges, which exclude legitimate yet effective settings. To help the industry move forward, dynamic performance specifications have been prepared. These include a control valve dynamic specification [10], and a controller dynamic specification [8]. Currently these specification documents are widely used in the pulp and paper industry, frequently for equipment purchase. As a result, most valve suppliers to the pulp and paper industry are actively using the valve specification as a performance guide to test their products [20]. In a number of cases new or improved control valve products have emerged. It is anticipated that future control valves will have combined backlash and stiction as low as 1% or lower, and time delays in fractions of a second. When these levels cannot be tolerated, mills are considering variable speed drive technology for pumps as an alternative to the control valve.

4.5 Linear Control Concepts — Lambda Tuning — Loop Performance Coordinated with Process Goals

In the pulp and paper industry, the term "Lambda Tuning" refers to a concept in which the designer/tuner specifies the loop performance by choosing the closed-loop time constant (called Lambda (λ)). The calculation of the parameters required by a specific control algorithm is determined by a "Lambda tuning rule," — a transformation of the closed-loop time constant, the process dynamics, and the control algorithm structure into the appropriate gains. Lambda tuning is considered a useful concept for designing and tuning control loops in continuous processes, especially when low product variability is the objective. The concept of specifying a desired performance for each loop is the key to coordinating the overall performance of a unit process with hundreds of control loops.

The idea of specifying a particular closed-loop performance is foreign to most people, who equate high bandwidth with performance, and hence would tune loops to be as fast as practically possible. The resulting closed-loop dynamics are a function only of the open-loop dynamics and the stability margins required by robustness considerations. Yet if every control loop in a paper machine application was independently tuned for maximum bandwidth, it is doubtful that the paper machine would run at all, let alone make saleable paper!

The beginnings of the Lambda tuning concept can be traced back to the work of Newton, Gould, and Kaiser [16], and the analytical design of linear feedback controls in which the concept of stochastic control systems for minimum bandwidth and minimization of mean-square error was put forward. Dahlin [6], applied the analytical design concept to controlling processes with time delay and first-order dynamics and used the notion of a user-selected bandwidth. This was done by specifying the desired closed-loop pole position, λ. Dahlin originated the expression, "Lambda Tuning," — meaning a single parametric dynamic specification for closed-loop performance. Later these ideas, generalized and extended by others [18], specified the control loop performance by choosing a closed-loop time constant, as opposed to a pole location. The evolution of the internal control concept (IMC) further extended these ideas to processes of arbitrary dynamics, while considering the associated problem of loop robustness [15]. More recently these concepts have been applied to the development of "tuning rules" for simple PI and PID controllers [5]. Finally, Lambda tuning has been adopted as the preferred controller tuning approach in the pulp and paper industry [17].

The mathematics of "Lambda tuning" is well established in the literature cited. But almost nothing is said about selecting the closed-loop time constant λ. Only the trade-off between performance and robustness [15] is recognized, and that robustness suffers as the loop performance specification is more aggressive. Yet it is precisely the ability to choose a specific closed-loop time constant for each loop that makes Lambda tuning so useful in the pulp and paper industry. This allows uniquely coordinated tuning of all of the loops which make up a unit process to enhance the manufacturing of uniform product. This idea is explored in the following blend chest design example.

4.5.1 Design Example — Blend Chest Dynamics using Lambda Tuning

For the now familiar blend chest example, let us establish the manufacturing objectives and translate these into control engineering objectives as desired closed-loop time constants for each loop.

Blend Chest Manufacturing Objectives

The manufacturing or papermaking objectives are:

1. provide a uniformly blended source of paper stock, at a desired blend ratio

2. insure that the stocks leaving both the hardwood and softwood chests are at uniform consistencies
3. insure uniform refining of both stocks, so that these fibers have uniformly developed bonding strength
4. insure that the blend chest level is maintained at the level set point, and never spills over or goes below, say 40%, while allowing the paper machine production rate to change by, say 20%

Uniform stock delivery from the hardwood and softwood chests is critically important for two reasons. First, the refiners depend on uniform stock delivery to insure that refining action is uniformly applied. Overrefined fibers provide higher bonding strength but reduce the drainage rate on the wire and cause high sheet moisture. Underrefined fibers cause loss of sheet strength, higher drainage rate, and lower moisture content in the sheet. Overrefined and underrefined fibers are discrete qualities of stock and do not "average-out."

Blend Chest Control Objectives:

1. Maintaining tight control over hardwood (NC104) and softwood (NC204) consistencies is certainly the most important objective, because it will insure constant and uniform fiber delivery. These two loops are then of prime importance. At the high density storage chests, the disturbance energy is normally very high and somewhat unpredictable. Hence, these loops should be tuned for maximum practical bandwidth. However, the consistency loops have time-delay dominant dynamics and, as a result, a high-frequency resonance (discussed later) slightly above the cutoff frequency. This high-frequency resonance typically occurs at about 0.2 radians/second (about 30 seconds per cycle, frequency depends on tuning) and should be minimized so as not to amplify process noise at this frequency and allow this resonance to propagate further down the paper machine line. The choice of the closed-loop time constant (λ), determines the extent of the resonance. For a λ two times the time delay, the resonance will be about +2 dB (AR = 1.26), and a λ equal to the time delay will cause a resonance of +3.6 dB (AR = 1.51). Because the noise in the resonant band is amplified by over 30%, this choice is too aggressive. The hardwood and softwood consistency loops with about 5 seconds of time delay, should be tuned for a λ of about 15 seconds.
2. The blend chest has a residence time of 20 minutes when full. At the normal operating point of 70% full, there are 14 minutes of stock residence time and an air space equivalent to 6 minutes. We are told that the paper machine production rate changes will be no more than 20%. A paper machine stoppage will cause the blend chest to overflow in 6 minutes. A 20% reduction in flow would cause the chest to overflow in 30 minutes. The process immediately down stream from the blend chest is the machine chest, which is itself level controlled. As a result the actual changes in blend chest outflow will be subject to the tuning of the machine chest level controller. The purpose of the blend chest is to provide a surge capacity. Fast tuning of the level will tightly couple disturbance in the outflow to the very sensitive upstream refiner and consistency control problems. Hence the level controller LC301, should be tuned as slowly as practically possible (minimum bandwidth), subject to the stipulation that the level must never make excursions greater than 30%. A tank level, with the inlet flows controlled by a Lambda tuned PI controller, has a load response to a step change in outflow in which the maximum excursion occurs in one closed-loop time constant. A good choice for the closed-loop time constant may then be 10 or 15 minutes. Either of these choices will maintain the level well within the required 30%. It is important to insure that the level controller never oscillates, because oscillation is damaging to all of the surrounding loops. Hence the tuning must insure that the closed-loop dominant poles remain on the negative real axis (discussed later).
3. The hardwood (FC105) and softwood (FC205) flows are the inner cascade loops for the level controller LC301, and their set points are adjusted via ratio stations which achieve the 70:30 blend ratio. To maintain a constant blend ratio under dynamic conditions, both flows must be tuned for the same absolute closed-loop time constant. Yet the hardwood and softwood lines probably differ, in pumps, pipe diameters, pipe lengths, valve sizes, valve types, flow meters, hence, open-loop dynamics. Any tuning method in which the closed-loop dynamics are a function only of the open-loop dynamics (e.g., Ziegler–Nichols, 5% overshoot, etc.) will produce different closed-loop dynamics for the hardwood and softwood flows. Only Lambda tuning can achieve the goal of identical closed-loop dynamics. The actual choice of closed-loop time constants for both flows is also critically important. Both of these flows can be tuned to closed-loop time constants under 10 seconds without too much trouble. However, too fast a time constant will disturb the consistency loops and the refiner operation through hydraulic coupling. This will be especially true when accounting for the nonlinear behaviour of the valves. Too slow a time constant will interfere with the operation of the level controller which requires the inner loops to be considerably faster than its closed-loop time constant. Analysis is required to determine the exact value of the flow time constant to insure adequate stability margins for the level control. However, for

a level λ of 15 minutes, it is likely that the flows can be tuned for λs in the two minute range. This choice will insure relatively light coupling between the flow and the consistency loops when tuned for closed-loop time constants of 15 seconds.

4. The blend chest consistency loop NC302 can be tuned in the same exact way as the other consistency loops. Alternatively, if a repeatable noise model structure for this loop can be identified by studying the noise power spectrum on manual, it may be possible to tune NC302 using minimize-variance principles. If the noise model is the fairly common integrating-moving-average (IMA) type (drifts plus noise) and if the corner frequency of the noise structure is slower than the frequency at which the consistency loop has significant resonance [12], λ can be chosen to match the corner frequency of the noise structure, thereby, approximately cancelling the low-frequency drifts and producing a "white" power spectrum.

Conclusion — Design of Blend Chest Dynamics with Lambda Tuning

The blend chest example illustrates six reasons for choosing specific closed-loop time constant values: 1) maximum nonresonant bandwidth for the loops considered the most important (hardwood and softwood consistencies), 2) minimum possible bandwidth for the least important loop (level), 3) equal closed-loop time constants for loops controlling parallel processes which must be maintained at a given ratio (hardwood and softwood flows), 4) a closed-loop time constant dictated by the dynamics of an upper cascade loop (flow loops), 5) a closed-loop time constant for loops of lesser importance sufficiently slower than adjacent coupled loops of greater importance (flows versus consistencies), and, finally, 6) choosing the closed-loop time constant to minimize variance through matching the regulator sensitivity function to the inverse of an IMA type noise structure by matching the regulator cut-off frequency to the IMA corner frequency.

4.5.2 Pulp and Paper Process Dynamics

Next, let us consider the types of process dynamics present in the pulp and paper industry. Most process dynamics can be described by one of the following two general transfer functions shown below:

$$G_P(s) = \frac{K_P(\beta s + 1)e^{-sT_d}}{s^a(\tau_1 s + 1)(\tau_2 s + 1)} \quad \text{or}$$

$$G_P(s) = \frac{K_P(\beta s + 1)e^{-sT_d}}{s^a(\tau_1^2 s^2 + 2\varsigma\tau_1 s + 1)} \tag{4.2}$$

where $a = 0$ or 1, β = lead time constant(positive or negative), τ_1, τ_2 = time constants ($\tau_1 \geq \tau_2$), ς = damping coeficient, and T_d = deadtime. Typical parameter values for pulp and paper process variables are listed in Table 4.3.

4.5.3 Lambda Tuning

Lambda tuning employs the general principles used in the Internal Model Control (IMC) concept which has the following requirements for a controller:

1. The controller should cancel the process dynamics, process poles with controller zeros, and process zeros with controller poles.
2. The controller should provide at least one free integrator (Type 1 loop) in the loop transfer function to insure that offsets from set point are canceled.
3. The controller must allow the speed of response to be specified (Lambda (λ) to be set by designer/tuner).

For the the process transfer function of Equation 4.2, these general principles call for PID control with a series filter (PID.F). For nonintegrating processes, this translates into the following controller transfer functions:

$$G_C(s) = \frac{(\tau_1 s + 1)(\tau_2 s + 1)}{K_P(s)(\lambda + T_d)(|\beta| s + 1)} \quad \text{or}$$

$$G_C(s) = \frac{(\tau_1^2 s^2 + 2\varsigma\tau_1 s + 1)}{K_P(s)(\lambda + T_d)(|\beta| s + 1)} \tag{4.3}$$

where closed-loop time constant = λ, setpoint response bandwidth (inverse sensitivity function) $\cong \frac{1}{\lambda}$, and load response bandwidth (sensitivity function) $\cong \frac{1}{(\lambda + T_d)}$.

For integrating processes, it is normally necessary to specify control loops of Type 2 form, because a controller integrator is usually desirable to overcome offsets due to load disturbances. The controller required for this task is typically of the form,

$$G_C(s) = \frac{[(2\lambda + T_d)s + 1](\tau_1 s + 1)}{K_P s(\lambda + T_d)^2(|\beta| s + 1)}. \tag{4.4}$$

Whereas the form of these controllers can be implemented in PID.F form, in most cases a PI controller will suffice, especially when the performance objectives are reasonably modest relative to the loop robustness limits. This gives the following advantages: the series filters need not be implemented (not a standard feature of current distributed control systems (DCSs), the resulting controller is within the training scope of the instrument technician, and the large actuation "spikes" caused by derivative control are avoided. The form of the PI controller contained in most DCSs is

$$G_C(s) = K_C \frac{(T_R s + 1)}{T_R s}. \tag{4.5}$$

Equating the controller gain (K_C) and reset (or integral) time (T_R) to the general process parameter values of Equations 4.3 and 4.4 yields the following tuning rules [5]. Consider the two most important cases of process dynamics, first-order plus deadtime, and integrating plus deadtime, which together represent 85% of all process dynamics in the pulp and paper industry. The tuning rules are listed in Table 4.4 below.

These two tuning rules form the basis for the bulk of the Lambda tuning for single loops in the pulp and paper industry. Whereas the choice of closed-loop time constant λ, provides the

TABLE 4.3 Typical Dynamics of Pulp and Paper Process Variables.

Process Variable	K_P	a	β (sec.)	τ_1 (sec.)	τ_2 (sec.)	ζ	T_d (sec.)
Stock flow	1.0#	0	-	3	-	-	0.2
Stock flow with air entrainment	1.0#	0	8	15	3	-	0.2
Stock flow, flexible pipe supports	1.0#	0	-	3	-	0.2	0.2
Stock consistency	−1.0#	0	-	5	-	-	5
Stock pressure	1.0#	0	-	3	-	-	0.1
Headbox total head with "hornbostle"	1.0#	0	-	3	-	0.2	-
Headbox total head with variable speed fan pump	1.0#	0	-	1	-	-	-
Paper machine basis weight	*	0	-	45	-	-	100
Paper machine moisture	*	0	-	120	-	-	120
Pulp dryer basis weight	*	0	-	10	-	-	300
Dryer steam pressure	0.005#	1	300	20	-	-	-
Chest level	0.005#	1	-	-	-	-	-
Chest level with controlled flows	0.005#	1	-	15!	-	-	-
Bleach plant chlorine gas flow	1.0#	0	-	2	-	-	0.1
Bleach plant pulp brightness	1.0#	0	-	20	-	-	20
Bleach plant steam mixer temperature	1.0#	0	-	20	5	-	-
Bleach plant oxygen gas flow	1.0#	0	-	0.05	-	-	-
Bleach plant D1 Stage 'J tube' brightness	1.0#	0	-	120	-	-	120
Digester Chip Bin Level	0.005#	1	-	-	-	-	200
Boiler feed water deaerator level	0.005#	1	-	-	-	-	20
Boiler steam drum level	0.005#	1	−30	30	-	-	-
Boiler master - steam header control with bark boiler	0.005#	1	-	-	-	-	300

* varies
\# nominal value, depends on equipment sizing
! tuning dependent

mechanism for coordinating the tuning of many control loops, this must always be done with control loop stability and robustness in mind. Furthermore, variability audit experience indicates that the tendency of control loops to resonate should be avoided at all costs. Hence, closed-loop poles must always be located on the negative real axis, and time-delay-induced resonance must be limited to manageable quantities, never more than, say, +3 dB. Such design objectives should significantly limit propagation of resonances within pulp and paper processes. Let us examine these issues further.

TABLE 4.4 PI Tuning Rules for First-Order and Integrating Dynamics.

	Process Dynamics	K_C	T_R
1)	$G_P(s) = \frac{K_P e^{-sT_d}}{(\tau s+1)}$	$\frac{\tau}{K_P(\lambda+T_d)}$	τ
2)	$G_P(s) = \frac{K_P e^{-sT_d}}{s}$	$\frac{2\lambda+T_d}{K_P(\lambda+T_d)^2}$	$2\lambda + T_d$

4.5.4 Choosing the Closed-loop Time Constant for Robustness and Resonance

First-Order Plus Deadtime

Consider a consistency control loop, as an example of a first-order plus deadtime process, with the following process parameters (time in seconds, ignore negative gain): $K_P = 1$, $\tau = 3$, and $T_d = 5$. By applying the Lambda tuning rule of Table 4.4, the loop transfer function becomes

$$G_L(s) = \frac{(\tau s+1)}{K_P(\lambda+T_d)s}\frac{K_P e^{-sT_d}}{(\tau s+1)} = \frac{e^{-sT_d}}{(\lambda+T_d)s}. \qquad (4.6)$$

As long as pole-zero cancellation has been achieved (or nearly so), the resulting dynamics depend only on T_d and the choice of λ. Let us consider what will happen as λ is varied as a function of T_d. Figure 4.13 shows the root locus plot for the cases $\lambda/T_d = 1, 2, 3,$ and 4. The deadtime is approximated by a first-order Pade' approximation (open-loop poles are *x*s, zeros are *o*s, and closed-loop poles are open squares).

Figure 4.13 shows the time-delay pole/zero pair of the Pade' approximation at ±0.4, together with the controller integrator pole at the origin. The most aggressive of the tuning choices, $\lambda/T_d = 1$, $\lambda = T_d = 5$ has a damping coefficient of about 0.7 and hence is just starting to become oscillatory. Figure 4.14 shows the sensitivity functions plotted for all four cases. The resonance for the case of $\lambda = 2T_d = 10$ is fairly acceptable at +2 dB, or an amplitude ratio of 1.26, hence noise amplification by 26%. From the viewpoint of robustness, this tuning produces a gain margin of 4.7 and a phase margin of 71°, a fairly acceptable result considering that changes in loop gain by a factor of two can be expected and changes of 50% in both time constant and deadtime can also be expected. This tuning is thus recommended for general use. On the other hand, a choice of $\lambda = T_d = 5$ is barely acceptable, having a resonance of +3.6 dB, (amplitude ratio of 1.51), a gain margin of only 3.1, and a phase margin of

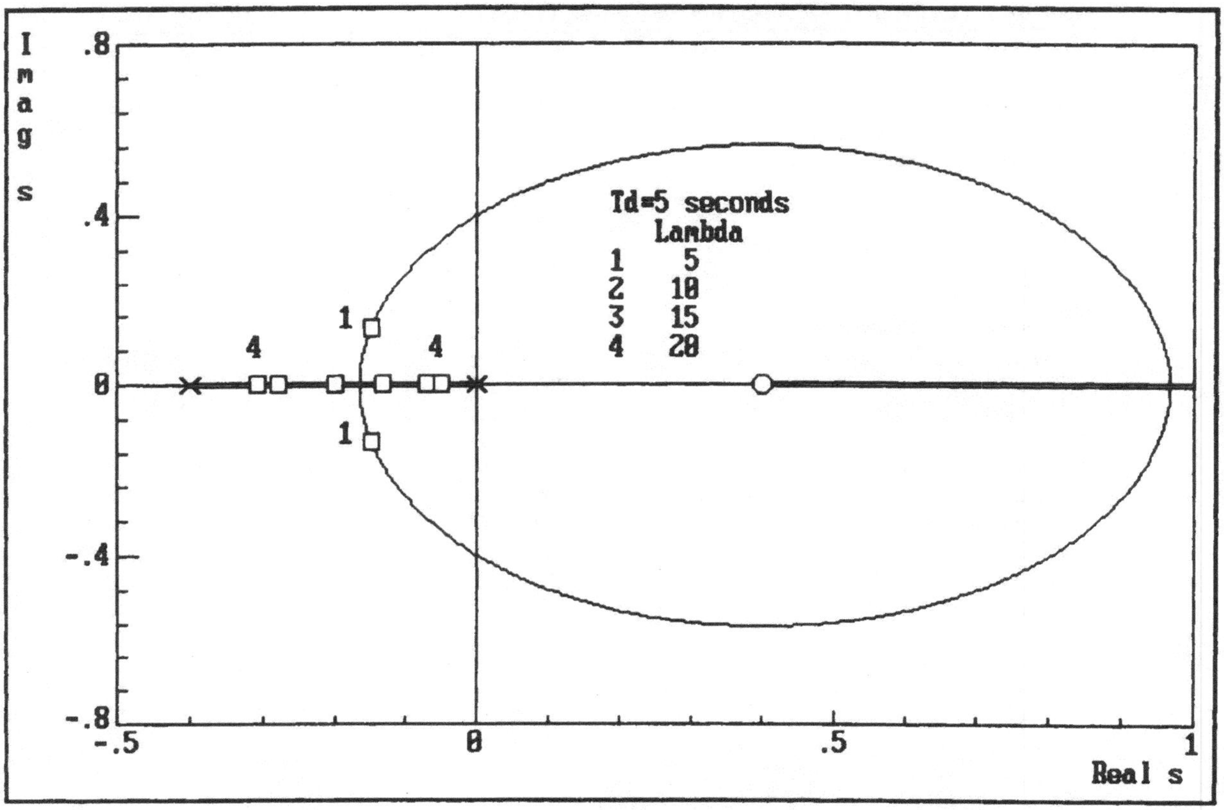

Figure 4.13 Consistency control loop with PI controller tuned for $\lambda/T_d = 1, 2, 3, 4$.

61°. A summary of these results is contained in Table 4.5. These values represent the limiting conditions for the tuning of such loops based on robustness and resonance. The actual choices of closed-loop time constants are likely to be made using process related considerations as outlined in the blend chest example.

Integrator Plus Deadtime

Now consider the integrator plus deadtime case. The loop transfer function is

$$G_L(s) = \frac{[(2\lambda + T_d)s + 1]}{K_P(\lambda + T_d)^2 s} \frac{K_P e^{-sT_d}}{s}$$

$$= \frac{[(2\lambda + T_d)s + 1]e^{-sT_d}}{(\lambda + T_d)^2 s^2} \quad (4.7)$$

Let us use a digester chip bin level example to illustrate this case. Wood chips are transported from the chip pile via a fixed speed conveyor belt to the chip bin. The transport delay is 200 seconds. Let us approximate this as 3 minutes for the sake of this example. The chip bin level controller adjusts the speed of a variable speed feeder which deposits the chips on the belt. The process parameters are (time in minutes): $K_P = 0.005$, and $T_d = 3$. From Equation 4.7 the loop transfer function is a function only of the deadtime and the closed-loop time constant. Let us consider two cases, $\lambda/T_d = 2$ and 3.

Figure 4.15 shows the root locus plot for the case of $\lambda = 2T_d =$ 6 minutes. The root locus plot shows the Pade' pole/zero pair approximation for deadtime at $\pm 2/3$, the reset zero at $-1/9$, and two poles at the origin. The closed poles are located on a root locus segment which moves quickly into the right half-plane (RHP), and the poles have a damping coefficient of only 0.69. Even more aggressive tuning speeds up the controller zero and increases the loop gain. This causes the root locus segment to bend towards the RHP earlier and reduce the damping coefficient.

Clearly this and faster tunings are far too aggressive. Figure 4.16 shows a root locus plot for $\lambda = 2T_d = 9$ minutes. The slower controller zero and lower loop gain cause the root locus to encircle the controller zero fully before breaking away again. Figure 4.16 shows an expanded plot of the region near the origin.

Even though there is still a branch containing a pair of poles which goes to the RHP, the system is quite robust to changes in loop gain. Changes in deadtime move the deadtime pole/zero pair closer to the origin as the deadtime lengthens, causing the eventual separation of the root locus branch as in Figure 4.15. From this analysis, it is clear that for reasons of robustness and resonance, the closed-loop time constant should not be made faster than three times the deadtime. Figure 4.17 shows the load response for the chip bin level control tuned for $\lambda = 3T_d = 9$ minutes, for a step change in outflow from the chip bin. The level initially sinks until the controller manages to increase the flow into the chest, to arrest the change for the first time. This occurs at 9.9 minutes in this case. For minimum-phase systems (no deadtime or RHP zeros), this point occurs exactly at λ. Otherwise the closed-loop time constant is a good approximation for the time when the load disturbance will be arrested. The time

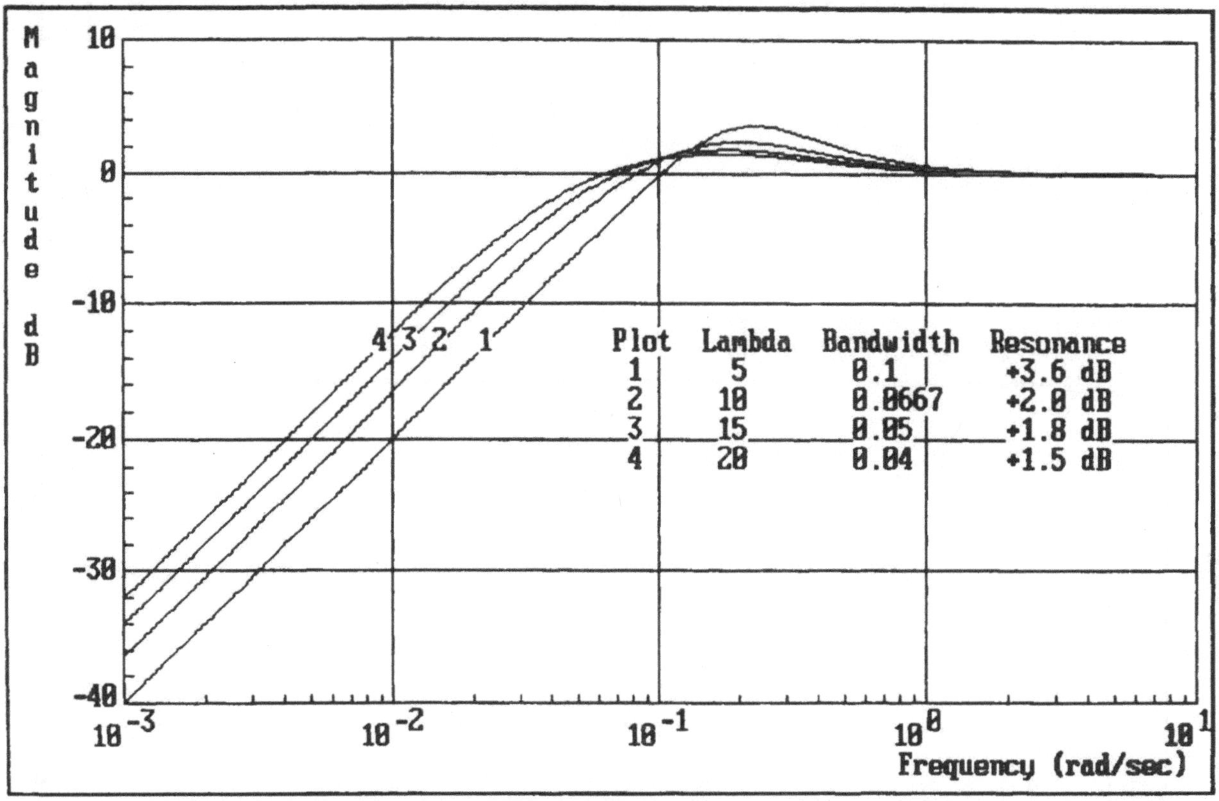

Figure 4.14 Consistency loop sensitivity functions for $\lambda / T_d = 1, 2, 3, 4$.

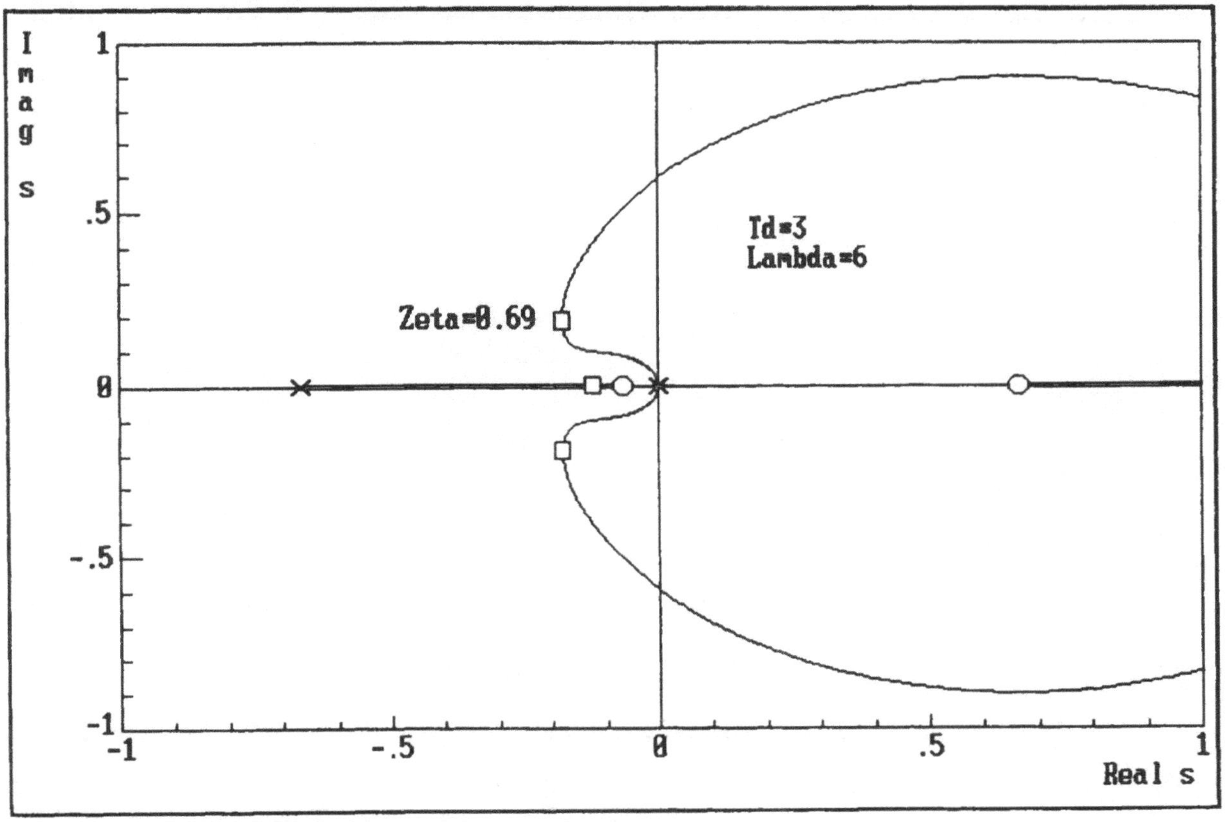

Figure 4.15 Digester chip bin level root locus for $\lambda = 2T_d$.

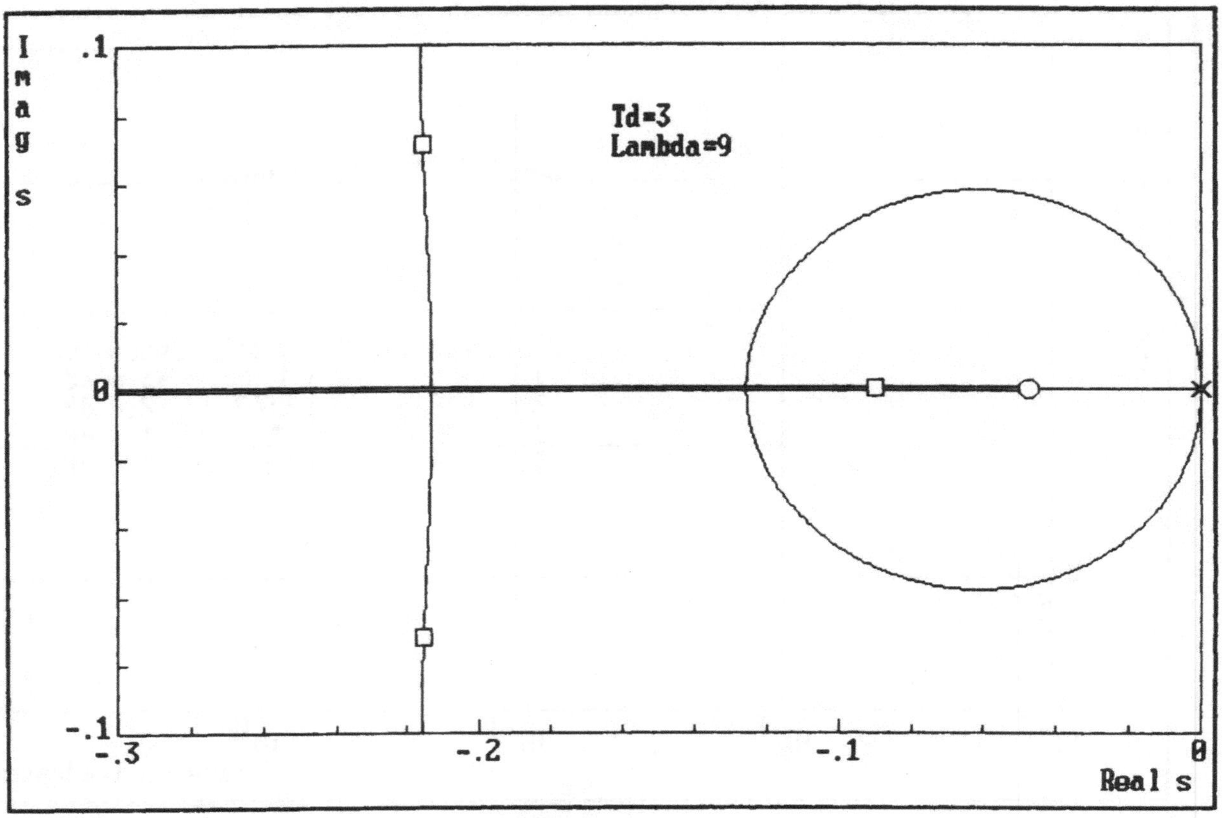

Figure 4.16 Digester chip bin level root locus for $\lambda = 3T_d$ expanded plot.

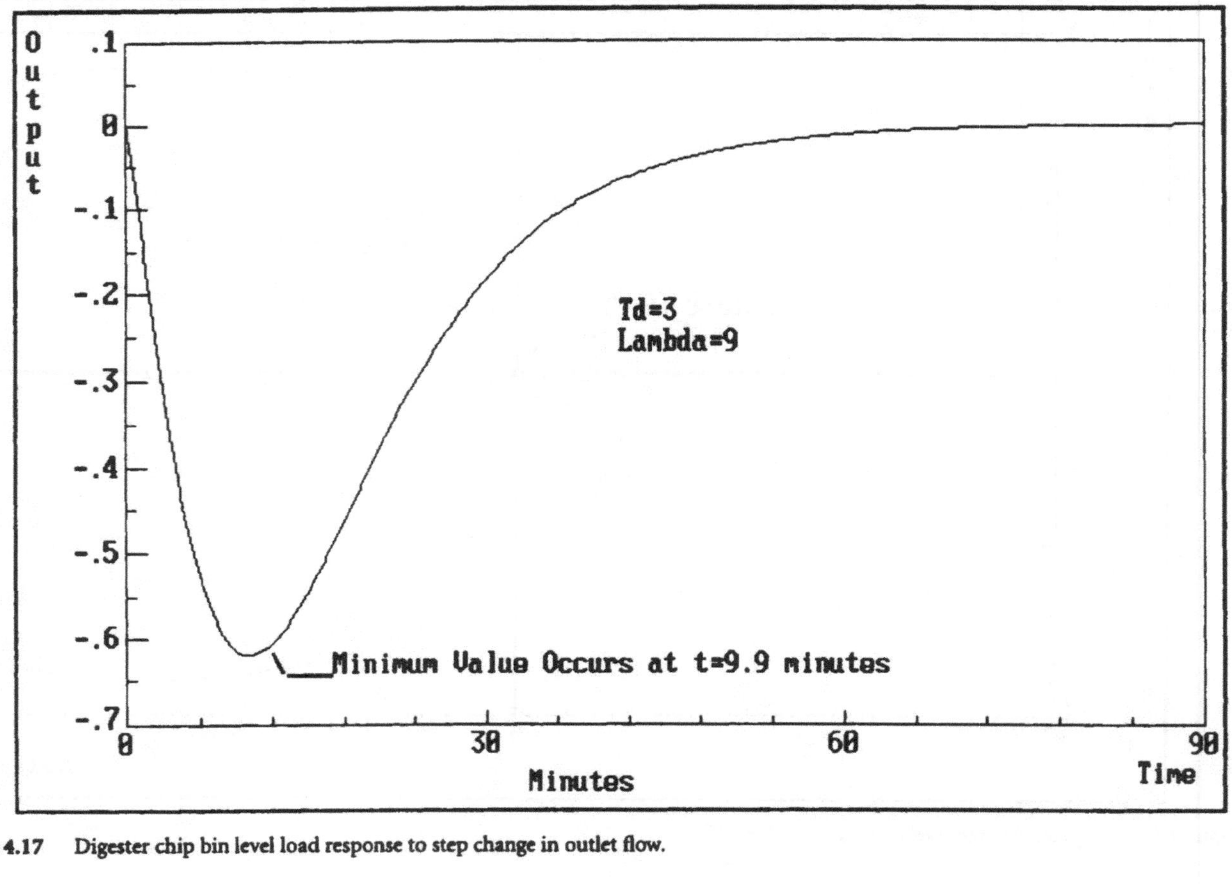

Figure 4.17 Digester chip bin level load response to step change in outlet flow.

TABLE 4.5 First-Order Plus Deadtime Lambda Tuning Performance/Robustness Trade-Off.

Tuning (λ/T_d) ratio	1	2	3	4
Closed-loop time constant λ (sec.)	5	10	15	20
Bandwidth $(1/(\lambda + T_d))$ radians/sec.	0.1	0.0667	0.05	0.04
Resonant Peak (dB)	+3.6	+2.0	+1.8	+1.5
Amplitude Ratio	1.51	1.26	1.23	1.19
Gain Margin	3.1	4.7	6.8	7.9
Phase Margin (°)	61	71	76	79

to correct fully for the whole change is approximately six times longer.

4.5.5 Multiple Control Objectives

A control loop can have multiple control objectives, such as

- good set point tracking,
- good rejection of major load disturbances, and
- minimizing process variability.

The tuning for each of these is unlikely to be the same. The first two relate to fairly large excursions, where speed of response is of the essence and process noise does not dominate. Minimizing process variability, on the other hand, is of major importance in the pulp and paper industry, and is essentially a regulation problem around fixed set points with noise dominant. There is a strong need to be able to accommodate the different tuning requirements (different λs) of these cases, all of which apply to the same loop at different times. One way to approach this problem might be to use logic within the control system, which would gain schedule based on the magnitude of the controller error signal. For errors below some threshold, say 3%, the controller would be tuned with λs suitable for rejecting noise over a certain bandwidth. This tuning is likely to be fairly conservative. On the other hand, when the controller error is above some threshold, say 10%, the gains could correspond to an aggressive λ chosen to correct the disturbance as quickly as practically possible. In addition, it is useful to consider the use of set point feedforward control to help differentiate between set point changes and major load changes.

4.5.6 Algorithms

SISO Control Using PI, PID, PID.F, Cascade, and Feedforward

The foregoing discussion has centered on the PI controller and SISO control (98% of the control loops in pulp and paper). The reasons for such wide use of single loop control is partly historical, as this is how the industry has evolved. There is also a good technical justification given the sheer complexity of a process with hundreds of control loops. The operator may arbitrarily place any of these in manual mode for operating reasons. Hence the concept of designing a workable MIMO system is almost unworkable given the complexity (note that the pilot of an aircraft cannot put the left aileron on manual, only the entire autopilot system). Furthermore, the concept of Lambda tuning allows the PI controller to provide an effective solution in most cases. PID control is needed when the controller must be able to cancel two process poles (e.g., the control of an underdamped second-order process, the aggressive control of a second-order overdamped process). There are times when it may be advantageous to include a series filter with PID algorithm (PID.F) (e.g., the control of a second-order process with a transfer function zero). The filter is then used to cancel the zero. There is extensive use in the pulp and paper industry of cascade control (e.g., basis weight and stock flow, paper moisture and steam pressure, bleach plant brightness and bleaching chemical flow, boiler steam drum level and feedwater flow). There is also quite frequent use of feedforward control (e.g., boiler steam drum feedforward from steam demand).

Deadtime Compensation

The pulp and paper industry uses deadtime compensator algorithms extensively (e.g., basis weight control, moisture control, Kappa number control, bleach plant brightness control). The common types are the Smith predictor [19], and the Dahlin algorithm [6]. It is interesting to note that the original work of Smith focused heavily on the set point response of the algorithm, for which the technique is well suited. However, almost all of the deadtime compensators in commission are employed as regulators, operating in steady state with the goal of disturbance rejection and minimizing variability. Although not a new discovery, Haggman and Bialkowski [12] have shown that the sensitivity function of these regulators, as well as deadtime compensators using IMC structure [15], is essentially the same as that of a Lambda tuned PI controller for a given bandwidth. There is no way of escaping the time-delay phase lag induced resonance ("water-bed" effect) as long as simple process output feedback is used. These algorithms offer no advantage for low-frequency attenuation or in providing less resonance when tuned for the same bandwidth. On the other hand, they are more complex and have a greater sensitivity to model mismatch than the PI controller. One exception is the Kalman filter-based deadtime compensator [2] which uses state feedback from an upstream state variable in the deadtime model. Because the feedback control structure is identical to that of a Smith predictor, the time-delay resonance is also present. However, the Kalman filter prevents the excitation of this resonance through the low-pass dynamics of the Kalman filter update cycle which has equal bandwidth to the regulator, thereby attenuating noise in the region of the primary time-delay resonance by some −10 dB.

Adaptive and MIMO Control

With the exception of gain scheduling, adaptive control has not achieved wide use in the pulp and paper industry, because its algorithms are too complex for the current technical capability in most mills. Whereas there are occasional reports of quite advanced work being done [such as [1], on MIMO adaptive control of Kamyr digester chip level using general predictive control (GPC)], in most cases the results of such advanced work are very difficult to maintain for long in an operational state in a mill environment. As for the much simpler commercially available self-tuning controllers, most of these perform some type of identification (e.g., the relay method) and then implement a simple adaptation scheme which often takes the form of automating a Ziegler–Nichols-like tuning. This usually results in a lightly damped closed-loop system which is quite unsuitable for the control objectives previously discussed (e.g., to prevent propagation of resonances). Finally, there is fairly wide use of simple decoupling schemes which depend mainly on static decouplers (e.g., basis weight and moisture, multiply headbox control of fan pumps).

4.6 Control Strategies for Uniform Manufacturing

The relationship between algorithms, control strategy, and variability was put into a concise perspective by Downs and Doss, [7], and some of their thoughts have been put into a pulp and paper context here. The control algorithm, however simple or complex, does not eliminate variability, but it causes the variability to be shifted to another place in the process. For instance, consistency control shifts variability from the stock consistency to the dilution valve and the dilution header, from the fiber stream to the dilution stream. The control algorithm and its tuning determines the efficiency with which this is done. The control strategy determines the pathway that the variability will follow. The control strategy can be defined as

1. defining control system objectives
2. selecting sensor types and their location
3. selecting actuator types and their location
4. deciding on input/output pairing for single loops
5. designing multivariable control where needed
6. designing process changes that make control more effective

The control strategy design hence determines where the variability will attempt to go. Let us revisit the paper machine blend chest example. Both the hardwood and softwood steams have consistency control, as these stocks leave their respective stock chests. Both of these consistency controls modulate dilution valves which take dilution water from the same header. As a result, they interact (see Figure 4.4). Hence variability will move from the hardwood consistency to the dilution header, to the softwood consistency, to the blend chest consistency, and back again. When these stock flows enter the blend chest, they will undergo mixing, which will act as a low-pass filter. Hence, the consistency loops have attenuated the low-frequency content, while causing interaction between each other at high frequency. Then, the high-frequency content will be attenuated by the mixing action of the chest. However, on leaving the blend chest, the blend chest consistency controller also draws dilution water from the same header. Clearly, this design will compromise almost everything that has been gained at the high-frequency end of the spectrum.

This example illustrates also that the control engineer, acting alone in the domain of the control algorithm, cannot achieve effective control over process variability. What is needed is an integration of process and control design.

4.6.1 Minimizing Variability — Integrated Process and Control Design

To design a process which produces low variability product, a true integration of the process and control design disciplines is required. The old way of designing the process in steady state, and adding the controls later, has produced the pulp and paper mills of today, which, as variability audits have already shown, are variable far in excess of potential. Control algorithm design follows a very general methodology and is largely based on linear dynamics. When thinking about control loop performance, the engineer pays no attention to the actual behavior of the process. For instance, the most important phenomena in the pulp and paper industry concern pulp slurries and the transport of fiber in two- or three-phase flow. The physics that govern these phenomena involve the principles of Bernoulli and Reynolds and are very nonlinear. The linear transfer function is a necessary abstraction to allow the control engineer to perform linear control design, the only analysis that can be done well. Yet in the final analysis, control only moves variability from one stream to another, where hopefully it will be less harmful. Yet the process design is not fixed. What about the strategy of creating new streams?

Integrated process and control design must take a broader view of control strategy design. Control is only a high-pass attenuation mechanism for variability. Process mixing and agitation provide low-pass attenuation of variability via mixing internal process streams. Yet both of these techniques only attenuate by so many dB. But the customer demands that the absolute variability of the product be within some specified limit to meet the manufacturing needs of his process. Surely, eliminating sources of variability is the best method to insure that no variability will be present which will need attenuation. These issues are in the domain of process design and control strategy design. Control strategy design does not lend itself to elegant and general analysis. Each process is different and must be understood in its specific detail. Nonlinear dynamic simulation offers a powerful tool to allow detailed analysis of performance trade-offs and is the only available method for investigating the impact of different design decisions on variability. Such decisions must question current process design practice.

From the blend chest example, it is clear that a number of process design issues compromise variability, and that integrated process and control design could lead to the following design alternatives:

1. Eliminate all stock flow control valves, and use variable frequency pump drives. This will eliminate all of the nonlinear problems of backlash and stiction.
2. Redesign the existing blend chest with two compartments, each separately agitated. This will convert the existing agitation from a first-order low pass filter to a second-order filter and provide a high-frequency attenuation asymptote at −40 dB/decade instead of −20 dB/decade.
3. Provide a separate dilution header for NC302, and use a variable frequency pump drive instead of the control valve. This will eliminate the high-frequency noise content of the existing header from disturbing NC302, and will also provide nearly linear control of the dilution water.
4. Replace the dilution header control valve of PC309 by a variable frequency pump drive. This will allow much faster tuning of this loop, substantially reducing the interaction between NC104 and NC204.

Each of these design alternatives should be evaluated using dynamic simulation before significant funds are committed. The alterations proposed above vary in capital cost from $10,000 to $1,000,000. Hence, the simulation must have high fidelity representing the phenomena of importance. In addition, network analysis techniques may determine how variability spectra propagate through a process and control strategy. Changes in process and control strategy design alter these pathways by creating new streams. The plant can be viewed as a network of connected nodes (process variables) with transmission paths (e.g., control loops, low pass process dynamics, etc.) which allow variability spectra to propagate. These ideas will need time to develop.

4.7 Conclusions

This chapter has attempted to provide a general overview of control engineering in the pulp and paper industry in the mid 1990s. There is a brief introduction to wood, pulp, paper products, and the unit processes. The results of variability audits were then presented to show how much potential exists in this industry to improve product uniformity, especially when there is an increasingly strong demand for uniform product. The concept of specifying the closed-loop performance of each control loop to match the process needs is presented as the Lambda tuning concept. The use of Lambda tuning is illustrated in a paper mill example, and the performance and robustness of tuned control loops are explored. There is a general review of various algorithms and their use in the industry. Finally, the concept of integrating control strategy and design of both the process and control is presented. This is seen as a new avenue of thought which promises to provide a design methodology for the pulp and paper mills of the future, which will be far more capable of efficiently manufacturing highly uniform product than today's mills.

4.8 Defining Terms

AR: Amplitude ratio.

Backlash: Hysteresis or lost motion in an actuator.

Basis weight: The paper property of mass per unit area (gsm, lbs/3000 sq. ft., etc.)

Bump test: Step test.

β: Process transfer function zero time constant.

C: ISA symbol for control.

Cellulose: Long chain polymer of glucose, the basic building block of wood fiber.

Chest: Tank.

Consistency: Mass percentage of solids or fiber content of a pulp or stock slurry.

CPPA: Canadian Pulp and Paper Association, 1155 Metcalfe St. Montreal, Canada, H3B 4T6.

DCS: Distributed control system.

Deadtime: Time delay.

F: ISA symbol for flow.

$G_C(s)$: Controller transfer function in the continuous (Laplace) domain.

$G_P(s)$: Process transfer function in the continuous (Laplace) domain.

gsm: Grams per square meter.

HD Chest: High Density chest, a large production capacity stock tank with consistency typically in the 10 to 15% range and with a dilution zone in the bottom.

Hemicellulose: Polymers of sugars other than glucose, a constituent of wood fiber.

IMA: Integrating moving average noise structure.

ISA: Instrument Society of America.

ISA tags: ISA tagging convention (e.g., FIC177 means Flow Indicating Controller No. 177).

K_C: Controller gain.

K_P: Process gain.

L: ISA symbol for level.

Lambda tuning: Tuning which requires the user to specify the desired closed-loop time constant, Lambda.

Lambda (λ): The desired closed-loop time constant, usually in seconds.

Lignin: Organic compound which binds the wood fiber structure together.

Limit cycle: A cycle induced in a control loop by nonlinear elements.

N: ISA symbol for consistency.

P: ISA symbol for pressure.

PI: Proportional-Integral controller.

PID: Proportional-Integral-Derivative controller.

PID.F: Proportional-Integral-Derivative controller with series filter.

Positioner: Control valve accessory which acts as a local pneumatic feedback servo.

Pulping: The process of removing individual fibers from solid wood.

Refiner: A machine with rotating plates used in pulping which disintegrates the wood chips into individual fibers through mechanical action, and in papermaking, to "fibrillate" the fibers to enhance bonding strength.

RHP: Right half-plane of the s-plane—the unstable region.

Standard PI form: $G_C(s) = K_C\left[1 + \frac{1}{T_R s}\right]$.

Stiction: Static friction in an actuator.

Stock: Pulp slurry.

T_d: Deadtime.

T_R: Controller reset or integral time/repeat.

τ_1, τ_2: process time constants ($\tau_1 \geq \tau_2$).

TAPPI: Tech. Asn. P & P Ind., P. O. Box 105133, Atlanta, GA, USA, 30348-5113.

References

[1] Allison, B. J., Dumont G. A., and Novak L. H., *Multi-Input Adaptive Control of Kamyr Digester Chip Level: Industrial Results and Practical Considerations*, CPPA Proc., Control Systems '90, Helsinki, Finland, 1990.

[2] Bialkowski, W. L., Application of Kalman Filters to the Regulation of Dead Time Processes, *IEEE Trans. Automat. Control*, AC-28, 3, 1983.

[3] Bialkowski, W. L., *Dreams Versus Reality: A View From Both Sides of the Gap*, Keynote Address, Control Systems '92, Whistler, British Columbia, 1992, published, *Pulp Paper Canada*, 94, 11, 1993.

[4] Bialkowski, W. L., Haggman, B. C., and Millette, S. K., Pulp and Paper Process Control Training Since 1984, *Pulp Paper Canada*, 95, 4, 1994.

[5] Chien, I-L. and Fruehauf, P. S., *Consider IMC Tuning to Improve Controller Performance, Hydrocarbon Proc.*, 1990.

[6] Dahlin, E. B., Designing and Tuning Digital Controllers, *Instrum. Control Syst.*, 41(6), 77, 1968.

[7] Downs, J. J. and Doss, J. E., *Present Status and Future Needs — a view from North American Industry*, Fourth International Conf. Chem. Proc. Control, Padre Island, Texas, 1991, 17–22.

[8] EnTech^{TM} — Automatic Controller Dynamic Specification (Version 1.0, 11/93) (EnTech Literature).

[9] EnTech^{TM} — Competency in Process Control-Industry Guidelines (Version 1.0, 3/94) (EnTech Literature).

[10] EnTech^{TM} — Control Valve Dynamic Specification (Version 2.1, 3/94) (EnTech Literature).

[11] EnTech^{TM} — Digital Measurement Dynamics - Industry Guidelines (Version 1.0, 8/94) (EnTech Literature).

[12] Haggman, B. C. and Bialkowski, W. L., Performance of Common Feedback Regulators for First-Order and Deadtime Dynamics. *Pulp Paper Canada*, 95, 4, 1994.

[13] Kaminaga, H., *One in a thousand*, Proc. CPPA Control Systems '94, Stockholm, Sweden, 1994.

[14] Kocurek, M. J., Series Ed., *Pulp and Paper Manufacture*, Joint (TAPPI, CPPA) Textbook Committee of the Pulp and Paper Industry, 1983 to 1993, Vol. 1 to 10.

[15] Morari, M. and Zafiriou, E., *Robust Process Control*, Prentice Hall, 1989.

[16] Newton, G. C., Gould, L. A., and Kaiser, J. F., *Analytical Design of Linear Feedback Controls*, John Wiley & Sons, 1957.

[17] Sell, N., Editor, Bialkowski, W. L., and Thomason, F. Y., contributors, *Process Control Fundamentals for the Pulp & Paper Industry*, TAPPI Textbook, to be published by TAPPI Press, 1995.

[18] Smith, C. A. and Corripio, A. B., *Principles and Practice of Automatic Process Control*, John Wiley & Sons, 1885.

[19] Smith, O. J. M., Closer Control of Loops with Dead Time, *Chem. Eng. Prog.*, 53(5), 217–219, 1957.

[20] Taylor, G., *The Role of Control Valves in Process Performance*, Proc. CPPA, Tech. Section, Canadian Pulp and Paper Assoc., Montreal, 1994.

[21] Ziegler, J. G. and Nichols, N. B., Optimum settings for automatic controllers, *Trans. ASME*, 759–768, 1942.

5

Control for Advanced Semiconductor Device Manufacturing: A Case History

T. Kailath, C. Schaper, Y. Cho, P. Gyugyi, S. Norman, P. Park, S. Boyd, G. Franklin, and K. Saraswat
Department of Electrical Engineering, Stanford University, Stanford, CA

M. Moslehi and C. Davis
Semiconductor Process and Design Center, Texas Instruments, Dallas, TX

5.1 Introduction 67
5.2 Modeling and Simulation 70
5.3 Performance Analysis 71
5.4 Models for Control 72
5.5 Control Design 76
5.6 Proof-of-Concept Testing 77
5.7 Technology Transfer to Industry 79
5.8 Conclusions 83
References 83

5.1 Introduction

Capital[1] costs for new integrated circuit (IC) fabrication lines are growing even more rapidly than had been expected even quite recently. Figure 5.1 was prepared in 1992, but a new Mitsubishi factory in Shoji, Japan, is reported to have cost $3 billion. Few companies can afford investments on this scale (and those that can perhaps prefer it that way). Moreover these factories are inflexible. New equipment and new standards, which account for roughly 3/4 of the total cost, are needed each time the device feature size is reduced, which has been happening about every 3 years. It takes about six years to bring a new technology on line. The very high development costs, the high operational costs (e.g., equipment down time is extremely expensive so maintenance is done on a regular schedule, whether it is needed or not), and the intense price competition compel a focus on high-volume low cost commodity lines, especially memories. Low volume, high product mix ASIC (application-specific integrated circuit) production does not fit well within the current manufacturing scenario.

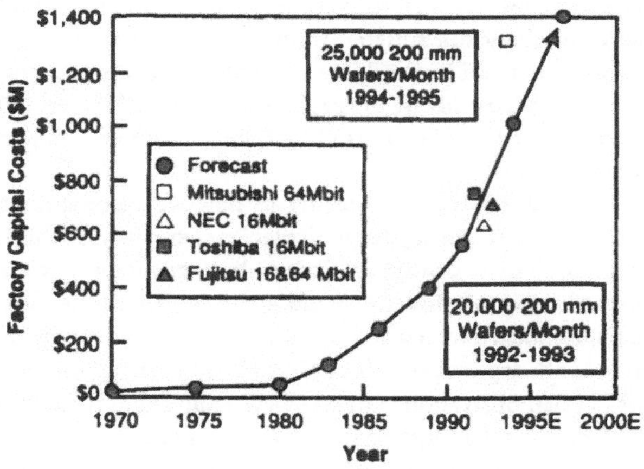

Figure 5.1 Capital cost for a new IC factory. (Source: *Texas Instruments Technical Journal*, 9(5), 8, 1992.)

In 1989, the Advanced Projects Research Agency(ARPA), Air Force Office of Scientific Research(AFOSR), and Texas Instruments (TI) joined in a $150 million cost-shared program called MMST (Microelectronics Manufacturing Science and Technology) to "establish and demonstrate (new) concepts for semiconductor device manufacture which will permit flexible, cost-effective manufacturing of application-specific logic integrated circuits in relatively low volume ... during the mid 1990s and beyond."

The approach taken by MMST was to seek fast cycle time by performing all single-wafer processing using highly instrumented flexible equipment with advanced process controls. The goal of the equipment design and operation was to quickly adapt the equipment trajectories to a wide variety of processing specifications and to quickly reduce the effects of manufacturing disturbances associated with small lot sizes (e.g., 1, 5 or 24 wafers) without the need for pilot wafers. Many other novel features

[1]This research was supported by the Advanced Research Projects Agency of the Department of Defense, under Contract F49620-93-1-0085 monitored by the Air Force Office of Scientific Research.

0-8493-0054-3/00/$0.00+$.50

were associated with MMST including a factory wide CIM (computer integrated manufacturing) computer system. The immediate target was a 1000-wafer demonstration (including demonstration of "bullet wafers" with three-day cycle times) of an all single-wafer factory by May 1993.

In order to achieve the MMST objectives, a flexible manufacturing tool was needed for the thermal processing steps associated with IC manufacturing. For a typical CMOS process flow, more than 15 different thermal processing steps are used, including chemical vapor deposition (CVD), annealing, and oxidation. The MMST program decided to investigate the use of Rapid Thermal Processing (RTP) tools to achieve these objectives.

TI awarded Professor K. Saraswat of Stanford's Center for Integrated Systems (CIS) a subcontract to study various aspects of RTP. About a year later, a group of us at Stanford's Information Systems Laboratory got involved in this project. Manufacturing was much in the news at that time. Professor L. Auslander, newly arrived at ARPA's Material Science Office, soon came to feel that the ideas and techniques of control, optimization, and signal processing needed to be more widely used in materials manufacturing and processing. He suggested that we explore these possibilities, and after some investigation, we decided to work with CIS on the problems of RTP.

RTP had been in the air for more than a decade, but for various reasons, its study was still in a research laboratory phase. Though there were several small companies making equipment for RTP, the technology still suffered from various limitations. One of these was an inability to achieve adequate temperature uniformity across the wafer during the rapid heating (e.g., 20°C to 1100°C in 20 seconds), hold (e.g., at 1100°C for 1–5 minutes), and rapid cooling phases.

This chapter is a case history of how we successfully tackled this problem, using the particular "systems-way-of-thinking" very familiar to control engineers, but seemingly not known or used in semiconductor manufacturing. In a little over two years, we started with simple idealized mathematical models and ended with deployment of a control system during the May, 1993, MMST demonstration. The system was applied to eight different RTP machines conducting thirteen different thermal operations, over a temperature range of 450°C to 1100°C and pressures ranging from 10^{-3} to 1 atmosphere.

Our first step was to analyze the performance of available commercial equipment. Generally, a bank of linear lamps was used to heat the wafer (see Figure 5.2).

The conventional wisdom was that a uniform energy flux to the wafer was needed to achieve uniform wafer temperature distribution. However, experimentally it had been seen that this still resulted in substantial temperature nonuniformities, which led to crystal slip and misprocessing. To improve performance, various heuristic strategies were used by the equipment manufacturers, e.g., modification of the reactor through the addition of guard rings near the wafer edge to reflect more energy to the edge, modification of the lamp design by using multiple lamps with a fixed power ratio, and various types of reflector geometries. However, these modifications turned out to be satisfactory only for a narrow range of conditions.

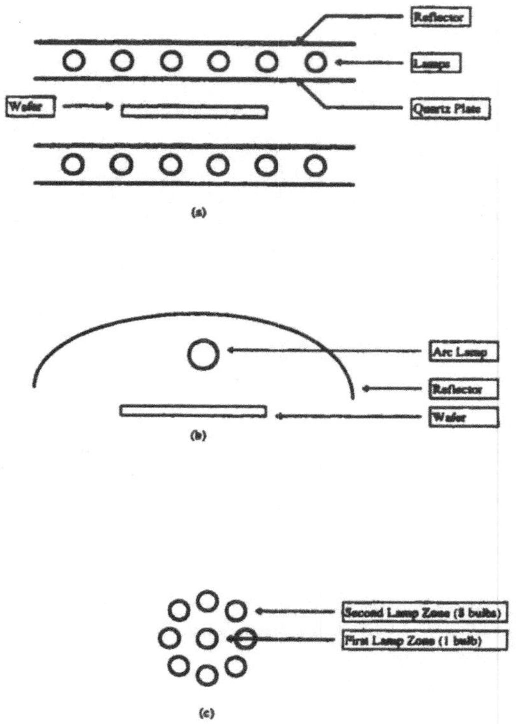

Figure 5.2 RTP lamp configurations: (a) bank of linear lamps, (b) single arc lamp, (c) two-zone lamp array.

The systems methodology suggests methods attempting to determine the performance limitations of RTP systems. To do this, we proceeded to develop a simple mathematical model, based on energy transfer relations that had been described in the literature. Computer simulations with this model indicated that conventional approaches trying to achieve uniform flux across the wafer would never work; there was always going to be a large temperature roll-off at the wafer edge (Figure 5.3). To improve

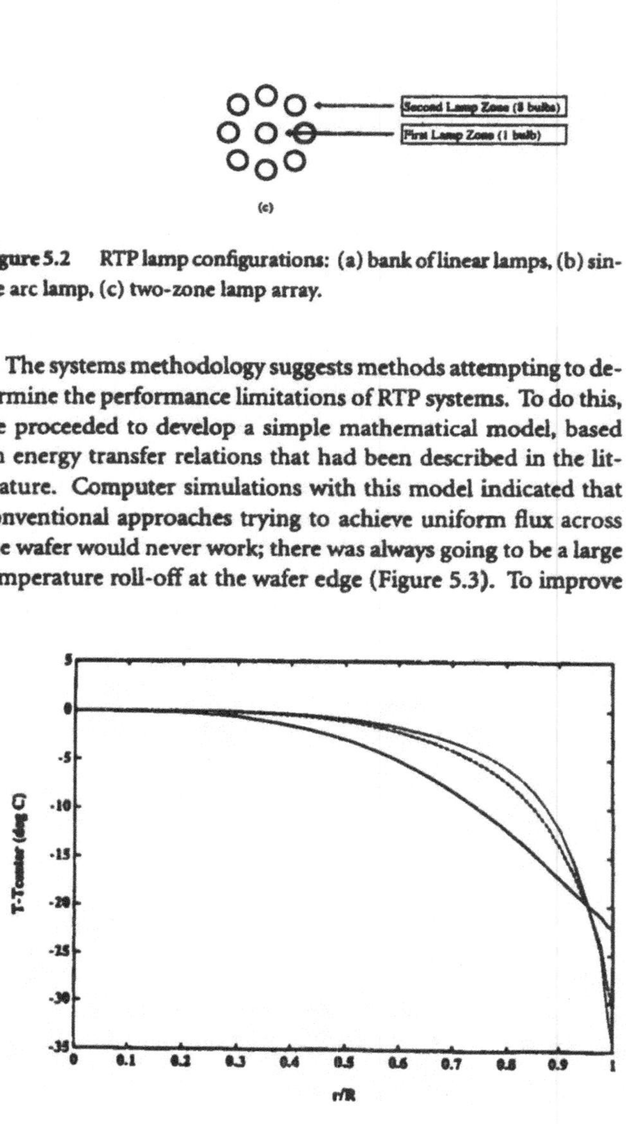

Figure 5.3 Nonuniformity in temperature induced by uniform energy flux impinging on the wafer top surface (center temperatures - solid line: 600°C; dashed line: 1000°C; dotted line: 1150°C.). R is the radius of the wafer, r is the radial distance from the center of the wafer.

performance, we decided to study the case where circularly symmetric rings of lamps were used to heat the wafer. With this configuration, two cases were considered: (1) a single power supply in a fixed power ratio, a strategy being used in the field and (2) independently controllable multiple power supplies (one for each ring of lamps). Both steady-state and dynamic studies indicated that it was necessary to use the (second) multivariable configuration to achieve wafer temperature uniformity within specifications. These modeling and analysis results are described in Sections 5.2 and 5.3, respectively.

The simulation results were presented to Texas Instruments, which had developed prototype RTP equipment for the MMST program with two concentric lamp zones, but operated in a scalar control mode using a fixed ratio between the two lamp zones. At our request, Texas Instruments modified the two zone lamp by adding a third zone and providing separate power supplies for each zone, allowing for multivariable control. The process engineers in the Center for Integrated Systems (CIS) at Stanford then evaluated the potential of multivariable control by their traditional so called "hand-tuning" methodology, which consists of having experienced operators determining the settings of the three lamp powers by manual iterative adjustment based on the results of test wafers. Good results were achieved (see Figure 5.4), but it took 7–8 hours and a large number of wafers before the procedure converged. Of course, it had to be repeated the next day because of unavoidable changes in the ambient conditions or operating conditions. Clearly, an "automatic" control strategy was required.

However, the physics-based equations used to simulate the RTP were much too detailed and contained far too many uncertain parameters for control design. The two main characteristics of the simulation model were (1) the relationship between the heating zones and the wafer temperature distribution and (2) the nonlinearities (T^4) of radiant heat transfer. Two approaches were used to obtain a reduced-order model. The first used the physical relations as a basis in deriving a lower-order approximate form. The resulting model captured the important aspects of the interactions and the nonlinearities, but had a simpler structure and fewer unknown parameters. The second approach viewed the RTP system as a black box. A novel model identification procedure was developed and applied to obtain a state-space model of the RTP system. In addition to identifying the dynamics of the process, these models were also studied to assess potential difficulties in performance and control design. For example, the models demonstrated that the system gain and time constants changed by a factor of 10 over the temperature range of interest. Also, the models were used to improve the condition number of the equipment via a change in reflector design. The development of control models is described in Section 5.4.

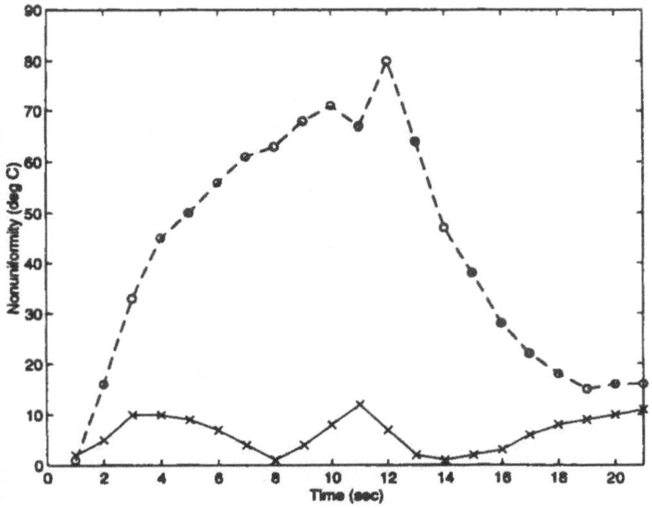

Figure 5.4 Temperature nonuniformity when the powers to the lamp were manually adjusted ("hand-tuning"). These nonuniformities correspond to a ramp and hold from nearly room temperature to 600°C at roughly 40°C/s. The upper curve (-o-) corresponds to scalar control (fixed power ratio to lamps). The lower curve (x-x) corresponds to multivariable control.

Using these models, a variety of control strategies was evaluated. The fundamental strategy was to use feedforward in combination with feedback control. Feedforward control was used to get close to the desired trajectory and feedback control was used to compensate for inevitable tracking errors. A feedback controller based on the Internal Model Control (IMC) design procedures was developed using the low-order physics-based model. An LQG feedback controller was developed using the black-box model. Gain scheduling was used to compensate for the nonlinearities. Optimization procedures were used to design the feedforward controller. Controller design is described in Section 5.5.

Our next step was to test the controller experimentally on the Stanford RTP system. After using step response and PRBS (Pseudo Random Binary Sequence) data to identify models of the process, the controllers were used to ramp up the wafer temperature from 20°C to 900°C at approximately 45°C/s, followed by a hold for 5 minutes at 900°C. For these experiments, the wafer temperature distribution was sensed by three thermocouples bonded to the wafer. The temperature nonuniformity present during the ramp was less than ±5°C from 400°C to the processing temperature and better than ±0.5°C on average during the hold. These proof-of-concept experiments are described in Section 5.6.

These results were presented to Texas Instruments, who were preparing their RTP systems for a 1000 wafer demonstration of the MMST concept. After upper level management review, it was decided that the Stanford temperature control system would be integrated within their RTP equipment. The technology transfer involved installing and testing the controller on eight different RTP machines conducting thirteen different thermal operations used in two full-flow 0.35 μm CMOS process technologies (see Figure 5.5 taken from an article appearing in a semiconductor manufacturing trade journal). More discussion concerning the

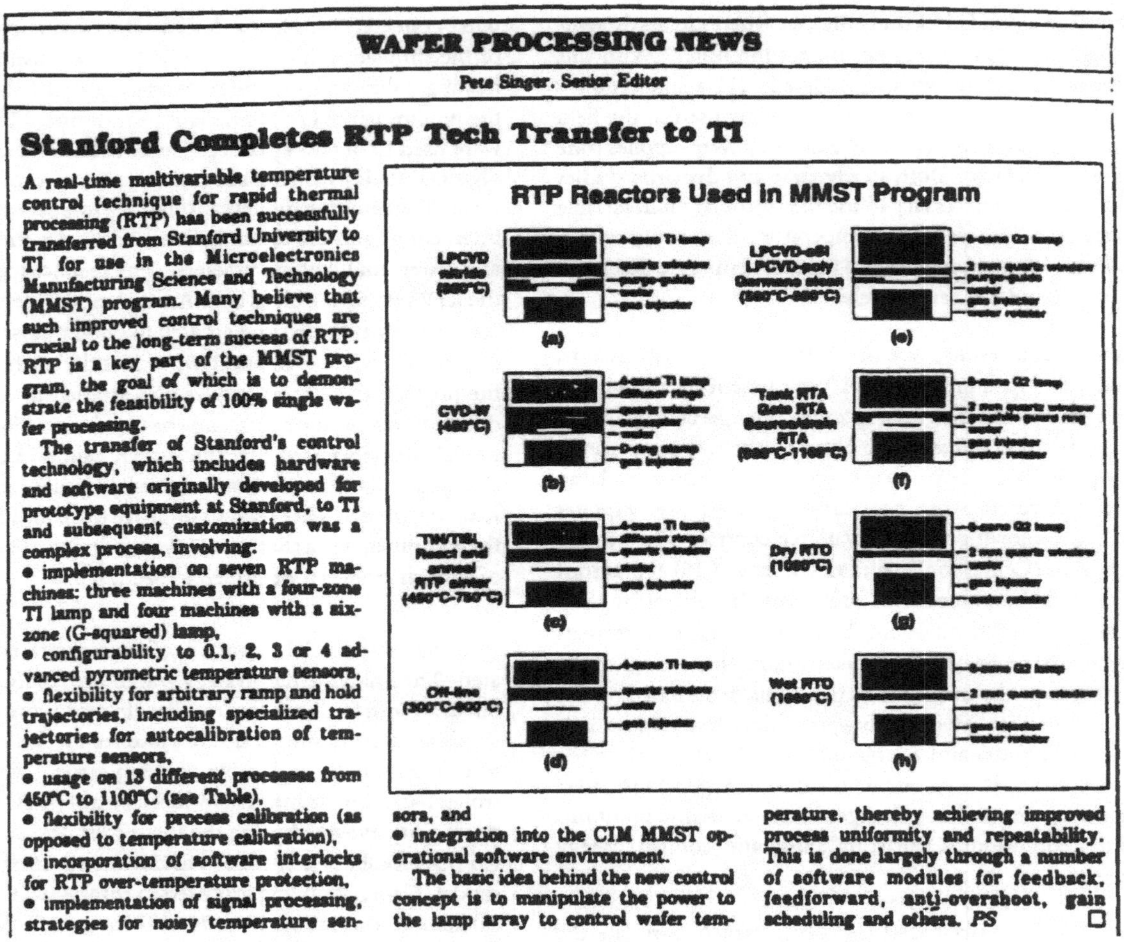

WAFER PROCESSING NEWS

Pete Singer, Senior Editor

Stanford Completes RTP Tech Transfer to TI

A real-time multivariable temperature control technique for rapid thermal processing (RTP) has been successfully transferred from Stanford University to TI for use in the Microelectronics Manufacturing Science and Technology (MMST) program. Many believe that such improved control techniques are crucial to the long-term success of RTP. RTP is a key part of the MMST program, the goal of which is to demonstrate the feasibility of 100% single wafer processing.

The transfer of Stanford's control technology, which includes hardware and software originally developed for prototype equipment at Stanford, to TI and subsequent customization was a complex process, involving:

- implementation on seven RTP machines: three machines with a four-zone TI lamp and four machines with a six-zone (G-squared) lamp,
- configurability to 0.1, 2, 3 or 4 advanced pyrometric temperature sensors,
- flexibility for arbitrary ramp and hold trajectories, including specialized trajectories for autocalibration of temperature sensors,
- usage on 13 different processes from 450°C to 1100°C (see Table),
- flexibility for process calibration (as opposed to temperature calibration),
- incorporation of software interlocks for RTP over-temperature protection,
- implementation of signal processing, strategies for noisy temperature sensors, and
- integration into the CIM MMST operational software environment.

The basic idea behind the new control concept is to manipulate the power to the lamp array to control wafer temperature, thereby achieving improved process uniformity and repeatability. This is done largely through a number of software modules for feedback, feedforward, anti-overshoot, gain scheduling and others. *PS*

Figure 5.5 Description of technology transfer in *Semiconductor International*, 16(7), 58, 1993.

technology transfer and results of the MMST demonstration is given in Section 5.7. Finally, some overview remarks are offered in Section 5.8.

5.2 Modeling and Simulation

Three alternative lamp configurations for rapidly heating a semiconductor wafer are shown in Figure 5.2. In Figure 5.2(a), linear lamps are arranged above and below the wafer. A single arc lamp is shown in Figure 5.2(b). Concentric rings of single bulbs are presented in Figure 5.2(c). These designs can be modified with guard rings around the wafer edge, specially designed reflectors, and diffusers placed on the quartz window. These additions allowed fine-tuning of the energy flux profile to the wafer to improve temperature uniformity.

To analyze the performance of these and related equipment designs, a simulator of the heat transfer effects was developed starting from physical relations for RTP available in the literature [1], [2]. The model was derived from a set of PDE's describing the radiative, conductive and convective energy transport effects. The basic expression is

$$\frac{1}{r}\frac{\partial}{\partial r}\left(kr\frac{\partial T}{\partial r}\right)+\frac{1}{r^2}\frac{\partial}{\partial \theta}\left(k\frac{\partial T}{\partial \theta}\right)+\frac{\partial}{\partial z}\left(k\frac{\partial T}{\partial z}\right)=\rho C_p\frac{\partial T}{\partial t} \tag{5.1}$$

where T is temperature, k is thermal conductivity, ρ is density, and C_p is specific heat. Both k and C_p are temperature dependent. The boundary conditions are given by

$$\begin{aligned} k\frac{\partial T}{\partial r} &= q_{edge}(\theta, z), r = R, \\ k\frac{\partial T}{\partial z} &= q_{bottom}(r, \theta), z = 0, \text{ and} \\ k\frac{\partial T}{\partial z} &= q_{top}(r, \theta), z = Z, \end{aligned}$$

where q_{edge}, q_{bottom}, and q_{top} are heat flow per unit area into the wafer edge, bottom, and top, respectively, via radiative and convective heat transfer mechanisms, Z is the thickness of the wafer, and R is the radius of the wafer. These terms coupled the effects of the lamp heating zones to the wafer.

Approximations were made to the general energy balance assuming axisymmetry and neglecting axial temperature gradients. The heating effects in RTP were developed by discretizing the wafer into concentric annular elements. Within each annular wafer element, the temperature was assumed uniform [2]. The resulting model was given by a set of nonlinear vector differential equations:

$$\begin{aligned} \mathbf{C}\dot{T} &= \mathbf{K}^{rad}T^4 + \mathbf{K}^{cond}T + \mathbf{K}^{conv}(T - T_{gas}) \\ &\quad + \mathbf{F}P + q^{wall} + q^{dist} \end{aligned} \qquad (5.2)$$

where

$$\begin{aligned} T &= [T_1\ T_2\ \ldots\ T_N]^T \\ T^4 &= \left[T_1^4\ T_2^4\ \ldots\ T_N^4\right]^T \\ P &= [P_1\ P_2\ \ldots\ P_M]^T \end{aligned}$$

where N denotes the number of wafer elements and M denotes the number of radiant heating zones; $\mathbf{K}^{rad}$ is a full matrix describing the radiation emission characteristics of the wafer, $\mathbf{K}^{cond}$ is a tridiagonal matrix describing the conductive heat transfer effects across the wafer, $\mathbf{K}^{conv}$ is a diagonal matrix describing the convective heat transfer effects from the wafer to the surrounding gas, $\mathbf{F}$ is a full matrix quantifying the fraction of energy leaving each lamp zone that radiates onto the wafer surface, q^{dist} is a vector of disturbances, q^{wall} is a vector of energy flux leaving the chamber walls and radiating onto the wafer surface, and $\mathbf{C}$ is a diagonal matrix relating the heat flux to temperature transients. More details can be found in [2] and [3].

5.3 Performance Analysis

We first used the model to analyze the case of uniform energy flux impinging on the wafer surface. In Figure 5.3, the temperature profile induced by a uniform input energy flux is shown for the cases where the center portion of the wafer was specified to be at either 600°C, 1000°C, or 1150°C. A roll-off in temperature is seen in the plots for all cases because the edge of the wafer required a different amount of energy flux than the interior due to differences in surface area. Conduction effects within the wafer helped to smooth the temperature profile. These results qualitatively agreed with those reported in the literature where, for example, sliplines at the wafer edge were seen because of the large temperature gradients induced by the uniform energy flux conditions.

We then analyzed the multiple concentric lamp zone arrangement of Figure 5.2(c) to assess the capability of achieving uniform temperature distribution during steady-state and transients. We considered each of four lamp zones to be manipulated independently. The optimal lamp powers were determined to minimize the peak temperature difference across the wafer at a steady-state condition,

$$\max_{0\le r\le R} \left|T^{ss}(r, P) - T^{set}\right| \qquad (5.3)$$

where T^{set} is the desired wafer temperature and $T^{ss}(r, P)$ is the steady-state temperature at radius r with the constant lamp power vector P, subject to the constraint that each entry P_j of P satisfies $0 \le P_j \le P_j^{max}$. Using the finite difference model, the objective function of Equation 5.3 was approximated as

$$\max_i \left|T_i^{ss}(P) - T^{set}\right| = \left\|T^{ss}(P) - T^{set}\right\|_\infty \qquad (5.4)$$

where $T_i^{ss}(P)$ is the steady-state temperature of element i with constant lamp power vector P and T^{set} is a vector with all entries equal to T^{set}. A two-step numerical optimization procedure was then employed in which two minimax error problems were solved to determine the set of lamp powers that minimize Equation 5.3 [4] and [2]. In Figure 5.6, the temperature deviation about the set points of 650°C, 1000°C, and 1150°C is shown. The deviation is less than ±1°C, much better than for the case of uniform energy flux.

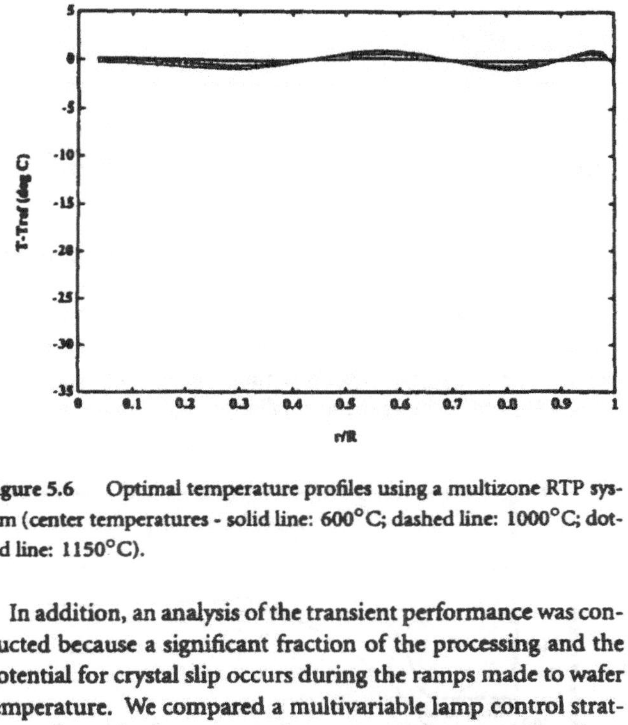

Figure 5.6 Optimal temperature profiles using a multizone RTP system (center temperatures - solid line: 600°C; dashed line: 1000°C; dotted line: 1150°C).

In addition, an analysis of the transient performance was conducted because a significant fraction of the processing and the potential for crystal slip occurs during the ramps made to wafer temperature. We compared a multivariable lamp control strategy and a scalar lamp control strategy. Industry, at that time, employed a scalar control strategy. For the scalar case, the lamps were held in a fixed ratio of power while total power was allowed to vary. We selected the optimization criterion of minimizing

$$\max_{t_o\le t\le t_f} \left\|T(t) - T^{ref}(t)\right\|_\infty \qquad (5.5)$$

which denotes the largest temperature error from the specified trajectory $T^{ref}(t)$ at any point on the wafer at any time between an initial time T_o and a final time t_f. The reference temperature trajectory was selected as a ramp from 600°C to 1150°C in 5 seconds. The optimization was carried out with respect to the power to the four lamp zones, in the case of the multilamp configuration, or to the total power for a fixed ratio that was optimal only at a 1000°C steady-state condition. The temperature at the center of the wafer matched the desired temperature trajectory almost exactly for both the multivariable and scalar control cases. However, the peak temperature difference across the wafer was much less for the multivariable case compared to the scalar (fixed-ratio) case as shown in Figure 5.7.

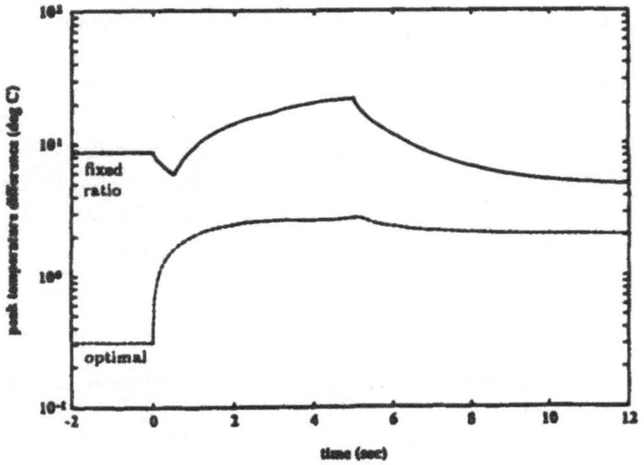

Figure 5.7 Peak temperature nonuniformity during ramp.

For the case of the fixed-ratio lamps, the peak temperature difference was more than 20°C during the transient and the multivariable case resulted in a temperature deviation of about 2°C. The simulator suggested that this nonuniformity in temperature for the fixed-ratio case would result in crystal slip as shown in Figure 5.8 which shows the normalized maximum resolved stress

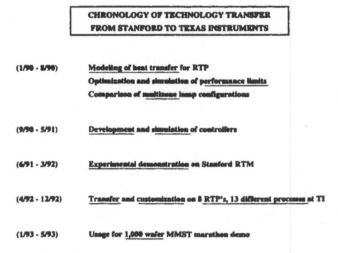

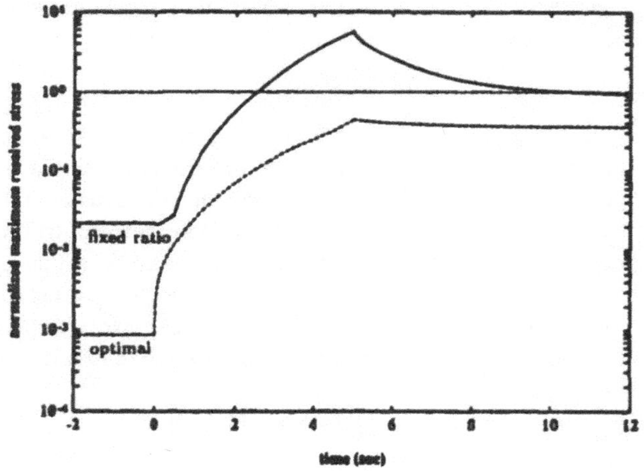

Figure 5.8 Normalized maximum resolved stress during ramp.

(based on simulation) as a function of time. No slip was present in the multivariable case. This analysis of the transient performance concluded that RTP systems configured with multiple independently controllable lamps can substantially outperform any existing scalar RTP system: for the same temperature schedule, much smaller stress and temperature variation across the wafer was achieved; and for the same specifications for stress and temperature variation across the wafer, much faster rise times can be achieved [2].

At the time of these simulations, prototype RTP equipment was being developed at Texas Instruments for implementation in the MMST program. TI had developed an RTP system with two concentric lamp zones. Their system at that time was operated in a scalar control mode with a fixed ratio between the two lamp zones. Upon presenting the above results, the two zone lamp was modified by adding a third zone and providing separate power supplies for each zone. This configuration allowed multivariable control. A resulting three-zone RTP lamp was then donated by TI to Stanford University. The chronology of this technology transfer is shown in Figure 5.9.

CHRONOLOGY OF TECHNOLOGY TRANSFER FROM STANFORD TO TEXAS INSTRUMENTS

(1/90 - 8/90)	Modeling of heat transfer for RTP Optimization and simulation of performance limits Comparison of multizone lamp configurations
(9/90 - 5/91)	Development and simulation of controllers
(6/91 - 3/92)	Experimental demonstration on Stanford RTM
(4/92 - 12/92)	Transfer and customization on 8 RTP's, 13 different processes at TI
(1/93 - 5/93)	Usage for 1,000 wafer MMST marathon demo

Figure 5.9 Chronology of the technology transfer to Texas Instruments.

A schematic of the Stanford RTP system and a picture of the three-zone arrangement are shown in Figures 5.10 and 5.11, respectively.

"Hand-tuning" procedures were used to evaluate the performance of the RTP equipment at Stanford quickly. In this approach, the power settings to the lamp were manually manipulated in real-time to achieve a desirable temperature response. In Figure 5.4, open-loop, hand-tuned results are shown for scalar control (i.e., fixed power ratio) and multivariable control as well as the error during the transient. Clearly, this comparison demonstrated that multivariable control was preferred to the scalar control method [5]. However, the hand-tuning approach was a trial and error procedure that was time-consuming and resulted in sub-optimal performance. An automated real-time control strategy is described in the following sections.

5.4 Models for Control

Two approaches were evaluated to develop a model for control design. In the first approach, the nonlinear physical model presented earlier was used to produce a reduced-order version. An energy balance equation on the ith annular element can be ex-

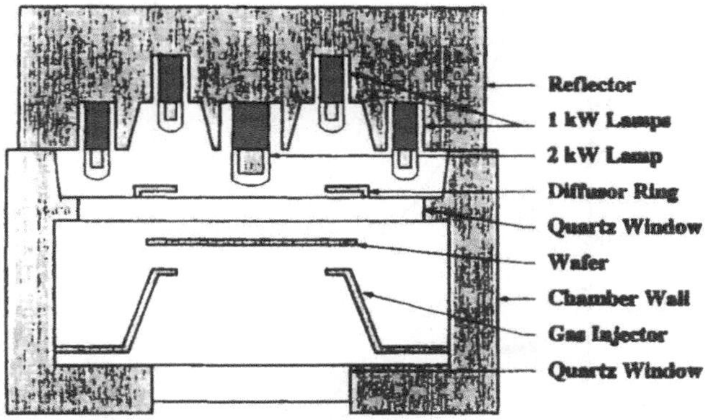

Figure 5.10 Schematic of the rapid thermal processor.

Figure 5.11 Picture of the Stanford three-zone RTM lamp.

pressed as [3] and [6]

$$\rho V_i C_p \frac{dT_i}{dt} = -\epsilon\sigma A_i \sum_{j=1}^{N} D_{i,j} T_j^4 - h_i A_i (T_i - T_{gas}) + q_i^{cond} + q_i^{wall} + q_i^{dist} + \epsilon \sum_{j=1}^{M} F_{i,j} P_j \quad (5.6)$$

where ρ is density, V_i is the volume of the annular element, C_p is heat capacity, T_i is temperature, ϵ is total emissivity, σ is the Stefan-Boltzmann constant, A_i is the surface area of the annular element, $D_{i,j}$ is a lumped parameter denoting the energy transfer due to reflections and emission, h_i is a convective heat transfer coefficient, q_i^{cond} is heat transfer due to conduction, $F_{i,j}$ is a view factor that represents the fraction of energy received by the ith annular element from the jth lamp zone, and P_j is the power from the jth lamp zone.

To develop a simpler model, the temperature distribution of the wafer was considered nearly uniform and much greater than that of the water-cooled chamber walls. With these approximations, q_i^{cond} and q_i^{wall} were negligible. In addition, the term accounting for radiative energy transport due to reflections can be simplified by analyzing the expansion,

$$\sum_{j=1}^{N} D_{i,j} T_j^4 = \sum_{j=1}^{N} D_{i,j} \left(T_i^4 + 4T_i^3\delta_{i,j} + 6T_i^2\delta_{i,j}^2 + 4T_i\delta_{i,j}^3 + \delta_{i,j}^4\right), \quad (5.7)$$

where $\delta_{i,j} = T_j - T_i$. After eliminating the terms involving $\delta_{i,j}$ (since $T_i >> \delta_{i,j}$), the resulting model was,

$$\rho V C_p \frac{dT_i}{dt} = -\epsilon\sigma A_i T_i^4 \sum_{j=1}^{N} D_{i,j} - h_i A_i (T_i - T_{ambient}) + \epsilon \sum_{j=1}^{M} F_{i,j} P_j. \quad (5.8)$$

It was noted that Equation 5.8 was interactive because each lamp zone affects the temperature of each annular element and noninteractive because the annular elements did not affect one another.

The nonlinear model given by Equation 5.8 was then linearized about an operating point $(\bar{T}_i, \bar{P}_i)$,

$$\rho V C_p \frac{d\tilde{T}_i}{dt} = -\left[4\epsilon\sigma A_i \bar{T}_i^3 \sum_{j=1}^{N} D_{i,j} + h_i A_i\right] \tilde{T}_i + \epsilon \sum_{j=1}^{M} F_{i,j} \tilde{P}_j, \quad (5.9)$$

where the deviation variables are defined as $\tilde{T}_i = T_i - \bar{T}_i$ and $\tilde{P}_i = P_i - \bar{P}_i$. This equation can be expressed more conveniently as

$$\tau_i \frac{d\tilde{T}_i}{dt} = -\tilde{T}_i + \sum_{j=1}^{M} K_{i,j} \tilde{P}_j, \quad (5.10)$$

where the gain and time-constant are given by

$$K_{i,j} = \frac{\epsilon F_{i,j}}{4\epsilon\sigma A_i \bar{T}_i^3 \sum_{j=1}^{N} D_{i,j} + h_i A_i} \quad (5.11)$$

$$\tau_i = \frac{\rho V C_p}{4\epsilon\sigma A_i \bar{T}_i^3 \sum_{j=1}^{N} D_{i,j} + h_i A_i} \quad (5.12)$$

From Equation 5.11, the gain decreases as $\bar{T}$ was increased. Larger changes in the lamp power were required at higher $\bar{T}$

to achieve an equivalent rise in temperature. In addition, from Equation 5.12, the time constant decreases as $\bar{T}$ is increased. Thus, the wafer temperature responded faster to changes in the lamp power at higher $\bar{T}$. The nonlinearities due to temperature were substantial, as the time constant and gain vary by a factor of 10 over the temperature range associated with RTP.

The identification scheme to estimate τ_i and K from experimental data is described in [7], [8]. A sequence of lamp power values was sent to the RTP system. This sequence was known as a recipe. The recipe was formulated so that reasonable spatial temperature uniformity was maintained at all instants in order to satisfy the approximation used in the development of the low-order model. The eigenvalues of the system were estimated at various temperature using a procedure employing the TLS ESPRIT algorithm [9]. After the eigenvalues were estimated, the amplitude of the step response was estimated. This was difficult because of the temperature drift induced by the window heating; however, a least-squares technique can be employed. The gain of the system and view factors were then identified using a least-squares algorithm again. The results are shown in Figure 5.12 and 5.13 for the estimation of the effects of temperature on the gain and time constant, respectively.

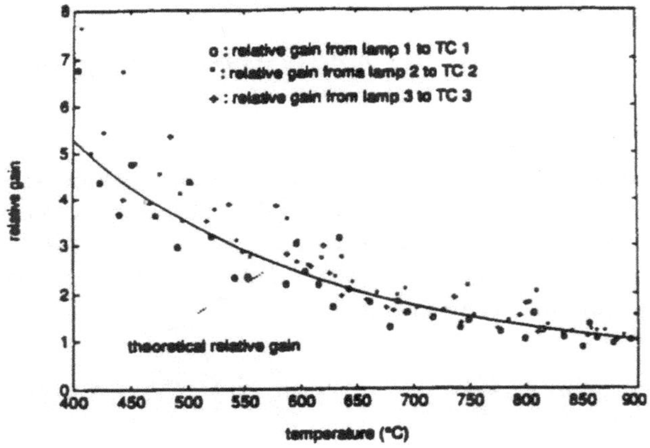

Figure 5.12 Gain of the system relative to that at 900°C and comparison with theory.

The model was expressed in discrete-time format for use in designing a control system. Using the zero-order hold to describe the input sequence, the discrete-time expression of Equation 5.10 was given by

$$\bar{T}(z) = \Gamma(z)\mathbf{K}\bar{P}(z) \tag{5.13}$$

where z denotes the z-transform,

$$\Gamma(z) = diag\left[\frac{(1-e^{-\Delta t/\tau_i})z^{-1}}{1-e^{-\Delta t/\tau_i}z^{-1}}\right] \tag{5.14}$$

and Δt denotes the sampling time. The system model was inherently stable since the poles lie within the unit circle for all operating temperatures.

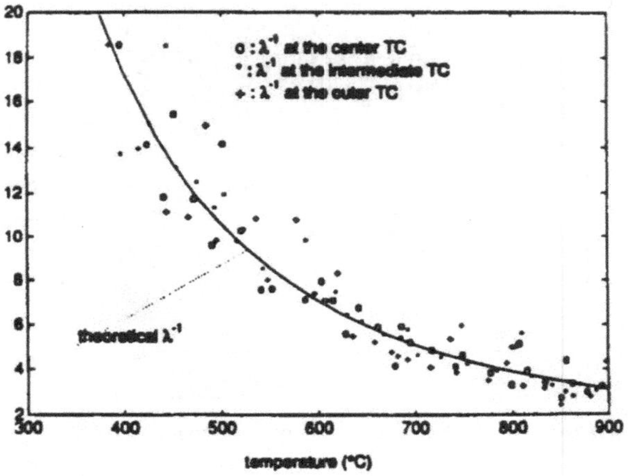

Figure 5.13 Time constant of the process model as a function of temperature and comparison with theory.

In order to obtain a complete description of the system, this relationship was combined with models describing sensor dynamics and lamp dynamics [4]. The sensor and lamp dynamics can be described by detailed models. However, for the purpose of model-based control system design, it was only necessary to approximate these dynamics as a simple time-delay relation,

$$\bar{T}_{m,i}(t) = \bar{T}_i(t-\theta). \tag{5.15}$$

The measured temperature at time t was denoted by $T_{m,i}(t)$, and the time delay was denoted by θ. The resulting model expressed in z-transform notation was given by

$$\bar{T}_m(z) = z^{-d}\Gamma(z)\mathbf{K}\bar{P}(z) \tag{5.16}$$

where $d = \theta/\Delta t$ was rounded to the nearest integer.

The power was supplied to the lamp filaments by sending a 0–10 volt signal from the computer control system to the power supplies that drive the lamps. The relation between the voltage signal applied to the supplies and the power sent to the lamps was not necessarily linear. Consequently, it was important to model that nonlinearity, if possible, so that it could be accounted for by the control system. It was possible to determine the nonlinearity with a transducer installed on the power supply to measure the average power sent to the lamps. By applying a known voltage to the power supplies and then recording the average power output, the desired relationship can be determined. This function can be described by a polynomial and then inverted to remove the nonlinearity from the loop because the model is linear with respect to radiative power from the lamps (see Equation 5.16). We noted that the average power to the lamps may not equal the radiative power from the lamps. The offset was due to heating losses within the bulb filament. However, this offset was implicitly incorporated in the model when the gain matrix, K, was determined from experimental data.

A second strategy that considered the RTP system as a black box was employed to identify a linear model of the process [11], [12]. Among numerous alternatives, an ARX model was used to describe the system,

$$T(t) - T_{ss} = \sum_{k=1}^{n_a} A_k(T(t-k) - T_{ss}) + \sum_{k=1}^{n_b} B_k(P(t-k) - P_{ss}) + n(t), \quad (5.17)$$

where T is an $l \times 1$ vector describing temperature, P is an $M \times 1$ vector describing percent of maximum zone power, A_k is an $l \times l$ matrix, B_k is an $l \times M$ matrix, and l is the number of sensors on the wafer measuring temperature. The steady-state temperature and power were denoted by T_{ss} and P_{ss}, respectively. Because the steady-state temperature was difficult to determine accurately because of drift, a slight modification to the model was made. Let $T_{ss} = \hat{T} + \Delta\hat{T}$. The least squares problem for model identification can then be formulated as

$$min_{A_i, B_i} \sum_{t=1}^{N} \left\| (T(t) - \hat{T}) - \sum_{k=1}^{n_a} A_k(T(t-k) - \hat{T}) - \sum_{k=1}^{n_b} B_k(P(t-k) - P_{ss}) - T_{bias} \right\|_2^2 \quad (5.18)$$

where $T_{bias} = (I - \sum_{k=1}^{n_a} A_k)\Delta\hat{T}$.

The strategy for estimating the unknown model parameters utilized PRBS (pseudo-random binary sequence) to excite the system to obtain the necessary input-output data. The mean temperature that this excitation produced is designated as $\hat{T}$. Some other issues that were accounted for during model identification included the use of a data subsampling method so that the ARX model could span a longer time interval and observe a larger change of the temperature. Subsampling was needed because the data collection rate was 10 Hz and over that interval the temperature changed very little. Consequently, the least-squares formulation, as an identification method, may contain inherent error sources due to the effect of the measurement sensor noise (which was presumed to be Gaussian distributed) and the quantization noise (which was presumed to be uniformly distributed with quantization level of 0.5°C).

With the black box approach, the model order needed to be selected. The criterion used to determine the appropriateness of the model was to be able to make the prediction error smaller than the quantization level using as small a number of ARX model parameters as possible. For our applications, this order was three A matrices and three B matrices with subsampling selected as four.

We are now going to show the value of the identified models by using them to study an important characteristic of the RTP system, its DC gain. For the ARX model with coefficients $\{A_i, B_i\}$, the identified DC gain was given by the formula

$$\text{DC gain} = \mathbf{D}_o = (I - \sum_{i=1}^{3} A_i)^{-1} \sum_{l=1}^{3} B_i$$

Substituting in the appropriate values for 700°C (with $l = 3$, $J = 3$),

$$\mathbf{D}_o = \begin{bmatrix} 2.07 & 4.41 & 4.50 \\ 1.11 & 4.78 & 4.91 \\ 0.73 & 5.08 & 5.51 \end{bmatrix}. \quad (5.19)$$

Note that the magnitude of the first column of $\mathbf{D}_o$ is smaller than those of the second and third columns, which was due to the difference in the maximum power of each lamp: the first (center) lamp has 2 kW maximum power, the second (intermediate) lamp 12 kW maximum power, and the third (outer) lamp 24 kW maximum power. Also note the similarity of the second and third column of $\mathbf{D}_o$, which says that the second lamp will affect the wafer temperature in a manner similar to the third lamp. As a result, we have effectively two lamps rather than three (recall we have physically three lamps), which may cause difficulties in maintaining temperature uniformity in the steady state because of an inadequate number of degrees of freedom. This conclusion will be more clearly seen from an SVD (Singular Value Decomposition) analysis, about which more will be said later. To increase the independence of the control effects of the two outside lamps, a baffle was installed to redistribute the light energy from the lamps to the wafer. The same identification technique described earlier was used to identify the RTP system model, and the DC gain was computed from the identified model with the result

$$\mathbf{D}_n = \begin{bmatrix} 2.02 & 5.27 & 3.97 \\ 1.19 & 5.53 & 4.01 \\ 0.83 & 5.11 & 5.15 \end{bmatrix}. \quad (5.20)$$

We can observe that the second column of the new DC gain matrix was no longer similar to the third column, as it was in Equation 5.19. As a result, the three lamps heated the wafer in different ways. The first (center) lamp heated mostly the center of the wafer, and the third (outer) lamp heated mostly the edge of the wafer. On the other hand, the second (intermediate) lamp heated the wafer overall, acting like a bulk heater. Of course, the second lamp heated the intermediate portion of the wafer more than the center and edge of the wafer, but the difference was not so significant.

Even if the idea of installing a baffle was partly motivated by the direct investigation of the DC gain matrix, it was in fact deduced from an SVD (Singular Value Decomposition) analysis of the DC gain matrix.

The SVD of $\mathbf{D}_o$ in Equation 5.19 is given by

$$u_1 = [0.54, 0.57, 0.63],\ u_2 = [-0.80, 0.13, 0.58],\ u_3 = [0.25, -0.81, 0.53] \quad (5.21)$$

$$v_1 = [0.18, 0.68, 0.71],\ v_2 = [-0.98, 0.04, 0.21],\ v_3 = [0.12, -0.73, 0.67] \quad (5.22)$$

$$\sigma_1 = 12.15, \sigma_2 = 1.12, \sigma_3 = 0.11 \quad (5.23)$$

From this, we can conclude that $[1, 1, 1]$ (u_1) is a strong output direction. Of course, u_1 is $[0.54, 0.57, 0.62]$ and is not exactly equal to $[1, 1, 1]$. However, $[0.54, 0.57, 0.62]$ was close to $[1, 1, 1]$ in terms of direction in a 3-dimensional coordinate system and was denoted as the [1, 1, 1] direction, here. Since [1, 1, 1] was the strong output direction, we can affect the wafer temperature in the [1, 1, 1] direction by a minimal input power change. This means that if we maintain the temperature uniformity at the reference temperature (700°C), we can maintain the uniformity near 700°C (say, at 710°C) with a small change in the input lamp power. The weak output direction (the vector u_3 — approximately [1, −1, 1]) says that it is difficult to increase the temperature of the center and outer portions of the wafer while cooling down the intermediate portion of the wafer, which was, more or less, expected. The gain of the weak direction (σ_3) was two orders of magnitude smaller than that of the strong direction (σ_1). This meant that there were effectively only two lamps in the RTP system in terms of controlling the temperature of the wafer, even if there were physically three lamps. This naturally led to the idea of redesigning the RTP chamber to get a better lamp illumination pattern. Installing a baffle (see [10] for more details) into the existing RTP system improved our situation as shown in the SVD analysis of the new DC gain $\mathbf{D}_n$ in Equation 5.20. The SVD of $\mathbf{D}_n$ was given by

$$u_1 = [0.56, 0.57, 0.60], \; u_2 = [-0.63, -0.17, 0.76],$$

$$u_3 = [0.53, -0.80, 0.26] \tag{5.24}$$

$$v_1 = [0.19, -0.72, 0.67], \; v_2 = [0.76, -0.3, -0.57],$$

$$v_3 = [0.63, 0.61, 0.48] \tag{5.25}$$

$$\sigma_1 = 12.14, \; \sigma_2 = 1.17, \; \sigma_3 = 0.52 \tag{5.26}$$

Compared to the SVD of the previous DC gain, the lowest singular value (σ_3) has been increased by a factor of 5, a significant improvement over the previous RTP system. In other words, only one-fifth of the power required to control the temperature in the weak direction, using the previous RTP system, was necessary for the same task with the new RTP system. As a result, we obtained three independent lamps by merely installing a baffle into the existing RTP system. Independence of the three lamps in the new RTP system was crucial in maintaining the temperature uniformity of the wafer.

5.5 Control Design

The general strategy of feedback combined with feedforward control was investigated for RTP control. In this strategy, a feedforward value of the lamp power was computed (in response to a change in the temperature set point) according to a predetermined relationship. This feedforward value was then added to a feedback value and the resultant lamp power was sent to the system. The concept behind this approach was that the feedforward power brings the temperature close to the desired temperature; the feedback value compensates for modeling errors and disturbances.

The feedback value can be determined with a variety of design techniques, two of which are described below. Several approaches were investigated to determine the feedforward value. One approach was based on replaying the lamp powers of previous runs. Another approach was based on a model-based optimization.

The physics-based model was employed to develop a controller using a variation of the classical Internal Model Control (IMC) design procedure [13], [6]. The IMC approach consisted of factoring the linearized form of the nonlinear low-order model (see Equation 5.16) as,

$$\tilde{\mathbf{G}}_p(z) = \tilde{\mathbf{G}}_p^+(z)\tilde{\mathbf{G}}_p^-(z) \tag{5.27}$$

where $\tilde{\mathbf{G}}_p^+(z)$ contains the time delay terms, z^{-d}, all right half-plane zeros, zeros that are close to $(-1, 0)$ on the unit disk, and has unity gain. The IMC controller is then obtained by

$$\mathbf{G}_c^*(z) = \tilde{\mathbf{G}}_p^-(z)^{-1}\mathbf{F}(z) \tag{5.28}$$

where $\mathbf{F}(z)$ is a matrix of filters used to tune the closed-loop performance and robustness and to obtain a physically realizable controller. The inversion $\tilde{\mathbf{G}}_p^-(z)^{-1}$ was relatively straightforward because the dynamics of the annular wafer elements of the linearized form of the nonlinear model were decoupled.

The tuning matrix, or IMC filter, $\mathbf{F}(z)$ was selected to satisfy several requirements of RTP. The first requirement was related to repeatability in which zero offset between the actual and desired trajectory was to be guaranteed at steady-state condition despite modeling error. The second requirement was related to uniformity in which the closed-loop dynamics of the wafer temperature should exhibit similar behavior. The third requirement was related to robustness and implementation in which the controller should be as low-order as possible. Other requirements were ease of operator usage and flexibility. One simple selection of $\mathbf{F}(z)$ that meets these requirements was given by the first-order filter

$$\mathbf{F}(z) = f(z)\mathbf{I}, \tag{5.29}$$

$$f(z) = \frac{1-\alpha}{1-\alpha z^{-1}}, \tag{5.30}$$

where α is a tuning parameter, the speed of response. This provided us with a simple controller with parameters that could be interpreted from a physical standpoint.

In this approach to control design, the nonlinear dependency of $\mathbf{K}$ and τ_i on temperature can be parameterized explicitly in the controller. Hence, a continuous gain-scheduling procedure can be applied. It was noted that, as temperature increased, the process gain decreased. Since the controller involved the inverse of the process model, the controller gain increased as temperature was increased. Consequently, the gain-scheduling provided consistent response over the entire temperature range. Thus, control design at one temperature should also apply at other temperatures.

In addition to the IMC approach, a multivariable feedback control law was determined by an LQG design which incorporated integral control action to reduce run-to-run variations [12], [14]. The controller needed to be designed carefully, because, in a nearly singular system such as the experimental RTP, actuator saturation and integrator windup can cause problems. To solve this problem partially, integral control was applied in only the (strongly) controllable directions in temperature error space, helping to prevent the controller from trying to remove uncontrollable disturbances.

For the LQG design, the black-box model was used. It can be expressed in the time domain as follows:

$$y_k^0 = CA^{k-1}Bu_0^0 + \ldots + CABu_{k-2}^0 + CBu_{k-1}^0,$$

and the resulting equations ordered from y_1^0 to y_N^0:

$$\begin{bmatrix} y_1^0 \\ y_2^0 \\ \vdots \\ y_N^0 \end{bmatrix} = \begin{bmatrix} CB & \cdots & 0 \\ CAB & \cdots & 0 \\ \vdots & \ddots & \vdots \\ CA^{N-1}B & \cdots & CB \end{bmatrix} \begin{bmatrix} u_0^0 \\ u_1^0 \\ \vdots \\ u_{N-1}^0 \end{bmatrix}.$$

These combined equations determine a linear relationship between the input and output of the system in the form $Y = HU$, where the notation Y and U is used to designate the $n_o N \times 1$ and $n_i N \times 1$ stacked vectors.

The identified model of the system was augmented with new states representing the integral of the error along the m easiest to control directions, defined as $\xi \overset{def}{=} [\ \xi_1\ \xi_2\ \ldots\ \xi_m\ ]^T$. The new system model was, then,

$$\begin{aligned} \begin{bmatrix} x_{k+1} \\ \xi_{k+1} \end{bmatrix} &= \begin{bmatrix} A & 0 \\ U_{1:m}^T C & I \end{bmatrix} \begin{bmatrix} x_k \\ \xi_k \end{bmatrix} + \begin{bmatrix} B \\ 0 \end{bmatrix} u, \\ y_k &= \begin{bmatrix} C & 0 \\ 0 & I \end{bmatrix} \begin{bmatrix} x_k \\ \xi_k \end{bmatrix}, \end{aligned}$$

where $U_{1:m}$ is the first m columns of U, the output matrix from the SVD of the open-loop transfer matrix $H(z)|_{z=1} = USV^T$, and represented the (easily) controllable subspace. The output y then consisted of thermocouple readings and the integrator states (in practice the integrator states were computed in software from the measured temperature errors). Weights for the integrator states were chosen to provide a good transient response. A complete description of the control design can be found in [14], [15].

The goal of our feedforward control was to design, in advance, a reference input trajectory which will cause the system to follow a predetermined reference output trajectory, assuming no noise or modeling error. The approach built upon the analysis done in [16] and expressed the open-loop trajectory generation problem as a convex optimization with linear constraints and a quadratic objective function [14], [15].

The input/output relationship of the system can be described by a linear matrix equation. This relationship was used to convert convex constraints on the output trajectory into convex constraints on the input trajectory.

The RTP system imposed a number of linear constraints. These were linear constraints imposed by the RTP hardware, specifically, the actuators had both saturation effects and a maximum rate of increase.

Saturation constraints were modeled as follows: P_k^{lm} is defined as the $n_i \times 1$ vector of steady-state powers at the point of linearization of the above model at time k. With an outer feedback control loop running, to insure that the feedback controller has some room to work (for example ±10% leeway), the total powers should be constrained to the operating range of $10 \leq P^{total} \leq 90$, which translates into a constraint on U of $(10 - P^{lm}) \leq U \leq (90 - P^{lm})$.

Maximum rates of increase (or slew rate limits) for our actuators were included also. These were due to the dynamics of the halogen bulbs in our lamp. We included this constraint as $u_{k+1}^0 - u_k^0 \leq 5$, which can be expressed in a matrix equation of the form $SU \leq 5$, where S has 1 on the main diagonal and −1 on the off-diagonal.

The quality of our optimized trajectory can be measured in two ways: minimized tracking error (following a reference trajectory) and minimized spatial nonuniformity across the wafer. Because the tracking error placed an upper bound on the nonuniformity error, we concentrate on it here. We define the desired trajectory Y^{ref} as relative to the same linearized starting point used for system identification. The tracking error E can be defined as $E = Y - Y^{ref}$, where E again denotes the stacked error vector.

We define our objective function to be a quadratic constraint on E as $F(x) = E^T E$, and expand

$$F(x) = U^T H^T HU - 2(Y^{ref})^T HU + (Y^{ref})^T Y^{ref}. \quad (5.31)$$

Software programs exist, such as the FORTRAN program LSSOL [17], which can take the convex constraints and produce a unique solution, if one exists.

After achieving successful results in simulation, the control system was implemented in a real-time computing environment linked to the actual RTP equipment. The computing environment included a VxWorks real-time operating system, SUN IPC workstation, VME I/O boards and a Motorola 68030 processor.

5.6 Proof-of-Concept Testing

The Stanford RTP system was used for multiprocessing applications in which sequential thermal process steps were performed within the same reactor. A schematic of the RTM is shown in Figure 5.10. A concentric three-zone 38-kW illuminator, constructed and donated by Texas Instruments, was used for wafer heating. The center zone consisted of a 2-kW bulb, the intermediate zone consisted of 12 1-kW bulbs and the outer zone consisted of 24 1-kW bulbs. A picture of the three-zone lamp is presented in Figure 5.11. The reflector was water and air cooled. An annular gold-plated stainless steel opaque ring was placed on the quartz window to provide improved compartmentalization between the intermediate and outer zones. This improvement was achieved by reducing the radiative energy from the outer zone impinging on the interior location of the wafer and from the intermediate zone impinging on the edge of the wafer. The RTM was used for 4-inch wafer processing. The wafer was manually loaded onto three supporting quartz pins of low thermal mass. The wafer was placed in the center of the chamber which was approximately 15 inches in diameter and 6 inches in height. Gas was injected via two jets. For the control experiments presented

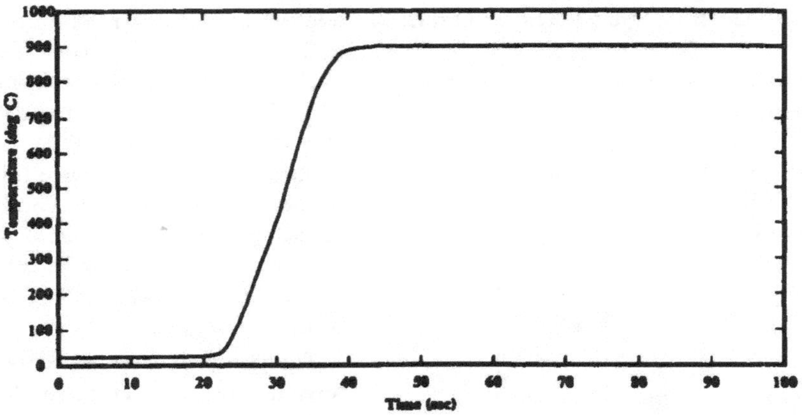

Figure 5.14 Temperature trajectory of the three sensors over the first 100 seconds and the 5 minute process hold.

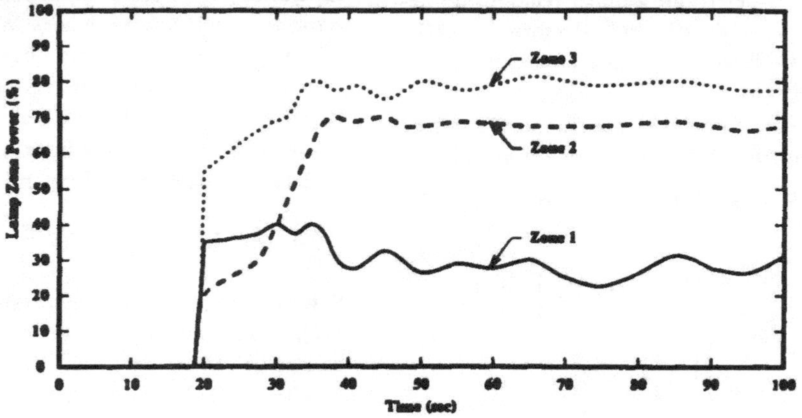

Figure 5.15 Powers of the three zones used to control temperature over the first 100 seconds.

below, temperature was measured with a thermocouple instrumented wafer. Three thermocouples were bonded to the wafer along a common diameter at radial positions of 0 inch (center), 1 inch, and 1 7/8 inches. The experiments were conducted in a N_2 environment at 1 atmosphere pressure.

The control algorithms were evaluated for control of temperature uniformity in achieving a ramp from room temperature to 900°C at a ramp rate of 45°C/s followed by a hold for 5 minutes at 1 atm pressure and 1000 sccm (cc/min gas at standard conditions) N_2 [4]. This trajectory typified low-temperature thermal oxidation or annealing operations. The ramp rate was selected to correspond to the performance limit (in terms of satisfying uniformity requirements) of the equipment. The control system utilized simultaneous IMC feedback and feedforward control. Gain scheduling was employed to compensate for the nonlinearities induced by radiative heating.

The wafer temperature for the desired trajectory over the first 100 seconds is plotted in Figure 5.14 for the center, middle, and edge locations where thermocouples are bonded to the wafer along a common diameter at radial positions of 0 inch (center), 1 inch, and 1 7/8 inches.

The ramp rate gradually increased to the specified 45°C/s and then decreased as the desired process hold temperature was approached. The corresponding lamp powers, that were manipulated by commands from the control system to achieve the desired temperature trajectory, are shown in Figure 5.15.

The time delay of the system can be seen by comparing the starting times of the lamp powers to the temperature response. Approximately, a two second delay existed in the beginning of the response. Of this delay, approximately 1.5 seconds was caused by a power surge interlock on the lamp power supplies which only functions when the lamp power is below 15% of the total power. The remaining delay was caused by the sensor and filament heating dynamics. In the power profile plot, the rate limiting of the lamp powers is seen. This rate-limiting strategy was employed as a safety precaution to prevent a large inrush current to the lamps. However, these interlocks prevented higher values of ramp rates from being achieved.

The nonuniformity of the controlled temperature trajectory was then analyzed. From the measurements of the entire five minute run (i.e., the first 100 seconds shown in Figure 5.14 along with an additional 400 second hold at 900°C not shown in the figure), the nonuniformity was computed by the peak-to-peak temperature error of the temperature measurements of the three thermocouples. The result is plotted in Figure 5.16. The maximum temperature nonuniformity of approximately 15°C occurred during the ramp around a mean temperature of 350°C. This nonuniformity occurred at a low temperature and does not effect processing or damage the wafer via slip. As the ramp progressed from this point, the nonuniformity decreased. The sig-

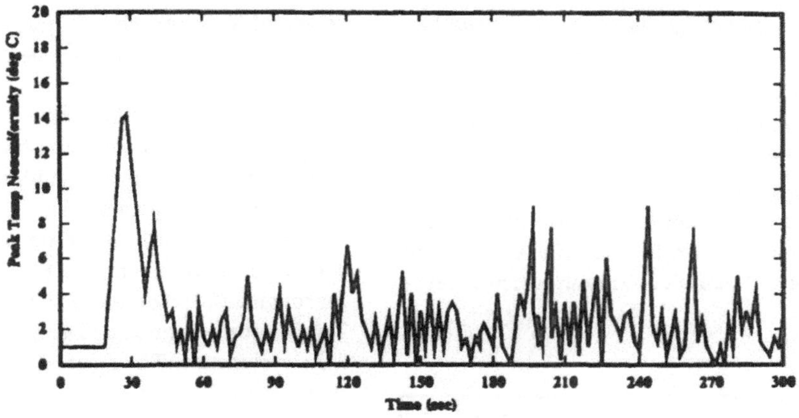

Figure 5.16 Temperature nonuniformity for the 5 minute run as measured by the three temperature sensors.

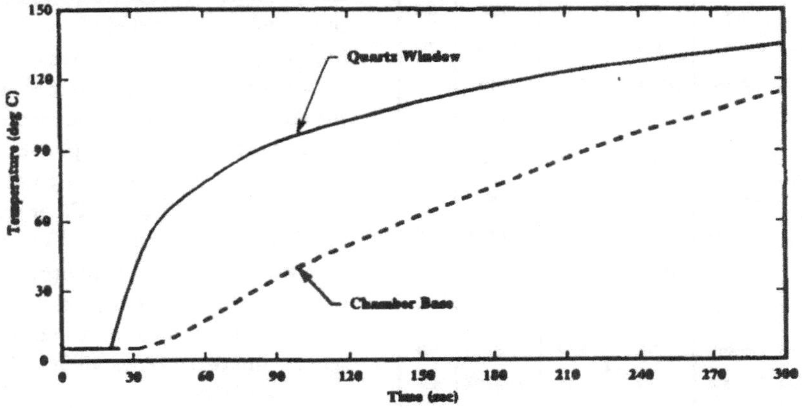

Figure 5.17 Temperatures of the quartz window and chamber base over the 5 minute run.

nificant sensor noise can be seen.

The capability of the controller to hold the wafer temperature at a desired process temperature despite the presence of dynamic heating from extraneous sources was then examined. As seen in Figure 5.14, the control system held the wafer temperature at the desired value of 900°C. Although the sensors were quite noisy and had resolution of 0.5°C, the wafer temperature averaged over the entire hold portion for the three sensors corresponded to 900.9°C, 900.7°C, and 900.8°C, respectively. This result was desired because the uniformity of the process parameters, such as film thickness and resistivity, generally depend on the integrated or averaged temperature over time. The capability of the control system to hold the wafer temperature at the desired value, albeit slightly higher, is demonstrated by plotting the dynamics of the quartz window and chamber base of the RTM in Figure 5.17.

The slow heating of these components of the RTM corresponded to slow disturbances to the wafer temperature. Because of the reduced gain of the controller to compensate for time delays, these disturbances impacted the closed-loop response by raising the temperature to a value slightly higher than the set point. However, without the feedback temperature control system, the wafer temperature would have drifted to a value more than 50°C higher than the set point as opposed to less than 1°C in the measured wafer temperature.

5.7 Technology Transfer to Industry

After demonstrating the prototype RTP equipment at Stanford, the multivariable control strategy (including hardware and software) was transferred to Texas Instruments for application in the MMST program. This transfer involved integration on eight RTP reactors: seven on-line and one off-line. These RTP systems were eventually to be used in a 1000 wafer demonstration of two full-flow sub-half-micron CMOS process technologies in the MMST program at TI [18], [19],[20],[21] and [22].

Although there were similarities between the RTP equipment at TI and the three-zone RTP system at Stanford, there were also substantial differences. Two types of illuminators were used for MMST, a four-zone system constructed at Texas Instruments for MMST applications and a six-zone system manufactured by G^2 Semiconductor Corp. Both systems utilized concentric zone heating; the TI system employed a circular arrangement of lamps, and the G^2 system used a hexagonal arrangement of lamps. A thick (roughly 15 mm) quartz window was used in the TI system to separate the lamps from the reaction chamber, and a thin (3 mm) quartz window was used in the G^2 system. Wafer rotation was not employed with the TI system but was used with the G^2 system. The rotation rate was approximately 20 rpm. Moreover, six-inch wafer processing took place using up to four pyrometers for feedback. Most reactors employed different configurations purge ring assemblies, guard rings, and susceptors (see Figure 5.5).

The on-line RTP reactors configured with the IMC controller were used for thirteen different thermal processes: LPCVD Nitride, LPCVD Tungsten, Silicide react, Silicide anneal, sinter, LPCVD polysilicon, LPCVD amorphous silicon, germane clean, dry RTO, wet RTO, source/drain anneal, gate anneal, and tank anneal. These processes ranged from 450° to 1100°C, from 1 to 650 torr pressure, and from 30 seconds to 5 minutes of processing time (see Figure 5.18).

RTP Process	Carrier Gas	T_{ph} [°C]	T_{pr} [°C]	t_{ph} [s]	t_{pr} [s]	P [Torr]
sinter	N_2	450	450	0	180	650
LPCVD-W	Ar/H_2	425-475	425-475	0	60-180	30
LPCVD-amor Si	Ar	450-500	500-560	5-15	60-180	15
LPCVD-poly	Ar	450-550	650	5-15	120-240	15
silicide react	N_2	450-500	650	5-15	180	1
silicide anneal	Ar	450-550	750	5-15	60	1
LPCVD-SiO_2	O_2	450-550	750	5-15	30-180	1-5
germane clean	H_2	450-550	650-750	5-15	120	15
LPCVD-Si_3N_4	NH_3	550-650	850	5-15	60-180	1-5
gate RTA	Ar	750-800	900	5-15	30	650
dry RTO	O_2	750-800	1000	5-15	120-180	650
wet RTO	O_2	750-800	1000	5-15	120-180	650
source/drain RTA	Ar	750-800	1000-1050	5-15	15-30	650
tank RTA	NH_3	750-800	1100	5-15	300	650

Figure 5.18 List of processes controlled during the MMST program. The preheat temperature (T_{ph}) and time (t_{ph}) and the process temperature (T_{pr}) and time (t_{pr}) are given. The carrier gases and operating pressures are also presented.

There were several challenges in customizing the temperature control system for operation in an all-RTP factory environment [13]. These challenges included substantial differences among the eight reactors and thirteen processes, operation in a prototyping development environment, ill-conditioned processing equipment, calibrated pyrometers required for temperature sensing, equipment reliability tied to control power trajectories, multiple lamp-zone/sensor configurations, detection of equipment failures, and numerous operational and communication modes.

Nonetheless, it was possible to develop a single computer control code with the flexibility of achieving all of the desired objectives. This was accomplished by developing a controller in a modular framework based on a standardized model of the process and equipment. The control structure remained the same while the model-based parameters of the controller differed from process to process and reactor to reactor. It was possible to read these parameters from a data file while holding the controller code and logic constant. Consequently, it was only necessary to maintain and modify a single computer control code for the entire RTP factory.

We present results here for an LPCVD-Nitride process that employed a TI illuminator and for an LPCVD-Poly process that employed a G^2 illuminator. Additional temperature and process control results are presented in [13] and [6].

The desired temperature trajectory for the LPCVD-Nitride process involved a ramp to 850°C and then a hold at 850°C for roughly 180 seconds in a SiH_4/NH_3 deposition environment. Temperature was measured using four radially distributed 3.3 μm InAs pyrometers. The center and edge pyrometers were actively used for real-time feedback control and the inner two pyrometers were used to monitor the temperature. The reasons for this analysis were: (1) repeatable results were possible using only two pyrometers, (2) an analysis of the benefits of using pyrometers for feedback could be assessed, and (3) fewer pyrometers were maintained during the marathon demonstration. In Figure 5.19, the center temperature measurement is shown for a 24-wafer lot process. The offsets in the plot during the ramps are merely due to differences in the starting points of the ramps.

During the hold at 850°C, the reactive gases were injected, and the deposition took place. The standard deviation (computed over the 24 runs) of the temperature measurements during the deposition time was analyzed. In Figure 5.20, the standard deviation of the four sensor measurements are shown. The controlled sensors improved repeatability over the monitored sensor locations by a factor of seven. A three-sigma interpretation shows roughly that the controlled sensors held temperature to within ±0.3°C and the monitored sensors were repeatable at ±2.0°C.

We analyzed the power trajectories to the lamp zones to evaluate the repeatability of the equipment. In Figure 5.21, the power to the center zone is presented for the 24 runs. The intermediate two zones were biased off the center and edge zones, respectively. From these results, it was clear that the lamp power decreased substantially during a nitride deposition run because the chamber and window heat more slowly than the wafer; because the chamber and window provide energy to the wafer, the necessary energy from the lamps to achieve a specified wafer temperature was less as the chamber and window heat up. In addition, we noted the chamber and window heating effect from run-to-run by observing the lowered lamp energy requirements as the lot processing progresses. These observations can be used in developing fault detection algorithms.

To study the capability of temperature control on the process parameter, we compared the thickness of the LPCVD poly process determined at the center for each wafer of a 24-wafer lot where multizone feedback temperature control was used and no real-time feedback temperature control (i.e., open-loop operation) was used. For the open-loop case, a predetermined lamp power trajectory was replayed for each wafer of the 24-wafer lot. The comparison is shown in Figure 5.22. It is clear that the multizone feedback control is much better than open-loop control. In some sense, this comparison is a worst case analysis since the lamp powers themselves for both cases had no control, not usual in industry. In our experiments, variations in line voltage

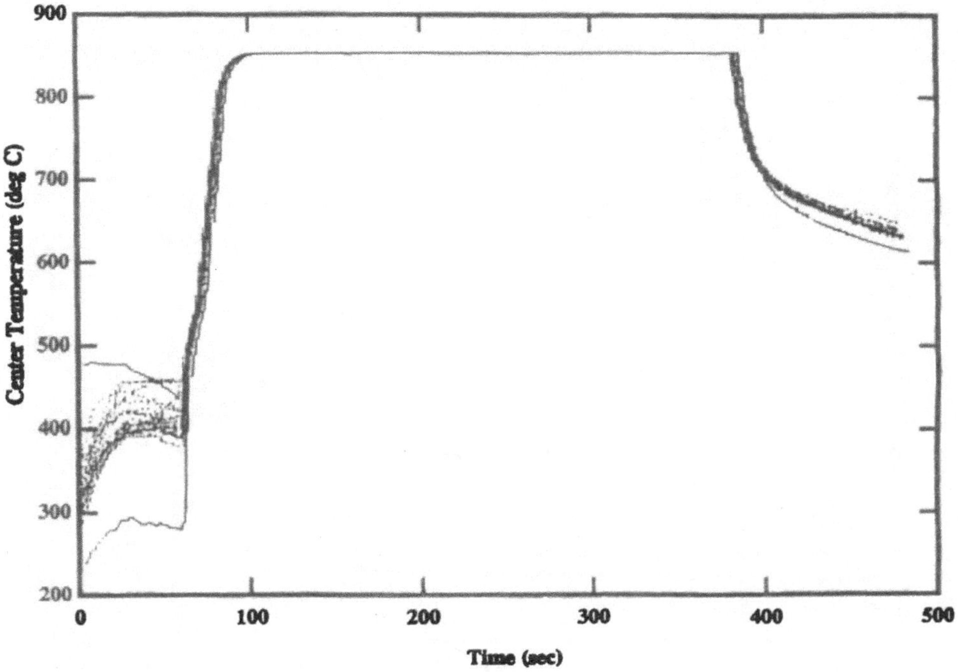

Figure 5.19 The center temperature of the LPCVD nitride process for a 24-wafer lot run.

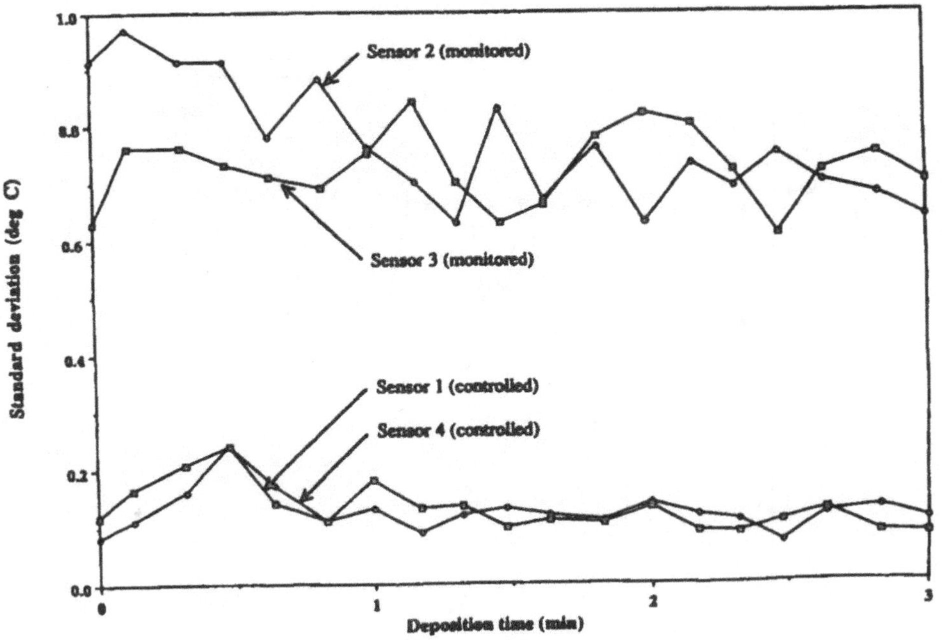

Figure 5.20 Standard deviation of temperature measurements during the three minute deposition step of the LPCVD nitride process for the 24-wafer lot run.

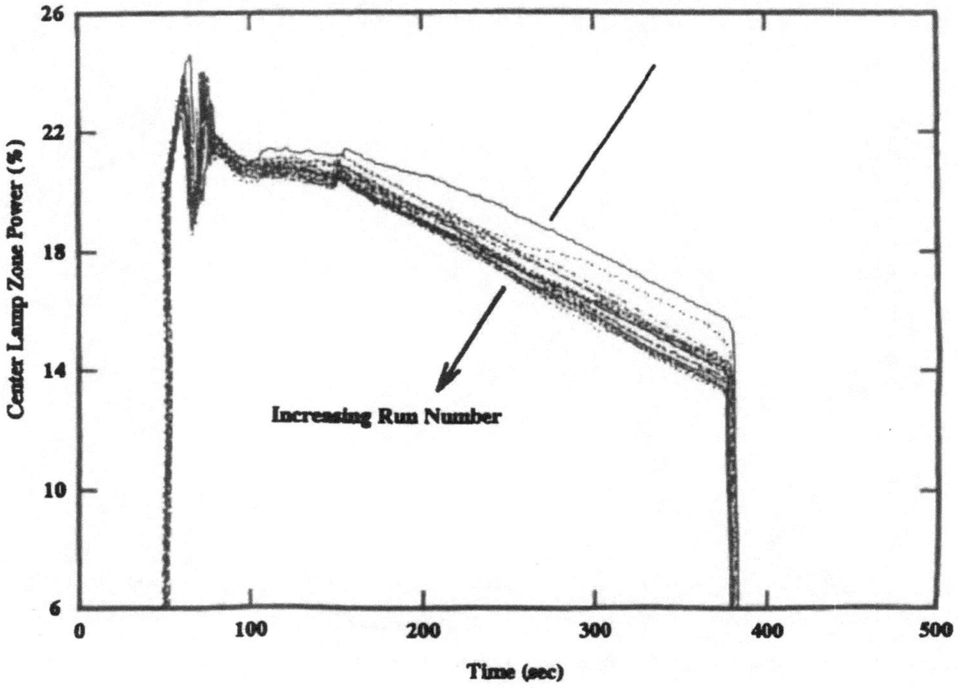

Figure 5.21 Power to the center zone for the 24-wafer lot run of the LPCVD nitride process.

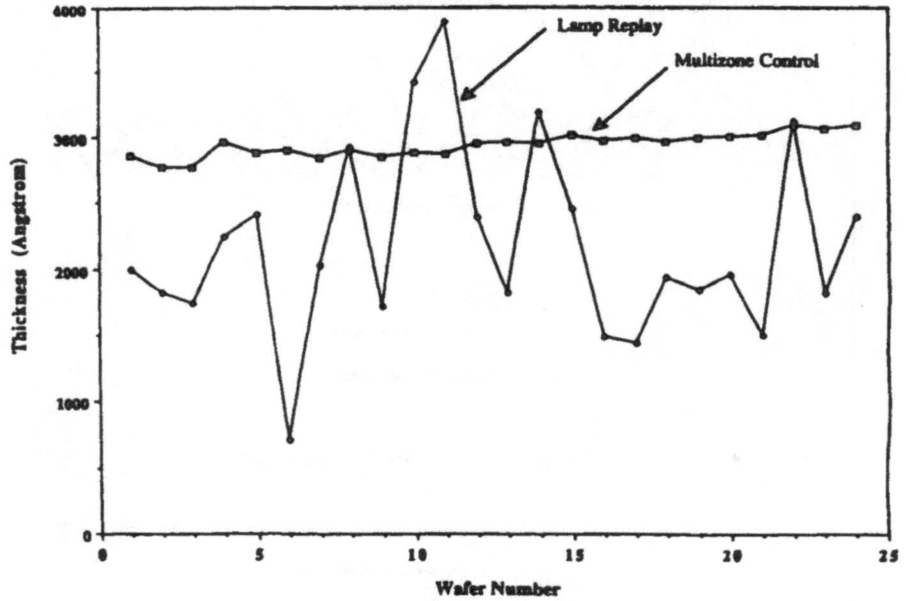

Figure 5.22 Comparison of multizone feedback control and open-loop lamp replay for LPCVD polysilicon process.

proceeded unfiltered through the reactor causing unprovoked fluctuations in the lamp power and inducing strong temperature effects. However, the feedback temperature control system can compensate somewhat for these fluctuations. For the open-loop case, these fluctuations pass on directly and result in unacceptable repeatability.

5.8 Conclusions

A systems approach has been used for a study in semiconductor manufacturing. This methodology has included developing models, analyzing alternative equipment designs from a control perspective, establishing model identification techniques to develop a model for control design, developing a real-time control system and embedding it within a control processor, proof-of-concept testing with a prototype system, and then transferring the control technology to industry. This application has shown the role that control methodologies can play in semiconductor device manufacturing.

References

[1] Lord, H., Thermal and stress analysis of semiconductor wafers in a rapid thermal processing oven, *IEEE Trans. Semicond. Manufact.*, 1, 141–150, 1988.

[2] Norman, S.A., *Wafer Temperature Control in Rapid Thermal Processing*, Ph.D. Thesis, Stanford University, 1992.

[3] Cho, Y., Schaper, C. and Kailath, T., Low order modeling and dynamic characterization of rapid thermal processing, *Appl. Phys. A: Solids and Surfaces*, A:54(4), 317–326, 1992.

[4] Norman, S.A., Schaper, C.D. and Boyd, S.P., Improvement of temperature uniformity in rapid thermal processing systems using multivariable control. In *Mater. Res. Soc. Proc.: Rapid Thermal and Integrated Processing*. Materials Research Society, 1991.

[5] Saraswat, K. and Apte, P., Rapid thermal processing uniformity using multivariable control of a circularly symmetric three-zone lamp, *IEEE Trans. on Semicond. Manufact.*, 5, 1992.

[6] Saraswat, K., Schaper, C., Moslehi, M. and Kailath, T., Modeling, identification, and control of rapid thermal processing, *J. Electrochem. Soc.*, 141(11), 3200–3209, 1994.

[7] Cho, Y., *Fast Subspace Based System Identification: Theory and Practice*, Ph.D. Thesis, Stanford University, CA, 1993.

[8] Cho, Y. and Kailath, T., Model identification in rapid thermal processing systems, *IEEE Trans. Semicond. Manufact.*, 6(3), 233–245, 1993.

[9] Roy, R., Paulraj, A. and Kailath, T., ESPRIT - a subspace rotation approach to estimation of parameters of cisoids in noise, *IEEE Trans. ASSP*, 34(5), 1340–1342, 1986.

[10] Schaper, C., Cho, Y., Park, P., Norman, S., Gyugyi, P., Hoffmann, G., Balemi, S., Boyd, S., Franklin, G., Kailath, T., and Sarawat, K., Dynamics and control of a rapid thermal multiprocessor. In *SPIE Conference on Rapid Thermal and Integrated Processing*, September 1991.

[11] Cho, Y.M., Xu, G., and Kailath, T., Fast recursive identification of state-space models via exploitation of displacement structure, *Automatica*, 30(1), 45–59, 1994.

[12] Gyugyi, P., Cho, Y., Franklin, G., and Kailath, T., Control of rapid thermal processing: A system theoretic approach. In *IFAC World Congress*, 1993.

[13] Saraswat, K., Schaper, C., Moslehi, M., and Kailath, T., Control of MMST RTP: Uniformity, repeatability, and integration for flexible manufacturing, *IEEE Trans. on Semicond. Manifact.*, 7(2), 202–219, 1994.

[14] Gyugyi, P., *Application of Model-Based Control to Rapid Thermal Processing Systems*. Ph.D. Thesis, Stanford University, 1993.

[15] Gyugyi, P.J., Cho, Y.M., Franklin, G., and Kailath, T., Convex optimization of wafer temperature trajectories for rapid thermal processing. In *The 2nd IEEE Conf. Control Appl.*, Vancouver, 1993.

[16] Norman, S.A., Optimization of transient temperature uniformity in RTP systems, *IEEE Trans. Electron Dev.*, January 1992.

[17] Gill, P.E., Hammarling, S.J., Murray, W., Saunders, M.A., and Wright, M.H., User's guide for LSSOL (Version 1.0): A FORTRAN package for constrained least-squares and convex quadratic programming, Tech. Rep. SOL 86-1, Operations Research Dept., Stanford University, Stanford, CA, 1986.

[18] Chatterjee, P. and Larrabee, G., Manufacturing for the gigabit age, *IEEE Trans. on VLSI Technology*, 1, 1993.

[19] Bowling, A., Davis, C., Moslehi, M., and Luttmer, J., Microeletronics manufacturing science and technology: Equipment and sensor technologies, *TI Technical J.*, 9, 1992.

[20] Davis, C., Moslehi, M., and Bowling, A., Microeletronics manufacturing science and technology: Single-wafer thermal processing and wafer cleaning, *TI Technical J.*, 9, 1992.

[21] Moslehi, M. et al., Single-wafer processing tools for agile semiconductor production, *Solid State Technol.*, 37(1), 35–45, 1994.

[22] Saraswat, K. et al., Rapid thermal multiprocessing for a programmable factory for adaptable manufacturing of ic's, *IEEE Trans. on Semicond. Manufact.*, 7(2), 159–175, 1994.

II

Mechanical Control Systems

6

Automotive Control Systems

J. A. Cook
Ford Motor Company, Scientific Research Laboratory, Control Systems Department, Dearborn, MI

J. W. Grizzle
Department of EECS, Control Systems Laboratory, University of Michigan, Ann Arbor, MI

J. Sun
Ford Motor Company, Scientific Research Laboratory, Control Systems Department, Dearborn, MI

M. K. Liubakka
Advanced Vehicle Technology, Ford Motor Company, Dearborn, MI

D.S. Rhode
Advanced Vehicle Technology, Ford Motor Company, Dearborn, MI

J. R. Winkelman
Advanced Vehicle Technology, Ford Motor Company, Dearborn, MI

P. V. Kokotović
ECE Department, University of California, Santa Barbara, CA

6.1 Engine Control 87
Introduction • Air-Fuel Ratio Control System Design • Idle Speed Control • Acknowledgments
References 100
6.2 Adaptive Automotive Speed Control 100
Introduction • Design Objectives • The Design Concept • Adaptive Controller Implementation
6.3 Performance in Test Vehicles 105
Conclusions
Appendix 107
References 111

6.1 Engine Control

J. A. Cook, *Ford Motor Company, Scientific Research Laboratory, Control Systems Department, Dearborn, MI*
J. W. Grizzle, *Department of EECS, Control Systems Laboratory, University of Michigan, Ann Arbor, MI*
J. Sun, *Ford Motor Company, Scientific Research Laboratory, Control Systems Department, Dearborn, MI*

6.1.1 Introduction

Automotive engine control systems must satisfy diverse and often conflicting requirements. These include regulating exhaust emissions to meet increasingly stringent standards without sacrificing good drivability; providing increased fuel economy to satisfy customer desires and to comply with Corporate Average Fuel Economy (CAFE) regulations; and delivering these performance objectives at low cost, with the minimum set of sensors and actuators. The dramatic evolution in vehicle electronic control systems over the past two decades is substantially in response to the first of these requirements. It is the capacity and flexibility of microprocessor-based digital control systems, introduced in the 1970s to address the problem of emission control, that have resulted in the improved function and added convenience, safety, and performance features that distinguish the modern automobile [8].

Although the problem of automotive engine control may encompass a number of different power plants, the one with which this chapter is concerned is the ubiquitous four-stroke cycle, spark ignition, internal combustion gasoline engine. Mechanically, this power plant has remained essentially the same since Nikolaus Otto built the first successful example in 1876. In automotive applications, it consists most often of four, six or eight cylinders wherein reciprocating pistons transmit power via a simple connecting rod and crankshaft mechanism to the wheels. Two complete revolutions of the crankshaft comprise the following sequence of operations.

0-8493-0054-3/00/$0.00+$.50

The initial 180 degrees of crankshaft revolution is the intake stroke, where the piston travels from top-dead-center (TDC) in the cylinder to bottom-dead-center (BDC). During this time an intake valve in the top of the cylinder is opened and a combustible mixture of air and fuel is drawn in from an intake manifold. Subsequent 180-degree increments of crankshaft revolution comprise the compression stroke, where the intake valve is closed and the mixture is compressed as the piston moves back to the top of the cylinder; the combustion stroke when, after the mixture is ignited by a spark plug, torque is generated at the crankshaft by the downward motion of the piston caused by the expanding gas; and finally, the exhaust stroke, when the piston moves back up in the cylinder, expelling the products of combustion through an exhaust valve.

Three fundamental control tasks affect emissions, performance, and fuel economy in the spark ignition engine: (1) air-fuel ratio (A/F) control, that is, providing the correct ratio of air and fuel for efficient combustion to the proper cylinder at the right time; (2) ignition control, which refers to firing the appropriate spark plug at the precise instant required; and (3) control of exhaust gas recirculation to the combustion process to reduce the formation of oxide of nitrogen (NOx) emissions.

Ignition Control

The spark plug is fired near the end of the compression stroke, as the piston approaches TDC. For any engine speed, the optimal time during the compression stroke for ignition to occur is the point at which the maximum brake torque (MBT) is generated. Spark timing significantly in advance of MBT risks damage from the piston moving against the expanding gas. As the ignition event is retarded from MBT, less combustion pressure is developed and more energy is lost to the exhaust stream.

Numerous methods exist for energizing the spark plugs. For most of automotive history, cam-activated breaker points were used to develop a high voltage in the secondary windings of an induction coil connected between the battery and a distributor. Inside the distributor, a rotating switch, synchronized with the crankshaft, connected the coil to the appropriate spark plug. In the early days of motoring, the ignition system control function was accomplished by the driver, who manipulated a lever located on the steering wheel to change ignition timing. A driver that neglected to retard the spark when attempting to start a hand-cranked Model T Ford could suffer a broken arm if he experienced "kickback." Failing to advance the spark properly while driving resulted in less than optimal fuel economy and power.

Before long, elaborate centrifugal and vacuum-driven distributor systems were developed to adjust spark timing with respect to engine speed and torque. The first digital electronic engine control systems accomplished ignition timing simply by mimicking the functionality of their mechanical predecessors. Modern electronic ignition systems sense crankshaft position to provide accurate cycle-time information and may use barometric pressure, engine coolant temperature, and throttle position along with engine speed and intake manifold pressure to schedule ignition events for the best fuel economy and drivability subject to emissions and spark knock constraints. Additionally, ignition timing may be used to modulate torque to improve transmission shift quality and in a feedback loop as one control variable to regulate engine idle speed. In modern engines, the electronic control module activates the induction coil in response to the sensed timing and operating point information and, in concert with dedicated ignition electronics, routes the high voltage to the correct spark plug.

One method of providing timing information to the control system is by using a magnetic proximity pickup and a toothed wheel driven from the crankshaft to generate a square wave signal indicating TDC for successive cylinders. A signature pulse of unique duration is often used to establish a reference from which absolute timing can be determined. During the past ten years there has been substantial research and development interest in using in-cylinder piezoelectric or piezoresistive combustion pressure sensors for closed-loop feedback control of individual cylinder spark timing to MBT or to the knock limit. The advantages of combustion-pressure-based ignition control are reduced calibration and increased robustness to variability in manufacturing, environment, fuel, and component aging. The cost is in an increased sensor set and additional computing power.

Exhaust Gas Recirculation

Exhaust gas recirculation (EGR) systems were introduced as early as 1973 to control (NOx) emissions. The principle of EGR is to reduce NOx formation during the combustion process by diluting the inducted air-fuel charge with inert exhaust gas. In electronically controlled EGR systems, this is accomplished using a metering orifice in the exhaust manifold to enable a portion of the exhaust gas to flow from the exhaust manifold through a vacuum-actuated EGR control valve and into the intake manifold. Feedback based on the difference between the desired and measured pressure drop across the metering orifice is employed to duty cycle modulate a vacuum regulator controlling the EGR valve pintle position. Because manifold pressure rate and engine torque are directly influenced by EGR, the dynamics of the system can have a significant effect on engine response and, ultimately, vehicle drivability. Such dynamics are dominated by the valve chamber filling response time to changes in the EGR duty cycle command.

The system can be represented as a pure transport delay associated with the time required to build up sufficient vacuum to overcome pintle shaft friction cascaded with first-order dynamics incorporating a time constant that is a function of engine exhaust flow rate. Typically, the EGR control algorithm is a simple proportional-integral (PI) or proportional-integral-derivative (PID) loop. Nonetheless, careful control design is required to provide good emission control without sacrificing vehicle performance. An unconventional method to accomplish NOx control by exhaust recirculation is to directly manipulate the timing of the intake and exhaust valves. Variable-cam-timing (VCT) engines have demonstrated NOx control using mechanical and hydraulic actuators to adjust valve timing and to affect the amount of internal EGR remaining in the cylinder after the exhaust stroke is completed. Early exhaust valve closing has the additional advantage that unburned hydrocarbons (HC) normally emitted to the exhaust stream are recycled through a second combustion event, reducing HC emissions as well. Although VCT engines eliminate the normal EGR system dynamics, the fundamentally multivariable nature of the resulting system presages a difficult engine control problem.

Air-Fuel Ratio Control

Historically, fuel control was accomplished by a carburetor that used a venturi arrangement and a simple float-and-valve mechanism to meter the proper amount of fuel to the engine. For special operating conditions, such as idle or acceleration, additional mechanical and vacuum circuitry was required to assure satisfactory engine operation and good drivability. The demise of the carburetor was occasioned by the advent of three-way catalytic converters (TWC) for emission control. These devices simultaneously convert oxidizing [HC and carbon monoxide (CO)] and reducing (NOx) species in the exhaust, but, as shown in Figure 6.1, require precise control of A/F to the stoichiometric value to be effective. Consequently, the electronic fuel system of

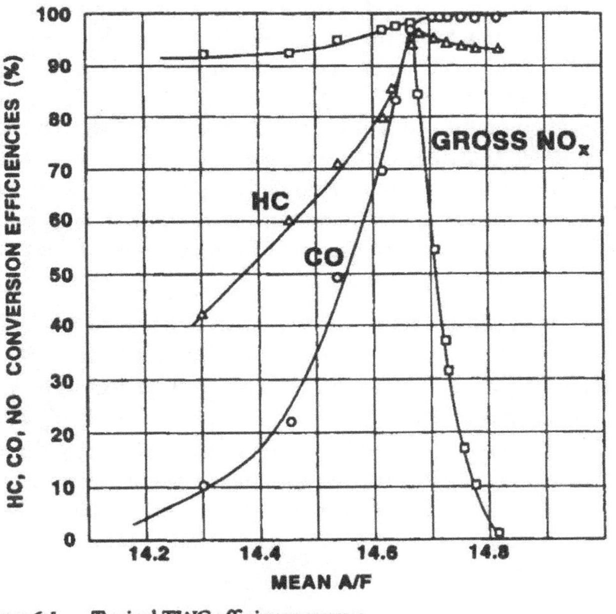

Figure 6.1 Typical TWC efficiency curves.

a modern spark ignition automobile engine employs individual fuel injectors located in the inlet manifold runners close to the intake valves to deliver accurately timed and metered fuel to all cylinders. The injectors are regulated by an A/F control system that has two primary components: a feedback portion, in which a signal related to A/F from an exhaust gas oxygen (EGO) sensor is fed back through a digital controller to regulate the pulse width command sent to the fuel injectors; and a feedforward portion, in which injector fuel flow is adjusted in response to a signal from an air flow meter.

The feedback, or closed-loop portion of the control system, is fully effective only under steady-state conditions and when the EGO sensor has attained the proper operating temperature. The feedforward, or open-loop portion of the control system, is particularly important when the engine is cold (before the closed-loop A/F control is operational) and during transient operation [when the significant delay between the injection of fuel (usually during the exhaust stroke, just before the intake valve opens) and the appearance of a signal at the EGO sensor (possibly long after the conclusion of the exhaust stroke) inhibits good control]. First, in Section 6.1.2, the open-loop A/F control problem is examined with emphasis on accounting for sensor dynamics. Then, the closed-loop problem is addressed from a modern control systems perspective, where individual cylinder control of A/F is accomplished using a single EGO sensor.

Idle Speed Control

In addition to these essential tasks of controlling ignition, A/F, and EGR, the typical on-board microprocessor performs many other diagnostic and control functions. These include electric fan control, purge control of the evaporative emissions canister, turbocharged engine wastegate control, overspeed control, electronic transmission shift scheduling and control, cruise control, and idle speed control (ISC). The ISC requirement is to maintain constant engine RPM at closed throttle while rejecting disturbances such as automatic transmission neutral-to-drive transition, air conditioner compressor engagement, and power steering lock-up.

The idle speed problem is a difficult one, especially for small engines at low speeds where marginal torque reserve is available for disturbance rejection. The problem is made more challenging by the fact that significant parameter variation can be expected over the substantial range of environmental conditions in which the engine must operate. Finally, the ISC design is subject not only to quantitative performance requirements, such as overshoot and settling time, but also to more subjective measures of performance, such as idle quality and the degree of noise and vibration communicated to the driver through the body structure. The ISC problem is addressed in Section 6.1.3.

6.1.2 Air-Fuel Ratio Control System Design

Due to the precipitous falloff of TWC efficiency away from stoichiometry, the primary objective of the A/F control system is to maintain the fuel metering in a stoichiometric proportion to the incoming air flow [the only exception to this occurs in heavy load situations where a rich mixture is required to avoid premature detonation (or knock) and to keep the TWC from overheating]. Variation in air flow commanded by the driver is treated as a disturbance to the system. A block diagram of the control structure is illustrated in Figure 6.2, and the two major subcomponents treated here are highlighted in bold outline. The first part of this section describes the development and implementation of a cylinder air charge estimator for predicting the air charge entering the cylinders downstream of the intake manifold plenum on the basis of available measurements of air mass flow rate upstream

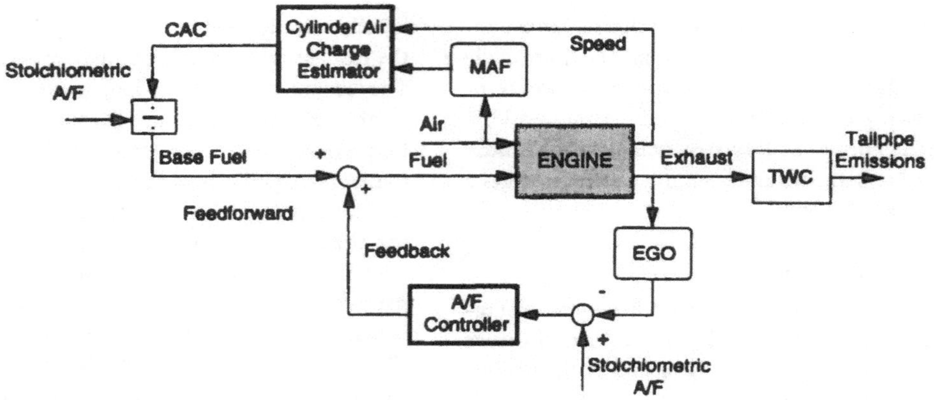

Figure 6.2 Basic A/F control loop showing major feedforward and feedback elements.

of the throttle. The air charge estimate is used to form the base fuel calculation, which is often then modified to account for any fuel-puddling dynamics and the delay associated with closed-valve fuel injection timing. Finally, a classical, time-invariant, single-input single-output (SISO) PI controller is normally used to correct for any persistent errors in the open-loop fuel calculation by adjusting the average A/F to perceived stoichiometry.

Even if the average A/F is controlled to stoichiometry, individual cylinders may be operating consistently rich or lean of the desired value. This cylinder-to-cylinder A/F maldistribution is due, in part, to injector variability. Consequently, fuel injectors are machined to close tolerances to avoid individual cylinder flow discrepancies, resulting in high cost per injector. However, even if the injectors are perfectly matched, maldistribution can arise from individual cylinders having different breathing characteristics due to a combination of factors, such as intake manifold configuration and valve characteristics. It is known that such A/F maldistribution can result in increased emissions due to shifts in the closed-loop A/F setpoint relative to the TWC [9]. The second half of this section describes the development of a nonclassical, periodically time-varying controller for tuning the A/F in each cylinder to eliminate this maldistribution.

Hardware Assumptions

The modeling and control methods presented here are applicable to just about any fuel-injected engine. For illustration purposes only, it is assumed that the engine is a port fuel-injected V8 with independent fuel control for each bank of cylinders. The cylinders are numbered one through four, starting from the front of the right bank, and five through eight, starting from the front of the left bank. The firing order of the engine is 1-3-7-2-6-5-4-8, which is not symmetric from bank to bank. Fuel injection is timed to occur on a closed valve prior to the intake stroke (induction event). For the purpose of closed-loop control, the engine is equipped with a switching type EGO sensor located at the confluence of the individual exhaust runners and just upstream of the catalytic converter. Such sensors typically incorporate a ZrO_2 ceramic thimble employing a platinum catalyst on the exterior surface to equilibrate the exhaust gas mixture. The interior surface of the sensor is exposed to the atmosphere. The output voltage is exponentially related to the ratio of O_2 partial pressures across the ceramic, and thus the sensor is essentially a switching device indicating by its state whether the exhaust gas is rich or lean of stoichiometry.

Cylinder Air Charge Computation

This section describes the development and implementation of an air charge estimator for an eight-cylinder engine. A very real practical problem is posed by the fact that the hot-wire anemometers currently used to measure mass air flow rate have relatively slow dynamics. Indeed, the time constant of this sensor is often on the order of an induction event for an engine speed of 1500 RPM and is only about four to five times faster than the dynamics of the intake manifold. Taking these dynamics into account in the air charge estimation algorithm can significantly improve the accuracy of the algorithm and have substantial benefits for reducing emissions.

Basic Model The air path of a typical engine is depicted in Figure 6.3. An associated lumped-parameter phenomenological model suitable for developing an on-line cylinder air charge estimator [2] is now described. Let P, V, T and m be the pressure in the intake manifold (psi), volume of the intake manifold and runners (liters), temperature (°R), and mass (lbm) of the air in the intake manifold, respectively. Invoking the ideal gas law, and assuming that the manifold air temperature is slowly varying, leads to

$$\frac{d}{dt}P = \frac{RT}{V}[MAF_a - Cyl(N, P, T_{EC}, T_i)], \tag{6.1}$$

where MAF_a is the actual mass air flow metered in by the throttle, R is the molar gas constant, $Cyl(N, P, T_{EC}, T_i)$ is the average instantaneous air flow pumped out of the intake manifold by the cylinders, as a function of engine speed, N (RPM), manifold pressure, engine coolant temperature, T_{EC} (^oR), and air inlet temperature, T_i (^oR). It is assumed that both MAF_a and $Cyl(N, P, T_{EC}, T_i)$ have units of lbm/s.

The dependence of the cylinder pumping or induction function on variations of the engine coolant and air inlet temperatures

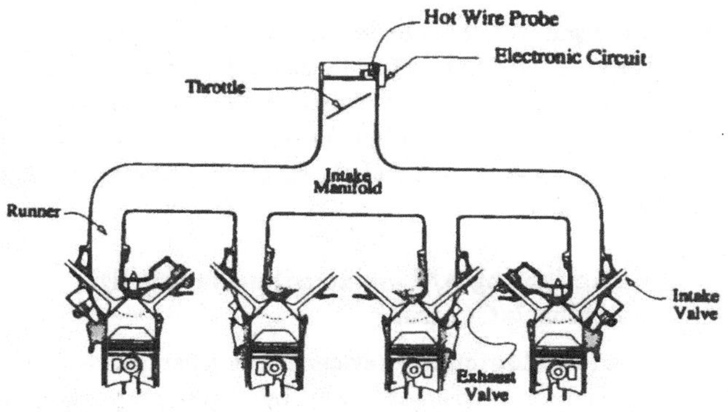

Figure 6.3 Schematic diagram of air path in engine.

is modeled empirically by [10], as

$$Cyl(N, P, T_{EC}, T_i) = Cyl(N, P)\sqrt{\frac{T_i}{T_i^{\text{mapping}}}}\;\frac{T_{EC}^{\text{mapping}} + 2460}{T_{EC} + 2460}, \qquad (6.2)$$

where the superscript "mapping" denotes the corresponding temperatures (^oR) at which the function $Cyl(N, P)$ is determined, based on engine mapping data. An explicit procedure for determining this function is explained in the next subsection.

Cylinder air charge per induction event, CAC, can be determined directly from Equation 6.1. In steady state, the integral of the mass flow rate of air pumped out of the intake manifold over two engine revolutions, divided by the number of cylinders, is the air charge per cylinder. Since engine speed is nearly constant over a single induction event, and the time in seconds for two engine revolutions is $\frac{120}{N}$, a good approximation of the inducted air charge on a per-cylinder basis is given by

$$CAC = \frac{120}{nN} Cyl(N, P, T_{EC}, T_i)\ \text{lbm}, \qquad (6.3)$$

where n is the number of cylinders.

The final element to be incorporated in the model is the mass air flow meter. The importance of including this was demonstrated in [2]. For the purpose of achieving rapid on-line computations, a simple first-order model is used

$$\gamma \frac{d}{dt} MAF_m + MAF_m = MAF_a, \qquad (6.4)$$

where MAF_m is the measured mass air flow and γ is the time constant of the air meter. Substituting the left-hand side of Equation 6.4 for MAF_a in Equation 6.1 yields

$$\frac{d}{dt}P = \frac{RT}{V}\left[\gamma \frac{d}{dt} MAF_m + MAF_m - Cyl(N, P, T_{EC}, T_i)\right] \qquad (6.5)$$

To eliminate the derivative of MAF_m in Equation 6.5, let $x = P - \gamma \frac{RT}{V} MAF_m$. This yields

$$\frac{d}{dt}x = \frac{RT}{V}\left[MAF_m - Cyl(N, x + \gamma \frac{RT}{V} MAF_m, T_{EC}, T_i)\right]. \qquad (6.6)$$

Cylinder air charge is then computed from Equation 6.3 as

$$CAC = \frac{120}{nN} Cyl(N, x + \gamma \frac{RT}{V} MAF_m, T_{EC}, T_i). \qquad (6.7)$$

Note that the effect of including the mass air flow meter's dynamics is to add a feedforward term involving the mass air flow rate to the cylinder air charge computation. When $\gamma = 0$, Equations 6.6 and 6.7 reduce to an estimator that ignores the air meter's dynamics or, equivalently, treats the sensor as being infinitely fast.

Determining Model Parameters The pumping function $Cyl(N, P)$ can be determined on the basis of steady-state engine mapping data. Equip the engine with a high-bandwidth manifold absolute pressure (MAP) sensor and exercise the engine over the full range of speed and load conditions while recording the steady-state value of the instantaneous mass air flow rate as a function of engine speed and manifold pressure. For this purpose, any external exhaust gas recirculation should be disabled. A typical data set should cover every 500 RPM of engine speed from 500 to 5,000 RPM and every half psi of manifold pressure from 3 psi to atmosphere. For the purpose of making these measurements, it is preferable to use a laminar air flow element as this, in addition, allows the calibration of the mass air flow meter to be verified. Record the engine coolant and air inlet temperatures for use in Equation 6.2. The function $Cyl(N, P)$ can be represented as a table lookup or as a polynomial regressed against the above mapping data. In either case, it is common to represent it in the following functional form:

$$Cyl(N, P) = \mu(N)P + \beta(N). \qquad (6.8)$$

The time constant of the air meter is best determined by installing the meter on a flow bench and applying step or rapid sinusoidal variations in air flow to the meter. Methods for fitting an approximate first-order model to the data can be found in any textbook on classical control. A typical value is $\gamma = 20$ ms. Though not highly recommended, a value for the time constant can be determined by on-vehicle calibration, if an accurate determination of the pumping function has been completed. This is explained at the end of the next subsection.

Model Discretization and Validation The estimator modeled by Equations 6.6 and 6.7 must be discretized for implementation. In engine models, an event-based sampling scheme is often used [7]. For illustration purposes, the discretization is carried out here for a V8; the modifications required for other configurations will be evident. Let k be the recursion index and let Δt_k be the elapsed time in seconds per 45 degrees of crank-angle advancement, or $\frac{1}{8}$ revolution; that is, $\Delta t_k = \frac{7.5}{N_k}$ s, where N_k is the current engine speed in RPM. Then Equation 6.6 can be Euler-integrated as

$$x_k = x_{k-1} + \Delta t_k \frac{RT_{k-1}}{V}\Big[MAF_{m,k-1} - Cyl(N_{k-1}, x_{k-1} + \gamma \frac{RT_{k-1}}{V} MAF_{m,k-1}, T_{EC}, T_i)\Big] \qquad (6.9)$$

The cylinder air charge is calculated by

$$CAC_k = 2\Delta t_k Cyl(N_k, x_k + \gamma \frac{RT_k}{V} MAF_{m,k}, T_{EC}, T_i) \tag{6.10}$$

and need be computed only once per 90 crank-angle degrees.

The accuracy of the cylinder air charge model can be easily validated on an engine dynamometer equipped to maintain constant engine speed. Apply very rapid throttle tip-ins and tip-outs, as in Figure 6.4, while holding the engine speed constant. If the model parameters have been properly determined, the calculated manifold pressure accurately tracks the measured manifold pressure. Figure 6.5 illustrates one such test at 1500 RPM. The dynamic

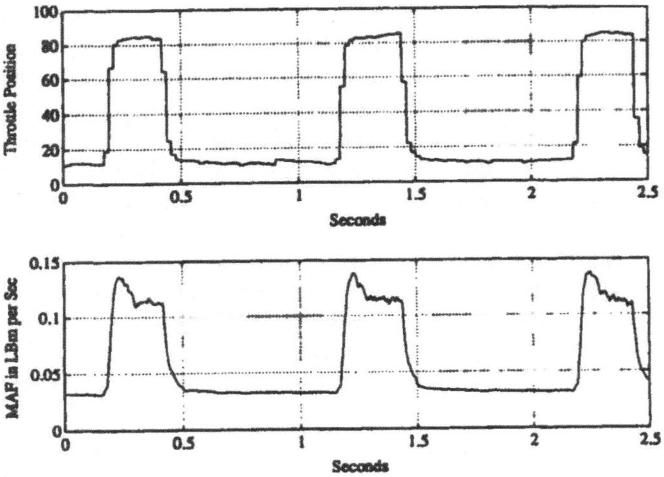

Figure 6.4 Engine operating conditions at nominal 1500 RPM.

responses of the measured and computed values match up quite well. There is some inaccuracy in the quasi-steady-state values at 12 psi; this corresponds to an error in the pumping function $Cyl(N, P)$ at high manifold pressures, so, in this operating condition, it should be reevaluated.

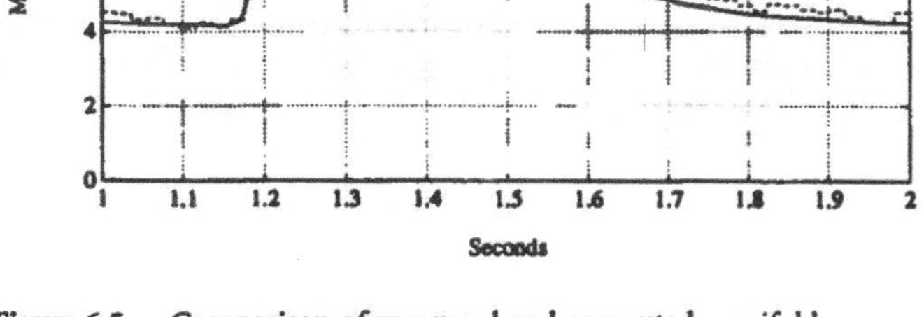

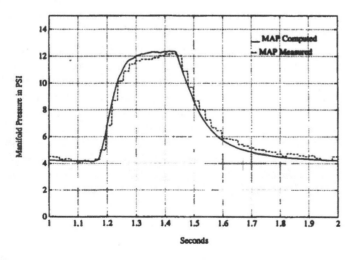

Figure 6.5 Comparison of measured and computed manifold pressure.

From Figure 6.5, it can be seen that the air meter time constant has been accurately identified in this test. If the value for γ in Equation 6.4 had been chosen too large, then the computed manifold pressure would be leading the measured manifold pressure; conversely, if γ were too small, the computed value would lag the measured value.

Eliminating A/F Maldistribution through Feedback Control

A/F maldistribution is evidenced by very rapid switching of the EGO sensor on an engine-event-by-engine-event basis. Such cylinder-to-cylinder A/F maldistribution can result in increased emissions due to shifts in the closed-loop A/F setpoint relative to the TWC [9]. The trivial control solution, which consists of placing individual EGO sensors in each exhaust runner and wrapping independent PI controllers around each injector-sensor pair, is not acceptable from an economic point of view; technically, there would also be problems due to the nonequilibrium condition of the exhaust gas immediately upon exiting the cylinder. This section details the development of a controller that eliminates A/F maldistribution on the basis of a single EGO sensor per engine bank. The controller is developed for the left bank of the engine.

Basic Model A control-oriented block diagram for the A/F system of the left bank of an eight-cylinder engine is depicted in Figure 6.6. The model evolves at an engine speed-

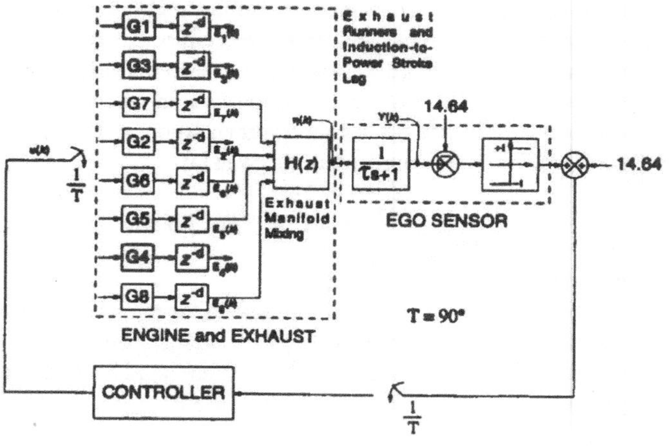

Figure 6.6 Control-oriented block diagram of A/F subsystem.

dependent sampling interval of 90 crank-angle degrees consistent with the eight-cylinder geometry (one exhaust event occurs every 90 degrees of crankshaft rotation). The engine is represented by injector gains G1 through G8 and pure delay z^{-d}, which accounts for the number of engine events that occur from the time that an individual cylinder's fuel injector pulse width is computed until the corresponding exhaust valve opens to admit the mixture to the exhaust manifold. The transfer function H(z) represents the mixing process of the exhaust gases from the individual exhaust ports to the EGO sensor location, including any transport delay. The switching type EGO sensor is represented by a first-order transfer function followed by a preload (switch) nonlinearity.

Note that only cylinders 5 through 8 are inputs to the mixing model, H(z). This is due to the fact that the separate banks of the V8 engine are controlled independently. The gains and delays for cylinders 1 through 4 correspond to the right bank of the engine and are included in the diagram only to represent the firing order. Note furthermore that cylinders 5 and 6 exhaust within 90 degrees of one another, whereas cylinders 7 and 8 exhaust 270 degrees apart. Since the exhaust stroke of cylinder 6 is not complete before the exhaust stroke of cylinder 5 commences, for any exhaust manifold configuration, there is mixing of the exhaust gases from these cylinders at the point where the EGO sensor samples the exhaust gas. We adopt the notation that the sampling index, k, is a multiple of 90 degrees, that is, $x(k)$ is the quantity x at $k \cdot 90$ degrees; moreover, if we are looking at a signal's value during a particular point of an engine cycle, then we will denote this by $x(8k+j)$, which is x at $(8k+j)\cdot 90$ degrees, or x at the j-th event of the k-th engine cycle. The initial time will be taken as $k=0$ at TDC of the compression stroke of cylinder 1. The basic model for a V6 or a four-cylinder engine is simpler; see [3].

A dynamic model of the exhaust manifold mixing is difficult to determine with current technology. This is because a linear A/F measurement is required, and, currently, such sensors have a slow dynamic response in comparison with the time duration of an individual cylinder event. Hence, standard system identification methods break down. In [6], a model structure for the mixing dynamics and an attendant model parameter identification procedure, compatible with existing laboratory sensors, is provided. This is outlined next.

The key assumption used to develop a mathematical model of the exhaust gas mixing is that once the exhaust gas from any particular cylinder reaches the EGO sensor, the exhaust of that cylinder from the previous cycle (two revolutions) has been completely evacuated from the exhaust manifold. It is further assumed that the transport lag from the exhaust port of any cylinder to the sensor location is less than two engine cycles. With these assumptions, and with reference to the timing diagram of Figure 6.7, a model for the exhaust-mixing dynamics may be expressed as relating the A/F at the sensor over one 720-crank-angle-degree period beginning at $(8k)$ as a linear combination of the A/Fs admitted to the exhaust manifold by cylinder 5 during the exhaust strokes occurring at times $(8k+7)$, $(8k-1)$, and $(8k-9)$; by cylinder 6 at $(8k+6)$, $(8k-2)$, and $(8k-10)$; by cylinder 7 at $(8k+4)$, $(8k-4)$, and $(8k-12)$; and by cylinder 8 at $(8k+1)$, $(8k-7)$, and $(8k-15)$. This relationship is given by

$$\begin{bmatrix} \eta(8k) \\ \eta(8k+1) \\ \vdots \\ \eta(8k+6) \\ \eta(8k+7) \end{bmatrix} \tag{6.11}$$

$$= \begin{bmatrix} a_5(1,0) & \dots & a_8(1,0) \\ \vdots & & \\ a_5(8,0) & \dots & a_8(8,0) \end{bmatrix} \begin{bmatrix} E_5(8k+7) \\ E_6(8k+6) \\ E_7(8k+4) \\ E_8(8k+1) \end{bmatrix}$$

$$+ \begin{bmatrix} a_5(1,1) & \dots & a_8(1,1) \\ \vdots & & \\ a_5(8,1) & \dots & a_8(8,1) \end{bmatrix} \begin{bmatrix} E_5(8k-1) \\ E_6(8k-2) \\ E_7(8k-4) \\ E_8(8k-7) \end{bmatrix}$$

$$+ \begin{bmatrix} a_5(1,2) & \dots & a_8(1,2) \\ \vdots & & \\ a_5(8,2) & \dots & a_8(8,2) \end{bmatrix} \begin{bmatrix} E_5(8k-9) \\ E_6(8k-10) \\ E_7(8k-12) \\ E_8(8k-15) \end{bmatrix}$$

where η is the actual A/F at the production sensor location, E_n is the exhaust gas A/F from cylinder n $(n = 5, 6, 7, 8)$, and a_n is the time-dependent fraction of the exhaust gas from cylinder n contributing to the A/F fuel ratio at the sensor. It follows from the key assumption that only 32 of the 96 coefficients in Equation 6.11 can be nonzero. Specifically, every triplet $\{a_n(k,0), a_n(k,1), a_n(k,2)\}$ has, at most, one nonzero element. This is exploited in the model parameter identification procedure.

Determining Model Parameters The pure delay z^{-d} is determined by the type of injection timing used (open or closed valve) and does not vary with engine speed or load. A typical value for closed-valve injection timing is $d = 8$. The time constant of the EGO sensor is normally provided by the manufacturer; if not, it can be estimated by installing it directly in the exhaust runner of one of the cylinders and controlling the fuel pulse width to cause a switch from rich to lean and then lean to rich. A typical average value of these two times is $\tau = 70$ ms.

The first step towards identifying the parameters in the exhaust-mixing model is to determine which one of the parameters $\{a_n(k,0), a_n(k,1), a_n(k,2)\}$ is the possibly nonzero element; this can be uniquely determined on the basis of the transport delay between the opening of the exhaust valve of each cylinder and the time of arrival of the corresponding exhaust gas pulse at the EGO sensor. The measurement of this delay is accomplished by installing the fast, switching type EGO sensor in the exhaust manifold in the production location and carefully balancing the A/F of each cylinder to the stoichiometric value. Then apply a step change in A/F to each cylinder and observe the time delay. The results of a typical test are given in [6]. The transport delays change as a function of engine speed and load and, thus, should be determined at several operating points. In addition, they may not be a multiple of 90 degrees. A practical method to account for these issues through a slightly more sophisticated sampling schedule is outlined in [3].

At steady state, Equation 6.11 reduces to

$$\begin{bmatrix} \eta(8k) \\ \eta(8k+1) \\ \vdots \\ \eta(8k+6) \\ \eta(8k+7) \end{bmatrix} = A_{mix} \begin{bmatrix} E_5(8k-1) \\ E_6(8k-2) \\ E_7(8k-4) \\ E_8(8k+1) \end{bmatrix} \tag{6.12}$$

where

$$A_{mix} = \begin{bmatrix} \sum_{j=0}^{2} a_5(1,j) & \dots & \sum_{j=0}^{2} a_8(1,j) \\ \vdots & \dots & \vdots \\ \sum_{j=0}^{2} a_5(8,j) & \dots & \sum_{j=0}^{2} a_8(1,j) \end{bmatrix} \tag{6.13}$$

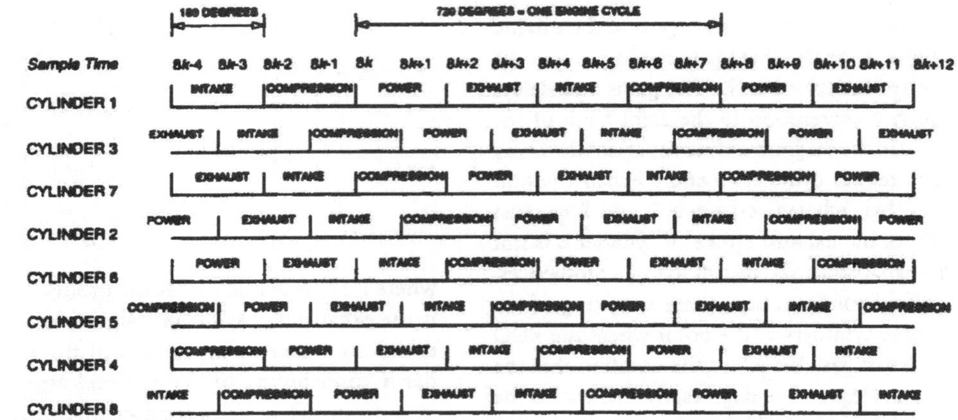

Figure 6.7 Timing diagram for eight-cylinder engine.

This leads to the second part of the parameter identification procedure in which the values of the summed coefficients of Equation 6.12 may be identified from "steady-state" experiments performed with a linear EGO sensor installed in the production location. Then, knowing which of the coefficients is nonzero, Equation 6.11 may be evaluated.

Install a linear EGO sensor in the production location. Then the measured A/F response, $y(k)$, to the sensor input, $\eta(k)$, is modeled by

$$\begin{aligned} w(k+1) &= \alpha w(k) + (1-\alpha)\eta(k) \\ y(k) &= w(k), \end{aligned} \tag{6.14}$$

where $\alpha = e^{-T/\tau_L}$, τ_L is the time constant of the linear EGO sensor, and T is the sampling interval, that is, the amount of time per 90 degrees of crank-angle advance. It follows [6] that the combined steady-state model of exhaust-mixing and linear EGO sensor dynamics is

$$Y = Q_s A_{mix} E \quad , \tag{6.15}$$

where

$$Y = \begin{bmatrix} y(8k) \\ y(8k+1) \\ \vdots \\ y(8k+6) \\ y(8k+7) \end{bmatrix}, \quad E = \begin{bmatrix} E_5(8k-1) \\ E_6(8k-2) \\ E_7(8k-4) \\ E_8(8k+1) \end{bmatrix},$$

and

$$Q_s = \frac{1-\alpha}{1-\alpha^8} \begin{bmatrix} \alpha^7 & \alpha^6 & \cdots & \alpha & 1 \\ \alpha^8 & \alpha^7 & \cdots & \alpha^2 & \alpha \\ \vdots & \vdots & \cdots & \vdots & \vdots \\ \alpha^{13} & \alpha^{12} & \cdots & \alpha^7 & \alpha^6 \\ \alpha^{14} & \alpha^{13} & \cdots & \alpha^8 & \alpha^7 \end{bmatrix} + \begin{bmatrix} 0 & 0 & \cdots & 0 & 0 \\ 1-\alpha & 0 & \cdots & 0 & 0 \\ \vdots & \vdots & \cdots & \vdots & \vdots \\ \alpha^5(1-\alpha) & \alpha^4(1-\alpha) & \cdots & 0 & 0 \\ \alpha^6(1-\alpha) & \alpha^5(1-\alpha) & \cdots & 1-\alpha & 0 \end{bmatrix}.$$

Next, carefully balance each of the cylinders to the stoichiometric A/F, then offset each cylinder, successively, 1 A/F rich and then 1 A/F lean to assess the effect on the A/F at the sensor location. At each condition, let the system reach steady state, then record Y and E over three to ten engine cycles, averaging the components of each vector in order to minimize the impact of noise and cycle-to-cycle variability in the combustion process. This provides input A/F vectors $\bar{E} = [\bar{E}^1 \ldots \bar{E}^8]$ and output A/F vectors $\bar{Z} = [\bar{Y}^1 \ldots \bar{Y}^8]$, where the overbar represents the averaged value. The least squares solution to Equation 6.15 is then given by

$$A_{mix} = Q_s^{-1} \bar{Z} \bar{E}^T (\bar{E}\bar{E}^T)^{-1} . \tag{6.16}$$

The identified coefficients of A_{mix} should satisfy two conditions: (1) the entries in the matrix lie in the interval [0, 1]; (2) the sum of the entries in any row of the matrix is unity. These conditions correspond to no chemical processes occurring in the exhaust system (which could modify the A/F) between the exhaust valve and the EGO sensor. Inherent nonlinearities in the "linear" EGO sensor or errors in the identification of its time constant often lead to violations of these conditions. In this case, the following fix is suggested. For each row of the matrix, identify the largest negative entry and subtract it from each entry so that all are non-negative; then scale the row so that its entries sum to one.

Assembling and Validating the State-Space Model A state-space model will be used for control law design. The combined dynamics of the A/F system from the fuel scheduler to the EGO sensor is shown in Figure 6.8. The coefficients $\kappa_1(k), \ldots, \kappa_{16}(k)$ arise from constructing a state-space representation of Equation 6.11 and, thus, are directly related to the $a_n(k, j)$. In particular, they are periodic, with period equal to one engine cycle. Figure 6.9 provides an example of these coefficients for the model identified in [6]. Assigning state variables as indicated, the state-space model can be expressed as

$$\begin{aligned} x(k+1) &= A(k)x(k) + B(k)u(k) \\ y(k) &= C(k)x(k) . \end{aligned} \tag{6.17}$$

This is an 8-periodic SISO system.

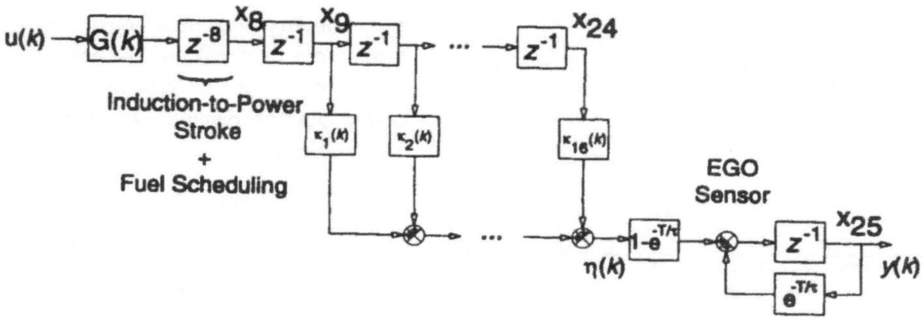

Figure 6.8 Periodically time-varying model of engine showing state variable assignments.

k	$\kappa_1(k)$	$\kappa_2(k)$	$\kappa_3(k)$	$\kappa_4(k)$	$\kappa_5(k)$	$\kappa_6(k)$	$\kappa_7(k)$	$\kappa_8(k)$	$\kappa_9(k)$	$\kappa_{10}(k)$	$\kappa_{11}(k)$	$\kappa_{12}(k)$	$\kappa_{13}(k)$	$\kappa_{14}(k)$	$\kappa_{15}(k)$	$\kappa_{16}(k)$
0	0	$a_8(1,1)$	0	$a_7(1,1)$	0	0	$a_8(1,1)$	0	$a_5(1,2)$	0	0	0	0	0	0	0
1	0	0	$a_8(2,1)$	0	$a_7(2,1)$	0	0	$a_8(2,1)$	0	$a_5(2,2)$	0	0	0	0	0	0
2	$a_8(3,0)$	0	0	$a_8(3,1)$	0	$a_7(3,1)$	0	0	0	0	$a_5(3,2)$	0	0	0	0	0
3	0	$a_8(4,0)$	0	0	$a_8(4,1)$	0	$a_7(4,1)$	0	0	0	0	$a_5(4,2)$	0	0	0	0
4	0	0	$a_8(5,0)$	0	0	$a_8(5,1)$	0	$a_7(5,1)$	0	0	0	0	$a_5(5,2)$	0	0	0
5	0	0	0	$a_8(6,0)$	0	0	$a_8(6,1)$	0	$a_7(6,1)$	0	0	0	0	$a_5(6,2)$	0	0
6	0	$a_7(7,0)$	0	0	$a_8(7,0)$	0	0	$a_8(7,1)$	0	0	0	0	0	0	$a_5(7,2)$	0
7	0	0	$a_7(8,0)$	0	0	$a_8(8,0)$	0	0	$a_8(8,1)$	0	0	0	0	0	0	$a_5(8,2)$

Figure 6.9 Time-dependent coefficients for Figure 6.8.

For control design, it is convenient to transform this system, via lifting [1], [5], to a linear, time-invariant multiple-input multiple-output (MIMO) system as follows. Let $\bar{x}(k) = x(8k)$, $Y(k) = [y(8k), \ldots, y(8k+7)]^T$, $U(k) = [u(8k), \ldots, u(8k+7)]^T$. Then

$$\begin{aligned} \bar{x}(k+1) &= \bar{A}\bar{x}(k) + \bar{B}U(k) \ , \\ Y(k) &= \bar{C}\bar{x}(k) + \bar{D}U(k) \end{aligned} \tag{6.18}$$

where

$$\begin{aligned} \bar{A} &= A(7)A(6)\cdots A(1)A(0), \\ \bar{B} &= [A(7)A(6)\cdots A(1)B(0) : A(7)A(6)\cdots \\ &\quad \cdots A(2)B(1) : \cdots : A(7)B(6) : B(7)] \ , \\ \bar{C} &= \begin{bmatrix} C(0) \\ C(1)A(0) \\ \vdots \\ C(6)A(5)\cdots A(0) \\ C(7)A(6)\cdots A(0) \end{bmatrix} , \end{aligned} \tag{6.19}$$

$$\bar{D} = \begin{bmatrix} 0 & \vdots & 0 & \cdots & 0 & \vdots & 0 \\ C(1)B(0) & \vdots & 0 & \cdots & 0 & \vdots & 0 \\ \vdots & \vdots & \vdots & \cdots & \vdots & \vdots & \vdots \\ C(6)A(5)\cdots A(1)B(0) & \vdots & C(6)A(5)\cdots A(2)B(1) & \cdots & 0 & \vdots & 0 \\ C(7)A(6)\cdots A(1)B(0) & \vdots & C(7)A(6)\cdots A(2)B(1) & \cdots & C(7)B(6) & \vdots & 0 \end{bmatrix} .$$

Normally, $\bar{D}$ is identically zero because the time delay separating the input from the sensor is greater than one engine cycle. Since only cylinders 5 through 8 are to be controlled, the $\bar{B}$ and $\bar{D}$ matrices may be reduced by eliminating the columns that correspond to the control variables for cylinders 1 through 4. This results in a system model with four inputs and eight outputs.

Additional data should be taken to validate the identified model of the A/F system. An example of the experimental and modeled response to a unit step input in A/F is shown in Figure 6.10.

Control Algorithm for ICAFC The first step is to check the feasibility of independently controlling the A/F in the four cylinders. This will be possible if and only if[1] the model of Equation 6.18, with all of the injector gains set to unity, has

[1] Since the model is asymptotically stable, it is automatically stabilizable and detectable.

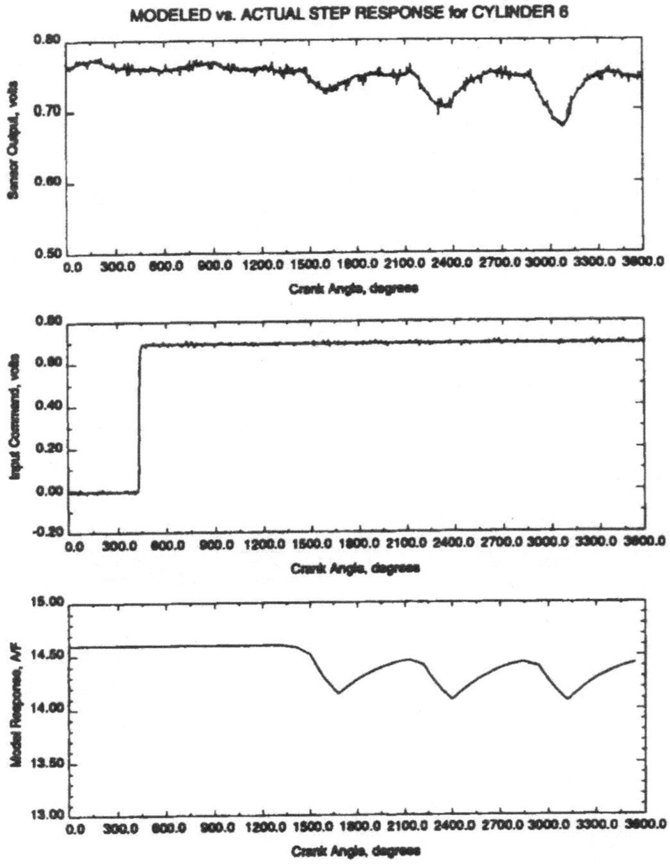

Figure 6.10 Comparison of actual and modeled step response for cylinder number 6.

"full rank[2] at dc" (no transmission zeros at 1). To evaluate this, compute the dc gain of the system

$$G_{dc} = \bar{C}(I - \bar{A})^{-1}\bar{B} + \bar{D}, \qquad (6.20)$$

then compute the singular value decomposition (SVD) of G_{dc}. For the regulation problem to be feasible, the ratio of the largest to the fourth largest singular values should be no larger than 4 or 5. If the ratio is too large, then a redesign of the hardware is necessary before proceeding to the next step [6].

In order to achieve individual set-point control on all cylinders, the system model needs to be augmented with four integrators. This can be done on the input side by

$$\begin{aligned} \bar{x}(k+1) &= \bar{A}\bar{x}(k) + \bar{B}U(k) \\ U(k+1) &= U(k) + V(k) \\ Y(k) &= \bar{C}\bar{x}(k) + \bar{D}U(k)\,, \end{aligned} \qquad (6.21)$$

where $V(k)$ is the new control variable; or on the output side. To do the latter, the four components of Y that are to be regulated to stoichiometry must be selected. One way to do this is to choose four components of Y on the basis of achieving the best numerically conditioned dc gain matrix when the other four output components are deleted. Denote the resulting reduced output by $\bar{Y}(k)$. Then integrators can be added as

$$\begin{aligned} \bar{x}(k+1) &= \bar{A}\bar{x}(k) + \bar{B}U(k) \\ W(k+1) &= W(k) + \Delta\bar{Y}_m(k) \\ Y(k) &= \bar{C}\bar{x}(k) + \bar{D}U(k)\,, \end{aligned} \qquad (6.22)$$

where $\Delta\bar{Y}_m$ is the error between the measured value of $\bar{Y}$ and the stoichiometric setpoint.

In either case, it is now very easy to design a stabilizing controller by a host of techniques presented in this handbook. For implementation purposes, the order of the resulting controller can normally be significantly lowered through the use of model reduction methods. Other issues dealing with implementation are discussed in [3], such as how to incorporate the switching aspect of the sensor into the final controller and how to properly schedule the computed control signals. Specific examples of such controllers eliminating A/F maldistribution are given in [3] and [6].

6.1.3 Idle Speed Control

Engine idle is one of the most frequently encountered operating conditions for city driving. The quality of ISC affects almost every aspect of vehicle performance such as fuel economy, emissions, drivability, etc. The ISC problem has been extensively studied, and a comprehensive overview of the subject can be found in [4].

The primary objective for ISC is to maintain the engine speed at a desired setpoint in the presence of various load disturbances. The key factors to be considered in its design include:

- **Engine speed setpoint.** To maximize fuel economy, the reference engine speed is scheduled at the minimum that yields acceptable combustion quality; accessory drive requirements; and noise, vibration, and harshness (NVH) properties. As the automotive industry strives to reduce fuel consumption by lowering the idle speed, the problems associated with the idle quality (such as allowable speed droop and recovery transient, combustion quality and engine vibration, etc.) tend to be magnified and thus put more stringent requirements on the performance of the control system.
- **Accessory load disturbances.** Typical loads in today's automobile include air conditioning, power steering, power windows, neutral-to-drive shift, alternator loads, etc. Their characteristics and range of operation determine the complexity of the control design and achievable performance.
- **Control authority and actuator limitations.** The control variables for ISC are air flow (regulated by the throttle or a bypass valve) and spark timing. Other variables, such as A/F, also affect engine operation,

[2]Physically, this corresponds to being able to use constant injector inputs to arbitrarily adjust the A/F in the individual cylinders.

but A/F is not considered as a control variable for ISC because it is the primary handle on emissions. The air bypass valve (or throttle) and spark timing are subject to constraints imposed by the hardware itself as well as other engine control design considerations. For example, in order to give spark enough control authority to respond to the load disturbances, it is necessary to retard it from MBT to provide appreciable torque reserve. On the other hand, there is a fuel penalty associated with the retarded spark, which, in theory, can be compensated by the lower idle speed allowed by the increased control authority of spark. The optimal trade-off, however, differs from engine to engine and needs to be evaluated by taking into consideration combustion quality and the ignition hardware constraints (the physical time required for arming the coil and firing the next spark imposes a limitation on the allowable spark advance increase between two consecutive events).

- **Available measurement.** Typically, only engine speed is used for ISC feedback. MAP, or inferred MAP, is also used in some designs. Accessory load sensors (such as the air conditioning switch, neutral-to-drive shift switch, power steering pressure sensor, etc.) are installed in many vehicles to provide information on load disturbances for feedforward control.
- **Variations in engine characteristics over the entire operating range.** The ISC design has to consider different operational and environmental conditions such as temperature, altitude, etc. To meet the performance objectives for a large fleet of vehicles throughout their entire engine life, the control system has to be robust enough to incorporate changes in the plant dynamics due to aging and unit-to-unit variability.

The selection of desired engine setpoint and spark retard is a sophisticated design trade-off process and is beyond the scope of this chapter. The control problem addressed here is the speed tracking problem, which can be formally stated as: *For a given desired engine speed setpoint, design a controller that, based on the measured engine speed, generates commands for the air bypass valve and spark timing to minimize engine speed variations from the setpoint in the presence of load disturbances.* A schematic control system diagram is shown in Figure 6.11.

Engine Models for ISC

An engine model that encompasses the most important characteristics and dynamics of engine idle operation is given in Figure 6.12. It uses the model structure developed in [7] and consists of the actuator characteristics, manifold filling dynamics, engine pumping characteristics, intake-to-power stroke delay, torque characteristics, and engine rotational dynamics (EGR is not considered at idle). The assumption of sonic flow through the throttle, generally satisfied at idle, has led to a much simplified model where the air flow across the throttle is only a function of the throttle position. The differential equations describing the overall dynamics are given by

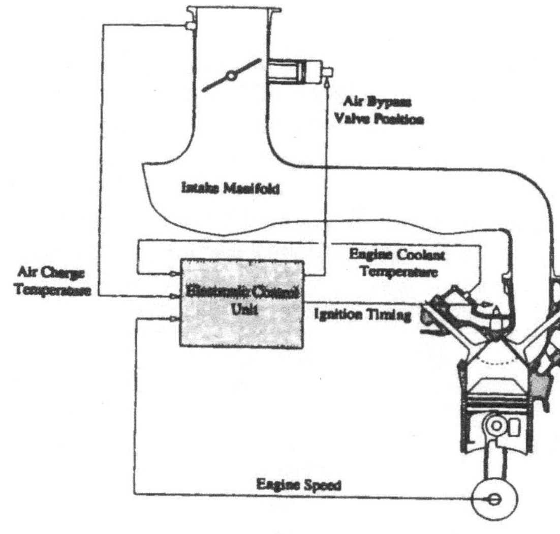

Figure 6.11 Sensor-actuator configuration for ISC.

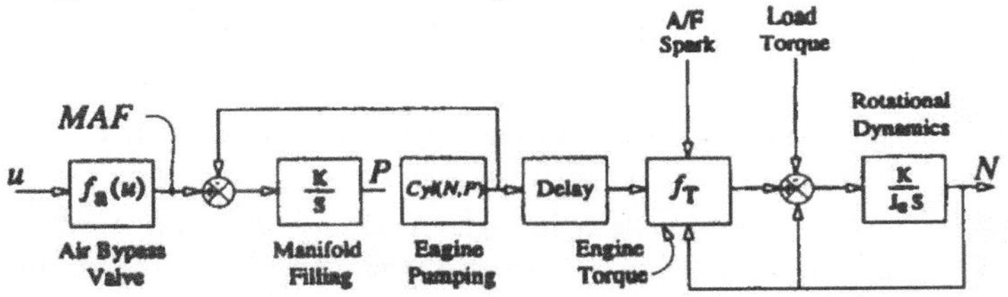

Figure 6.12 Nonlinear engine model.

$$
\begin{array}{rcll}
MAF & = & f_a(u) & \\
\dot{P} & = & K_m(MAF - \dot{m}) & \\
\dot{m} & = & Cyl(N, P) & (6.23) \\
J_e\dot{N} & = & T_q - T_L, & \\
T_q(t) & = & f_T(\dot{m}(t-\sigma), N(t), r(t-\sigma), \delta(t)) & \\
\text{where } u & = & \text{duty cycle for the air bypass valve} & \\
r & = & \text{A/F} & \\
\delta & = & \text{spark timing in terms of crank-angle} & \\
 & & \text{degrees before TDC} & \\
T_L & = & \text{load torque} &
\end{array}
$$

J_e and K_m in Equation 6.23 are two engine-dependent constants, where J_e represents the engine rotational inertia, and K_m is a function of the gas constant, air temperature, manifold volume, etc. Both J_e and K_m can be determined from engine design specifications and given nominal operating conditions. The time delay σ in the engine model equals approximately 180 degrees of crank-angle advance and, thus, is a speed-dependent parameter. This is one reason that models for ISC often use crank-angle instead of time as the independent variable. Additionally, most engine control activities are event driven and synchronized with crank position; the use of $\frac{dM}{d\theta}$, $\frac{dP}{d\theta}$ instead of $\frac{dM}{dt}$, $\frac{dP}{dt}$, respectively, tends to have a linearizing effect on the pumping and torque generation blocks.

Performing a standard linearization procedure results in the linear model shown in Figure 6.13, with inputs Δu, $\Delta\delta$, ΔT_L (the change of the bypass valve duty cycle, spark, and load torque from their nominal values, respectively) and output ΔN (the deviation of the idle speed from the setpoint). The time delay in the continuous-time feedback loop usually complicates the control design and analysis tasks. In a discrete-time representation, however, the time delay in Figure 6.13 corresponds to a rational transfer function z^{-n} where n is an integer that depends on the sampling scheme and the number of cylinders. It is generally more convenient to accomplish the controller design using a discrete-time model.

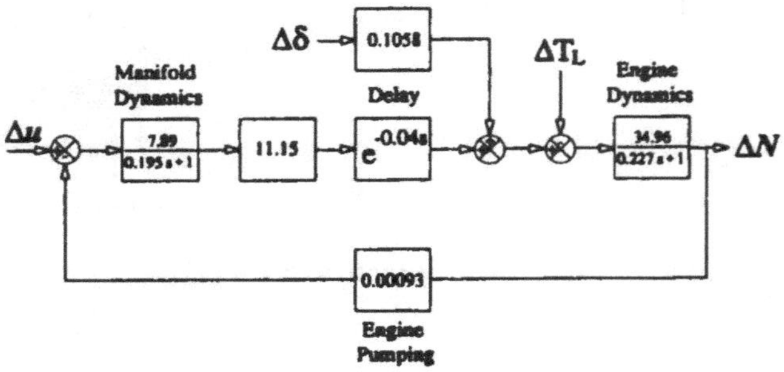

Figure 6.13 Linearized model for typical eight-cylinder engine.

Determining Model Parameters

In the engine model of Equation 6.23, the nonlinear algebraic functions f_a, f_T, Cyl describe characteristics of the air bypass valve, torque generation, and engine pumping blocks. These functions can be obtained by regressing engine dynamometer test data, using least squares or other curve-fitting algorithms. The torque generation function is developed on the engine dynamometer by individually sweeping ignition timing, A/F, and mass flow rate (regulated by throttle or air bypass valve position) over their expected values across the idle operating speed range. For a typical eight-cylinder engine, the torque regression is given by

$$\begin{aligned} T_q &= f_T(MAF, N, r, \delta) \\ &= -28.198 + 128.38MAF - 0.196N \\ &\quad + 13.845r - 0.306\delta - 5.669MAF^2 \\ &\quad + 7.39 \times 10^{-5}N^2 - 0.6257r^2 - 0.0257\delta^2 \\ &\quad - 0.0379MAF \cdot N + 0.2843MAF \cdot r \\ &\quad - 0.2483MAF \cdot \delta - 0.00059N \cdot r \\ &\quad + 0.00067N \cdot \delta + 0.0931r \cdot \delta \end{aligned}$$

The steady-state speed-torque relation to spark advance is illustrated in Figure 6.14.

For choked (i.e., sonic) flow, the bypass valve's static relationship is developed simply by exercising the actuator over its operating envelope and measuring either mass airflow using a

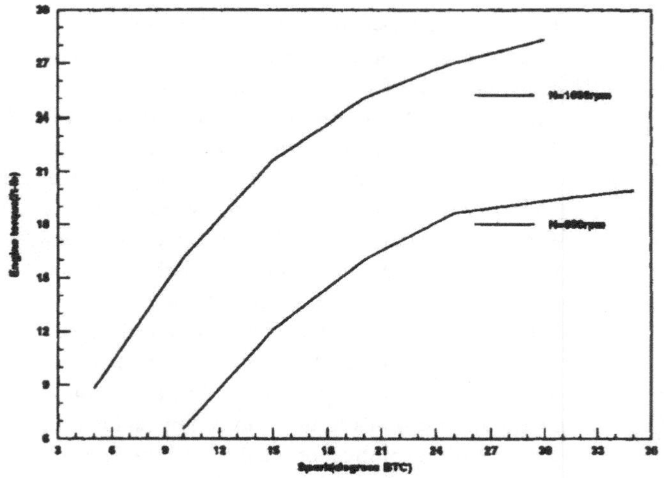

Figure 6.14 Spark-torque relation for different engine speeds (with A/F=14.64).

hot-wire anemometer or volume flow by measuring the pressure drop across a calibrated laminar flow element. The dynamic elements in the linearized idle speed model can be obtained by linearizing the model in Section 6.1.2, or they can be estimated by evaluating (small) step response data from a dynamometer. In particular, the intake manifold time constant can be validated by constant-speed, sonic-flow throttle step tests, using a sufficiently high-bandwidth sensor to measure manifold pressure.

ISC Controller Design

The ISC problem lends itself to the application of various control design techniques. Many different design methodologies, ranging from classical (such as PID) to modern (such as LQG, H_∞, adaptive, etc.) and nonconventional (such as neural networks and fuzzy logic) designs have been discussed and implemented [4]. The mathematical models described previously and commercially available software tools can be used to design different control strategies, depending on the implementor's preference and experience. A general ISC system with both feedforward and feedback components is shown in the block diagram of Figure 6.15.

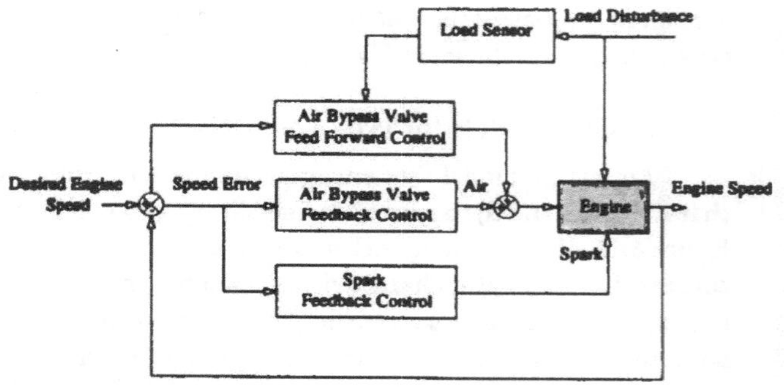

Figure 6.15 General ISC system with feedforward and feedback elements.

Feedforward Control Design Feedforward control is considered as an effective mechanism to reject load disturbances, especially for small engines. When a disturbance is measured (most disturbance sensors used in vehicles are on-off type), control signals can be generated in an attempt to counteract its effect. A typical ISC strategy has feedforward only for the air bypass valve control, and the feedforward is designed based on static engine mapping data. For example, if an air conditioning switch sensor is installed, an extra amount of air will be scheduled to prevent engine speed droop when the air conditioning compressor is engaged. The amount of feedforward control can be determined as follows. At the steady state, since $\dot{P}$, $\dot{N} \approx 0$, the available engine torque to balance the load torque is related to the mass air flow and engine speed through

$$T_q = f_T(MAF, N, r, \delta).$$

By estimating the load torque presented to the engine by the measured disturbance, one can calculate, for fixed A/F and spark, the amount of air that is needed to maintain the engine speed at the fixed setpoint. The feedforward control can be applied either as a multiplier or an adder to the control signal.

Feedforward control introduces extra cost due to the added sensor and software complexity; thus, it should be used only when necessary. Most importantly, it should not be used to replace the role of feedback in rejecting disturbances, since it does not address the problems of operating condition variations, miscalibration, etc.

Feedback Design Feedback design for ISC can be pursued in many different ways. Two philosophically different approaches are used in developing the control strategy. One is the SISO approach, which treats the air and spark control as separate entities and designs one loop at a time. When the SISO approach is used, the natural separation of time scale in the air and spark dynamics (spark has a fast response compared to air flow, which has a time lag due to the manifold dynamics and intake-to-power delay) suggests that the spark control be closed first as an inner loop. Then the air control, as an outer loop, is designed by including the spark feedback as part of the plant dynamics. Another approach is to treat the ISC as a multiple-input single-output (MISO) or, when the manifold pressure is to be controlled, a MIMO problem. Many control strategies, such as LQ-optimal control and H_∞ have been developed within the MIMO framework. This approach generally leads to a coordinated air and spark control strategy and improved performance.

Despite the rich literature on ISC featuring different control design methodologies, PID control in combination with static feedforward design is still viewed as the control structure of choice in the automotive industry. In many cases, controllers designed using advanced theory, such as H_∞ and LQR, are ultimately implemented in the PID format to reduce complexity and to append engineering content to design parameters. A typical production ISC feedback strategy has a PID for the air bypass valve (or throttle) control and a simple proportional feedback for the spark control. This control configuration is dictated by the following requirements: (1) at steady state, the spark should return to its nominal value independent of the load disturbances; (2) zero steady-state error has to be attained for step disturbances.

Calibration of ISC

Control system development in the automotive industry has been traditionally an empirical process with heavy reliance on manual tuning. As engine control systems have become more complex because of increased functionality, the old-fashioned trial-and-error approach has proved inadequate to achieve optimum performance for interactively connected systems. The trends in today's automotive control system development are in favor of more model-based design and systematic calibration. Tools introduced for performing systematic in-vehicle calibration include dynamic optimization packages, which are used to search for optimal parameters based on a large amount of vehicle data, and controller fine-tuning techniques. Given the reality that most ISC strategies implemented in vehicles are of PID type, we discuss two PID tuning approaches that have proved effective in ISC calibration.

The first method is based on the sensitivity functions of the engine speed with respect to the controller parameters. Let K be a generic controller parameter (possibly vector valued), and suppose that we want to minimize a performance cost function $J(\Delta N)$ [a commonly used function for J is $J = (\Delta N)^2$] by adjusting K. Viewing ΔN as a function of K and noting that $\frac{\partial \Delta N}{\partial K} = \frac{\partial N}{\partial K}$, we have

$$\begin{aligned} J(K + \Delta K) &\approx J(K) + 2\Delta N \frac{\partial N}{\partial K}\Delta K \\ &\quad + (\Delta K)^{\mathsf{T}} \left(\frac{\partial N}{\partial K}\right)^{\mathsf{T}} \frac{\partial N}{\partial K}\Delta K. \end{aligned}$$

According to Newton's method, ΔK, which minimizes J, is given by

$$\Delta K = -\left[\left(\frac{\partial N}{\partial K}\right)^{\mathsf{T}} \frac{\partial N}{\partial K}\right]^{-1} \left(\frac{\partial N}{\partial K}\right)^{\mathsf{T}} \Delta N. \tag{6.24}$$

By measuring the sensitivity function $\frac{\partial N}{\partial K}$, we can use a simple gradient method or Equation 6.24 to iteratively minimize the cost function J. The controller gains for the air and spark loops can be adjusted simultaneously. The advantages of the method are that the sensitivity functions are easy to generate. For the ISC calibration, the sensitivity functions of N with respect to PID controller parameters can be obtained by measuring the signal at the sensitivity points, as illustrated in Figure 6.16.

It should be pointed out that this offline tuning principle can be used to develop an on-line adaptive PID control scheme (referred to as the M.I.T. rule in the adaptive control literature). The sensitivity function method can also be used to investigate the robustness of the ISC system with respect to key plant parameters by evaluating $\frac{\partial N}{\partial K_p}$ where K_p is the plant parameter vector.

The second method is the well-known Ziegler-Nichols PID tuning method. It gives a set of heuristic rules for selecting the optimal PID gains. For the ISC applications, modifications have

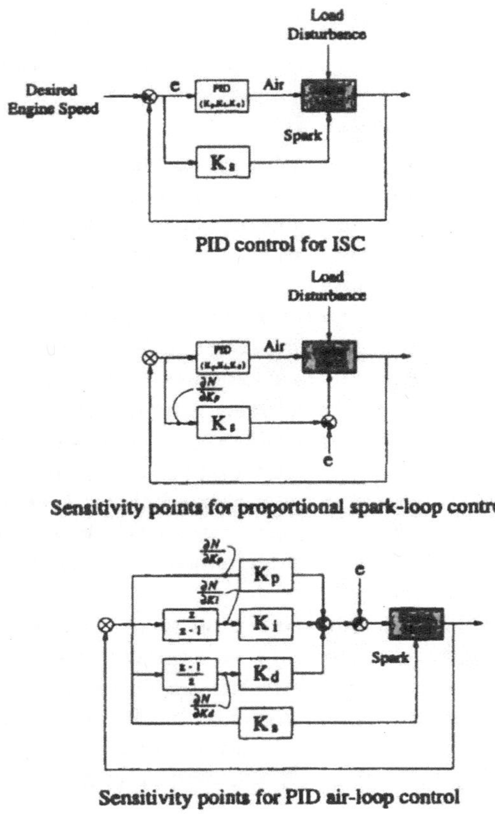

Figure 6.16 Sensitivity points for calibrating PID idle speed controller.

to be introduced to accommodate the time delay and other constraints. Generally, the Ziegler-Nichols sensitivity method is used to calibrate the PID air feedback loop after the proportional gain for the spark is fixed.

6.1.4 Acknowledgments

This work was supported in part by the National Science Foundation under contract NSF ECS-92-13551.

The authors also acknowledge their many colleagues at Ford Motor Company and the University of Michigan who contributed to the work described in this chapter, with special thanks to Dr. Paul Moraal of Ford.

References

[1] Buescher, K.L., Representation, Analysis, and Design of Multirate Discrete-Time Control Systems, Master's thesis, Department of Electrical and Computer Engineering, University of Illinois, Urbana-Champaign, 1988.

[2] Grizzle, J.W., Cook, J.A., and Milam, W.P., Improved cylinder air charge estimation for transient air fuel ratio control, in *Proc. 1994 Am. Control Conf.*, Baltimore, MD, June 1994, 1568–1573.

[3] Grizzle, J.W., Dobbins, K.L., and Cook, J.A., Individual cylinder air fuel ratio control with a single EGO sensor, *IEEE Trans. Vehicular Technol.*, 40(1), 280–286, February 1991.

[4] Hrovat, D. and Powers, W.F., Modeling and Control of Automotive Power Trains, in *Control and Dynamic Systems*, Vol. 37, Academic Press, New York, 1990, 33–64.

[5] Khargonekar, P.P., Poolla, K., and Tannenbaum, A., Robust control of linear time-invariant plants using periodic compensation, *IEEE Trans. Autom. Control*, 30(11), 1088–1096, 1985.

[6] Moraal, P.E., Cook, J.A., and Grizzle, J.W., Single sensor individual cylinder air-fuel ratio control of an eight cylinder engine with exhaust gas mixing, in *Proc. 1993 Am. Control Conf.*, San Francisco, CA, June 1993, 1761–1767.

[7] Powell, B.K. and Cook, J.A., Nonlinear low frequency phenomenological engine modeling and analysis, in *Proc. 1987 Am. Control Conf.*, Minneapolis, MN, June 1987, 332–340.

[8] Powers, W.F., Customers and controls, *IEEE Control Syst. Mag.*, 13(1), February 1993, 10–14.

[9] Shulman, M.A. and Hamburg, D.R., Non-ideal properties of Z_rO_2 and T_iO_2 exhaust gas oxygen sensors, SAE Tech. Paper Series, No. 800018, 1980.

[10] Taylor, C.F., The Internal Combustion Engine in Theory and Practice, Vol. 1: Thermodynamics, Fluid Flow, Performance, MIT Press, Cambridge, MA, 1980, 187.

6.2 Adaptive Automotive Speed Control

M. K. Liubakka, Advanced Vehicle Technology, Ford Motor Company, Dearborn, MI

D.S. Rhode, Advanced Vehicle Technology, Ford Motor Company, Dearborn, MI

J. R. Winkelman, Advanced Vehicle Technology, Ford Motor Company, Dearborn, MI

P. V. Kokotović, ECE Department, University of California, Santa Barbara, CA

6.2.1 Introduction

One of the main goals for an automobile speed control [3] (cruise control) system is to provide acceptable performance over a wide range of vehicle lines and operating conditions. Ideally, this is to be achieved with one control module, without recalibration for

[3]©1993 IEEE. Reprinted, with permission, from *IEEE Transactions on Automatic Control;* Volume 38, Number 7, Pages 1011–1020; July 1993.

different vehicle lines. For commonly used proportional feedback controllers, no single controller gain is adequate for all vehicles and all operating conditions. Such simple controllers no longer have the level of performance expected by customers.

The complexity of speed control algorithms has increased through the years to meet the more stringent performance requirements. The earliest systems simply held the throttle in a fixed position [1]. In the late 1950s speed control systems with feedback appeared [2]. These used proportional (P) feedback of the speed error, with the gain typically chosen so that 6 to 10 mph of error would pull full throttle. The next enhancement was proportional control with an integral preset or bias input (PI) [3]. This helped to minimize steady-state error as well as speed droop when the system was initialized. Only with the recent availability of inexpensive microprocessors have more sophisticated control strategies been implemented. Proportional-integral-derivative (PID) controllers, optimal LQ regulators, Kalman filters, fuzzy logic, and adaptive algorithms have all been tried [4]-[10].

Still, it is hard to beat the performance of a well-tuned PI controller for speed control. The problem is how to keep the PI controller well tuned, since both the system and operating conditions vary greatly. The optimal speed control gains are dependent on:

- Vehicle parameters (engine, transmission, weight, etc.)
- Vehicle speed
- Torque disturbances (road slope, wind, etc.)

Gain scheduling over vehicle speed is not a viable option because the vehicle parameters are not constant and torque disturbances are not measurable. Much testing and calibration work has been done to tune PI gains for a controller that works across more than one car line, but as new vehicles are added, retuning is often necessary. For example, with a PI speed control, low-power cars generally need higher gains than high-power cars. This suggests a need for adaptation to vehicle parameters. For an individual car, the best performance on flat roads is achieved with low-integral gain, while rolling hill terrain requires high-integral gain. This suggests a need for adaptation to disturbances.

Our goal was to build an adaptive controller that outperforms its fixed-gain competitors, yet retains their simplicity and robustness. This goal has been achieved with a slow-adaptation design using a sensitivity-based gradient algorithm. This algorithm, driven by the vehicle response to unmeasured load torque disturbances, adjusts the proportional and integral gains, K_p and K_i, respectively, to minimize a quadratic cost functional. Through simulations and experiments a single cost functional was found that, when minimized, resulted in satisfactory speed control performance for each vehicle and all operating conditions. Adaptive minimization of this cost functional improved the performance of every tested vehicle over varying road terrain (flat, rolling hills, steep grades, etc.). This is not possible with a fixed-gain controller.

Our optimization type adaptive design has several advantages. The slow adaption of only two adjustable parameters is simple and makes use of knowledge already acquired about the vehicle. The main requirement for slow adaptation is the existence of a fixed-gain controller that provides the desired performance when properly tuned. Since the PI control meets this requirement and is well understood, the design and implementation of the adaptive control with good robustness properties become fairly easy tasks. With only two adjustable parameters, all but perfectly flat roads provide sufficient excitation for parameter convergence and local robustness. These properties are strengthened by the sensitivity filter design and speed-dependent initialization.

The main idea of the adaptive algorithm employed here comes from a sensitivity approach proposed in the 1960s [11] but soon abandoned because of its instabilities in fast adaptation. Under the ideal model-matching conditions, such instabilities do not occur in more complex schemes developed in the 1970s to 1980s. However, the ideal model-matching requires twice as many adjustable parameters as the dynamic order of the plant. If the design is based on a reduced-order model, the resulting unmodeled dynamics may cause instability and robust redesigns are required. This difficulty motivated our renewed interest in *a sensitivity-based approach in which both the controller structure and the adjustable parameters are free to be chosen independently of the plant order*. Such an approach would be suitable for adaptive tuning of simple controllers to higher-order plants if a verifiable condition for its stability could be found. For this purpose we employ the "pseudogradient condition," recently derived by Rhode [12], [13] using the averaging results of [14]-[16]. A brief outline of this derivation is given in the appendix. From the known bounds on vehicle parameters and torque disturbances, we evaluate, in the frequency domain, a "phase-uncertainty envelope." Then we design a sensitivity filter to guarantee that the pseudogradient condition is satisfied at all points encompassed by the envelope.

6.2.2 Design Objectives

The automobile speed control is simpler than many other automotive control problems: engine efficiency and emissions, active suspension, four-wheel steering, to name only a few. It is, therefore, required that the solution to the speed control problem be simple. However, this simple solution must also satisfy a set of challenging performance and functional requirements. A typical list of these is as follows:

- *Performance requirements*
  - Speed tracking ability for low-frequency commands.
  - Torque disturbance attenuation for low frequencies, with zero steady-state error for large grades (within the capabilities of the vehicle power train).
  - Smooth and minimal throttle movement.
  - Robustness of the above properties over a wide range of operating conditions.

- *Functional requirements*
  - Universality: the same control module must meet the performance requirements for different vehicle lines without recalibration.
  - Simplicity: design concepts and diagrams should be understandable to automotive engineers with basic control background.

The dynamics that are relevant for this design problem are organized in the form of a generic vehicle model in Figure 6.17.

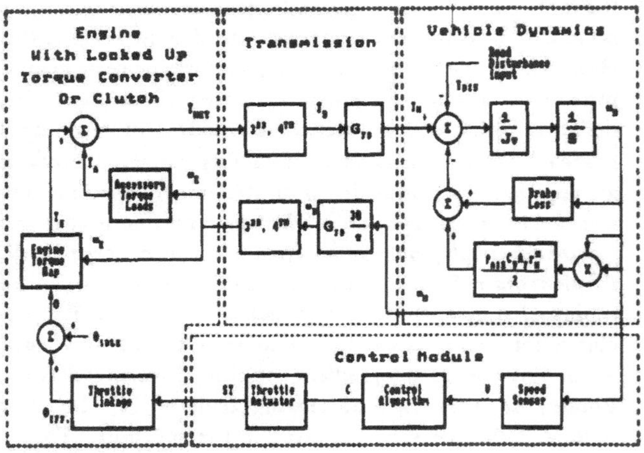

Figure 6.17 Vehicle model for speed control. Different vehicle lines are represented by different structures and parameters of individual blocks.

To represent vehicles of different lines (Escort, Mustang, Taurus, etc.) individual blocks will contain different parameters and, possibly, slightly different structures (e.g., manual or automatic transmission). Although the first-order vehicle dynamics are dominant, there are other blocks with significant higher-frequency dynamics. Nonlinearities such as dead zone, saturation, multiple gear ratios, and backlash are also present. In conjunction with large variations of static gain (e.g., low- or high-power engine) these nonlinearities may cause limit cycles, which should be suppressed, especially if noticeable to the driver.

There are two inputs: the speed set-point y_{set} and the road load disturbance torque T_{dis}. While the response to set-point changes should be as specified, the most important performance requirement is the accommodation of the torque disturbance. The steady-state error caused by a constant torque disturbance (e.g., constant-slope road) must be zero. For other types of roads (e.g., rolling hills) a low-frequency specification of disturbance accommodation is defined.

This illustrative review of design objectives, which is by no means complete, suffices to motivate the speed control design presented in Sections 6.2.3 and 6.2.4. Typical results with test vehicles presented in Section 6.3 show how the above requirements have been satisfied.

6.2.3 The Design Concept

The choice of a design concept for mass production differs substantially from an academic study of competing theories, in this case, numerous ways to design an adaptive scheme. With physical constraints, necessary safety nets, and diagnostics, the implementation of an analytically conceived algorithm may appear similar to an "expert," "fuzzy," or "intelligent" system. Innovative terminologies respond to personal tastes and market pressures, but the origins of most successful control designs are often traced to some fundamental concepts. The most enduring among these are PI control and gradient type algorithms. Recent theoretical results on conditions for stability of such algorithms reduce the necessary ad hoc fixes required to assure reliable performance. They are an excellent starting point for many practical adaptive designs and can be expanded by additional nonlinear compensators and least-square modifications of the algorithm.

For our design, a PI controller is suggested by the zero steady-state error requirement, as well as by earlier speed control designs. In the final design, a simple nonlinear compensator was added, but is not discussed in this text. The decision to adaptively tune the PI controller gains K_p and K_i was reached after it was confirmed that a controller with gain scheduling based on speed cannot satisfy performance requirements for all vehicle lines under all road load conditions. Adaptive control is chosen to eliminate the need for costly recalibration and to satisfy the universal functionality requirement.

The remaining choice was that of a parameter adaptation algorithm. Based on the data about the vehicle lines and the fact that the torque disturbance is not available for measurement, the choice was made of an optimization-based algorithm. A reference-model approach was not followed because no single model can specify the desired performance for the wide range of dynamics and disturbances. On the other hand, through simulation studies and experience with earlier designs, a *single quadratic cost functional* was constructed whose minimization led to an acceptable performance for each vehicle and each operating condition. For a given vehicle subjected to a given torque disturbance, the minimization of the cost functional generates an optimal pair of the PI controller gains K_p and K_i. In this sense, the choice of a single cost functional represents an implicit map from the set of vehicles and operating conditions to an admissible region in the parameter plane (K_p, K_i). This region was chosen to be a rectangle with preassigned bounds.

The task of parameter adaptation was to minimize the selected quadratic cost functional for each unknown vehicle and each unmeasured disturbance. A possible candidate was an indirect adaptive scheme with an estimator of the unknown vehicle and disturbance model parameters and an on-line LQ optimization algorithm. In this particular system, the frequency content in the disturbance was significantly faster than the plant dynamics. This resulted in difficulties in estimating the disturbance. After some experimentation, this scheme was abandoned in favor of a simpler sensitivity-based scheme, which more directly led to adaptive minimization of the cost functional and made better use of the knowledge acquired during its construction.

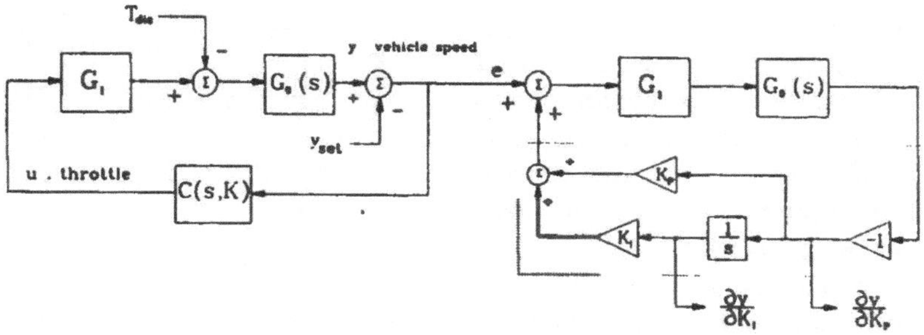

Figure 6.18 The system and its copy generate sensitivity functions for optimization of the PI controller parameters K_p and K_i.

The sensitivity-based approach to parameter optimization exploits the remarkable sensitivity property of linear systems: *the sensitivity function (i.e., partial derivative) of any signal in the system with respect to any constant system parameter can be obtained from a particular cascade connection of the system and its copy.* For a linearized version of the vehicle model in Figure 6.17, the sensitivities of the vehicle speed error $e = y - y_{set}$ with respect to the PI controller parameters K_p and K_i are obtained as in Figure 6.18, where $G_0(s)$ and $G_1(s)$ represent the vehicle and power train dynamics, respectively; $C(s, K) = -K_p - \frac{K_i}{s}$; and the control variable u is throttle position. This result can be derived by differentiation of

$$e(s, K) = \frac{1}{1 + C(s, K)G_1(s)G_0(s)} [y_{set} - G_0(s)T_{dis}] \tag{6.25}$$

with respect to $K = [K_p, K_i]$, namely,

$$\frac{\partial e}{\partial K} = \frac{\partial C}{\partial K} \frac{G_1(s)G_0(s)}{1 + C(s, K)G_1(s)G_0(s)} e(s, K) \tag{6.26}$$

where $\frac{\partial C}{\partial K} = \left(-1, -\frac{1}{s}\right)$. Expressions analogous to Equation 6.26 can be obtained for the control sensitivities $\frac{\partial u}{\partial K}$. Our cost functional also uses a high-pass filter $F(s)$ to penalize higher frequencies in u; that is, $\tilde{u}(s, K) = F(s)u(s, K)$. The sensitivities of $\tilde{u}$ are obtained simply as $\frac{\partial \tilde{u}}{\partial K} = F(s)\frac{\partial u}{\partial K}$.

When the sensitivity functions are available, a continuous-gradient algorithm for the PI controller parameters is

$$\begin{aligned} \frac{dK_p}{dt} &= -\epsilon\left(\beta_1 \tilde{u}\frac{\partial \tilde{u}}{\partial K_p} + \beta_2 e \frac{\partial e}{\partial K_p}\right) \\ \frac{dK_i}{dt} &= -\epsilon\left(\beta_1 \tilde{u}\frac{\partial \tilde{u}}{\partial K_i} + \beta_2 e \frac{\partial e}{\partial K_i}\right) \end{aligned} \tag{6.27}$$

where the adaptation speed determined by ϵ must be kept sufficiently small so that the averaging assumption (K_p and K_i are constant) is approximately satisfied. With ϵ small, the method of averaging [14]-[16] is applicable to Equation 6.27 and proves that, as $t \to \infty$, the parameters K_p and K_i converge to an ϵ neighborhood of the values that minimize the quadratic cost functional

$$J = \int_0^\infty \left(\beta_1 \tilde{u}^2 + \beta_2 e^2\right) dt. \tag{6.28}$$

With a choice of the weighting coefficients, β_1 and β_2 (to be discussed later), our cost functional is Equation 6.28. Thus, Equation 6.27 is a convergent algorithm that can be used to minimize this functional when the system is known, so that its copy can be employed to generate the sensitivities needed in Equation 6.27. In fact, our computational procedure for finding a cost functional good for all vehicle lines and operating conditions made use of this algorithm.

Unfortunately, when the vehicle parameters are unknown, the exact-gradient algorithm of Equation 6.27 cannot be used because a copy of the system is not available. In other words, an algorithm employing exact sensitivities is not suitable for adaptive control. A practical escape from this difficulty is to generate some approximations of the sensitivity functions

$$\psi_1 \approx \frac{\partial \tilde{u}}{\partial K_p}, \ \psi_2 \approx \frac{\partial \tilde{u}}{\partial K_i}, \ \psi_3 \approx \frac{\partial e}{\partial K_p}, \ \psi_4 \approx \frac{\partial e}{\partial K_i} \tag{6.29}$$

and to employ them in a "pseudogradient" algorithm

$$\begin{aligned} \frac{dK_p}{dt} &= -\epsilon\left(\beta_1 \tilde{u}\psi_1 + \beta_2 e \psi_3\right) \\ \frac{dK_i}{dt} &= -\epsilon\left(\beta_1 \tilde{u}\psi_2 + \beta_2 e \psi_4\right) . \end{aligned} \tag{6.30}$$

A filter used to generate ψ_1, ψ_2, ψ_3, and ψ_4 is called a *pseudosensitivity filter*. The fundamental problem in the design of the pseudosensitivity filter is to guarantee not only that the algorithm of Equation 6.30 converges, but also that the values to which it converges are close to those that minimize the chosen cost functional.

6.2.4 Adaptive Controller Implementation

The adaptive speed control algorithm presented here is fairly simple and easy to implement, but care must be taken when choosing its free parameters and designing pseudosensitivity filters. This section discusses the procedure used to achieve a robust system and to provide the desired speed control performance.

Pseudosensitivity Filter

While testing the adaptive algorithm it becomes obvious that the gains K_p and K_i and the vehicle parameters vary greatly for operating conditions and vehicles. This makes it impossible to implement the exact sensitivity filters for the gradient algorithm of Equation 6.27. Our approach is to generate a "pseudogradient" approximation of $\partial J/\partial P$, satisfying the stability and convergence conditions summarized in the appendix.

In the appendix, the two main requirements for stability and convergence are: a persistently exciting (PE) input condition and a "pseudogradient condition," which, in our case, is a *phase condition on the nominal sensitivity filters.* Since we are using a reduced-order controller with only two adjustable parameters, the PE condition is easily met by the changing road loads. Road disturbances have an approximate frequency spectrum centered about zero that drops off with the square of frequency. This meets the PE condition for adapting two gains, K_p and K_i.

To satisfy the pseudogradient condition, the phase of the pseudosensitivity filter must be within ±90° of the phase of the actual sensitivity at the dominant frequencies. To help guarantee this for a wide range of vehicles and operating conditions, we varied the system parameters in the detailed vehicle model to generate an envelope of possible exact sensitivities. Then the pseudosensitivity filter was chosen near the center of this envelope. An important element of the phase condition is the fact that it is a condition on the sum of frequencies; that is, the phase condition is most important in the range of frequencies where there are dominant dynamics. If the pseudosensitivity filters do not meet the phase conditions at frequencies where there is little dominant spectral content, the algorithm may still be convergent, provided the phase conditions are strongly met in the region of dominant dynamics. Thus, the algorithm possesses a robustness property.

Figure 6.19 shows the gain and phase plots for the pseudosensitivity filter $\partial y/\partial K_p$. The other three sensitivities are left out for brevity. Figure 6.20 shows the ±90° phase boundary (solid lines) along with exact sensitivity phase angles (dashed lines) as vehicle inertia, engine power, and the speed control gains are varied over their full range. From this plot it is seen that the chosen pseudosensitivity filter meets the pseudogradient condition along with some safety margin to accommodate unmodeled dynamics.

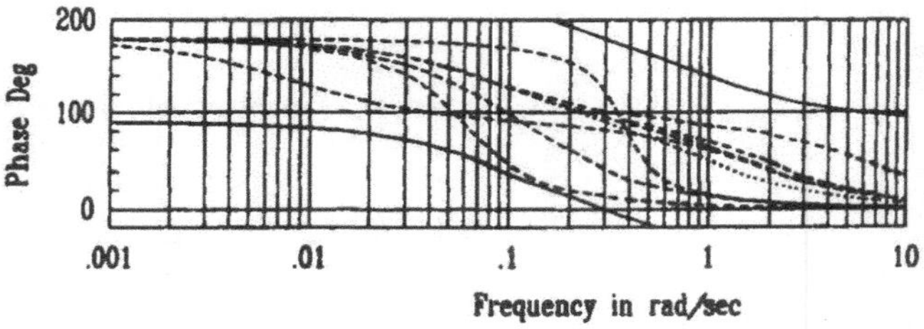

Figure 6.20 Envelope of possible exact sensitivity phase angles. The solid curves mark the ±90° boundaries, and the dashed curves denote the limits of plant variation.

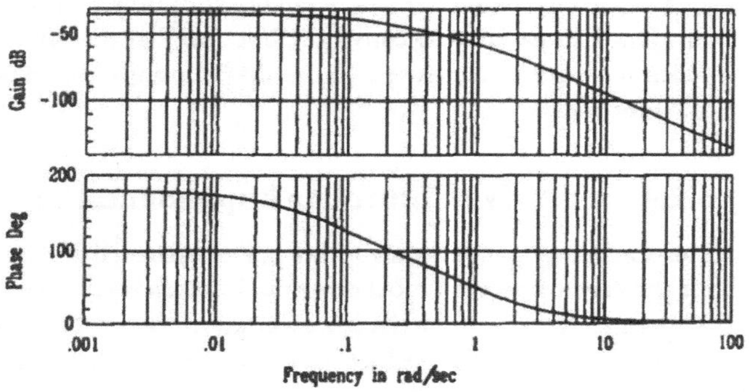

Figure 6.19 Gain and phase plot of pseudosensitivity $\partial y/\partial K_p$.

Choice of ϵ, β_1, and β_2

For implementation, the first parameters that must be chosen are the adaptation gain ϵ and the weightings in the cost functional, β_1 and β_2. The adaptation gain ϵ determines the speed of adaptation and should be chosen based on the slowest dynamics of the system. For speed control, the dominant dynamics result from the vehicle inertia and have a time constant on the order of 30 to 50 seconds. To avoid interaction between the adaptive controller states and those of the plant, the adaptation should be approximately an order of magnitude slower than the plant. As shown in [14]-[16], this allows one to use the frozen parameter system and averaging to analyze stability of the adaptive system. The adaptation law takes up to several minutes to converge, depending on initial conditions and the road load disturbances.

The two extreme choices of βs are (1) $\beta_1 = 0, \beta_2 = k$ and (2) $\beta_1 = k, \beta_2 = 0$ where $k > 0$. For the first extreme, $\beta_1 = 0$, we are penalizing only speed error, and the adaptation will tune to an unacceptable high-gain controller. High gain will cause too much throttle movement, resulting in nonsmooth behavior as felt by the passengers, and the system will be less robust from a stability point of view. For the second case, $\beta_2 = 0$, the adaptation will try to keep a fixed throttle angle and will allow large speed errors. Obviously, some middle values for the weightings are desired. An increase of the ratio β_1/β_2 reduces unnecessary throttle movement, while to improve tracking and transient speed errors we need to decrease this ratio.

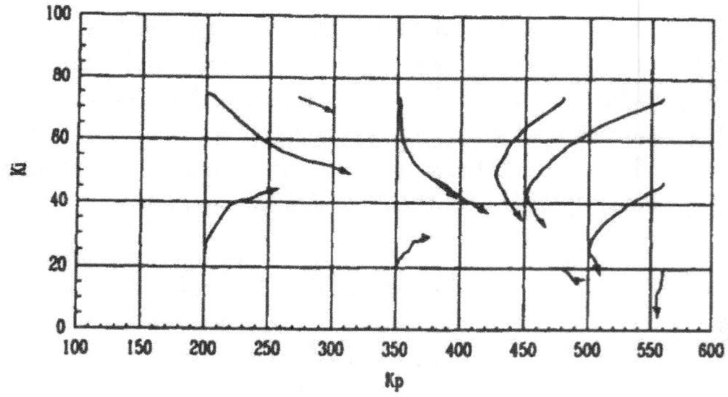

Figure 6.21 Gain trajectories on flat ground at 30 mph.

The choice of the weightings was based on experience with tuning standard PI speed controllers. Much subjective testing has been performed on vehicles to obtain the best fixed gains for the PI controller when the vehicle parameters are known. With this information, simulations were run on a detailed vehicle model with a small ϵ, and the βs were varied until the adaptation converged to approximately the gains obtained from subjective testing. The cost functional weights β_1 and β_2 are a different parameterization of the controller tuning problem. For development engineers, who may not be control system engineers, β_1

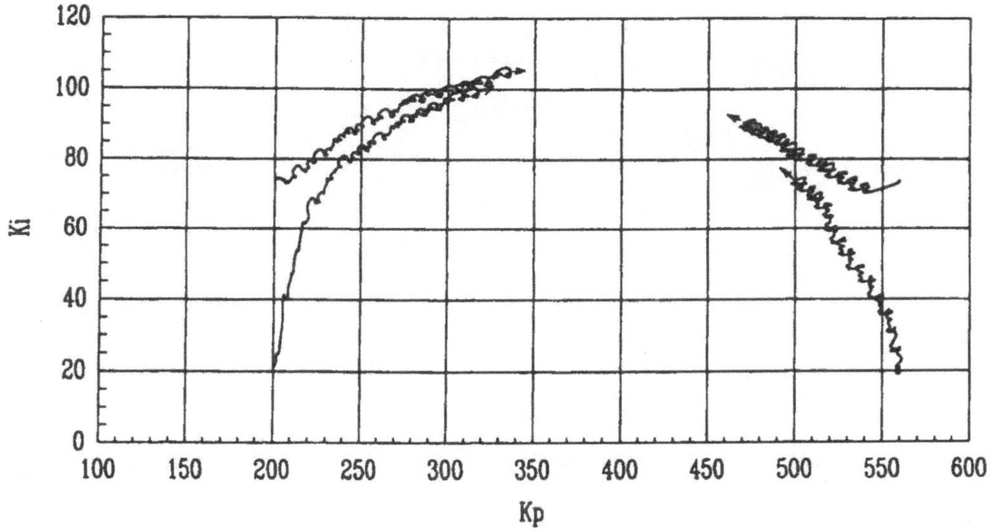

Figure 6.22 Gain trajectories on low-frequency hills at 30 mph.

and β_2 represent a pair of tunable parameters that relate directly to customer needs. This allows for a broader range of engineering inputs into the tuning process.

As examples of adaptive controller performance, simulation results in Figures 6.21 and 6.22 show the trajectories of the gains for a vehicle on two different road terrains. Much can be learned about the behavior of the adaptive algorithm from these types of simulations. In general, K_p varies proportionally with vehicle speed and K_i varies with road load. The integral gain K_i tends toward low values for small disturbances or for disturbances too fast for the vehicle to respond to. This can be seen in Figure 6.21. Conversely, K_i tends toward high values for large or slowly varying road disturbances, as can be seen in Figure 6.22.

Modifications For Improved Robustness

Additional steps have been taken to ensure robust performance in the automotive environment. First, the adaptation is turned off if the vehicle is operating in regions where the modeling assumptions are violated. These include operation at closed throttle or near wide-open throttle, during start-up transients, and when the set speed is changing. When the adaptation is turned off, the gains are frozen, but the sensitivities are still computed.

Care has been taken to avoid parameter drift due to noise and modeling imperfections. Two common ways to reduce drift are projection and a dead band on error. Projection, which limits the range over which the gains may adapt, is more attractive given the *a priori* knowledge of reasonable gains for the speed control system. To minimize computation, a simple projection is used, constraining the tuned gains to a predetermined set as shown in Figure 6.23.

There are other unmeasurable disturbances that can affect performance, such as the driver overriding speed control by use of the throttle. This condition, which can cause large gain changes, cannot be detected immediately because throttle position is not measured. To minimize these unwanted gain changes, the rate at which the gains can adapt is limited. If the adaptation adjusts more quickly than a predetermined rate, the adaptation gain ϵ is lowered, limiting the rate of change of the gains K_p and K_i.

The final parameters to choose are the initial guesses for K_p and K_i. Since the adaptation is fairly slow, a poor choice of initial gains can cause poor start-up performance. For quick convergence, the initial controller gains are scheduled with vehicle speed at the points A, B, and C. Figure 6.23 shows these initial gains along with the range of possible tuning. The proportional relationship between vehicle speed and the optimal controller gains can be seen.

6.3 Performance in Test Vehicles

The adaptive algorithm discussed in this chapter has been tried in a number of vehicles with excellent results. The adaptive control has worked well in every car tested so far and makes significant performance improvements in vehicles that have poor performance with the fixed-gain PI control. For vehicles where speed control already performs well, improvements from the adaptive algorithm are still significant at low speeds or at small throttle angles. Many vehicles with conventional speed control systems exhibit a limit cycle or surge condition at low speeds or on down slopes [7]. This is due in part to nonlinearities such as the idle stop, which limits throttle movement. In such vehicles, the adaptive controller reduces the magnitude and lengthens the period of the limit cycle, thus improving performance.

The first set of data, Figures 6.24 to 6.26, is for a vehicle that had a low-speed surge. Here the limit cycle is very noticeable in the data as well as while driving the car. The high frequency (0.2 Hz) of the limit cycle is what makes it noticeable to the driver. The adaptive controller greatly improves performance by decreasing the amplitude and frequency to the point where the driver cannot feel the limit cycle. Looking at the control gains during this test, it is seen that the gains initially decrease to reduce the limit cycle, then they continue to adjust in small amounts for the varying

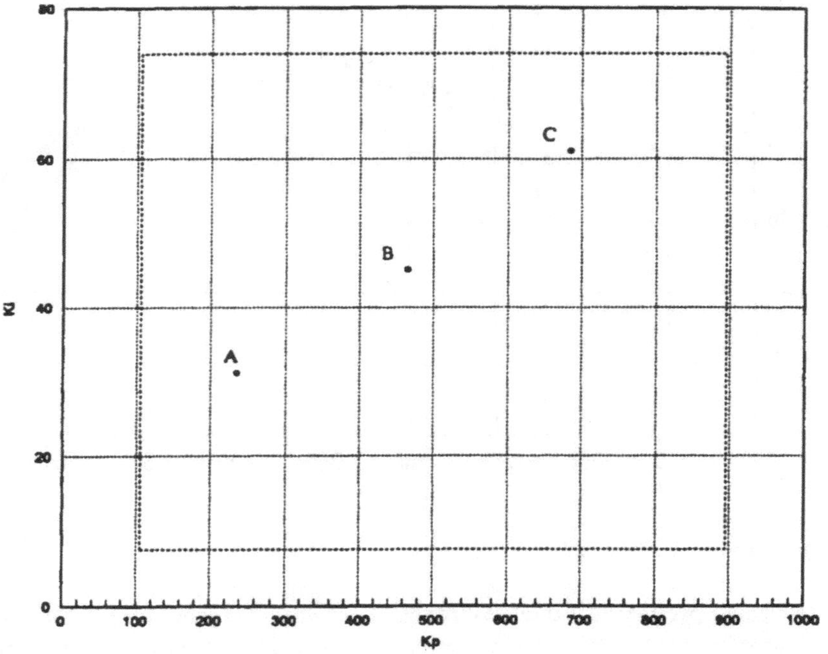

Figure 6.23 Limits and initial conditions for control gains: (A) $y_{set} < 35$ mph; (B) 35 mph $\leq y_{set} < 50$ mph; (C) $y_{set} \geq 50$ mph.

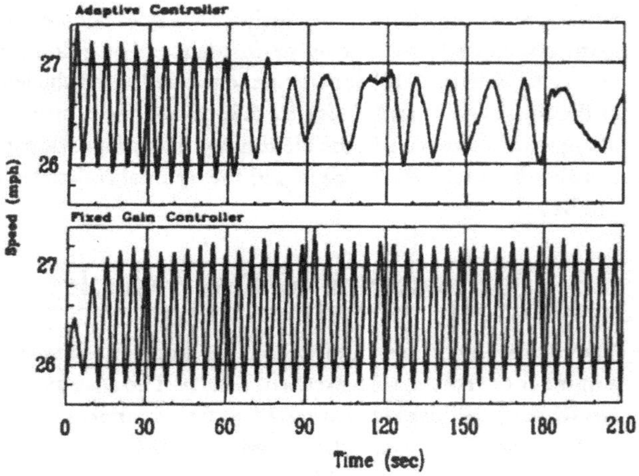

Figure 6.24 Vehicle with low speed limit cycle: adaptive controller; top, bottom, fixed gain controller.

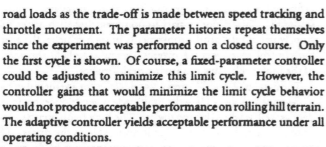

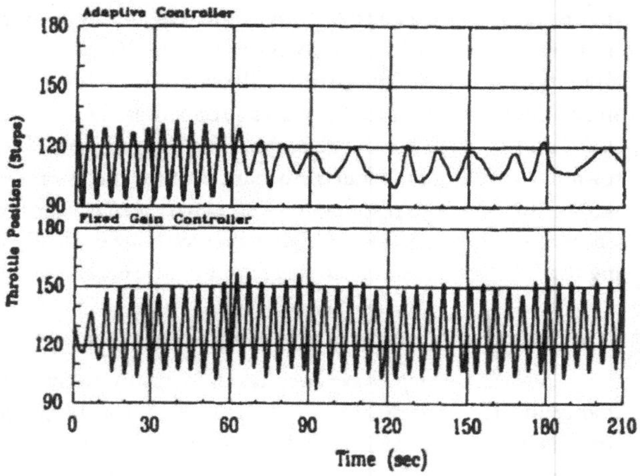

Figure 6.25 Vehicle with low-speed limit cycle: top, adaptive controller; bottom, fixed-gain controller.

road loads as the trade-off is made between speed tracking and throttle movement. The parameter histories repeat themselves since the experiment was performed on a closed course. Only the first cycle is shown. Of course, a fixed-parameter controller could be adjusted to minimize this limit cycle. However, the controller gains that would minimize the limit cycle behavior would not produce acceptable performance on rolling hill terrain. The adaptive controller yields acceptable performance under all operating conditions.

The next three Figures show the gains for three different vehicles traveling over the same road at 40 mph. As expected, even though the initial gains are acceptable, each vehicle tunes to a different set of optimal gains for this road. One thing to notice in Figure 6.27 is the initial increase in K_i. This occurred because the integral preset of the control did not match this vehicle, causing an initial overshoot or droop in speed. To the adaptive algorithm, this offset looks like a large road disturbance. Again, since these tests were performed on a closed course, the gains eventually become periodic.

Figure 6.30 to 6.32 show a vehicle driving on a freeway which passes below the grade level to pass under surface roads. An approximate road profile for a section of this road is shown in Figure 6.33. The dips are where the surface streets cross over the freeway. Because the road disturbances are of such high frequency, the adaptive controller cannot greatly improve the tracking ability of the vehicle, but much of the high-frequency limit cycle has been removed. For this disturbance, the proportional gain tunes up and the integral gain tunes down. The integral

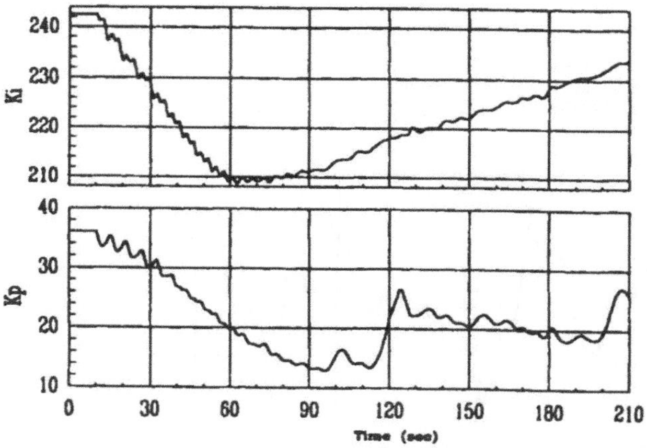

Figure 6.26 Controller gains.

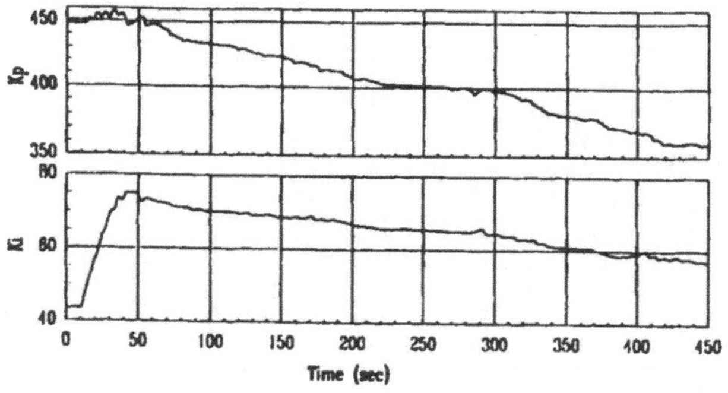

Figure 6.27 Test car A, 40 mph.

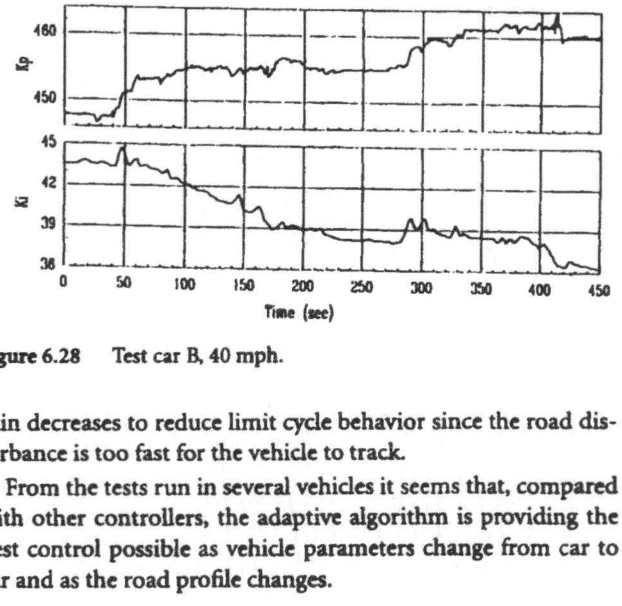

Figure 6.28 Test car B, 40 mph.

gain decreases to reduce limit cycle behavior since the road disturbance is too fast for the vehicle to track.

From the tests run in several vehicles it seems that, compared with other controllers, the adaptive algorithm is providing the best control possible as vehicle parameters change from car to car and as the road profile changes.

6.3.1 Conclusions

This chapter has presented an adaptive algorithm to adjust the gains of a vehicle speed control system. By continuously adjusting the PI control gains, speed control performance can be optimized for each vehicle and operating condition. This helps to design a single speed control module without additional calibration or sacrifices in performance for certain car lines. It also allows im-

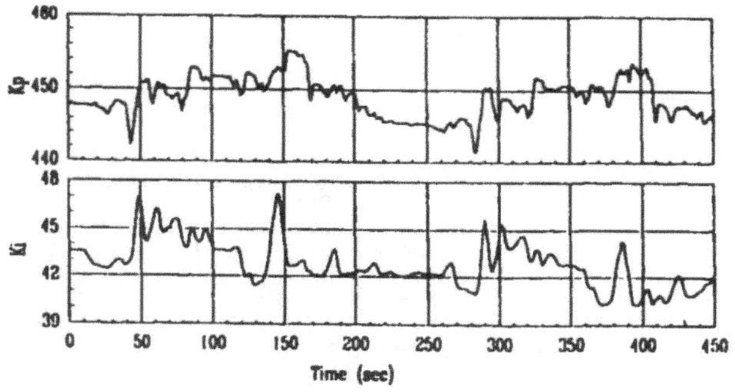

Figure 6.29 Test car C, 40 mph.

proved performance for changing road conditions not possible with a fixed-gain control or other types of adaptive control.

The results of initial vehicle testing confirm the performance improvements and robustness of the adaptive controller. Vehicle speed control is not the only automotive application of adaptive control at Ford Motor Company. The adaptive control technique presented in this chapter has also been applied to solve other problems of electronically controlled automotive systems.

Appendix

The pseudogradient adaptive approach used in this application relies upon the properties of slow adaptation. Such gradient tuning algorithms can be described by

$$\dot{K}(t) = -\epsilon\Psi(t, K)e(t, K), \quad K \in \Re^m \tag{6.31}$$

where K is a vector of tunable parameters, $e(t, K)$ is an error signal, and $\Psi(t, K)$ is the regressor, which is an approximation of the sensitivity of the output with respect to the parameters $\frac{\partial y(t,K)}{\partial K}$. In this derivation, $e(t, K)$ is defined as the difference between the output $y(t, K)$ and a desired response $y_m(t)$,

$$e(t, K) \equiv y(t, K) - y_m(t), \tag{6.32}$$

so that the pseudogradient update law of Equation 6.31 aims to minimize the average of $e(t, K)^2$. However, the same procedure is readily modified to minimize a weighted sum of $e(t, K)^2$ and $u(t, K)^2$ as in this speed control application. Since slow adaptation is used only for performance improvement, and not for stabilization, we constrain the vector K of adjustable controller parameters, $k_1, \ldots, k_m$, to remain in a set $\bar{K}$ such that with constant K ($\epsilon = 0$), the resulting linear system is stable. Our main tool in this analysis is an integral manifold, the so-called *slow manifold* [14], that separates the fast linear plant and controller states from the slow parameter dynamics. To assure the existence of this manifold, we assume that $\Psi(t, K)$ and $e(t, K)$ are differentiable with respect to K and the input to the system, $r(t)$, is a uniformly bounded, almost periodic function of time. Then applying the averaging theorem of Bogolubov, the stability of the adaptive algorithm is determined from the response of the linear system with constant parameters and the average update law. The following analysis is restricted to initial conditions near this slow manifold.

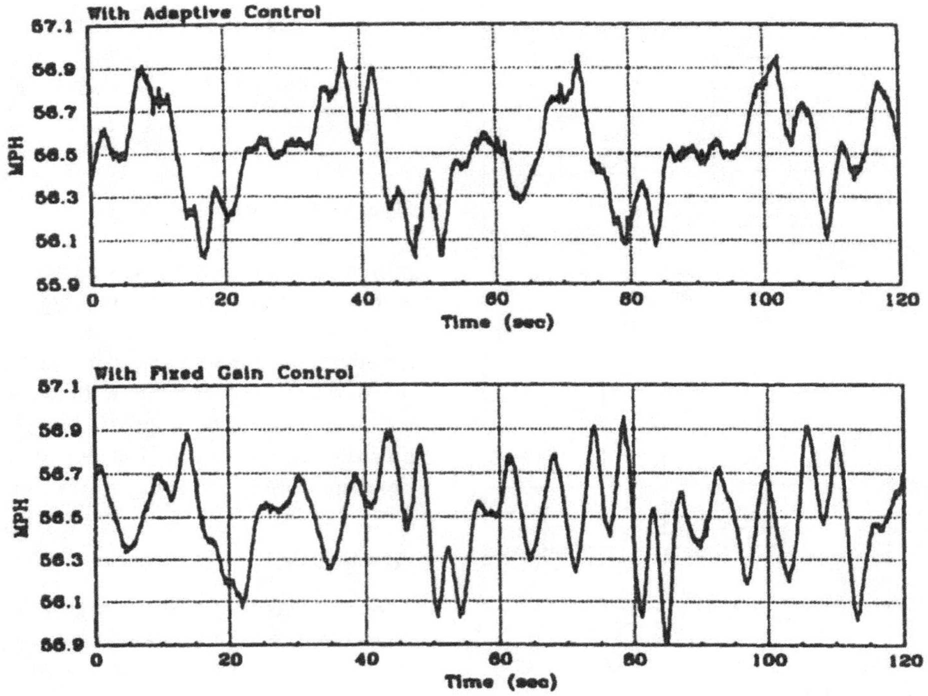

Figure 6.30 Vehicle with high-frequency road disturbance: top, with adaptive control; bottom, with fixed-gain control.

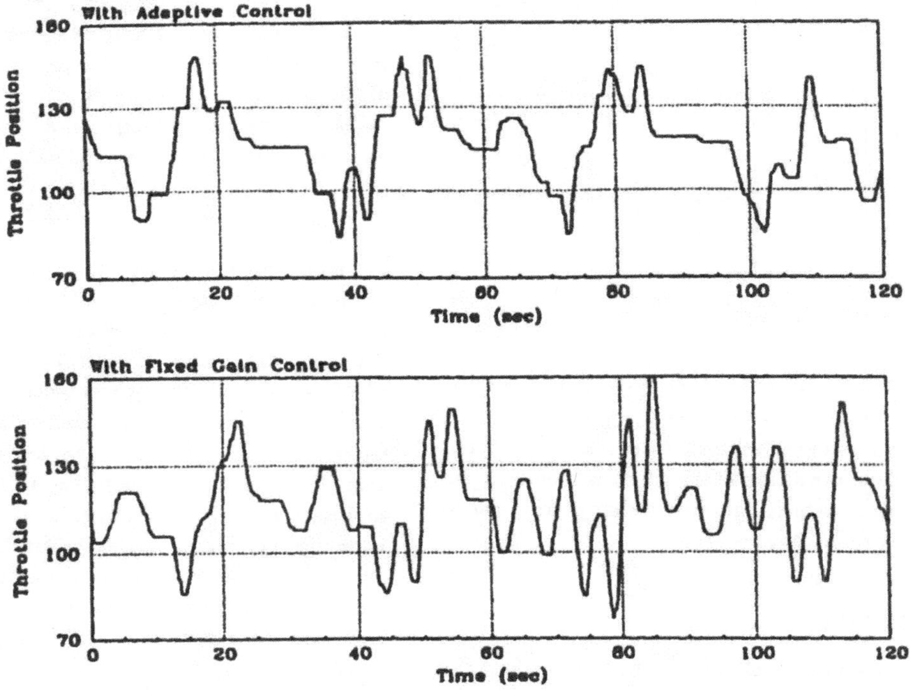

Figure 6.31 Vehicle with high-frequency road disturbance: top, with adaptive control; bottom, with fixed-gain control.

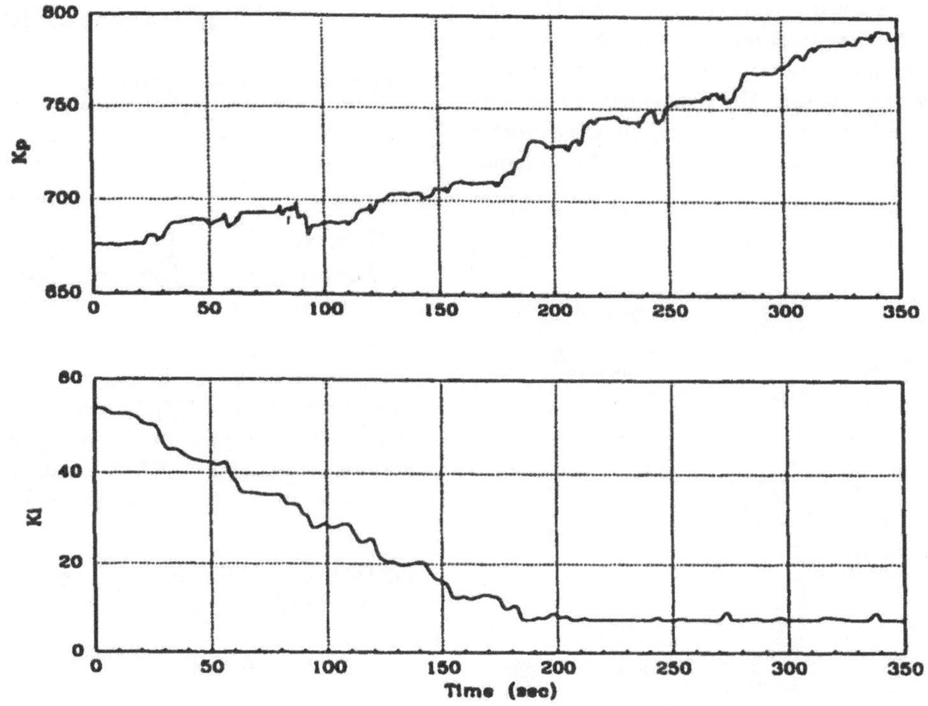

Figure 6.32 Gains with high-frequency disturbances.

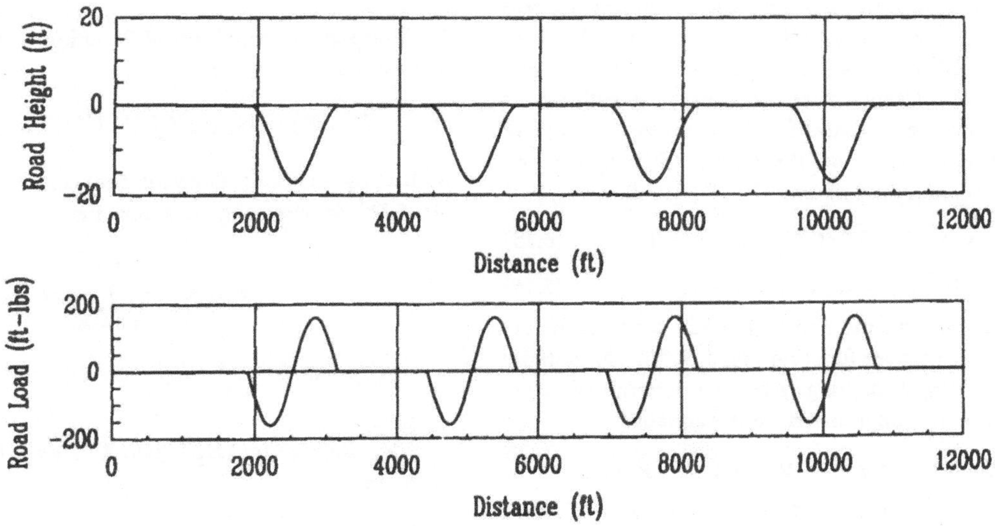

Figure 6.33 Approximate freeway profile.

In the pseudogradient approach used here, the regressor $\Psi(t, K)$ is an approximation of the sensitivities of the error with respect to parameters $\frac{\partial e(t,K)}{\partial K}$. We assume that the adjustable controller parameters K feed a common summing junction as shown in Figure 6.34. In this case, if the system in Figure 6.34 was known, these sensitivities would be obtained as in Figure 6.35, where $H_\Sigma(s, K)$ is the scalar transfer function from the summing junction input to the system output as shown in Figure 6.36. However, this transfer function is unknown, and we approximate it by the sensitivity filter $H_\Psi(s)$. The pseudogradient stability condition to be derived here specifies a bound on the allowable mismatch between $H_\Sigma(s, K^*)$ and $H_\Psi(s)$, where K^* is such that

$$\lim_{T\to\infty} \frac{1}{T}\int_t^{t+T} \Psi(\tau, K^*)e(\tau, K^*)d\tau = \quad (6.33)$$

$$[\Psi(t, K^*)e(t, K^*)]_{\text{ave}} = 0. \quad (6.34)$$

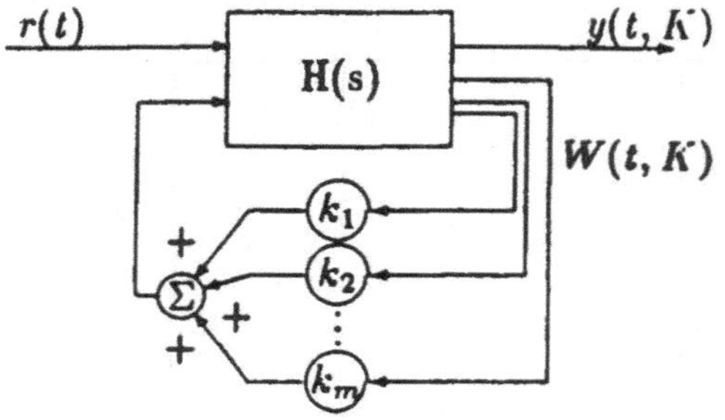

Figure 6.34 Adjustable linear system.

With reduced-order controllers, it is unrealistic to assume that the error $e(t, K)$ can be eliminated entirely for any value of controller parameters and plant variations. The remaining nonzero error is called the *tuned error*, $e^*(t) = e(t, K^*)$. The error $e(t, K)$ is comprised of $e^*(t)$ and a term caused by the parameter error $\tilde{K} = K - K^*$. Using a mixed (t, s) notation with signals as functions of time and transfer functions denoted as Laplace transforms, the error $e(t, K)$ and the regressor $\Psi(t, K)$ are both obtained from the measured signals $W(t, K)$ as follows:

$$e(t, K) = H_\Sigma(s, K^*)[\tilde{K}^T(t)W(t, K)] + e^*(t) \quad (6.35)$$

$$\Psi(t, K) = H_\Psi(s)[W(t, K)]. \quad (6.36)$$

The error expression illustrated in Figure 6.37 contains the exact sensitivity filter transfer function $H_\Sigma(s, K^*)$. The stability properties of this slowly adapting system are determined by examination of the average parameter update equation

$$\begin{aligned}\dot{\tilde{K}}(t) &= -\epsilon\{\Psi(t, K)V(t, K)^T\}_{\text{ave}}\tilde{K}(t) \\ &\quad -\epsilon\Delta - \epsilon\{\Psi(t, K)e^*(t)\}_{\text{ave}}. \quad (6.37)\end{aligned}$$

where we have again used $H_\Sigma(s, K^*)$ in the definition of the signal

$$V(t, K) \equiv H_\Sigma(s, K^*)[W(t, K)]. \quad (6.38)$$

The average swapping term Δ in Equation 6.37 is

$$\begin{aligned}\Delta &= \{H_\Sigma(s, K^*)[W(t, K)^T\tilde{K}(t)] \\ &\quad - H_\Sigma(s, K^*)[W(t, K)^T]\tilde{K}(t)\}_{\text{ave}} \quad (6.39)\end{aligned}$$

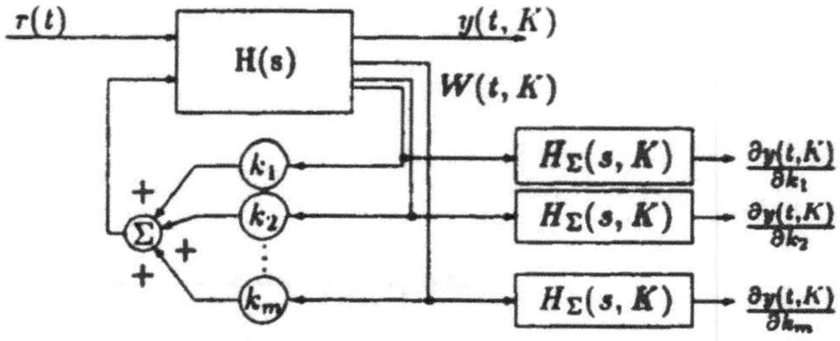

Figure 6.35 Sensitivity system.

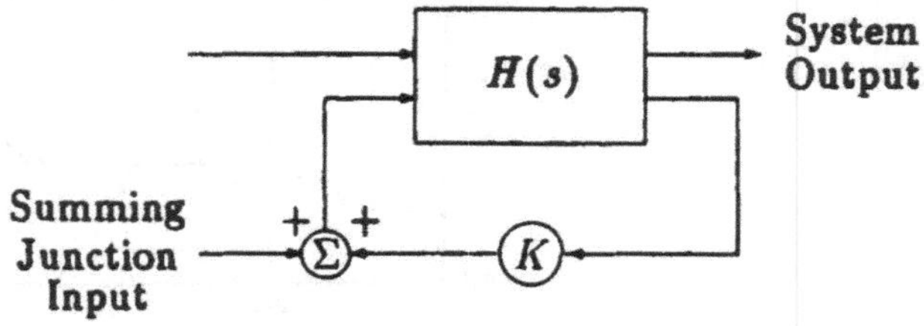

Figure 6.36 $H_\Sigma(s, K)$.

As shown in [14] and [15], for slow adaptation this term is small and does not affect the stability of Equation 6.37 in a neighborhood of the equilibrium K^*. The stability of this equilibrium, in the case when $e^*(t) = 0$, is established as follows.

LEMMA 6.1 [15] Consider the average update system (Equation 6.37) with $\Delta(t) = 0$, $e(t, K^*) = 0$, and $K = K^*$. If

$$\text{Re}\lambda\{\Psi(t, K)V^T(t, K)\}_{\text{ave}} > 0 \quad (6.40)$$

then there exists an $\epsilon^* > 0$ such that $\forall\epsilon \in (0, \epsilon^*]$ the equilibrium $\tilde{K} = 0$ is uniformly asymptotically stable.

To interpret the condition of Equation 6.40, we represent the regressor as

$$\Psi(t, K) = H_\Psi(s)H_\Sigma^{-1}(s, K^*)[V(t, K)] \quad (6.41)$$

so that the matrix in Equation 6.40 can be visualized as in Figure 6.38. Representing the almost periodic signal $V(t, K)$ as

$$V(t, K) = \sum_{i=-\infty}^{+\infty} v_i e^{j\omega_i t} \quad (6.42)$$

a sufficient condition for Equation 6.40 to hold is

$$\sum_{i=-\infty}^{+\infty} \text{Re}[H_\Psi(j\omega_i)H_\Sigma^{-1}(j\omega_i, K^*)]\text{Re}[v_i\bar{v}_i^T] > 0. \quad (6.43)$$

If the signal, $V(t)$, possesses an autocovariance, $R_V(z)$, it is PE if and only if $R_V(0) > 0$ [17]. It follows that $V(t, K)$ is PE if and only if $\sum_{i=-\infty}^{\infty} R_e[v_i\bar{v}_i^T] > 0$. Clearly, if the sensitivity filter is exact, $H_\Psi(j\omega) = H_\Sigma(j\omega, K^*)$, and $V(t, K)$ is PE, then the sufficient stability condition in Equation 6.43 holds. When $H_\Psi(s)$ cannot be made exact, this stability condition is still satisfied when the pseudosensitivity filter is chosen such that $\text{Re}[H_\Psi(j\omega)H_\Sigma^{-1}(j\omega, K^*)] > 0$ for dominant frequencies, that

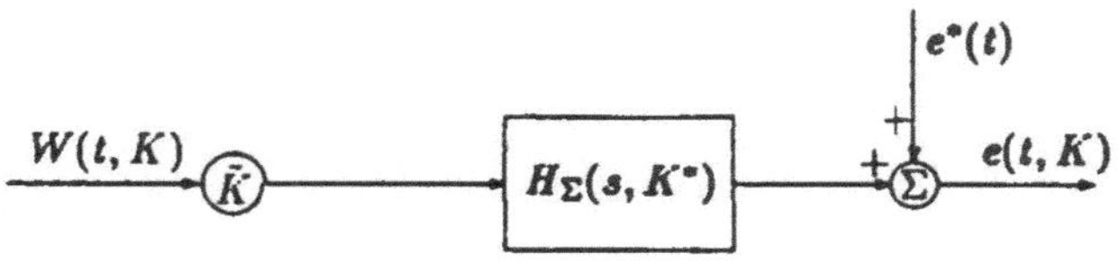

Figure 6.37 Error model.

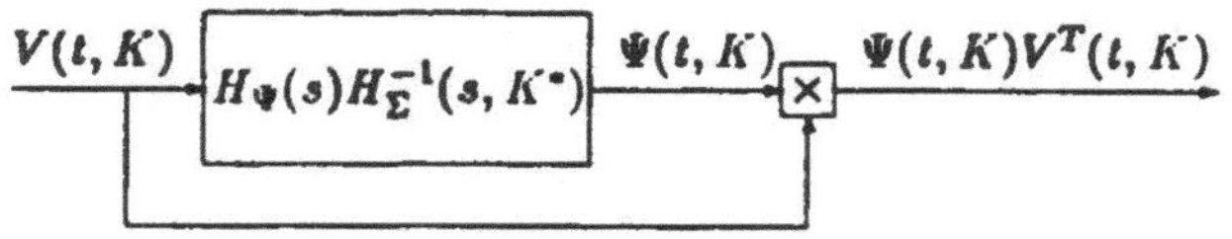

Figure 6.38 Feedback matrix.

is, the frequencies where $v_i \bar{v}_i^T$ is large. This condition serves as a guide for designing the pseudosensitivity filter $H_\Psi(s)$. As $H_\Sigma(s, K^*)$ is unknown beforehand, we use the *a priori* information about the closed-loop system to design a filter $H_\Psi(s)$ such that in the dominant frequency range the following *pseudogradient condition* is satisfied for all plant and controller parameters of interest:

$$-90^o < \angle\{H_\Psi(\jmath\omega)H_\Sigma^{-1}(\jmath\omega, K^*)\} < 90^o \quad (6.44)$$

When Equation 6.44 is satisfied, then the equilibrium K^* of the average update law of Equation 6.37 with $\Delta \equiv 0, e^*(t) \equiv 0$ is uniformly asymptotically stable. By Bogolubov's theorem and slow manifold analysis [14], this implies the local stability property of the actual adaptive system, provided $e^*(t) \neq 0$ is sufficiently small. Since, in this approach, both the controller structure and the number of adjustable parameters are free to be chosen independently of plant order, there is no guarantee that $e^*(t)$ will be sufficiently small. Although conservative bounds for $e^*(t)$ may be calculated [12], in practice, since the design objective of the controller and pseudogradient adaptive law is to minimize the average of $e(t, K)^2$, $e^*(t)$ is typically small.

References

[1] Ball, J.T., Approaches and Trends in Automatic Speed Controls, SAE Tech. Paper #670195, 1967.

[2] Follmer, W.C., Electronic Speed Control, SAE Tech. Paper #740022, 1974.

[3] Sobolak, S.J., Simulation of the Ford Vehicle Speed Control System, SAE Tech. Paper #820777, 1982.

[4] Nakamura, K., Ochiai, T., and Tanigawa, K., Application of microprocessor to cruise control system, *Proc. IEEE Workshop Automot. Appl. Microprocessors*, 37–44, 1982.

[5] Chaudhure, B., Schwabel, R.J., and Voelkle, L.H., Speed Control Integrated into the Powertrain Computer, SAE Tech. Paper #860480, 1986.

[6] Tabe, T., Takeuchi, H., Tsujii, M., and Ohba, M., Vehicle speed control system using modern control theory, *Proc. 1986 Int. Conf. Industrial Electron., Control Instrum.*, 1, 365–370, 1986.

[7] Uriuhara, M., Hattori, T., and Morida, S., Development of Automatic Cruising Using Fuzzy Control System, *J. SAE Jpn.*, 42(2), 224–229, 1988.

[8] Abate, M. and Dosio, N., Use of Fuzzy Logic for Engine Idle Speed Control, SAE Tech. Paper #900594, 1990.

[9] Tsujii, T., Takeuchi, H., Oda, K., and Ohba, M., Application of self-tuning to automotive cruise control, *Proc. Am. Control Conf.*, 1843–1848, 1990.

[10] Hong, G. and Collings, N., Application of Self-Tuning Control, SAE Tech. Paper #900593, 1990.

[11] Kokotovic, P.V., Method of sensitivity points in the investigation and optimization of linear control systems, *Automation Remote Control*, 25, 1670–1676, 1964.

[12] Rhode, D.S., Sensitivity Methods and Slow Adaptation, Ph.D. thesis, University of Illinois at Urbana-Champaign, 1990.

[13] Rhode, D.S. and Kokotovic, P.V., Parameter Convergence conditions independent of plant order, in *Proc. Am. Control Conf.*, 981–986, 1990.

[14] Riedle, B.D. and Kokotovic, P.V., Integral manifolds of slow adaptation, *IEEE Trans. Autom. Control*, 31, 316–323, 1986.

[15] Kokotovic, P.V., Riedle, B.D., and Praly, L., On a stability criterion for continuous slow adaptation, *Sys. Control Lett.*, 6, 7–14, 1985.

[16] Anderson, B.D.O., Bitmead, R.R., Johnson, C.R., Jr., Kokotovic, P.V., Kosut, R.L., Mareels, I., Praly, L., and Riedle, B.D., *Stability of Adaptive Systems: Passivity and Averaging Analysis*, MIT Press, Cambridge, MA, 1986.

[17] Boyd S. and Sastry, S.S., Necessary and sufficient conditions for parameter convergence in adaptive control, *Automatica*, 22(6), 629–639, 1986.

7

Aerospace Controls

7.1 Flight Control of Piloted Aircraft 113
Introduction • Flight Mechanics • Nonlinear Dynamics • Actuators • Flight Control Requirements • Dynamic Analysis • Conventional Flight Control • Time-Scale Separation • Actuator Saturation Mitigation in Tracking Control • Nonlinear Inner Loop Design • Flight Control of Piloted Aircraft

References 128

Further Reading 129

7.2 Spacecraft Attitude Control 129
Introduction • Modeling • Spacecraft Attitude Sensors and Control Actuators • Spacecraft Rotational Kinematics • Spacecraft Rotational Dynamics • Linearized Spacecraft Rotational Equations • Control Specifications and Objectives • Spacecraft Control Problem: Linear Control Law Based on Linearized Spacecraft Equations • Spacecraft Control Problem: Bang-Bang Control Law Based on Linearized Spacecraft Equations • Spacecraft Control Problem: Nonlinear Control Law Based on Nonlinear Spacecraft Equations • Spacecraft Control Problem: Attitude Control in Circular Orbit • Other Spacecraft Control Problems and Control Methodologies • Defining Terms

References 141

Further Reading 141

7.3 Control of Flexible Space Structures 142
Introduction • Single-Body Flexible Spacecraft • Multibody Flexible Space Systems • Summary

References 152

7.4 Line-of-Sight Pointing and Stabilization Control System 152
Introduction • Overall System Performance Objectives • Physical System Description and Modeling • Controller Design • Performance Achieved • Concluding Remarks • Defining Terms

References 164

M. Pachter
Department of Electrical and Computer Engineering, Air Force Institute of Technology, Wright-Patterson AFB, OH

C. H. Houpis
Department of Electrical and Computer Engineering, Air Force Institute of Technology, Wright-Patterson AFB, OH

Vincent T. Coppola
Department of Aerospace Engineering, The University of Michigan, Ann Arbor, MI

N. Harris McClamroch
Department of Aerospace Engineering, The University of Michigan, Ann Arbor, MI

S. M. Joshi and A. G. Kelkar
NASA Langley Research Center

David Haessig
GEC-Marconi Systems Corporation, Wayne, NJ

7.1 Flight Control of Piloted Aircraft

M. Pachter, Department of Electrical and Computer Engineering, Air Force Institute of Technology, Wright-Patterson AFB, OH
C. H. Houpis, Department of Electrical and Computer Engineering, Air Force Institute of Technology, Wright-Patterson AFB, OH

7.1.1 Introduction

Modern Flight Control Systems (FCS) consist of (1) aerodynamic control surfaces and/or the engines' nozzles, (2) actuators, (3) sensors, (4) a sampler and ZOH device, and (5) compensators. The first four components of an FCS are hardware elements, whereas the controller (the digital implementation of the compensator) is an *algorithm* executed in real time in the on-board digital computer. In this chapter the design of the compensation/controller element/algorithm of the FCS, for a given aircraft, after the actuators, sensors and samplers have been chosen, is addressed. The way in which control theory is applied to the FCS's controller design is the main focus of this article. An advanced and comprehensive perspective on flight control is presented. The emphasis is on maneuvering flight control. Thus, attention is given to the process of setting up the flight control problem from its inception. Flight Mechanics is used to obtain a rigorous formulation of the nonlinear dynamic model of the controlled "plant;" the linearization of the latter yields Linear Time Invariant (LTI) models routinely used for controller design. Also, it is important to remember that the *pilot* will be closing an additional outer feedback loop. This transforms the FCS design problem from one of meeting flying quality specifications into one of meeting handling quality specifications.

0-8493-0054-3/00/$0.00+$.50

The essence of flight control is the design of an FCS for *maneuvering* flight. Hence, we chose not to dwell on outer-loop control associated with autopilot design and, instead, the focus is on the challenging problems of maneuvering flight control and the design of an inner-loop FCS, and pilot-in-the-loop issues. By its very nature, maneuvering flight entails large state variable excursions, which forces us to address nonlinearity and cross-coupling.

To turn, pilots will bank their aircraft and pull g's. It is thus realized that the critical control problem in maneuvering flight is the stability axis, or, velocity vector, roll maneuver. During a velocity vector roll both the lateral/directional *and* the pitch channels are controlled simultaneously. Hence, this maneuver epitomizes maneuvering flight control, for it brings into the foreground the pitch and lateral/directional channels' cross-coupling, nonlinearity, time-scale separation, tracking control design, actuator saturation concerns, and pilot-in-the-loop issues—all addressed in this chapter. Moreover, when additional simplifying assumptions apply, the velocity vector roll maneuver is general enough to serve as the starting point for derivating the classical LTI aircraft model, where the longitudinal and lateral/directional flight control channels are decoupled. Evidently, the design of a FCS for velocity vector rolls is a vehicle for exploring the important aspects of maneuvering flight control. Hence, this article's leitmotif is the high Angle Of Attack (AOA) velocity vector roll maneuver.

Since this chapter is configured around maneuvering flight, and because pilots employ a high AOA and bank in order to turn, the kinematics of maneuvering flight and high AOA velocity vector rolls are now illustrated. Thus, should the aircraft roll about its $x-$body axis, say, as a result of adverse yaw, then at the point of attainment of a bank angle of 90°, the AOA will have been totally converted into sideslip angle, as illustrated in Figure 7.1e. In this case, the aircraft's nose won't be pointed in the right direction. During a velocity vector roll the aircraft rotates about an axis aligned with its velocity vector, as illustrated in Figures 7.1a–7.1d. The operational significance of this maneuver is obvious, for it allows the pilot to slew quickly and point the aircraft's nose using a fast roll maneuver, without pulling g's, i.e., without increasing the normal acceleration, and turning. This maneuver is also a critical element of the close air combat S maneuver, where one would like to roll the loaded airframe, rather than first unload, roll and pull gs. Thus, in this chapter, maneuvering flight control of modern fighter aircraft, akin to an F-16 derivative, is considered.

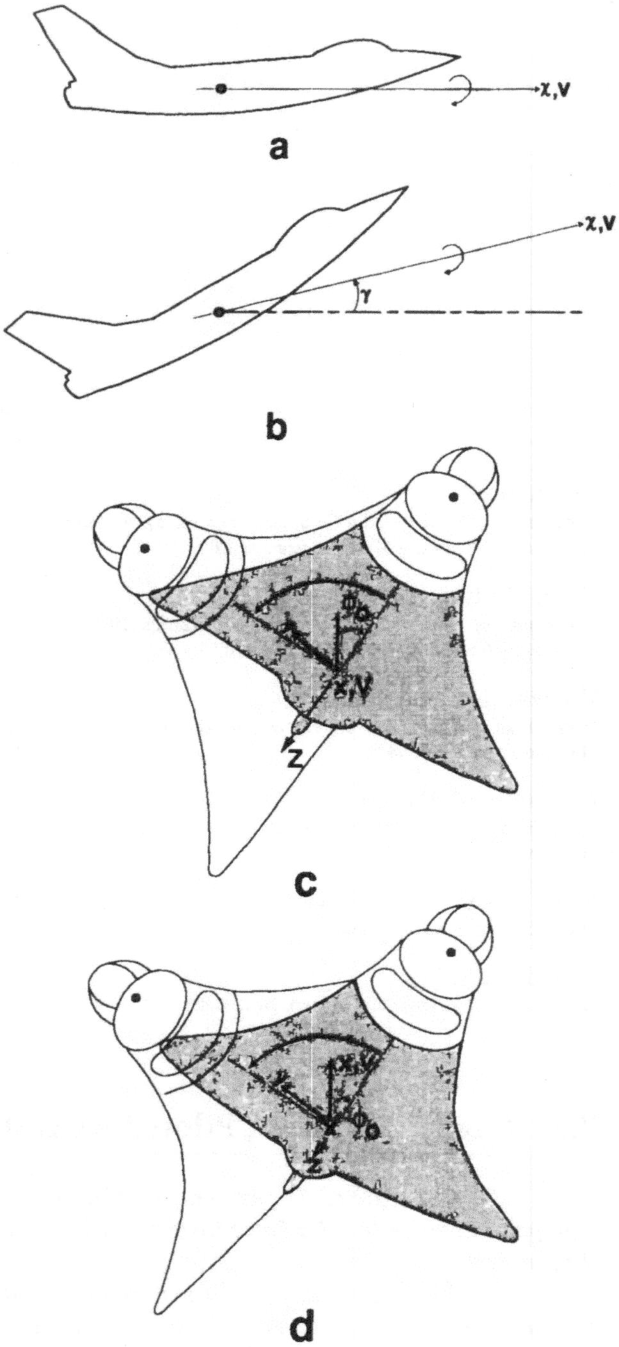

Figure 7.1 Initial and final flight configurations for high AOA maneuvers; V is always the aircraft velocity vector. a) level flight. $\psi_0 = \theta_0 = \phi_0 = 0$; $\psi_f = \theta_f = 0, \phi_f = \phi$ b) climbing flight. $\psi_0 = 0, \theta_0 = \gamma, \phi_0 = 0$; $\psi_f = 0, \theta_f = \gamma, \phi_f = \phi$ c) planar S maneuver. $\psi_0 = \theta_0 = 0, \phi_0 = -\phi_0$; $\psi_f = \theta_f = 0, \phi_f = \phi_0$ d) climbing S maneuver. $\psi_0 = 0, \theta_0 = \gamma, \phi_0 = -\phi_0$; $\psi_f = 0, \theta_f = \gamma, \phi_f = \phi_0$

7.1.2 Flight Mechanics

Proper application of the existing control theoretic methods is contingent on a thorough understanding of the "plant." Hence, a careful derivation of the aircraft, ("plant") model required in FCS design is given. To account properly for the nonlinearities affecting the control system in maneuvering flight, the plant model must be rigorously derived from the fundamental nine state equations of motion [1] and [2]. The Euler equations for a rigid body yield the following equations of motion. The *Force*

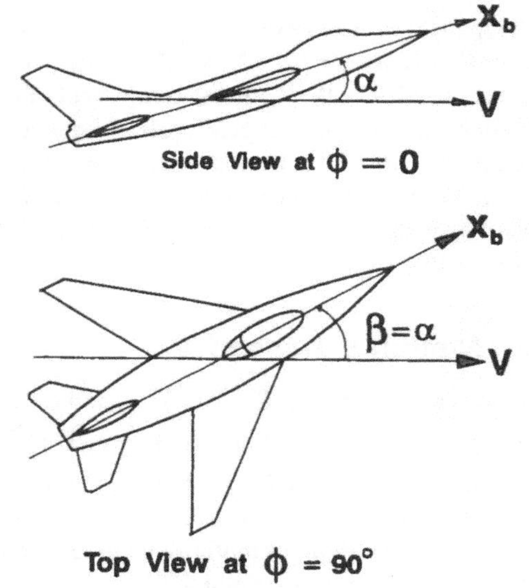

e

Figure 7.1 *(Cont.)* Initial and final flight configurations for high AOA maneuvers; V is always the aircraft velocity vector. e) roll about x-body axis.

Equations are

$$\dot{U} = VR - WQ + \frac{1}{m}\sum F_x, \quad U(0) = U_0 = \bar{U}, \ 0 \le t \le t_f, \tag{7.1}$$

$$\dot{V} = WP - UR + \frac{1}{m}\sum F_y, \ V(0) = V_0, \tag{7.2}$$

and

$$\dot{W} = UQ - VP + \frac{1}{m}\sum F_z, \ W(0) = 0, \tag{7.3}$$

where m is the mass of the aircraft and U, V, and W are the components of the aircraft's velocity vector, resolved in the respective x, y, and z body axes; similarly, P, Q, and R are the components of the aircraft's rotational speed, resolved in the x, y, and z body axes (see Figure 7.2). The origin of the body axes is collocated with the aircraft's CG. The initial condition in Equation 7.3 is consistent with the use of stability axes. If body axes were used instead, then the initial condition would be $W(0) = W_0$. The *Moment Equations* are

$$\begin{aligned}\dot{P} = {} & \frac{I_{xz}}{D}(I_x - I_y + I_z)PQ \\ & + \frac{1}{D}(I_y I_z - I_z^2 - I_{xz}^2)QR \\ & + \frac{I_z}{D}\sum L + \frac{I_{xz}}{D}\sum N, \\ & P(0) = P_0,\end{aligned} \tag{7.4}$$

$$\begin{aligned}\dot{Q} = {} & \frac{I_z - I_x}{I_y}PR + \frac{I_{xz}}{I_y}(R^2 - P^2) + \frac{1}{I_y}\sum M, \\ & Q(0) = Q_0,\end{aligned} \tag{7.5}$$

and

$$\begin{aligned}\dot{R} = {} & \frac{1}{D}(I_x^2 - I_x I_y + I_{xz}^2)PQ \\ & + \frac{I_{xz}}{D}(I_y - I_x - I_z)QR \\ & + \frac{I_{xz}}{D}\sum L + \frac{I_x}{D}\sum N, \\ & R(0) = R_0,\end{aligned} \tag{7.6}$$

where I denotes the aircraft's inertia tensor and the definition $D = I_x I_z - I_{xz}^2$ is used. Also note that the modified "moment equations" above are obtained by transforming the classical Euler equations for the dynamics of a rigid body into state-space form.

The *Kinematic Equations* are concerned with the propagation of the Euler angles, which describe the attitude of the aircraft in inertial space: The Euler angles specify the orientation of the aircraft's body axes triad, and rotations are measured with reference to a right handed inertial frame whose z-axis points toward the center of the earth. The body axes are initially aligned with the inertial frame and the orientation of the body axes triad is determined by three *consecutive* rotations of ψ, θ and ϕ radians about the respective z, y, and x body axes. When the three consecutive rotations of the body axes frame is performed in the specified order above, ψ, ϕ, and θ are referred to as (3,2,1) Euler angles, the convention adhered to in this chapter. The Euler angles, which are needed to resolve the force of gravity into the body axes, are determined by the aircraft's angular rates P, Q, and R, according to the following equations:

$$\begin{aligned}\dot{\theta} = {} & Q\cos\phi - R\sin\phi, \\ & \theta(0) = \theta_0, \ \theta(t_f) = \theta_f,\end{aligned} \tag{7.7}$$

$$\begin{aligned}\dot{\phi} = {} & P + Q\sin\phi\tan\theta + R\cos\phi\tan\theta, \\ & \phi(0) = \phi_0, \ \phi(t_f) = \phi_f,\end{aligned} \tag{7.8}$$

$$\begin{aligned}\dot{\psi} = {} & Q\frac{\sin\phi}{\cos\theta} + R\frac{\cos\phi}{\cos\theta}, \\ & \psi(0) = \psi_0, \ \psi(t_f) = \psi_f,\end{aligned} \tag{7.9}$$

The contribution of the force of gravity, the contributions of the aerodynamic and propulsive forces and moments acting on the aircraft, and the control input contributions, are contained in the force (F) and moment (M) summations in Equations 7.1 – 7.6. The aerodynamic forces and moments are produced by the aircraft's relative motion with respect to the air flow, and are proportional to the air density ρ and the square of the airspeed $\bar{U}$. In addition, the aerodynamic forces and moments are determined by the orientation angles with respect to the relative wind, viz., the AOA α and the sideslip angle β. The aerodynamic angles are depicted in Figure 7.2, where the aerodynamic forces and moments are also shown.

The standard [4] nondimensional aerodynamic force and moment coefficients, designated by the letter C, are introduced. The dynamic pressure $\bar{q} = \frac{1}{2}\rho(U^2 + V^2 + W^2) = \frac{1}{2}\rho\bar{U}^2$, and the wing's area S, are used to nondimensionalize the aerodynamic forces; the additional parameters, c, the wing's mean aerodynamic chord, or, b, the wing's span, are used to nondimensionalize the aerodynamic moments. Moreover, two subscript levels are used. The first subscript of C designates the pertinent aerodynamic force or moment component, and the second subscript

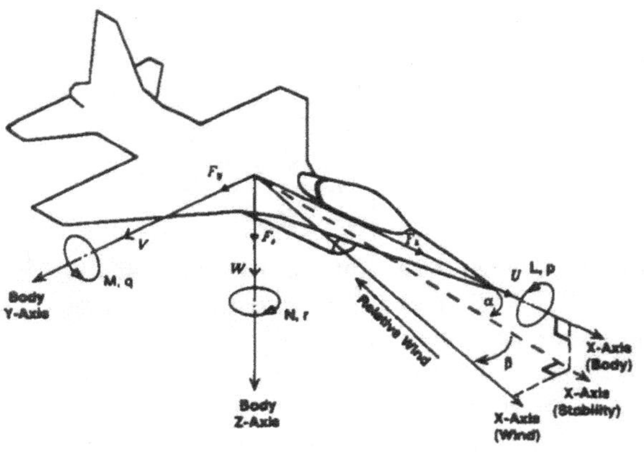

Figure 7.2 Diagram depicting axes definitions, componentes of velocity vector of the aircraft, angular rates of the aircraft, and aerodynamic forces and moments.

pertains to a specific component of the state vector. For example, the first stability derivative in Equation 7.10 is C_{l_p}, and it yields the aircraft's roll rate contribution to the nondimensional rolling moment coefficient C_l. Thus, the aerodynamic derivatives are "influence coefficients." Using three generalized control inputs, aileron, elevator, and rudder deflections, δ_a, δ_e, and δ_r, respectively, these force and moment summation equations are

$$
\begin{aligned}
\sum L &= \bar{q}Sb\,(\frac{b}{2\bar{U}}C_{l_p}P + \frac{b}{2\bar{U}}C_{l_r}R + C_{l_\beta}\beta + C_{l_{\delta_a}}\delta_a \\
&\quad + C_{l_{\delta_r}}\delta_r), && (7.10)\\
\sum M &= \bar{q}Sc\,(C_{m_0} + C_{m_\alpha}\alpha + \frac{c}{2\bar{U}}C_{m_q}Q \\
&\quad + \frac{c}{2\bar{U}}C_{m_{\dot{\alpha}}}\dot{\alpha} + C_{m_{\delta_e}}\delta_e), && (7.11)\\
\sum N &= \bar{q}Sb\,(\frac{b}{2\bar{U}}C_{n_p}P + \frac{b}{2\bar{U}}C_{n_r}R + C_{n_\beta}\beta \\
&\quad + C_{n_{\delta_a}}\delta_a + C_{n_{\delta_r}}\delta_r), && (7.12)\\
\sum F_y &= mg\,sin\phi\,cos\theta + \bar{q}S(\frac{b}{2\bar{U}}C_{y_p}P \\
&\quad + \frac{b}{2\bar{U}}C_{y_r}R + C_{y_\beta}\beta + C_{y_{\delta_r}}\delta_r), \\
\sum F_z &= mg\,cos\phi\,cos\theta + T_z + \bar{q}S(-C_{z_\alpha}\alpha_{0L} \\
&\quad + C_{z_\alpha}\alpha + \frac{c}{2\bar{U}}C_{z_q}Q + \frac{c}{2\bar{U}}C_{z_{\dot{\alpha}}}\dot{\alpha} \\
&\quad + C_{z_{\delta_e}}\delta_e). && (7.13)
\end{aligned}
$$

In Equation 7.13, T_z is the z-axis component of the thrust (usually < 0) and α_{0L} (< 0) is the zero lift Angle Of Attack (AOA) referenced from the x-stability axis. Hence, α_{0L} is determined by the choice of stability axes. The same is true for the C_{m_0} stabilty derivative in Equation 7.11. The C_{m_0} stability derivative used here is $C'_{m0} - C_{m_\alpha}\alpha_{0L}$, where C'_{m_0} pertains to AOA measurements referenced to the aircraft's zero lift plane. Both C_{m_0} and α_{0L} are defined with reference to a nominal elevator setting ($\delta_e = 0$).

The x-axis velocity component is assumed constant throughout the short time horizon of interest in inner-loop flight control work. Hence, the thrust (control) setting is not included in inner-loop flight control work and Equation 7.1 is not used. The pertinent Equations are 7.2 – 7.13. The thrust control setting is also not included in the T_z thrust component. Furthermore, the heading angle ψ does not play a role in Equations 7.2 – 7.8 and Equations 7.10 – 7.13. Hence, one need not consider Equation 7.9 and thus the pertinent velocity vector roll dynamics are described by the seven DOF system of Equations 7.2 – 7.8 and Equations 7.10 – 7.13.

7.1.3 Nonlinear Dynamics

The seven states are P, Q, R, $\frac{V}{U}$, $\frac{W}{U}$, θ, and ϕ. The aerodynamic angles are defined as follows: $\alpha = tan^{-1}(\frac{W}{U})$, $\beta = tan^{-1}(\frac{V}{U})$. In Equations 7.2 – 7.6 and Equations 7.10 – 7.13, consolidated stability and control derivatives (which combine like terms in the equations of motion) are used, and the following nonlinear dynamics are obtained [2]:

$$
\begin{aligned}
\frac{d}{dt}(\frac{V}{\bar{U}}) &= (\frac{W}{\bar{U}})P + (C_{y_r} - 1)R + \frac{g}{\bar{U}}\,sin\phi\,cos\theta \\
&\quad + C_{y_p}P + C_{y_\beta}\beta + C_{y_{\delta_r}}\delta_r\,, \\
&\quad 0 \le t\,,\ V(0) = V_0 && (7.14)\\
\frac{d}{dt}(\frac{W}{\bar{U}}) &= (1 + C_{z_q})Q - (\frac{V}{\bar{U}})P + \frac{g}{\bar{U}}\,cos\phi\,cos\theta \\
&\quad + C_{T_z} - C_{z_\alpha}\alpha_{0L} + C_{z_\alpha}\alpha + C_{z_{\dot{\alpha}}}\dot{\alpha} + C_{z_{\delta_e}}\delta_e, \\
&\quad W(0) = 0 && (7.15)\\
\frac{dP}{dt} &= C_{l_{pq}}PQ + C_{l_{qr}}QR + C_{l_p}P + C_{l_r}R \\
&\quad + C_{l_\beta}\beta + C_{l_{\delta_a}}\delta_a + C_{l_{\delta_r}}\delta_r\,, \\
&\quad P(0) = P_0 && (7.16)\\
\frac{dQ}{dt} &= C_{m_{pr}}PR + C_{m_{r^2}}R^2 - C_{m_{p^2}}P^2 + C_{m_0} \\
&\quad + C_{m_\alpha}\alpha + C_{m_{\dot{\alpha}}}\dot{\alpha} + C_{m_q}Q + C_{m_\delta}\delta_e\,, \\
&\quad Q(0) = Q_0 && (7.17)\\
\frac{dR}{dt} &= C_{n_{pq}}PQ - C_{l_{pq}}QR + C_{n_p}P + C_{n_r}R \\
&\quad + C_{n_\beta}\beta + C_{n_{\delta_a}}\delta_a + C_{n_{\delta_r}}\delta_r\,, \\
&\quad R(0) = R_0 && (7.18)
\end{aligned}
$$

It is assumed that stability axes are used. Furthermore, the particular stability axes used are chosen at time $t = 0$. Also, in Equations 7.15 and 7.17, $\dot{\alpha} = (\frac{\dot{W}}{U})/(1 + \frac{W^2}{U^2})$. The kinematic Equations 7.7 and 7.8 are also included in the nonlinear dynamical system.

Trim Analysis

The LHS of the seven differential Equations 7.2 – 7.6 (or Equations 7.14 – 7.18) and Equations 7.7 and 7.8, is set equal to zero to compute the so-called "trim" values of the aircraft states and controls. Trim conditions are now considered where the aircraft's angular rates P_0, Q_0, and R_0, its sideslip velocity component V_0, and the respective pitch and bank Euler angles θ_0, and ϕ_0, are constant. The pertinent seven states are V, W, P, Q, R, θ, and ϕ, and the (generalized) controls are δ_a, δ_e, and δ_r. In the sequel, the trim controls and states are denoted with the subscript 0, with the exception of the trim value of U, which is barred. For fixed control trim settings, an initial trim condition (or nominal trajectory) is established. Also, when some of the trim states are specified, the remaining states and control trim settings can be obtained, provided that the total number of specified variables (controls and/or states) is three.

Using generalized stability and control derivatives, the following algebraic *trim equations* are obtained from Equations 7.14, 7.16, and 7.18:

$$\frac{g}{\bar{U}}cos\theta_0\, sin\phi_0 + C_{y_p}P_0 + (C_{y_r} - 1)R_0 + C_{y_\beta}\beta_0 = -C_{y_{\delta_r}}\delta_{r_0}, \tag{7.19}$$

$$C_{l_{pq}}P_0Q_0 + C_{l_{qr}}Q_0R_0 + C_{l_p}P_0 + C_{l_r}R_0 + C_{l_\beta}\beta_0 = -C_{l_{\delta_a}}\delta_{a_0} - C_{l_{\delta_r}}\delta_{r_0}, \tag{7.20}$$

and

$$C_{n_{pq}}P_0Q_0 - C_{l_{pq}}Q_0R_0 + C_{n_p}P_0 + C_{n_r}R_0 + C_{n_\beta}\beta_0 = -C_{n_{\delta_a}}\delta_{a_0} - C_{n_{\delta_r}}\delta_{r_0}. \tag{7.21}$$

In Equations 7.19 – 7.21 the unknowns are P_0, Q_0, R_0, β_0, δ_{a_0}, and δ_{r_0}. In addition, and as will be shown in the sequel, Equations 7.7 and 7.8 yield $\theta_0 = \theta_0(P_0, Q_0, R_0)$ and $\phi_0 = \phi_0(P_0, Q_0, R_0)$. Hence, we have obtained three equations in six unknowns. This then requires the specification of three trim variables, whereupon the remaining three trim variables are obtained from the above three trim equations. For example, if the equilibrium angular rates P_0, Q_0, and R_0 are specified then the required control trim settings δ_{a_0} and δ_{r_0}, and the trim sideslip angle β_0, can be calculated. The calculation then entails the solution of a set of three linear equations in three unknowns. The system matrix that needs to be inverted obtained from Equations 7.19 – 7.21 is

$$M = \begin{bmatrix} C_{y_\beta} & 0 & C_{y_{\delta_r}} \\ C_{l_\beta} & C_{l_{\delta_a}} & C_{l_{\delta_r}} \\ C_{n_\beta} & C_{n_{\delta_a}} & C_{n_{\delta_r}} \end{bmatrix}.$$

Obviously, the M matrix must be nonsingular for a trim solution with constant angular rates to be feasible, i.e., the condition $C_{y_\beta}C_{l_{\delta_a}}C_{n_{\delta_r}} + C_{y_{\delta_r}}C_{l_\beta}C_{n_{\delta_a}} \neq C_{y_\beta}C_{l_{\delta_r}}C_{n_{\delta_a}} + C_{y_{\delta_r}}C_{l_{\delta_a}}C_{n_\beta}$ must hold.

From the remaining Equations 7.15 and 7.17, the two equations which determine the zero lift angle included between the velocity vector and the aircraft's zero lift plane, α_{0L}, and the trim elevator setting, δ_{e_0}, are obtained:

$$\frac{g}{\bar{U}}cos\theta_0\, cos\phi_0 + C_{T_z} - C_{z_\alpha}\alpha_{0L} - P_0 tan\beta_0 + (1 + C_{z_q})Q_0 + C_{z_{\delta_e}}\delta_{e_0} = 0, \tag{7.22}$$

and

$$C_{m_0} + C_{m_{\delta_e}}\delta_{e_0} + C_{m_q}Q_0 + C_{m_{pr}}P_0R_0 + C_{m_{r^2}}R_0^2 - C_{m_{p^2}}P_0^2 = 0. \tag{7.23}$$

Once the angular rates and the sideslip angle β_0 have been established, the δ_{e_0} and α_{0L} unknowns are determined from the linear Equations 7.22 and 7.23. Obviously, δ_{e_0} must be feasible, i.e., $-\delta_{e_{max}} \le \delta_{e_0} \le \delta_{e_{max}}$.

An analysis of Equations 7.7 and 7.8 is now undertaken. Recall that the trim trajectory is flown at a constant pitch angle. If both Q_0 and R_0 are not 0, then the bank angle ϕ_0 must be constant. The trim pitch and bank angles are obtained from the Euler angles Equations 7.7 and 7.8. Thus, if $R_0 \neq 0$, then $\phi_0 = tan^{-1}(\frac{Q_0}{R_0})$, and $\theta_0 = -tan^{-1}(\frac{P_0}{R_0}cos\phi_0)$. Hence, if $P_0 = 0$, then $\theta_0 = 0$ and this trim condition entails a level turn, illustrated in Figure 7.1c. In the general case where $P_0 \neq 0$, a steady spiral climb or a steady corkscrew descent, as illustrated in Figure 7.1d, is being considered. Moreover, for the above flight maneuvers, Equation 7.9 yields the elegant heading rate result $\dot{\psi} = \sqrt{P_0^2 + Q_0^2 + R_0^2}$.

Obviously, not every pair of prespecified pitch and bank trim angles, θ_0 and ϕ_0, respectively, is feasible. Indeed, the solution of the equations above must satisfy $-\delta_{a_{max}} \le \delta_{a_0} \le \delta_{a_{max}}$ and $-\delta_{r_{max}} \le \delta_{r_0} \le \delta_{r_{max}}$. In the very special case where the bank angle $\phi_0 = 0$, a symmetrical flight condition ensues with $\beta_0 = \delta_{a_0} = \delta_{r_0} = 0$. A steady climb in the pitch plane is then considered, as illustrated in Figure 7.1b. Finally, α_{0L} and δ_{e_0} are then determined from Equations 7.22 and 7.23, where $\phi_0 = P_0 = Q_0 = R_0 = 0$, i.e., $\frac{g}{U}cos\theta_0 + C_{T_z} - C_{z_\alpha}\alpha_{0L} + C_{z_{\delta_e}}\delta_{e_0} = 0$, *and* $C'_{m_0} - C_{m_\alpha}\alpha_{0L} + C_{m_{\delta_e}}\delta_{e_0} = 0$.

The trim conditions identified above represent interesting flight phases. For example, and as shown in Figure 7.1a, one can initiate a velocity vector roll in trimmed and level flight, where $P_0 = Q_0 = R_0 = \theta_0 = \phi_0 = \beta_0 = \delta_{a_0} = \delta_{r_0} = 0$, and one can subsequently arrest the velocity vector roll at a new trim flight condition which entails a level turn, shown in Figure 7.1c, and where Q_0 and R_0 are prespecified, $P_0 = \theta_0 = 0$, and $\phi_0 = tan^{-1}(\frac{Q_0}{R_0})$. The new bank angle satisfies the equation $sin\phi_0 = \frac{Q_0}{\sqrt{Q_0^2+R_0^2}} = \frac{Q_0}{\dot{\psi}}$. The required trim β_0, δ_{a_0}, and δ_{r_0} is determined by solving the linear trim equations. Since the trim bank angle changes, this will require a change in the trim elevator setting and in α_{0L}. The latter directly translates into a change in AOA. To account for the difference in the required AOA for trim before and after the maneuver is accomplished, an $\alpha = \alpha_{0L_2} - \alpha_{0L_1}$ must be commanded. To establish the new trim condition, a new trim sideslip angle $\beta = \beta_0$ must also be commanded.

Similarly, a level and symmetric S maneuver will be initiated in a level and trimmed turn, say, to the right, and the end trim condition will be a level and trimmed turn to the left. In this case, the new trim variables are $Q_0 \leftarrow Q_0$, $-R_0 \leftarrow R_0$, $P_0 = \theta_0 = 0$, $\pi - \phi_0 \leftarrow \phi_0$. Evidently, R_0 changes, hence a new trim sideslip angle must be established. A similar analysis can be conducted for climbing turns, corkscrew descents and nonplanar S maneuvers, where $\theta_0 \neq 0$.

7.1.4 Actuators

The control effectors' actuator dynamics play an important role in flight control. Typical fighter aircraft actuator Transfer Functions (TFs) are given below.

Elevator dynamics:

$$\frac{\delta_e(s)}{\delta_{e_c}(s)} = \frac{2138\,(s^2+11.27s+6872)}{s^4+154.1s^3+1612.8s^2+495589s+14692336} \tag{7.24}$$

The Bode plot of the elevator actuator TF is shown in Figure 7.3. Note the phase lag at high frequency. High frequency operation is brought about by a high loop gain, which, in turn, is specified for robustness purposes. Thus, robust (flight) control mandates the use of high-order actuator models, which, in turn, complicates the control design problem. In addition, the following nonlinearities play an important role in actuator modeling and significantly impact the control system's design and performance. These nonlinearities are of saturation type and entail 1) Maximum elevator deflection of, e.g., ±22° and 2) Maximum elevator deflection rate of ±60°/sec. The latter significantly impacts the aircraft's maneuverability.

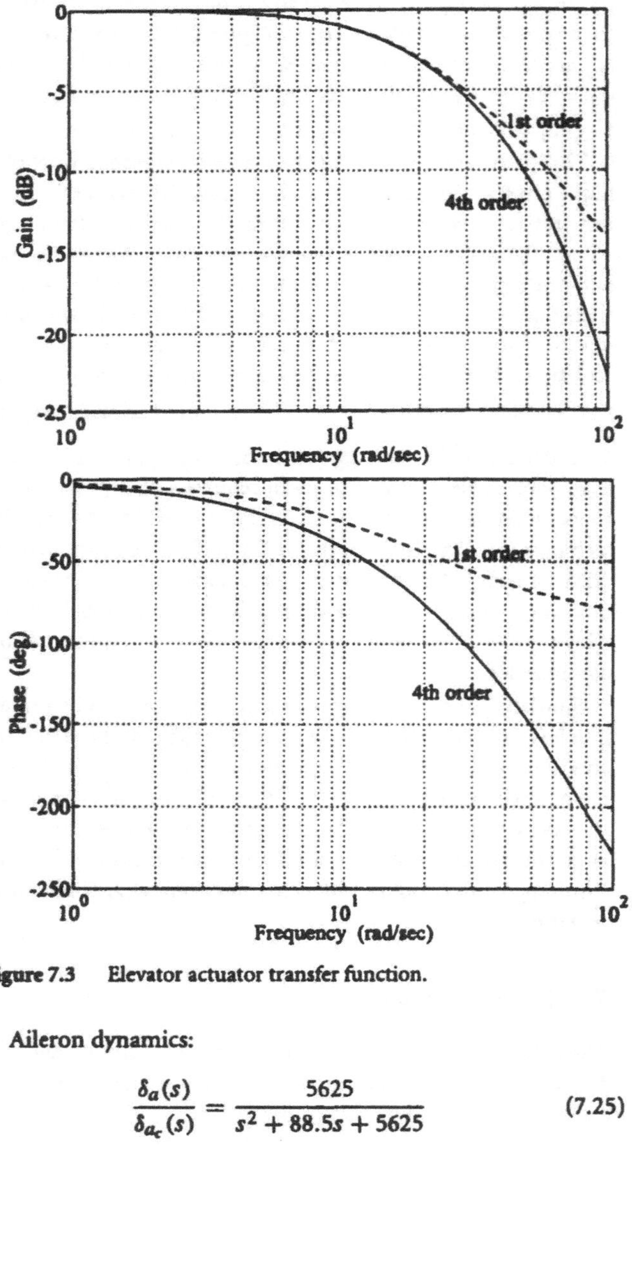

Figure 7.3 Elevator actuator transfer function.

Aileron dynamics:

$$\frac{\delta_a(s)}{\delta_{a_c}(s)} = \frac{5625}{s^2+88.5s+5625} \tag{7.25}$$

The aileron's saturation limits are similar to the elevator's.

Rudder dynamics:

$$\frac{\delta_r(s)}{\delta_{r_c}(s)} = \frac{5148}{s^2+99.36s+5148} \tag{7.26}$$

The maximum rudder deflection is ±30° and its maximum deflection rate is ±60°/sec..

Multiaxes Thrust Vectoring (MATV) actuator dynamics:

The MATV actuator entails a thrust vectoring nozzle. Its deflection in pitch and yaw is governed by the following transfer function:

$$\frac{\delta_{TV}(s)}{\delta_{TV_c}(s)} = \frac{400}{s^2+26s+400} \tag{7.27}$$

Note the reduced bandwidth of the MATV actuator. In addition, the dynamic range of the MATV actuator is ±20° and its maximum deflection rate is limited to ±45°/sec.

Weighting Matrix

The three generalized control inputs δ_e, δ_a, and δ_r are responsible for producing the required control moments about the aircraft's x, y, and z axes. Modern combat aircraft are, however, equipped with redundant control effectors. The latter are of aerodynamic type, and, in addition, Multiaxes Thrust Vectoring (MATV) is employed. Hence, a weighting matrix W for control authority apportionment is used to transform the three generalized command inputs into the six command signals δ_{ar}, δ_{er}, δ_{ep}, δ_{TVp}, δ_{ry}, and δ_{TVy} that control the six physical actuators. In industrial flight control circles the weighting matrix is also referred to as the "Mixer." In our flight control problem, W is a 6×3 matrix.

It is noteworthy that a weighting matrix is required even in the case of aircraft equipped with basic control effectors only, in which case the generalized controls are also the actual physical controls and the weighting matrix is 3×3. The flight control concept of an aileron rudder interconnect will mitigate some of the adverse yaw effects encountered at high AOAs. Thus, a roll command would not only command an aileron deflection, but also generate a rudder command to remove adverse yaw produced by the aileron deflection. The appropriate weighting matrix coefficient is based on the ratio of the applicable control derivatives because the latter are indicative of the yaw moment control power of the respective effector. Hence, the weighting matrix

$$W_1 = \begin{bmatrix} 1 & 0 & 0 \\ 0 & 1 & 0 \\ 0 & \frac{C_{n_{\delta_a}}}{C_{n_{\delta_r}}} & 0 \end{bmatrix} \text{ is oftentimes employed,}$$

$$\text{and } \begin{bmatrix} \delta_e \\ \delta_a \\ \delta_r \end{bmatrix} \leftarrow W_1 \begin{bmatrix} \delta_e \\ \delta_a \\ \delta_r \end{bmatrix}.$$

When the number of physical effectors exceeds the number of generalized controls, the control authority in each axis needs to be adequately apportioned among the physical effectors, as shown in Figure 7.4. Note that the six physical control effectors are partitioned into three distinct groups which correspond to the three control axes of the aircraft. There are two physical effectors in each group. The physical effectors δ_{ep} and δ_{TVp} are in the δ_e group, the physical effectors δ_{ar} and δ_{er} are in the δ_a group, and the physical effectors δ_{ry} and δ_{TVy} are in the δ_r group. Now, the underlying control concept is to have the control effectors in a group saturate simultaneously. Hence, scale factors are used in the weighting matrix to account for dissimilar maximum control deflections within the control effectors in each group. For example, a maximal aileron command should command the roll ailerons to their maximum deflection of 20 degrees, but the roll elevators to their preassigned maximum roll control authority of 5 degrees only. Thus, scaling is achieved by multiplying the generalized control command by the ratio of the respective maximum actuator limits in its group. Hence,

$$W_2 = \begin{bmatrix} 1 & 0 & 0 \\ \frac{\gamma}{\delta_{ep_{max}}} & 0 & 0 \\ 0 & 1 & 0 \\ 0 & \frac{\delta_{er_{max}}}{\delta_{ar_{max}}} & 0 \\ 0 & 0 & 1 \\ 0 & 0 & \frac{\gamma}{\delta_{ry_{max}}} \end{bmatrix}$$

Finally, the weighting matrices W_1 and W_2 are combined to form the weighting matrix $W = W_2 W_1$.

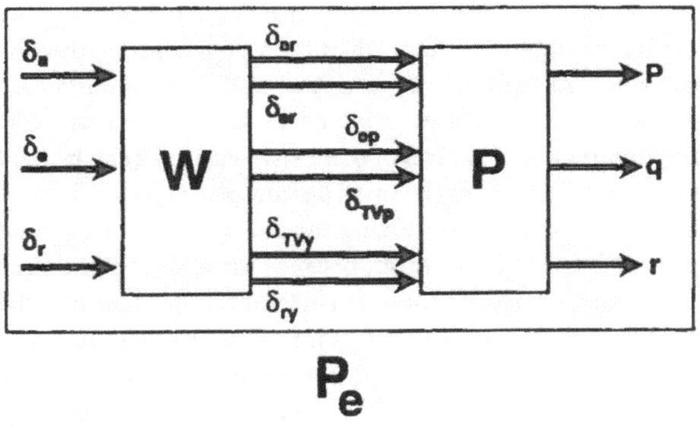

Figure 7.4 Mixer.

A more detailed analysis also considers the relative magnitudes of the control moments and control forces generated by the two effectors in each control group, which requires knowledge of the control derivatives. The control theoretic problem of optimal control authority apportionment among the various control effectors is currently a research topic in flight control.

7.1.5 Flight Control Requirements

The Flight Control System constitutes the pilot/aircraft interface and its primary mission is to enable the pilot/aircraft to accomplish a prespecified task. This entails the following:

1. Accommodate *high* α (velocity vector roll) *maneuvers.*
2. Meet handling/flying qualities specifications over the entire flight envelope and for all aircraft configurations, including tight tracking and accurate pointing for Fire Control.
3. Above and beyond item 1, obtain optimal performance. Pilot "gets what he wants."
4. Minimize number of placards, the number of limitations and restrictions the pilot must adhere to → Pilot can fly his aircraft "in abandon."

Specifications

Mil-Std 1797A [3] defines the flying quality specifications, Level 1 the best. The specifications (airworthiness criteria) are given as a mixture of time and frequency domain specifications. The longitudinal channel uses time-domain specifications based on the pitch rate q response to a step input command calculated from the two-degrees-of-freedom model given by the fast, Short Period (SP), approximation. The time-domain specifications are based on two straight lines drawn on the q response shown in Figure 7.5. To meet Level 1 flying quality specifications, the equivalent time delay (t_1) must be less than 0.12 sec, the transient peak ratio ($\frac{\Delta q_2}{\Delta q_1}$) less than 0.30, and the effective rise time ($\Delta t = t_2 - t_1$) between $9/\bar{U}$ and $500/\bar{U}$, where $\bar{U}$ is the true airspeed (ft/s).

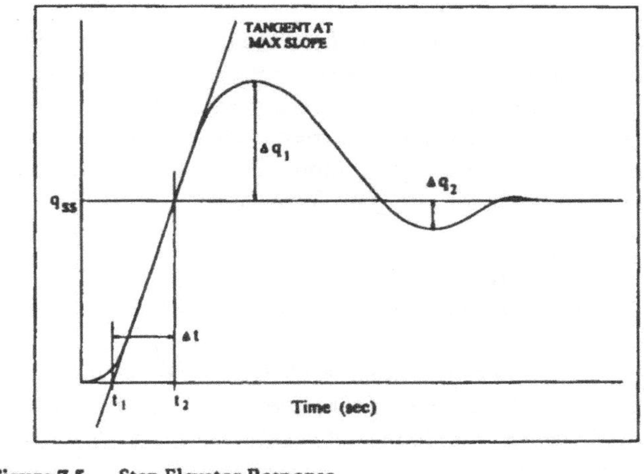

Figure 7.5 Step Elevator Response.

For the lateral/directional channel, the flying quality specifications apply to simultaneously matching the Bode plots of the final system with those of the equivalent fourth-order transfer functions given by

$$\frac{\phi(s)}{\delta_{rstk}(s)} = \frac{K_\phi(s^2 + 2\zeta_\phi\omega_\phi s + \omega_\phi^2)\, e^{-\tau_{ep}s}}{(s + \frac{1}{T_R})(s + \frac{1}{T_s})(s^2 + 2\zeta_d\omega_d s + \omega_d^2)}, \text{ and}$$

$$\frac{\beta(s)}{\delta_{rud}(s)} = \frac{(A_3 s^3 + A_2 s^2 + + A_1 s + A_0)\, e^{-\tau_{e\beta}s}}{(s + \frac{1}{T_R})(s + \frac{1}{T_s})(s^2 + 2\zeta_d\omega_d s + \omega_d^2)}.$$

To meet Level 1 flying qualities, the Roll Mode time constant (T_R) must be less than 1.0 sec, the Dutch Roll Mode damping

(ζ_d) greater than 0.40, the Dutch Roll Mode natural frequency (ω_d) greater than 1 rad/sec, the roll rate time delay (τ_{ep}) less than 0.1 sec, and the time to double of the Spiral Mode ($-ln2\ T_S$) greater than 12 sec. Disturbance rejection specifications in the lateral/directional channel are based on control surface usage and are variable throughout the flight envelope.

In addition, the open-loop phase margin angle γ of both the longitudinal and lateral/directional designs must be greater than 30°, and the open-loop gain margin a must be greater than 6dB. To determine phase and gain margins, the open-loop transmissions from stick inputs to the required outputs in each loop are examined. Finally, the phase margin frequency (cutoff frequency $= \omega_\phi$) should be less than 30 rad/sec to prevent deleterious interaction with the bending modes of the aircraft.

Concerning cross-coupling disturbance bounds: One requirement is that a sustained 10° sideslip will use less than 75% of the available roll axis control power (aileron). There is also a complicated specification for the amount of sideslip β allowed for a particular roll angle. If one simplifies the requirements in a conservative fashion over the majority of the flight envelope, a roll command of 1°/sec should result in less than 0.022° of β, but, at low speeds, β is allowed to increase to 0.067°. The maximal β allowed in any roll maneuver is 6°.

Evidently, a "robust" set of specifications is not available because the flying qualities depend on flight condition throughout the flight envelope. The lack of uniform performance bounds over the entire envelope illustrates a shortcoming of the current robust control paradigms. Moreover, the specifications above must be met in the face of the following.

A. Constraints
Practical limitations on achievable performance (including robustness) are imposed by
(1) actuator rate limits (e.g., ca. 60°/sec), (2) actuator saturation (e.g., ca. 22°), (3) sensor noise/sensor quality, (4) low frequency elastic modes, and (5) highest possible sampling rate for digital implementation.

B. Environment
The FCS needs to accommodate the following:
(1) "plant" variability/uncertainty in the extended flight envelope, and aircraft configurations possibilities. Thus, in the real world, *structured* uncertainty needs to be addressed. Also, accommodation of inflight aerodynamic control surfaces failures/battle damage is required, (2) atmospheric disturbances/ turbulence suppression, (3) unstable aircraft (for reduced trim drag and improved cruise performance); guard against adverse interaction with actuator saturation, for the latter can cause loss of control, i.e., departure, (4) control authority allocation for trim, maneuverability, additional channel control, stabilization. Example: In the current F-16, the apportionment of control authority is 10, 7, 4, and 2 degrees of elevator deflection, respectively, (5) optimal control redundancy management, and (6) pilot-in-the-loop.

7.1.6 Dynamic Analysis

The linearization of the state equations of motion about trimmed flight conditions is the first step performed in classical flight control system design. The simple trim condition shown in Figure 7.1a, which entails symmetric and level, or climbing, flight, and where $P_0 = Q_0 = R_0 = \phi_0 = 0$ and $\beta_0 = \delta_{r_0} = \delta_{a_0} = 0$, is considered.

Linearization is based on the small perturbations hypothesis, which predicates that only small excursions in the state variables about a given trim condition occur [1], [2], [5]. For conventional aircraft with an (x,z) plane of symmetry, and for small roll rates, an additional benefit accrues: Linearization brings about the decoupling of the aircraft's pitch and lateral/directional dynamics, thus significantly simplifying the flight control problem. Unfortunately, the velocity vector roll is a large amplitude maneuver consisting of a high roll rate and large excursions in Euler angles. Thus, the velocity vector roll maneuver could entail a high (stability axis) roll rate P. According to the flying quality specifications in Section 7.1.5, a maximal roll rate is desired, hence P is not necessarily small. In addition, during a velocity vector roll maneuver, large excursions in the bank angle occur. Hence, for maneuvering flight, the standard linearization approach must be modified. All of the system states, other than the roll rate and the Euler angle ϕ, can still be represented as small perturbation quantities (denoted by lower case variables), and the squares and products of these state variables can be neglected as is typically done in the small perturbations based linearization process. All terms containing the roll rate P and bank angle ϕ are not linearized, however, because these variables do not represent small perturbations. Furthermore, the elevator deflection δ_e is now measured with reference to the trim elevator setting of δ_{e_0}. Therefore, α_{0L} does not feature in Equation 7.15, nor does C_{m_0} feature in Equation 7.17. Furthermore, the gravity term in Equation 7.15 is modified to account for the trim condition properly. Moreover, the FCS strives to maintain q and r small, and our choice of stabilty axes yields the velocity vector roll's end conditions $\theta_f = \theta_0$. Hence, it is assumed that, throughout the velocity vector roll, the pitch angle perturbation $\theta \approx 0$. In addition, the stabilty axes choice lets us approximate the AOA as $\alpha = tan^{-1}(\frac{w}{U}) \approx \frac{w}{U}$ and, at this particular trim condition ($\beta_0 = 0$), the sideslip angle $\beta = tan^{-1}(\frac{v}{U}) \approx \frac{v}{U}$. Thus, the "slow" dynamics in Equations 7.8, 7.15 and 7.14 are reduced to

$$\dot{\phi} = P + (q\ sin\phi + r\ cos\phi)tan\theta_0, \tag{7.28}$$

$$\dot{\alpha} = q - P\beta + \frac{g}{U}(cos\theta_0 cos\phi - sin\theta_0\ \theta\ cos\phi - cos\theta_0) + C_{z_\alpha}\alpha + C_{z_q}q + C_{z_{\dot{\alpha}}}\dot{\alpha} + C_{z_{\delta_e}}\delta_e, \tag{7.29}$$

and

$$\begin{aligned}\dot{\beta} &= P\alpha - r + \frac{g}{U}(\cos\theta_0 \sin\phi - \sin\theta_0\, \theta \sin\phi) + C_{y_p} P \\ &\quad + C_{y_r} r + C_{y\beta}\beta + C_{y_{\delta_r}}\delta_r\,, \qquad (7.30)\end{aligned}$$

where θ now denotes the perturbation in the pitch angle. Equations 7.16, 7.17 and 7.18 yield the fast dynamics

$$\begin{aligned}\dot{P} &= C_{l_p} P + C_{l_{pq}} Pq + C_{l_r} r + C_{l_\beta}\beta + C_{l_{\delta_{ar}}}\delta_{ar} \\ &\quad + C_{l_{\delta_{er}}}\delta_{er} + C_{l_{\delta_{ry}}}\delta_{ry} + C_{l_{\delta_{TVy}}}\delta_{TVy}, \qquad (7.31)\\ \dot{q} &= C_{m_{pr}} Pr - C_{m_{p^2}} P^2 + C_{m_\alpha}\alpha + C_{m_q} q \\ &\quad + C_{m_{\dot\alpha}}\dot{\alpha} + C_{m_{\delta_{ep}}}\delta_{ep} + C_{m_{\delta_{TVp}}}\delta_{TVp}, \qquad (7.32)\\ \dot{r} &= C_{n_{pq}} Pq + C_{n_p} P + C_{n_r} r + C_{n_\beta}\beta + C_{n_{\delta_{ar}}}\delta_{ar} \\ &\quad + C_{n_{\delta_{er}}}\delta_{er} + C_{n_{\delta_{ry}}}\delta_{ry} + C_{n_{\delta_{TVy}}}\delta_{TVy}, \qquad (7.33)\end{aligned}$$

and

$$\dot{\theta} = q\cos\phi - r\sin\phi. \qquad (7.34)$$

Hence, the dynamics of maneuvering flight and of the velocity vector roll maneuver entail a seven DOF model which is specified by Equations 7.28 – 7.34. The three generalized control effectors' inputs δ_e, δ_a and δ_r appearing in the six DOF equations represent the six physical control effector inputs, i.e., the pitch elevator deflection δ_{e_p}, the deflection in pitch of the thrust vectoring engine nozzle δ_{TVp}, the aileron deflection δ_{ar}, the roll elevator deflection δ_{er} (we here refer to the differential tails), the rudder deflection δ_{ry}, and the deflection in yaw of the thrust vectoring engine nozzle δ_{TVy}.

When the initial trim condition is straight and level flight, $\theta_0 = 0$. Then a further simplification of the Equations of motion 7.28 – 7.30 is possible:

$$\begin{aligned}\dot{\phi} &= P, \qquad (7.35)\\ \dot{\alpha} &= (1 + C_{z_q})q - P\beta + (\frac{g}{U})(\cos\phi - 1) + C_{z_{\dot\alpha}}\dot{\alpha} + C_{z_\alpha}\alpha \\ &\quad + C_{z_{\delta_{ep}}}\delta_{ep} + C_{z_{\delta_{TVp}}}\delta_{TVp}, \text{ and} \qquad (7.36)\\ \dot{\beta} &= P\alpha + (\frac{g}{U})\sin\phi + C_{y_p} P + (C_{y_r} - 1)r + C_{y_\beta}\beta \\ &\quad + C_{y_{\delta_{ar}}}\delta_{ar} + C_{y_{\delta_{er}}}\delta_{er} + C_{y_{\delta_{ry}}}\delta_{ry} \\ &\quad + C_{y_{\delta TVy}}\delta_{TVy}. \qquad (7.37)\end{aligned}$$

Equations 7.35 – 7.37 are used in conjunction with the "fast" dynamics in Equations 7.31 – 7.34 to complete the description of the aircraft dynamics for an initial straight and level trim.

7.1.7 Conventional Flight Control

It is remarkable that the very same assumption of trimmed *level* flight renders in the pitch channel the classical second-order Short Period (SP) approximation. Thus, $P = \phi = 0$ yields the celebrated linear SP pitch dynamics

$$\begin{aligned}\dot{\alpha} &= C_{z_\alpha}\alpha + (1 + C_{z_q})q + C_{z_{\dot\alpha}}\dot{\alpha} + C_{z_{\delta_e}}\delta_e \qquad (7.38)\\ \dot{q} &= C_{m_\alpha}\alpha + C_{m_q} q + C_{m_{\dot\alpha}}\dot{\alpha} + C_{m_{\delta_e}}\delta_e \qquad (7.39)\end{aligned}$$

In the more general case of trimmed and symmetric *climbing* flight in the vertical plane, the longitudinal SP dynamics are of third order [2], [6]:

$$\begin{aligned}\dot{\alpha} &= C_{z_\alpha}\alpha + (1 + C_{z_q})q - \frac{g}{U}\sin\theta_0\,\theta + C_{z_{\dot\alpha}}\dot{\alpha} + C_{z_{\delta_e}}\delta_e \\ \dot{q} &= C_{m_\alpha}\alpha + C_{m_q} q + C_{m_{\dot\alpha}}\dot{\alpha} + C_{m_{\delta_e}}\delta_e \\ \dot{\theta} &= q\end{aligned}$$

The third-order dynamics depend on the pitch angle parameter θ_0. At elevated pitch angles the SP/Phugoid time-scale separation's validity is questionable.

Furthermore, applying the small perturbations hypothesis to the additional variables P and ϕ in Equations 7.28, 7.30, 7.31, and 7.33, allows us to neglect the terms which contain perturbation products. This decouples the lateral/directional channel from the pitch channel. Thus, the conventional lateral/directional "plant" is

$$\begin{aligned}\dot{p} &= C_{l_p} p + C_{l_r} r + C_{l_\beta}\beta + C_{l_{\delta_a}}\delta_a + C_{l_{\delta_r}}\delta_r \qquad (7.40)\\ \dot{r} &= C_{n_p} p + C_{n_r} r + C_{n_\beta}\beta + C_{n_{\delta_a}}\delta_a \\ &\quad + C_{n_{\delta_r}}\delta_r \qquad (7.41)\\ \dot{\beta} &= C_{y_p} p + (C_{y_r} - 1)r + C_{y_\beta}\beta + \frac{g}{U}\phi \\ &\quad + C_{y_{\delta_r}}\delta_r \qquad (7.42)\\ \dot{\phi} &= p \qquad (7.43)\end{aligned}$$

In conclusion, the horizontal flight assumption $\theta_0 = 0$ reduces the complexity of both the pitch channel and the velocity vector roll control problems.

The stability and control derivatives in Equations 7.38, 7.39, and Equations 7.40 – 7.43 depend on flight condition. Therefore, in most aircraft implementations, gain scheduling is used in the controller. Hence, low order controllers are used. A simple first-order controller whose gain and time constant are scheduled on the dynamic pressure $\bar{q}$ will do the job.

In modern approaches to flight control, robust controllers that do not need gain scheduling are sought. A full envelope controller for the F-16 VISTA that meets the flying quality specifications of Section 7.1.5 has been synthesized using the QFT robust control design method [7]. The pitch channel's inner and outer loop compensators, G_1 and G_2, and the prefilter F, are

$$\begin{aligned}G_1(s) &= \frac{3.1(s+13)(s+17)}{s(s+100)}, \\ G_2(s) &= \frac{30s^4 + 1356.6s^3 + 57136s^2 + 177000s + 280000}{s^4 + 109.8s^3 + 1029s^2 + 4900s}, \\ F(s) &= \frac{16}{s^2 + 6s + 16}.\end{aligned}$$

7.1.8 Time-Scale Separation

Time-scale separation plays an important role in flight control. The fast states P, q, and r which represent the angular rates are dominant during maneuvering flight where the velocity vector roll is initiated and arrested. The perturbations α in AOA and in the sideslip angle β are maintained at near zero in the short time periods of the transition regimes. Therefore, the six DOF dynamical model obtained when $\theta_0 = 0$ can be decoupled into "slow" dynamics associated with the α, β and ϕ variables and the "fast" dynamics of the P, q, and r angular rates. Two primary regimes of the roll maneuver are identified, viz., the transition and free stream regions shown in Figure 7.6. The dynamic transition regions represent the initiation and arrest of the velocity vector roll, which occur in a relatively short amount of time (on the order of one second). In these transition regions the fast states of the aircraft, viz., the angular rates, are dominant. During these initial and/or terminal phases of the velocity vector roll maneuver a desired roll rate needs to be established, viz., the controller's modus operandi entails tracking. In the free stream regime, the established roll rate is maintained through the desired bank angle, while the perturbations in AOA and sideslip angle are regulated to zero. Thus, in the "free stream" regime, the controller's function entails regulation.

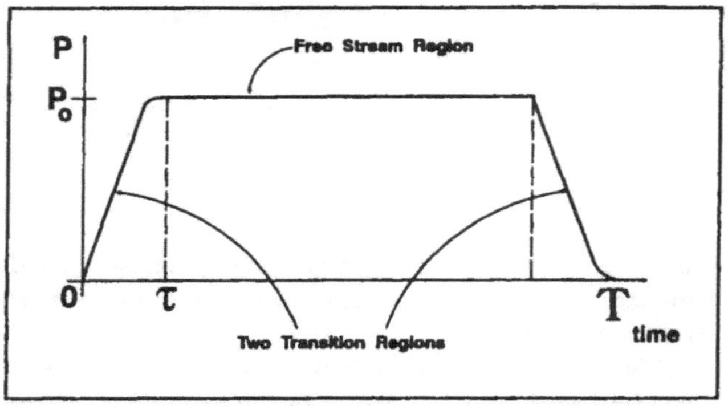

Figure 7.6 Time-scale separation.

The design of the FCS for maneuvering flight and for large amplitude velocity vector roll maneuvers at high AOAs hinges on this discernible time-scale separation, and is now outlined. The time-scale separation leads to the two-loop control structure shown in Figure 7.7. The inner loop of the FCS consists of a three-axis rate-commanded control system which controls the fast states of the system. This control function is essential in the transition regions of the velocity vector roll maneuver. The bandwidth of the inner loop is relatively high. The outer loop serves to regulate to zero the relatively slower aerodynamic angles, viz., the AOA and sideslip angle perturbations which accrue during the velocity vector roll, in particular, during the transition regions. The outer loop is therefore applicable to both the transition and free stream regimes of the velocity vector roll.

Time-Domain Approach

Maneuvering flight control entails tracking control, as opposed to operation on autopilot, where regulation is required. Tracking control in particular in the presence of actuator saturation requires a time-domain approach. Therefore, the objective is the derivation of a receding horizon/model predictive tracking control law, in the case where both fast and slow state variables feature in the plant model. An optimal control sequence for the whole planning horizon is obtained and its first element is applied to the plant, following which the control sequence is reevaluated over a one time step displaced horizon. Thus, feedback action is obtained. A procedure for synthesizing an optimal control sequence for solving a fixed horizon optimal control problem, which exploits the time-scale separation inherent in the dynamical model, is employed. The time-scale separation decomposes the receding horizon control problem into two nested, but lower dimensional, receding horizon control optimization problems. This yields an efficient control signal generation algorithm. The outer loop optimal control problem's horizon is longer than the inner loop optimal control problem's horizon, and the discretization time step in the inner loop is smaller than in the outer loop. Thus, to apply model predictive tracking control to dynamical systems with time-scale separation, the classical nested feedback control loop structure of the frequency domain is transcribed into the time domain.

The plant model for the inner loop's "fast" control system entails reduced order dynamics and hence the "fast" state x_1 consists of the angular rates P, Q, and R. In the inner loop the control signal is $[\delta_e, \delta_a, \delta_r]^T$. The reference variables are the desired angular rates, namely P_c, Q_c, and R_c. The latter are supplied by the slower outer loop. The on-line optimization performed in the inner loop yields the optimal control signals δ_e^*, δ_a^*, and δ_r^*.

The slow variables are passed into the inner loop from the outer loop. Thus, in the inner loop the slow variables $x_2 = (\alpha, \beta, \theta, \phi)$ are fixed, and if linearization is employed, their perturbations are set to 0. In the slower, outer loop, only the "slow" states α, β, θ, and ϕ are employed. In the outer loop the control variables are the "fast" angular rates P, Q, and R. Their optimal values are determined over the outer loop optimization horizon, and they are subsequently used as the reference signal in the fast inner loop. The outer loop's reference variables are commanded by the pilot. For example, at low $\bar{q}$ the pilot's commands are Q_c, P_c, β_c.

7.1.9 Actuator Saturation Mitigation in Tracking Control

The classical *regulator* paradigm applies to outer loop flight control, where autopilot design problems are addressed. More challenging maneuvering, or inner loop, flight control entails the on-line solution of *tracking* control problems. The design of an advanced pitch channel FCS which accommodates actuator saturation is presented.

Linear control law synthesis methods, e.g., LQR, H_∞, and QFT, tend to ignore "hard" control constraints, viz., actuator saturation. Unfortunately, good tracking requires the use of high gain in the linear design, and consequently actuator saturation effects are pronounced in tracking control. Hence, in this section, the basic problems of tracking control and mitigating actuator saturation effects are synergetically addressed. The emphasis is on real-time synthesis of the control law, so that implementation in a feedback flight control system is possible. Thus, a hybrid optimization approach is used for mitigating actuator saturation effects in a high gain situation. Specifically, model following control is considered. The Linear Quadratic Regulator (LQR) and

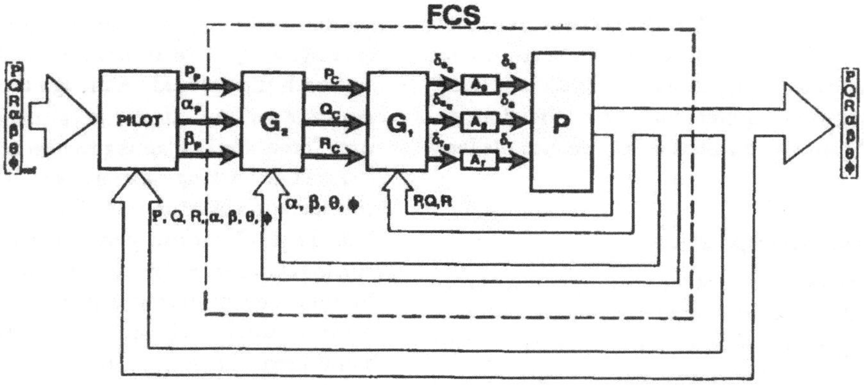

Figure 7.7 Flight control system.

Linear Programming (LP) optimization paradigms are combined [8] so that a tracking control law that does not violate the actuator constraints is synthesized in real time. Furthermore, the multiscale/multigrid approach presented in Section 7.1.8 is applied. Attention is confined to the pitch channel only and the conventionally linearized plant Equations 7.38 and 7.39 is used.

Saturation Problem

Actuator saturation reduces the benefits of feedback and degrades the tracking performance. Actuator saturation mitigation is of paramount importance in the case where feedback control is employed to stabilize open-loop unstable aircraft, e.g., the F-16: Prolonged saturation of the input signal to the plant is tantamount to opening the feedback control loop. Hence, in the event where the controller *continuously* outputs infeasible control signals which will saturate the plant[1], instability and departure will follow. At the same time, conservative approaches, which rely on small signal levels to avoid saturation, invariably yield inferior (sluggish) control performance. Hence, the following realization is crucial: The cause of saturation—precipitated instability is the continuous output by the linear controller of *infeasible* control signals. Conversely, if the controller sends only *feasible* control signals to the actuator, including controls on the boundary of the control constraint set, instability won't occur. In this case, in the event where the fed back measurement tells the controller that a reduction in control effort is called for, the latter can be instantaneously accomplished. Thus, a "windup" situation, typical in linear controllers, does not occur and no 'anti-windup" measures are required. Therefore, the feedback loop has not been opened, and the onset of instability is precluded. Obviously, a *nonlinear* control strategy is needed to generate *feasible* control signals. To command feasible control signals u, one must compromise on the tracking performance which is achievable with unconstrained controls.

[1] "Plant" here means the original plant and the actuator.

Tracking Control

The model following a state feedback control system is illustrated schematically in Figure 7.8. The pitch channel, modeled by Equations 7.38 and 7.39, is considered. The controlled variable is pitch rate q. The model M outputs the reference signal r, which is the pilot's desired pitch rate q_C. P represents the aircraft's dynamics, A the actuator dynamics, and G the controller. K_i and K_o are adaptive gains in the feedback and feedforward loops, respectively.

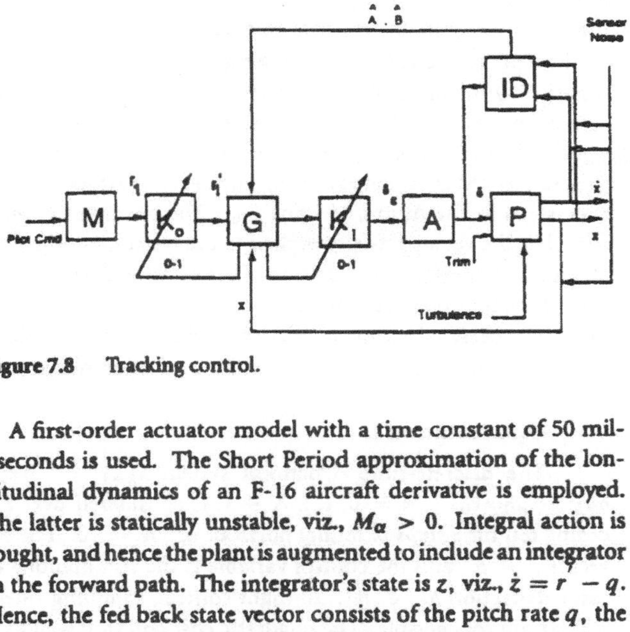

Figure 7.8 Tracking control.

A first-order actuator model with a time constant of 50 milliseconds is used. The Short Period approximation of the longitudinal dynamics of an F-16 aircraft derivative is employed. The latter is statically unstable, viz., $M_\alpha > 0$. Integral action is sought, and hence the plant is augmented to include an integrator in the forward path. The integrator's state is z, viz., $\dot{z} = r - q$. Hence, the fed back state vector consists of the pitch rate q, the AOA α, the elevator deflection δ_e, and the integrator state z.

The command signal u= δ_{e_c} at the actuator's input consists of two parts, a fed back signal which is a linear function of the state and a signal which is a linear function of the pilot's input to the model and hence of the reference signal r. Therefore, actuator saturation will be precipitated by either a high amplitude excursion of the state, caused by a disturbance applied to the plant, or by a high amplitude reference signal when the flight control system is being driven hard by the pilot. Now, it is most desirable to track the reference model output driven directly by the pilot. However, because saturation requires the generation by the controller of feasible control signals, tracking performance must be relaxed. Hence, saturation mitigation requires that $r_1' \neq r_1$, therefore, reduce the feedforward gain K_o, or, alternatively, reduce the loop gain K_i.

Reference Signal Extrapolation

The optimal control solution of the tracking control problem (as opposed to the regulator problem) requires advance knowledge of the exogenous reference signal. Therefore, the reference signal r must be extrapolated over the optimization time horizon. However, long range extrapolation is dangerous. Hence, receding horizon optimal control provides a natural framework for tracking control.

Even though past reference signals are employed for extrapolation, some of which (r') are divorced from the pilot's reference signal input (r), a good way must be found to incorporate the pilot's extrapolated (future) reference signal demand into the extrapolated reference signal being supplied to the LQR algorithm. Hence, the extrapolation of the pilot's signal and its incorporation into the reference signal provided to the LQR algorithm provides good lead and helps enhance tracking performance. At the same time, the use of the past reference signal output by the controller (and not necessarily by the pilot-driven model) will not require abrupt changes in the current elevator position and this will ameliorate the elevator rate saturation problem.

Since the receding horizon tracking controller requires an estimate of q for the future, the latter is generated as a parabolic interpolation of r_0', the yet to be determined adjusted reference signal r_1', and $\hat{r}_{10}$; $\hat{r}_{10}$ is generated by parabolically interpolating the pilot issued commands r_{-1}, r_0, and r_1, and extrapolating ten time steps in the future, yielding $\hat{r}_{10}$. The ten point r' vector (which is the vector actually applied to the system, not the r vector) is parameterized as a function of the yet to be determined r_1'. Therefore, the objective of the controller's LP algorithm is to find the reference signal at time now, r_1', that will not cause a violation of the actuator limits over the ten-point prediction horizon.

Optimization

The discrete-time dynamical system $x_{k+1} = Ax_k + bu_k$, $x_0 \equiv x_0$, $k = 0, 1, ...N-1$, is considered over a receding ten time steps planning horizon, i.e., $N = 10$. The state vector $x \in R^4$ and the control variable is the commanded elevator deflection $\delta_{e_c} \in R^1$. In flight control, sampling rates of 100 Hz are conceivable. The quadratic cost criterion weighs the tracking error $r' - q$, the control effort (commanded elevator deflection), and control rate effort, over the ten time steps planning horizon which is 0.1 sec.

At the time instant $k = 0$, the future control sequence $u_0, u_1, ...u_{N-1}$ is planned based on the initial state information x_0 and the reference signal sequence $r_1, r_2, ...r_N$ given ahead of time. Now, the actual reference signal received at time $k = 0$ is r_1; the sequence $r_2, ...r_N$ is generated by polynomially (quadratically) extrapolating the reference signal forward in time, based on the knowledge of past reference signals. The extrapolated reference signals $r_2, ...r_N$ are linear in r_1. An open-loop optimization approach is employed. Although the complete optimal control sequence u_0, $u_1, ...u_{N-1}$ is calculated, only u_0 is exercised at time "now," thus obtaining feedback action.

An LQR tracking control law is first synthesized without due consideration of saturation. The objective here is to obtain good "small signal" tracking performance. If saturation does not occur, the LQR control law will be exercised. Saturation is addressed by using the explicit formula that relates the applied reference signal sequence $r_1', ...r_N'$ and the LQR algorithm-generated optimal control sequence $u_0, \ldots, u_N$. Thus, the dependence of the optimal control sequence on the initial state and on the reference signal r_1' is transparent:

$$[\, u_0 \quad u_1 \quad . \quad . \quad . \quad u_{N-1} \,]^T = a(p)r_1' + b(p)x_0 + c(p) \tag{7.44}$$

where the vectors a, b and c $\in R^N$. The vector $p \in R^3$ represents the parameters (weights) of the LQR algorithm, viz., $p^T = [\frac{R}{Q_P}, \frac{R_R}{Q_P}, \frac{Q_L}{Q_P}]$. We are mainly concerned with actuator rate saturation and hence the important parameter is $\frac{R_R}{Q_P}$.

Next, the LP problem is formulated to enforce the nonsaturation of the command signal to the actuator and of the command rate signal. The following $2N$ (= 20) inequalities must hold:

$$\begin{aligned}
-def_{\max} \le |\; & u_0 & | \le def_{\max}, \\
-defr_{\max} \le |\; & \tfrac{u_0 - u_{-1}}{\Delta T} & | \le defr_{\max}, \\
-def_{\max} \le |\; & u_1 & | \le def_{\max}, \\
-defr_{\max} \le |\; & \tfrac{u_1 - u_0}{\Delta T} & | \le defr_{\max}, \\
& \text{and} & \\
& \vdots & \\
-def_{\max} \le |\; & u_{N-1} & | \le def_{\max}, \\
-defr_{\max} \le |\; & \tfrac{u_{N-1} - u_{N-2}}{\Delta T} & | \le defr_{\max}.
\end{aligned}$$

The control u_{-1} is available from the previous window. The def_{max} (22°) and $defr_{max}$ (60°/sec) bounds are the maximal elevator deflection and elevator deflection rate, respectively. The sampling rate is 100 Hz and hence ΔT=0.01 sec.

The 2N inequalities above yield a set of 2N constraints in the currently applied reference signal r_1'. Thus,

$$g_i(\frac{R_R}{Q_P}) \le r_1' \le f_i(\frac{R_R}{Q_P}), i = 1, 2, ...2N. \tag{7.45}$$

Next, $f(\frac{R_R}{Q_P}) = min_i[f_i]$, $g(\frac{R_R}{Q_P}) = max_i[g_i]$ is defined, and the case where $g \le f$ is considered. The set of 2N inequalities is consistent and a feasible solution r_1' exists. Then, given the pilot-injected reference signal r_1, a feasible solution is obtained, which satisfies the 2N inequalities above, and which is as close as possible to r_1. In other words,

$$r_1' = \begin{cases} r_1, & if\; g < r_1 < f \\ f, & if\; r_1 \ge f \\ g, & if\; r_1 \le g \end{cases} \tag{7.46}$$

Finally, at time k=0, set $r_1 := r_1'$. Thus, the action of the LP algorithm is equivalent to lowering the feedforward gain K_o. In the case where the set of inequalities is inconsistent (f<g) and a solution does not exist, the loop gain K_i needs to be reduced. This is accomplished by increasing (say doubling) the LQR algorithm's R_R penalty parameter, and repeating the optimization process.

Implementation

The implementation of the advanced tracking control law developed in the preceding sections is discussed. The aircraft model represents the conventionally linearized second-order pitch plane dynamics (short period) of an F-16 aircraft derivative at the $M = 0.7$ and $h = 10,000$ ft flight condition. The bare aircraft model represented by Equations 7.38 and 7.39 is given in state-space form in Equation 7.47, where Z_α, Z_q, Z_δ, M_α, M_q, M_δ are the partial derivatives of normal force (Z) and pitching moment (M) with respect to AOA, pitch rate and stabilator deflection.

$$\begin{bmatrix} \dot{\alpha} \\ \dot{q} \end{bmatrix} = \begin{bmatrix} Z_\alpha & Z_q \\ M_\alpha & M_q \end{bmatrix} \begin{bmatrix} \alpha \\ q \end{bmatrix} + \begin{bmatrix} Z_\delta \\ M_\delta \end{bmatrix} \delta \tag{7.47}$$

The stability and control derivatives are defined in Table 7.1. The dynamics entail the algebraic manipulation of the equations of motion to eliminate the $\dot{\alpha}$ derivatives.

TABLE 7.1 Nominal Model.

	δ	α	q
Z	−0.1770	−1.1500	0.9937
M	−19.5000	3.7240	−1.2600

Failure or damage to an aerodynamic control surface results in a loss of control authority, which, in turn, exacerbates the actuator saturation problem. A failure is modeled as a 50% loss of horizontal tail area which occurs 5 sec into the "flight." The failure affects *all* of the short period stability and control derivatives. The short period model's parameters, after failure, are shown in Table 7.2. The reference model has the same structure as the plant model. A first-order actuator with bandwidth of 20 rad/sec is included. The parameters of M are shown in Table 7.3. Only the pitch rate output of the model is tracked. The actuator output is rate limited to ±1 rad/sec and deflection limited to ±0.37 rad.

TABLE 7.2 Failure Model.

	δ	α	q
Z	−0.0885	−1.11	0.9968
M	−9.7500	5.58	−0.8400

The block diagram for the complete system is shown in Figure 7.8. The system consists of a reference model, a receding horizon LQ controller which also includes a LP algorithm, and the plant model. Integral action is included in the flight control system and is mechanized by augmenting the dynamics, viz., $z_{k+1} = z_k + (q - r')$. Thus, although the bare aircraft model has two states, with the actuator and integrator included, there are four states: α, q, δ, and z.

TABLE 7.3 Reference Model.

	δ	α	q
Z	−0.1770	−1.2693	0.9531
M	−19.5000	−9.4176	−5.7307

A parameter estimation module is also included, rendering the flight control system adaptive and reconfigurable.

The simulation is performed with both the receding horizon LQ/LP controller and the System Identification module in the loop. The pilot commands are 1.0 sec duration pitch rate pulses of 0.2 rad/sec magnitude, with polarities of +, −, +, at times 0.0, 3.0, and 6 seconds, respectively. At 5.0 seconds, a loss of one horizontal stabilator is simulated. The failure causes a change in the trim condition of the aircraft, which biases the pitch acceleration ($\dot{q}$) by $-0.21\ rad/sec^2$.

Figure 7.9 is a comparison plot of the performance of two controllers designed to prevent actuator saturation. REF is the desired pitch rate produced by a model of the desired pilot input to pitch rate response. The GAIN curve represents an inner loop gain limiter (0-1) denoted as K_i in Figure 7.10. The K_i value is computed from driving a rate and position limited actuator model and a linear actuator model with the command vector from the receding horizon controller. The ratio of the limited and unlimited value at each time step is computed; the smallest ratio is K_i (see Figure 7.10). The PROG curve is from an LP augmented solution that determined the largest magnitude of the input that can be applied without saturating the actuator. This is equivalent to outer-loop gain K_o attenuation. As can be seen, both PROG and GAIN track very well. At 5.0 seconds the failure occurs with a slightly larger perturbation for PROG than for GAIN. PROG has a slightly greater overshoot after the failure, primarily due to the reduced equivalent actuator rate.

7.1.10 Nonlinear Inner Loop Design

During high AOA maneuvers the pitch and lateral/directional channels are coupled and nonlinear effects are important. *Robust* control can be brought to bear on *nonlinear* control problems. In this Section is shown how QFT and the concept of structured plant parameter uncertainty accommodate the nonlinearities in the "fast" inner loop. The plant is specified by Equations 7.31 to 7.33 with $\alpha = \dot{\alpha} = \beta = 0$. Based on available data, a maximum bound of approximately 30°/sec can be placed on the magnitude of P for a 30 degree AOA velocity vector roll. Structured uncertainty is modeled with a set of LTI plants which describe the uncertain plant over the range of uncertainty. Therefore, intro-

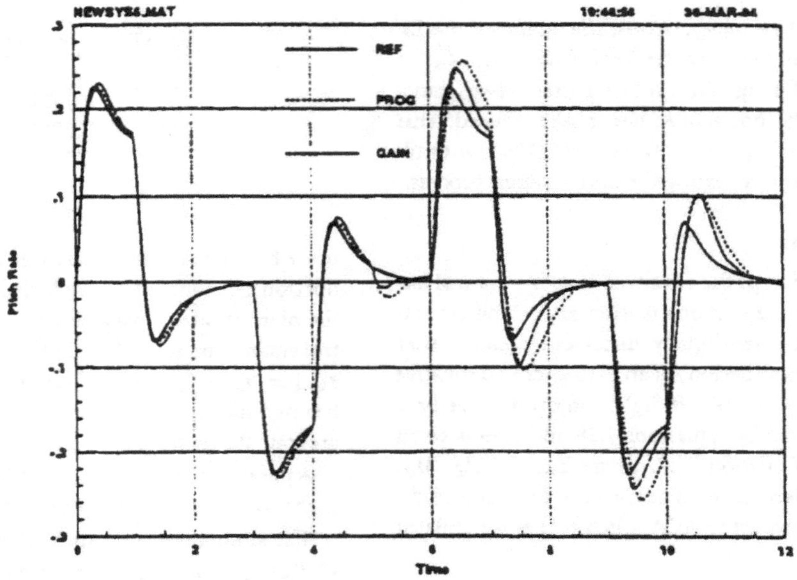

Figure 7.9 Tracking performance.

ducing a roll parameter P_p which varies over the range of 0 to 30°/sec in place of the state variable P in the nonlinear terms, allows the roll rate to be treated as structured uncertainty in the plant. Note that the roll rate parameter replaces one of the state variables in the P^2 term, viz., $P^2 \approx P_p P$. With the introduction of the roll rate parameter, a set of LTI plants with three states which model the nonlinear three DOFs plant are obtained. The generalized control vector is also three dimensional. The state and input vectors are

$$\begin{aligned} x &= [\, P \quad q \quad r \,]^T , \\ u &= [\, \delta_{ar} \quad \delta_{er} \quad \delta_{ep} \quad \delta_{TVp} \quad \delta_{TVy} \quad \delta_{ry} \,]^T , \end{aligned}$$

where $\delta_{ar}, \delta_{er}, \delta_{ep}, \delta_{TVp}, \delta_{TVy}$, and δ_{ry} are the differential (roll) ailerons, differential (pitch) elevators, collective pitch elevators, pitch thrust vectoring, yaw thrust vectoring, and rudder deflections, respectively.

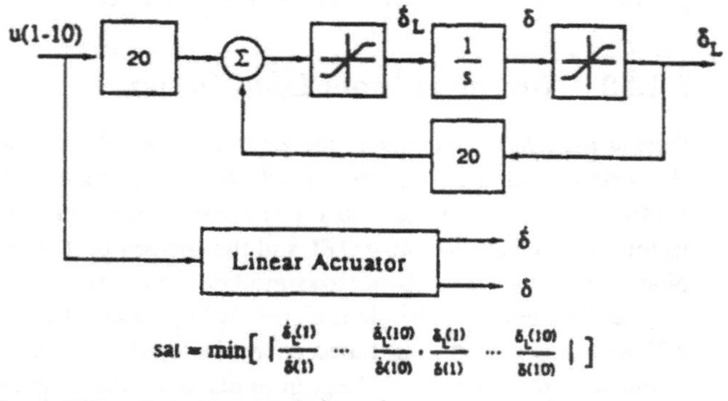

Figure 7.10 Actuator antiwindup scheme.

The linear, but parameter-dependent plant dynamics, and input matrices, are

$$A = \begin{bmatrix} C_{l_p} & C_{l_{pq}} P_p & C_{l_r} \\ -C_{m_{p^2}} P_p & C_{m_q} & C_{m_{pr}} P_p \\ C_{n_p} & C_{n_{pq}} P_p & C_{n_r} \end{bmatrix},$$

$$B = \begin{bmatrix} C_{l_{\delta_{ar}}} & C_{l_{\delta_{er}}} & 0 & 0 & C_{l_{\delta_{TVy}}} & C_{l_{\delta_{ry}}} \\ 0 & 0 & C_{m_{\delta_{ep}}} & C_{m_{\delta_{TVp}}} & 0 & 0 \\ C_{n_{\delta_{ar}}} & C_{n_{\delta_{er}}} & 0 & 0 & C_{n_{\delta_{TVy}}} & C_{n_{\delta_{ry}}} \end{bmatrix}$$

The low $\bar{q}$ corner of the flight envelope is considered and four values of the roll rate parameter ($P_p = 0,\ 8,\ 16,\ 24$) deg/sec are used to represent the variation in roll rate during the roll onset or roll arrest phases [9], [2]. Thus, structured uncertainty is used to account for the nonlinearities introduced by the velocity vector roll maneuver. Finally, the QFT linear robust control design method is used to accommodate the structured uncertainty and design for the flying quality specifications of Section 7.1.5, thus obtaining a linear controller for a nonlinear plant. The weighting matrix

$$W = \begin{bmatrix} 0.2433 & 0 & 0 \\ 0.1396 & 0 & 0 \\ 0 & 0.0838 & 0 \\ 0 & 0.0698 & 0 \\ -0.03913 & 0 & 0.1745 \\ -0.11813 & 0 & 0.5236 \end{bmatrix}$$

is used, and the respective roll, pitch and yaw channels controllers in the G_1 block, in pole-zero format, are

$$\begin{aligned} G_{1_{11}} &= \frac{205.128(-13)(-24 \pm j18)}{(0)(-50)(-60)} , \\ G_{1_{22}} &= \frac{1225(-5.5)(-17.1 \pm j22.8)(-32)}{(0)(-57 \pm j18.735)(-65)} , \\ G_{1_{33}} &= \frac{500(-6)(-14 \pm j14.2829)}{(0)(-50)(-50)} \end{aligned}$$

7.1.11 Flight Control of Piloted Aircraft

The pilot is closing an additional outer loop about the flight control system, as illustrated in Figure 7.6. This brings into the picture the human operator's weakness, namely, the pilot introduces a transport delay into the augmented flight control system. This problem is anticipated and is factored into the FCS design using the following methods.

Neal–Smith Criterion

This is a longitudinal FCS design criterion which includes control system dynamics used in preliminary design. The target tracking mission is considered and the aircraft's handling qualities are predicted.

The expected pilot workload is quantified in terms of the required generation of lead or lag by the pilot. These numbers have been correlated with qualitative Pilot Ratings (PR) of the aircraft, according to the Cooper–Harper chart [10], [3], and [4]. Hence, this design method uses a synthetic pilot model in the outer control loop. The augmented "plant" model includes the airframe and FCS dynamics. Thus, the plant model represents the transfer function from stick force F_s to pitch rate q. The augmented longitudinal FCS with the pilot in the loop is illustrated in Figure 7.11.

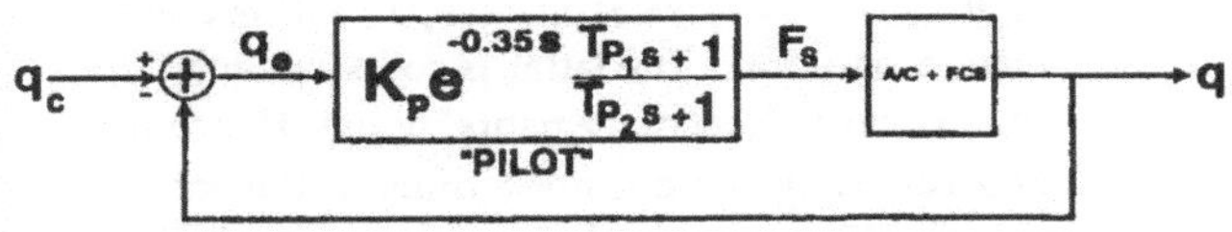

Figure 7.11 "Paper" pilot.

The "pilot" is required to solve the following parameter optimization problem: Choose a high gain K_P so that the closed-loop system's bandwidth is $\geq$ 3.5 rad/sec, and at the same time try to set the lead and lag parameters, τ_{p_1}, and τ_{p_2}, so that the peak closed-loop response $|\frac{Q(j\omega)}{Q_c(j\omega)}|_{max}$ is minimized and the droop at low frequencies is reduced. The closed-loop bandwidth is specified by the frequency where the closed-loop system's phase angle becomes $-90°$. Hence, the pilot chooses the gain and lead and lag parameters K_P, τ_{p_1}, and τ_{p_2} so that the closed-loop system's frequency response is as illustrated in Figure 7.12. Next, the angle

$$\angle_{PC} \equiv \angle \frac{\tau_{p_1}s + 1}{\tau_{p_2}s + 1}|_{s=j\omega,\omega=-BW}$$

is read off the chart. This angle is a measure of pilot compensation required.

If $\angle_{PC} > 0$, the model predicts that the pilot will need to apply lead compensation to obtain adequate handling qualities. If however $\angle_{PC} < 0$, the model predicts that lag compensation must be applied by the pilot. Thus, the aircraft's handling qualities in pitch are predicted from the chart shown in Figure 7.13, where PR corresponds to the predicted Cooper–Harper pilot rating.

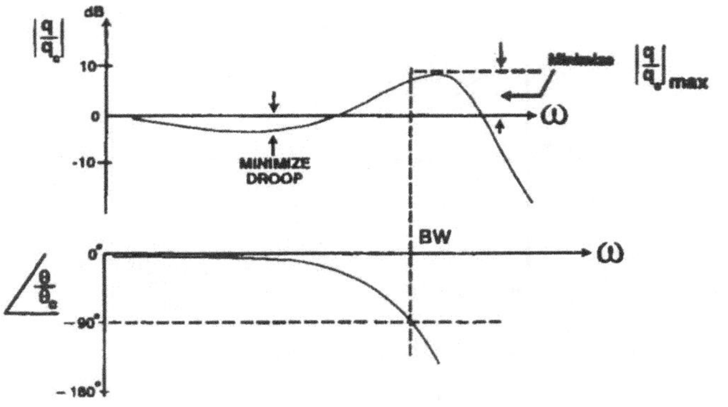

Figure 7.12 Closed-Loop frequency response.

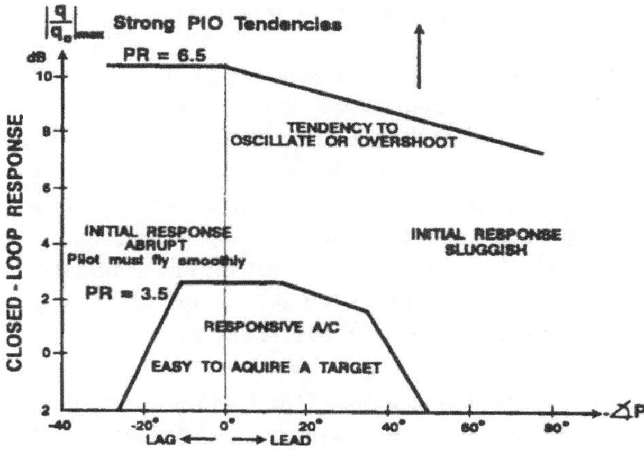

Figure 7.13 Pilot rating prediction chart.

Pilot-Induced Oscillations

Pilot-Induced Oscillations (PIOs) are a major concern in FCS design. PIOs are brought about by the following factors: (1) the RHP zero of the pilot model, (2) an airframe with high ω_{sp} (as is the case at low altitude and high $\bar{q}$ flight) and low damping ζ_{sp}, (3) excessive lag in the FCS (e.g., space shuttle in landing flare flight condition), (4) high-order compensators in digital FCSs (an effective transport delay effect is created), (5) FCS nonlinearities (interaction of nonlinearity and transport delay will cause oscillations), (6) stick force gradient too low (e.g., F-4), and/or "leading stick" (e.g., F-4, T-38 aircraft), and (7) high stress mission, where the pilot applies "high gain."

The Ralph–Smith criterion [11] has been developed for use during the FCS design cycle, to predict the possible onset of PIOs. Specifically, it is recognized that the pilot's transport delay "eats up" phase. Hence, the PIO tendency is predicted from the open-loop $\frac{n_z}{F_s}$ transfer function, and a sufficiently large Phase Margin (PM) is required for PIO prevention.

The Ralph–Smith criterion requires that the resonant frequency ω_R be determined (see Figure 7.14). The following condition must be satisfied: $PM \geq 14.3\ \omega_R$, in which case the aircraft/FCS is not PIO prone.

Pilot Role

It is of utmost importance to realize that (fighter) pilots play a crucial role when putting their aircraft through aggres-

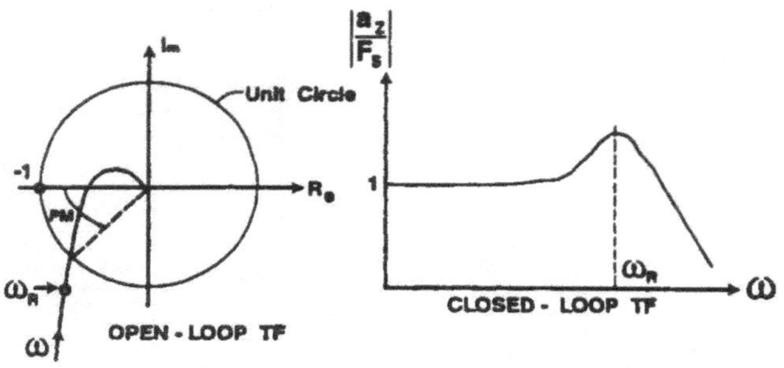

Figure 7.14 Ralph-Smith PIO criterion.

sive maneuvers. When the pilot/aircraft man/machine system is investigated, the following are most pertinent:

1. In Handling Qualities work, the problem of which variable the pilot actually controls is addressed. This is dictated by the prevailing flight condition. In the pitch channel and at low dynamic pressures $\bar{q}$, the controlled variables are α or Q; at high dynamic pressures, the controlled variables are $\dot{\alpha}$, n_z, or $C^* \equiv n_z + \bar{U}q$. In the lateral channel, the pilot commands roll rate P_C. In the directional channel and at low $\bar{q}$, the pilot controls the sideslip angle β, and, at high $\bar{q}$, the pilot controls $\dot{\beta}$ or n_y.
2. The classical analyses of "pilot in the loop" effects are usually confined to single channel flight control. The discussion centers on the deleterious effects of the $e^{-\tau s}$ transport delay introduced into the flight control system by the pilot. The transport delay-caused phase lag can destabilize the augmented FCS system, which includes the pilot in its outer loop. Hence, a linear stability analysis of the augmented flight control system is performed, as outlined in Sections 7.1.11 and 7.1.11. Thus, a very narrow and a very specialized approach is pursued. Unfortunately, this somewhat superficial investigation of control with a "pilot in the loop" only amplifies the obvious and unavoidable drawbacks of manual control of flight vehicles. It is, however, our firm belief that the above rash conclusion is seriously flawed. In simple, "one-dimensional," and highly structured scenarios it might indeed be hard to justify the insertion of a human operator into the control loop, for it is precisely in such environments that automatic machines outperform the pilot. However, in unstructured environments, (automatic) machines have a hard time beating the control prowess of humans. Human operators excel at high level tasks, where a degree of perception is required. This important facet of "pilot in the loop" operation is further amplified in items 3 and 4 in the sequel.
3. A subtle aspect of the pilot's work during maneuvering flight entails the following. The pilot compensates for deficiencies in the existing control law. These are caused by discrepancies between the simplified plant model used in control law design and the actual nonlinear dynamics of the aircraft. Obviously, some of this "modeling error" is being accommodated by the "benefits of feedback" and the balance, we believe, is being relegated to "pilot workload." However, the FCS designer's job is to strive to reduce the pilot's workload as much as possible.
4. The analysis of different trim conditions in Section 7.1.3 indicates that, in order to perform operationally meaningful maneuvers, not only the roll rate P, but also Q, R, α and β need to be controlled. Thus, control laws for the three control axes are simultaneously synthesized in the outer control loop which includes the pilot. Furthermore, since perfect decoupling is not realistically achievable, it is the pilot who synergetically works the FCS's three control channels to perform the required maneuvers. In other words, the pilot performs the nontrivial task of *multivariable control.* Although the inclusion of a pilot in the flight control loop comes not without a price—a transport time lag is being introduced—the benefits far outweigh this inherent drawback of a human operator, for the pilot brings to the table the intelligent faculty of on-line multivariable control synthesis. Indeed, high AOA and maneuvering flight entails a degree of on-line perception and pattern recognition. The latter is colloquially referred to by pilots as "seat of the pants" flying. Hence stick-and-rudder prowess is not a thing of the past, and the pilot plays a vital role in maneuvering flight.

References

[1] Blakelock, J.H., *Automatic Control of Aircraft and Missiles,* John Wiley & Sons, New York, 1991.

[2] Pachter, M., *Modern Flight Control,* AFIT Lecture Notes, 1995, obtainable from the author at AFIT/ENG, 2950 P Street, Wright Patterson AFB, OH 45433-7765.

[3] MIL-STD-1797A: Flying Qualities of Piloted Aircraft, US Air Force, Feb. 1991.

[4] Roskam, J., *Airplane Flight Dynamics, Part 1,* Roskam Aviation and Engineering Corp., Lawrence, Kansas, 1979.

[5] Etkin, B., *Dynamics of Flight: Stability and Control,* John Wiley & Sons, New York, 1982.

[6] Stevens, B.L. and Lewis, F.L., *Aircraft Control and Simulation,* John Wiley & Sons, New York, 1992.

[7] Reynolds, O.R., Pachter, M., and Houpis, C.H., *Full Envelope Flight Control System Design Using QFT,* Proceedings of the American Control Conference, pp 350–354, June 1994, Baltimore, MD; to appear in the ALAA Journal of Guidance, Control and Dynamics.

[8] Chandler, P.R., Mears, M., and Pachter, M., *A Hybrid LQR/LP Approach for Addressing Actuator Saturation*

in Feedback Control, Proceedings of the Conference on Decision and Control, pp 3860–3867, 1994, Orlando, FL.

[9] Boyum, K.E., Pachter, M., and Houpis, C.H., *High Angle Of Attack Velocity Vector Rolls*, Proceedings of the 13th IFAC Symposium on Automatic Control in Aerospace, pp 51–57, 1994, Palo Alto, CA, and Control Engineering Practice, 3(8), 1087–1093, 1995.

[10] Neal, T.P. and Smith, R.E., An In-Flight Investigation to Develop Control System Design Criteria for Fighter Airplanes, AFFDL-TR-70-74, Vols. 1 and 2, Air Force Flight Dynamics Laboratory, Wright Patterson AFB, 1970.

[11] Chalk, C.R., Neal, T.P., and Harris, T.M., Background Information and User Guide for MIL-F-8785 B - Military Specifications and Flying Qualities of Piloted Airplanes, AFFDL-TR-69-72, Air Force Flight Dynamics Laboratory, Wright Patterson AFB, August 1969.

Further Reading

a) Monographs

1. C. D. Perkins and R. E. Hage, "Airplane Performance, Stability and Control," Wiley, New York, 1949.
2. "Dynamics of the Airframe," Northrop Corporation, 1952.
3. W. R. Kolk, "Modern Flight Dynamics," Prentice Hall, 1961.
4. B. Etkin, "Dynamics of Atmospheric Flight," Wiley, 1972.
5. D. McRuer, I. Ahkenas and D. Graham, "Aircraft Dynamics and Automatic Control," Princeton University Press, Princeton, NJ, 1973.
6. J. Roskam, "Airplane Flight Dynamics, Part 2," Roskam Aviation, 1979.
7. A. W. Babister, "Aircraft Dynamic Stability and Response," Pergamon Press, 1980.
8. R. C. Nelson, "Flight Stability and Automatic Control," McGraw-Hill, 1989, Second Edition.
9. D. McLean, "Automatic Flight Control Systems," Prentice Hall, 1990.
10. E. H. Pallett and S. Coyle, "Automatic Flight Control," Blackwell, 1993, Fourth Edition.
11. A. E. Bryson, "Control of Spacecraft and Aircraft," Princeton University Press, Princeton, NJ, 1994.
12. "Special Issue: Aircraft Flight Control," International Journal of Control, Vol. 59, No 1, January 1994.

b) The reader is encouraged to consult the bibliography listed in References [8], [9] and [10] in the text.

7.2 Spacecraft Attitude Control

Vincent T. Coppola, *Department of Aerospace Engineering, The University of Michigan, Ann Arbor, MI*

N. Harris McClamroch, *Department of Aerospace Engineering, The University of Michigan, Ann Arbor, MI*

7.2.1 Introduction

The purpose of this chapter is to provide an introductory account of spacecraft attitude control and related control problems. Attention is given to spacecraft kinematics and dynamics, the control objectives, and the sensor and control actuation characteristics. These factors are combined to develop specific feedback control laws for achieving the control objectives. In particular, we emphasize the interplay between the spacecraft kinematics and dynamics and the control law design, since we believe that the particular attributes of the spacecraft attitude control problem should be exploited in any control law design. We do not consider specific spacecraft designs or implementations of control laws using specific sensor and actuation hardware.

Several different rotational control problems are considered. These include the problem of transferring the spacecraft to a desired, possibly time-variable, angular velocity and maintaining satisfaction of this condition without concern for the orientation. A special case is the problem of transferring the spacecraft to rest. A related but different problem is to bring the spacecraft to a desired, possibly time-variable, orientation and to maintain satisfaction of this condition. A special case is the problem of bringing the spacecraft to a constant orientation. We also consider the spin stabilization problem, where the spacecraft is desired to have a constant spin rate about a specified axis of rotation. All of these control problems are considered, and we demonstrate that there is a common framework for their study.

Our approach is to provide a careful development of the spacecraft dynamics and kinematics equations and to indicate the assumptions under which they are valid. We then provide several general control laws, stated in terms of certain gain constants. No attempt has been made to provide algorithms for selecting these gains, but standard optimal control and robust control approaches can usually be used. Only general principles are mentioned that indicate how these control laws are obtained; the key is that they result in closed-loop systems that can be studied using elementary methods to guarantee asymptotic stability or some related asymptotic property.

7.2.2 Modeling

The motion of a spacecraft consists of its orbital motion, governed by translational equations, and its attitude motion, governed by rotational equations. An inertial frame is chosen to be at the center of mass of the orbited body and nonrotating with respect to some reference (e.g., the polar axis of the earth or the fixed stars). Body axes x, y, z are chosen as a reference frame fixed in the spacecraft body with origin at its center of mass. For spacecraft in circular orbits, a local horizontal-vertical reference frame is used to measure radial pointing. The local vertical is defined to be radial from the center of mass of the orbited body with the local horizontal aligned with the spacecraft's velocity.

The spacecraft is modeled as a rigid body. The rotational inertia matrix is assumed to be constant with respect to the x, y, z axes. This rigidity assumption is rather strong considering that many real spacecraft show some degree of flexibility and/or internal motion, for example, caused by fuel slosh. These nonrigidity effects may be sometimes modeled as disturbances of the rigid spacecraft.

We assume that gravitational attraction is the dominant force experienced by the spacecraft. Since the environmental forces (e.g., solar radiation, magnetic forces) are very weak, the spacecraft's orbital motion is well modeled as an ideal two-body, purely Keplerian orbit (i.e., a circle or ellipse), at least for short time periods. Thus, the translational motion is assumed decoupled from the rotational motion.

The rotational motion of the spacecraft responds to control moments and moments arising from environmental effects such as gravity gradients. In contrast to the translational equations, the dominant moment for rotational motion is not environmental, but is usually the control moment. In such cases, all environmental influences are considered as disturbances.

7.2.3 Spacecraft Attitude Sensors and Control Actuators

There are many different attitude sensors and actuators used in controlling spacecraft. Sensors provide indirect measurements of orientation or rate; models of the sensor (and sometimes of the spacecraft itself) can be used to compute orientation and rate from available measurements. Actuators are used to control the moments applied to influence the spacecraft rotational motion in some desired way. Certain sensors and actuators applicable for one spacecraft mission may be inappropriate for another, depending on the spacecraft characteristics and the mission requirements. A discussion of these issues can be found in [5].

Several types of sensors require knowledge of the spacecraft orbital motion. These include sun sensors, horizon sensors, and star sensors, which measure orientation angles. The measurement may be taken using optical telescopes, infrared radiation, or radar. Knowledge of star (and possibly planet or moon) positions may also be required to determine inertial orientation. Magnetometers provide a measurement of orientation based upon the magnetic field. The limited ability to provide highly accurate models of the magnetosphere limit the accuracy of these devices.

Gyroscopes mounted within the spacecraft can provide measurements of both angular rate and orientation. Rate gyros measure angular rate; integrating gyros provide a measure of angular orientation. The gyroscope rotor spins about its symmetry axis at a constant rate in an inertially fixed direction. The rotor is supported by one or two gimbals to the spacecraft. The gimbals move as the spacecraft rotates about the rotor. Angular position and rate are measured by the movement of the gimbals. The model of the gyroscope often ignores the inertia of the gimbals and the friction in the bearings; these effects cause the direction of the gyro to drift over time.

An inertial measurement unit (IMU) consists of three mutually perpendicular gyroscopes mounted to a platform, either in gimbals or strapped down. Using feedback of the gyro signals, motors apply moments to the gyros to make the angular velocity of the platform zero. The platform then becomes an inertial reference.

The Global Positioning System (GPS) allows for very precise measurements of the orbital position of earth satellites. Two GPS receivers, located sufficiently far apart on the satellite, can be used to derive attitude information based on the phase shift of the signals from the GPS satellites. The orientation and angular rates are computed based upon a model of the rotational dynamics of the satellite.

Gas-jet thrusters are commonly employed as actuators for spacecraft attitude control. At least 12 thrust chambers are needed to provide three-axis rotational control. Each axis is controlled by two pairs of thrusters: one pair provides clockwise moment; the other pair, counterclockwise. A thruster pair consists of two thrusters that operate simultaneously at the same thrust level but in opposite directions. They create no net force on the spacecraft but do create a moment about an axis perpendicular to the plane containing the thrust directions. Thrusters may operate continuously or in full-on, full-off modes. Although the spacecraft loses mass during thruster firings, it is often negligible compared to the mass of the spacecraft and is ignored in the equations of motion.

Another important class of actuators used for attitude control are reaction wheel devices. Typically, balanced reaction wheels are mounted on the spacecraft so that their rotational axes are rigidly attached to the spacecraft. As a reaction wheel is spun up by an electric motor rigidly attached to the spacecraft, there is a reaction moment on the spacecraft. These three reaction moments provide the control moments on the spacecraft. In some cases, the effects of the electric motor dynamics are significant; these dynamics are ignored in this chapter.

7.2.4 Spacecraft Rotational Kinematics

The orientation or attitude of a rigid spacecraft can be expressed by a 3×3 rotation matrix R [3], [4]. Since a body-fixed x, y, z frame is rigidly attached to the spacecraft, the orientation of the spacecraft is the orientation of the body frame expressed with respect to a reference frame X, Y, Z. The columns of the rotation matrix are the components of the three standard basis vectors of the X, Y, Z frame expressed in terms of the three body-fixed standard basis vectors of the x, y, z frame. It can be shown that rotation matrices are necessarily orthogonal matrices; that is, they have the property that

$$RR^T = I \text{ and } R^T R = I, \tag{7.48}$$

where I is the 3×3 identity matrix, and $\det(R) = 1$. Upon differentiating with respect to time, we obtain

$$\frac{d}{dt}(RR^T) = \dot{R}R^T + R\dot{R}^T = 0 \tag{7.49}$$

Consequently,

$$(\dot{R}R^T) = -(\dot{R}R^T)^T \tag{7.50}$$

is a skew-symmetric matrix. It can be shown that this skew-symmetric matrix can be expressed in terms of the components of the angular velocity vector in the body-fixed coordinate frame $(\omega_x, \omega_y, \omega_z)$ as

$$\dot{R}R^T = S(\omega) \tag{7.51}$$

where

$$S(\omega) = \begin{bmatrix} 0 & \omega_z & -\omega_y \\ -\omega_z & 0 & \omega_x \\ \omega_y & -\omega_x & 0 \end{bmatrix}. \tag{7.52}$$

Consequently, the spacecraft attitude is described by the kinematics equation

$$\dot{R} = S(\omega)R. \tag{7.53}$$

This linear matrix differential equation describes the spacecraft attitude time dependence. If the angular velocity vector is a given vector function of time, and if an initial attitude is specified by a rotation matrix, then the matrix differential Equation 7.53 can be integrated using the specified initial data to obtain the subsequent spacecraft attitude. The solution of this matrix differential equation must necessarily be an orthogonal matrix at all instants of time. It should be noted that this matrix differential equation can also be written as nine scalar linear differential equations, but if these scalar equations are integrated, say, numerically, care must be taken to guarantee that the resulting solution satisfies the orthogonality property. In other words, the nine scalar entries in a rotation matrix are not independent.

The description of the spacecraft attitude in terms of a rotation matrix is conceptually natural and elegant. However, use of rotation matrices directly presents difficulties in computations and in physical interpretation. Consequently, other descriptions of attitude have been developed. These descriptions can be seen as specific parameterizations of rotation matrices using fewer than nine parameters. The most common parameterization involves the use of three Euler angle parameters.

Although there are various definitions of Euler angles, we introduce the most common definition (the 3-2-1 definition) that is widely used in spacecraft analyses. The three Euler angles are denoted by Ψ, θ, and ϕ, and are referred to as the yaw angle, the pitch angle, and the roll angle, respectively. A general spacecraft orientation defined by a rotation matrix can be achieved by a sequence of three elementary rotations, beginning with the body-fixed coordinate frame coincident with the reference coordinate frame, defined as follows:

1. A rotation of the spacecraft about the body-fixed z axis by a yaw angle Ψ
2. A rotation of the spacecraft about the body-fixed y axis by a pitch angle θ
3. A rotation of the spacecraft about the body-fixed x axis by a roll angle ϕ

It can be shown that the rotation matrix can be expressed in terms of the three Euler angle parameters Ψ, θ, ϕ according to the relationship [4]

$$R = \begin{bmatrix} \cos\theta\cos\Psi & \cos\theta\sin\Psi & -\sin\theta \\ (-\cos\phi\sin\Psi + \sin\phi\sin\theta\cos\Psi) & (\cos\phi\cos\Psi + \sin\phi\sin\theta\sin\Psi & \sin\phi\cos\theta \\ (\sin\phi\sin\Psi + \cos\phi\sin\theta\cos\Psi) & (-\sin\phi\cos\Psi + \cos\phi\sin\theta\sin\Psi) & \cos\phi\cos\theta \end{bmatrix}. \tag{7.54}$$

The components of a vector, expressed in terms of the x, y, z frame, are the product of the rotation matrix and the components of that same vector, expressed in terms of the X, Y, Z frame.

One of the deficiencies in the use of the Euler angles is that they do not provide a global parameterization of rotation matrices. In particular, the Euler angles are restricted to the range

$$\begin{aligned} -\pi &< \Psi < \pi, \\ -\tfrac{\pi}{2} &< \theta < \tfrac{\pi}{2}, \\ -\pi &< \phi < \pi, \end{aligned} \tag{7.55}$$

in order to avoid singularities in the above representation. This limitation is serious in some attitude control problems and motivates the use of other attitude representations.

Other attitude representations that are often used include axis-angle variables and quaternion or Euler parameters. The latter representations are globally defined but involve the use of four parameters rather than three. Attitude control problems can be formulated and solved using these alternative attitude parameterizations. In this chapter, we make use only of the Euler angle approach; however, we are careful to point out the difficulties that can arise by using Euler angles.

The spacecraft kinematic equations relate the components of angular velocity vector to the rates of change of the Euler angles. It can also be shown [4] that the angular velocity components in the body-fixed coordinate frame can be expressed in terms of the rates of change of the Euler angles as

$$\begin{aligned} \omega_x &= -\dot{\Psi}\sin\theta + \dot{\phi}, \\ \omega_y &= \dot{\Psi}\cos\theta\sin\phi + \dot{\theta}\cos\phi, \\ \omega_z &= \dot{\Psi}\cos\theta\cos\phi - \dot{\theta}\sin\phi. \end{aligned} \tag{7.56}$$

Conversely, the rates of change of the Euler angles can be expressed in terms of the components of the angular velocity vector as

$$\begin{aligned} \dot{\Psi} &= \omega_y \sec\theta\sin\phi + \omega_z\sec\theta\cos\phi, \\ \dot{\theta} &= \omega_y\cos\phi - \omega_z\sin\phi, \\ \dot{\phi} &= \omega_x + \omega_y\tan\theta\sin\phi + \omega_z\tan\theta\cos\phi. \end{aligned} \tag{7.57}$$

These equations are referred to as the spacecraft kinematics equations; they are used subsequently in the development of attitude control laws.

7.2.5 Spacecraft Rotational Dynamics

We consider the rotational dynamics of a rigid spacecraft. The body-fixed coordinates are chosen to be coincident with the principal axes of the spacecraft. We first consider the case where three pairs of thrusters are employed for attitude control. We next consider the case where three reaction wheels are employed for attitude control; under appropriate assumptions, the controlled spacecraft dynamics for the two cases are identical. The uncontrolled spacecraft dynamics are briefly studied.

We first present a model for the rotational dynamics of a rigid spacecraft controlled by thrusters as shown in Figure 7.15.

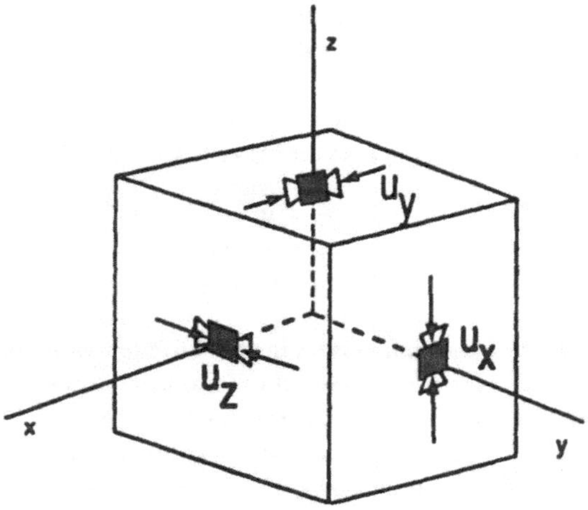

Figure 7.15 Spacecraft with gas-jet thrusters.

These dynamic equations are most naturally expressed in body-fixed coordinates. We further assume that the body-fixed coordinate frame is selected to be coincident with the spacecraft principal axes and that there are three pairs of thrusters that produce attitude control moments about these three principal axes. The resulting spacecraft dynamics are given by

$$\begin{aligned}
I_{xx}\dot{\omega}_x &= \left(I_{yy}-I_{zz}\right)\omega_y\omega_z + u_x + T_x, \\
I_{yy}\dot{\omega}_y &= \left(I_{zz}-I_{xx}\right)\omega_z\omega_x + u_y + T_y, \\
I_{zz}\dot{\omega}_z &= \left(I_{xx}-I_{yy}\right)\omega_x\omega_y + u_z + T_z,
\end{aligned} \tag{7.58}$$

where $(\omega_x, \omega_y, \omega_z)$ are the components of the spacecraft angular velocity vector, expressed in body-fixed coordinates; (u_x, u_y, u_z) are the components of the control moments due to the thrusters about the spacecraft principal axes; and (T_x, T_y, T_z) are the components of the external disturbance moment vector, expressed in body-fixed coordinates. The terms I_{xx}, I_{yy}, I_{zz} are the constant moments of inertia of the spacecraft with respect to its principal axes.

We now present a model for the rotational dynamics of a rigid spacecraft controlled by reaction wheels as shown in Figure 7.16.

As before, we express the rotational dynamics in body-fixed coordinates. We further assume that the body-fixed coordinate frame is selected to be coincident with the spacecraft principal axes and that there are three balanced reaction wheels that pro-

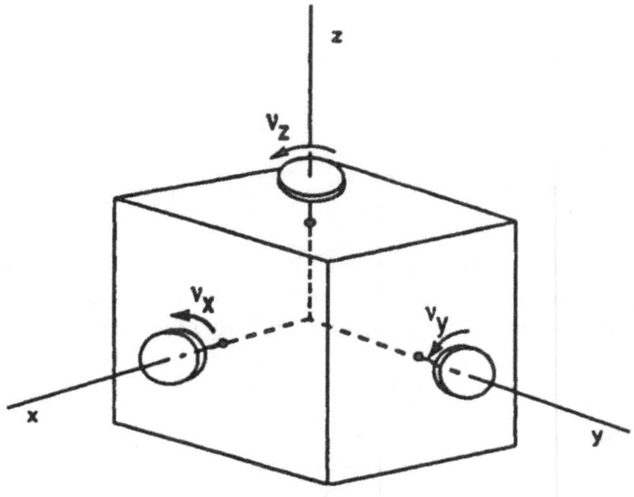

Figure 7.16 Spacecraft with three reaction wheels.

duce attitude control moments; the rotation axes of the reaction wheels are the spacecraft principal axes. The resulting spacecraft dynamics are given by [2]

$$\begin{aligned}
I_{xx}\dot{\omega}_x &= \left(I_{yy}-I_{zz}\right)\omega_y\omega_z \\
&\quad + J_y\omega_z(\omega_y+\nu_y) - J_z\omega_y(\omega_z+\nu_z) \\
&\quad + u_x + T_x, \\
I_{yy}\dot{\omega}_y &= \left(I_{zz}-I_{xx}\right)\omega_z\omega_x \\
&\quad + J_z\omega_x(\omega_z+\nu_z) - J_x\omega_z(\omega_x+\nu_x) \\
&\quad + u_y + T_y, \\
I_{zz}\dot{\omega}_z &= \left(I_{xx}-I_{yy}\right)\omega_x\omega_y \\
&\quad + J_x\omega_y(\omega_x+\nu_x) - J_y\omega_x(\omega_y+\nu_y) \\
&\quad + u_z + T_z,
\end{aligned} \tag{7.59}$$

and the dynamics of the reactions wheels are given by

$$\begin{aligned}
J_x(\dot{\omega}_x+\dot{\nu}_x) &= -u_x, \\
J_y(\dot{\omega}_y+\dot{\nu}_y) &= -u_y, \\
J_z(\dot{\omega}_z+\dot{\nu}_z) &= -u_z,
\end{aligned} \tag{7.60}$$

where $(\omega_x, \omega_y, \omega_z)$ are the components of the spacecraft angular velocity vector, expressed in body-fixed coordinates; (ν_x, ν_y, ν_z) are the relative angular velocities of the reaction wheels with respect to their respective axes of rotation (the spacecraft principal axes); (u_x, u_y, u_z) are the control moments developed by the electric motors rigidly mounted on the spacecraft with shafts aligned with the spacecraft principal axes connected to the respective reaction wheels; and (T_x, T_y, T_z) are the components of the external disturbance moment vector, expressed in body-fixed coordinates. Letting $(\mathsf{I}_{xx}, \mathsf{I}_{yy}, \mathsf{I}_{zz})$ denote the principal moments of inertia of the spacecraft, $I_{xx} = \mathsf{I}_{xx} - J_x$, $I_{yy} = \mathsf{I}_{yy} - J_y$, $I_{zz} = \mathsf{I}_{zz} - J_z$, where (J_x, J_y, J_z) are the (polar) moments of inertia of the reaction wheels.

Our interest is in attitude control of the spacecraft, so that Equation 7.60 for the reaction wheels does not play a central role. Note that if the terms in Equation 7.59 that explicitly couple the spacecraft dynamics and the reaction wheel dynamics are assumed to be small so that they can be ignored, then Equation 7.59 reduces formally to Equation 7.58.

Our subsequent control development makes use of Equation 7.58, or equivalently, the simplified form of Equation 7.59, so that the development applies to spacecraft attitude control using either thrusters or reaction wheels. In the latter case, it should be kept in mind that the coupling terms in Equation 7.59 are ignored as small. If these terms in Equation 7.59 are in fact not small, suitable modifications can be made to our subsequent development to incorporate the effect of these additional terms. One conceptually simple approach is to introduce *a priori* feedback loops that eliminate these coupling terms.

Disturbance moments have been included in the above equations, but it is usual to ignore external disturbances in the control design. Consequently, we subsequently make use of the controlled spacecraft dynamics given by

$$\begin{aligned} I_{xx}\dot{\omega}_x &= (I_{yy} - I_{zz})\,\omega_y\omega_z + u_x, \\ I_{yy}\dot{\omega}_y &= (I_{zz} - I_{xx})\,\omega_z\omega_x + u_y, \\ I_{zz}\dot{\omega}_z &= (I_{xx} - I_{yy})\,\omega_x\omega_y + u_z, \end{aligned} \tag{7.61}$$

where it is assumed that the disturbance moment vector $(T_x, T_y, T_z) = (0, 0, 0)$.

If there are no control or disturbance moments on the spacecraft, then the rotational dynamics are described by the Euler equations

$$\begin{aligned} I_{xx}\dot{\omega}_x &= (I_{yy} - I_{zz})\,\omega_y\omega_z, \\ I_{yy}\dot{\omega}_y &= (I_{zz} - I_{xx})\,\omega_z\omega_x, \\ I_{zz}\dot{\omega}_z &= (I_{xx} - I_{yy})\,\omega_x\omega_y. \end{aligned} \tag{7.62}$$

These equations govern the uncontrolled angular velocity of the spacecraft. Both the angular momentum H and the rotational kinetic energy T given by

$$\begin{aligned} 2T &= I_{xx}\omega_x^2 + I_{yy}\omega_y^2 + I_{zz}\omega_z^2, \\ H^2 &= (I_{xx}\omega_x)^2 + (I_{yy}\omega_y)^2 + (I_{zz}\omega_z)^2, \end{aligned} \tag{7.63}$$

are constants of the motion. These constants describe two ellipsoids in the $(\omega_x, \omega_y, \omega_z)$ space with the actual motion constrained to lie on their intersection. Since the intersection occurs as closed one-dimensional curves (or points), the angular velocity components $(\omega_x, \omega_y, \omega_z)$ are periodic time functions. Point intersections occur when two of the three components are zero: the body is said to be in a simple spin about the third body axis. Moreover, the spin axis is aligned with the (inertially fixed) angular momentum vector in the simple spin case.

Further analysis shows that simple spins about either the minor or major axis (i.e., the axes with the least and greatest moment of inertia, respectively) in Equation 7.62 are Lyapunov stable, while a simple spin about the intermediate axis is unstable. However, the minor axis spin occurs when the kinetic energy is maximum for a given magnitude of the angular momentum; thus, if dissipation effects are considered, spin about the minor axis becomes unstable. This observation leads to the major-axis rule: An energy-dissipating rigid body eventually arrives at a simple spin about its major axis. Overlooking this rule had devastating consequences for the U.S.' first orbital satellite *Explorer I*.

The orientation of the body with respect to inertial space is described in terms of the Euler angle responses Ψ, θ, ϕ [3]. This is best understood when the inertial directions are chosen so that the angular momentum vector lies along the inertial Z axis. Then the x axis is said to be precessing about the angular momentum vector at the rate $d\Psi/dt$ (which is always positive) and nutating at the rate $d\theta/dt$ (which is periodic and single signed). The angle $(\pi/2 - \theta)$ is called the coning or nutation angle. The motion of the x axis in inertial space is periodic in time; however, in general, the orientation of the body is not periodic because the spin angle ϕ is not commensurate with periodic motion of the x axis.

Although simple spins about major/minor axes are Lyapunov stable to perturbations in the angular velocities, neither is stable with respect to perturbations in orientation since the angular momentum vector cannot resist angular rate perturbations about itself [4]. However, it can resist perturbations orthogonal to itself. Hence, both the major and minor axes are said to be linearly directionally stable (when dissipation is ignored) since small orientation perturbations of the simple spin do not cause large deviations in the direction of the spin axis. That is, the direction of the spin axis is linearly stable.

7.2.6 Linearized Spacecraft Rotational Equations

The spacecraft kinematics and dynamics are described by nonlinear differential Equations 7.57 and 7.61. In this section, we develop linear differential equations that serve as good approximations for the spacecraft kinematics and dynamics in many instances.

We first consider a *linearization of the spacecraft equations near a rest solution.* Suppose that there are no external moments applied to the spacecraft; we note that the spacecraft can remain at rest in any fixed orientation. This corresponds to a constant attitude and zero angular velocity. If the reference coordinate frame defines this constant attitude, then it is easily seen that

$$R = I, \;\; \omega = 0$$

satisfy the spacecraft kinematics and dynamics given by Equations 7.53 and 7.61. Equivalently, this reference attitude corresponds to the Euler angles being identically zero

$$(\Psi, \theta, \phi) = (0, 0, 0)$$

and it is easily seen that the kinematics Equation 7.56 is trivially satisfied.

Now if we assume that applied control moments do not perturb the spacecraft too far from its rest solution, then we can assume that the components of the spacecraft angular velocity vector are small; thus, we ignore the product terms to obtain the linearized approximation to the spacecraft dynamics

$$\begin{aligned} I_{xx}\dot{\omega}_x &= u_x, \\ I_{yy}\dot{\omega}_y &= u_y, \\ I_{zz}\dot{\omega}_z &= u_z. \end{aligned} \tag{7.64}$$

Similarly, by assuming that the Euler angles and their rates of change are small, we make the standard small angle assumptions to obtain the linearized approximation to the spacecraft kinematics equations

$$\begin{aligned} \dot{\phi} &= \omega_x, \\ \dot{\theta} &= \omega_y, \\ \dot{\Psi} &= \omega_z, \end{aligned} \tag{7.65}$$

Thus, the linearized dynamics and kinematics equations can be combined in a decoupled form that approximates the rolling, pitching, and yawing motions, namely,

$$\begin{aligned} I_{xx}\ddot{\phi} &= u_x, \\ I_{yy}\ddot{\theta} &= u_y, \\ I_{zz}\ddot{\Psi} &= u_z, \end{aligned} \tag{7.66}$$

We next present a *linearized model for the spacecraft equations near a constant spin solution.* Suppose that there is no control moment applied to the spacecraft. Another solution for the spacecraft motion, in addition to the rest solution, corresponds to a simple spin at constant rate of rotation about an inertially fixed axis. To be specific, we assume that the spacecraft has a constant spin rate Ω about the inertially fixed Z axis; the corresponding solution of the spacecraft kinematics Equation 7.57 and the spacecraft dynamics Equation 7.61 is given by

$$(\Psi, \theta, \phi) = (\Omega t, 0, 0), (\omega_x, \omega_y, \omega_z) = (0, 0, \Omega).$$

Now if we assume that the applied control moments do not perturb the spacecraft motion too far from this constant spin solution, then we can assume that the angular velocity components ω_x, ω_y, and $\omega_z - \Omega$ are small. Thus, we can ignore the products of small terms to obtain the linearized approximation to the spacecraft dynamics

$$\begin{aligned} I_{xx}\dot{\omega}_x &= (I_{yy} - I_{zz})\,\Omega\omega_y + u_x, \\ I_{yy}\dot{\omega}_y &= (I_{zz} - I_{xx})\,\Omega\omega_x + u_y, \\ I_{zz}\dot{\omega}_z &= u_z. \end{aligned} \tag{7.67}$$

Similarly, by assuming that the Euler angles θ and ϕ (but not Ψ) are small, we obtain the linearized approximation to the spacecraft kinematics

$$\begin{aligned} \dot{\Psi} &= \omega_z, \\ \dot{\theta} &= \omega_y - \Omega\phi, \\ \dot{\phi} &= \omega_x + \Omega\theta. \end{aligned} \tag{7.68}$$

Thus, the linearized dynamics and kinematics equations can be combined into

$$\begin{aligned} I_{xx}(\ddot{\phi} - \Omega\dot{\theta}) &= (I_{yy} - I_{zz})\Omega(\dot{\theta} + \Omega\phi) + u_x, \\ I_{yy}(\ddot{\theta} + \Omega\dot{\phi}) &= (I_{zz} - I_{xx})\Omega(\dot{\phi} - \Omega\theta) + u_y, \\ I_{zz}\ddot{\Psi} &= u_z, \end{aligned} \tag{7.69}$$

which makes clear that the yawing motion (corresponding to the motion about the spin axis) is decoupled from the pitching and rolling motion.

There is little loss of generality in assuming that the nominal spin axis is the inertially fixed Z axis; this is simply a matter of defining the inertial reference frame in a suitable way. This choice has been made since it leads to a simple formulation in terms of the Euler angles.

7.2.7 Control Specifications and Objectives

There are a number of possible rotational control objectives that can be formulated. In this section, several common problems are described. In all cases, our interest, for practical reasons, is to use feedback control. We use the control actuators to generate control moments that cause a desired spacecraft rotational motion; operation of these control actuators depends on feedback of the spacecraft rotational motion variables, which are obtained from the sensors. Most commonly, these sensors provide instantaneous measurements of the angular velocity vector (with respect to the body-fixed coordinate frame) and the orientation, or equivalently, the Euler angles.

The use of feedback control is natural for spacecraft rotational control applications; the same control law can be used to achieve and maintain satisfaction of the control objective in spite of certain types of model uncertainties and external disturbances. The desirable features of feedback are best understood in classical control theory, but the benefits of feedback can also be obtained for the fundamentally nonlinear rotational control problems that we consider here. Conversely, feedback, if applied unwisely, can do great harm. Consequently, we give considerable attention to a careful description of the feedback control laws and to justification for those control laws via specification of formal closed-loop stability properties. This motivates our use of mathematical models and careful analysis and design, thereby avoiding the potential difficulties associated with *ad hoc* or trial-and-error based control design.

We first consider spacecraft control problems where the control objectives are specified in terms of the spacecraft angular velocity. A general angular velocity control objective is that the spacecraft angular velocity vector exactly track a specified angular velocity vector in the sense that the angular velocity error be brought to zero and maintained at zero. If $(\omega_{xd}, \omega_{yd}, \omega_{zd})$ denotes the desired, possibly time-variable, angular velocity vector, this control objective is described by the asymptotic condition that

$$(\omega_x, \omega_y, \omega_z) \to (\omega_{xd}, \omega_{yd}, \omega_{zd}) \text{ as } t \to \infty.$$

One special case of this general control objective corresponds to bringing the spacecraft angular velocity vector to zero and maintaining it at zero. In this case, the desired angular velocity vector is zero, so the control objective is described by the asymptotic condition that

$$(\omega_x, \omega_y, \omega_z) \to (0, 0, 0) \text{ as } t \to \infty.$$

Another example of this general control objective corresponds to specifying that the spacecraft spin about a specified axis fixed to the spacecraft with a given angular velocity. For example, if $(\omega_{xd}, \omega_{yd}, \omega_{zd}) = (0, 0, \Omega)$, the control objective is that

$$(\omega_x, \omega_y, \omega_z) \rightarrow (0, 0, \Omega) \text{ as } t \rightarrow \infty.$$

so that the spacecraft asymptotically spins about its body-fixed z axis with a constant angular velocity Ω.

In all of the above cases, the spacecraft orientation is of no concern. Consequently, control laws that achieve these control objectives make use solely of the spacecraft dynamics.

We next consider spacecraft control problems where the control objectives are specified in terms of the spacecraft attitude. A more general class of control objectives involves the spacecraft attitude as well as the spacecraft angular velocity. In particular, suppose that it is desired that the spacecraft orientation exactly track a desired orientation in the sense that the attitude error is brought to zero and maintained at zero. The specified orientation may be time variable. If the desired orientation is described in terms of a rotation matrix function of time R_d, then the control objective is described by the asymptotic condition that

$$R \rightarrow R_d \text{ as } t \rightarrow \infty.$$

If we assume that R_d is parameterized by the desired Euler angle functions $(\Psi_d, \theta_d, \phi_d)$, then the control objective can be described by the asymptotic condition that

$$(\Psi, \theta, \phi) \rightarrow (\Psi_d, \theta_d, \phi_d) \text{ as } t \rightarrow \infty.$$

Since the specified matrix R_d is a rotation matrix, it follows that there is a corresponding angular velocity vector $\omega_d = (\omega_{xd}, \omega_{yd}, \omega_{zd})$ that satisfies

$$\dot{R}_d = S(\omega_d)R_d.$$

Consequently, it follows that the spacecraft angular velocity vector must also satisfy the asymptotic condition

$$(\omega_x, \omega_y, \omega_z) \rightarrow (\omega_{xd}, \omega_{yd}, \omega_{zd}) \text{ as } t \rightarrow \infty.$$

An important special case of the general attitude control objective is to bring the spacecraft attitude to a specified constant orientation and to maintain it at that orientation. Let R_d denote the desired constant spacecraft orientation; then the control objective is described by the asymptotic condition that

$$R \rightarrow R_d \text{ as } t \rightarrow \infty.$$

Letting R_d be parameterized by the desired constant Euler angles $(\Psi_d, \theta_d, \phi_d)$, the control objective can be described by the asymptotic condition that

$$(\Psi, \theta, \phi) \rightarrow (\Psi_d, \theta_d, \phi_d) \text{ as } t \rightarrow \infty.$$

Since the desired orientation is constant, it follows that the angular velocity vector must be brought to zero and maintained at zero; that is,

$$(\omega_x, \omega_y, \omega_z) \rightarrow (0, 0, 0) \text{ as } t \rightarrow \infty.$$

If the reference coordinate frame is selected to define the desired spacecraft orientation, then $R_d = I$, the 3 by 3 identity matrix, and $(\Psi_d, \theta_d, \phi_d) = (0, 0, 0)$ and the control objective is

$$R \rightarrow I \text{ as } t \rightarrow \infty,$$

or equivalently,

$$(\Psi, \theta, \phi) \rightarrow (0, 0, 0) \text{ as } t \rightarrow \infty.$$

Control laws that achieve the above control objectives should make use of both the spacecraft kinematics and the spacecraft dynamics. Special cases of such control objectives are given by the requirements that the spacecraft always point at the earth's center, a fixed star, or another spacecraft.

Another control objective corresponds to specifying that the spacecraft spin about a specified axis that is inertially fixed with a given angular velocity Ω. If the reference coordinate frame is chosen so that it is an inertial frame and its positive Z axis is the spin axis, then it follows that the control objective can be described by the asymptotic condition that

$$(\theta, \phi) \rightarrow (0, 0) \text{ as } t \rightarrow \infty,$$
$$(\omega_x, \omega_y, \omega_z) \rightarrow (0, 0, \Omega) \text{ as } t \rightarrow \infty,$$

which guarantees that the spacecraft tends to a constant angular velocity about its body-fixed z axis and that its z axis is aligned with the inertial Z axis. Control laws that achieve the above control objectives should make use of both the spacecraft kinematics and the spacecraft dynamics.

An essential part of any control design process is performance specifications. When the spacecraft control inputs are adjusted automatically according to a specified feedback control law, the resulting system is a closed-loop system. In terms of control law design, performance specifications are naturally imposed on the closed-loop system. Since the spacecraft kinematics and dynamics are necessarily nonlinear in their most general form, closedloop specifications must be given in somewhat nonstandard form. There are no uniformly accepted technical performance specifications for this class of control problems, but it is generally accepted that the closed loop should exhibit rapid transient response, good steady-state accuracy, good robustness to parameter uncertainties, a large domain of attraction, and the ability to reject certain classes of external disturbances.

These conceptual control objectives can be quantified, at least if the rotational motions of the closed loop are sufficiently small so that the dynamics and kinematics are adequately described by linear models. In such case, good transient response depends on the eigenvalues or characteristic roots. Desired closed-loop properties can be specified if there are uncertainties in the models and external disturbances of certain classes. Control design to achieve performance specifications, such as steady-state accuracy, robustness, and disturbance rejection, have been extensively treated in the theoretical literature, for both linear and nonlinear control systems, and are not explicitly studied here. We note that if there are persistent disturbances, then there may be nonzero steady-state errors for the control laws that we subsequently propose; those control laws can easily be modified to include integral error terms to improve the steady-state accuracy. Examples of relatively simple specifications of the closed loop are illustrated in the subsequent sections that deal with design of feedback control laws for several spacecraft rotational control problems.

7.2.8 Spacecraft Control Problem: Linear Control Law Based on Linearized Spacecraft Equations

In this section, we assume that the spacecraft dynamics and kinematics can be described in terms of the linearized Equations 7.64 and 7.65. We assume the control moments can be adjusted to any specified level; hence, we consider the use of linear control laws.

We first consider *control of the spacecraft angular velocity*. It is assumed that the spacecraft dynamics are described by the linearized Equation 7.64; a control law (u_x, u_y, u_z) is desired in feedback form, so that the resulting closed-loop system satisfies the asymptotic condition

$$(\omega_x, \omega_y, \omega_z) \rightarrow (\omega_{xd}, \omega_{yd}, \omega_{zd}) \text{ as } t \rightarrow \infty,$$

and has desired closed-loop properties. Since the spacecraft dynamics are linear and uncoupled first-order equations, a standard linear control approach can be used to design a linear feedback control of the form

$$\begin{aligned} u_x &= -c_x(\omega_x - \omega_{xd}) + I_{xx}\dot{\omega}_{xd}, \\ u_y &= -c_y(\omega_y - \omega_{yd}) + I_{yy}\dot{\omega}_{yd}, \\ u_z &= -c_z(\omega_z - \omega_{zd}) + I_{zz}\dot{\omega}_{zd}, \end{aligned} \tag{7.70}$$

where the control gains c_x, c_y, c_z are chosen as positive constants. Based on the linearized equations, introduce the variables

$$\delta_x = \omega_x - \omega_{xd}, \quad \delta_y = \omega_y - \omega_{yd}, \quad \delta_z = \omega_z - \omega_{zd}, \tag{7.71}$$

so that the closed-loop equations are

$$\begin{aligned} \dot{\delta}_x + \left(\frac{c_x}{I_{xx}}\right)\delta_x &= 0, \\ \dot{\delta}_y + \left(\frac{c_y}{I_{yy}}\right)\delta_y &= 0, \\ \dot{\delta}_z + \left(\frac{c_z}{I_{zz}}\right)\delta_z &= 0. \end{aligned} \tag{7.72}$$

Hence, the angular velocity errors in roll rate, pitch rate, and yaw rate are brought to zero asymptotically, at least for sufficiently small perturbations. The values of the control gains can be chosen to provide specified closed-loop time constants. Consequently, the control law is guaranteed to achieve exact tracking asymptotically, with speed of response that is determined by the values of the control gains.

A simple special case of the above control law corresponds to the choice that $(\omega_{xd}, \omega_{yd}, \omega_{zd}) = (0, 0, 0)$; i.e., the spacecraft is to be brought to rest. In simplified form, the control law becomes

$$\begin{aligned} u_x &= -c_x\omega_x, \\ u_y &= -c_y\omega_y, \\ u_z &= -c_z\omega_z, \end{aligned} \tag{7.73}$$

and it is guaranteed to bring the spacecraft asymptotically to rest.

It is important to note that the preceding analysis holds only for sufficiently small perturbations from rest. For large perturbations in the angular velocity of the spacecraft, the above control laws may not have the desired properties. An analysis of the closed-loop system, using the nonlinear dynamics Equation 7.61, is required to determine the domain of perturbations for which closed-loop stabilization is achieved. Such an analysis is beyond the scope of the present chapter, but we note that the stability domain depends on both the desired control objective and the control gains that are selected.

We next consider *control of the spacecraft angular attitude*. It is assumed that the spacecraft dynamics and kinematics are described by the linearized Equation 7.66 and a control law (u_x, u_y, u_z) is desired in feedback form, so that the resulting closed-loop system satisfies the attitude control conditions

$$\begin{aligned} (\Psi, \theta, \phi) &\rightarrow (\Psi_d, \theta_d, \phi_d) \text{ as } t \rightarrow \infty, \\ (\omega_x, \omega_y, \omega_z) &\rightarrow (\omega_{xd}, \omega_{yd}, \omega_{zd}) \text{ as } t \rightarrow \infty, \end{aligned}$$

where the kinematic conditions of Equation 7.56 are assumed to be satisfied, and the closed-loop system has desired closed-loop properties. Since the equations are linear and uncoupled second-order equations, a standard linear control approach can be used to obtain linear feedback control laws of the form

$$\begin{aligned} u_x &= -c_x(\omega_x - \omega_{xd}) - k_x(\phi - \phi_d) + I_{xx}\ddot{\phi}_d, \\ u_y &= -c_y(\omega_y - \omega_{yd}) - k_y(\theta - \theta_d) + I_{yy}\ddot{\theta}_d, \\ u_z &= -c_z(\omega_z - \omega_{zd}) - k_z(\Psi - \Psi_d) + I_{zz}\ddot{\Psi}_d. \end{aligned} \tag{7.74}$$

Recall that the angular velocity components are the rates of change of the Euler angles according to the linearized relations of Equation 7.65, so that the control law can equivalently be expressed in terms of the Euler angles and their rates of change. Based on the linearized equations, introduce the variables

$$\varepsilon_x = \phi - \phi_d, \quad \varepsilon_y = \theta - \theta_d, \quad \varepsilon_z = \Psi - \Psi_d, \tag{7.75}$$

so that the closed-loop equations are

$$\begin{aligned} I_{xx}\ddot{\varepsilon}_x + c_x\dot{\varepsilon}_x + k_x\varepsilon_x &= 0, \\ I_{yy}\ddot{\varepsilon}_y + c_y\dot{\varepsilon}_y + k_y\varepsilon_y &= 0, \\ I_{zz}\ddot{\varepsilon}_z + c_z\dot{\varepsilon}_z + k_z\varepsilon_z &= 0, \end{aligned} \tag{7.76}$$

Hence, the spacecraft attitude errors in roll angle, pitch angle, and yaw angle are brought to zero asymptotically, at least for sufficiently small perturbations. The values of the control gains can be chosen to provide specified closed-loop natural frequencies and damping ratios. Consequently, the above control laws are guaranteed to bring the spacecraft to the desired attitude asymptotically, with speed of response that is determined by the values of the control gains.

A simple special case of the above control law corresponds to the choice that $(\Psi_d, \theta_d, \phi_d)$ is a constant and $(\omega_{xd}, \omega_{yd}, \omega_{zd}) = (0, 0, 0)$. The simplified control law is of the form

$$\begin{aligned} u_x &= -c_x\omega_x - k_x(\phi - \phi_d), \\ u_y &= -c_y\omega_y - k_y(\theta - \theta_d), \\ u_z &= -c_z\omega_z - k_z(\Psi - \Psi_d), \end{aligned} \tag{7.77}$$

and it is guaranteed to bring the spacecraft asymptotically to the desired constant attitude and to maintain it in the desired attitude. The resulting control law is referred to as an attitude stabilization control law. Further details are available in [1].

Again, it is important to note that the preceding analysis holds only for sufficiently small perturbations in the spacecraft angular velocity and attitude. For large perturbations in the angular velocity and attitude of the spacecraft, the control law may not have the desired properties. An analysis of the closed-loop system, using the nonlinear dynamics Equation 7.61 and the nonlinear kinematic Equation 7.57, is required to determine the domain of perturbations for which closed-loop stabilization is achieved. Such an analysis is beyond the scope of the present chapter, but we note that the stability domain depends on the desired attitude and angular velocity and the control gains that are selected.

We now consider *spin stabilization of the spacecraft about a specified inertial axis*. A control law (u_x, u_y, u_z) is desired in feedback form, so that the resulting closed-loop system satisfies the asymptotic control objective

$$(\theta, \phi) \to (0, 0) \text{ as } t \to \infty,$$
$$(\omega_x, \omega_y, \omega_z) \to (0, 0, \Omega) \text{ as } t \to \infty,$$

corresponding to an asymptotically constant spin rate Ω about the body-fixed z axis, which is aligned with the inertial Z axis, and the closed loop has desired properties. It is assumed that the spacecraft dynamics and kinematics are described by the linearized Equations 7.67 and 7.68. Since the spacecraft equations are linear and uncoupled, a standard linear control approach can be used. Consider the linear feedback control law of the form

$$\begin{aligned} u_x &= -(I_{yy} - I_{zz})\Omega(\dot{\theta} + \Omega\phi) - I_{xx}\Omega\dot{\theta} - c_x\dot{\phi} - k_x\phi, \\ u_y &= -(I_{zz} - I_{xx})\Omega(\dot{\phi} - \Omega\theta) + I_{yy}\Omega\dot{\phi} - c_y\dot{\theta} - k_y\theta, \\ u_z &= -c_z(\omega_z - \Omega). \end{aligned} \quad (7.78)$$

Recall that the angular velocity components are related to the rates of change of the Euler angles according to the linearized relations of Equation 7.68, so that the control law can be expressed either in terms of the rates of change of the Euler angles (as in Equation 7.78) or in terms of the components of the angular velocity vector.

Based on the linearized equations, introduce the variables

$$\varepsilon_x = \phi, \quad \varepsilon_y = \theta, \quad \delta_z = \omega_z - \Omega, \quad (7.79)$$

so that the closed-loop equations are

$$\begin{aligned} I_{xx}\ddot{\varepsilon}_x + c_x\dot{\varepsilon}_x + k_x\varepsilon_x &= 0, \\ I_{yy}\ddot{\varepsilon}_y + c_y\dot{\varepsilon}_y + k_y\varepsilon_y &= 0, \\ I_{zz}\dot{\delta}_z + c_z\delta_z &= 0. \end{aligned} \quad (7.80)$$

Hence, the spacecraft attitude errors in pitch angle and roll angle are brought to zero asymptotically and the yaw rate is brought to the value Ω asymptotically, at least for sufficiently small perturbations. The values of the control gains can be chosen to provide specified closed-loop transient responses. Consequently, the above control law is guaranteed to bring the spacecraft to the desired spin rate about the specified axis of rotation, asymptotically, with speed of response that is determined by the values of the control gains. The control law of Equation 7.78 has the desirable property that it requires feedback of only the pitch and roll angles, which characterize, in this case, the errors in the instantaneous spin axis; feedback of the yaw angle is not required.

Again, it is important to note that the preceding analysis holds only for sufficiently small perturbations in the spacecraft angular velocity and attitude. For large perturbations in the angular velocity and attitude of the spacecraft, the above control law may not have the desired properties. An analysis of the closed-loop system, using the nonlinear dynamics Equation 7.61 and the nonlinear kinematic Equation 7.57, is required to determine the domain of perturbations for which closed-loop stabilization is achieved. Such an analysis is beyond the scope of the present chapter, but we note that the stability domain depends on the desired spin rate and the control gains that are selected.

7.2.9 Spacecraft Control Problem: Bang-Bang Control Law Based on Linearized Spacecraft Equations

As developed previously, it is assumed that the spacecraft kinematics and dynamics are described by linear equations obtained by linearizing about the rest solution. The linearized spacecraft dynamics are described by Equation 7.64, and the linearized spacecraft kinematics are described by Equation 7.65.

Since certain types of thrusters are most easily operated in an on-off mode, the control moments produced by each pair of thrusters may be limited to fixed values (of either sign) or to zero. We now impose these control constraints on the development of the control law by requiring that each of the control moment components can take only the values $\{-U, 0, +U\}$ at each instant of time.

We first present a simple control law that *stabilizes the spacecraft to rest*, ignoring the spacecraft attitude. The simplest control law, satisfying the imposed constraints, that stabilizes the spacecraft to rest is given by

$$\begin{aligned} u_x &= -U\,\text{sgn}(\omega_x), \\ u_y &= -U\,\text{sgn}(\omega_y), \\ u_z &= -U\,\text{sgn}(\omega_z), \end{aligned} \quad (7.81)$$

where the signum function is the discontinuous function defined by

$$\text{sgn}(\sigma) = \begin{bmatrix} -1 & \text{if} & \sigma < 0, \\ 0 & \text{if} & \sigma = 0, \\ 1 & \text{if} & \sigma > 0. \end{bmatrix} \quad (7.82)$$

It is easily shown that the resulting closed-loop system has the property that

$$(\omega_x, \omega_y, \omega_z) \to (0, 0, 0) \text{ as } t \to \infty,$$

at least for sufficiently small perturbations in the spacecraft angular velocity vector.

There are necessarily errors in measuring the angular velocity vector, and it can be shown that the above control law can be improved by using the modified feedback control law

$$\begin{aligned} u_x &= -U \operatorname{dez}(\omega_x, \varepsilon), \\ u_y &= -U \operatorname{dez}(\omega_y, \varepsilon), \\ u_z &= -U \operatorname{dez}(\omega_z, \varepsilon), \end{aligned} \qquad (7.83)$$

where the dead-zone function is the discontinuous function defined by

$$\operatorname{dez}(\sigma, \varepsilon) = \begin{bmatrix} -1 & \text{if} & \sigma < -\varepsilon, \\ 0 & \text{if} & -\varepsilon < \sigma < \varepsilon, \\ 1 & \text{if} & \sigma > \varepsilon, \end{bmatrix} \qquad (7.84)$$

and ε is a positive constant, the dead-zone width. The resulting closed-loop system has the property that

$$(\omega_x, \omega_y, \omega_z) \to S \text{ as } t \to \infty,$$

where S is an open set containing (0,0,0); the (maximum) diameter of S has the property that it goes to zero as ε goes to zero. Thus, the dead-zone parameter ε can be selected appropriately, so that the angular velocity vector is maintained small while the closed loop is not excessively sensitive to measurement errors in the angular velocity vector.

We now present a simple control law that *stabilizes the spacecraft to a fixed attitude*, which is given by the Euler angles being all zero. The simplest control law, satisfying the imposed constraints, is of the form

$$\begin{aligned} u_x &= -U \operatorname{sgn}(\phi + \tau_x \dot{\phi}), \\ u_y &= -U \operatorname{sgn}(\theta + \tau_y \dot{\theta}), \\ u_z &= -U \operatorname{sgn}(\Psi + \tau_z \dot{\Psi}), \end{aligned} \qquad (7.85)$$

where τ_x, τ_y, τ_z are positive constants and the linear arguments in the control laws of Equation 7.85 are the switching functions. Consequently, the closed-loop system is decoupled into independent closed loops for the roll angle, the pitch angle, and the yaw angle. By analyzing each of these closed-loop systems, it can be shown that

$$(\Psi, \theta, \phi) \to (0, 0, 0) \text{ as } t \to \infty,$$

at least for sufficiently small perturbations in the spacecraft angular velocity vector and attitude. It should be noted that a part of the solution involves a chattering solution where the switching functions are identically zero over a finite time interval.

There are necessarily errors in measuring the angular velocity vector and the attitude, and it can be shown that the above control law can be improved by using the modified feedback control law

$$\begin{aligned} u_x &= -U \operatorname{dez}(\phi + \tau_x \dot{\phi}, \varepsilon), \\ u_y &= -U \operatorname{dez}(\theta + \tau_y \dot{\theta}, \varepsilon), \\ u_z &= -U \operatorname{dez}(\Psi + \tau_z \dot{\Psi}, \varepsilon), \end{aligned} \qquad (7.86)$$

where ε is a positive constant, the dead-zone width. The resulting closed-loop system has the property that

$$(\Psi, \theta, \phi) \to S \text{ as } t \to \infty,$$

where S is an open set containing (0, 0, 0); in fact, the spacecraft attitude is asymptotically periodic with a maximum amplitude that tends to zero as ε tends to zero. Thus, the dead-zone parameter ε can be selected appropriately, so that the angular velocity vector and the attitude errors are maintained small while the closed loop is not excessively sensitive to measurement errors in the angular velocity vector or the attitude errors. Further details are available in [1].

7.2.10 Spacecraft Control Problem: Nonlinear Control Law Based on Nonlinear Spacecraft Equations

In this section, we present nonlinear control laws that guarantee that the closed-loop equations are exactly linear; this approach can be viewed as using feedback both to cancel out the nonlinear terms and then to add in linear terms that result in good closed-loop linear characteristics. This approach is often referred to as feedback linearization or dynamic inversion.

We first consider *control of the spacecraft angular velocity*. It is assumed that the spacecraft dynamics are described by the nonlinear Equation 7.61 and a control law (u_x, u_y, u_z) is desired in feedback form, so that the resulting closed-loop system is linear and satisfies the asymptotic condition

$$(\omega_x, \omega_y, \omega_z) \to (\omega_{xd}, \omega_{yd}, \omega_{zd}) \text{ as } t \to \infty.$$

Control laws that accomplish these objectives are given in the form

$$\begin{aligned} u_x &= -(I_{yy} - I_{zz})\omega_y\omega_z - c_x(\omega_x - \omega_{xd}) + I_{xx}\dot{\omega}_{xd}, \\ u_y &= -(I_{zz} - I_{xx})\omega_z\omega_x - c_y(\omega_y - \omega_{yd}) + I_{yy}\dot{\omega}_{yd}, \\ u_z &= -(I_{xx} - I_{yy})\omega_x\omega_y - c_z(\omega_z - \omega_{zd}) + I_{zz}\dot{\omega}_{zd}, \end{aligned} \qquad (7.87)$$

where the control gains c_x, c_y, c_z are chosen as positive constants. If the variables

$$\delta_x = \omega_x - \omega_{xd}, \quad \delta_y = \omega_y - \omega_{yd}, \quad \delta_z = \omega_z - \omega_{zd}, \qquad (7.88)$$

are introduced, the closed-loop equations are

$$\begin{aligned} \dot{\delta}_x + \left(\frac{c_x}{I_{xx}}\right)\delta_x &= 0, \\ \dot{\delta}_y + \left(\frac{c_y}{I_{yy}}\right)\delta_y &= 0, \\ \dot{\delta}_z + \left(\frac{c_z}{I_{zz}}\right)\delta_z &= 0. \end{aligned} \qquad (7.89)$$

Hence, the angular velocity errors in roll rate, pitch rate, and yaw rate are brought to zero asymptotically. The values of the control gains can be chosen to provide specified closed-loop time constants. Consequently, the above control laws are guaranteed to achieve exact tracking asymptotically, with speed of response that is determined by the values of the control gains.

A simple special case of the above control law corresponds to the choice that $(\omega_{xd}, \omega_{yd}, \omega_{zd}) = (0, 0, 0)$. Thus, the simplified control law is given by

$$\begin{aligned} u_x &= -(I_{yy} - I_{zz})\omega_y\omega_z - c_x\omega_x, \\ u_y &= -(I_{zz} - I_{xx})\omega_z\omega_x - c_y\omega_y, \\ u_z &= -(I_{xx} - I_{yy})\omega_x\omega_y - c_z\omega_z, \end{aligned} \tag{7.90}$$

and it is guaranteed to bring the spacecraft asymptotically to rest.

The preceding analysis holds globally, that is, for all possible perturbations in the angular velocity vector, since the closed-loop system is exactly linear and asymptotically stable. This is in sharp contrast with the development that was based on the linearized equations. Thus, the family of control laws given here provides excellent closed-loop properties. The price of such good performance is the relative complexity of the control laws and the associated difficulty in practical implementation.

We now consider *spin stabilization of the spacecraft about a specified inertial axis.* It is assumed that the spacecraft dynamics are described by the nonlinear Equation 7.61, and the kinematics are described by the nonlinear Equation 7.57. A control law (u_x, u_y, u_z) is desired in feedback form, so that the resulting closed-loop system is linear and satisfies the control objective

$$\begin{aligned} &(\theta, \phi) \to (0, 0) \text{ as } t \to \infty, \\ &(\omega_x, \omega_y, \omega_z) \to (0, 0, \Omega) \text{ as } t \to \infty. \end{aligned}$$

This corresponds to control of the spacecraft so that it asymptotically spins about its body-fixed z axis at a spin rate Ω, and this spin axis is aligned with the inertially fixed Z axis.

In order to obtain closed-loop equations that are exactly linear, differentiate each of the first two equations in Equation 7.61, substitute for the time derivatives of the angular velocities from Equation 7.61; then select the feedback control law to cancel out all nonlinear terms in the resulting equations and add in desired linear terms. After considerable algebra, this feedback linearization approach leads to the following nonlinear control law

$$\begin{aligned} u_x &= -I_{xx} f_x - c_x\dot{\phi} - k_x\phi, \\ u_y &= -[I_{yy} f_y + c_y\dot{\theta} + k_y\theta] \sec\phi, \\ u_z &= (I_{yy} - I_{xx})\omega_x\omega_y - c_z(\omega_z - \Omega), \end{aligned} \tag{7.91}$$

where

$$\begin{aligned} f_y &= \frac{c_z}{I_{zz}}(\omega_z - \Omega)\sin\phi + \frac{(I_{zz} - I_{xx})}{I_{yy}}\omega_x\omega_z\cos\phi \\ &\quad - \omega_y\dot{\phi}\sin\phi - \omega_z\dot{\phi}\cos\phi, \\ f_x &= \frac{(I_{yy} - I_{zz})}{I_{xx}}\omega_y\omega_z \\ &\quad + \frac{(I_{zz} - I_{xx})}{I_{yy}}\omega_x\omega_z\sin\phi\tan\theta \\ &\quad - [f_y + \frac{c_y}{I_{yy}}\dot{\theta} + \frac{k_y}{I_{yy}}\theta]\tan\phi\tan\theta \\ &\quad - \frac{c_z}{I_{zz}}(\omega_z - \Omega)\cos\phi\tan\theta \\ &\quad + \omega_y\frac{d}{dt}[\sin\phi\tan\theta] + \omega_z\frac{d}{dt}[\cos\phi\tan\theta]. \end{aligned} \tag{7.92}$$

Note that the control law can be expressed either in terms of the rates of change of the Euler angles or in terms of the components of the angular velocity vector; the above expression involves a mixture of both. In addition, the control law can be seen not to depend on the yaw angle.

If we introduce the variables

$$\varepsilon_x = \phi, \quad \varepsilon_y = \theta, \quad \delta_x = \omega_z - \Omega, \tag{7.93}$$

it can be shown (after substantial algebra) that the closed-loop equations are the linear decoupled equations

$$\begin{aligned} I_{xx}\ddot{\varepsilon}_x + c_x\dot{\varepsilon}_x + k_x\varepsilon_x &= 0, \\ I_{yy}\ddot{\varepsilon}_y + c_y\dot{\varepsilon}_y + k_y\varepsilon_y &= 0, \\ I_{zz}\dot{\delta}_z + c_z\delta_z &= 0. \end{aligned} \tag{7.94}$$

Hence, the spacecraft attitude errors in pitch angle and roll angle are brought to zero asymptotically and the yaw rate is brought to the value Ω asymptotically. The values of the control gains can be chosen to provide specified closed-loop responses. Consequently, the above control laws are guaranteed to bring the spacecraft to the desired spin rate about the specified axis of rotation with speed of response that is determined by the values of the control gains.

The preceding analysis holds nearly globally, that is, for all possible perturbations in the angular velocity vector and for all possible perturbations in the Euler angles, excepting the singular values, since the closed-loop system of Equation 7.94 is exactly linear. This is in sharp contrast with the previous development that was based on linearized equations. It can easily be seen that the control law of Equation 7.78, obtained using the linearized approximation, also results from a linearization of the nonlinear control law of Equation 7.91 obtained in this section. Thus, the control law given by Equation 7.91 provides excellent closed-loop properties. The price of such good performance is the relative complexity of the control law of Equation 7.91 and the associated difficulty in its practical implementation.

7.2.11 Spacecraft Control Problem: Attitude Control in Circular Orbit

An important case of interest is that of a spacecraft in a circular orbit. In such a case, it is natural to describe the orientation of the spacecraft not with respect to an inertially fixed coordinate frame, but rather with respect to a locally horizontal-vertical coordinate frame as reference, defined so that the X axis of this frame is tangent to the circular orbit in the direction of the orbital motion, the Z axis of this frame is directed radially at the center of attraction, and the Y axis completes a right-hand orthogonal frame. Let the constant orbital angular velocity of the locally horizontal coordinate frame be

$$\Omega = \sqrt{\frac{g}{R}} \tag{7.95}$$

where g is the local acceleration of gravity and R is the orbital radius; the direction of the orbital angular velocity vector is along the negative Y axis.

The nonlinear dynamics for a spacecraft in circular orbit can be described in terms of the body-fixed angular velocity components. However, moment terms arise from the gravitational forces on the spacecraft when the spacecraft is modeled as a finite body rather than a point mass. These gravity gradient moments can be expressed in terms of the orbital angular velocity and the Euler angles, which describe the orientation of the body-fixed frame with respect to the locally horizontal-vertical coordinate frame. The complete nonlinear dynamics equations are not presented here due to their complexity.

The linearized expressions for the spacecraft dynamics about the constant angular velocity solution $(0, -\Omega, 0)$ are obtained by introducing the perturbation variables

$$\delta_x = \omega_x, \quad \delta_y = \omega_y + \Omega, \quad \delta_z = \omega_z; \tag{7.96}$$

the resulting linearized dynamics equations are

$$\begin{aligned} I_{xx}\dot{\delta}_x &= \Omega(I_{zz} - I_{yy})\delta_z + T_x, \\ I_{yy}\dot{\delta}_y &= T_y, \\ I_{zz}\dot{\delta}_z &= \Omega(I_{yy} - I_{xx})\delta_x + T_z, \end{aligned} \tag{7.97}$$

The external moments on the spacecraft are given by

$$\begin{aligned} T_x &= 3\Omega^2(I_{zz} - I_{yy})\phi + u_x, \\ T_y &= 3\Omega^2(I_{xx} - I_{zz})\theta + u_y, \\ T_z &= u_z, \end{aligned} \tag{7.98}$$

where (u_x, u_y, u_z) denotes the control moments, and the other terms describe the linearized gravity gradient moments on the spacecraft. Thus, the linearized spacecraft dynamics are given by

$$\begin{aligned} I_{xx}\dot{\delta}_x &= \Omega(I_{zz} - I_{yy})\delta_z + 3\Omega^2(I_{zz} - I_{yy})\phi + u_x, \\ I_{yy}\dot{\delta}_y &= -3\Omega^2(I_{xx} - I_{zz})\theta + u_y, \\ I_{zz}\dot{\delta}_z &= \Omega(I_{yy} - I_{xx})\delta_x + u_z. \end{aligned} \tag{7.99}$$

The linearized kinematics equations are given by

$$\begin{aligned} \dot{\phi} &= \delta_x + \Omega\Psi, \\ \dot{\theta} &= \delta_y, \\ \dot{\Psi} &= \delta_z - \Omega\phi, \end{aligned} \tag{7.100}$$

where the extra terms arise since the locally horizontal-vertical coordinate frame has a constant angular velocity of $-\Omega$ about the Y axis. Thus, the linearized dynamics and kinematics equations can be combined in a form that makes clear that the pitching motion is decoupled from the rolling and yawing motions

$$I_{yy}\ddot{\theta} = -3\Omega^2(I_{xx} - I_{zz})\theta + u_y, \tag{7.101}$$

but the rolling and yawing motions are coupled

$$\begin{aligned} I_{xx}\ddot{\phi} &= \Omega(I_{xx} + I_{zz} - I_{yy})\dot{\Psi} \\ &\quad + 4\Omega^2(I_{zz} - I_{yy})\phi + u_x, \\ I_{zz}\ddot{\Psi} &= \Omega(I_{yy} - I_{xx} - I_{zz})\dot{\phi} - \Omega^2(I_{yy} - I_{xx})\Psi + u_z. \end{aligned} \tag{7.102}$$

A control law (u_x, u_y, u_z) is desired in feedback form, so that the resulting closed-loop system satisfies the asymptotic attitude conditions

$$\begin{aligned} (\Psi, \theta, \phi) &\to (0, 0, 0) \text{ as } t \to \infty, \\ (\delta_x, \delta_y, \delta_z) &\to (0, 0, 0) \text{ as } t \to \infty, \end{aligned}$$

which guarantees that

$$(\omega_x, \omega_y, \omega_z) \to (0, -\Omega, 0) \text{ as } t \to \infty;$$

that is, the spacecraft has an asymptotically constant spin rate consistent with the orbital angular velocity as desired.

Since the preceding spacecraft equations are linear, a standard linear control approach can be used to obtain linear feedback control laws for the pitching motion and for the rolling and yawing motion of the form

$$\begin{aligned} u_x &= -\Omega(I_{xx} + I_{zz} - I_{yy})\dot{\Psi} - c_x\dot{\phi} \\ &\quad - \left[4\Omega^2(I_{zz} - I_{yy}) + k_x\right]\phi, \\ u_y &= -c_y\dot{\theta} - [3\Omega^2(I_{xx} - I_{zz}) + k_y]\theta, \\ u_z &= -c_z\dot{\Psi} - [-\Omega^2(I_{yy} - I_{xx}) + k_z]\Psi \\ &\quad + \Omega(I_{xx} + I_{zz} - I_{yy})\dot{\phi}. \end{aligned} \tag{7.103}$$

The control law can be expressed either in terms of the rates of change of the Euler angles or in terms of the components of the angular velocity vector.

If we introduce the perturbation variables

$$\varepsilon_x = \phi, \quad \varepsilon_y = \theta, \quad \varepsilon_z = \Psi, \tag{7.104}$$

the resulting closed-loop system is described by

$$\begin{aligned} I_{xx}\ddot{\varepsilon}_x + c_x\dot{\varepsilon}_x + k_x\varepsilon_x &= 0, \\ I_{yy}\ddot{\varepsilon}_y + c_y\dot{\varepsilon}_y + k_y\varepsilon_y &= 0, \\ I_{zz}\ddot{\varepsilon}_z + c_z\dot{\varepsilon}_z + k_z\varepsilon_z &= 0. \end{aligned} \tag{7.105}$$

If the gains are chosen so that the pitching motion is asymptotically stable and the rolling and yawing motion is asymptotically stable, then the spacecraft attitude errors are automatically brought to zero asymptotically, at least for sufficiently small perturbations of the orientation from the locally horizontal-vertical reference. The values of the control gains can be chosen to provide specified closed-loop response properties. Consequently, the above control laws are guaranteed to bring the spacecraft to the desired attitude with speed of response that is determined by the values of the control gains.

We again note that the preceding analysis holds only for sufficiently small attitude perturbations from the local horizontal-vertical reference. For large perturbations, the above control laws may not have the desired properties. An analysis of the closed-loop system, using the nonlinear dynamics and kinematics equations, is required to determine the domain of perturbations for which closed-loop stabilization is achieved. Such an analysis is beyond the scope of the present chapter.

7.2.12 Other Spacecraft Control Problems and Control Methodologies

Our treatment of spacecraft attitude control has been limited, both by the class of rotational control problems considered and by the assumptions that have been made. In this section, we briefly indicate other classes of problems for which results are available in the published literature.

Throughout our development, specific control laws have been developed using orientation representations expressed in terms of Euler angles. As we have indicated, the Euler angles are not global representations for orientation. Other orientation representations, including quaternions and Euler axis-angle variables, have been studied and various control laws have been developed using these representations.

Several of our control approaches have been based on use of the linearized spacecraft kinematics and dynamics equations. We have also suggested a nonlinear control approach, feedback linearization, to develop several classes of feedback control laws for the nonlinear kinematics and dynamics equations. Other types of control approaches have been studied for spacecraft reorientation, including optimal control and pulse-width-modulated control schemes.

We should also mention other classes of spacecraft attitude control problems. Our approach has assumed use of pairs of gas-jet thrusters or reaction wheels modeled in the simplest way. Other assumptions lead to somewhat different models and, hence, somewhat different control problems. In particular, we mention that control problems in the case there are only two, rather than three, control moments have recently been studied.

A key assumption throughout our development is that the spacecraft is a rigid body. There are important spacecraft designs where this assumption is not satisfied, and the resulting attitude control problems are somewhat different from what has been considered here; usually these control problems are even more challenging. Examples of such control problems occur when nutation dampers or control moment gyros are used for attitude control. Dual-spin spacecraft and multibody spacecraft are examples where there is relative motion between spacecraft components that must be taken into account in the design of attitude control systems. Numerous modern spacecraft, due to weight constraints, consist of flexible components. Attitude control of flexible spacecraft is a very important and widely studied subject.

7.2.13 Defining Terms

Attitude: The orientation of the spacecraft with respect to some reference frame.

Body axes: A reference frame fixed in the spacecraft body and rotating with it.

Euler angles: A sequence of angle rotations that are used to parametrize a rotation matrix.

Pitch: The second angle in the 3-2-1 Euler angle sequence. For small rotation angles, the pitch angle is the rotation angle about the spacecraft y axis.

Roll: The third angle in the 3-2-1 Euler angle sequence. For small rotation angles, the roll angle is the rotation angle about the spacecraft x axis.

Rotation matrix: A matrix of direction cosines relating unit vectors of two different coordinate frames.

Simple spin: A spacecraft spinning about a body axis whose direction remains inertially fixed.

Yaw: The first rotation angle in the 3-2-1 Euler angle sequence. For small rotation angles, the yaw angle is the rotation angle about the spacecraft z axis.

References

[1] Bryson, A.E., *Control of Spacecraft and Aircraft*, Princeton University Press, Princeton, NJ, 1994.

[2] Crouch, P.E., Spacecraft attitude control and stabilization: applications of geometric control theory to rigid body models, *IEEE Trans. Autom. Control*, 29(4), 321–331, 1984.

[3] Greenwood, D.T., *Principles of Dynamics*, 2nd ed., Prentice Hall, Englewood Cliffs, NJ, 1988.

[4] Hughes, P.C., *Spacecraft Attitude Dynamics*, Wiley, New York, 1986.

[5] Wertz, J.R., Ed., *Spacecraft Attitude Determination and Control*, Kluwer, Dordrecht, Netherlands, 1978.

Further Reading

The most complete reference on the control of spacecraft is *Spacecraft Attitude Determination and Control*, a handbook edited by J. R. Wertz [5]. It has been reprinted often and is available from Kluwer Academic Publishers.

Two introductory textbooks on spacecraft dynamics are Wiesel, W.E. 1989, *Spaceflight Dynamics*, McGraw-Hill, New York, and Thomson, W.T. 1986, *Introduction to Space Dynamics*, Dover Publications (paperback), NY, 1986.

A more comprehensive treatment can be found in *Spacecraft Attitude Dynamics* by P.C. Hughes [4].

A discussion of the orbital motion of spacecraft can be found in Bate, Mueller, White, 1971, *Fundamentals of Astrodynamics*, Dover Publications, New York, and Danby, J.M.A. 1988, *Fundamentals of Celestial Mechanics*, 2nd ed., Willmann-Bell Inc., Richmond, VA.

7.3 Control of Flexible Space Structures

S. M. Joshi and A. G. Kelkar, NASA *Langley Research Center*

7.3.1 Introduction[2]

A number of near-term space missions as well as future mission concepts will require flexible space structures (FSS) in low Earth and geostationary orbits. Examples of near-term missions include multipayload space platforms, such as Earth observing systems and space-based manipulators for on-orbit assembly and satellite servicing. Examples of future space mission concepts include mobile satellite communication systems, solar power satellites, and large optical reflectors, which would require large antennas, platforms, and solar arrays. The dimensions of such structures would range from 50 meters (m) to several kilometers (km). Because of their relatively light weight and, in some cases, expansive sizes, such structures tend to have low-frequency, lightly damped structural (elastic) modes. The natural frequencies of the elastic modes are generally closely spaced, and some natural frequencies may be lower than the controller bandwidth. In addition, the elastic mode characteristics are not known accurately. For these reasons, control systems design for flexible space structures is a challenging problem.

Depending on their missions, flexible spacecraft can be roughly categorized as single-body spacecraft and multibody spacecraft. Two of the most important control problems for single-body FSS are (1) fine-pointing of FSS in space with the required precision in attitude (represented by three Euler angles) and shape, and (2) large-angle maneuvering ("slewing") of the FSS to orient to a different target. The performance requirements for both of these problems are usually very high. For example, for a certain mobile communication system concept, a 122-meter diameter space antenna will have to be pointed with an accuracy of 0.03 degree root mean square (RMS). The requirements for other missions vary, but some are expected to be even more stringent, on the order of 0.01 arc-second. In some applications, it would be necessary to maneuver the FSS quickly through large angles to acquire a new target on the Earth in minimum time and with minimum fuel expenditure, while keeping the elastic motion and accompanying stresses within acceptable limits. Once the target is acquired, the FSS must point to it with the required precision.

For multibody spacecraft with articulated appendages, the main control problems are (1) fine-pointing of some of the appendages to their respective targets, (2) rotating some of the appendages to follow prescribed periodic scanning profiles, and (3) changing the orientation of some of the appendages through large angles. For example, a multipayload platform would have the first two requirements, while a multilink manipulator would have the third requirement to reach a new end-effector position.

The important feature that distinguishes FSS from conventional older generation spacecraft is their highly prominent structural flexibility which results in special dynamic characteristics. Detailed literature surveys on dynamics and control of FSS may be found in [11], [20].

The organization of this chapter is as follows. The problem of fine-pointing control of single-body spacecraft is considered in Section 7.3.2. This problem not only represents an important class of missions, but also permits analysis in the linear, time-invariant (LTI) setting. The basic linearized mathematical model of single-body FSS is presented, and the problems encountered in FSS control systems design are discussed. Two types of controller design methods, model-based controllers and passivity-based controllers, are presented. The highly desirable robustness characteristics of passivity-based controllers are summarized. Section 7.3.3 addresses another important class of missions, namely, multibody FSS. A generic nonlinear mathematical model of a multibody flexible space system is presented and passivity-based robust controllers are discussed.

7.3.2 Single-Body Flexible Spacecraft

Linearized Mathematical Model

Simple structures, such as uniform beams or plates, can be effectively modeled by infinite-dimensional systems (see [12]). In some cases, approximate infinite-dimensional models have been proposed for more complex structures such as trusses [2]. However, most of the realistic FSS are highly complex, not amenable to infinite-dimensional modeling. The standard practice is to use finite-dimensional mathematical models generated by using the finite-element method [19]. The basic approach of this method is dividing a continuous system into a number of elements using fictitious dividing lines, and applying the Lagrangian formulation to determine the forces at the points of intersection as functions of the applied forces. Suppose there are r force actuators and p torque actuators distributed throughout the structure. The ith force actuator produces the 3×1 force vector $f_i = (f_{xi}, f_{yi}, f_{zi})^T$, along the X, Y, and Z axes of a body-fixed coordinate system centered at the nominal center of mass (c.m.). Similarly, the ith torque actuator produces the torque vector $T_i = (T_{xi}, T_{yi}, T_{zi})^T$. Then the linearized equations of motion can be written as follows [12]:

rigid-body translation:

$$M\ddot{z} = \sum_{i=1}^{r} f_i, \tag{7.106}$$

rigid-body rotation:

$$J\ddot{\alpha} = \sum_{i=1}^{r} R_i \times f_i + \sum_{i=1}^{p} T_i, \quad \text{and} \tag{7.107}$$

elastic motion:

$$\ddot{q} + D\dot{q} + \Lambda q = \sum_{i=1}^{r} \Delta_i^T f_i + \sum_{i=1}^{p} \Phi_i^T T_i, \tag{7.108}$$

[2]This article is based on the work performed for the U.S. Government. The responsibility for the contents rests with the authors.

where M is the mass, z is the 3×1 position of the c.m., R_i is the location of f_i on the FSS, J is the 3×3 moment-of-inertia matrix, α is the attitude vector consisting of the three Euler rotation angles (ϕ, θ, ψ), and $q = (q_1, q_2, \ldots, q_{nq})^T$ is the $n_q \times 1$ modal amplitude vector for the n_q elastic modes. ("$\times$" in Equation 7.107 denotes the vector cross-product.) In general, the number of modes (n_q) necessary to characterize an FSS adequately is quite large, on the order of 100–1000. Δ_i^T, Φ_i^T are the $n_q \times 3$ translational and rotational mode shape matrices at the ith actuator location. The rows of Δ_i^T, Φ_i^T represent the X, Y, Z components of the translational and rotational mode shapes at the location of actuator i.

$$\Lambda = diag(\omega_1^2, \omega_2^2, \ldots, \omega_{n_q}^2) \tag{7.109}$$

where ω_k is the natural frequency of the kth elastic mode, and D is an $n_q \times n_q$ matrix representing the inherent damping in the elastic modes:

$$D = 2\,diag\,(\rho_1\omega_1, \rho_2\omega_2, ..., \rho_{nq}\omega_{nq}). \tag{7.110}$$

The inherent damping ratios (ρ_is) are typically on the order of 0.001–0.01. The finite-element method cannot model inherent damping. This proportional damping term is customarily added after an undamped finite-element model is obtained.

The translational and rotational positions z_p and y_p, at a location with coordinate vector R, are given by

$$z_p = z - R \times \alpha + \overline{\Delta} q \tag{7.111}$$

$$\text{and} \quad y_p = \alpha + \overline{\Phi} q, \tag{7.112}$$

where the ith columns of $\overline{\Delta}$ and $\overline{\Phi}$ represent the ith $3 \times n_q$ translational and rotational mode shapes at that location.

Figure 7.17 shows the mode-shape plots for a finite-element model of a completely free, 100 ft. $\times$ 100 ft. $\times$ 0.1 in. aluminum plate. The plots were obtained by computing the elastic displacements at many locations on the plate, resulting from nonzero values of individual modal amplitudes q_i. Figure 7.18 shows the mode-shape plots for the 122-m diameter, hoop/column antenna concept [22], which consists of a deployable central mast attached to a deployable hoop by cables held in tension.

Controllability and Observability

Precision attitude control is usually accomplished by torque actuators. Therefore, in the following material, only rotational equations of motion are considered. No force actuators are used. With $f_i = 0$ and denoting $\xi = [\alpha^T, q_1, q_2, \ldots q_{nq}]^T$, Equations 7.107 and 7.108 can be written as

$$\tilde{A}\ddot{\xi} + \tilde{B}\dot{\xi} + \tilde{C}\xi = \Gamma^T u \tag{7.113}$$

$$\tilde{A} = diag(J, I_{n_q}), \tilde{B} = diag(0_3, D), \tilde{C} = diag(0_3, \Lambda) \tag{7.114}$$

$$\Gamma^T = \begin{bmatrix} I_{3\times3} \quad I_{3\times3} \ldots I_{3\times3} \\ \Phi^T \end{bmatrix}_{n_1 \times m} \tag{7.115}$$

where $\Phi^T = \left[\Phi_1^T, \Phi_2^T, \ldots \Phi_p^T\right]$, $n_1 = n_q + 3$, and $m = 3p$. The system can be written in the state-space form as

$$\dot{x} = Ax + Bu \tag{7.116}$$

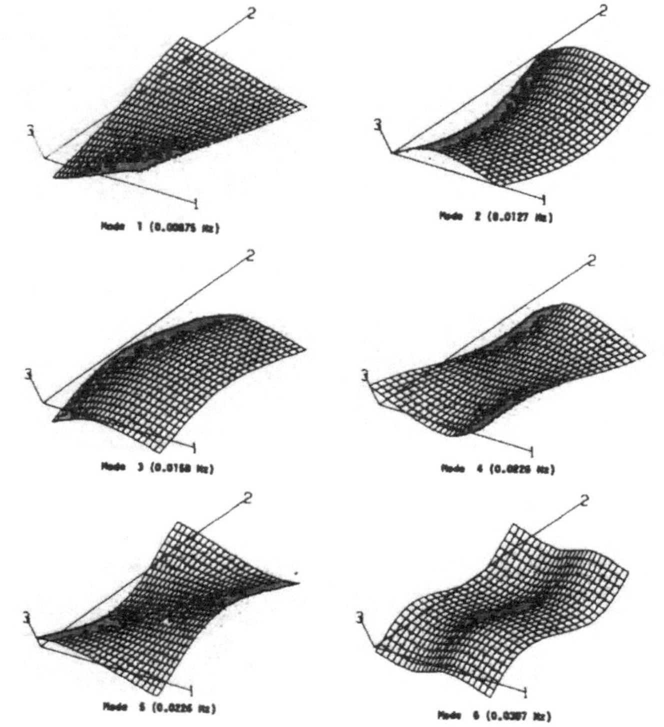

Figure 7.17 Mode-shape plots for a completely free plate.

where $x = (\alpha^T, \dot{\alpha}^T, q_1, \dot{q}_1, \ldots, q_{nq}, \dot{q}_{nq})^T$ is the n-dimensional state vector ($n = 2n_1$) and u is the $m \times 1$ control vector consisting of applied torques.

$$A = diag(A_{rb}, A_1, A_2, ..., A_{nq}) \tag{7.117}$$

$$A_{rb} = \begin{bmatrix} 0_3 & I_3 \\ 0_3 & 0_3 \end{bmatrix}; A_i = \begin{bmatrix} 0 & 1 \\ -\omega_i^2 & -2\rho_i\omega_i \end{bmatrix} \tag{7.118}$$

(0_k and I_k denote the $k \times k$ null and identity matrices, respectively.)

$$B = \left[B_{rb}^T, : 0_{m\times1}, \phi_1 : 0_{m\times1}, \phi_2 : \ldots : 0_{m\times1}, \phi_{nq}\right]^T \tag{7.119}$$

$$B_{rb} = \begin{bmatrix} 0_{3\times m} \\ J^{-1}, J^{-1} \ldots J^{-1} \end{bmatrix}_{6\times m} \tag{7.120}$$

where ϕ_k^T represents the kth row of Φ^T.

If a three-axis attitude sensor is placed at a location where the rotational mode shapes are given by the rows of the $n_q \times 3$ matrix Ψ^T, the sensed attitude (ignoring noise) would be

$$y_p = \alpha + \Psi q = \left[I_3, \Psi^T\right]\xi. \tag{7.121}$$

The conditions for controllability are given below.

Controllability Conditions

The system given by Equation 7.116 is controllable if, and only if, (iff) the following conditions are satisfied:

1. Rows of Φ^T corresponding to each distinct (in frequency) elastic mode have at least one nonzero entry.

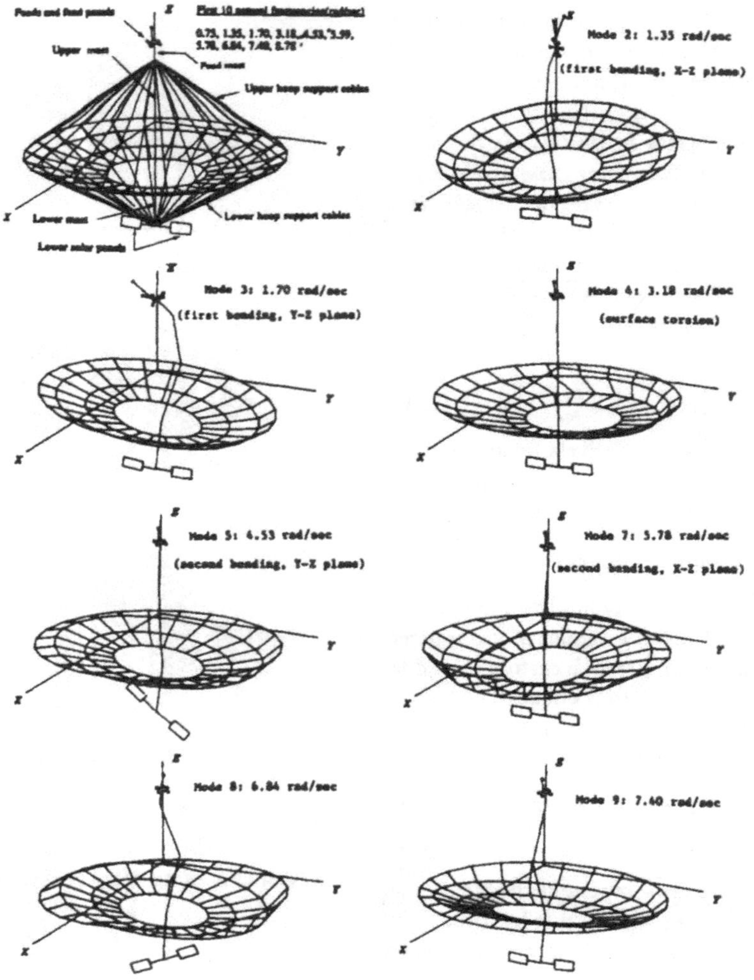

Figure 7.18 Typical mode shapes of hoop/column antenna.

2. If there are ν elastic modes with the same natural frequency $\overline{\omega}$, the corresponding rows of Φ^T form a linearly independent set.

A proof of this result can be found in [12]. Condition (1) would be satisfied iff the rotational mode shape (X, Y, or Z component) for each mode is nonzero at the location of at least one actuator. Condition (2) needs to be tested only when there is more than one elastic mode with the same natural frequency, which can typically occur in the case of symmetric structures.

Similar necessary and sufficient conditions can be obtained analogously for observability. It should be noted that the rigid-body modes are not observable using attitude-rate sensors alone without attitude sensors. However, a three-axis attitude sensor can be sufficient for observability even if no rate sensors are used.

Problems in Controller Design for FSS

Precision attitude control requires controlling the rigid rotational modes and suppressing the elastic vibration. The objectives of the controller are

1. **fast transient response:** Quickly damp out the pointing errors resulting from step disturbances such as thermal distortion resulting from entering or leaving Earth's shadow or nonzero initial conditions, resulting from the completion of a large-angle attitude maneuver.
2. **disturbance rejection:** Maintain the attitude as close as possible to the desired attitude in the presence of noise and disturbances.

The first objective translates into the closed-loop bandwidth requirement, and the second translates into minimizing the RMS pointing error. In addition, the elastic motion must be very small, i.e., the RMS shape distortions must be below prescribed limits. For applications such as large communications antennas, the typical bandwidth requirement is 0.1 rad/sec., with at most a 4 sec. time constant for all of the elastic modes (closed loop). Typical allowable RMS errors are 0.03 degrees pointing error, and 6-mm surface distortion.

The problems encountered in designing an attitude controller are

1. An adequate model of an FSS is of high order because it contains a large number of elastic modes; however, a practically implementable controller has to be of sufficiently low order.
2. The inherent energy dissipation (damping) is very small.
3. The elastic frequencies are low and closely spaced.
4. The parameters (frequencies, damping ratios, and mode shapes) are not known accurately.

The simplest controller design approach would be truncating the model beyond a certain number of modes and designing a reduced-order controller. This approach is routinely used for controlling relatively rigid conventional spacecraft, wherein only the rigid modes are retained in the design model. Second-order filters are included in the loop to attenuate the contribution of the elastic modes. This approach is not generally effective for FSS because the elastic modes are much more prominent. Figure 7.19 shows the effect of using a truncated design model. When constructing a control loop around the "controlled" modes, an unintentional feedback loop is also constructed around the truncated modes, which can make the closed-loop system unstable. The inadvertent excitation of the truncated modes by the input and the unwanted contribution of the truncated modes to the sensed output were aptly termed by Balas [4] as "control spillover" and "observation spillover," respectively. The spillover terms may cause performance degradation and even instability, leading to catastrophic failure.

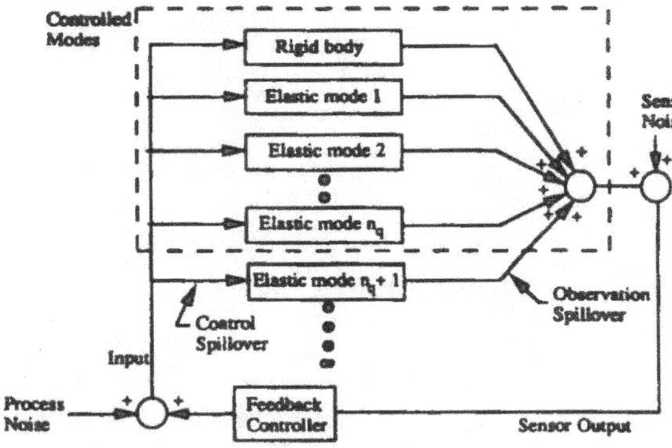

Figure 7.19 Control and observation spillover.

In addition to the truncation problem, the designer also lacks accurate knowledge of the parameters. Finite-element models give reasonably accurate estimates of the frequencies and mode shapes only for the first few modes and can provide no estimates of inherent damping ratios. Premission ground testing for parameter estimation is not generally possible because many FSS are not designed to withstand the gravitational force (while deployed), and because the test facilities required, such as a vacuum chamber, would be excessively large. Another consideration in controller design is that the actuators and sensors have nonlinearities, and finite response times. In view of these problems, the attitude controller must be a "robust" one, that is, it must at least maintain stability, and perhaps performance, despite modeling errors, uncertainties, nonlinearities, and component failures. The next two sections present linear controller design methods for the attitude control problem.

Model-Based Controllers

Consider the nth order state space model of an FSS in the presence of process noise and measurement noise,

$$\dot{x} = Ax + Bu + v, \quad y = Cx + Du + w, \tag{7.122}$$

where $v(t)$ and $w(t)$ are, respectively, the $n \times 1$ and $l \times 1$ process noise and sensor noise vectors. v and w are assumed to be mutually uncorrelated, zero-mean, Gaussian white noise processes with covariance intensity matrices V and W. A linear quadratic Gaussian (LQG) controller can be designed to minimize

$$J = \lim_{t_f \to \infty} \frac{1}{t_f} \mathcal{E} \int_0^{t_f} [x^T(t)Qx(t) + u^T(t)Ru(t)]dt \tag{7.123}$$

where "$\mathcal{E}$" denotes the expectation operator and $Q = Q^T \geq 0$, $R = R^T > 0$ are the state and control weighting matrices. The resulting nth order controller consists of a Kalman–Bucy filter (KBF) in tandem with a linear quadratic regulator (LQR) with the form,

$$\dot{\hat{x}} = (A + BG - KC)\hat{x} + Ky, \quad u = G\hat{x}, \tag{7.124}$$

where $\hat{x}$ is the controller state vector, and $G_{m \times n}$ and $K_{n \times l}$ are the LQR and KBF gain matrices. Any controller using an observer and state estimate feedback has the same mathematical structure. The order of the controller is n, the same as that of the plant. An adequate model of an FSS typically consists of several hundred elastic modes. However, to be practically implementable, the controller must be of sufficiently low order. A reduced-order controller design can be obtained in two ways, either by using a reduced-order "design model" of the plant or by obtaining a reduced-order approximation to a high-order controller. The former method is used more widely than the latter because high-order controller design relies on the knowledge of the high frequency mode parameters, which is usually inaccurate.

A number of model-order reduction methods have been developed during the past few years. The most important of these include the singular perturbation method, the balanced truncation method, and the optimal Hankel norm method (see [7] for a detailed description). In the singular perturbation method, higher frequency modes are approximated by their quasi-static representation. The balanced truncation method uses a similarity transformation that makes the controllability and observability matrices equal and diagonal. A reduced-order model is then obtained by retaining the most controllable and observable state variables. The optimal Hankel norm approximation method aims to minimize the Hankel norm of the approximation error and can yield a smaller error than the balanced truncation method. A disadvantage of the balanced truncation and the Hankel norm methods is that the resulting (transformed) state variables are mutually coupled and do not correspond to individual modes, resulting in the loss of physical insight. A disadvantage of the singular perturbation and Hankel norm methods is that they can yield non-strictly proper reduced-order models. An alternate method of overcoming these difficulties is to rank the elastic modes according their contributions to the overall transfer function, in the sense of H_2, H_∞, or $\mathcal{L}_1$ norms [9]. The highest ranked modes are then retained in the design model. This method retains the physical significance of the modes and also yields a strictly proper model. Note that the rigid-body modes must always be included in the design model, no matter which order-reduction method is used. A model-based controller can then be designed based on the reduced-order design model.

LQG Controller

An LQG controller designed for the reduced-order design model is guaranteed to stabilize the nominal design model. However, it may not stabilize the full-order plant because of the control and observation spillovers. Some time-domain methods for designing spillover-tolerant, reduced-order LQG controllers are discussed in [12]. These methods basically attempt to reduce the norms of spillover terms $\|B_t G\|$ and $\|KC_t\|$, where B_t and C_t denote the input and observation matrices corresponding to the truncated modes. Lyapunov-based sufficient conditions for stability are derived in terms of upper bounds on the spillover norms and are used as guidelines in spillover reduction. The controllers obtained by these methods are generally quite conservative and also require the knowledge of the truncated mode parameters to ensure stability.

Another approach to LQG controller design is the application of multivariable frequency-domain methods, wherein the truncated modes are represented as an additive uncertainty term $\Delta P(s)$ that appears in parallel with the design model (i.e., nominal plant) transfer function $P(s)$, as shown in Figure 7.20.

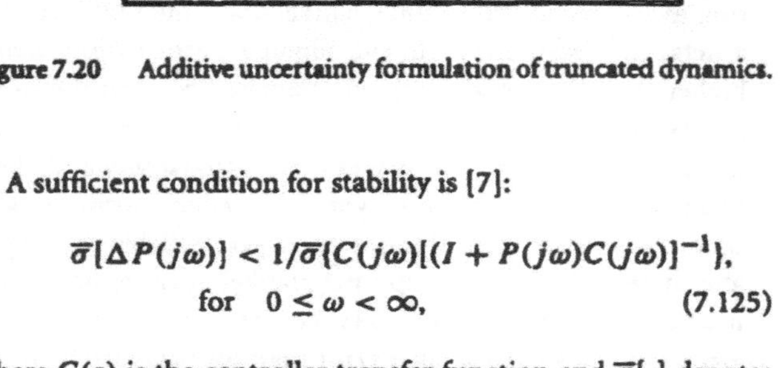

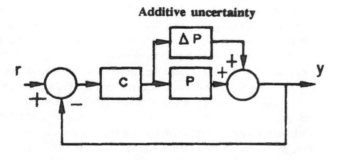

Figure 7.20 Additive uncertainty formulation of truncated dynamics.

A sufficient condition for stability is [7]:

$$\overline{\sigma}[\Delta P(j\omega)] < 1/\overline{\sigma}\{C(j\omega)[(I + P(j\omega)C(j\omega)]^{-1}\},$$
$$\text{for} \quad 0 \leq \omega < \infty, \qquad (7.125)$$

where $C(s)$ is the controller transfer function and $\overline{\sigma}[.]$ denotes the largest singular value. An upper bound on $\overline{\sigma}[\Delta P(j\omega)]$ can be obtained from (crude knowledge of) the truncated mode parameters to generate an "uncertainty envelope." Figure 7.21 shows the stability test Equation 7.125 for the 122-m hoop/column antenna where the design model consists of the three rigid rotational modes and the first three elastic modes.

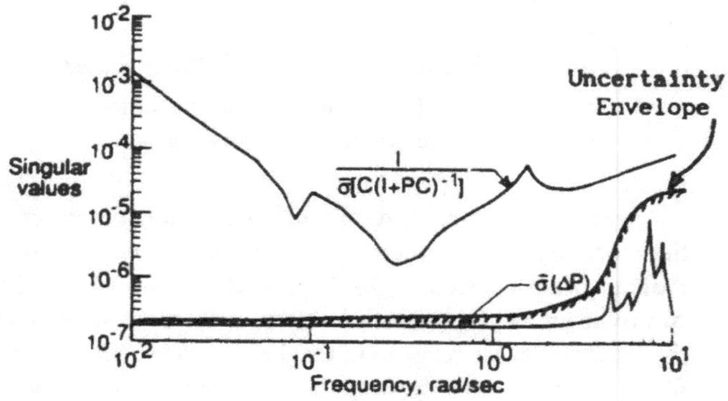

Figure 7.21 Stability test for additive uncertainty.

A measure of the nominal closed-loop performance is given by the bandwidth of the closed loop transfer function, $G_{cl} = PC(I + PC)^{-1}$, shown in Figure 7.22 for the hoop/column antenna.

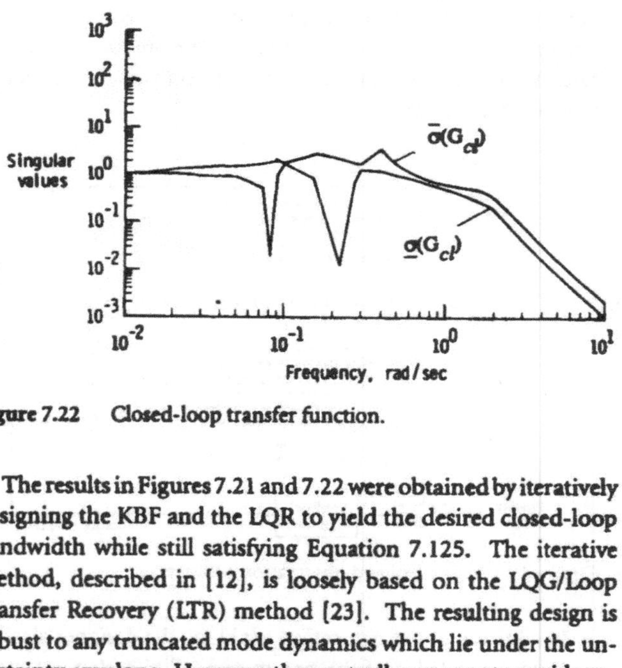

Figure 7.22 Closed-loop transfer function.

The results in Figures 7.21 and 7.22 were obtained by iteratively designing the KBF and the LQR to yield the desired closed-loop bandwidth while still satisfying Equation 7.125. The iterative method, described in [12], is loosely based on the LQG/Loop Transfer Recovery (LTR) method [23]. The resulting design is robust to any truncated mode dynamics which lie under the uncertainty envelope. However, the controller may not provide robustness to parametric uncertainties in the design model. A small uncertainty in the natural frequencies of the design model (i.e., the "controlled modes") can cause closed-loop instability because the very small open-loop damping ratios cause very sharp peaks in the frequency response, so that a small error in the natural frequency produces a large error peak in the frequency response [12].

H_∞- and μ-Synthesis Methods

The H_∞ controller design method [7] represents a systematic method for obtaining the desired performance as well as robustness to truncated mode dynamics. A typical design objective is minimizing the H_∞ norm of the frequency-weighted transfer function from the disturbance inputs (e.g., sensor and actuator noise) to the controlled variables, while insuring stability in the presence of truncated modes. An example of the application of the H_∞ method to FSS control is given in [17]. The problem also can be formulated to include parametric uncertainties represented as unstructured uncertainty. However, the resulting controller design is usually very conservative and provides inadequate performance.

The structured singular value method [6], [21] also known as the "μ-synthesis" method, can overcome the conservatism of the H_∞ method. In this method, the parametric uncertainties are individually "extracted" from the system block diagram and arranged as a diagonal block that forms a feedback connection with the nominal closed-loop system. The controller design problem is formulated as one of H_∞-norm minimization subject to a constraint on the structured singular value of an appropriate transfer function. The μ-synthesis problem can also be formulated to provide robust performance, i.e., the performance specifications must be satisfied despite model uncertainties. An application of the μ-synthesis method to FSS control is presented in [3].

The next section presents a class of controllers that can circumvent the problems due to spillovers and parametric uncertainties.

Passivity-Based Controllers

Consider the case where an attitude sensor and a rate sensor are collocated with each of the p torque actuators. Then the $m \times 1$ ($m = 3p$) sensed attitude and rate vectors y_p and y_r are given by

$$y_p = \Gamma\xi, \quad y_r = \Gamma\dot{\xi} \tag{7.126}$$

The transfer function from $U(s)$ to $Y_p(s)$ is $G(s) = G'(s)/s$, where $G'(s)$ is given by

$$G'(s) = \frac{\mathcal{I}J^{-1}\mathcal{I}^T}{s} + \sum_{1}^{n_q} \frac{\phi_i\phi_i^T s}{s^2 + 2\rho_i\omega_i s + \omega_i^2} \tag{7.127}$$

where the $m \times 3$ matrix $\mathcal{I} = [I_3, I_3, \ldots I_3]^T$, and ϕ_i^T denotes the ith row of Φ^T. The entries of ϕ_i^T represent the rotational mode shapes for the ith mode. An important consequence of collocation of actuators and sensors is that the operator from u to y_r is passive [5], or equivalently, $G'(s)$ is Positive-Real as defined below.

DEFINITION 7.1 A rational matrix-valued function $T(s)$ of the complex variable s is said to be positive-real if all of its elements are analytic in $Re[s] > 0$, and $T(s) + T^T(s^*) \geq 0$ in $Re[s] > 0$, where $*$ denotes the complex conjugate.

Scalar positive-real (PR) functions have a relative degree (i.e., the difference between the degrees of the denominator and numerator polynomials) of -1, 0, or 1 [24]. PR matrices have no transmission zeros or poles in the open right-half of the complex plane, and the poles on the imaginary axis are simple and have nonnegative definite residues. It can be shown that it is sufficient to check for positive semidefiniteness of $T(s)$ only on the imaginary axis ($s = j\omega$, $0 \leq \omega < \infty$), i.e., the condition becomes $T(j\omega) + T^*(j\omega) \geq 0$, where $*$ denotes complex conjugate transpose. Suppose (A, B, C, D) is an nth order minimal realization of $T(s)$. From [1], a necessary and sufficient condition for $T(s)$ to be positive real is that an $n \times n$ symmetric positive definite matrix P and matrices W and L exist so that

$$\begin{aligned} A^TP + PA &= -LL^T \\ C &= B^TP + W^TL \\ W^TW &= D + D^T \end{aligned} \tag{7.128}$$

This result is also generally known in the literature as the Kalman–Yakubovich lemma. Positive realness of $G'(s)$ gives rise to a large class of robustly stabilizing controllers, called dissipative controllers. Such controllers can be divided into static dissipative and dynamic dissipative controllers.

Static Dissipative Controllers.

Consider the proportional-plus-derivative control law,

$$u = -G_p y_p - G_r y_r, \tag{7.129}$$

where G_p and G_r are symmetric positive-definite matrices. The closed-loop equation then becomes

$$\bar{A}\ddot{\xi} + (\bar{B} + \Gamma^T G_r \Gamma)\dot{\xi} + (\bar{C} + \Gamma^T G_p \Gamma)\xi = 0. \tag{7.130}$$

It can be shown that $(\bar{B} + \Gamma^T G_r \Gamma)$ and $(\bar{C} + \Gamma^T G_p \Gamma)$ are positive-definite matrices, and that this control law stabilizes the plant $G(s)$, i.e., the closed-loop system of Equation 7.130 is asymptotically stable (see [12]). The closed-loop stability is not affected by the number of truncated modes or the knowledge of the parametric values, that is, the stability is *robust*. The only requirements are that the actuators and sensors be collocated and that the feedback gains be positive definite. Furthermore, if G_p, G_r are diagonal, then the robust stability holds even when the actuators and sensors have certain types of nonlinear gains.

Stability in the Presence of Actuator and Sensor Nonlinearities

Suppose that G_p, G_r are diagonal, and that

1. the actuator nonlinearities, $\psi_{ai}(\nu)$, are monotonically nondecreasing and belong to the $(0, \infty)$ sector, i.e., $\psi_{ai}(0) = 0$, and $\nu\psi_{ai}(\nu) > 0$ for $\nu \neq 0$.
2. the attitude and rate sensor nonlinearities, $\psi_{si}(\nu)$, belong to the $(0, \infty)$ sector.

Then the closed-loop system with the static dissipative control law is globally asymptotically stable.

A proof of this result can be obtained by slightly modifying the results in [12]. Examples of permissible nonlinearities are shown in Figure 7.23. It can be seen that actuator and sensor saturation are permissible nonlinearities which will not destroy the robust stability property.

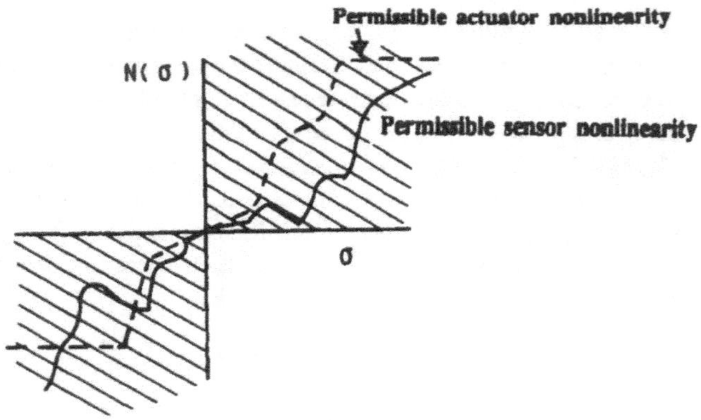

Figure 7.23 Permissible actuator and sensor nonlinearities.

Some methods for designing static dissipative controllers are discussed in [12]. In particular, the static dissipative control law minimizes the quadratic performance function,

$$\mathcal{J} = \int_0^{\infty} \Big[y_p^T G_p G_r^{-1} G_p y_p + y_r^T G_r y_r + 2\dot{q}^T D\dot{q} + 2y_p^T G_p G_r^{-1} u + u^T G_r^{-1} u \Big] dt.$$

This performance function can be used as a basis for controller design. Another approach for selecting gains is to minimize the norms of the differences between the actual and desired values of the closed-loop coefficient matrices, $(\bar{B} + \Gamma^T G_r \Gamma)$ and $(\bar{C} + \Gamma^T G_p \Gamma)$ (see [12]).

The performance of static dissipative controllers is inherently limited because of their restrictive structure. Furthermore, direct output feedback allows the addition of unfiltered sensor noise to the input. These difficulties can be overcome by using dynamic dissipative controllers, which also offer more design freedom and potentially superior performance.

Dynamic Dissipative Controllers

The positive realness of $G'(s)$ also permits robust stabilization of $G(s)$ by a class of dynamic compensators. The following definition is needed to define this class of controllers.

DEFINITION 7.2 A rational matrix-valued function $T(s)$ of the complex variable s is said to be marginally strictly positive-real (MSPR) if $T(s)$ is PR, and $T(j\omega) + T^*(j\omega) > 0$ for $\omega \in (-\infty, \infty)$.

This definition of MSPR matrices is a weaker version of the definitions of "strictly positive-real (SPR)" matrices which have appeared in the literature [18]. The main difference is that MSPR matrices can have poles on the imaginary axis. It has been shown in [13] that the negative feedback connection of a PR system and an MSPR system is stable, i.e., the composite system consisting of minimal realizations of the two systems is asymptotically stable (or equivalently, one system "stabilizes" the other system).

Consider an $m \times m$ controller transfer function matrix $\mathcal{K}(s)$ which uses the position sensor output $y_p(t)$ to generate the input $u(t)$. The following sufficient condition for stability is proved in [14].

Stability with Dynamic Dissipative Controller

Suppose that the controller transfer function $\mathcal{K}(s)$ has no transmission zeros at the origin and that $C(s) = \mathcal{K}(s)/s$ is MSPR. Then $\mathcal{K}(s)$ stabilizes the plant $G(s)$.

This condition can also be stated in terms of $[A_k, B_k, C_k, D_k]$, a minimal realization of $\mathcal{K}(s)$ [14]. The stability property depends only on the positive realness of $G'(s)$, which is a consequence of actuator-sensor collocation. Therefore, just as in the case of the static dissipative controller, the stability property holds regardless of truncated modes or the knowledge of the parameters. Controllers which satisfy the above property are said to belong to the class called "dynamic dissipative" controllers.

The condition that $\mathcal{K}(s)/s$ be MSPR is generally difficult to test. However, if $\mathcal{K}(s)$ is restricted to be diagonal, i.e., $\mathcal{K}(s) = diag[\mathcal{K}_1(s), \ldots, \mathcal{K}_m(s)]$, the condition is easier to check. For example, for the diagonal case, let

$$\mathcal{K}_i(s) = k_i \frac{s^2 + \beta_{1i}s + \beta_{0i}}{s^2 + \alpha_{1i}s + \alpha_{0i}}. \tag{7.131}$$

It is straightforward to show that $\mathcal{K}(s)/s$ is MSPR if, and only if, (for $i = 1, \ldots, m$), $k_i, \alpha_{0i}, \alpha_{1i}, \beta_{0i}, \beta_{1i}$ are positive,

$$\alpha_{1i} - \beta_{1i} > 0 \tag{7.132}$$

$$\alpha_{1i}\beta_{0i} - \alpha_{0i}\beta_{1i} > 0 \tag{7.133}$$

For higher order $\mathcal{K}_i$s, the conditions on the polynomial coefficients are harder to obtain. One systematic procedure for obtaining such conditions for higher order controllers is the application of Sturm's theorem [24]. Symbolic manipulation codes can then be used to derive explicit inequalities similar to Equations 7.132 and 7.133. Using such inequalities as constraints,the controller design problem can be posed as a constrained optimization problem which minimizes a given performance function. An example design of a dynamic dissipative controller for the hoop/column antenna concept is presented in [14], wherein diagonal $\mathcal{K}(s)$ is assumed. For the case of fully populated $\mathcal{K}(s)$, however, there are no straightforward methods and it remains an area of future research.

The preceding stability result for dynamic dissipative controllers can be used to show that the robust stability property of the static dissipative controller is maintained even when the actuators have a finite bandwidth:

1. For the static dissipative controller (Equation 7.129), suppose that G_p and G_r are diagonal with positive entries (denoted by subscript i), and that actuators represented by the transfer function $G_{Ai}(s) = \frac{k_i}{(s+a_i)}$ are present in the ith control channel. Then the closed-loop system is asymptotically stable if $G_{ri} > G_{pi}/a_i$ (for $i = 1, \ldots, m$).
2. Suppose that the static dissipative controller also includes the feedback of the acceleration $y_a (= \Gamma\ddot{\xi})$, that is,

$$u = -G_p y_p - G_r y_r - G_a y_a$$

where G_p, G_r, and G_a are diagonal with positive entries. Suppose that the actuator dynamics for the ith input channel are given by $G_{Ai}(s) = k_i/(s^2 + \mu_i s + \nu_i)$, with k_i, μ_i, ν_i positive. Then the closed-loop system is asymptotically stable if

$$\frac{G_{ri}}{G_{ai}} \leq \mu_i < \frac{G_{ri}}{G_{pi}} \quad (i = 1, ..., m).$$

Because of the requirement that $\mathcal{K}(s)/s$ be MSPR, the controller $\mathcal{K}(s)$ is not strictly proper. From a practical viewpoint, it is sometimes desirable to have a strictly proper controller because it attenuates sensor noise as well as high-frequency disturbances. Furthermore, the most common types of controllers, which include the LQG as well as the observer/pole placement controllers, are strictly proper (they have a first-order rolloff). It is possible to realize $\mathcal{K}(s)$ as a strictly proper controller wherein both y_p and y_r are utilized for feedback. Let $[A_k, B_k, C_k, D_k]$ be a minimal realization of $\mathcal{K}(s)$ where C_k is of full rank.

Strictly Proper Dissipative Controller

The plant with y_p and y_r as outputs is stabilized by the controller given by

$$\dot{x}_k = A_k x_k + [\, B_k - A_k L \quad L \,] \begin{bmatrix} y_p \\ y_r \end{bmatrix}, \tag{7.134}$$

$$y_k = C_k x_k, \tag{7.135}$$

where the $n_k \times m (n_k \geq m)$ matrix L is a solution of

$$D_k - C_k L = 0. \tag{7.136}$$

Equation 7.136 represents m^2 equations in mn_k unknowns. If $m < n_k$ (i.e., the compensator order is greater than the number of plant inputs), there are many possible solutions for L. The solution which minimizes the Frobenius norm of L is

$$L = C_k^T (C_k C_k^T)^{-1} D_k. \tag{7.137}$$

If $m = n_k$, Equation 7.136 gives the unique solution, $L = C_k^{-1} D_k$.

The next section addresses another important class of systems, namely, multibody flexible systems, which are described by nonlinear mathematical models.

7.3.3 Multibody Flexible Space Systems

This section considers the problem of controlling a class of nonlinear multibody flexible space systems consisting of a flexible central body with a number of articulated appendages. A complete nonlinear rotational dynamic model of a multibody flexible spacecraft is considered. It is assumed that the model configuration consists of a branched geometry, i.e., it has a central flexible body to which various flexible appendage bodies are attached (Figure 7.24). Each branch by itself can be a serial chain of structures. The actuators and sensors are assumed to be collocated. The global asymptotic stability of such systems is established using a nonlinear feedback control law. In many applications, the central body has a large mass and moments of inertia as compared to any other appendage bodies. As a result, the motion of the central body is small and can be assumed to be in the linear range. For this special case, the robust stability results are given for linear static as well as dynamic dissipative compensators. The effects of realistic nonlinearities in the actuators and sensors are also considered.

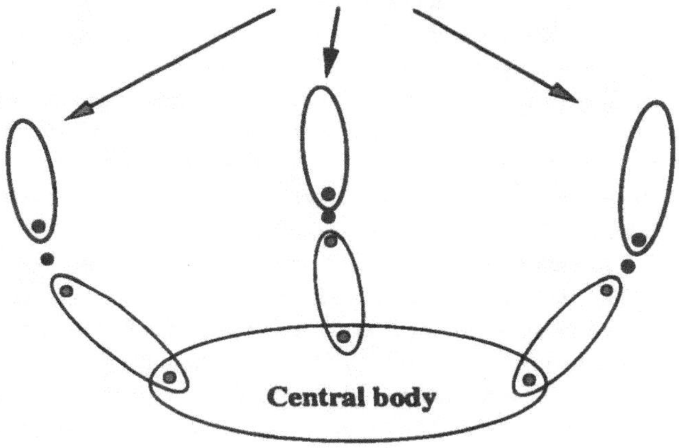

Figure 7.24 Multibody flexible system.

Mathematical Model

Equations of Motion

Consider a spacecraft (shown in Figure 7.24) consisting of a central flexible body and a chain of $(k-3)$ flexible links. (Although a single chain is considered here, all the results are also valid when several chains are present.) Using the Lagrangian formulation, the following equations of motion can be obtained (see [15] for details):

$$M(p)\ddot{p} + C(p, \dot{p})\dot{p} + D\dot{p} + Kp = B^T u \tag{7.138}$$

where $\dot{p} = \{\omega^T, \dot{\theta}^T, \dot{q}^T\}^T$; ω is the 3×1 inertial angular velocity vector (in body-fixed coordinates) for the central body; $\theta = (\theta_1, \theta_2, ..., \theta_{(k-3)})^T$, where θ_i denotes the joint angle for the ith joint expressed in body-fixed coordinates; q is the $(n-k)$ vector of flexible degrees of freedom (modal amplitudes); $p = (\gamma^T, \theta^T, q^T)^T$, and $\dot{\gamma} = \omega$; $M(p) = M^T(p) > 0$ is the configuration-dependent, mass-inertia matrix, and $\tilde{K}$ is the symmetric positive-definite stiffness matrix related to the flexible degrees of freedom. $C(p, \dot{p})$ corresponds to Coriolis and centrifugal forces; D is the symmetric, positive semidefinite damping matrix; $B = [\, I_{k\times k} \quad 0_{k\times(n-k)} \,]$ is the control influence matrix and u is the k-vector of applied torques. The first three components of u represent the attitude control torques (about the X-, Y-, and Z- axes) applied to the central body, and the remaining components are the torques applied at the $(k-3)$ joints. K and D are symmetric, positive-semidefinite stiffness and damping matrices,

$$K = \begin{bmatrix} 0_{k\times k} & 0_{k\times(n-k)} \\ 0_{(n-k)\times k} & \tilde{K}_{(n-k)\times(n-k)} \end{bmatrix}$$
$$D = \begin{bmatrix} 0_{k\times k} & 0_{k\times(n-k)} \\ 0_{(n-k)\times k} & \tilde{D}_{(n-k)\times(n-k)} \end{bmatrix} \quad (7.139)$$

where $\tilde{K}$ and $\tilde{D}$ are symmetric positive definite. The angular measurements for the central body are Euler angles (not the vector γ), whereas the remaining angular measurements between bodies are relative angles. In deriving equations of motion, it is assumed that the elastic displacements are small enough to be in the linear range and that the total displacement can be obtained by the principle of superposing rigid and elastic motions. One important inherent property (which will be called "Property $\mathcal{S}$") of such systems crucial to the stability results is given next.

Property $\mathcal{S}$: For the system represented by Equation 7.138, the matrix $(\frac{1}{2}\dot{M} - C)$ is skew-symmetric.

The justification of this property can be found in [15].

The central body attitude (Euler angle) vector η is given by $E(\eta)\dot{\eta} = \omega$, where $E(\eta)$ is a 3×3 transformation matrix [8]. The sensor outputs consist of three central body Euler angles, the $(k-3)$ joint angles, and the angular rates, i.e., the sensors are collocated with the torque actuators. The sensor outputs are then given by

$$y_p = B\hat{p} \quad \textit{and} \quad y_r = B\dot{p} \quad (7.140)$$

where $\hat{p} = (\eta^T, \theta^T, q^T)^T$ in which η is the Euler angle vector for the central body. $y_p = (\eta^T, \theta^T)^T$ and $y_r = (\omega^T, \dot{\theta}^T)^T$ are measured angular position and rate vectors, respectively. It is assumed that the body rate measurements, ω, are available from rate gyros. Using Property $\mathcal{S}$, it can be proved that the operator from u to y_r is passive [14].

Quaternion as a Measure of Attitude

The orientation of a free-floating body can be minimally represented by a three-dimensional orientation vector. However, this representation is not unique. One minimal representation commonly used to represent the attitude is Euler angles. The 3×1 Euler angle vector η is given by $E(\eta)\dot{\eta} = \omega$, where $E(\eta)$ is a 3×3 transformation matrix. $E(\eta)$ becomes singular for certain values of η. However, the limitations imposed on the allowable orientations due to this singularity are purely mathematical without physical significance. The problem of singularity in three-parameter representation of attitude has been studied in detail in the literature. An effective way of overcoming the singularity problem is to use the quaternion formulation (see [10]).

The unit quaternion α is defined as follows.

$$\alpha = \{\overline{\alpha}^T, \alpha_4\}^T, \quad \overline{\alpha} = \begin{bmatrix} \hat{\alpha}_1 \\ \hat{\alpha}_2 \\ \hat{\alpha}_3 \end{bmatrix} sin(\tfrac{\phi}{2}),$$
$$\alpha_4 = cos(\tfrac{\phi}{2}). \quad (7.141)$$

$\hat{\alpha} = (\hat{\alpha}_1, \hat{\alpha}_2, \hat{\alpha}_3)^T$ is the unit vector along the eigen-axis of rotation and ϕ is the magnitude of rotation. The quaternion is also subjected to the norm constraint,

$$\overline{\alpha}^T\overline{\alpha} + \alpha_4^2 = 1. \quad (7.142)$$

It can be also shown [10] that the quaternion obeys the following kinematic differential equations:

$$\dot{\overline{\alpha}} = \tfrac{1}{2}(\omega \times \overline{\alpha} + \alpha_4\omega), \quad \text{and} \quad (7.143)$$
$$\dot{\alpha}_4 = -\tfrac{1}{2}\omega^T\overline{\alpha}. \quad (7.144)$$

The quaternion representation can be effectively used for the central body attitude. The quaternion can be computed [10] using Euler angle measurements (Equation 7.140). The open-loop system, given by Equations 7.138, 7.143, and 7.144, has multiple equilibrium solutions: $(\overline{\alpha}_{ss}{}^T, \alpha_{4ss}, \theta_{ss}^T)^T$ where the subscript 'ss' denotes the steady-state value (the steady-state value of q is zero). Defining $\beta = (\alpha_4 - 1)$ and denoting $\dot{p} = v$, Equations 7.138, 7.143, and 7.144 can be rewritten as

$$M\dot{v} + Cv + Dv + \tilde{K}q = B^T u \quad (7.145)$$
$$\begin{bmatrix} \dot{\theta} \\ \dot{q} \end{bmatrix} = \begin{bmatrix} 0_{(n-3)\times3} & I_{(n-3)} \end{bmatrix} v \quad (7.146)$$
$$\dot{\overline{\alpha}} = \tfrac{1}{2}(\omega \times \overline{\alpha} + (\beta+1)\omega) \quad (7.147)$$
$$\dot{\beta} = -\tfrac{1}{2}\omega^T\overline{\alpha} \quad (7.148)$$

In Equation 7.145 the matrices M and C are functions of p, and $(p, \dot{p})$, respectively. It should be noted that the first three elements of p associated with the orientation of the central body can be fully described by the unit quaternion. The system represented by Equations 7.145–7.148 can be expressed in the state-space form as follows:

$$\dot{x} = f(x, u) \quad (7.149)$$

where $x = (\overline{\alpha}^T, \beta, \theta^T, q^T, v^T)^T$. Note that the dimension of x is $(2n+1)$, which is one more than the dimension of the system in Equation 7.138. However, one constraint (Equation 7.142) is now present. It can be verified from Equations 7.143 and 7.144 that the constraint (Equation 7.142) is satisfied for all $t > 0$ if it is satisfied at $t = 0$.

A Nonlinear Feedback Control Law

Consider the dissipative control law u, given by

$$u = -G_p\tilde{p} - G_r y_r \quad (7.150)$$

where $\tilde{p} = \{\overline{\alpha}^T, \theta^T\}^T$. Matrices G_p and G_r are symmetric positive-definite $(k \times k)$ matrices and G_p is given by

$$G_p = \begin{bmatrix} (1 + \frac{(\beta+1)}{2})G_{p1} & 0_{3\times(k-3)} \\ 0_{(k-3)\times3} & G_{p2(k-3)\times(k-3)} \end{bmatrix}. \quad (7.151)$$

Note that Equations 7.150 and 7.151 represent a nonlinear control law. If G_p and G_r satisfy certain conditions, this control law renders, the time rate of change of the system's energy negative along all trajectories; i.e., it is a 'dissipative' control law.

The closed-loop equilibrium solution can be obtained by equating all the derivatives to zero in Equations 7.138, 7.147, and 7.148. After some algebraic manipulations, there appear to be two equilibrium points in the state space. However, it can be shown [14] that they refer to a single equilibrium point.

If the objective of the control law is to transfer the state of the system from one orientation (equilibrium) position to another without loss of generality, the target orientation can be defined as zero and the initial orientation, given by $[\bar{\alpha}(0), \alpha_4(0), \theta(0)]$, can always be defined so that $|\theta_i(0)| \leq \pi$, $0 \leq \alpha_4(0) \leq 1$ (corresponding to $|\phi| \leq \pi$) and $[\bar{\alpha}(0), \alpha_4(0)]$ satisfy Equation 7.142.

The following stability result is proved in [14].

Stability Result: Suppose that $G_{p2_{(k-3)\times(k-3)}}$ and $G_{r(k\times k)}$ are symmetric and positive definite, and $G_{p1} = \mu I_3$, where $\mu > 0$. Then, the closed-loop system given by Equations 7.149–7.151 is globally asymptotically stable (g.a.s.).

This result states that the control law in Equation 7.150 stabilizes the nonlinear system despite unmodeled elastic mode dynamics and parametric errors, that is, a multibody spacecraft can be brought from any initial state to the desired final equilibrium state. The result also applies to a particular case, namely, single-body FSS, that is, this control law can bring a rotating spacecraft to rest or perform robustly stable, large-angle, rest-to-rest maneuvers. A generalization of the control law in Equations 7.150–7.151 to the case with fully populated G_{p1} matrix is given in [16].

The next section considers a special case of multibody systems.

Systems in Attitude-Hold Configuration

Consider a special case where the central body attitude motion is small. This can occur in many realistic situations. For example, in the case of a space-station-based or shuttle-based manipulator, the moments of inertia of the base (central body) are much larger than that of any manipulator link or payload. In such cases the rotational motion of the base can be assumed to be in the linear region, although the payloads (or links) attached to it can undergo large rotational and translational motions and nonlinear dynamic loading due to Coriolis and centripetal accelerations. For this case, the attitude of the central body is simply γ, the integral of the inertial angular velocity ω, and the use of quaternions is not necessary. The equations of motion (Equation 7.138) can now be expressed in the state-space form as

$$\dot{\bar{x}} = \begin{bmatrix} 0 & I \\ -M^{-1}K & -M^{-1}(C+D) \end{bmatrix} \bar{x} + \begin{bmatrix} 0 \\ M^{-1}B^T \end{bmatrix} u \quad (7.152)$$

where $\bar{x} = \{p^T, \dot{p}^T\}^T$, and $p = \{\gamma^T, \theta^T, q^T\}^T$. Note that M and C are functions of $\bar{x}$, and hence the system is nonlinear.

Stability with Dissipative Controllers

The static dissipative control law u is given by

$$u = -\overline{G}_p y_p - G_r y_r \quad (7.153)$$

where $\overline{G}_p$ and G_r are constant symmetric positive definite $(k \times k)$ matrices,

$$y_p = Bp \quad and \quad y_r = B\dot{p}. \quad (7.154)$$

Where y_p and y_r are measured angular position and rate vectors. The following result is proved in [14].

Stability Result: Suppose that $\overline{G}_{p_{k\times k}}$ and $G_{r_{k\times k}}$ are symmetric and positive definite. Then the closed-loop system given by Equations 7.152, 7.153, and 7.154 is globally asymptotically stable.

The significance of the two stability results presented in this section is that any nonlinear multibody system belonging to these classes can be robustly stabilized with the dissipative control laws given. In the case of manipulators, this means that one can accomplish any terminal angular position (of the links) from any initial position with guaranteed asymptotic stability. Furthermore, for the static dissipative case, the stability result also holds when the actuators and sensors have nonlinearities. In particular, the stability result of the Section on page 147 for the linear, single-body FSS also extends to the case of nonlinear flexible multibody systems in attitude-hold configuration, that is, the closed-loop system with the static dissipative controller is globally asymptotically stable in the presence of monotonically nondecreasing $(0, \infty)$-sector actuator nonlinearities, and $(0, \infty)$-sector sensor nonlinearities. This result is proved in [14].

For the more general case where the central body motion is not in the linear range, the robust stability in the presence of actuator/sensor nonlinearities cannot be easily extended because the stabilizing control law (Equation 7.150) is nonlinear.

The robust stability with dynamic dissipative controllers (Section 7.3.2) can also be extended to the multibody case (in attitude-hold configuration). As stated previously, the advantages of using dynamic dissipative controllers include higher performance, more design freedom, and better noise attenuation.

Consider the system given by Equation 7.152 with the sensor outputs given by Equation 7.154. As in the linear case, consider a $k \times k$ controller $\mathcal{K}(s)$ which uses the angular position vector $y_p(t)$ to produce the input $u(t)$. The closed-loop system consisting of nonlinear plant (Equation 7.152) and the controller $\mathcal{K}(s)$ is shown in Figure 7.25. $\mathcal{K}(s)$ is said to stabilize the nonlinear plant if the closed-loop system is globally asymptotically stable (with $\mathcal{K}(s)$ represented by its minimal realization). The conditions under which $\mathcal{K}(s)$ stabilizes the nonlinear plant are the same as the linear, single-body case, discussed in the Section on page 148, that is, the closed-loop system in Figure 7.25 is g.a.s. if $\mathcal{K}(s)$ has no transmission zeros at $s = 0$, and $C(s) = \frac{\mathcal{K}(s)}{s}$ is MSPR. A proof of this result is given in [14]. The proof does not make any assumptions regarding the model order or the knowledge of the parametric values. Hence, the stability is robust to modeling errors and parametric uncertainties. As shown in Section 7.3.2, this controller can also be realized as a strictly proper controller that utilizes the feedback of both y_p and y_r.

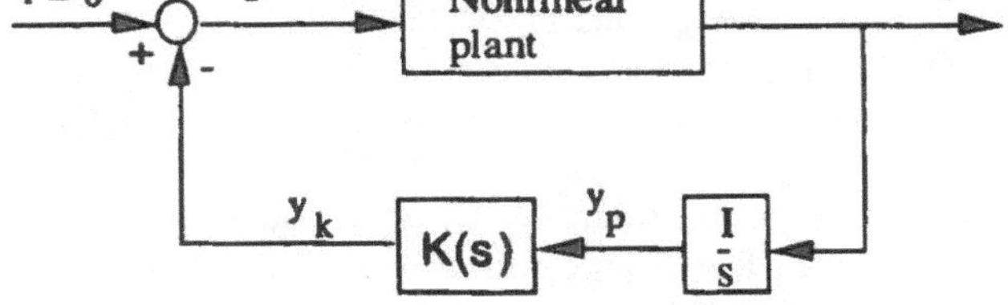

Figure 7.25 Stabilization with dynamic dissipative controller.

7.3.4 Summary

This chapter has provided an overview of various control issues for flexible space structures (FSS), which are classified into single-body FSS or multibody FSS. The first problem considered is fine attitude pointing and vibration suppression of single-body FSS, which can be formulated in the linear, time-invariant setting. For this problem, two types of controllers, model-based and passivity-based ("dissipative"), are discussed. The robust stability properties of dissipative controllers are highlighted and design techniques are discussed. For the case of multibody FSS, a generic nonlinear mathematical model is presented and is shown to have a certain passivity property. Nonlinear and linear dissipative control laws which provide robust global asymptotic stability are presented.

References

[1] Anderson, B. D. O., A System Theory Criterion for Positive-Real Matrices, *SIAM J. Control*, 5, 1967.

[2] Balakrishnan, A. V., Combined Structures-Controls Optimization of Lattice Trusses, *Computer Methods Appl. Mech. Eng.*, 94, 1992.

[3] Balas, G. J. and Doyle, J. C., Control of Lightly Damped Flexible Modes in the Controller Crossover Region, *J. Guid. Control Dyn.*, 17(2), 370–377, 1994.

[4] Balas, M. J., Trends in Large Space Structures Control Theory: Fondest Hopes, Wildest Dreams, *IEEE Trans. Automat. Control*, AC-27(3), 1982.

[5] Desoer, C. A. and Vidyasagar, M., *Feedback Systems: Input-Output Properties*, Academic, New York, 1975.

[6] Doyle, J. C., Analysis of Feedback Systems With Structured Uncertainty, *IEEE Proc.*, 129D(6), 1982.

[7] Green, M. and Limebeer, D. J. N., *Linear Robust Control*, Prentice Hall, Englewood Cliffs, NJ, 1995.

[8] Greenwood, D. T., *Principles of Dynamics*, Prentice Hall, Englewood Cliffs, NJ, 1988.

[9] Gupta, S., *State Space Characterization and Robust Stabilization of Dissipative Systems*, D.Sc. Thesis, George Washington University, 1994.

[10] Haug, E. G., *Computer-Aided Kinematics and Dynamics of Mechanical Systems*, Allyn and Bacon Series in Engineering, 1989.

[11] Hyland, D. C., Junkins, J. L., and Longman, R. W., Active Control Technology for Large Space Structures, *J. Guid. Control Dyn.*, 16(5), 1993.

[12] Joshi, S. M., *Control of Large Flexible Space Structures, Lecture Notes in Control and Information Sciences*, Springer, Berlin, 1989, Vol. 131.

[13] Joshi, S. M. and Gupta, S., Robust Stabilization of Marginally Stable Positive-Real Systems. NASA TM-109136, 1994.

[14] Joshi, S. M., Kelkar, A. G., and Maghami, P. G., A Class of Stabilizing Controllers for Flexible Multibody Systems. NASA TP-3494, 1995.

[15] Kelkar, A. G., Mathematical Modeling of a Class of Multibody Flexible Space Structures. NASA TM-109166, 1994.

[16] Kelkar, A. G. and Joshi, S. M., *Global Stabilization of Multibody Spacecraft Using Quaternion-Based Nonlinear Control Law*, Proc. Am. Control Conf., Seattle, Washington, 1995.

[17] Lim, K. B., Maghami, P. G., and Joshi, S. M., Comparison of Controller Designs for an Experimental Flexible Structure, *IEEE Control Syst. Mag.*, 3, 1992.

[18] Lozano-Leal, R. and Joshi, S. M., Strictly Positive Real Functions Revisited, *IEEE Trans. Automat. Control*, 35(11), 1243–1245, 1990.

[19] Meirovitch, L., *Methods of Analytical Dynamics*, MacGraw-Hill, New York, 1970.

[20] Nurre, G. S., Ryan, R. S., Scofield, H. N., and Sims, J. L., Dynamics and Control of Large Space Structures, *J. Guid. Control Dyn.*, 7(5), 1984.

[21] Packard, A. K., Doyle, J. C., and Balas, G. J., Linear Multivariable Robust Control With a μ-Perspective, *ASME J. Dyn. Meas. & Control*, 115(2(B)), 426–438, 1993.

[22] Russell, R. A., Campbell, T. G., and Freeland, R. E., A Technology Development Program for Large Space Antenna, NASA TM-81902, 1980.

[23] Stein, G. and Athans, M., The LQG/LTR Procedure for Multivariable Feedback Control Design, *IEEE Trans. Automat. Control*, 32(2), 105–114, 1987.

[24] Van Valkenberg, M. E., *Introduction to Modern Network Synthesis*, John Wiley & Sons, New York, 1965.

7.4 Line-of-Sight Pointing and Stabilization Control System

David Haessig, GEC-Marconi Systems Corporation, Wayne, NJ

7.4.1 Introduction

To gain some insight into the functions of line-of-sight pointing and stabilization, consider the human visual system. It is quite apparent that we can control the direction of our eyesight, but few are aware that they have also been equipped with a stabilization capability. The vestibulo-occular reflex is a physiological mechanism which acts to fix the direction of our eyes inertially when the head is bobbling about for whatever reason [1]. (Try shaking your head rapidly while viewing this text and note that you can fix your eyesight on a particular word.) Its importance becomes obvious when you imagine life without it. Catching a football while running would be next to impossible. Reading would be difficult while traveling in a car over rough road, and driving dangerous. Our vision would appear jittery and blurry. This line-of-sight pointing and stabilization mechanism clearly improves and expands our capabilities.

Similarly, many man-made systems have been improved by adding pointing and stabilization functionality. In the night vision systems used extensively during the Persian Gulf War, image clarity and resolution, and their precision strike capability, were enhanced by the pointing and stabilization devices they contained.

The Falcon Eye System [11] was developed in the late 1980s to build upon the strengths of the night vision systems developed earlier and used extensively during Desert Storm. The Falcon Eye differs from previous night vision systems developed for fighter aircraft in that it is head-steerable. It provides a high degree of night situational awareness by allowing the pilot to look in any direction including directly above the aircraft. This complicates the control system design problem because not only must the system isolate the line of sight from image blurring angular vibration, it also must simultaneously track pilot head motion. Nevertheless, this system has been successfully designed, built, extensively flight tested on a General Dynamics F-16 Fighting Falcon, and shown to markedly improve a pilot's ability to find and attack fixed or mobile targets at night.

The Falcon Eye System is pictured in Figures 7.26 and 7.27. It consists of (1) a helmet-mounted display with dual optical combiners suspended directly in front of the pilot's eyes to permit a merging of Flir (forward looking infrared) imagery with external light, (2) a head angular position sensor consisting of a magnetic sensor attached to the helmet and coupled to another attached to the underside of the canopy, and (3) a Flir sensor providing three-axis control of the line-of-sight orientation and including a two-axis fine stabilization assembly. The infrared image captured by this sensor is relayed to the pilot's helmet-mounted display, providing him with a realistic Flir image of the outside world. The helmet orientation serves as a commanded reference that the Flir line of sight must follow.

7.4.2 Overall System Performance Objectives

The system's primary design goal was to provide a night vision capability that works and feels as much like natural daytime vision as possible. This means the pilot must be able to turn his head to look in any direction. Also, the Flir scene cannot degrade due to the high frequency angular vibration present in the airframe where the equipment is mounted.

The Falcon Eye Flir attempts to achieve these effects by serving as a buffer between the vehicle and the outside world, which must (1) track the pilot's head with sufficient accuracy so that lag or registration errors between the Flir scene and the outside world scene are imperceptible and (2) isolate the Flir sensor from vehicle angular vibration which can cause the image to appear blurry (in a single image frame) and jittery (jumping around from frame to frame).

The pilot can select either a 1:1 or a 5.6:1 level of scene magnification. Switching to the magnified view results in a reduction in the size of the scene from 22 × 30 degrees to the very narrow 4 × 4.5 degrees, hence the name *narrow field of view*. When in narrow field-of-view mode, the scene resolution is much finer, 250 μrads, and therefore the stabilization requirements are tighter. This chapter focuses on the more difficult task of designing a controller for the stabilization assembly in narrow field-of-view mode.

Control System Structure

The Falcon Eye's field of regard is equal to that of the pilot's. Consequently, the system must precisely control the orientation of the line of sight over a very large range of angles. It is difficult to combine fine stabilization and large field-of-regard capabilities in a single mechanical control effector. Therefore, these tasks are separated and given to two different parts of the overall system (see Figure 7.28). A coarse gimbal set accomplishes the task of achieving a large field of regard. A fine stabilization assembly attached to the inner gimbal acts as a vernier which helps to track commands and acts to isolate the line of sight from angular vibration, those generated by the vehicle and those generated within the coarse gimbal set. The coarse positioning gimbal system tracks pilot head position commands, but will fall short because of its limited dynamic response and because of the effects of torque disturbances within the gimbal drive systems. These torque disturbances include stiction, motor torque ripple and cogging, bearing torque variations, and many others. The stabilization system assists in tracking the commanded head position by (1) isolating the line-of-sight from vehicle vibration and (2) reducing the residual tracking error left by the gimbal servos. The reason for sending the command rates rather than the command angles to the stabilization system will be covered in the description of the stabilization system controller.

Control System Performance Requirements

Ideally the Flir sensor's line of sight should perfectly track the pilot's line of sight. The control system must therefore limit the difference between these two pointing directions. (Recognize that the pilot's line of sight is defined here by the orientation of his head, not that of his eyes. A two-dimensional scene appears in the display and he is free to direct his eyes at any particular point in that scene, much like natural vision.)

Two quantities precisely define the pilot's line of sight in inertial space: vehicle attitude relative to inertial space θ_a and pilot head attitude relative to the vehicle ϕ_p. (θ is used for inertial variables and ϕ for relative variables.) This is depicted in one dimension in Figure 7.29, i.e., $\theta_p = \phi_p + \theta_a$.

The Flir, although connected to the aircraft, is physically separated from the pilot by a nonrigid structure which can exhibit vibrational motion relative to the pilot's location. This relative angular motion will be referred to as ϕ_v. The Flir's line of sight θ_f equals the vehicle attitude θ_a plus vehicle vibration ϕ_v, plus the Flir line of sight relative to the vehicle ϕ_f, i.e., $\theta_f = \phi_f + \phi_v + \theta_a$. The difference between the two is the error,

$$e = \theta_f - \theta_p = \phi_f - \phi_p + \phi_v. \tag{7.155}$$

As Equation 7.155 indicates, for this error to equal zero, the Flir must track pilot head motion and move antiphase to vehicle vibration. Head orientation is therefore accurately described as a commanded reference input, and vehicle vibration as a disturbance.

Figure 7.26 The Falcon Eye Flir, consisting of a helmet-mounted display, a head position sensor, a Flir optical system, and the line-of-sight position control system partially visible just forward of the canopy. (By courtesy of Texas Instruments, Defense Systems & Electronics Group.).

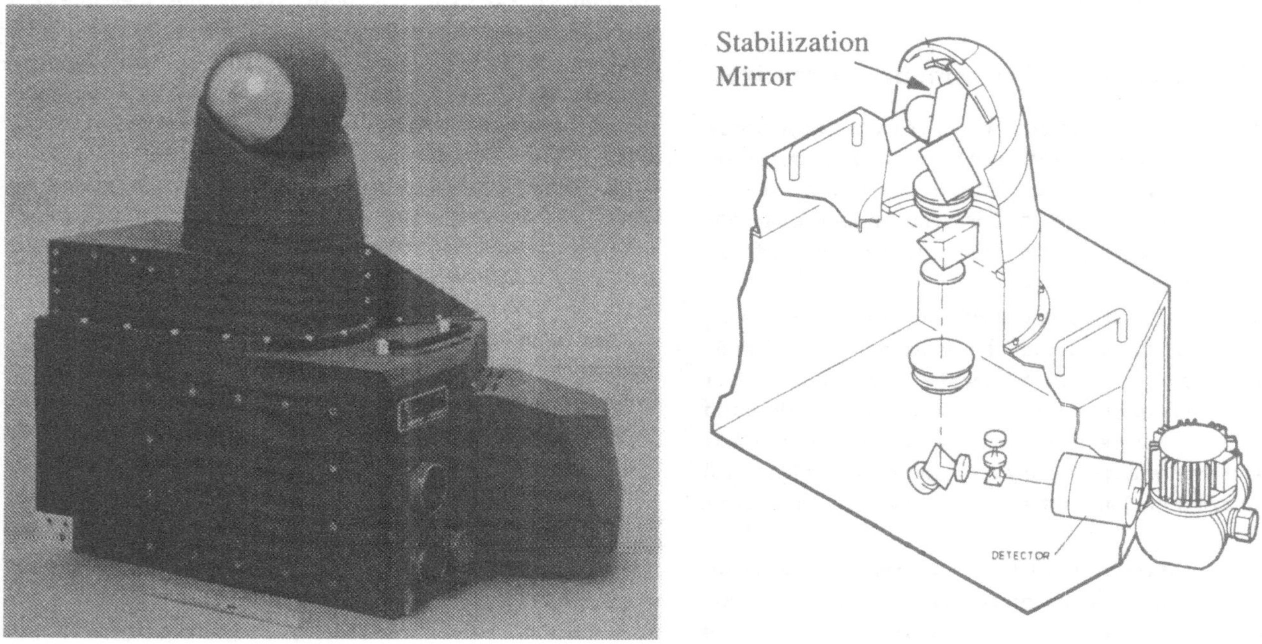

Figure 7.27 The Falcon Eye gimbals follow pilot head motion. An image stabilization assembly attached inside the ball helps to follow the pilot's head and isolates the line of sight from angular vibration.

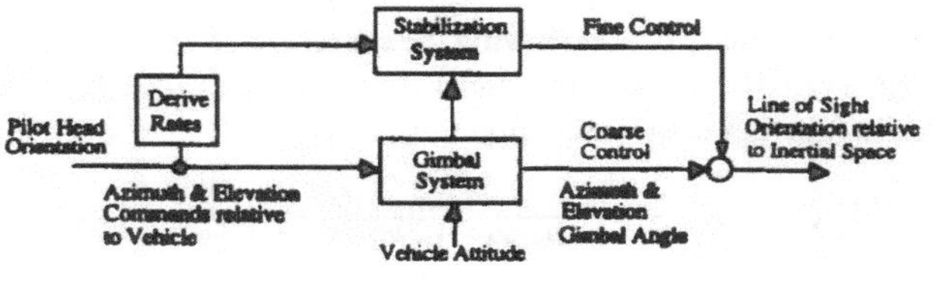

Figure 7.28 Partitioning of the control responsibility: the gimbals track commands; the stabilization system helps to track commands and stabilizes the image.

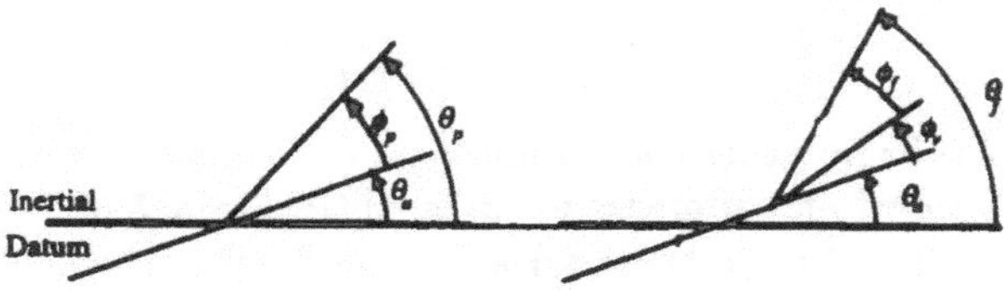

Figure 7.29 Tracking of the pilot's line of sight involves isolation from vehicle vibration ϕ_v, tracking of vehicle maneuvering θ_a, and tracking of pilot head motion relative to the vehicle ϕ_p.

The ideal response to various inputs to the system is shown in Figure 7.30. The output is the Flir line of sight angle relative to inertial space. The inputs consist of vehicle angle relative to inertial space at the Flir ($\theta_v = \theta_a + \phi_v$), pilot head commanded angle relative to the vehicle, and a third input not mentioned thus far, the tracking error due to torque disturbances generated within the gimbals. The goal is to point the Flir sensor's line of sight continuously in the same direction as the pilot's. Therefore, pilot head commands are ideally tracked perfectly, as the upper plot indicates. However, because these commands are relative to the vehicle, the vehicle maneuver motion θ_a, which is generally low frequency, must also be tracked, as the lower plot depicts. At higher frequencies, vehicle angular vibration ϕ_v injects an input that affects the Flir sensor but not the pilot. So, as the lower plot indicates, the ideal response to vehicle vibration is complete attenuation. Between these frequency regimes there is a transition region. In this application, maneuver tracking must occur for frequencies up to about 0.3 Hz. The transition occurs between 0.3 and about 3 Hz, and good isolation performance is needed in the 3 to 30 Hz range. Beyond that, because there is little vibration energy at the higher frequencies, the transfer function from vehicle motion to line-of-sight motion can increase without significant penalty.

The center curve of Figure 7.30 is the ideal system response to torque disturbances generated within the gimbals. Ideally, internally generated tracking errors should be completely attenuated by the stabilization assembly. Thus the magnitude of this transfer function is zero, as shown. This is implied by the perfect command tracking ideal response depicted in the upper plot. However, it is shown separately to distinguish and clearly identify these two characteristics.

The ideal performance requirements discussed thus far serve primarily to guide the specification of realistic, quantifiable performance requirements defining allowable responses to specific

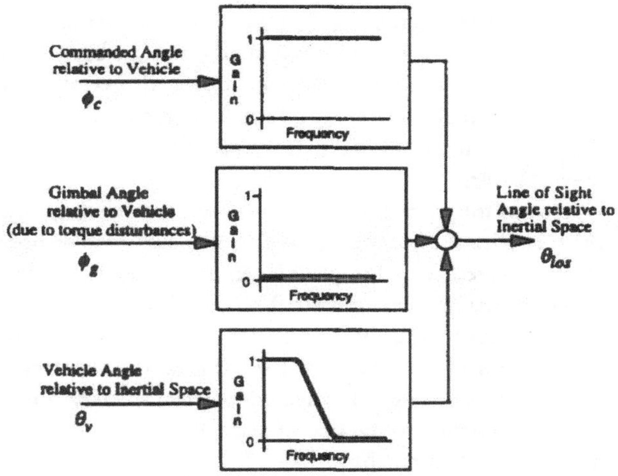

Figure 7.30 The response of an ideal line-of-sight control system to the various inputs.

input environments. For the Falcon Eye these fall into 4 areas: (1) stabilization error, the high frequency line-of-sight motion caused by vehicle vibration or other disturbances, (2) tracking smoothness, the tracking error due to internally generated torque disturbances, (3) command tracking, the error in tracking specifically defined head motion commands, and (4) maneuver tracking, the tracking error that occurs in response to specific maneuvers. In this chapter our evaluation of performance will cover the first two in detail.

Stabilization Requirement Vehicle angular vibration produces line-of-sight motion at higher frequencies (above 5 Hz) that can degrade the resolution of the image. Resolution is not impaired by vibration, but is limited by the size of the pixel associated with the infrared (IR) detector, if the line of sight *rms* motion due to vibration is less than 1/4 the size of the pixel, which in the Falcon Eye is 250 μrads in narrow field of view. The image stabilization assembly must allow no more than 62 μrads rms of line-of-sight motion when subjected to vehicle angular vibration.

Tracking Smoothness Requirement Testing has indicated that a pilot can see and is bothered by a jump or jitter in the image at intermediate frequencies (2 to 10 Hz) when they approach or exceed the size of one pixel, i.e., 250 μrads. Therefore, a smoothness of tracking requirement is applied which limits the peak-to-peak tracking error to less then 250 μrads in the 2 to 10 Hz range when following constant-rate head motion.

Command and Maneuver Tracking Requirements In the next section we will see that the stabilization system must provide command and maneuver tracking bandwidths of greater than 30 and 0.3 Hz, respectively.

7.4.3 Physical System Description and Modeling

The stabilization assembly, pictured in Figure 7.31, is rigidly attached to the inner gimbal of the coarse positioning system (see Figure 7.27). It produces small changes in the line-of-sight direction by changing the angular position of the mirror relative to the base structure. The mirror is attached to the base by a flexure hinge, a solid but flexible structure which permits angular motion only about two-axes and applies a small restoring spring torque. Rotation is limited to ± 5 μrads ($\pm 1/3°$), by hardstops under the mirror.

Between the mirror and the base there is a two-axis torquer which applies the control signals. A two-axis angular position sensor, referred to as the pickoff sensor, measures the angular position of the mirror relative to the base. A two-axis angular rate sensor attached rigidly in the base senses angular rate, including the vibration disturbance that the stabilization assembly must counter. The rate sensor is oriented to detect any rotation about axes perpendicular to the line-of-sight, because only those alter the line-of-sight direction.

While operating within the hardstops, each axis of the mirror is governed by the dynamic equation,

$$J_m\ddot{\theta}_m + \beta_m(\dot{\theta}_m - \dot{\theta}_b) + K_m(\theta_m - \theta_b) = \tau_m, \quad (7.156)$$

where θ_m and θ_b are the mirror and base inertial angular positions, respectively, and τ_m is the control torque. Other parameters are defined in Table 7.4. Unmodeled nonlinear dynamics such as centripetal, coriolis, and friction torques are small and neglected.

The variable that we are interested in controlling is the line-of-sight angle θ_{los}. This variable cannot be sensed directly but is related to other variables that are sensed. Consider the situation where the mirror is fixed to the base so that $\phi_m = \theta_m - \theta_b = 0$. Then the line of sight follows the base exactly: $\theta_{los} = \theta_b$. If the mirror rotates relative to the base and the base remains stationary, the line of sight moves λ times farther due to optical effects (λ is 2 for a typical mirror). Thus

$$\theta_{los} = \theta_b + \lambda\phi_m. \quad (7.157)$$

Perfect inertial stabilization would be achieved by driving the mirror position ϕ_m to $-\theta_b/\lambda$. However, not everything in θ_b is to be rejected, only vehicle vibration, not vehicle maneuvers.

Exogenous Input Environments

Mathematical descriptions of the input command and disturbance environments (exogenous inputs) are used in evaluating and designing the control system compensators. Models of the disturbance inputs serve as a basis for the structure of the design model and become embedded within the controller. The resulting designs were evaluated, analytically and through simulation, using these models to define or generate the inputs.

Vehicle Vibration Vehicle angular vibration at the location where the Flir equipment is mounted was expected to be equal or less than that defined by the solid line in the spectrum of Figure 7.32. It has an *rms* angular acceleration and angular position of 4.4 rad/sec^2 and 560 μrads, respectively. The bulk of its energy is concentrated around 10 Hz and it extends from 5 to 200 Hz. Angular motion contributed beyond 200 Hz is negligible in magnitude. Angular motion below 5 Hz falls in the jitter range and does not have a blurring effect.

A stationary process having the spectrum of Figure 7.32 can be generated by passing unit variance white noise through a shaping filter having the transfer function,

$$V_{\ddot{\theta}}(s) = K_v \frac{s^2(s/\alpha_1+1)(s/\alpha_2+1)}{(s/\alpha_3+1)(s/\alpha_4+1)(s/\alpha_5+1)(s^2/\Omega^2+2\zeta s/\Omega+1)}, \quad (7.158)$$

where $K_v = 2.22 \times 10^{-4}$, $\alpha_1 = 94$ rad/sec, $\alpha_2 = 377$ rad/sec, $\alpha_3 = 56.5$ rad/sec, $\alpha_4 = 188$ rad/sec, $\alpha_5 = 942$ rad/sec, $\Omega = 62.8$ rad/sec, and $\zeta = 0.5$. Twice integrating yields angular position,

$$V_\theta(s) = \frac{1}{s^2} V_{\ddot{\theta}}(s). \quad (7.159)$$

The image stabilization requirement of 65 μrads rms must be achieved when subjected to this input (see Figure 7.36).

Vehicle Maneuvering A precise mathematical definition of all input environments is not always available to the control system designer, or if one is available, it may not accurately represent the input environment that will exist when the equipment is in use. It may be necessary, as was the case here, to estimate the environment. The F-16 is capable of very high dynamic maneuvering. However, a pilot would not be doing those types of maneuvers when using this equipment in the 5.6:1 magnified, narrow field-of-view mode. (It is described to be like flying through a straw.) Under these conditions, maneuvers would be held to a minimum. Therefore, an assumption was made that the bulk of maneuvering motion energy, when in narrow field of view, would be concentrated at low frequencies, between D.C. and 0.3 Hz. This consideration will impact the design of the LQG (linear quadratic guassian) controller. By selecting the filter weighting matrices, the poles of the Kalman filter associated with the maneuver state will be positioned at locations having a magnitude of 0.3 Hz.

Pilot Head Motion Like maneuver motion, a good mathematical description of pilot head motion (when flying through the straw) was not available. One would think that the narrow field-of-view conditions would result in very little head motion, however, the pilots will glance down at their instruments, producing periods of moderately high rates of motion that must be tracked. Our assumption was that this head motion frequency content would extend out to about 3 or 4 Hz. Tight tracking of those inputs require bandwidths of 30 Hz or greater (well beyond the capabilities of the coarse positioning servo).

Self-Induced Torque Disturbances There are many sources of internally generated error present in the typical coarse line-of-sight pointing system. These include motor cogging (also called slot-lock or detent), torque ripple, stiction of seals and bearings, bearing retainer friction, bearing disturbance torques, etc. Friction or stiction in the gimbals causes a hang-off error whenever the gimbals change direction or whenever motion is initiated after a period of rest. Friction estimation and compensation techniques, described in [2], [8], as well as in this handbook in Section 9.1 (Friction Compensation), are particularly important in systems that do not include a fine stabilization assembly.

Tracking error due to cogging and torque ripple is of particular concern. Both are periodic functions of motor shaft angle

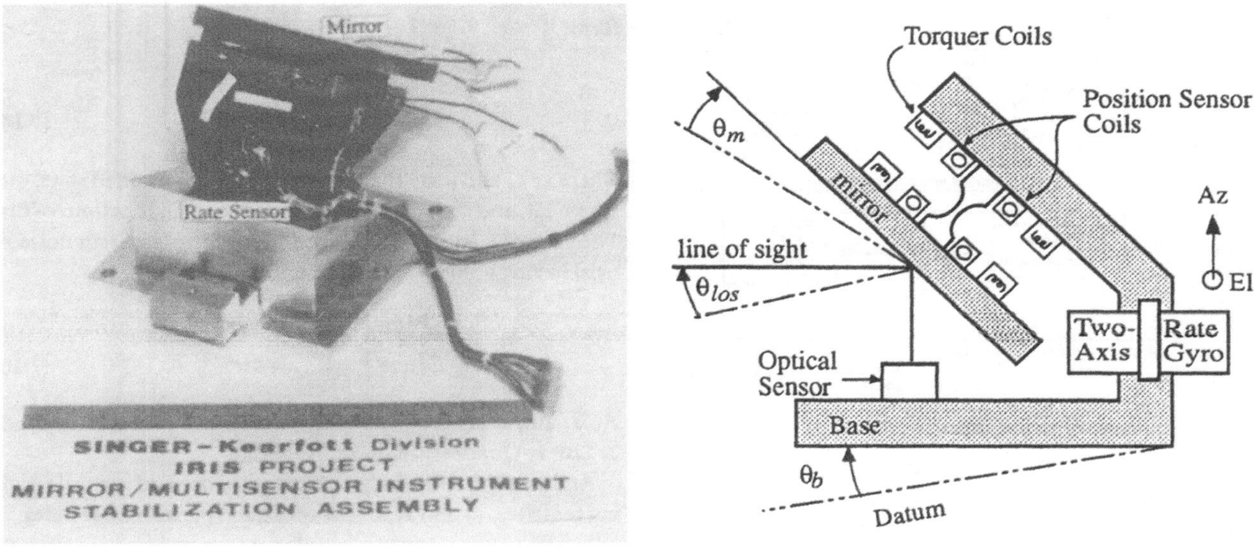

Figure 7.31 The two-axis stabilization assembly, consisting of a mirror, a two-axis angular rate Multisensor (Trademark of GEC Systems), mirror angular position sensors and torquers. Changes in line-of-sight direction are effected by tilting the mirror.

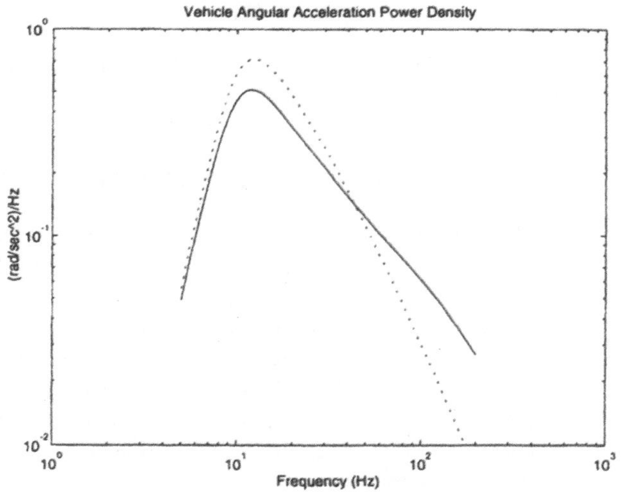

Figure 7.32 When subjected to this angular vibration environment (solid) having an amplitude of 560 μrads rms, the image stabilization assembly cannot permit more than 65 μrads rms of line-of-sight motion. An approximation (dashed) is used for controller design.

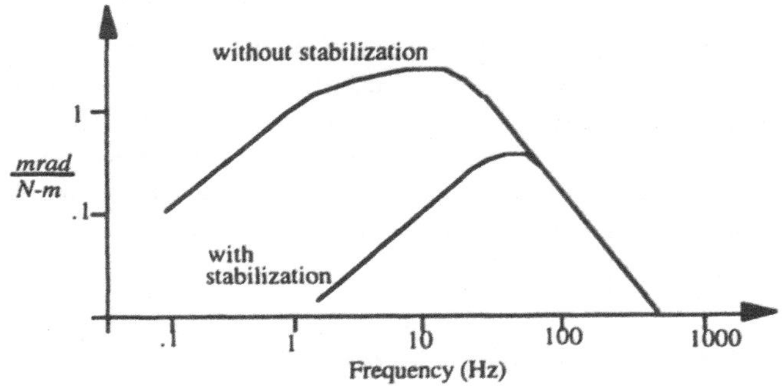

Figure 7.33 The stabilization system helps attenuate tracking error caused by internally generated torque disturbances.

which produce a noticeable jumping of the image when the pilot is steadily turning his head. The coarse positioning system will have a certain response to these disturbances as defined by their Torque Disturbance Sensitivity function, which is typically shaped like the curve in Figure 7.33. In the Falcon Eye, torque disturbances produce a worst-case gimbal tracking error of about 3 mrad peak to peak, far in excess of the 250 μrad limit imposed by the IR detector resolution. The stabilization system therefore must attenuate those tracking errors by more than a factor of 12 for frequencies up to 10 Hz.

Stabilization Control System Structure and Modeling

The architecture of the overall control system is shown in some detail in Figure 7.34. It represents a single axis of the two-axis system (coordinate transformations not shown). A discussion of the coordinate systems relating gimbals to mirror stabilization assembly is not provided because it will not help to understand this control system design effort. All that one needs to recognize is that coordinate frames used in designing the stabilization system controllers are attached locally to the base of the stabilization assembly, and that, in the actual implementation, transformations from the gimbal to the stabilization assembly coordinate frames, and back, were employed.

Linear Truth Model A linear model of one axis of the stabilization control system is given in Figure 7.35. It defines the stabilization system's response to the three inputs discussed in conjunction with Figure 7.30 and will be used in evaluating system performance with the compensators to be derived. Transfer functions from those three inputs, ϕ_g, ϕ_c, and θ_v, to the line of

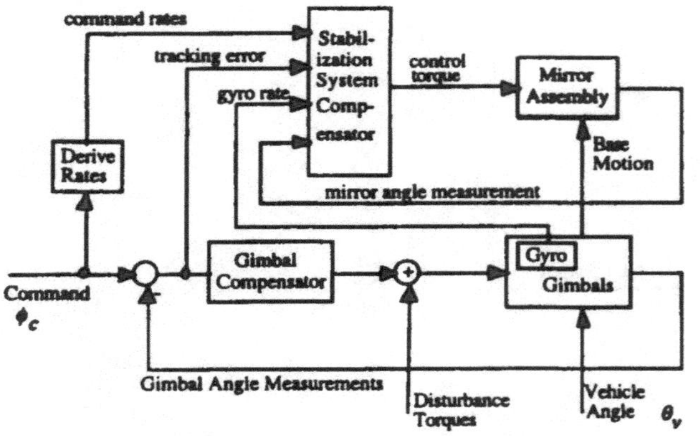

Figure 7.34 Structure controller system.

sight angle θ_{los} are to be referred to as $T_g(s)$, $T_c(s)$, and $T_v(s)$, respectively. The system parameters of this model appear in Table 7.4. Units N-m refer to Newton-meters.

Although implemented digitally, the compensator is shown here in continuous-time form. It was possible to design and analyze the compensator in the continuous-time domain because the fastest time constants of the closed-loop system were significantly less (e.g., 4 times) than the sample frequency (2700 Hz). Nevertheless, delays due to processing and the sample and holding (S/H) of data, and other performance limiting effects were included in the model. These include a 2 kHz, three-pole, low-pass Butterworth antialias filter $A(s)$, and a 1 kHz, three-pole low-pass Butterworth reconstruction filter $R(s)$ on the control output from the D/A converter.

7.4.4 Controller Design

Design Procedure

Our objective was to develop a compensator that achieves a performance approaching the ideal performance described with Figure 7.30. This was accomplished using the tools of Linear Quadratic Gaussian (LQG) optimal control theory. The design procedure, briefly, is as follows. It begins with the development of a design model capturing the essential features of the plant and exogenous inputs. This model serves as a basis for the design of a steady-state LQ regulator and Kalman filter, which combine to form a linear time-invariant dynamic compensator. The designer adjusts the regulator and filter weighting matrices (Q, R, V, and W) until a design is produced which achieves satisfactory system performance. This is described in greater detail below.

1. Design Model Definition

The first step in the design process entails defining a linear state-space model of the process to be controlled. It becomes imbedded within the controller, therefore the simplest model that adequately describes the plant and exogenous disturbance inputs should be used. In this application an adequate description is one which will enable the controller, given the measured inputs, to distinguish the various exogenous inputs from one another, and thereby to effect the proper control action.

The design model will be represented in the usual state-space form:

$$\begin{aligned} \dot{x} &= Ax + Bu + v, \\ y &= Cx + w, \end{aligned} \tag{7.160}$$

where x, y, and u are the appropriately dimensioned state, measurement, and control vectors, respectively, all functions of time. The vectors v and w are independent, white Gaussian noise signals, having autocovariance matrices,

$$\begin{aligned} E\{vv'\} &= V\delta(t), \quad \text{and} \\ E\{ww'\} &= W\delta(t). \end{aligned} \tag{7.161}$$

A, B, and C are constant coefficient system matrices.

2. Linear Quadratic Regulator Design

At this stage of the design process the state and control weighting matrices, Q and R, of the quadratic performance index,

$$J = \int_0^{\infty}(x'Qx + u'Ru)dt, \tag{7.162}$$

are selected, and the control matrix G of the linear control law

$$u = -Gx \tag{7.163}$$

is derived in accordance with $G = R^{-1}B'M$, where M is the solution to the Control Algebraic Riccati Equation,

$$0 = MA + A'M - MBR^{-1}B'M + Q. \tag{7.164}$$

The coefficients of Q, the state weighting, are defined in accordance with the error signal to be minimized. This error will be some linear combination of state variables.

The control weighting R is adjusted to achieve some other criterion. It usually suffices to weight each control independently. Then all of the off-diagonal coefficients in R are zero. The diagonal elements are typically chosen to achieve a bandwidth criterion, as was the case in this application, described below.

3. Kalman Filter Design

The state vector x of the linear control law (Equation 7.163) is typically not completely known. Some states will not be measured directly, and those measured may be corrupted by noise. The steady-state Kalman filter

$$\begin{aligned} \dot{\hat{x}} &= A\hat{x} + B\hat{u} + K(y - C\hat{x}), \\ \hat{u} &= -G\hat{x}, \end{aligned} \tag{7.165}$$

is enlisted to provide an estimate $\hat{x}$ of the actual state x. The designer specifies the coefficients in the weighting matrices, V and W, which, along with the system matrices A and C and the Filter Algebraic Riccati Equation,

$$0 = AP + PA' - PC'W^{-1}CP + V, \tag{7.166}$$

define the filter gain $K = PC'W^{-1}$.

It is important to recognize that we are not designing a Kalman filter for accurately estimating the system state, but we are using the Kalman filter as a synthesis tool for designing a compensator.

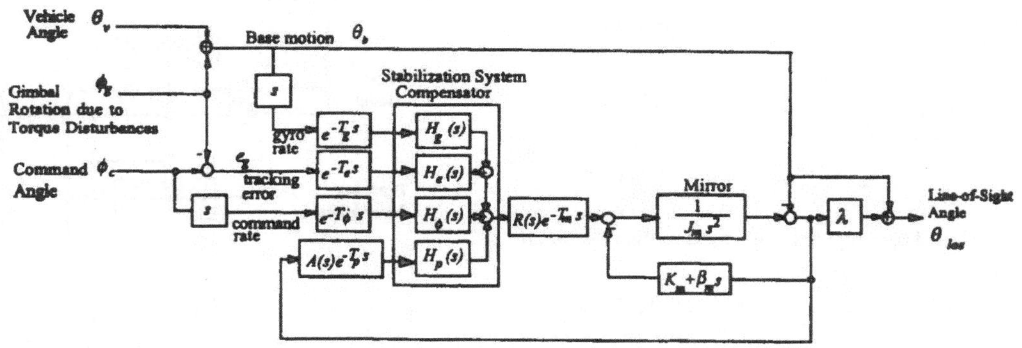

Figure 7.35 Linear model used in evaluating stabilization system performance.

TABLE 7.4 Stabilization System Actuator and Error Source Parameters

Symbol	Value	Units	Parameter name
J_m	3.5×10^{-5}	N-m-s^2	mirror inertia
β_m	4×10^{-3}	N-m-s	flexure damping
K_m	.096	N-m/rad	flexure stiffness
T_g	1.75	msec	gyro signal latency
T_e	85	μsec	tracking error processing delay
T_ϕ	40	μsec	command delay
T_p	90	μsec	pickoff measurement processing delay
T_m	240	μsec	stab. compensator processing & S/H delay
λ	1.386	-	optical scale factor

The matrices, V and W, are not viewed as descriptions of actual noise present in the system, but as design parameters, i.e., knobs, that the designer adjusts to shape the dynamic response of the filter and to achieve design goals.

To simplify the task of finding an acceptable design, one would like to work with the smallest set of parameters that provides the needed design freedom. To this end we assume that the state and measurement noise vectors, v and w, are not only independent of each other, but that the individual elements of v and w are independent. Then, only the diagonal elements of V and W are nonzero and are employed as adjustable design parameters. (There are cases when the off-diagonal elements must be adjusted to achieve certain design characteristics. An example is given in [4].)

The filter and regulator combine to form a dynamic compensator having the matrix transfer function,

$$u(s) = H(s)y(s), \tag{7.167}$$

where

$$H(s) = -G(sI - A + BG + KC)^{-1}K. \tag{7.168}$$

4. Evaluation of Performance

At this stage we verify that the design goals have been satisfied. In this application, this involved evaluating the performance predicted by the linear model of the system (Figure 7.35) with the newly derived controller. If the requirements are met, a more thorough evaluation can be performed by simulating the operation of this controller in a detailed truth model of the system. This is a model that includes all of the important dynamic and nonlinear effects, the coordinate transformations, Coriolis and centripetal couplings, the gimbals, friction, backlash, vehicle and pilot head attitude data, etc. If satisfactory performance is achieved there, the design is complete. If not, a judgment is made as to what can be changed to improve performance, and the design process is repeated. This entails 1) going back to Steps 2 and 3, modifying one or more weighting matrices, Q, R, V, and W, and generating a new compensator based on the same design model, or (2) going back to Step 1, altering the design model to incorporate some effect now understood to be important, and repeating the design. This continues until satisfactory performance with an adequate controller is achieved.

In this application the stabilization performance was computed in accordance with the model of Figure 7.36, where $V_\theta(j\omega)$ is the shaping filter (see Equation 7.158) defining worst-case angular motion, and $T_v(j\omega)$ is the stabilization system's response to vehicle angular motion inputs.

The resulting line-of-sight motion due to angular vibration in the frequency range over which the stabilization requirement applies (5 to 200 Hz) is given by

$$\sigma_{los}^2 = \frac{1}{2\pi}\int_{2\pi 5}^{2\pi 200} |V_\theta(j\omega)T_v(j\omega)|^2\,\sigma_n^2 d\omega \tag{7.169}$$

where $\sigma_n^2 = 1$.

The response of the complete system to internal torque disturbances was generated in accordance with the model in Figure 7.37

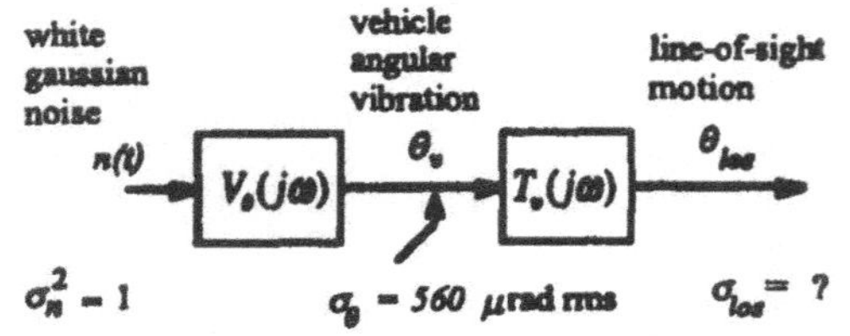

Figure 7.36 Model used to compute line-of-sight motion due to angular vehicle vibration.

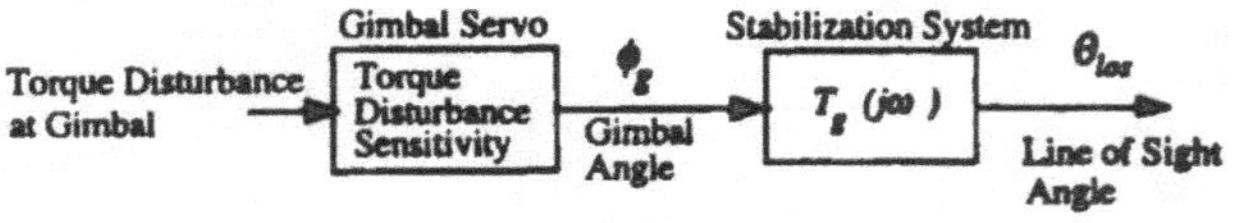

Figure 7.37 Model used to compute line of sight motion due to torque disturbances.

where $T_g(j\omega)$ is the stabilization system's response to gimbal angular input (whose ideal value is depicted by the center curve of Figure 7.30). Because the gimbal servos are not described, this check will entail verifying that the stabilization system provides adequate isolation (i.e., > 12 in the 2 to 10 Hz range) from gimbal angular position errors due to torque disturbances.

Design Model Development

The process of developing a design model is commonly one requiring several iterations. One reason for this is that it is highly unlikely, except in simple cases, that the designer will anticipate all of the important characteristics that should be in the design model in the first attempt. Secondly, it pays to begin with a simple model which may or may not be adequate, then to increase complexity until satisfactory performance is achieved, rather than beginning immediately with what may be an unnecessarily complicated, high-order model.

The design models developed for this application all include the stabilization mirror dynamics (Equation 7.156) and some form of base motion model. (Recall that base motion is defined as motion about a local coordinate system attached to the base to which the mirror is mounted.) Since the controller must isolate the line of sight from vehicle vibration, the base motion model must include a model of that disturbance. The actual vibration model, defined by the shaping filter of Equation 7.158 is fifth-order and probably more complicated than need be. A third-order approximation was developed for the purpose of the controller design. This model, when excited by white noise, yields a signal whose second derivative has the spectrum shown as a dashed line in Figure 7.32. It has an angular position output given by the transfer function,

$$V_\theta'(s) = \frac{1}{(s/\Omega_0+1)(s^2/\Omega_1^2+2\zeta_1 s/\Omega_1+1)}, \qquad (7.170)$$

where $\Omega_0 = 126$ rad/sec, $\Omega_1 = 62.8$ rad/sec, and $\zeta_1 = 0.5$. This model is part of the controller design model block diagram shown in Figure 7.38, depicting the design model in its final form after several stages of development. Those stages are described below.

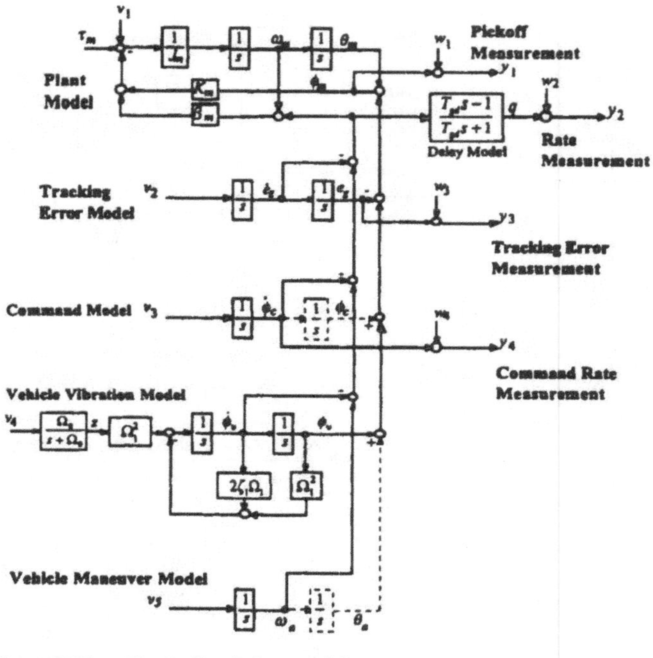

Figure 7.38 Controller design model.

The first design model developed for this application involved a base motion model consisting of vehicle vibration only. The resulting controller did a fine job in isolating the line of sight from vibration. However, it knew nothing of vehicle maneuvers or of gimbal motion, and interpreted those large motions, when tested in simulation, as vibration to be rejected, driving the mirror into its stops. To remedy this, models of vehicle maneuvering and commanded gimbal motion were added to the design model. Both are modeled as double integrators because white noise through a double integrator produces a signal having characteristics expected of maneuver and head motion — infinite DC gain needed to represent constant turning rates and a rolling off with increased frequency. An important distinction between the command and maneuver signals is that the command is a known input while the maneuver signal is not. The command rate is therefore included as a measured output in the design model.

A controller developed with this design model worked, in that it did not drive the mirror to the stops. However, stabilization performance was not quite adequate (85 µrads rms, relative to the required 62.) Performance was being limited by the pure delay present in the gyro signal measurement. Improved stabilization performance was achieved by adding a model of the delay to the design model. A first-order Pade' approximation to the delay incorporated in the design model increased the order of the compensator by one, but also improved stabilization performance sufficiently (to 55 µrads rms), as described below.

Although stabilization performance was now adequate, the tracking smoothness performance was not. Improved ability to isolate the line of sight from gimbal motion due to internally generated torque disturbances was achieved by adding the gimbal servo tracking error e_g, as an input to the stabilization system compensator. Two additional integrator states were added to allow including the tracking error as an input to the compensator, and therefore as an output of the design model. The resulting design model in its final form is shown in Figure 7.38.

Two of the integrators in the design model are drawn with dashed lines. The regulator gains on those states turn out to be zero. Consequently, they have no effect on the output of the controller and can be eliminated from the design model without effect. The model and the resulting compensator are therefore tenth order.

You will note that the command angle ϕ_c is no longer present as a state in the design model. And even if it were, it could not be used as a reference input to the stabilization system because the gain on that state is zero. The command rate is therefore used as the reference input defining pilot head motion.

The ten-element state vector and four-element measurement vector are ordered as follows:

$$x = \begin{bmatrix} \omega_m \\ \phi_m \\ \dot{e}_g \\ e_g \\ \dot{\phi}_c \\ z \\ \dot{\phi}_v \\ \phi_v \\ \omega_a \\ q \end{bmatrix} = \begin{bmatrix} \text{mirror inertial rate} \\ \text{mirror angular position} \\ \text{gimbal tracking error rate} \\ \text{gimbal tracking error} \\ \text{command rate} \\ \text{vehicle vibration filter state} \\ \text{vehicle vibration angular rate} \\ \text{vehicle vibration angular position} \\ \text{vehicle maneuver angular rate} \\ \text{gyro delay model variable} \end{bmatrix}$$

$$y = \begin{bmatrix} \phi_m \\ \omega_g \\ e_g \\ \dot{\phi}_c \end{bmatrix} = \begin{bmatrix} \text{mirror pickoff angle} \\ \text{gyro rate} \\ \text{gimbal tracking error} \\ \text{command rate} \end{bmatrix}$$

The system parameters contained in the design model are given in Table 7.5. The units of these parameters were changed from seconds to deciseconds (ds), adjusting the scaling of the system matrices to eliminate numerical problems which were precluding the solution of the Control and Filter Algebraic Riccati Equations. The controller gain and filter gain matrices were derived for this scaled system, and then scaled back to units of seconds prior to their usage.

The corresponding system matrices are

$$A = \begin{bmatrix} -.0133 & -.272 & .0133 & 0 & .0133 & 0 & .0133 & 0 & .0133 & 0 \\ 1 & 0 & 1 & 0 & -1 & 0 & -1 & 0 & -1 & 0 \\ 0 & 0 & \varepsilon & -.01 & 0 & 0 & 0 & 0 & 0 & 0 \\ 0 & 0 & 1 & 0 & 0 & 0 & 0 & 0 & 0 & 0 \\ 0 & 0 & 0 & 0 & \varepsilon & 0 & 0 & 0 & 0 & 0 \\ 0 & 0 & 0 & 0 & 0 & -1.26 & 0 & 0 & 0 & 0 \\ 0 & 0 & 0 & 0 & 0 & .394 & -.628 & -.394 & 0 & 0 \\ 0 & 0 & 0 & 0 & 0 & 0 & 1 & 0 & 0 & 0 \\ 0 & 0 & 0 & 0 & 0 & 0 & 0 & 0 & \varepsilon & 0 \\ 0 & 0 & -1 & 0 & 1 & 0 & 1 & 0 & 1 & -11.4 \end{bmatrix}$$

$$B = \begin{bmatrix} 2.82 & 0 & 0 & 0 & 0 & 0 & 0 & 0 & 0 & 0 \end{bmatrix}'$$

$$C = \begin{bmatrix} 0 & 1 & 0 & 0 & 0 & 0 & 0 & 0 & 0 & 0 \\ 0 & 0 & 1 & 0 & -1 & 0 & -1 & 0 & -1 & 22.8 \\ 0 & 0 & 0 & 1 & 0 & 0 & 0 & 0 & 0 & 0 \\ 0 & 0 & 0 & 0 & 1 & 0 & 0 & 0 & 0 & 0 \end{bmatrix}$$

The small negative entries ($\varepsilon = -1 \times 10^{-5}$) on the diagonal of A and the small spring term ($-.01$) at A_{34}, were also added to eliminate the numerical problems precluding the convergence of the Riccati equation solver.

LQ Regulator Design

A quadratic performance integral was formed which contains both the error e that the regulator must drive to zero and the control action τ_m :

$$J = \int_0^{\infty} (e^2 + \rho \tau_m^2) dt \tag{7.171}$$

Because there is only a single control variable, the mirror torque τ_m, the control weighting, ρ, is scalar.

An expression for the error e was derived from the tracking error expression given by Equation 7.155. The Flir line of sight relative to the vehicle ϕ_f is composed of the gimbal relative angle ϕ_g and the line of sight angle relative to the gimbals due to mirror rotation $\lambda\phi_m$:

$$\phi_f = \phi_g + \lambda\phi_m. \tag{7.172}$$

To include gimbal tracking error in this expression, we express the gimbal angle in terms of the command and gimbal tracking error:

$$\phi_g = \phi_c - e_g. \tag{7.173}$$

We also assume that the pilot's head attitude relative to the vehicle ϕ_p of Equation 7.155 equals the command angle ϕ_c (i.e., latency and noise in the head tracker data is ignored). Combining this and Equations 7.155, 7.172, and 7.173 yields the error equation,

$$e = \phi_v - e_g + \lambda\phi_m. \tag{7.174}$$

The state weighting matrix Q is formed to penalize this error expression. To compute Q easily, we express e in terms of the state

TABLE 7.5 Design Model Scaled System Parameters.

Parameter	Value	Units
Mirror inertia J_m	3.54×10^{-1}	N-m-ds^2
Mirror hinge damping β_m	4.0×10^{-3}	N-m-ds
Mirror hinge stiffness K_m	.096	N-m/rad
Gyro delay parameter T_{gd}	1/2(.175)	ds
Shaping filter parameter Ω_0	1.26	rad/ds
Shaping filter parameter Ω_1	.628	rad/ds
Shaping filter parameter ζ_1	0.5	—

vector $x : e = Lx$. With optical scale factor $\lambda = 1.386$ (which applies in narrow field-of-view mode) this vector L becomes

$$L = \begin{bmatrix} 0 & 1.386 & 0 & -1 & 0 & 0 & 0 & 1 & 0 & 0 \end{bmatrix}, \tag{7.175}$$

and with this the state weighting Q is computed as follows:

$$Q = L'L + \varepsilon I_{10} \tag{7.176}$$

It was necessary to add small positive terms ($\varepsilon = 1 \times 10^{-5}$) on the diagonal to makes Q slightly more positive definite, because, without them, the Ricatti Equation Solver halted when checking and mistaking Q for a negative-definite matrix.

Only a single parameter, the control weighting ρ of the matrix $R = [\rho]$, must be selected to define completely all of the matrices entering into the Control Algebraic Ricatti Equation. As ρ is decreased, the eigenvalues of the controlled plant, eig$(sI - A + BG)$, associated with the mirror dynamics, move to the left and approach asymptotes at 45° above and 45° below the negative real axis. The other eigenvalues, those associated with the exogenous inputs, do not move because those modes are uncontrollable. The coefficient ρ was adjusted to place the mirror eigenvalues at a location having a magnitude of about 100 Hz. This enables the regulator to respond adequately to exogenous inputs having lesser frequency content, and yet does not move those eigenvalues into a frequency regime where unmodeled phase lag in the system begins to have a destabilizing effect.

With $R = [0.125]$ the controller matrix G (for the scaled system), computed via solution of the Control Algebraic Ricatti Equation, is

$$\begin{aligned} G = \; & [1.6424 \quad 3.8249 \quad .46390 \quad -2.8286 \quad -1.6423 \\ & -.034147 \quad -.38867 \quad 2.8672 \quad -1.6423 \quad 0] \end{aligned}$$

Filter Design

There are a total of nine nonzero coefficients in V and W, all on the diagonal, which the designer must specify. To reduce the number of design parameters to a manageable number, the coefficients in W were set to fixed values proportional to the actual noise variances of the associated measurements. The tracking error and command rate signals are not actual measurements but are digital signals fed forward from the gimbal servoes, so their variances were set to values consistent with their quantization levels. The elements of W were, therefore, set to values proportional to the pickoff, gyro rate, tracking error, and command rate noise variances, respectively, which yielded

$$W = \text{diag}[1 \quad 5 \quad .1 \quad 16.7] \tag{7.177}$$

The five nonzero diagonal coefficients of V,

$$V = \text{diag}[v_1 \quad 0 \quad v_2 \quad 0 \quad v_3 \quad v_4 \quad 0 \quad 0 \quad v_5 \quad 0], \tag{7.178}$$

were selected by performing a manual search over the five-dimensional parameter space spanned by the plant noise variances v_1 through v_5. At selected points within that space, control system compensators were computed and the performance of the stabilization system with that compensator was evaluated. This included examining how closely its performance approached the ideal (Figure 7.30), and the calculations necessary to determine if the stabilization and tracking smoothness requirements were met. This manual search did not proceed blindly, however. It was possible to predict how adjustments made to specific plant noise levels should alter closed-loop performance. For example, what would you expect the compensator to do if you tell it there is more noise driving the vibration portion of the design model by increasing v_4? It should increase the degree of isolation against vehicle vibration. And in fact it does that, as shown in Figure 7.39. As v_4 is increased, the transfer function from vehicle angular motion to line-of-sight motion deepens.

This manual search settled upon

$$V = \text{diag}[500 \quad 0 \quad 100 \quad 0 \quad 100 \quad 10^5 \quad 0 \quad 0 \quad .01 \quad 0]$$

to which corresponds the Kalman gain matrix:

$$K = \begin{bmatrix} 1.9548e{+}01 & 4.5483e{+}00 & -1.2313e{+}00 & 1.0944e{-}01 \\ 7.6554e{+}00 & -1.1539e{+}00 & 7.0985e{-}01 & -5.3122e{-}02 \\ 1.0308e{+}00 & -7.1132e{-}01 & 3.1038e{+}01 & 2.1484e{-}02 \\ 7.0985e{-}02 & -9.7569e{-}02 & 7.8452e{+}00 & 4.1423e{-}03 \\ -8.8714e{-}01 & 5.9042e{-}01 & 6.9177e{-}01 & 2.4153e{+}00 \\ -9.5811e{+}01 & 7.3508e{+}01 & 9.7882e{+}01 & -5.0307e{+}00 \\ -1.1170e{+}01 & 1.2785e{+}01 & 2.2769e{+}01 & -1.5445e{+}00 \\ 1.9951e{+}00 & -2.2092e{-}01 & 1.8483e{+}00 & -5.8257e{-}01 \\ -4.3322e{-}02 & 3.8739e{-}02 & 5.9208e{-}02 & -3.8896e{-}03 \\ -8.2898e{-}01 & 1.0424e{+}00 & -5.4377e{-}01 & 4.4834e{-}02 \end{bmatrix}.$$

This matrix and the G matrix given above were unscaled, from units of deciseconds to seconds, before use.

7.4.5 Performance Achieved

Here we examine how closely the stabilization system's response approaches the ideal response defined in Figure 7.30. Transfer functions $T_v(s)$, $T_c(s)$, $T_g(s)$, from vehicle angle, command angle, and gimbal angle, respectively, to the line-of-sight angle, were

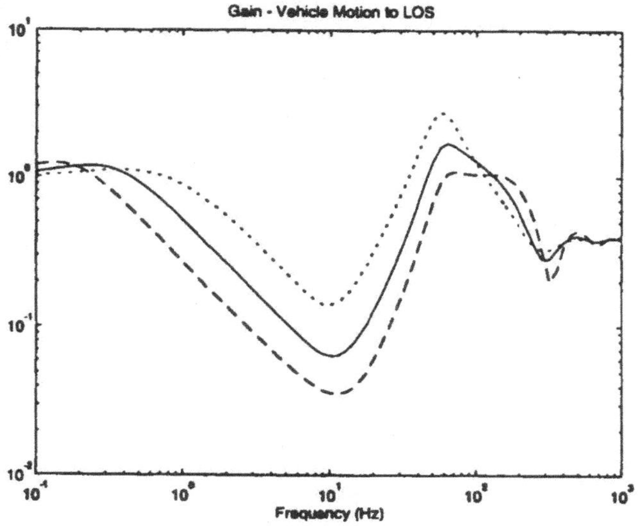

Figure 7.39 As noise variance v_4 driving the vibration part of the plant is increased, the level of line-of-sight (LOS) stabilization against vibration improves, as shown here with $v_4 = 10^4$ (dotted), 10^5 (solid), and 10^6 (dashed). All other elements in V, W, Q, and R were not changed.

derived in accordance with the linear model of Figure 7.35. The magnitudes of these transfer functions are plotted in Figure 7.40.

Clearly these plots approach the ideal response curves given in Figure 7.30, primarily in the low to middle frequency regimes. In the upper plot, command tracking is unity (the ideal) to a frequency of 10 Hz, and a command tracking bandwidth of nearly 100 Hz is provided, which exceeds the Command Tracking Requirement defined earlier. The controller contributes a measure of peaking in the response to reduce phase lag at lower frequencies where tight command tracking is needed.

As shown in the middle plot, the stabilization system achieves exceptionally good isolation from internally generated gimbal tracking errors.

Finally, in the lower plot, which concerns the system's response to vehicle angular motion, there is adequate maneuver tracking with a bandwidth of just over 0.3 Hz, and there is a transition from maneuver tracking at low frequencies to isolation from high frequency vibration.

Stabilization Performance The adequacy of the stabilization performance was verified by numerically evaluating Equation 7.169 with $T_v(j\omega)$ given by the lower plot of Figure 7.40. This results in

$$\sigma_{los} = 55\ \mu\text{rads rms}$$

which meets the 62 μrads requirement.

Smoothness of Tracking Performance The center plot of Figure 7.40 clearly indicates that this requirement (greater than a factor of $12x$, or 22 dB of attenuation in the 2 to 10 Hz range) is met by a large margin.

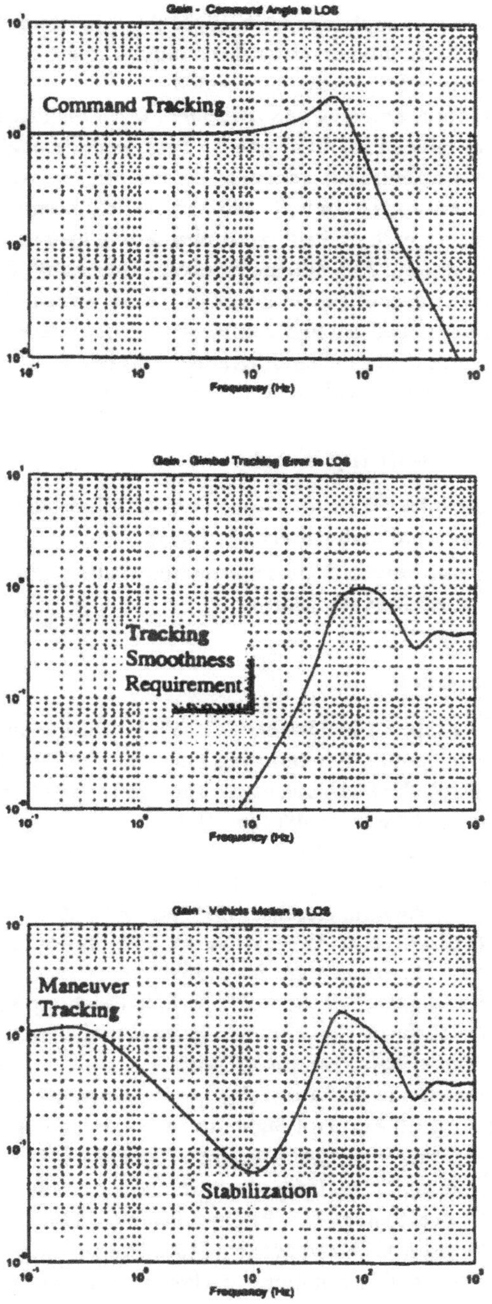

Figure 7.40 Response of the line-of-sight (LOS) stabilization system to the various inputs (compare to Figure 7.30).

7.4.6 Concluding Remarks

This chapter describes the design of a controller for the line-of-sight pointing and stabilization system contained within the Falcon Eye, a head-steered night vision system developed for the F16 Falcon. The chapter begins by discussing the purpose and performance objectives of the overall night vision system. This is followed by a description of the electro-optical subsystem used to point and stabilize the line-of-sight, i.e., the plant, and of the disturbance inputs. The chapter then discusses the performance requirements applying specifically to this subsystem. Finally, the design, using modern control techniques, of the controller that met those requirements is presented.

Although only one was described, seven distinct controllers were developed for this system. Four were developed for the two-axis stabilization assembly — one per axis with different controllers applied in wide and narrow field-of-view modes — and one for each of the three gimbal servos, azimuth, elevation, and derotation. All were extensively and successfully flight tested, as was the overall Falcon Eye night vision system. In fact, test pilots developed enough confidence in the system to use it at night when taking off, landing, and during low level flight (~ 200 ft above ground) with the Falcon Eye as their only visual source.

7.4.7 Defining Terms

Flir: An infrared imaging system typically directed to look forward in the direction of travel.

Exogenous inputs: Inputs arising from states that are unaffected by the control.

References

[1] Benson, A.J. and Barnes, G.R., Vision During Angular Oscillation: The Dynamic Interaction of Visual and Vestibular Mechanisms, *Aviation, Space, and Environmental Medicine*, 49(*1Sec.II*), 1978.

[2] Friedland, B. and Park, Y.J., On Adaptive Friction Compensation, *Proc. 30th IEEE Conf. Decision Control*, Brighton, England, 1991, pp. 2899–2903.

[3] Fuchs, C., et al., Two-axis Mirror Stabilization Assembly, U.S. Patent 4 881 800, 1989.

[4] Haessig, D., Selection of LQG/LTR Weighting Matrices through Constrained Optimization, *Proc. ACC*, Seattle, WA, 1995, pp. 458–460.

[5] Haessig, D. A., Mirror Positioning Assembly for Stabilizing the Line-of-Sight in a Two-Axis Line-of-Sight Pointing System, U.S. Patent 5 220 456, 1993.

[6] Haessig, D. A., Image Stabilization Assembly for an Optical System, U.S. Patent 5 307 206, 1994.

[7] Johnson, C.D. and Masten, M.K., Fundamental Concepts and Limitations in Precision Pointing and Tracking Problems, Acquisition, Tracking, and Pointing VII; *SPIE Proc.*, 1993, Vol. 1950, pp. 58–72.

[8] Maqueira, B. and Masten, M.K., Adaptive Friction Compensation for Line-of-Sight Pointing and Stabilization, *Proc. ACC*, 1993, pp. 1942–46.

[9] Masten, M.K.,, Electromechanical Systems for Optical Target Tracking Sensors. In *Multitarget-Multisensor Tracking: Advanced Applications*, Artech House, Norwood, MA, 1990, pp 321–360.

[10] Masten, M.K. and Sebesta, H.R., Line-of-Sight Stabilization/Tracking Systems: An Overview, *Proc. ACC*, 2, 1987, pp. 1477–1482.

[11] Scott, W.B., Falcon Eye Flir, GEC Helmet Aid F-16 Mission Flexibility, *Aviation Week*, April 17, 1989.

8

Control of Robots and Manipulators

Mark W. Spong
The Coordinated Science Laboratory, University of Illinois at Urbana-Champaign

Joris De Schutter
Katholieke Universiteit Leuven, Department of Mechanical Engineering, Leuven, Belgium

Herman Bruyninckx
Katholieke Universiteit Leuven, Department of Mechanical Engineering, Leuven, Belgium

John Ting-Yung Wen
Department of Electrical, Computer, and Systems Engineering, Rensselaer Polytechnic Institute

8.1 Motion Control of Robot Manipulators 165
Introduction • Dynamics • PD Control • Feedback Linearization • Robust and Adaptive Control • Time-Optimal Control • Repetitive and Learning Control
References 177
8.2 Force Control of Robot Manipulators 177
Introduction • System Equations • Hybrid Force/Position Control • Adaptive Control • Impedance Control
References 184
8.3 Control of Nonholonomic Systems 185
Introduction • Test of Nonholonomy • Nonholonomic Path Planning Problem • Stabilization
References 194

8.1 Motion Control of Robot Manipulators

Mark W. Spong, *The Coordinated Science Laboratory, University of Illinois at Urbana-Champaign*

8.1.1 Introduction

The design of intelligent, autonomous machines to perform tasks that are dull, repetitive, hazardous, or that require skill, strength, or dexterity beyond the capability of humans is the ultimate goal of robotics research. Examples of such tasks include manufacturing, excavation, construction, undersea, space, and planetary exploration, toxic waste cleanup, and robotic-assisted surgery. The field of robotics is highly interdisciplinary, requiring the integration of control theory with computer science, mechanics, and electronics.

The term *robot* has been applied to a wide variety of mechanical devices, from children's toys to guided missiles. An important class of robots are the manipulator arms, such as the PUMA robot shown in Figure 8.1 These manipulators are used primarily in materials handling, welding, assembly, spray painting, grinding, deburring, and other manufacturing applications. This chapter discusses the motion control of such manipulators.

Robot manipulators are basically multi-degree-of-freedom positioning devices. The robot, as the "plant to be controlled," is a multi-input/multi-output, highly coupled, nonlinear mechatronic system. The main challenges in the motion control problem are the complexity of the dynamics and uncertainties, both

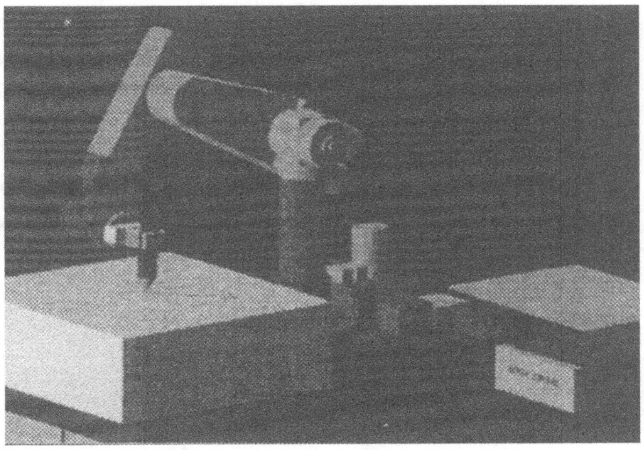

Figure 8.1 PUMA robot manipulator.

parametric and dynamic. Parametric uncertainties arise from imprecise knowledge of kinematic parameters and inertia parameters, while dynamic uncertainties arise from joint and link flexibility, actuator dynamics, friction, sensor noise, and unknown environment dynamics.

There are a number of excellent survey articles on the control of robot manipulators from an elementary viewpoint [2], [4], [6]. These articles present the basic ideas of independent joint control using linearized models and classical transfer function analysis. That material is not repeated here. Instead, we survey more recent and more advanced material that summarizes the work of many researchers from about 1985 to the present. Many of the ideas presented here are found in the papers reprinted in

0-8493-0054-3/00/$0.00+$.50

[10], which is recommended to the reader who wishes additional details. The present chapter is accessible to anyone having an elementary knowledge of Lagrangian dynamics and the state-space theory of dynamical systems, including the basics of Lyapunov stability theory. The textbooks by Asada and Slotine [1] and Spong and Vidyasagar [11] may be consulted for the background necessary to follow the material presented here.

Configuration Space and Task Space

We consider a robot manipulator with **n-links** interconnected by **joints** into an **open kinematic chain** as shown in Figure 8.2. For simplicity of exposition we assume that all joints are rotational, or **revolute**. Most of the discussion in this chapter remains valid for robots with sliding or **prismatic** joints.

The joint variables, $q_1, \ldots, q_n$, are the relative angles between the links; for example, q_i is the angle between link i and link $i-1$. A vector $q = (q_1, \ldots, q_n)^T$, with each $q_i \in [0, 2\pi)$ is called a **configuration**. The set of all possible configurations is called **configuration space** or **joint space**, which we denote as $\mathcal{C}$. The configuration space is an n-dimensional torus, $\mathcal{T}^n = S^1 \times \cdots \times S^1$, where S^1 is the unit circle.

The **task space**, or **end-effector space**, is the space of all positions and orientations (called **poses**) of the end-effector (usually a gripper or tool). If a coordinate frame, called the **base frame** or **world frame**, is established at the base of the robot and a second frame, called the **end-effector frame** or **task frame**, is attached to the end-effector, then the end-effector position is given by a vector $x \in \mathbf{R}^3$ specifying the coordinates of the origin of the task frame in the base frame, and the end-effector orientation is given by a 3×3 matrix R whose columns are the direction cosines of the coordinate axes of the task frame in the base frame. This **orientation matrix**, R, is orthogonal, i.e. $R^T R = I$, and satisfies $\det(R) = +1$. The set of all such 3×3 orientation matrices forms the **special orthogonal group**, $SO(3)$. The task space is then isomorphic to the **special Euclidean group**, $SE(3) = \mathbf{R}^3 \times SO(3)$. Elements of $SE(3)$ are called rigid motions [11].

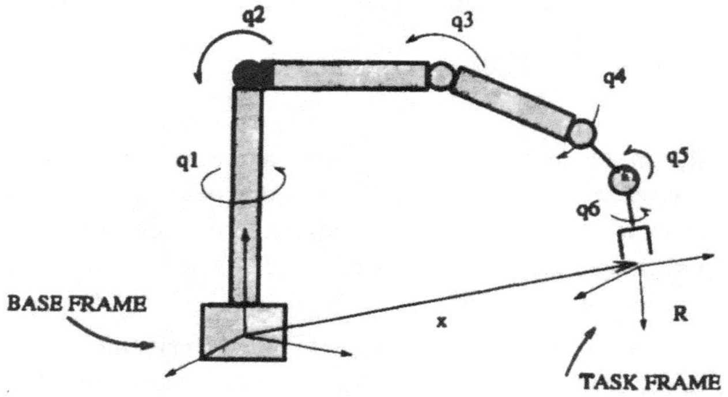

Figure 8.2 Configuration space and task space variables.

Kinematics

Kinematics refers to the geometric relationship between the motion of the robot in joint space and the motion of the end-effector in task space without consideration of the forces that produce the motion. The **forward kinematics problem** is to determine the mapping

$$X_0 = \begin{bmatrix} x(q) \\ R(q) \end{bmatrix} = f_0(q) : \mathcal{T}^n \to SE(3) \tag{8.1}$$

from configuration space to task space, which gives the end-effector pose in terms of the joint configuration. The **inverse kinematics problem** is to determine the inverse of this mapping, i.e., the joint configuration as a function of the end-effector pose. The forward kinematics map is many-to-one, so that several joint space configurations may give rise to the same end-effector pose. This means that the forward kinematics always has a unique pose for each configuration, while the inverse kinematics has multiple solutions, in general.

The kinematics problem is compounded by the difficulty of parameterizing the rotation group, $SO(3)$. It is well-known that there does not exist a minimal set of coordinates to "cover" $SO(3)$, i.e., a single set of three variables to represent all orientations in $SO(3)$ uniquely. The most common representations used are Euler angles and quaternions. Representational singularities, which are points at which the representation fails to be unique, give rise to a number of computational difficulties in motion planning and control.

Given a minimal representation for $SO(3)$, for example, a set of Euler angles ϕ, θ, ψ, the forward kinematics mapping may also be defined by a function

$$X_1 = \begin{bmatrix} x(q) \\ o(q) \end{bmatrix} = f_1(\cdot) : \mathcal{T}^n \to \mathbf{R}^6 \tag{8.2}$$

where $x(q) \in \mathbf{R}^3$ gives the Cartesian position of the end-effector and $o(q) = [\phi(q), \theta(q), \psi(q)]^T$ represents the orientation of the end-effector. The nonuniqueness of the inverse kinematics in this case includes multiplicities due to the particular representation of $SO(3)$ in addition to multiplicities intrinsic to the geometric structure of the manipulator.

The **velocity kinematics** is the relationship between the joint velocities and the end-effector velocities. If the mapping f_0 from Equation 8.1 is used to represent the forward kinematics, then the velocity kinematics is given by

$$V = \begin{pmatrix} v \\ \omega \end{pmatrix} = J_0(q)\dot{q} \tag{8.3}$$

where $J_0(q)$ is a $6 \times n$ matrix, called the **manipulator Jacobian**, and $V^T = (v^T, \omega^T)$ represents the linear and angular velocity of the end-effector. The vector $v \in \mathbf{R}^3$ is just $\frac{d}{dt}x(q)$, where $x(q)$ is the end-effector position vector from Equation 8.1. It is a little more difficult to see how the angular velocity vector ω is computed since the end-effector orientation in Equation 8.1 is specified by a matrix $R \in SO(3)$. If $\omega = (\omega_x, \omega_y, \omega_z)^T$ is a vector in $\mathbf{R}^3$, we may define a skew-symmetric matrix, $S(\omega)$, according to

$$S(\omega) = \begin{bmatrix} 0 & -\omega_z & \omega_y \\ \omega_z & 0 & -\omega_x \\ -\omega_y & \omega_x & 0 \end{bmatrix}. \tag{8.4}$$

The set of all skew-symmetric matrices is denoted by $so(3)$. Now, if $R(t)$ belongs to $SO(3)$ for all t, it can be shown that

$$\dot{R} = S(\omega(t))R, \tag{8.5}$$

for a unique vector $\omega(t)$ [11]. The vector $\omega(t)$ thus defined is the angular velocity of the end-effector frame relative to the base frame.

Singularities in the Jacobian $J_0(q)$, i.e., configurations where the Jacobian loses rank, are important configurations to identify for any manipulator. At singular configurations the manipulator loses one or more degrees-of-freedom. In many applications, it is important to place robots in work cells and to plan their motion in such a way that singular configurations are avoided.

If the mapping f_1 is used to represent the forward kinematics, then the velocity kinematics is written as

$$\dot{X}_1 = J_1(q)\dot{q} \tag{8.6}$$

where $J_1(q) = \partial f_1/\partial q$ is the 6×6 Jacobian of the function f_1. Singularities in J_1 include the representational singularities in addition to the manipulator singularities present in J_0. In the sequel, we use J to denote either the matrix J_0 or J_1.

Trajectories

Because of the complexity of both the kinematics and the dynamics of the manipulator and of the task to be carried out, the motion control problem is generally decomposed into three stages: **motion planning, trajectory generation,** and **trajectory tracking**. In the motion planning stage, desired paths are generated in the task space, $SE(3)$, without timing information, i.e., without specifying velocity or acceleration along the paths. Of primary concern is the generation of collision-free paths in the workspace. In the trajectory generation stage, the desired position, velocity, and acceleration of the manipulator along the path, as a function of time or as a function of arc length along the path, are computed. The trajectory planner may parameterize the end-effector path directly in task space, either as a curve in $SE(3)$ or as a curve in $\mathbb{R}^6$ using a particular minimal representation for $SO(3)$, or it may compute a trajectory for the individual joints of the manipulator as a curve in the configuration space $\mathcal{C}$.

In order to compute a joint space trajectory, the given end-effector path must be transformed into a joint space path via the inverse kinematics mapping. Because of the difficulty of computing this mapping on-line, the usual approach is to compute a discrete set of joint vectors along the end-effector path and to perform an interpolation in joint space among these points in order to complete the joint space trajectory. Common approaches to trajectory interpolation include polynomial spline interpolation, using trapezoidal velocity trajectories or cubic polynomial trajectories, as well as trajectories generated by reference models.

The computed reference trajectory is then presented to the controller, whose function is to cause the robot to track the given trajectory as closely as possible. A typical architecture for the robot control problem is illustrated in the block diagram of Figure 8.3. This chapter is concerned mainly with the design of the tracking controller, assuming that the path and trajectory have been precomputed. For background on the motion planning and trajectory generation problems the reader is referred to [10].

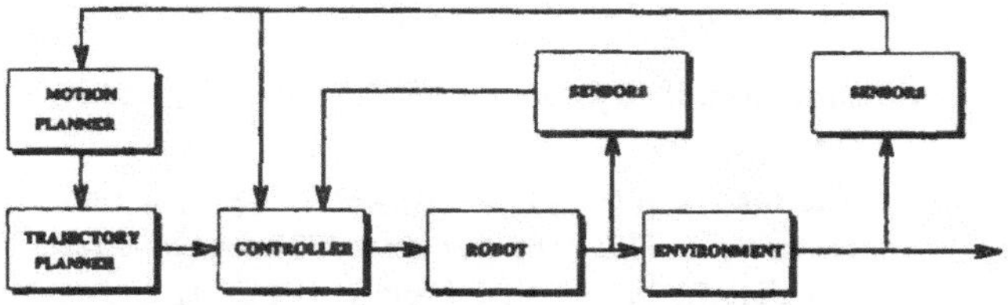

Figure 8.3 Block diagram of the robot control problem.

8.1.2 Dynamics

The dynamics of n-link manipulators are conveniently described by Lagrangian dynamics. In the Lagrangian approach, the joint variables, $q = (q_1, \ldots, q_n)^T$, serve as a suitable set of generalized coordinates. The **kinetic energy** of the manipulator is given by a symmetric, positive definite quadratic form,

$$\mathcal{K} = \frac{1}{2}\sum_{i,j=1}^{n} d_{ij}(q)\dot{q}_i\dot{q}_j = \frac{1}{2}\dot{q}^T D(q)\dot{q} \tag{8.7}$$

where $D(q)$ is the **inertia matrix** of the robot. Let $\mathcal{P} : \mathcal{C} \to \mathbb{R}$ be a continuously differentiable function (called the **potential energy**). For a rigid robot, the potential energy is due to gravity alone while for a flexible robot the potential energy also contains elastic potential energy. We define the function $\mathcal{L} = \mathcal{K} - \mathcal{P}$, which is called the **Lagrangian**. The dynamics of the manipulator are then described by Lagrange's equations [11],

$$\frac{d}{dt}\frac{\partial \mathcal{L}}{\partial \dot{q}_k} - \frac{\partial \mathcal{L}}{\partial q_k} = \tau_k, \quad k = 1, \ldots, n \tag{8.8}$$

where $\tau_1, \ldots, \tau_n$ represent input generalized forces. In local coordinates, Lagrange's equations can be written as

$$\sum_{j=1}^{n} d_{kj}(q)\ddot{q}_j + \sum_{i,j=1}^{n} \Gamma_{ijk}(q)\dot{q}_i\dot{q}_j + \phi_k(q) = \tau_k, \quad k = 1, \ldots, n \tag{8.9}$$

where

$$\Gamma_{ijk} = \frac{1}{2}\left\{\frac{\partial d_{kj}}{\partial q_i} + \frac{\partial d_{ki}}{\partial q_j} - \frac{\partial d_{ij}}{\partial q_k}\right\} \tag{8.10}$$

are known as **Christoffel symbols of the first kind**, and

$$\phi_k = \frac{\partial \mathcal{P}}{\partial q_k}. \tag{8.11}$$

In matrix form we can write Lagrange's Equation 8.9 as

$$D(q)\ddot{q} + C(q, \dot{q})\dot{q} + g(q) = \tau \tag{8.12}$$

Properties of the Lagrangian Dynamics

The Lagrangian dynamics given in Equation 8.12 possess a number of important properties that facilitate analysis and control system design. Among these are

1. The inertia matrix $D(q)$ is symmetric and positive definite, and there exist scalars, $\mu_1(q)$ and $\mu_2(q)$, such that

$$\mu_1(q)I \leq D(q) \leq \mu_2(q)I. \quad (8.13)$$

Moreover, if all joints are revolute, then μ_1 and μ_2 are constants.

2. The matrix $W(q, \dot{q}) = \dot{D}(q) - 2C(q, \dot{q})$ is skew symmetric. This property is easily shown by direct calculation. The kj-th component of $\dot{D}(q)$ is given by the chain rule as

$$\dot{d}_{kj} = \sum_{i=1}^{n} \frac{\partial d_{kj}}{\partial q_i} \dot{q}_i, \quad (8.14)$$

and the kj-th component of the matrix $C(q, \dot{q})$ is given as

$$\begin{aligned} c_{kj} &= \sum_{i=1}^{n} \Gamma_{ijk} \dot{q}_i \quad (8.15) \\ &= \frac{1}{2} \sum_{i=1}^{n} \left\{ \frac{\partial d_{kj}}{\partial q_i} + \frac{\partial d_{ki}}{\partial q_j} - \frac{\partial d_{ij}}{\partial q_k} \right\} \dot{q}_i \end{aligned}$$

Therefore the kj component of the matrix $W(q, \dot{q})$ is given by

$$\begin{aligned} w_{kj} &= \dot{d}_{kj} - 2c_{kj} \\ &= \sum_{i=1}^{n} \left[\frac{\partial d_{kj}}{\partial q_i} - \left\{ \frac{\partial d_{kj}}{\partial q_i} + \frac{\partial d_{ki}}{\partial q_j} - \frac{\partial d_{ij}}{\partial q_k} \right\} \right] \dot{q}_i \\ &= \sum_{i=1}^{n} \left[\frac{\partial d_{ij}}{\partial q_k} - \frac{\partial d_{ki}}{\partial q_j} \right] \dot{q}_i. \end{aligned}$$

Skew symmetry of $W(q, \dot{q})$ now follows by symmetry of the inertia matrix $D(q)$. Strongly related to the skew symmetry property is the so-called **passivity property**.

3. The mapping $\tau \rightarrow \dot{q}$ is passive; i.e., there exists $\beta \geq 0$ such that

$$\int_0^T \dot{q}^T(u)\tau(u)du \geq -\beta. \quad (8.16)$$

To show this property, let H be the total energy of the system

$$H = \frac{1}{2}\dot{q}^T D(q)\dot{q} + \mathcal{P}(q). \quad (8.17)$$

Then the change in energy, $\dot{H}$, satisfies

$$\dot{H} = \frac{1}{2}\dot{q}^T \dot{D}(q)\dot{q} + \dot{q}^T [D(q)\ddot{q} + g(q)] \quad (8.18)$$

since $g(q)^T$ is the gradient of $\mathcal{P}$. Substituting Equation 8.12 into Equation 8.18 yields

$$\dot{H} = \dot{q}^T\tau + \frac{1}{2}\dot{q}^T\{\dot{D}(q) - 2C(q, \dot{q})\}\dot{q} = \dot{q}^T\tau, \quad (8.19)$$

by skew symmetry of $\dot{D} - 2C$. Integrating both sides of Equation 8.19 with respect to time gives

$$\int_0^T \dot{q}^T(u)\tau(u)du = H(T) - H(0) \geq -H(0), \quad (8.20)$$

since the total energy $H(T)$ is non-negative, and passivity follows with $\beta = H(0)$.

4. Rigid robot manipulators are fully actuated; i.e., there is an independent control input for each degree-of-freedom. By contrast, robots possessing joint or link flexibility are no longer fully actuated and the control problems are more difficult, in general.

5. The equations of motion given in Equation (8.12) are linear in the inertia parameters. In other words, there is a constant vector $\theta \in \mathbf{R}^p$ and a function $Y(q, \dot{q}, \ddot{q}) \in \mathbf{R}^{n \times p}$ such that

$$D(q)\ddot{q} + C(q, \dot{q})\dot{q} + g(q) = Y(q, \dot{q}, \ddot{q})\theta = \tau. \quad (8.21)$$

The function $Y(q, \dot{q}, \ddot{q})$ is called the **regressor**. The parameter vector θ is comprised of link masses, moments of inertia, and the like, in various combinations. The dimension of the parameter space is not unique, and the search for the parameterization that minimizes the dimension of the parameter space is an important problem. Historically, the appearance of the passivity and linear parameterization properties in the early 1980s marked watershed events in robotics research. Using these properties, researchers have been able to prove elegant global convergence and stability results for robust and adaptive control. We detail some of these results in the following.

Additional Dynamics

So far, we have discussed only rigid body dynamics. Other important contributions to the dynamic description of manipulators include the dynamics of the actuators, joint and link flexibility, friction, noise, and disturbances. In addition, whenever the manipulator is in contact with the environment, the complete dynamic description includes the dynamics of the environment and the coupling forces between the environment and the manipulator. Modeling all of these effects produces an enormously complicated model. The key in robot control system design is to model the most dominant dynamic effects for the particular manipulator under consideration and to design the controller so that it is insensitive or robust to the neglected dynamics.

Friction and joint elasticity are dominant in geared manipulators such as those equipped with harmonic drives. Actuator dynamics are important in many manipulators, while noise is present in potentiometers and tachometers used as joint position and velocity sensors. For very long or very lightweight robots, particularly in space robots, the link flexibility becomes an important consideration.

Actuator Inertia and Friction The simplest modification to the rigid robot model given in Equation 8.12 is the inclusion of the actuator inertia and joint friction. The actuator inertia is specified by an $n \times n$ diagonal matrix

$$I = \text{diag}(I_1 r_1^2, \ldots, I_n r_n^2), \tag{8.22}$$

where I_i and r_i are the actuator inertia and gear ratio, respectively, of the i-th joint. The friction is specified by a vector, $f(q, \dot{q})$, and may contain only viscous friction, $B\dot{q}$, or it may include more complex models that include static friction. Defining $M(q) = D(q) + I$, we may modify the dynamics to include these additional terms as

$$M(q)\ddot{q} + C(q, \dot{q})\dot{q} + g(q) + f(q, \dot{q}) = \tau. \tag{8.23}$$

As can be seen, the inclusion of the actuator inertias and friction does not change the order of the equations. For simplicity of notation we ignore the friction terms $f(q, \dot{q})$ in the subsequent development.

Environment Forces Whenever the robot is in contact with the environment, additional forces are produced by the robot/environment interaction. Let F_e denote the force acting on the robot due to contact with the environment. It is easy to show, using the **principle of virtual work** [11], that a joint torque $\tau_e = J^T(q)F_e$ results, where $J(q)$ is the manipulator Jacobian. Thus, Equation 8.23 may be further modified as

$$M(q)\ddot{q} + C(q, \dot{q})\dot{q} + g(q) + J^T(q)F_e = \tau \tag{8.24}$$

to incorporate the external forces due to environment interaction. The problem of force control is not considered further in this chapter, but is treated in detail in Chapter 10.2.

Actuator Dynamics If the joints are actuated with permanent magnet dc motors, we may write the actuator dynamics as

$$L\frac{di}{dt} + Ri = V - K_b\dot{q}, \tag{8.25}$$

where i, V are vectors representing the armature currents and voltages, and L, R, K_b are matrices representing, respectively, the armature inductances, armature resistances, and back emf constants. Since the joint torque τ and the armature current i are related by $\tau = K_m i$, where K_m is the torque constant of the motor, we may write the complete system of Equations 8.23 to 8.25 as

$$\begin{aligned} M(q)\ddot{q} + C(q, \dot{q})\dot{q} + g(q) &= K_m i && (8.26)\\ L\frac{di}{dt} + Ri &= V - K_b\dot{q} && (8.27) \end{aligned}$$

The inclusion of these actuator dynamics increases the dimension of the state-space from $2n$ to $3n$. Other types of actuators, such as induction motors or hydraulic actuators, may introduce more-complicated dynamics and increase the system order further.

Joint Elasticity Joint elasticity, due to elasticity in the motor shaft and gears, is an important effect to model in many robots. If the joints are elastic, then the number of degrees-of-freedom is twice the number for the rigid robot, since the joint angles and motor shaft angles are no longer simply related by the gear ratio, but are now independent generalized coordinates. If we represent the joint angles by q_1 and the motor shaft angles by q_2 and model the joint elasticity by linear springs at the joints, then we may write the dynamics as

$$\begin{aligned} D(q_1)\ddot{q}_1 + C(q_1, \dot{q}_1)\dot{q}_1 + g(q_1) + K(q_1 - q_2) &= 0 && (8.28)\\ I\ddot{q}_2 + K(q_2 - q_1) &= \tau && (8.29) \end{aligned}$$

where I is the actuator inertia matrix and K is a diagonal matrix of joint stiffness constants. The model given in Equations 8.28 and 8.29 is derived under the assumptions that the inertia of the rotor is symmetric about its axis of rotation and that the inertia of the rotor axes, other than the motor shaft, may be neglected. For models of flexible joint robots that include these additional effects, the reader is referred to the article by DeLuca in [10, p. 98]. It is easy to show that the flexible joint model given in Equations 8.28 and 8.29 defines a passive mapping from motor torque τ to motor velocity $\dot{q}_2$, but does not define a passive mapping from τ to the link velocity $\dot{q}_1$. This is related to the classical problem of collocation of sensors and actuators and has important consequences in the control system design.

Singular Perturbation Models

In practice the armature inductances in L in Equation 8.25 are quite small, whereas the joint stiffness constants in K in Equations 8.28 and 8.29 are quite large relative to the inertia parameters in the rigid model given in Equation 8.23. This means that the rigid robot model (Equation 8.23) may be viewed as a **singular perturbation** [3] of both the system with actuator dynamics (Equations 8.23 to 8.25) and of the flexible joint system (Equations 8.28 and 8.29). In the case of actuator dynamics, suppose that all entries L_i/R_i in the diagonal matrix $R^{-1}L$ are equal to $\epsilon << 1$ for simplicity and write Equations 8.26 and 8.27 as

$$\begin{aligned} M(q)\ddot{q} + C(q, \dot{q})\dot{q} + g(q) &= K_m i && (8.30)\\ \epsilon\frac{di}{dt} + i &= R^{-1}(V - K_b\dot{q}) && (8.31) \end{aligned}$$

In the case of joint elasticity, define $z = K(q_2 - q_1)$ and suppose that all the joint stiffness constants are equal to $1/\epsilon^2$ for simplicity. It is easy to show that Equations 8.28 and 8.29 may be written as

$$\begin{aligned} D(q_1)\ddot{q}_1 + C(q_1, \dot{q}_1)\dot{q}_1 + g(q_1) &= z && (8.32)\\ \epsilon^2 I\ddot{z} + z &= \tau - I\ddot{q}_1 && (8.33) \end{aligned}$$

In both cases above, the rigid model given by Equation 8.23 is recovered in the limit as $\epsilon \to 0$. Singular perturbation techniques are thus of great value in designing and analyzing control laws for manipulators.

8.1.3 PD Control

In light of the complexity of the dynamics of n-link robots, it is quite remarkable to discover that very simple control laws can be used in a number of cases. For example, a simple independent joint proportional-derivative (PD) control can achieve global asymptotic stability for the set-point tracking in the absence of gravity. This fact is a fundamental consequence of the passivity property discussed earlier. To see this, consider the dynamic equations in the absence of friction

$$M(q)\ddot{q} + C(q,\dot{q})\dot{q} = u \tag{8.34}$$

where $u = \tau - g(q)$. Let

$$u = K_p(q^d - q) - K_d\dot{q} \tag{8.35}$$

be an independent joint PD control, where q^d represents a constant reference set-point, and K_p and K_d are positive, diagonal matrices of proportional and derivative gains, respectively. Consider the Lyapunov function candidate

$$V = \frac{1}{2}\dot{q}^T M(q)\dot{q} + \frac{1}{2}(q^d - q)^T K_p(q^d - q). \tag{8.36}$$

Then a simple calculation using the skew symmetry property shows that

$$\dot{V} = -\dot{q}^T K_d\dot{q} \leq 0. \tag{8.37}$$

LaSalle's invariance principle [5] can now be used to show that the equilibrium state $q = q^d$, $\dot{q} = 0$ is globally asymptotically stable. Thus, if gravity is absent or is compensated as

$$\tau = g(q) + u, \tag{8.38}$$

then a simple independent joint PD control achieves global asymptotic stability. Unfortunately, these results do not extend to the case of time-varying reference trajectories since they rely on LaSalle's Theorem.

Other researchers have investigated proportional-integral-derivative (PID) control and adaptive gravity compensation in order to overcome the requirement that the gravity parameters in Equation 8.38 be known exactly. It has also been shown that asymptotic tracking may be achieved using only the reference set-point q^d in the gravity compensation, i.e.,

$$\tau = g(q^d) + u, \tag{8.39}$$

provided the PD gains are chosen suitably, which simplifies the on-line computational requirements. These latter results require the addition of cross terms to the Lyapunov function (Equation 8.36) in order to show asymptotic stability.

In practice, however, the input given by Equation 8.35 or Equation 8.38 may result in input saturation because of the large gains that are usually required and the large initial tracking error. One of the primary reasons for trajectory generation, such as trapezoidal velocity profiles, in the first place is to help achieve proper scaling of the amplifier inputs.

Nevertheless, these simple PD results provide theoretical justification for the widespread use of such controllers in commercial industrial manipulators. In fact, it is not too difficult to show the same result for a flexible joint robot in the absence of gravity, provided the PD control is implemented using the motor variables. This is because the flexible joint robot dynamics defines a passive mapping from motor torque τ to the motor velocity $\dot{q}_2$.

8.1.4 Feedback Linearization

The notion of **feedback linearization** of nonlinear systems is a relatively recent idea in control theory, whose practical realization has been made possible by the rapid development of microprocessor technology. The basic idea of feedback linearization control is to transform a given nonlinear system into a linear system by use of a nonlinear coordinate transformation and nonlinear feedback. Feedback linearization is a useful paradigm because it allows the extensive body of knowledge from linear systems to be brought to bear to design controllers for nonlinear systems. The roots of feedback linearization in robotics predate the general theoretical development by nearly a decade, going back to the early notion of feedforward-computed torque [8].

In the robotics context, feedback linearization is known as **inverse dynamics**. The idea is to exactly compensate all of the coupling nonlinearities in the Lagrangian dynamics in a first stage so that a second-stage compensator may be designed based on a linear and decoupled plant. Any number of techniques may be used in the second stage. The feedback linearization may be accomplished with respect to the joint space coordinates or with respect to the task space coordinates. Feedback linearization may also be used as a basis for force control, such as hybrid control and impedance control.

Joint Space Inverse Dynamics

We first present the main ideas in joint space where they are easiest to understand. The control architecture we use is important as a basis for later developments. Thus, given the plant model

$$M(q)\ddot{q} + C(q,\dot{q})\dot{q} + g(q) = \tau, \tag{8.40}$$

we compute the nonlinear feedback control law

$$\tau = M(q)a_q + C(q,\dot{q})\dot{q} + g(q) \tag{8.41}$$

where $a_q \in \mathbb{R}^n$ is, as yet, undetermined. Since the inertia matrix $M(q)$ is invertible for all q, the closed-loop system reduces to the decoupled **double integrator**

$$\ddot{q} = a_q. \tag{8.42}$$

Given a joint space trajectory, $q^d(t)$, an obvious choice for the outer loop term a_q is as a **PD plus feedforward acceleration** control

$$a_q = \ddot{q}^d + K_P\left(q^d - q\right) + K_d\left(\dot{q}^d - \dot{q}\right) \tag{8.43}$$

Substituting Equation 8.43 into Equation 8.42 and defining

$$\tilde{q} = q - q^d \quad , \tag{8.44}$$

we have the linear and decoupled closed-loop system

$$\ddot{\tilde{q}} + K_d\dot{\tilde{q}} + K_P\tilde{q} = 0. \tag{8.45}$$

We can implement the joint space inverse dynamics in a so-called **inner loop/outer loop** architecture as shown in Figure 8.4.

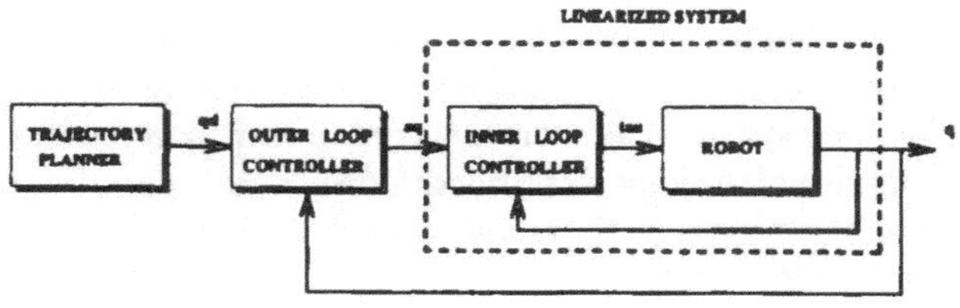

Figure 8.4 Inner loop/outer loop architecture.

The computation of the nonlinear terms in Equation 8.41 is performed in the inner loop, perhaps with a dedicated microprocessor to obtain high computation speed. The computation of the additional term a_q is performed in the outer loop. This separation of the inner loop and outer loop terms is important for several reasons. The structure of the inner loop control is fixed by Lagrange's equations. What control engineers traditionally think of as **control system design** is contained primarily in the outer loop. The outer loop control given in Equation 8.43 is merely the simplest choice of outer loop control and achieves asymptotic tracking of joint space trajectories in the ideal case of perfect knowledge of the model given by Equation 8.40. However, one has complete freedom to modify the outer loop control to achieve various other goals without the need to modify the dedicated inner loop control. For example, additional compensation terms may be included in the outer loop to enhance the robustness to parametric uncertainty, unmodeled dynamics, and external disturbances. The outer loop control may also be modified to achieve other goals, such as tracking of task space trajectories instead of joint space trajectories, regulating both motion and force, etc. The inner loop/outer loop architecture thus unifies many robot control strategies from the literature.

Task Space Inverse Dynamics

As a first illustration of the importance of the inner loop/outer loop paradigm, we show that tracking in task space can be achieved by modifying our choice of outer loop control a_q in Equation 8.42 while leaving the inner loop control unchanged. Let $X \in R^6$ represent the end-effector pose using any minimal representation of $SO(3)$. Since X is a function of the joint variables $q \in C$, we have

$$\begin{aligned} \dot{X} &= J(q)\dot{q} && (8.46) \\ \ddot{X} &= J(q)\ddot{q} + \dot{J}(q)\dot{q}. && (8.47) \end{aligned}$$

where $J = J_1$ is the Jacobian defined in Section 8.1.1. Given the double integrator (Equation 8.42) in joint space we see that if a_q is chosen as

$$a_q = J^{-1}\left\{a_X - \dot{J}\dot{q}\right\}, \qquad (8.48)$$

then we have a double integrator model in task space coordinates

$$\ddot{X} = a_X. \qquad (8.49)$$

Given a task space trajectory $X^d(t)$, we may choose a_X as

$$a_X = \ddot{X}^d + K_p(X^d - X) + K_d(\dot{X}^d - \dot{X}) \qquad (8.50)$$

so that the Cartesian space tracking error, $\tilde{X} = X - X^d$, satisfies

$$\ddot{\tilde{X}} + K_d\dot{\tilde{X}} + K_p\tilde{X} = 0. \qquad (8.51)$$

Therefore, a modification of the outer loop control achieves a linear and decoupled system directly in the task space coordinates, without the need to compute a joint trajectory and without the need to modify the nonlinear inner loop control.

Note that we have used a minimal representation for the orientation of the end-effector in order to specify a trajectory $X \in \mathbf{R}^6$. In general, if the end-effector coordinates are given in $SE(3)$, then the Jacobian J in the preceding formulation is the Jacobian J_0 defined in Section 8.1.1. In this case

$$V = \begin{pmatrix} v \\ \omega \end{pmatrix} = \begin{pmatrix} \dot{x} \\ \omega \end{pmatrix} = J(q)\dot{q} \qquad (8.52)$$

and the outer loop control

$$a_q = J^{-1}(q)\{\begin{pmatrix} a_x \\ a_\omega \end{pmatrix} - \dot{J}(q)\dot{q}\} \qquad (8.53)$$

applied to Equation 8.42 results in the system

$$\begin{aligned} \ddot{x} &= a_x \in \mathbf{R}^3 && (8.54) \\ \dot{\omega} &= a_\omega \in \mathbf{R}^3 && (8.55) \\ \dot{R} &= S(\omega)R, \; R \in SO(3), \; S \in so(3). && (8.56) \end{aligned}$$

Although, in this case, the dynamics have not been linearized to a double integrator, the outer loop terms a_v and a_ω may still be used to achieve global tracking of end-effector trajectories in $SE(3)$. In both cases, we see that nonsingularity of the Jacobian is necessary to implement the outer loop control.

The inverse dynamics control approach has been proposed in a number of different guises, such as **resolved acceleration control** and **operational space control**. These seemingly distinct approaches have all been shown to be equivalent and may be incorporated into the general framework.

It turns out that both the robot model including actuator dynamics given in Equations 8.26 and 8.27 and the model including joint flexibility given in Equations 8.28 and 8.29 are also feedback linearizable using a nonlinear coordinate transformation and nonlinear feedback. For the system with actuator dynamics given in Equations 8.26 and 8.27, if one chooses as state variables, the link positions, velocities, and accelerations, a nonlinear feedback control exists to reduce the system to a decoupled set of third-order integrators.

The coordinates in which the flexible joint system given in Equations 8.28 and 8.29 may be exactly linearized are the link positions, velocities, accelerations, and jerks; and a decoupled set of fourth-order integrators is achievable by suitable nonlinear feedback. See the article by Spong [p. 105] for details.

Composite Control

Although global feedback linearization in the case of actuator dynamics or joint flexibility is possible in theory, in practice it is difficult to achieve, mainly because the coordinate transformation is a function of the system parameters and, hence, sensitive to uncertainty. Also, the large differences in magnitude among the parameters, e.g., between the joint stiffness and the link inertia, may make the computation of the control ill-conditioned and the performance of the system poor. Using the singular perturbation models derived earlier, so-called **composite control** laws may produce better designs. We illustrate this idea using the flexible joint model given by Equations 8.32 and 8.33. The corresponding result for the system with actuator dynamics is similar.

Given the system of Equations 8.32 and 8.33, we define two related systems, the **quasi-steady-state system** and the **boundary layer system**, as follows. The quasi-steady-state system is the reduced-order system calculated by setting $\epsilon = 0$ in Equation 8.33,

$$\bar{z} = \bar{\tau} - I\ddot{\bar{q}}_1, \tag{8.57}$$

where the overbar indicates quantities computed at $\epsilon = 0$, and eliminating $\bar{z}$ in Equation 8.32. It can easily be shown that the quasi-steady-state system thus derived is equal to

$$M(\bar{q}_1)\ddot{\bar{q}}_1 + C(\bar{q}_1, \dot{\bar{q}}_1)\dot{\bar{q}}_1 + g(\bar{q}_1) = \bar{\tau}. \tag{8.58}$$

The quasi-steady-state is thus identical to the rigid robot model in terms of $\bar{q}_1$.

The boundary layer system represents the dynamics of $\eta = z - \bar{z}$, computed in the **fast time scale**, $\sigma = t/\epsilon$. This system can be shown to be

$$\frac{d^2}{d\sigma^2}\eta + (I^{-1} + D^{-1})\eta = \tau_f, \tag{8.59}$$

where $\tau_f = \tau - \bar{\tau}$. A **composite control** for the flexible joint system (Equations 8.32 and 8.33) is a control law of the form

$$\tau = \bar{\tau}(q_1, \dot{q}_2, t) + \tau_f(\eta, \dot{\eta}) \tag{8.60}$$

where $\bar{\tau}$ is designed for the quasi-steady-state system and τ_f is designed for the boundary layer system. The significant features of this approach are that

1. The control design is based on reduced-order systems, which is generally easier than designing a controller for the full-order system.
2. The quasi-steady-state system is just the rigid robot model, so existing controllers designed for rigid robots may be used without modification.

If τ_f is designed to render the boundary layer system asymptotically stable in the fast time scale and $\bar{\tau}$ is any control that achieves asymptotic tracking for the rigid system, it follows from Tichonov's Theorem [3] that

$$q_1(t) = \bar{q}_1(t) + O(\epsilon) \tag{8.61}$$

$$z(t) = \bar{z}(t) + \eta(\sigma) + O(\epsilon) \tag{8.62}$$

uniformly on a time interval $[0, 1/\sigma]$. Stronger results showing asymptotic tracking may also be achieved, depending on the particular control chosen for the quasi-steady-state system. Roughly speaking, the response of the system with joint flexibility using the composite control given by Equation 8.60 is nearly the same as the response of the rigid robot model using only $\bar{\tau}$. An important consequence is that the inverse dynamics control for the rigid robot may be used for the flexible joint robot, provided a correction term is superimposed on it to stabilize the boundary layer system. The additional control term τ_f represents the damping of the joint oscillations. Other "rigid" controllers may also be used, such as the robust and adaptive controllers detailed in the next section.

8.1.5 Robust and Adaptive Control

The feedback linearization approach of the previous section exploits important structural properties of robot dynamics. However, the practical implementation of such controllers requires consideration of various sources of uncertainties such as modeling errors, computation errors, external disturbances, unknown loads, and noise. Robust and adaptive control are concerned with the problem of maintaining precise tracking under uncertainty. We distinguish robust from adaptive control in the sense that an adaptive algorithm typically incorporates some sort of on-line parameter estimation scheme while a robust, nonadaptive scheme does not.

Robust Feedback Linearization

A number of techniques from linear and nonlinear control theory have been applied to the problem of robust feedback linearization for manipulators. Chief among these are sliding modes, Lyapunov's second method, and the method of stable factorizations. Given the dynamic equations

$$M(q)\ddot{q} + C(q, \dot{q})\dot{q} + g(q) = \tau, \tag{8.63}$$

the control input is chosen as

$$\tau = \hat{M}(q)a_q + \hat{C}(q, \dot{q})\dot{q} + \hat{g}(q) \tag{8.64}$$

where $\hat{(\cdot)}$ represents the computed or nominal value of $(\cdot)$ and indicates that the theoretically exact feedback linearization cannot be achieved in practice due to the uncertainties in the system. The error or mismatch $\tilde{(\cdot)} = (\cdot) - \hat{(\cdot)}$ is a measure of one's knowledge of the system dynamics. The outer loop control term a_q may be used to compensate for the resulting perturbation terms.

If we set

$$a_q = \ddot{q}^d + K_d(\dot{q}^d - \dot{q}) + K_p(q^d - q) + \delta a \tag{8.65}$$

and substitute Equations 8.64 and 8.65 into Equation 8.63 we obtain, after some algebra,

$$\ddot{\tilde{q}} + K_d\dot{\tilde{q}} + K_p\tilde{q} = \delta a + \eta(q, \dot{q}, \delta a, t) \tag{8.66}$$

where

$$\eta = M^{-1}\{\tilde{M}(\ddot{q}^d + K_d\dot{\tilde{q}} + K_p\tilde{q} + \delta a) + \tilde{C}\dot{q} + \tilde{g}\}. \tag{8.67}$$

In state-space we may write the system given by Equation 8.66 as

$$\dot{x} = Ax + B\{\delta a + \eta\} \tag{8.68}$$

where

$$x = \begin{pmatrix} \tilde{q} \\ \dot{\tilde{q}} \end{pmatrix}; \; A = \begin{bmatrix} 0 & I \\ -K_p & -K_d \end{bmatrix}; \; B = \begin{bmatrix} 0 \\ I \end{bmatrix}. \tag{8.69}$$

The approach is now to search for a time-varying scalar bound, $\rho(x,t) \geq 0$, on the uncertainty η, i.e.,

$$||\eta|| \leq \rho(x,t) \tag{8.70}$$

and to design the additional input term δa to guarantee asymptotic stability or, at least, ultimate boundedness of the state trajectory $x(t)$ in Equation 8.68. The bound ρ is difficult to compute, both because of the complexity of the perturbation terms in η and because the uncertainty η is itself a function of δa.

The sliding mode theory of variable structure systems has been extensively applied to the design of δa in Equation 8.68. The simplest such sliding mode control results from choosing the components δa_i of δa according to

$$\delta a_i = \rho_i(x,t)\mathrm{sgn}(s_i), \quad i = 1, \ldots, n \tag{8.71}$$

where ρ_i is a bound on the i-th component of η, $s_i = \dot{\tilde{q}}_i + \lambda_i \tilde{q}_i$ represents a sliding surface in the state-space, and sgn($\cdot$) is the signum function

$$\mathrm{sgn}(s_i) = \begin{cases} +1 & \text{if} \quad s_i > 0 \\ -1 & \text{if} \quad s_i < 0 \end{cases} \tag{8.72}$$

An alternative but similar approach is the so-called theory of guaranteed stability of uncertain systems, based on Lyapunov's second method. Since K_p and K_d are chosen in Equation 8.68 so that A is a Hurwitz matrix, for any $Q > 0$ there exists a unique symmetric $P > 0$ satisfying the Lyapunov equation,

$$A^T P + PA = -Q. \tag{8.73}$$

Using the matrix P, the outer loop term δa may be chosen as

$$\delta a = \begin{cases} -\rho(x,t)\frac{B^T Px}{||B^T Px||} & ; \text{ if } \quad ||B^T Px|| \neq 0 \\ 0 & ; \text{ if } \quad ||B^T Px|| = 0 \end{cases} \tag{8.74}$$

The Lyapunov function $V = x^T Px$ can be used to show that $\dot{V}$ is negative definite along solution trajectories of the system given by Equation 8.68.

In both the sliding mode approach and the guaranteed stability approach, problems arise in showing the existence of solutions to the closed-loop differential equations because the control signal δa is discontinuous in the state x. In practice, a chattering control signal results due to nonzero switching delays. There have been many refinements and extensions to the above approaches to robust feedback linearization, mainly to simplify the computation of the uncertainty bounds and to smooth the chattering in the control signal [10]

The method of stable factorizations has also been applied to the robust feedback linearization problem. In this approach a linear, dynamic compensator $C(s)$ is used to generate δa to stabilize the perturbed system. Since A is a Hurwitz matrix, the Youla-parameterization may be used to generate the entire class, Ω, of stabilizing compensators for the unperturbed system, i.e., Equation 8.68 with $\eta = 0$. Given bounds on the uncertainty, the Small Gain Theorem is used to generate a sufficient condition for stability of the perturbed system, and the design problem is to determine a particular compensator, $C(s)$, from the class of stabilizing compensators Ω that satisfies this sufficient condition. The interesting feature of this problem is that the perturbation terms appearing in Equation 8.68 are finite in the L_∞ norm, but not necessarily in the L_2 norm sense. This means that standard H_∞ design methods fail for this problem. For this reason, the robust manipulator control problem was influential in the development of the L_1 optimal control field. Further details on these various outer loop designs may be found in [10].

Adaptive Feedback Linearization

Once the linear parameterization property for manipulators became widely known in the mid-1980s, the first globally convergent adaptive control results began to appear. These first results were based on the inverse dynamics or feedback linearization approach discussed earlier. Consider the plant (Equation 8.63) and control (Equation 8.64) as previously, but now suppose that the parameters appearing in Equation 8.64 are not fixed as in the robust control approach, but are time-varying estimates of the true parameters. Substituting Equation 8.64 into Equation 8.63 and setting

$$a_q = \ddot{q}^d + K_d(\dot{q}^d - \dot{q}) + K_p(q^d - q), \tag{8.75}$$

it can be shown, after some algebra, that

$$\ddot{\tilde{q}} + K_d\dot{\tilde{q}} + K_p\tilde{q} = \hat{M}^{-1}Y(q,\dot{q},\ddot{q})\tilde{\theta} \tag{8.76}$$

where Y is the regressor function, and $\tilde{\theta} = \hat{\theta} - \theta$, and $\hat{\theta}$ is the estimate of the parameter vector θ. In state-space we write the system given by Equation 8.76 as

$$\dot{x} = Ax + B\Phi\tilde{\theta} \tag{8.77}$$

where

$$x = \begin{pmatrix} \tilde{q} \\ \dot{\tilde{q}} \end{pmatrix}; \; A = \begin{bmatrix} 0 & I \\ -K_p & -K_d \end{bmatrix};$$

$$B = \begin{bmatrix} 0 \\ I \end{bmatrix}; \; \Phi = \hat{M}^{-1}Y(q,\dot{q},\ddot{q}) \tag{8.78}$$

with K_p and K_d chosen so that A is a Hurwitz matrix. Suppose that an output function $y = Cx$ is defined for Equation 8.77 in such a way that the transfer function $C(sI - A)^{-1}B$ is strictly positive real (SPR). It can be shown using the passivity theorem that, for $Q > 0$, there exists a symmetric, positive definite matrix P satisfying

$$A^T P + PA = -Q \tag{8.79}$$

$$PB = C^T \tag{8.80}$$

If the parameter update law is chosen as

$$\dot{\hat{\theta}} = -\Gamma^{-1}\Phi^T Cx \tag{8.81}$$

where $\Gamma = \Gamma^T > 0$, then global convergence to zero of the tracking error with all internal signals remaining bounded can be shown using the Lyapunov function

$$V = x^T Px + \frac{1}{2}\tilde{\theta}^T \Gamma \tilde{\theta}. \tag{8.82}$$

Furthermore, the estimated parameters converge to the true parameters, provided the reference trajectory satisfies the condition of **persistency of excitation**,

$$\alpha I \leq \int_{t_0}^{t_0+T} Y^T(q^d, \dot{q}^d, \ddot{q}^d)Y(q^d, \dot{q}^d, \ddot{q}^d)dt \leq \beta I \tag{8.83}$$

for all t_0, where α, β, and T are positive constants.

In order to implement this adaptive feedback linearization scheme, however, one notes that the acceleration $\ddot{q}$ is needed in the parameter update law and that $\hat{M}$ must be guaranteed to be invertible, possibly by the use of projection in the parameter space. Later work was devoted to overcome these two drawbacks to this scheme, by using so-called indirect approaches based on a (filtered) prediction error.

Passivity-Based Approaches

By exploiting the passivity of the rigid robot dynamics it is possible to derive more elegant robust and adaptive control algorithms for manipulators, which are, at the same time, simpler to design. In the passivity-based approach we modify the inner loop control as

$$\tau = \hat{M}(q)a + \hat{C}(q,\dot{q})v + \hat{g}(q) - Kr \tag{8.84}$$

where v, a, and r are given as

$$\begin{aligned} v &= \dot{q}^d - \Lambda\tilde{q} \\ a &= \dot{v} = \ddot{q}^d - \Lambda\dot{\tilde{q}} \\ r &= \dot{q}^d - v = \dot{\tilde{q}} + \Lambda\tilde{q} \end{aligned}$$

with K, Λ diagonal matrices of positive gains. In terms of the linear parameterization of the robot dynamics, the control given in Equation 8.84 becomes

$$\tau = Y(q,\dot{q},a,v)\hat{\theta} - Kr \tag{8.85}$$

and the combination of Equation 8.84 with Equation 8.63 yields

$$M(q)\dot{r} + C(q,\dot{q})r + Kr = Y\tilde{\theta}. \tag{8.86}$$

Note that, unlike the inverse dynamics control given in Equation 8.41, the modified inner loop control of Equation 8.63 does not achieve a linear, decoupled system, even in the known parameter case, $\hat{\theta} = \theta$. However, the advantage achieved is that the regressor Y in Equation 8.86 does not contain the acceleration $\ddot{q}$, nor is the inverse of the estimated inertia matrix required.

Passivity Based Robust Control In the robust approach, the term $\hat{\theta}$ in Equation 8.85 is chosen as

$$\hat{\theta} = \theta_0 + u \tag{8.87}$$

where θ_0 is a fixed nominal parameter vector and u is an additional control term. The system given in Equation 8.86 then becomes

$$M(q)\dot{r} + C(q,\dot{q})r + Kr = Y(a,v,q,\dot{q})(\tilde{\theta} + u) \tag{8.88}$$

where $\tilde{\theta} = \theta_0 - \theta$ is a constant vector and represents the parametric uncertainty in the system. If the uncertainty can be bounded by finding a non-negative constant $\rho \geq 0$ such that

$$\|\tilde{\theta}\| = \|\theta_0 - \theta\| \leq \rho, \tag{8.89}$$

then the additional term u can be designed to guarantee stable tracking according to the expression

$$u = \begin{cases} -\rho\frac{Y^T r}{\|Y^T r\|} & ; \text{ if } \|Y^T r\| \neq 0 \\ 0 & ; \text{ if } \|Y^T r\| = 0 \end{cases} \tag{8.90}$$

The Lyapunov function

$$V = \frac{1}{2}r^T M(q)r + \tilde{q}\Lambda K\tilde{q} \tag{8.91}$$

can be used to show asymptotic stability of the tracking error. Note that $\tilde{\theta}$ is constant and so is not a state vector as in adaptive control. Comparing this approach with the approach in Section 8.1.5 we see that finding a constant bound ρ for the constant vector $\tilde{\theta}$ is much simpler than finding a time-varying bound for η in Equation 8.67. The bound ρ in this case depends only on the inertia parameters of the manipulator, while $\rho(x,t)$ in Equation 8.70 depends on the manipulator state vector and the reference trajectory and, in addition, requires some assumptions on the estimated inertia matrix $\hat{M}(q)$.

Various refinements of this approach are possible. By replacing the discontinuous control law with the continuous control

$$u = -\rho\frac{Y^T r}{\|Y^T r\| + \gamma_1 e^{-\gamma_2 t}} \tag{8.92}$$

not only is the problem of existence of solutions due to discontinuities in the control eliminated, but also the tracking errors can be shown to be globally exponentially stable. It is also possible to introduce an estimation algorithm to estimate the uncertainty bound ρ so that no *a priori* information of the uncertainty is needed.

Passivity-Based Adaptive Control In the adaptive approach, the vector $\hat{\theta}$ in Equation 8.86 is now taken to be a time-varying estimate of the true parameter vector θ. Instead of adding an additional control term, as in the robust approach, we introduce a parameter update law for $\hat{\theta}$. Combining the control law given by Equation 8.84 with Equation 8.63 yields

$$M(q)\dot{r} + C(q,\dot{q})r + Kr = Y\tilde{\theta}. \tag{8.93}$$

The parameter estimate $\hat{\theta}$ may be computed using standard methods, such as gradient or least squares. For example, using the gradient update law

$$\dot{\hat{\theta}} = -\Gamma^{-1} Y^T(q, \dot{q}, a, v) r \tag{8.94}$$

together with the Lyapunov function

$$V = \frac{1}{2} r^T M(q) r + \tilde{q}^T \Lambda K \tilde{q} + \frac{1}{2} \tilde{\theta}^T \Gamma \tilde{\theta} \tag{8.95}$$

results in global convergence of the tracking errors to zero and boundedness of the parameter estimates.

A number of important refinements to this basic result are possible. By using the reference trajectory instead of the measured joint variables in both the control and update laws, i.e.,

$$\tau = Y(q^d, \dot{q}^d, \ddot{q}^d)\hat{\theta} - K_p \dot{\tilde{q}} - K_d \tilde{q} \tag{8.96}$$

with

$$\dot{\hat{\theta}} = -\Gamma^{-1} Y^T(q^d, \dot{q}^d, \ddot{q}^d) r, \tag{8.97}$$

it is possible to show exponential stability of the tracking error in the known parameter case, asymptotic stability in the adaptive case, and convergence of the parameter estimation error to zero under persistence of excitation of the reference trajectory.

8.1.6 Time-Optimal Control

For many applications, such as palletizing, there is a direct correlation between the speed of the robot manipulator and cycle time. For these applications, making the robot work faster translates directly into an increase in productivity. Since the input torques to the robot are limited by the capability of the actuators as

$$\tau^{min}(t) \leq \tau(t) \leq \tau^{max}(t), \tag{8.98}$$

it is natural to consider the problem of time-optimal control. In many applications, such as seam tracking, the geometric path of the end-effector is constrained in the task space. In such cases, it is useful to produce time-optimal trajectories, i.e., time-optimal parameterizations of the geometric path that can be presented to the feedback controller. For this reason, most of the research into the time-optimal control of manipulators has gone into the problem of generating a minimum time trajectory along a given path in task space. Several algorithms are now available to compute such time-optimal trajectories.

Formulation of the Time-Optimal Control Problem

Consider an end-effector path $p(s) \in \mathbf{R}^6$ parameterized by arc length s along the path. If the manipulator is constrained to follow this path, then

$$p(s) = f(q) \tag{8.99}$$

where $f(q)$ is the forward kinematics map. We may also use the inverse kinematics map to write

$$q = f^{-1}(p(s)) = e(s) \in C \tag{8.100}$$

Either the end-effector path (Equation 8.99) or the joint space path (Equation 8.100) may be used in the subsequent development. We illustrate the formulation with the joint space path (Equation 8.100) and refer the reader to [10] for further details.

The final time t_f may be specified as

$$t_f = \int_0^{t_f} dt = \int_0^{s_{max}} ds/\dot{s}, \tag{8.101}$$

which suggests that, in order to minimize the final time, the velocity $\dot{s}$ along the path should be maximized. It turns out that the optimal solution in this case is bang-bang; i.e., the acceleration $\ddot{s}$ is either maximum or minimum along the path. Since the trajectory is parameterized by the scalar s, phase plane techniques may be used to calculate the maximum acceleration $\ddot{s}$, as we shall see. From Equation 8.100 we may compute

$$\dot{q} = e_s(s)\dot{s} \tag{8.102}$$

$$\ddot{q} = e_s(s)\ddot{s} + \dot{e}_s\dot{s} \tag{8.103}$$

where e_s is the Jacobian of the mapping given by Equation 8.100. Therefore, once the optimal solution $s(t)$ is computed, the above expressions can be used to determine the optimal joint space trajectory. Substituting the expressions for $\dot{q}$ and $\ddot{q}$ into the manipulator dynamics (Equation 8.63) leads to a set of second-order equations in the scalar arc length parameter s,

$$a_i(s)\ddot{s} + b_i(s)\dot{s}^2 + c_i(s) = \tau, \quad i = 1, \ldots, n. \tag{8.104}$$

Given the bounds on the joint actuator torques from Equation 8.98 bounds on $\ddot{s}$ can be determined by substituting Equation 8.104 into Equation 8.98

$$\tau_i^{min} \leq a_i(s)\ddot{s} + b_i(s)\dot{s}^2 + c_i(s) \leq \tau_i^{max} \quad i = 1, \ldots, n, \tag{8.105}$$

which can be written as a set of n-constraints on $\ddot{s}$ as

$$\alpha_i(s, \dot{s}) \leq \ddot{s} \leq \beta_i(s, \dot{s}) \tag{8.106}$$

where $\alpha_i = (\tau_i^{\alpha} - b_i\dot{s}^2 - c_i)/a_i$

$\beta_i = (\tau_i^{\beta} - b_i\dot{s}^2 - c_i)/a_i$

with $\tau_i^{\alpha} = \tau_i^{min}$ and $\tau_i^{\beta} = \tau_i^{max}$ if $a_i > 0$

$\tau_i^{\alpha} = \tau_i^{max}$ and $\tau_i^{\beta} = \tau_i^{min}$ if $a_i < 0$

Thus, the bounds on the scalar $\ddot{s}$ are determined as

$$\alpha(s, \dot{s}) \leq \ddot{s} \leq \beta(s, \dot{s}) \tag{8.107}$$

where

$$\alpha(s, \dot{s}) = \max\{\alpha_i(s, \dot{s})\}\ ; \quad \beta(s, \dot{s}) = \min\{\beta_i(s, \dot{s})\} \tag{8.108}$$

Since the solution is known to be bang-bang, the optimal control is determined by finding the times, or positions, at which $\ddot{s}$ switches between

$$\ddot{s} = \beta(s, \dot{s})$$

and

$$\ddot{s} = \alpha(s, \dot{s}).$$

These switching times may be found by constructing switching curves in the phase plane s-$\dot{s}$ corresponding to $\alpha(s, \dot{s}) = \beta(s, \dot{s})$. Various methods for constructing this switching curve are found in the references contained in [10, part 6]. A typical minimum-time solution is shown in Figure 8.5.

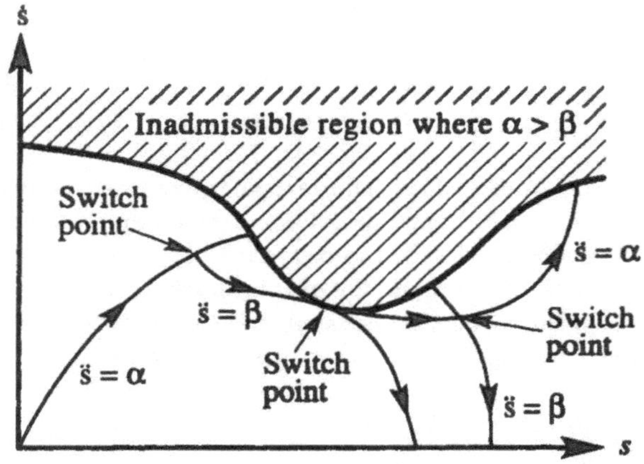

Figure 8.5 Minimum-time trajectory in the s-$\dot{s}$ plane.

8.1.7 Repetitive and Learning Control

Since many robotic applications, such as pick-and-place operations, painting, and circuit board assembly, involve repetitive motions, it is natural to consider using data gathered in previous cycles to try to improve the performance of the manipulator in subsequent cycles. This is the basic idea of **repetitive control** or **learning control**.

Consider the rigid robot model given by Equation 8.23 and suppose one is given a desired output trajectory on a finite time interval, $y_d(t)$, $0 \leq t \leq T$, which may represent a joint space trajectory or a task space trajectory. The reference trajectory $y_d(t)$ is used in repeated trials of the manipulator, assuming either that the trajectory is periodic, $y_d(T) = y_d(0)$ (repetitive control), or that the robot is reinitialized to lie on the desired trajectory at the beginning of each trial (learning control). Hereafter we use the term *learning control* to mean either repetitive or learning control.

Let $\tau_k(t)$ be the input torque during the k-th cycle, which produces an output $y_k(t)$, $0 \leq t \leq T$. The input/output pair $[\tau_k(t), y_k(t)]$ may be stored and utilized in the $k+1$-st cycle. The initial control input $\tau_0(t)$ can be any control input that produces a stable output, such as a PD control.

The learning control problem is to determine a recursive control

$$\tau_{k+1}(t) = F(\tau_k(t), \Delta y_k(t)), \quad 0 \leq t \leq T, \tag{8.109}$$

where $\Delta y_k(t) = y_k(t) - y_d(t)$, such that $||\Delta y_k|| \to 0$ as $k \to \infty$ in some suitably defined function space norm, $||\cdot||$. Such learning control algorithms are attractive because accurate models of the dynamics need not be known *a priori*.

Several approaches have been used to generate a suitable learning law F and to prove convergence of the output error. A P-type learning law is one of the form

$$\tau_{k+1}(t) = \tau_k(t) - \Phi\Delta y_k(t), \tag{8.110}$$

so called because the correction term to the input torque at each iteration is proportional to the error Δy_k. A D-type learning law is one of form

$$\tau_{k+1}(t) = \tau_k(t) - \Gamma\frac{d}{dt}\Delta y_k(t) \tag{8.111}$$

A more general PID type learning algorithm takes the form

$$\tau_{k+1}(t) = \tau_k(t) - \Gamma\frac{d}{dt}\Delta y_k(t) - \Phi\Delta y_k(t) - \int \psi\Delta y_k(u)du. \tag{8.112}$$

Convergence of $y_k(t)$ to $y_d(t)$ has been proved under various assumptions on the system. The earliest results considered the robot dynamics linearized around the desired trajectory and proved convergence for the linear time-varying system that results. Later results proved convergence based on the complete Lagrangian model.

Passivity has been shown to play a fundamental role in the convergence and robustness of learning algorithms. Given a joint space trajectory $q_d(t)$, let $\tau_d(t)$ be defined by the inverse dynamics, i.e.,

$$\tau_d(t) = M(q_d(t))\ddot{q}_d(t) + C(q_d(t), \dot{q}_d(t))\dot{q}_d(t) + g(q_d(t)). \tag{8.113}$$

The function $\tau_d(t)$ need not be computed; it is sufficient to know that it exists. Consider the P-type learning control law given by Equation 8.110 and subtract $\tau_d(t)$ from both sides to obtain

$$\Delta\tau_{k+1} = \Delta\tau_k - \Phi\Delta y_k \tag{8.114}$$

where $\Delta\tau_k = \tau_k - \tau_d$. It follows that

$$\Delta\tau_{k+1}^T\Phi^{-1}\Delta\tau_{k+1} = \Delta\tau_k^T\Phi^{-1}\Delta\tau_k + \Delta y_k^T\Phi\Delta y_k - 2\Delta y_k^T\Delta\tau_k. \tag{8.115}$$

Multiplying both sides by $e^{-\lambda t}$ and integrating over $[0, T]$ it can be shown that

$$||\Delta\tau_{k+1}||^2 \leq ||\Delta\tau_k||^2 - \beta||\Phi\Delta y_k||^2 \tag{8.116}$$

provided there exist positive constants λ and β such that

$$\int_0^T e^{-\lambda t}\Delta y_k^T\Delta\tau_k(t)dt \geq \frac{1+\beta}{2}||\Phi\Delta y_k||^2 \tag{8.117}$$

for all k. Equation 8.117 defines a passivity relationship of the exponentially weighted error dynamics. It follows from Equation 8.116 that $\Delta y_k \to 0$ in the L_2 norm.

References

[1] Asada, H. and Slotine, J-J. E., *Robot Analysis and Control*, John Wiley & Sons, Inc., New York, 1986.

[2] Dorf, R.C., Ed., *International Encyclopedia of Robotics: Applications and Automation*, John Wiley & Sons, Inc., 1988.

[3] Kokotović, P.V., Khalil, H.K., and O'Reilly, J., *Singular Perturbation Methods in Control: Analysis and Design*, Academic Press, Inc., London, 1986.

[4] Luh, J.Y.S., Conventional controller design for industrial robots: a tutorial, *IEEE Trans. Syst., Man, Cybern.*, 13(3), 298–316, May/June 1983.

[5] Khalil, H., *Nonlinear Systems*, Macmillan Press, New York, 1992.

[6] Nof, S.Y., Ed., *Handbook of Industrial Robotics*, John Wiley & Sons, Inc., New York, 1985.

[7] Ortega, R., and Spong, M.W., Adaptive control of rigid robots: a tutorial, in Proc. IEEE Conf. Decision Control, 1575–1584, Austin, TX, 1988.

[8] Paul, R.C., Modeling, Trajectory Calculation, and Servoing of a Computer Controlled Arm, Stanford A.I. Lab, A.I. Memo 177, Stanford, CA, November 1972.

[9] Spong, M.W., On the robust control of robot manipulators, *IEEE Trans. Autom. Control*, 37, 1782–1786, November 1992.

[10] Spong, M.W., Lewis, F., and Abdallah, C., *Robot Control: Dynamics, Motion Planning, and Analysis*, IEEE Press, 1992.

[11] Spong, M.W. and Vidyasagar, M., *Robot Dynamics and Control*, John Wiley & Sons, Inc., New York, 1989.

8.2 Force Control of Robot Manipulators

Joris De Schutter, Katholieke Universiteit Leuven, Department of Mechanical Engineering, Leuven, Belgium

Herman Bruyninckx, Katholieke Universiteit Leuven, Department of Mechanical Engineering, Leuven, Belgium

8.2.1 Introduction

Robots of the first generations were conceived as "open-loop" positioning devices; they operate with little or no feedback at all from the process in which they participate. For industrial assembly environments, this implies that all parts or subassemblies have to be prepositioned with a high accuracy, which requires expensive and rather inflexible peripheral equipment. Providing robots with sensing capabilities can reduce these accuracy requirements considerably. In particular, for industrial assembly, force feedback is extremely useful. But also for other tasks, in which a tool held by the robot has to make controlled contact with a work piece, as in deburring, polishing, or cleaning, it is not a good idea to rely fully on the positioning accuracy of the robot, and force feedback or force control becomes mandatory.

Force feedback is classified into two categories. In passive force feedback, the trajectory of the robot end-effector is modified by the interaction forces due to the inherent compliance of the robot; the compliance may be due to the structural compliance of links, joints, and end-effector or to the compliance of the position servo. In passive force feedback there is no actual force measurement, and the preprogrammed trajectory of the end-effector is never changed at execution time. On the other hand, in active force feedback, the interaction forces are measured, fed back to the controller, and used to modify, or even generate on-line, the desired trajectory of the robot end-effector.

Up till now force feedback applications in industrial environments have been mainly passive, for obvious reasons. Passive force feedback requires neither a force sensor nor a modified programming and control system and is therefore simple and cheap. In addition, it operates very fast. The Remote Center Compliance (RCC), developed by Draper Lab [1] and now widely available in various forms, is a well-known example of passive force feedback. It consists of a compliant end-effector that is designed and optimized for peg-into-hole assembly operations.

However, compared to active force feedback, passive force feedback has several disadvantages. It lacks flexibility, since for every robot task a special-purpose compliant end-effector has to be designed and mounted. Also, it can deal only with small errors of position and orientation. Finally, since no forces are measured, it can neither detect nor cope with error conditions involving excessive contact forces, and it cannot guarantee that high contact forces will never occur.

Clearly, although active force feedback has an answer to all of these issues, it is usually slower, more expensive, and more sophisticated than purely passive force feedback. Apart from a force sensor, it also requires an adapted programming and control system. In addition, it has been shown [2] that, in order to obtain a reasonable task execution speed and disturbance rejection capability, active force feedback has to be used in combination with passive force feedback.

In this chapter the design of active force controllers is discussed. More detailed reviews are given in [7] and [10].

In order to apply active force control, the following components are needed: force measurement, task specification, and control.

Force measurement. For a general force-controlled task, six force components are required to provide complete contact force information: three translational force components and three torques. Very often, a force sensor is mounted at the robot wrist, but other possibilities exist. The force signals may be obtained using strain measurements, which results in a stiff sensor, or using deformation measurements (e.g., optically), which results in a compliant sensor. The latter approach has an advantage if additional passive compliance is desired.

Task specification. For robot tasks involving constrained motion, the user has to specify more than just the desired motion as in the case of motion in free space. In addition, he has to specify how the robot has to interact with the external constraints. Basically there are two approaches, which are explained in this chapter. In the *hybrid force/position control* approach, the user specifies both desired motion and, explicitly, desired contact forces in two mutually independent subspaces. In the second approach, called *impedance control*, the user specifies how the robot has to comply with the external constraints; i.e., he specifies the dynamic relationship between contact forces and executed motion. In the hybrid approach, there is a clear separation between task specification and control; in the impedance approach, task specification and control are closely linked.

Control. The aim of the force control system is to make the actual contact forces, as measured by the sensor, equal to the desired contact forces, given by the task specification. This is called low-level or setpoint control, which is the main topic of this chapter. However, by interpreting the velocities actually executed by the robot, as well as the measured contact forces, much can be learned about the actual geometry of the constraints. This is the key for more high-level or adaptive control.

The chapter is organized as follows. Section 8.2.2 derives the system equations for control and task specification purposes. It also contains examples of hybrid force/position task specification. Section 8.2.3 describes methods for hybrid control of end-effector motion and contact force. Section 8.2.4 shows how to adapt the hybrid controller to the actual contact geometry. Section 8.2.5 briefly describes the impedance approach.

8.2.2 System Equations

Chapter 8.1 derived the general dynamic equations for a robot arm constrained by a contact with the environment:

$$M(q)\ddot{q} + C(q,\dot{q})\dot{q} + g(q) + J^T F_e = \tau. \qquad (8.118)$$

$F_e^T = \left(f^T \; m^T\right)$ is a six-vector representing Cartesian forces (f) and torques (m) occurring in the contact between robot and environment. $q = [q_1 \ldots q_n]^T$ are the n joint angles of the manipulator. $M(q)$ is the inertia matrix of the manipulator, expressed in joint space form. $C(q,\dot{q})\dot{q}$ and $g(q)$ are the velocity- and gravity-dependent terms, respectively. τ is the vector of joint torques. J is the manipulator's Jacobian matrix that transforms joint velocities $\dot{q}$ to the Cartesian linear and angular end-effector velocities represented by the six-vector $V^T = \left(v^T \; \omega^T\right)$:

$$V = J\dot{q}. \qquad (8.119)$$

The Cartesian contact force F_e is determined both by Equation 8.118 and by the dynamics of the environment. The dynamic model of the environment contains two aspects: (1) the kinematics of the contact geometry; i.e., in what directions is the manipulator's motion constrained by the environment; and (2) the relationship between force applied to the environment and the deformation of the constraint surface. We consider two cases: (1) the robot and the environment are perfectly rigid; and (2) the environment behaves as a mass-spring-damper system. The first model is most appropriate for:

- Motion specification purposes: The user has to specify the desired motion of the robot, as well as the desired contact forces. This is easier within a Cartesian, rigid, and purely geometric constraint model.
- Theoretical purposes: A perfectly rigid interaction between robot and environment is an interesting *ideal* limit case, to which all other control approaches can be compared [9].

The soft environment model corresponds better to most of the real situations. Usually, the model of the robot-environment interaction is simplified by assuming that all compliance in the system—including that of the manipulator, its servo, the force sensor, and the tool—is localized in the environment.

Rigid Environment

Two cases are considered: (1) the constraints are formulated in joint or configuration space; and (2) the constraints are formulated in Cartesian space.

Configuration Space Formulation Assume that the kinematic constraints imposed by the environment are expressed in configuration space by n_f algebraic equations

$$\psi_q(q) = 0. \qquad (8.120)$$

This assumes the constraints are rigid, bilateral, and holonomic. Assume also that these n_f constraints are mutually independent; then they take away n_f motion degrees of freedom from the manipulator.

The constrained robot dynamics are now derived from the unconstrained robot dynamics by the classical technique of incorporating the constraints into the Lagrangian function (see Chapter 8.1). For the unconstrained system, the Lagrangian is the difference between the system's kinetic energy, $\mathcal{K}$, and its potential energy, $\mathcal{P}$. Each of the constraint equations ψ_{qj}, $j = 1,\ldots,n_f$, should be identically satisfied. Lagrange's approach to satisfying both the dynamical equations of the system and the constraint requirements was to define the extended Lagrangian

$$\mathcal{L} = \mathcal{K} - \mathcal{P} - \sum_{j=1}^{n_f} \lambda_j \psi_{qj}(q).$$

$\lambda = [\lambda_1 \; \ldots \; \lambda_{n_f}]^T$ is a vector of Lagrange multipliers. The solution to the Lagrangian equations

$$\frac{d}{dt}\left(\frac{\partial \mathcal{L}}{\partial \dot{q}_k}\right) - \frac{\partial \mathcal{L}}{\partial q_k} = \tau_k, \; k = 1,\ldots,n$$

then results in

$$M(q)\ddot{q} + C(q,\dot{q})\dot{q} + g(q) + J_{\psi_q}^T(q)\lambda = \tau. \qquad (8.121)$$

$J_{\psi_q}(q)$ is the $n_f \times n$ Jacobian matrix (i.e., the matrix of partial derivatives with respect to the joint angles q) of the constraint function $\psi_q(q)$. J_{ψ_q} is of full rank, n_f, because all constraints are assumed to be independent. (If not, the constraints represent a so-called *hyperstatic* situation.) $J_{\psi_q}^T(q)\lambda$ represents the ideal contact forces, i.e., without contact friction.

Cartesian Space Formulation The kinematic constraints in Cartesian space are easily derived from the configuration space results, as long as the manipulator's Jacobian matrix J is square (i.e., $n = 6$) and nonsingular. In that case, Equation 8.121 is equivalent to [J^T denotes J transpose and $J^{-T} = (J^T)^{-1}$]

$$M(q)\ddot{q} + C(q,\dot{q})\dot{q} + g(q) + J^T\left(J^{-T}J^T_{\psi_q}(q)\right)\lambda = \tau.$$

Denote $J_{\psi_q}(q)J^{-1}$ by $J_{\psi_x}(q)$. This is an $n_f \times 6$ matrix. Then

$$M(q)\ddot{q} + C(q,\dot{q})\dot{q} + g(q) + J^T J^T_{\psi_x}(q)\lambda = \tau. \qquad (8.122)$$

Comparison with Equation 8.118 shows that

$$F_e = J^T_{\psi_x}\lambda. \qquad (8.123)$$

This means that the ideal Cartesian reaction forces F_e belong to an n_f dimensional vector space, spanned by the full rank matrix $J^T_{\psi_x}$. Let S_f denote a basis of this vector space, i.e., a set of n_f independent contact forces that can be generated by the constraints:

$$\forall F_e, \exists\phi = [\phi_1 \ \ldots \ \phi_{n_f}]^T \in \mathbb{R}^{n_f} : F_e = S_f\,\phi. \qquad (8.124)$$

The coordinate vector ϕ contains dimensionless scalars; all physical dimensions of Cartesian force are contained in the basis S_f.

The time derivative of Equation 8.120 yields

$$0 = \frac{d\psi_q(q)}{dt} = \frac{\partial\psi_q(q)}{\partial q}\dot{q} = J_{\psi_q}\dot{q}. \qquad (8.125)$$

Hence, using Equation 8.119 yields

$$\left(J_{\psi_q}J^{-1}\right)(J\dot{q}) = J_{\psi_x}V = 0. \qquad (8.126)$$

Combining Equations 8.123 and 8.126 gives the kinematic *reciprocity* relationship between the ideal Cartesian reaction forces F_e (spanning the so-called *force-controlled* subspace) and the Cartesian manipulator velocities V that obey the constraints (spanning the *velocity-controlled* subspace):

$$V^T F_e = F_e^T V = 0. \qquad (8.127)$$

This means that the velocity-controlled subspace is the n_x dimensional ($n_x = 6 - n_f$) reciprocal complement of the force-controlled subspace. It can be given a basis S_x, such that

$$\forall V, \exists\chi = [\chi_1 \ \ldots \ \chi_{n_x}]^T \in \mathbb{R}^{n_x} : V = S_x\,\chi. \qquad (8.128)$$

Again, χ is a vector with physically dimensionless scalars; the columns of S_x have the physical dimensions of a Cartesian velocity. From Equation 8.127 it follows that

$$S_x^T S_f = 0. \qquad (8.129)$$

Task Specification The task specification module specifies force and velocity setpoints, F^d and V^d, respectively. In order to be consistent with the constraints, these setpoints must lie in the force- and velocity-controlled directions, respectively. Hence, the instantaneous task description corresponds to specifying the vectors ϕ^d and χ^d:

$$F^d = S_f\,\phi^d, \quad V^d = S_x\,\chi^d. \qquad (8.130)$$

These equations are invariant with respect to the choice of reference frame and with respect to a change in the physical units. However, the great majority of tasks have a set of orthogonal reference frames in which the task specification becomes very easy and intuitive. Such a frame is called a *task frame* or *compliance frame*, [6]. Figures 8.6 and 8.7 show two examples.

Inserting a round peg in a round hole. The goal of this task is to push the peg into the hole while avoiding wedging and jamming. The peg behaves as a cylindrical joint; hence, it has two degrees of motion freedom ($n_x = 2$) while the force-controlled subspace is of rank four ($n_f = 4$). Hence, the task can be achieved by the following four force setpoints and two velocity setpoints in the task frame depicted in Figure 8.6:

1. A nonzero velocity in the Z direction
2. Zero forces in the X and Y directions
3. Zero torques about the X and Y directions
4. An arbitrary angular velocity about the Z direction

The task continues until a "large" reaction force in the Z direction is measured. This indicates that the peg has hit the bottom of the hole.

Sliding a block over a planar surface. The goal of this task is to slide the block over the surface without generating too large reaction forces and without breaking the contact. There are three velocity-controlled directions and three force-controlled directions ($n_x = n_f = 3$). Hence, the task can be achieved by the following setpoints in the task frame depicted in Figure 8.7:

1. A nonzero force in the Z direction
2. A nonzero velocity in the X direction
3. A zero velocity in the Y direction
4. A zero angular velocity about the Z direction
5. Zero torques about the X and Y directions

Soft Environment

For the fully constrained case, i.e., all end-effector degrees of freedom are constrained by the environment, the dynamic equation of the robot-environment interaction is given by

$$F_e = M_e\,\Delta a + C_e\,\Delta V + K_e\,\Delta X. \qquad (8.131)$$

$\Delta V = V - V_e$ is the Cartesian deformation velocity of the soft environment, with V_e the velocity of the environment. $\Delta a = (d\Delta V)/(dt)$ is the deformation acceleration, and $\Delta X = \int \Delta V dt$ is the deformation. Note that velocity is taken here as the basic motion input, since the difference $X - X_e$ of two-position

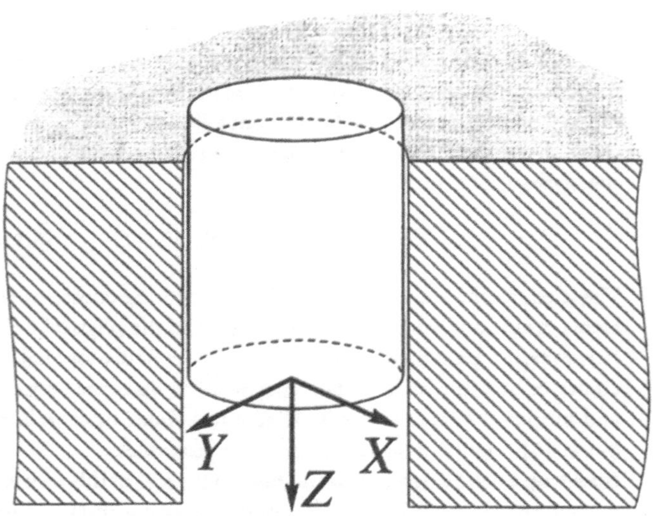

Figure 8.6 Peg-in-hole.

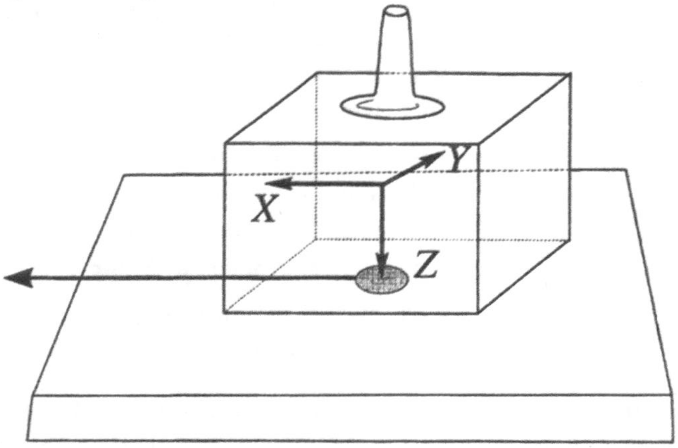

Figure 8.7 Sliding a block.

six-vectors is not well defined. Here X represents position and orientation of the robot end-effector, and X_e represents position and orientation of the environment. M_e is a positive definite inertia matrix; C_e and K_e are positive semidefinite damping and stiffness matrices. If there is sufficient passive compliance in the system, Equation 8.131 is approximated by

$$F_e = K_e \, \Delta X. \tag{8.132}$$

This case of soft environment is considered subsequently.

For partially constrained motion, the contact kinematics influence the dynamics of the robot-environment interaction. Moreover, in the case of a soft environment the measured velocity V does not completely belong to the ideal velocity subspace, defined for a rigid environment, because the environment can deform. Similarly, the measured force F_e does not completely belong to the ideal force subspace, also due to friction along the contact surface.

Hence, for control purposes the measured quantities V and F_e have to be projected onto the corresponding modeled subspaces. Algebraically, these projections are performed by projection matrices $P_x = S_x S_x^\dagger$ and $P_f = S_f S_f^\dagger$ [3]. $S_x^\dagger$ and $S_f^\dagger$ are (weighted) pseudo inverses, defined as

$$\begin{aligned} S_x^\dagger &= \left(S_x^T K_e S_x\right)^{-1} S_x^T K_e, \\ S_f^\dagger &= \left(S_f^T K_e^{-1} S_f\right)^{-1} S_f^T K_e^{-1}. \end{aligned} \tag{8.133}$$

With Equation 8.129 this yields

$$S_x^\dagger K_e^{-1} S_f = 0, \quad S_f^\dagger K_e S_x = 0. \tag{8.134}$$

The projection matrices decompose every Cartesian force F_e into a *force of constraint* $P_f F_e$, which is fully taken up by the constraint, and a *force of motion* $(I_6 - P_f)F_e$, which generates motion along the velocity-controlled directions. (I_6 is the 6×6 unity matrix.) Similarly, the Cartesian velocity V consists of a *velocity of freedom* part $P_x V$ in the velocity-controlled directions and a *velocity of constraint* part $(I_6 - P_x)V$, which deforms the environment. Within the velocity- and force-controlled subspaces, the measured velocity V and the measured force F_e correspond, respectively, to the following coordinate vectors with respect to the bases S_x and S_f:

$$\chi = S_x^\dagger V, \quad \phi = S_f^\dagger F_e. \tag{8.135}$$

Note that in the limit case of a rigid (and frictionless) environment, V and F_e do lie in the ideal velocity- and force-controlled subspaces. As a result, the projections of V and F_e onto the velocity- and force-controlled subspace, respectively, coincide with V and F_e. Hence, Equation 8.135 always gives the same results, whatever weighting matrices are chosen in Equation 8.133.

Using the projection matrices P_f and P_x, the ideal contact force of Equation 8.132 is rewritten for the partially constrained case as

$$F_e = P_f K_e (I_6 - P_x) \Delta X. \tag{8.136}$$

With Equation 8.134 this reduces to

$$F_e = P_f K_e \Delta X. \tag{8.137}$$

8.2.3 Hybrid Force/Position Control

The aim of hybrid control is to split up simultaneous control of both end-effector motion and contact force into two separate and decoupled subproblems [5]. Three different hybrid control approaches are presented here. In the first two, *acceleration-resolved* approaches, the control signals generated in the force and velocity subspaces are transformed to joint space signals, necessary to drive the robot joints, in terms of accelerations. Both the cases of a rigid and of a soft environment are considered. In the third, *velocity-resolved* approach, this transformation is performed in terms of velocities. Similarly as described in Chapter 8.1, all three approaches use a combination of an inner and an outer controller: the inner controller compensates the dynamics of the robot arm and may be model based; the outer controller is purely error driven.

We consider only the case of a constant contact geometry, so

$$\dot{S}_x = \ddot{S}_x = 0; \quad \dot{S}_f = \ddot{S}_f = 0. \tag{8.138}$$

Acceleratio-Resolved Control: Case of Rigid Environment

The system equations are given by Equation 8.118, where F_e is an ideal constraint force as in Equation 8.124. The inner loop controller is given by

$$\tau = M(q)a_q + C(q,\dot{q})\dot{q} + g(q) + J^T F^d, \tag{8.139}$$

where F^d is the desired force, specified as in Equation 8.130. a_q is a desired joint space acceleration, which is related to a_x, the desired Cartesian space acceleration resulting from the outer loop controller:

$$a_q = J^{-1}(a_x - \dot{J}\dot{q}). \tag{8.140}$$

The closed-loop dynamics of the system with its inner loop controller are derived as follows. Substitute Equation 8.140 in Equation 8.139, then substitute the result in Equation 8.118. Using the derivative of Equation 8.119,

$$J\ddot{q} + \dot{J}\dot{q} = A, \tag{8.141}$$

where A is the Cartesian acceleration of the end-effector, this results in

$$A + JM^{-1}J^T F_e = a_x + JM^{-1}J^T F^d. \tag{8.142}$$

In the case of a rigid environment, the Cartesian acceleration A is given by the derivative of Equation 8.128:

$$A = S_x \dot{\chi}. \tag{8.143}$$

The outer loop controller generates an acceleration in the motion-controlled subspace only:

$$a_x = S_x\left(\dot{\chi}^d + K_{dx}\tilde{\chi} + K_{px}\tilde{\chi}_\Delta\right), \tag{8.144}$$

where $\tilde{\chi} = \chi^d - \chi$; χ^d specifies the desired end-effector velocity as in Equation 8.130; χ represents the coordinates of the measured velocity, and is given by Equation 8.135; $\tilde{\chi}_\Delta$ represents the time integral of $\tilde{\chi}$; K_{dx} and K_{px} are control gain matrices with dimensions $n_x \times n_x$ and physical units $\frac{1}{time}$ and $\frac{1}{(time)^2}$, respectively.

Substituting the outer loop controller of Equation 8.144 into Equation 8.142 results in the closed-loop equation. This closed-loop equation is split up into two independent parts corresponding to the force- and velocity-controlled subspaces. First, premultiply Equation 8.142 with $S_x^\dagger K_e^{-1}J^{-T}MJ^{-1}$. This eliminates the terms containing F_e and F^d, because of Equation 8.134, and results in

$$S_x^\dagger K_e^{-1}J^{-T}MJ^{-1}A = S_x^\dagger K_e^{-1}J^{-T}MJ^{-1}a_x. \tag{8.145}$$

Since $S_x^\dagger K_e^{-1}J^{-T}MJ^{-1}S_x$ is a $n_x \times n_x$ nonsingular matrix, and with Equations 8.143 and 8.144, this reduces to

$$\dot{\tilde{\chi}} + K_{dx}\tilde{\chi} + K_{px}\tilde{\chi}_\Delta = 0. \tag{8.146}$$

Choosing diagonal gain matrices K_{dx} and K_{px} decouples the velocity control.

Similarly, premultiply Equation 8.142 with $S_f^\dagger K_e$. This eliminates the terms containing A and a_x, because of Equation 8.134, and results in

$$S_f^\dagger K_e JM^{-1}J^T F_e = S_f^\dagger K_e JM^{-1}J^T F^d. \tag{8.147}$$

Since $S_f^\dagger K_e JM^{-1}J^T S_f$ is an $n_f \times n_f$ nonsingular matrix, and with Equations 8.124 and 8.130, this reduces to

$$\phi = \phi^d, \tag{8.148}$$

or

$$\tilde{\phi} = 0, \tag{8.149}$$

with $\tilde{\phi} = \phi^d - \phi$. This proves the complete decoupling between velocity- and force-controlled subspaces. In the velocity-controlled subspace, the dynamics are assigned by choosing appropriate matrices K_{dx} and K_{px}. The force control is very sensitive to disturbance forces, since it contains no feedback. Suppose a disturbance force F_{dist} acts on the end-effector, e.g., due to modeling errors in the inner loop controller. This changes Equation 8.149 to

$$\tilde{\phi} = \phi_{dist}, \tag{8.150}$$

where $\phi_{dist} = S_f^\dagger F_{dist}$. This effect is compensated for by modifying F^d in Equation 8.139 to include feedback, e.g., $F^d + S_f K_{pf}(\phi^d - \phi)$, with K_{pf} a dimensionless $n_f \times n_f$ (diagonal) force control gain matrix. Using this feedback control law, Equation 8.149 results in

$$\tilde{\phi} = (I_{n_f} + K_{pf})^{-1}\phi_{dist}. \tag{8.151}$$

Acceleration Resolved Control: Case of Soft Environment

The system equations are given by Equation 8.118 for the robot and by Equation 8.137 for the robot-environment interaction. The inner loop controller is taken as

$$\tau = M(q)a_q + C(q,\dot{q})\dot{q} + g(q) + J^T F_e, \tag{8.152}$$

where F_e is the measured contact force, which is supposed to correspond to the real contact force. a_q is again given by Equation 8.140. The closed-loop dynamics of the system with its inner loop controller result in

$$A = a_x. \tag{8.153}$$

The outer loop controller contains two terms:

$$a_x = a_{xx} + a_{xf}. \tag{8.154}$$

a_{xx} corresponds to an *acceleration of freedom* and is given by Equation 8.144. On the other hand, a_{xf} corresponds to an *acceleration of constraint* and corresponds to

$$a_{xf} = K_e^{-1}S_f\left(\ddot{\phi}^d + K_{df}\dot{\tilde{\phi}} + K_{pf}\tilde{\phi}\right). \tag{8.155}$$

K_{df} and K_{pf} are (diagonal) control gain matrices with dimensions $n_f \times n_f$ and physical units $\frac{1}{time}$ and $\frac{1}{(time)^2}$ respectively.

Substituting Equation 8.154 in Equation 8.153 leads to the closed-loop dynamic equation. The closed-loop equation is split up into two independent parts corresponding to the force- and velocity-controlled subspaces. First, premultiply Equation 8.153 with $S_x^{\dagger}$. This eliminates the acceleration of constraint a_{xf}, because of Equation 8.134, and results in

$$S_x^{\dagger} A = S_x^{\dagger} a_{xx}. \tag{8.156}$$

With $\ddot{\chi} = S_x^{\dagger} A$, and with Equation 8.144, this reduces to Equation 8.146. Similarly, premultiply Equation 8.153 with $S_f^{\dagger} K_e$. In the case of a stationary environment, i.e., X_e is constant, the second derivative of Equation 8.137 is given by

$$\ddot{F}_e = P_f K_e A. \tag{8.157}$$

With Equation 8.157, with Equation 8.135, and Equation 8.134 this leads to:

$$\ddot{\tilde{\phi}} + K_{df}\dot{\tilde{\phi}} + K_{pf}\tilde{\phi} = 0. \tag{8.158}$$

Hence, there is complete decoupling between velocity- and force-controlled subspaces. Suppose a disturbance force F_{dist} acts on the end-effector. The influence on the force loop dynamics is derived as:

$$\ddot{\tilde{\phi}} + K_{df}\dot{\tilde{\phi}} + K_{pf}\tilde{\phi} = S_f^{\dagger} K_e J M^{-1} J^T F_{dist}. \tag{8.159}$$

Hence, as in the case of a rigid environment, disturbance forces directly affect the force loop; their effect is proportional to the contact stiffness. As a result, accurate force control is much easier in the case of soft contact. This is achieved by adding extra compliance in the robot end-effector.

Remarks.

1. Usually, $\dot{\phi}^d = \ddot{\phi}^d = 0$.
2. Usually, the measured force signal is rather noisy. Therefore, feedback of $\dot{\phi} = S_f^{\dagger} \dot{F}_e$ in the outer loop controller of Equation 8.155 is often replaced by $S_f^{\dagger} K_e J \dot{q}$, where the joint velocities $\dot{q}$ are measured using tachometers. In the case of a stationary environment, both signals are equivalent, and hence they result in the same closed-loop dynamics.
3. If in the inner controller of Equation 8.152 the contact force is compensated with the desired force F^d instead of the measured force F_e, the force loop dynamics become Equation 8.159, with $F_{dist} = F_e - F^d$. Hence, the dynamics of the different force coordinates are coupled.

Velocity-Resolved Control

The model-based inner loop controllers of Equations 8.139 and 8.152 are too advanced for implementation in current industrial robot controllers. The main problems are their computational complexity and the nonavailability of accurate inertial parameters of each robot link. These parameters are necessary to calculate $M(q)$, $C(q, \dot{q})$, and $g(q)$.

Instead, the state of practice consists of using a set of independent, i.e., completely decoupled, velocity controllers for each robot joint as the inner controller. Usually such a velocity controller is an analog proportional-integral (PI) type controller that controls the voltage of the joint actuator based on feedback of the joint velocity. This velocity is either obtained from a tachometer or derived by differentiating the position measurement. Such an analog velocity loop can be made very high bandwidth. Because of their high bandwidth, the independent joint controllers are able to decouple the robot dynamics in Equation 8.118 to a large extent, and they are able to suppress the effect of the environment forces to a large extent, especially if the contact is sufficiently compliant. Other disturbance forces are suppressed in the same way by the inner velocity controller before they affect the dynamics of the outer loop. This property has made this approach, in combination with a compliant end-effector, very popular for practical applications.

This practical approach is considered below. The closed-loop dynamics of the system with its high bandwidth inner velocity controller is approximated by

$$\dot{q} = \dot{q}^d, \tag{8.160}$$

or, using Equation 8.119, in the Cartesian space:

$$V = v_x. \tag{8.161}$$

v_x is the control signal generated by the outer loop controller. The outer loop controller contains two terms:

$$v_x = v_{xx} + v_{xf}. \tag{8.162}$$

v_{xx} corresponds to a *velocity of freedom* and is given by

$$v_x = S_x \left(\chi^d + K_{px} \tilde{\chi}_\Delta \right). \tag{8.163}$$

On the other hand, v_{xf} corresponds to a *velocity of constraint* and corresponds to

$$v_{xf} = K_e^{-1} S_f \left(\dot{\phi}^d + K_{pf} \tilde{\phi} \right). \tag{8.164}$$

Both K_{px} and K_{pf} have units $\frac{1}{time}$.

Substituting Equation 8.162 in Equation 8.161 leads to the closed-loop dynamic equation. The closed-loop equation is split up into two independent parts corresponding to the force- and velocity-controlled subspaces. First, premultiply Equation 8.161 with $S_x^{\dagger}$. This eliminates the velocity of constraint v_{xf}, because of Equation 8.134, and results in:

$$\tilde{\chi} + K_{px} \tilde{\chi}_\Delta = 0. \tag{8.165}$$

Similarly, premultiply Equation 8.161 with $S_f^{\dagger} K_e$. In the case of a stationary environment, i.e., X_e is constant, the derivative of Equation 8.137 is given by

$$\dot{F}_e = P_f K_e V. \tag{8.166}$$

This leads to

$$\dot{\tilde{\phi}} + K_{pf}\tilde{\phi} = 0. \tag{8.167}$$

Hence, there is complete decoupling between velocity- and force-controlled subspaces.

8.2.4 Adaptive Control

The hybrid control approach just presented explicitly relies on a decomposition of the Cartesian space into force- and velocity-controlled directions. The control laws implicitly assume that accurate models of both subspaces are available all the time. On the other hand, most practical implementations turn out to be rather robust against modeling errors. For example, for the two tasks discussed in Section 8.2.2—i.e., *peg-in-hole* and *sliding a block*—the initial relative position between the manipulated object and the environment may contain errors. As a matter of fact, to cope reliably with these situations is exactly why force control is used! The robustness of the force controller increases if it can continuously adapt its model of the force- and velocity-controlled subspaces. In this chapter, we consider only geometric parameters (i.e., the S_x and S_f subspaces), not the dynamical parameters of the manipulator and/or the environment.

The previous sections relied on the assumption that $\dot{S}_x = 0$ and $\dot{S}_f = 0$. If this assumption is not valid, the controller must follow (or *track*) the constraint's time variance by: (1) using feedforward (motion) information from the constraint equations (if known and available); (2) estimating the changes in S_x and S_f from the motion and/or force measurements. The adaptation involves two steps: (1) to identify the errors between the current constraint model and the currently measured contact situation; and (2) to feed back these identified errors to the constraint model. Figures 8.8 and 8.9 illustrate two examples of error identification.

Two-Dimensional Contour Following

The orientation of the contact normal changes if the environment is not planar. Hence, an error $\Delta\alpha$ appears. This error angle can be estimated with either the velocity or the force measurements only (Figure 8.8):

1. Velocity based: The $X_t Y_t$ frame is the modeled task frame, while $X_0 Y_0$ indicates the real task frame. Hence, the executed velocity V, which is tangential to the real contour, does not completely lie along the X_t axis, but has a small component V_{yt} along the Y_t axis. The orientation error $\Delta\alpha$ is approximated by the arc tangent of the ratio V_{yt}/V_{xt}.
2. Force based: The measured (ideal) reaction force F does not lie completely along the modeled normal direction (Y_t), but has a component F_{xt} along X_t. The orientation error $\Delta\alpha$ is approximated by the arc tangent of the ratio F_{xt}/F_{yt}.

The velocity-based approach is disturbed by mechanical compliance in the system; the force-based approach is disturbed by friction.

Sliding an Edge Over an Edge

This task has two geometric uncertainty parameters when the robot moves the object over the environment: (1) an uncertainty Δx of the position of the task frame's origin along the contacting edge of the object; and (2) an uncertainty $\Delta\alpha$ in the frame's orientation about the same edge (Figure 8.9). Identification equations for these uncertainties are

$$\begin{cases} \Delta\alpha = \arctan \frac{v_z}{v_y} \\ \Delta x = \frac{v_z}{\omega_y} \end{cases} \quad \text{(velocity based)}, \qquad (8.168)$$

$$\begin{cases} \Delta x = -\frac{f_z}{m_y} \\ \Delta\alpha = -\arctan \frac{f_z}{f_y} \end{cases} \quad \text{(force based)}. \qquad (8.169)$$

The results from the force- and/or velocity-based identification should be fed back to the model. Moreover, in the contour-following case, the identified orientation error can be converted into an error χ_Δ, such that the setpoint control laws of Equation 8.144 or Equation 8.163 make the robot track changes in the contact normal.

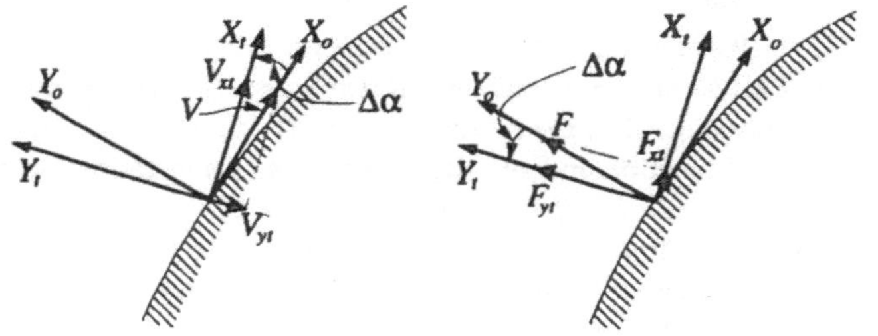

Figure 8.8 Estimation of orientation error.

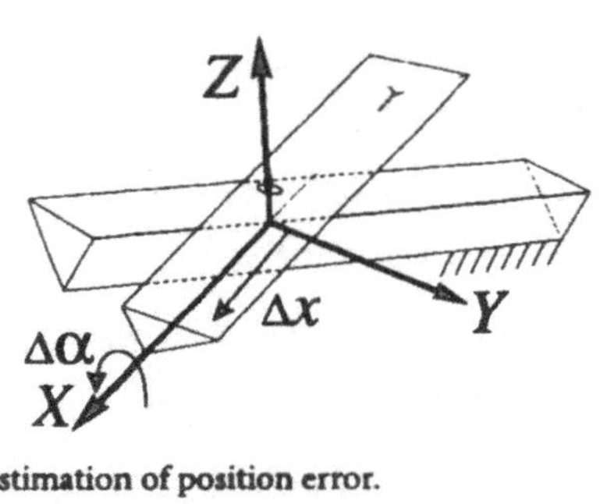

Figure 8.9 Estimation of position error.

8.2.5 Impedance Control

The previous sections focused on a *model-based* approach towards force control: the controller relies on an explicit geometric model of the force- and velocity-controlled directions. However, an alternative approach exists, called *impedance control*. It differs from the hybrid approach both in task specification and in control.

Task Specification

Hybrid control specifies desired motion *and* force trajectories; impedance control [4] specifies (1) a desired *motion* trajectory and (2) a desired *dynamic relationship* between the deviations from this desired trajectory, induced by the contact with the environment, and the forces exerted by the environment:

$$F = -M_c\,\dot{V} + C_c\,\tilde{V} + K_c \int \tilde{V} dt. \qquad (8.170)$$

$\tilde{V}$ is the Cartesian error velocity, i.e., the difference between the prescribed velocity V^d and the measured velocity V; $\dot{V}$ is the Cartesian acceleration; M_c, C_c, and K_c are user-defined inertia, damping, and stiffness matrices, respectively.

Compared to hybrid force/position control, the apparent advantage of impedance control is that no explicit knowledge of the constraint kinematics is required. However, in order to obtain a satisfactory dynamic behavior, the inertia, damping, and stiffness matrices have to be tuned for a particular task. Hence, they embody implicit knowledge of the task geometry, and hence task specification and control are intimately linked.

Control

For control purposes the dynamic relationship of Equation 8.170 can be interpreted in two ways. It is the model of an *impedance*; i.e., the robot reacts to the "deformations" of its planned position and velocity trajectories by generating forces. Special cases are *stiffness control* [8], where $M_c = C_c = 0$, and *damping control*, where $M_c = K_c = 0$. However, Equation 8.170 can also be interpreted in the other way as an *admittance*; i.e., the robot reacts to the constraint forces by deviating from its planned trajectory.

Impedance Control In essence, a stiffness or damping controller is a proportional-derivative (PD) position controller, with position and velocity feedback gains adjusted in order to obtain the desired impedance. Consider a PD joint position controller (see Chapter 8.1):

$$\begin{aligned} \tau &= K_{pq}(q^d - q) + K_{dq}(\dot{q}^d - \dot{q}) && (8.171) \\ &= K_{pq}\tilde{q} + K_{dq}\dot{\tilde{q}}. && (8.172) \end{aligned}$$

This compares to Equation 8.170, with $M_c = 0$ by writing

$$\begin{aligned} \tau &= J^T F && (8.173) \\ &= J^T\left(C_c\,\tilde{V} + K_c \int \tilde{V} dt\right) && (8.174) \\ &= J^T\left(C_c\,J\dot{\tilde{q}} + K_c J\tilde{q}\right). && (8.175) \end{aligned}$$

Hence, in order to obtain the desired stiffness or damping behavior, the PD gain matrices have to be chosen as

$$K_{pq} = J^T K_c J; \quad K_{dq} = J^T C_c J. \qquad (8.176)$$

Note that the gain matrices are position dependent due to the position dependence of the Jacobian J.

Admittance Control In this case, the measured constraint force F_e is used to modify the robot trajectory, given by $X^d(t)$ and $V^d(t)$. In *stiffness control* (or rather, inverse stiffness, or compliance control), a modified position X^m is commanded:

$$X^m = X^d - K_c^{-1} F_e. \qquad (8.177)$$

In *damping control* (or rather, inverse damping control), a modified velocity V^m is commanded:

$$V^m = V^d - C_c^{-1} F_e. \qquad (8.178)$$

Besides introducing damping and stiffness, the most general case of admittance control changes the apparent inertia of the robot manipulator. In this case a desired acceleration A^m is solved from Equation 8.170:

$$A^m = M_c^{-1}\left(-F_e + C_c\,\tilde{V} + K_c \int \tilde{V} dt\right). \qquad (8.179)$$

X^m, V^m, and A^m are applied to the motion controller (see Chapter 8.1) in which the constraint forces may be compensated for by the measured forces as an extra term. For example, in the general admittance case, the control torques are

$$\tau = M(q)a_q + C(q, \dot{q})\dot{q} + g(q) + J^T F_e, \qquad (8.180)$$

where a_q is given by Equation 8.140, in which $a_x = A^m$; and A_m is given by Equation 8.179.

References

[1] DeFazio, T. L., Seltzer, D.S., Whitney, D. E., The instrumented remote center compliance, *Industrial Robot*, 11(4), 238–242, 1984.

[2] De Schutter, J. and Van Brussel, H., Compliant robot motion, *Int. J. Robotics Res.*, 7(4), 3–33, 1988.

[3] Doty, K. L., Melchiorri, C., and Bonivento, C., A theory of generalized inverses applied to robotics, *Int. J. Robotics Res.*, 12(1), 1–19, 1993.

[4] Hogan, N., Impedance control: an approach to manipulation, I-III, *Trans. ASME, J. Dynamic Syst., Meas., Control*, 117, 1–24, 1985.

[5] Khatib, O., A Unified approach for motion and force control of robot manipulators: the operational space formulation, *IEEE J. Robotics Autom.*, 3(1), 43–53, 1987.

[6] Mason, M. T., Compliance and force control for computer controlled manipulators, *IEEE Trans. Syst., Man, Cybern.*, 11(6), 418–432, 1981.

[7] Patarinski, S. and Botev, R., Robot force control, a review, *Mechatronics*, 3(4), 377–398, 1993.

[8] Salisbury, J. K., Active stiffness control of a manipulator in cartesian coordinates, *19th IEEE Conf. Decision Control*, 95–100, 1980.

[9] Wang D. and McClamroch, N. H., Position and force control for constrained manipulator motion: Lyapunov's direct method, *IEEE Trans. Robotics Autom.*, 9(3), 308–313, 1993.

[10] Whitney, D. E., Historic perspective and state of the art in robot force control, *Int. J. Robotics Res.*, 6(1), 3–14, 1987.

8.3 Control of Nonholonomic Systems

John Ting-Yung Wen, Department of Electrical, Computer, and Systems Engineering, Rensselaer Polytechnic Institute

8.3.1 Introduction

When the generalized velocity of a mechanical system satisfies an equality condition that cannot be written as an equivalent condition on the generalized position, the system is called a nonholonomic system. Nonholonomic conditions may arise from constraints, such as pure rolling of a wheel, or from physical conservation laws, such as the conservation of angular momentum of a free floating body.

Nonholonomic systems pose a particular challenge from the control point of view, as any one who has tried to parallel park a car in a tight space can attest. The basic problem involves finding a path that connects an initial configuration to the final configuration and satisfies all the holonomic and nonholonomic conditions for the system. Both open-loop and closed-loop solutions are of interest: open loop solution is useful for off-line path generation and closed-loop solution is needed for real-time control.

Nonholonomic systems typically arise in the following classes of systems:

1. No-slip constraint
   Consider a single wheel rolling on a flat plane (see Figure 8.10). The no slippage (or pure rolling) contact condition means that the linear velocity at the contact point is zero. Let $\vec{\omega}$ and $\vec{v}$, respectively, denote the angular and linear velocity of the body frame attached to the center of the wheel. Then the no slippage condition at the contact point can be written as

   $$\vec{v} - \ell\vec{\omega} \times \vec{z} = 0. \qquad (8.181)$$

   We will see later that part of this constraint is nonintegrable (i.e., not reducible to a position constraint) and, therefore, nonholonomic.

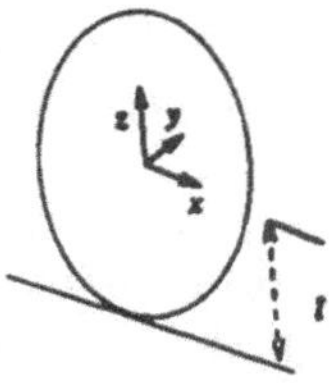

Figure 8.10 A wheel with no-slip contact.

   In modeling the grasping of an object by a robot hand, the so-called soft finger contact model is sometimes used. In this model, the finger is not allowed to rotate about the local normal, $\vec{z} \cdot \vec{\omega} = 0$, but is free to rotate about the local x and y axes. This velocity constraint is nonintegrable.
   The dynamic equations of wheeled vehicles and finger grasping are of similar forms. There are two sets of equations of motion, one for the unconstrained vehicle or finger, the other for the ground (stationary) or the payload. These two sets of equations are coupled by the constraint force/torque that keeps the vehicle on the ground with no wheel slippage or fingers on the object with no rotation about the local normal axis. These equations can be summarized in the following form:

   $$\begin{aligned} &(a) \quad M(q)\ddot{q} + C(q,\dot{q})\dot{q} + g(q) = u - J^T f, \\ &(b) \quad M_c\alpha_c + b_c + k_c = A^T f, \\ &(c) \quad Hf = 0, \qquad (8.182) \\ &(d) \quad v^+ = J\dot{\theta},\ v^- = Av_c = v^+ + H^T W, \text{ and} \\ &(e) \quad \alpha^+ = J\ddot{\theta} + \dot{J}\dot{\phi}\ \alpha^- = A\alpha_c + a \\ &\qquad = \alpha^+ + H^T\dot{W} + \dot{H}^T W. \end{aligned}$$

   Equations 8.182a and 8.182b are the equations of motion of the fingers and the payload, respectively, f is the constraint force related to the vehicle or fingers via the Jacobian transpose J^T, α_c denotes the payload acceleration, b_c and k_c are the Coriolis and gravity forces on the payload, H is a full row rank matrix whose null space specifies the directions where motion at the contact is allowed (these are also the directions with no constraint forces), v^+ and v^- are the velocity at the two sides of the contacts. Similarly, α^+ and α^- denote accelerations and W parameterizes the admissible velocity across the contact. The velocity constraint is specified in Equation 8.182d; premultiplying Equation 8.182d by the annihilator of H^T, denoted by $\hat{H}^T$,

   $$\hat{H}^T(J\dot{\theta} - Av_c) = 0. \qquad (8.183)$$

   In the single wheel case in Figure 8.10,

   $$\begin{aligned} v_c &= 0 \ , \quad \dot{\theta} = \begin{bmatrix} \vec{\omega} \\ \vec{v} \end{bmatrix} \\ J &= \begin{bmatrix} I & 0 \\ \ell\vec{z}\times & I \end{bmatrix}, \quad H = [I \ , \ 0], \\ \hat{H}^T &= [0 \ , \ I]. \qquad (8.184) \end{aligned}$$

   The velocity constraint Equation 8.183 is then the same as Equation 8.181.

2. Conservation of angular momentum
   In a Lagrangian system, if a subset of the generalized coordinates q_u does not appear in the mass matrix $M(q)$, they are called the cyclic coordinates. In this case, the Lagrangian equation associated with q_u is

   $$\frac{d}{dt}\left(\frac{\partial L}{\partial \dot{q}_{u_i}}\right) = \frac{\partial L}{\partial q_{u_i}} = 0. \qquad (8.185)$$

After integration, we obtain the conservation of generalized momentum condition associated with the cyclic coordinates.

As an example, consider a free floating multibody system with no external torque (such as a robot attached to a floating platform in space or an astronaut unassisted by the jet pack, as shown in Figure 8.11).

Figure 8.11 Examples of free floating multibody systems.

The equation of motion for such systems is

$$\begin{bmatrix} M(q) & M_1(q) \\ M_1^T(q) & M_b(q) \end{bmatrix} \begin{bmatrix} \ddot{q} \\ \dot{\omega} \end{bmatrix} + \begin{bmatrix} C_{11}(q,\dot{q}) & C_{12}(q,\dot{q},\omega) \\ C_{21}(q,\dot{q},\omega) & C_{22}(q,\dot{q},\omega) \end{bmatrix} \begin{bmatrix} \dot{q} \\ \omega \end{bmatrix} = \begin{bmatrix} u \\ 0 \end{bmatrix}, \tag{8.186}$$

where ω is the angular velocity of the multibody system about the center of mass. This is a special case of the situation described above with $L = \frac{1}{2}\dot{q}^T M(q)\dot{q} + \dot{q}^T M_1(q)\omega + \frac{1}{2}\omega^T M_b(q)\omega$. Identifying $\dot{q}_u$ with ω, Equation 8.185 becomes

$$M_1^T(q)\dot{q} + M_b(q)\omega = 0 \tag{8.187}$$

which is a nonintegrable condition.

3. Underactuated mechanical system
An underactuated mechanical system is one that does not have all of its degrees of freedom independently actuated. The nonintegrable condition can arise in terms of velocity, as we have seen above, or in terms of acceleration which cannot be integrated to a velocity condition. The latter case is called the *second-order nonholonomic condition* [1].
First-Order condition: Consider a rigid spacecraft with less than three independent torques.

$$I\dot{\omega} + \omega \times I\omega = Bu \tag{8.188}$$

where B is a full column rank matrix with rank less than three. Let $\hat{B}$ be the annihilator of B, i.e., $\hat{B}B = 0$. Then premultiplying Equation 8.188 by $\hat{B}$ gives $\frac{d}{dt}(\hat{B}I\omega) = 0$. Assuming the initial velocity is zero, then we arrive at a nonintegrable velocity constraint,

$$\hat{B}I\omega = 0.$$

Second-Order condition: Consider a robot with some of the joints unactuated. The general dynamic equation can be written as

$$M(q)\ddot{q} + C(q,\dot{q})\dot{q} + g(q) = \begin{bmatrix} u \\ 0 \end{bmatrix}. \tag{8.189}$$

By premultiplying by $\hat{B} = [0 \; I]$ which annihilates the input vector, we obtain a condition involving the acceleration,

$$\hat{B}(M(q)\ddot{q} + C(q,\dot{q})\dot{q} + g(q)) = 0. \tag{8.190}$$

It can be shown that this equation is integrable to a velocity condition, $h(q,\dot{q},t) = 0$, if, and only if, the following conditions hold [1]:

(a) the gravitational torque for the unactuated variables, $g_u(q) = \hat{B}g(q)$, is a constant and
(b) the mass matrix $M(q)$ does not depend on the unactuated coordinates, $q_u = \hat{B}q$.

This implies that any earthbound robots with nonplanar, articulated, underactuated degrees of freedom would satisfy a nonintegrable second-order constraint because $g_u(q)$ would not be constant.

The control problem associated with a nonholonomic system can be posed based on the kinematics alone (with an ideal dynamic controller assumed) or the full dynamical model.

In the kinematics case, nonholonomic conditions are linear in the velocity, v,

$$\Omega(q)v = 0. \tag{8.191}$$

Assuming that the rank of $\Omega(q)$ is constant over q, then Equation 8.191 can be equivalently stated as,

$$v = f(q)u \tag{8.192}$$

where the columns of $f(q)$ form a basis of the null space of $\Omega(q)$. Equation 8.192 can be regarded as a control problem with u as the control variable and the configuration variable, q, as the state if $v = \dot{q}$. If v is nonintegrable (as is the case for the angular velocity), there would be an additional kinematic equation $\dot{q} = h(q)v$ (such as the attitude kinematic equation); the control problem then becomes $\dot{q} = h(q)f(q)u$. Note that in either case, the right hand side of the differential equation does not contain a term dependent only in q. Such systems are called *driftless systems*.

Solving the control problem associated with the kinematic Equation 8.192 produces a feasible path. To actually follow the path, a real-time controller is needed to produce the required force or torque. This procedure of decomposing path planning and path following is common in industrial robot motion control. Alternatively, one can also consider the control of the full dynamical system directly. In other words, consider Equation 8.182 for the rolling constraint case, or Equations 8.186, 8.188 or 8.189 for the underactuated case, with u as the control input. In the rolling constraint case, the contact force also needs to be controlled, similar to a robot performing a contact task. Otherwise, slippage or even loss of contact may result (e.g., witness occasional truck rollovers on highway exit ramps). The dynamical equations also differ from the kinematic problem Equation 8.192 in a fundamental way: a control-independent term, called the drift term, is present in the dynamics. In contrast to driftless systems, there is no known general global controllability condition for such systems. However, the presence of the drift term sometimes simplifies the problem by rendering the linearized system locally controllable.

This chapter focus mainly on the kinematic control problem. In addition to the many research papers already published on this subject, excellent summaries of the current state of research can be found in [2, 3].

In the remainder of this chapter, we address the following aspects of the kinematic control of a nonholonomic system:

1. Determination of Nonholonomy. Given a set of constraints, how does one classify them as holonomic or nonholonomic?
2. Controllability. Given a nonholonomic system, does a path exist that connects an initial configuration to the desired final configuration?
3. Path Planning. Given a controllable nonholonomic system, how does one construct a path that connects an initial configuration to the desired final configuration?
4. Stabilizability. Given a nonholonomic system, can one construct a stabilizing feedback controller, and if it is possible, how does one do so?
5. Output stabilizability. Given a nonholonomic system, can one construct a feedback controller that drives a specified output to the desired target while maintaining the boundedness of all the states, and, if it is possible, how does one do so?

We shall use a simple example to illustrate various concepts and results throughout this section. Consider a unicycle with a fat wheel, i.e., it cannot fall (see Figure 8.12). For this system,

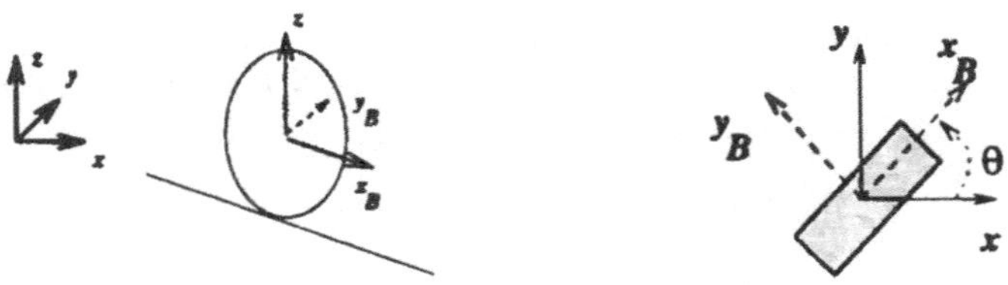

Figure 8.12 Unicycle model and coordinate definition.

there are four constraints:

$$\begin{aligned} \vec{x}_B \cdot \vec{\omega} &= 0, \\ \text{and } \vec{v} - \vec{\omega} \times \ell \vec{z} &= 0. \end{aligned} \tag{8.193}$$

The first equation specifies the no-tilt constraint and the second equation is the no-slip constraint.

8.3.2 Test of Nonholonomy

As motivated in the previous section, consider a set of constraints in the following form:

$$\Omega(q)\dot{q} = 0 \tag{8.194}$$

where $q \in \mathcal{R}^n$ is the configuration variable, $\dot{q}$ is the velocity, and $\Omega(q) \in \mathcal{R}^{\ell \times n}$ specifies the constraint directions.

The complete integrability of the velocity condition in Equation 8.194 means that $\Omega(q)$ is the Jacobian of some function, $h(q) \in \mathcal{R}^{\ell}$, i.e.,

$$\frac{\partial h}{\partial q} = \Omega(q). \tag{8.195}$$

In this case, Equation 8.194 can be written as an equivalent holonomic condition, $h(q) = c$, where c is some constant vector. Equation 8.194 may be only partially integrable, which means that some of the rows of $\Omega(q)$, say, $\Omega_{k+1}, \ldots, \Omega_\ell$, satisfy

$$\frac{\partial h_i}{\partial q} = \Omega_i \quad , \quad i = k+1, \ldots, \ell. \tag{8.196}$$

for some scalar functions $h_i(q)$. Substituting Equation 8.196 in Equation 8.194, we have $\ell - k$ integrable constraints

$$\frac{\partial h_i}{\partial q}\dot{q} = 0, \tag{8.197}$$

which can be equivalently written as $h_i(q) = c_i$ for some constants c_i. If $\ell - k$ is the maximum number of such $h_i(q)$ functions, the remaining k constraints are then nonholonomic.

To determine if the constraint Equation 8.194 is integrable, we can apply the Frobenius theorem. We first need some definitions:

DEFINITION 8.1

1. A *vector field* is a smooth mapping from the configuration space to the tangent space.
2. A *distribution* is the subspace generated by a collection of vector fields. The *dimension* of a distribution is the dimension of any basis of the distribution.
3. The *Lie bracket* between two vector fields, f and g, is defined as
$$[f, g] \triangleq \frac{\partial g}{\partial q} f(q) - \frac{\partial f}{\partial q} g(q).$$
4. An *involutive distribution* is a distribution that is closed with respect to the Lie bracket, that is, if f, g belong to a distribution Δ, then $[f, g]$ also belongs to Δ.
5. A distribution, Δ, with constant dimension m, consisting of vector fields in $\mathcal{R}^n$ is *integrable* if $n - m$ functions, $h_1, \ldots, h_{n-m}$ exist so that the Lie derivative of h_i along each vector field $f \in \Delta$ is zero, that is,
$$L_f h_i(q) \triangleq \frac{\partial h_i}{\partial q} \cdot f(q) = 0.$$

6. The *involutive closure* of a distribution Δ is the smallest involutive distribution that contains Δ.

The Frobenius theorem can be simply stated:

THEOREM 8.1 *A distribution is integrable if, and only if, it is involutive.*

To apply the Frobenius theorem to Equation 8.194, first observe that $\dot{q}$ must be within the null space of $\Omega(q)$ denoted by Δ. Suppose the constraints are independent throughout the configuration space, then the dimension of Δ is $n - \ell$; let a basis of Δ be $g_1(q), \ldots, g_{n-\ell}(q)$:

$$\Delta = \text{span}\{g_1(q), \ldots, g_{n-\ell}(q)\}.$$

Let $\overline{\Delta}$ be the involutive closure of Δ. Suppose the dimension of $\overline{\Delta}$ is constant, $n - \ell + k$. Since $\overline{\Delta}$ is involutive by definition, from the Frobenius theorem, $\overline{\Delta}$ is integrable. This means that functions $h_i, i = 1, \ldots, \ell - k$ exist, so that $\frac{\partial h_i}{\partial q}$ annihilates $\overline{\Delta}$:

$$\frac{\partial h_i}{\partial q} f = 0 \tag{8.198}$$

for all $f \in \overline{\Delta} \supset \Delta$. Since $\dot{q} \in \Delta$, it follows that Equation 8.196 is satisfied for all $\dot{q}$. Hence, among the ℓ constraints given by Equation 8.194, $\ell - k$ are holonomic (obtained from the annihilator of $\overline{\Delta}$) and k are nonholonomic. Geometrically, this means that the flows of the system lie on a $n - \ell + k$ dimensional manifold given by h_i =constant, $i = 1, \ldots, \ell - k$.

To illustrate the above discussion, consider the unicycle example presented at the end of Section 8.3.1. First write the constraints Equation 8.193 in the same form as Equation 8.191

$$\begin{bmatrix} \vec{x}_B \cdot & 0 \\ \ell \vec{z} \times & I \end{bmatrix} \begin{bmatrix} \vec{\omega} \\ \vec{v} \end{bmatrix} = 0.$$

This implies that

$$\begin{bmatrix} \vec{\omega} \\ \vec{v} \end{bmatrix} \in \text{span} \left\{ \begin{bmatrix} \vec{y}_B \\ \ell \vec{x}_B \end{bmatrix}, \begin{bmatrix} \vec{z} \\ 0 \end{bmatrix} \right\}.$$

Represent the top portion of each vector field in the body coordinates, $\vec{y}_B = [0, 1, 0]^T$, $\vec{z} = [0, 0, 1]^T$, and the bottom portion in the world coordinates, $\vec{x}_B = [\ell c_\theta, \ell s_\theta, 0]^T$, $c_\theta = \cos\theta$ and $s_\theta = \sin\theta$, θ is the steering angle. We have

$$\Delta = \text{span} \left\{ \begin{bmatrix} 0 \\ 1 \\ 0 \\ \ell c_\theta \\ \ell s_\theta \\ 0 \end{bmatrix}, \begin{bmatrix} 0 \\ 0 \\ 1 \\ 0 \\ 0 \\ 0 \end{bmatrix} \right\}. \tag{8.199}$$

The involutive closure of Δ can be computed by taking repeated Lie brackets:

$$\overline{\Delta} = \text{span} \left\{ \begin{bmatrix} 0 \\ 1 \\ 0 \\ \ell c_\theta \\ \ell s_\theta \\ 0 \end{bmatrix}, \begin{bmatrix} 0 \\ 0 \\ 1 \\ 0 \\ 0 \\ 0 \end{bmatrix}, \begin{bmatrix} 0 \\ 0 \\ 0 \\ -\ell s_\theta \\ \ell c_\theta \\ 0 \end{bmatrix}, \begin{bmatrix} 0 \\ 0 \\ 0 \\ \ell c_\theta \\ \ell s_\theta \\ 0 \end{bmatrix} \right\} \tag{8.200}$$

which is of constant dimension four. The annihilator of $\overline{\Delta}$ is

$$\overline{\Delta}^{\perp} = \text{span}\{[1, 0, 0, 0, 0, 0], [0, 0, 0, 0, 0, 1]\}.$$

From the Frobenius theorem, the annihilator of $\overline{\Delta}$ is integrable. Indeed, the corresponding holonomic constraints are what one could have obtained by inspection:

$$z = \text{constant}, \quad \psi = \text{roll angle} = 0.$$

Eliminating the holonomic constraints results in a common form of the kinematic equation for unicycle,

$$\begin{bmatrix} \dot{x} \\ \dot{y} \\ \dot{\theta} \\ \dot{\phi} \end{bmatrix} = \begin{bmatrix} \ell c_\theta \\ \ell s_\theta \\ 0 \\ 1 \end{bmatrix} u_1 + \begin{bmatrix} 0 \\ 0 \\ 1 \\ 0 \end{bmatrix} u_2 \tag{8.201}$$

where $u_1 = \omega_{y_B}$ and $u_2 = \omega_z$. Because the exact wheel rotational angle is frequently inconsequential, the ϕ equation is often omitted. In that case, the kinematic Equation 8.201 becomes

$$\begin{bmatrix} \dot{x} \\ \dot{y} \\ \dot{\theta} \end{bmatrix} = \begin{bmatrix} \ell c_\theta \\ \ell s_\theta \\ 0 \end{bmatrix} u_1 + \begin{bmatrix} 0 \\ 0 \\ 1 \end{bmatrix} u_2. \tag{8.202}$$

We shall refer to the system described by either Equation 8.201 or Equation 8.202 as the unicycle problem.

8.3.3 Nonholonomic Path Planning Problem

The nonholonomic path planning problem, also called nonholonomic motion planning, involves finding a path connecting specified configurations that satisfies the nonholonomic condition as in Equation 8.194. As discussed in Section 8.3.1, this problem can be written as an equivalent nonlinear control problem:

> *Given the system*
>
> $$\dot{q} = f(q)u \ , \ q \in \mathcal{R}^n \ , u \in \mathcal{R}^m, \tag{8.203}$$
>
> *and initial and desired final configurations,* $q(0) = q_0$ *and* q_f, *find* $\underline{u} = \{u(t) : t \in [0, 1]\}$ *so that the solution of Equation* 8.203 *satisfies* $q(1) = q_f$.

The terminal time has been normalized to 1. In Equation 8.203, $f(q)$ is a full rank matrix whose columns span the null space of $\Omega(q)$ in Equation 8.194, and u parameterizes the degree of freedom in the velocity space. By construction, $f(q)$ is necessarily a tall matrix, that is, an $n \times m$ matrix with $n > m$.

Controllability

For $u = 0$, every q in $\mathcal{R}^n$ is an equilibrium. The linearized system about any equilibrium q^* is

$$\frac{d}{dt}(q - q^*) = f(q^*)u. \tag{8.204}$$

Because $f(q)$ is tall, this linear time-invariant system is not controllable (the controllability matrix, $[f(q^*) : 0 : \ldots : 0]$, has maximum rank m). This is intuitively plausible; as the nonholonomic condition restricts the flows in the tangent space, the system can locally only move in directions compatible with the nonholonomic condition, contradicting the controllability requirement. However, the system may still be controllable globally.

For a driftless system, such as a nonholonomic system described by Equation 8.203, the controllability can be ascertained through the following sufficient condition (sometimes called *Chow's theorem*):

THEOREM 8.2 The system given by Equation 8.203 is controllable if the involutive closure of the columns of $f(q)$ is of constant rank n for all q.

The involutive closure of a set of vector fields is in general called the *Lie algebra* generated by these vector fields. In the context of control systems where the vector fields are the columns of the input matrix $f(q)$, the Lie algebra is called the *control Lie algebra.*

For systems with drift terms, the above full rank condition is only sufficient for local accessibility. For a linear time-invariant system, this condition simply reduces to the usual controllability rank condition. This theorem is nonconstructive, however. The path planning problem basically deals with finding a specific control input to steer the system from a given initial condition to a given final condition, once the controllability rank condition is satisfied.

Since the involutive closure of the null space of the constraints is just the control Lie algebra of the corresponding nonholonomic control system, the control system (with the holonomic constraints removed) is globally controllable as long as the constraints remain independent for all configurations. For the unicycle problem given by Equation 8.201, the control Lie algebra is $\overline{\Delta}$ in Equation 8.200 (with z and ψ coordinates removed). Because the dimension of $\overline{\Delta}$ and the state-space dimension are both equal to four, the system is globally controllable.

An alternate way to view Equation 8.203 is to regard it as a nonlinear mapping of the input function $\underline{u}$ to the final state $q(1)$:

$$q(1) = F(q_0, \underline{u}). \tag{8.205}$$

Given q_0 and $\underline{u}$, denote the solution of Equation 8.203 by

$$q(t) = \phi_{\underline{u}}(t; q_0). \tag{8.206}$$

Then, $F(q_0, \underline{u}) = \phi_{\underline{u}}(1; q_0)$. In general, the analytic expression for F is impossible to obtain.

By definition, global controllability means that $F(q_0, \cdot)$ is an onto mapping for every q_0. For a given $\underline{u}$, $\nabla_{\underline{u}} F(q_0, \underline{u})$ corresponds to the system linearized about a trajectory $\underline{q} = \{q(t) : t \in [0, 1]\}$ which is generated by $\underline{u}$:

$$\delta\dot{q} = A(t)\delta q + B(t)\delta u, \quad \delta q(0) = 0, \tag{8.207}$$

where, $A(t) \triangleq [\frac{\partial f}{\partial q_1}(q(t))u(t) \; : \; \cdots \; : \; \frac{\partial f}{\partial q_n}(q(t))u(t)]$, and $B(t) \triangleq f(q(t))$. Since $\delta q(0) = 0$, the solution to this equation is,

$$\delta q(1) = \int_0^1 \Phi(1, s)B(s)\delta u(s)\, ds \tag{8.208}$$

where Φ is the state transition matrix of the linearized system. It follows that

$$\left(\nabla_{\underline{u}} F(q_0, \underline{u})\right) \underline{v} = \int_0^1 \Phi(1, s)B(s)v(s)\, ds. \tag{8.209}$$

Controllability of the system in Equation 8.207 implies that for any final state $\delta q(1) \in \mathcal{R}^n$, a control δu exists which drives the linear system from $\delta q(0) = 0$ to $\delta q(1)$. This is equivalent to the operator $\nabla_{\underline{u}} F$ being onto (equivalently, the null space of the adjoint operator, $[\nabla_{\underline{u}} F]^*$, being zero). In the case that $\underline{u} = 0$, $\nabla_{\underline{u}} F$ reduces to the linear time-invariant system Equation 8.204. In this case, $\nabla_{\underline{u}} F$ cannot be of full rank because the linearized system is not controllable.

Path Planning Algorithms

Steering with Cyclic Input In Equation 8.203, because $f(q)$ is full rank for all q, there is a coordinate transformation so that $f(q)$ becomes $\begin{bmatrix} I \\ f_1(q) \end{bmatrix}$. In other words, the inputs are simply the velocities of m configuration variables. For example, in the unicycle problem described by Equation 8.201, u_1 and u_2 are equal to $\dot{\theta}$ and $\dot{\phi}$. The subspace corresponding to these variables is called the *base space* (also called the *shape space*). A cyclic motion in the base space returns the base variables to their starting point, but the configuration variables would have a net change (called the geometric phase) as shown in Figure 8.13. In the unicycle case, cyclic motions in θ and ϕ result in the following net changes in the x and y coordinates:

$$\begin{aligned} x(T) - x(0) &= \int_0^T \cos\theta\dot{\phi}\, dt = \oint \cos\theta\, d\phi; \\ y(T) - y(0) &= \int_0^T \sin\theta\dot{\phi}\, dt = \oint \sin\theta\, d\phi. \end{aligned} \tag{8.210}$$

By Green's theorem, they can be written as surface integrals

$$\begin{aligned} x(T) - x(0) &= \iint_S -\sin\theta\, d\theta\, d\phi; \quad \text{and} \\ y(T) - y(0) &= \iint_S \cos\theta\, d\theta\, d\phi \end{aligned} \tag{8.211}$$

where S is the surface enclosed by the closed contour in the (ϕ, θ) space.

A general strategy for path planning would then consist of two steps: first drive the base variables to the desired final location, then appropriately choose a closed contour in the base space to achieve the desired change in the configuration variables without affecting the base variables. This idea has served as the basis of many path planning algorithms.

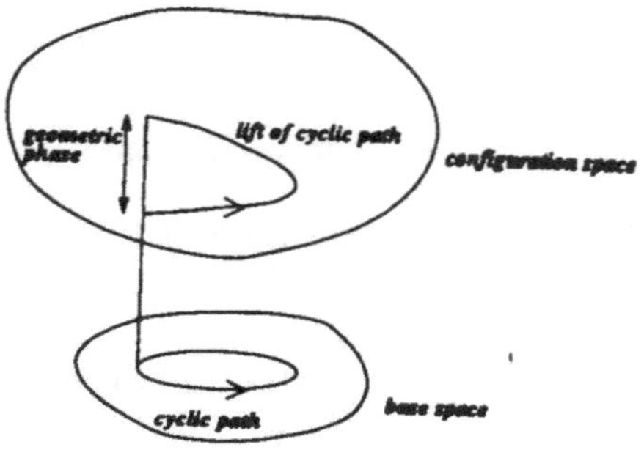

Figure 8.13 Geometric phase.

To illustrate this procedure for path planning for the unicycle example, assume that the base variables, ϕ and θ, have reached their target values. We choose them to be sinusoids with integral frequencies, so that at $t = 1$, they return to their initial values:

$$u_1 = a_1 \cos(4\pi t), \text{ and } u_2 = a_2 \cos(2\pi t). \tag{8.212}$$

By direct integration,

$$\phi = \frac{a_1}{4\pi} \sin(4\pi t), \text{ and } \theta = \frac{a_2}{2\pi} \sin(2\pi t). \tag{8.213}$$

For several values of a_1 and a_2, the closed contours in the $\phi - \theta$ plane given by Equation 8.213 are as shown in Figure 8.14. The net changes in x and y over the period $[0, 1]$ are given by the surface integrals Equation 8.211 over the area enclosed by the contours. To achieve the desired values for x and y, the two equations can be numerically solved for a_1 and a_2.

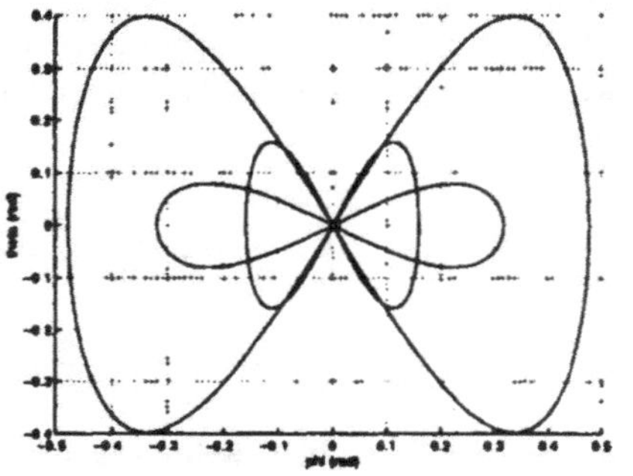

Figure 8.14 Closed contour in base space.

This procedure can also be performed directly in the time domain. For the chosen sinusoidal inputs, the changes in x and y are

$$\Delta x = \int_0^1 a_1 \cos(4\pi t) \cos\left(\frac{a_2}{2\pi} \sin(2\pi t)\right) dt \tag{8.214}$$

$$\Delta y = \int_0^1 a_1 \cos(4\pi t) \sin\left(\frac{a_2}{2\pi} \sin(2\pi t)\right) dt. \tag{8.215}$$

Using Fourier series expansion for even functions,

$$\cos\left(\frac{a_2}{2\pi} \sin(2\pi t)\right) = \sum_{k=0}^{\infty} \alpha_k \cos(2\pi k t)$$

$$\sin\left(\frac{a_2}{2\pi} \sin(2\pi t)\right) = \sum_{k=0}^{\infty} \beta_k \cos(2\pi k t).$$

After the integration, we obtain

$$\Delta x = \frac{1}{2} a_1 \alpha_1 \quad , \quad \Delta y = \frac{1}{2} a_1 \beta_1. \tag{8.216}$$

Because α_1 and β_1 depend on a_2, given the desired motion in x and y, Equation 8.216 results in a one-dimensional line search for a_2:

$$\frac{\Delta x}{\alpha_1(a_2)} - \frac{\Delta y}{\beta_1(a_2)} = 0. \tag{8.217}$$

Once a_2 is found (there may be multiple solutions), a_1 can be found from Equation 8.216.

The above procedure of using sinusoidal inputs for path planning can be generalized to systems in the following canonical form (written for systems with two inputs), called the ***chain form***:

$$\begin{bmatrix} \dot q_1 \\ \dot q_2 \\ \dot q_3 \\ \dot q_4 \\ \vdots \\ \dot q_n \end{bmatrix} = \begin{bmatrix} u_1 \\ u_2 \\ q_2 u_1 \\ q_3 u_1 \\ \vdots \\ q_{n-1} u_1. \end{bmatrix} \tag{8.218}$$

For example, the unicycle problem Equation 8.202 can be converted to the chain form by defining

$$\begin{aligned} q_1 &= \theta \\ q_2 &= c_\theta x + s_\theta y \\ q_3 &= s_\theta x - c_\theta y. \end{aligned}$$

Then

$$\begin{aligned} \dot q_1 &= u_1 \\ \dot q_2 &= -q_3 u_1 + \ell u_2 \\ \dot q_3 &= q_2 u_1. \end{aligned}$$

By defining the right hand side of the $\dot q_2$ equation as the new u_2, the system is now in the chain form.

For a general chain system, consider the sinusoidal inputs,

$$u_1 = a \sin(2\pi t), \quad u_2 = b \cos(2\pi k t) \tag{8.219}$$

It follows that, for $i < k + 2$, $q_i(t)$ consists of sinusoids with period 1; therefore, $q_i(1) = q_i(0)$. The net change in q_{k+2} can be computed as

$$q_{k+2}(1) - q_{k+2}(0) = \left(\frac{a}{4\pi}\right)^k \frac{b}{k!}. \tag{8.220}$$

The parameters a and b can then be chosen so that q_{k+2} is driven to the desired value in [0, 1] without affecting all of the states preceding it. A steering algorithm will then consist of the following steps:

1. Drive q_1 and q_2 to the desired values.
2. For each q_{k+2}, $k = 1, \ldots, n-2$, drive q_{k+2} to its desired values by using the sinusoidal input Equation 8.219 with a and b determined from Equation 8.220.

Many systems can be converted to the chain form, e.g., kinematic car, space robot etc. In [4], a general procedure is provided to transform a given system to the chain form. There are also some systems that cannot be transformed to the chain form, e.g., a ball rolling on a flat plate.

Optimal Control Another approach to nonholonomic path planning is optimal control. Consider the following two-input, three-state chain system (we have shown that the unicycle can be converted to this form):

$$\dot{q} = \begin{bmatrix} u_1 \\ u_2 \\ q_2 u_1 \end{bmatrix}, \quad q(0) = q_0. \tag{8.221}$$

The inputs u_i are to be chosen to drive $q(t)$ from q_0 to $q(1) = 0$ while minimizing the input energy:

$$J = \int_0^1 \frac{1}{2} \| u(t) \|^2 \, dt.$$

The Hamiltonian associated with this optimal control problem is

$$H(q, u, \lambda) = \frac{1}{2} \| u \|^2 + \lambda^T f(q) u \tag{8.222}$$

where λ is the co-state vector. From the Maximum Principle, the optimal control can be found by minimizing H with respect to u:

$$u_1 = -(\lambda_1 + \lambda_3 q_2), \quad u_2 = -\lambda_2. \tag{8.223}$$

The co-state satisfies

$$\dot{\lambda} = -\frac{\partial H}{\partial q} = \begin{bmatrix} 0 \\ -\lambda_3 u_1 \\ 0 \end{bmatrix}. \tag{8.224}$$

Differentiating the optimal control in Equation 8.223,

$$\dot{u}_1 = c u_2, \text{ and } \dot{u}_2 = -c u_1 \tag{8.225}$$

where c is a constant ($c = -\lambda_3$). This implies that u_1 and u_2 are sinusoids:

$$\begin{aligned} u_1(t) &= -a \cos ct + u_1(0), \text{ and} \\ u_2(t) &= a \sin ct + u_2(0). \end{aligned} \tag{8.226}$$

Substituting in the equation of motion Equation 8.221 and choosing $c = 2\pi$,

$$\begin{aligned} q_1(1) &= u_1(0) + q_1(0) \\ q_2(1) &= u_2(0) + q_2(0) \\ q_3(1) &= \frac{a^2}{4\pi} + \frac{u_1(0)a}{2\pi} + q_2(0)u_1(0) + \frac{u_1(0)u_2(0)}{2}. \end{aligned} \tag{8.227}$$

The requirement on the zero final state can be used to solve for the constants in the control:

$$\begin{aligned} u_1(0) &= -q_1(0) \\ u_2(0) &= -q_2(0) \\ a &= q_1(0) + \sqrt{q_1^2(0) + 2\pi q_1(0) q_2(0)}. \end{aligned} \tag{8.228}$$

If the expression within the square root is negative, then the constant c should be chosen as -2π to render it positive.

The optimization approach described above can be generalized to certain higher order systems, but, in general, the optimal control for nonholonomic systems is more complicated. One can also try finding an optimal solution numerically; this would, in general, entail the solving a two-point boundary value problem. A nonlinear programming approach based on the following Ritz approximation of the input function space has also been proposed:

$$u(t) = \sum_{k=0}^{N} \alpha_k \psi_k(t) \tag{8.229}$$

where ψ_k's are chosen to be independent orthonormal functions (such as the Fourier basis) and α_k's are constant vectors parameterizing the input function. The minimum input energy criterion can then be combined with a final state penalty term, resulting in the following optimization criterion:

$$\begin{aligned} J &= \gamma \| q_f - q(1) \|^2 + \int_0^1 \| u(t) \|^2 \, dt \\ &= \gamma \| q_f - q(1) \|^2 + \sum_{k=0}^{N} \| \alpha_k \|^2. \end{aligned} \tag{8.230}$$

The optimal α_k's can be solved numerically by using nonlinear programming. The penalty weighting γ can be iteratively increased to enforce the final state constraint. The problem is not necessarily convex. Consequently, as in any nonlinear programming problem, only local convergence can be asserted. In next section, we will describe a similar approach without the control penalty term in J. As a result of this modification, a stronger convergence condition can be established.

Path Space Iterative Approach As shown in the beginning of Section 8.3.3, the differential equation governing the nonholonomic motion Equation 8.203 can be written as a nonlinear operator relating an input function, $\underline{u}$, to a path, $\underline{q}$. By writing the final state error as

$$y = q_f - F(q_0, \underline{u}) \tag{8.231}$$

the path planning problem can be regarded as a nonlinear least-squares problem. Global controllability means, that for any q_f, there is at least one solution $\underline{u}$. Many numerical algorithms exist for the solution of this problem. In general, the solution involves lifting a path connecting the initial y to the desired $y = 0$ to the $\underline{u}$ space (see Figure 8.15). Let $\underline{u}(0)$ be the first guess of the input function and $y(0)$ be the corresponding final state error as given by Equation 8.231. The goal is to modify $\underline{u}$ iteratively so that

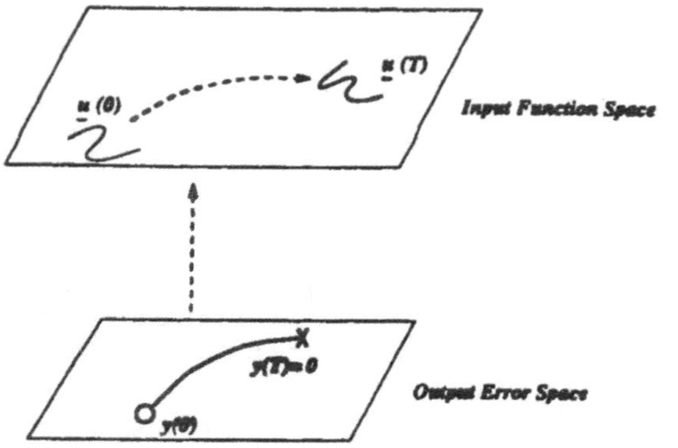

Figure 8.15 Path planning by lifting a path in output error space.

y converges to 0 asymptotically. To this end, choose a path in the error space connecting $y(0)$ to 0, call it $y_d(\tau)$, where τ is the iteration variable. The derivative of $y(\tau)$ is

$$\frac{dy}{d\tau} = -\nabla_{\underline{u}} F(q_0, \underline{u}) \frac{d\underline{u}}{d\tau}. \tag{8.232}$$

If $\nabla_{\underline{u}} F(q_0, \underline{u})$ is full rank, then we can choose the following update rule for $\underline{u}(\tau)$ to force y to follow y_d:

$$\begin{aligned} \frac{d\underline{u}}{d\tau} &= -\left[\nabla_{\underline{u}} F(q_0, \underline{u})\right]^+ \\ &\quad \left[-\alpha((q_f - F(q_0, \underline{u})) - y_d) + \frac{dy_d}{d\tau}\right] \end{aligned} \tag{8.233}$$

where $\alpha > 0$ and $\left[\nabla_{\underline{u}} F(q_0, \underline{u})\right]^+$ denotes the Moore-Penrose pseudo-inverse of $\nabla_{\underline{u}} F(\underline{u})$. This is essentially the continuous version of Newton's method. Equation 8.233 is an initial value problem in $\underline{u}$ with a chosen $\underline{u}(0)$. With $\underline{u}$ discretized by a finite dimensional approximation (e.g., using Fourier basis as in Equation 8.229), it can be solved numerically by an ordinary differential equation solver.

As discussed in Section 8.3.3, the gradient of F, $\nabla_{\underline{u}} F(q_0, \underline{u})$, can be computed from the system Equation 8.203 linearized about the path corresponding to $\underline{u}$. A sufficient condition for the convergence of the iterative algorithm Equation 8.233 is that $\nabla_{\underline{u}} F(q_0, \underline{u}(\tau))$ is full rank for all τ, or equivalently, the time varying linearized system Equation 8.207, generated by linearizing Equation 8.203 about $\underline{u}(\tau)$, is controllable. For controllable systems without drift, it has been shown in [5] that this full rank condition is true generically (i.e., for almost all $\underline{u}$ in the C_∞ topology).

In the cases where $\nabla_{\underline{u}} F(q_0, \underline{u})$ loses rank (possibly causing the algorithm to get stuck), a *generic loop* (see Figure 8.16) can be appended to the singular control causing the composite control to be nonsingular and thus allowing the algorithm to continue its progress toward a solution. A generic loop can be described as follows: For some small time interval $[0, T/2]$, generate a nonsingular control $\underline{v}_a$ (which can be randomly chosen, due to the genericity property). Then let $\underline{v}$ be the control on $[0, T]$ consisting of $\underline{v}_a$ on $[0, T/2]$ and $-\underline{v}_a$ on $[T/2, T]$. Because nonholonomic systems have no drift term, it follows that the system makes a "loop" starting at $q(1) = F(q_0, \underline{u})$ ending once again at the same point. Appending $\underline{v}$ to $\underline{u}$ and renormalizing the time interval to $[0, 1]$ yields a nonsingular control which does not change y. The algorithm is therefore guaranteed to converge to any arbitrary neighborhood of the desired final configuration.

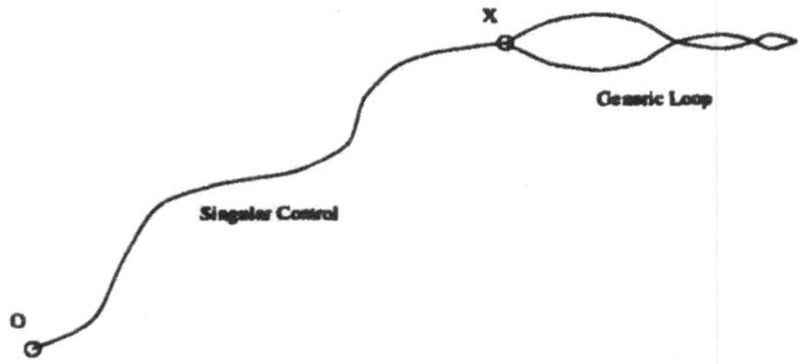

Figure 8.16 Generic loop.

This algorithm has been extended to include inequality constraints such as joint limits, collision avoidance, etc. [6], by using an exterior penalty function approach. Consider state inequality constraints given by

$$c(\underline{q}) \leq 0 \tag{8.234}$$

where $\underline{q}$ is the complete path in the configuration space, $c(\cdot)$ is a vector, and the inequality is interpreted in the componentwise sense. The state trajectory, $\underline{q}$, can be related to the input function $\underline{u}$ through a nonlinear operator (which is typically not possible to find analytically),

$$\underline{q} = \mathcal{F}(q_0, \underline{u}). \tag{8.235}$$

The inequality constraint Equation 8.234 can then be expressed in terms of $\underline{u}$:

$$c(\mathcal{F}(q_0, \underline{u})) \leq 0. \tag{8.236}$$

Inequality constraints in optimization problems are typically handled through penalty functions. There are two types, interior and exterior penalty functions. An interior penalty function sets up barriers at the boundary of the inequality constraints. As the height of the barrier increases, the corresponding path becomes closer to being feasible. If the optimization problem, in our case, the feasible path problem, can be solved for each finite barrier, then convergence to the optimal solution is assured as the barrier height tends to infinity. In the exterior penalty function approach, the ith inequality constraint is converted to an equality constraint by using an exterior penalty function,

$$z_i(\underline{u}) = \gamma_i \sum_{j=1}^{N} g(c_i(\mathcal{F}_j(q_0, \underline{u}))) \tag{8.237}$$

where $\gamma_i > 0$, c_i is the ith constraint, $\mathcal{F}_j$ denotes the jth discretized time point where the constraint is checked, and g is a continuous scalar function with the property that g is equal to zero, when c_i is less than or equal to zero, and is greater than zero and monotonic when c_i is greater than zero. The same iterative approach presented for the equality-only case can now be applied to the composite constraint vector:

$$\psi(\underline{u}) = \begin{bmatrix} y(\underline{u}) \\ z(\underline{u}) \end{bmatrix}. \tag{8.238}$$

For a certain class of convex polyhedral constraints, the generic full rank condition for the augmented problem still holds. This approach has been successfully applied to many complex examples, such as cars with multiple trailers, subject to a variety of collision avoidance and joint limits constraints [6].

8.3.4 Stabilization

State Stabilization

Stabilizability means the existence of a feedback controller that will render the closed-loop system asymptotically stable about an equilibrium point. For linear systems, controllability implies stabilizability. It would be of great value if this were true for special classes of nonlinear systems such as the nonholonomic systems considered in this article (where controllability can be checked through a rank condition on the control Lie algebra). It was shown by [7] that this assertion is not true in general. For a general nonlinear system $\dot{q} = f_0(q, u)$, with equilibrium at q_0, $f_0(q_0, 0) = 0$ and $f_0(\cdot, \cdot)$ continuous in a neighborhood of $(q_0, 0)$, a necessary condition for the existence of a continuous time-invariant control law, that renders $(q_0, 0)$ asymptotically stable, is that f maps any neighborhood of $(q_0, 0)$ to a neighborhood of 0. For a nonholonomic system described by Equation 8.203, $f_0(q, u) = f(q)u$. Then the range of $\{f_0(q, u) : (q, u)$ in a neighborhood of $(q_0, 0)\}$ is equal to the span of the columns of $f(q)$ which is of dimension m (number of inputs). Because a neighborhood about the zero state is n dimensional, the necessary condition above is not satisfied unless $m \geq n$.

There are two approaches to deal with the lack of a continuous time-invariant stabilizing feedback. The first is to relax the continuity requirement to allow piecewise smooth control laws; the second is to relax the time invariance requirement and allow a time-varying feedback.

In either approach, an obvious starting point is to begin with an initial feasible path obtained by using any one of the open-loop methods discussed in Section 8.3.3 and then to apply a feedback to stabilize the system around the path. Given an initial feasible path, if the nonlinear kinematic model linearized about the path is controllable, almost always true as mentioned in Section 8.3.3, a time-varying stabilizing controller can be constructed by using standard techniques. The resulting system will then be locally asymptotically stable.

Consider the unicycle problem Equation 8.202 as an example. Suppose an open-loop trajectory, $\{(x^*(t), y^*(t), \theta^*(t), t \in [0, 1]\}$, and the corresponding input, $\{u_1^*(t), u_2^*(t), t \in [0, 1]\}$, are already generated by using any of the methods discussed in Section 8.3.3. The system equation can be linearized about this path:

$$\begin{aligned} \delta\dot{x} &= -(\sin(\theta^*(t))\, u_1^*(t))\, \delta\theta + \cos(\theta^*(t))\, \delta u_1, \\ \delta\dot{y} &= (\cos(\theta^*(t)) u_1^*(t))\, \delta\theta + \sin(\theta^*(t))\, \delta u_1, \\ \text{and } \delta\dot{\theta} &= \delta u_2. \end{aligned} \tag{8.239}$$

This is a linear time-varying system. It can be easily verified that as long as u_1^* is not identically zero, the system is controllable. One can then construct a time-varying stabilizing feedback to keep the system on the planned open-loop path.

Stabilizing control laws can also be directly constructed without first finding a feasible open-loop path. In [8], it was shown that all nonholonomic systems can be feedback stabilized with a smooth periodic controller. For specific classes of systems, such as mobile robots in [9], or, more generally, the so-called power systems as in [10], explicit constructive procedures for such controllers have been demonstrated.

We will again use the unicycle example to illustrate the basic idea of constructing a time-varying stabilizing feedback by using a time-dependent coordinate transformation so that the equation of motion contains a time-varying drift term. Define a new variable z by

$$z = \theta + k(t, x, y) \tag{8.240}$$

where k is a function that will be specified later. Differentiating z,

$$\dot{z} = v \triangleq u_2 + \frac{\partial k}{\partial t} + (\frac{\partial k}{\partial x}\cos\theta + \frac{\partial k}{\partial y}\sin\theta)u_1.$$

Consider a quadratic Lyapunov function candidate $V = \frac{1}{2}(x^2 + y^2 + z^2)$. The derivative along the solution trajectory is

$$\dot{V} = (x\cos\theta + y\sin\theta)u_1 + zv.$$

By choosing

$$\begin{aligned} u_1 &= -\alpha_1(x\cos\theta + y\sin\theta), \\ v &= -\alpha_2 z, \quad \alpha_1, \alpha_2 > 0, \end{aligned} \tag{8.241}$$

which means

$$\begin{aligned} u_2 = &-\frac{\partial k}{\partial t} - (\frac{\partial k}{\partial x}\cos\theta + \frac{\partial k}{\partial y}\sin\theta)u_1 \\ &- \alpha_2(\theta + k(t, x, y)), \end{aligned}$$

we obtain a negative semidefinite $\dot{V} = -\alpha_1(x\cos\theta + y\sin\theta)^2 - \alpha_2 z^2$. This implies that, as $t \to \infty$, $z \to 0$ and $x\cos\theta + y\sin\theta \to 0$. Substituting in the definition of z, we get $\theta(t) \to -k(t, x(t), y(t))$. From the other asymptotic condition, $\theta(t)$ also converges to $-\tan^{-1}\left(\frac{x(t)}{y(t)}\right)$. As $\dot{x}$ and $\dot{y}$ asymptotically vanish, $x(t)$ and $y(t)$, and therefore, $\theta(t)$, tend to constants. Equating the two asymptotic expressions for $\theta(t)$, we conclude that $k(t, x(t), y(t))$ converges to a constant. By suitably choosing $k(t, x, y)$, e.g., $k(t, x, y) = (x^2 + y^2)\sin(t)$, the only condition under which $k(t, x, y)$ can converge to a constant is that $x^2 + y^2$ converges to zero, which in turn implies that $\theta(t) \to 0$. In contrast to the indirect approach (i.e., using a linear time varying control law to stabilize a system about a planned open-loop path), this control law is globally stabilizing.

Output Stabilization

In certain cases, it may only be necessary to control the state to a certain manifold rather than to a particular configuration.

For example, in the case of a robot manipulator on a free floating mobile base, it may only be necessary to control the tip of the manipulator so that it can perform useful tasks. In this case, a smooth output stabilizing controller can frequently be found.

Suppose the output of interest is

$$y = g(q)\ ,\ \ y \in \mathcal{R}^p \qquad (8.242)$$

and $p < n$. At a particular configuration, q,

$$\dot{y} = \nabla_q g(q)\, f(q)\, u. \qquad (8.243)$$

Define $K(q) = \nabla_q g(q)\, f(q)$. If $K(q)$ is onto, i.e., $p \leq m$ and $K(q)$ is full rank, then the system is locally output controllable (there is a u that can move y arbitrarily within a small enough ball) though it is not locally state controllable.

The output stabilization problem involves finding a feedback controller u (possibly dependent on the full state) to drive y to a set point, y_d. Provided that $K(q)$ is of full row rank, an output stabilizing controller can be easily found:

$$u = -QK^T(q)(y - y_d)\quad ,\quad Q > 0. \qquad (8.244)$$

Therefore, y is governed by

$$\dot{y} = -K(q)QK^T(q)(y - y_d). \qquad (8.245)$$

Under the full row rank assumption on $K(q)$, $K(q)QK^T(q)$ is positive definite, which implies that y converges to y_d asymptotically. In general, either y converges to y_d or q converges to a singular configuration of $K(q)$ (where $K(q)$ loses row rank) and $(y - y_d)$ converges to the null space of $K^T(q)$.

We will again use the unicycle problem as an illustration. Suppose the output of interest is (x, θ) and the goal is to drive (x, θ) to (x_d, θ_d) where θ_d is not a multiple of $\frac{\pi}{2}$. By choosing the control law,

$$u_1 = -\alpha_1 \cos\theta (x - x_d),\ \ u_2 = -\alpha_2(\theta - \theta_d). \qquad (8.246)$$

The closed-loop system for the output is

$$\dot{x} = -\alpha_1(\cos^2\theta)(x - x_d),\ \ \dot{\theta} = -\alpha_2(\theta - \theta_d). \qquad (8.247)$$

The closed-loop system contains a singularity at $\theta = \frac{\pi}{2}$, but, if $\theta_d \neq \frac{\pi}{2}$, this singularity will not be attractive. The output stabilization of (x, θ) can be concatenated with other output stabilizing controllers, with other choices of outputs, to obtain full state stabilization. For example, once x is driven to zero, θ can be independently driven to zero (with $u_1 = 0$), and, finally, y can be driven to zero without affecting x and θ. These stages can be combined together as a piecewise smooth state stabilizing feedback controller.

Consider a space robot on a platform as another example. Suppose the output of interest is the end effector coordinate, y. The singular configurations in this case are called the dynamic singularities. The output velocity is related to the joint velocity and center of mass angular velocity by the kinematic Jacobians,

$$\dot{y} = J(q)\dot{q} + J_b(q)\omega.$$

As discussed in Section 8.3.1, the nonholonomic nature of the problem follows from the conservation of the angular momentum Equation 8.187:

$$M_1^T(q)\dot{q} + M_b(q)\omega = 0.$$

Eliminating ω,

$$\dot{y} = (J(q) - J_b M_b^{-1} M_1^T(q))\dot{q}.$$

The effective Jacobian, $K(q) = J(q) - J_b M_b^{-1} M_1^T(q)$, sometimes called the dynamic Jacobian, now contains inertia parameters (hence the modifier "dynamic") in contrast to a terrestrial robot Jacobian which only depends on the kinematic parameters. If the dimension of q is at least as large as the dimension of y, the output can be effectively controlled provided that the dynamic Jacobian does not lose rank (i.e., q is away from the dynamic singularities).

References

[1] Oriolo, G. and Nakamura, Y., Control of mechanical systems with second-order nonholonomic constraints: Underactuated manipulators, *Proc. 30th IEEE Conference on Decision and Control*, 2398–2403, Brighton, England, 1991.

[2] Li, Z. and Canny, J.F., Eds., *Nonholonomic motion planning*, Kluwer Academic, Boston, MA, 1993.

[3] Sastry, S.S., Murray, R.M., and Li, Z., *A Mathematical Introduction to Robotic Manipulation*, CRC Press, Boca Raton, FL, 1993.

[4] Murray, R.M. and Sastry, S.S., Nonholonomic motion planning – steering using sinusoids, *IEEE Trans. Automat. Control*, 38, 700–716, 1993.

[5] Lin, Y. and Sontag, E.D., Universal formula for stabilization with bounded controls, *Syst. Control Lett.*, 16(6), 393–397, 1991.

[6] Divelbiss, A. and Wen, J.T., Nonholonomic motion planning with inequality constraints, *Proc. IEEE Int. Conf. Robotics Automat.*, San Diego, CA, 1994.

[7] Brockett, R.W., Asymptotic stability and feedback stabilization, in *Differential Geometric Control Theory*, Brockett, R.W., Millman, R.S., and Sussmann, J.J, Eds., Birkhauser, 1983, vol. 27, 181–208.

[8] Coron, J.-M., Global assymptotic stabilization for controllable systems without drift, *Math. Control, Signals, Syst.*, 5(3), 1992.

[9] Samson, C. and Ait-Abderrahim, K., Feedback control of a nonholonomic wheeled cart in Cartesian space, in *Proc. IEEE Robotics Automat. Conf.*, Sacramento, CA, 1991.

[10] Teel, A., Murray, R., and Walsh, G., Nonholonomic control systems: From steering to stabilization with sinusoids, *Proc. 31th IEEE Conf. Dec. Control*, Tucson, AZ, 1992.

9

Miscellaneous Mechanical Control Systems

Brian Armstrong
Department of Electrical Engineering and Computer Science, University of Wisconsin—Milwaukee, Milwaukee, WI

Carlos Canudas de Wit
Laboratoire d'Automatique de Grenoble, ENSIEG, Grenoble, France

Jacob Tal
Galil Motion Control, Inc.

Thomas R. Kurfess
The George W. Woodruff School of Mechanical Engineering, The Georgia Institute of Technology, Atlanta, GA

Hodge Jenkins
The George W. Woodruff School of Mechanical Engineering, The Georgia Institute of Technology, Atlanta, GA

Maarten Steinbuch
Philips Research Laboratories, Eindhoven, The Netherlands

Gerrit Schootstra
Philips Research Laboratories, Eindhoven, The Netherlands

Okko H. Bosgra
Mechanical Engineering Systems and Control Group, Delft University of Technology, Delft, The Netherlands

9.1 Friction Modeling and Compensation 195
Introduction • Friction Modeling • Simulation • Off-Line Friction Parameter Identification • Friction Compensation • Conclusion
References 208
9.2 Motion Control Systems 208
Introduction • System Elements and Operation • Stability Analysis Example • Motion Profiling • Tools for Motion Coordination • Design Example – Glue Dispensing
References 212
9.3 Ultra-High Precision Control 212
Introduction • System Description • System Identification and Modeling • Control Design • Example: Diamond turning machine • Conclusions • Defining Terms
References 230
9.4 Robust Control of a Compact Disc Mechanism 231
Introduction • Compact Disc Mechanism • Modeling • Performance Specification • μ-synthesis for the CD Player • Implementation Results • Conclusions
References 237

9.1 Friction Modeling and Compensation

Brian Armstrong, Department of Electrical Engineering and Computer Science, University of Wisconsin—Milwaukee, Milwaukee, WI
Carlos Canudas de Wit, Laboratoire d'Automatique de Grenoble, ENSIEG, Grenoble, France

9.1.1 Introduction

The successful implementation of friction compensation brings together aspects of servo control theory, tribology (the science of friction), machine design, and lubrication engineering. The design of friction compensation is thus intrinsically an interdisciplinary challenge. Surveys of contributions from the diverse fields that are important for friction modeling, motion analysis, and compensation are presented in [2], [10].

The challenge to good control posed by friction is often thought of as being stick-slip, which is an alternation between sliding and sticking due to static friction. Stick-slip is most common when integral control is used and can prevent a machine from ever reaching its intended goal position. However, other forms of frictional disturbance can be of equal or greater importance. The tracking error introduced by friction into multi-axis motion is an example. This error, called ***quadrature glitch*** is illustrated in Figure 9.1. The two-axis machine fails to accurately track the desired circular contour because as one axis goes through zero velocity, it is arrested for a moment by static friction while the other axis continues to move.

Even when frictional disturbances are eliminated, friction in mechanical servos may still impact cost and performance. As outlined in Section 9.1.5 , lubrication and hardware modification are two commonly used approaches to improving closed-loop performance. These approaches are grouped under problem avoidance, and a survey of engineers in industry [2] suggests that they are the most commonly used techniques for eliminating frictional disturbances. But these techniques have a cost that

0-8493-0054-3/00/$0.00+$.50

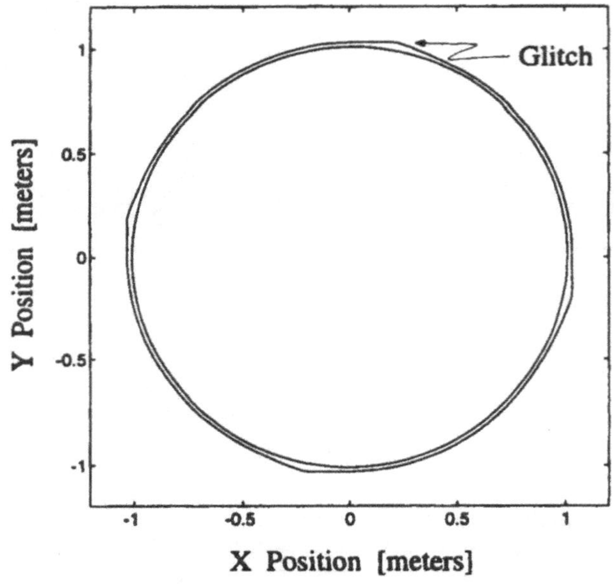

Figure 9.1 An example of quadrature glitch in 2-axis motion: X-Y position trace showing quadrature glitch. The desired trajectory is shown with 2% radial reduction to highlight the tracking error.

may be hidden if more effective servo compensation of friction is not considered. For example, special materials, called *friction materials* (see Section 9.1.5) are often used on the machine tool slideways to eliminate stick-slip. These materials have a high coulomb friction but a relatively lower static friction, and thus the machine slideway is not prone to stick-slip. But the higher coulomb friction of the slideway introduces increased power and energy requirements, which increases the initial costs for a larger drive and energy costs throughout the lifetime of the machine. More effective friction compensation by feedback control could provide higher performance at lower cost.

9.1.2 Friction Modeling

The simplest friction model has the instantaneous friction force, $F_f(t)$, expressed as a function of instantaneous sliding velocity, $v(t)$. Such a model may include coulomb, viscous, and/or static friction terms, which are described in Section 9.1.2. Typical examples are shown in Figure 9.2. The components of this simple model are given in Equations 9.1 to 9.3.

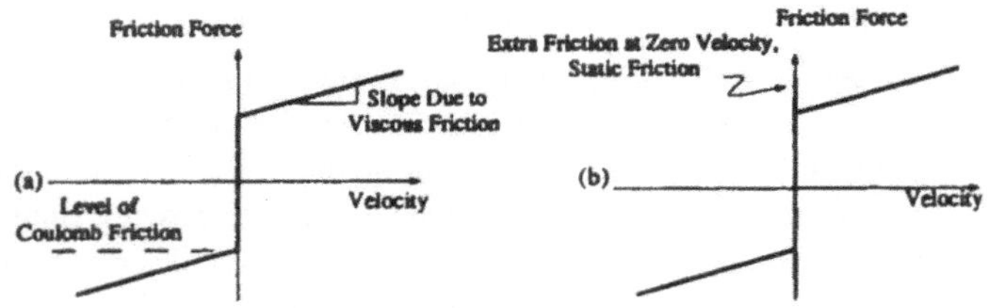

Figure 9.2 Overly simplistic, but nonetheless common, friction models: (a) coulomb + viscous friction model; (b) static + coulomb + viscous friction model.

Glossary of Terms and Effects

Coulomb friction: A force of constant magnitude, acting in the direction opposite to motion.

$$\text{When } v(t) \neq 0: \quad F_f(t) = -F_c \operatorname{sgn}(v(t)) \tag{9.1}$$

Viscous friction: A force proportional to velocity.

$$\text{When } v(t) \neq 0: \quad F_f(t) = -F_v v(t) \tag{9.2}$$

Static friction: Is not truly a force of friction, as it is neither dissipative nor a consequence of sliding. Static friction is a force of constraint [12].

$$\begin{aligned} \text{When } v(t) &= 0: \\ F_f &= \begin{cases} u(t); & |u(t)| \leq (F_c + F_s) \\ (F_c + F_s)\operatorname{sgn}(u(t)); & \text{otherwise} \end{cases} \end{aligned} \tag{9.3}$$

where $u(t)$ is the externally applied force.

For many servo applications, the simplest friction model–instantaneous friction as a function (any function) of instantaneous sliding velocity–is inadequate to accurately predict the interaction of control parameters and frictional disturbances to motion. In addition to coulomb, viscous, and static friction, dynamic friction must be considered. The state of the art in tribology does not yet provide a friction model derived from first principles. Four friction phenomena, however, are consistently observed in lubricated machines [10]:

1. **Stribeck friction or the Stribeck curve** is the negatively sloped and nonlinear friction-velocity characteristic occurring at low velocities for most lubricated and some dry contacts. The Stribeck curve is illustrated in Figure 9.3. The negatively sloped portion of the curve is an important contributor to stick-slip. Because of dynamic friction effects, instantaneous friction is not simply a function of instantaneous velocity. When velocity is steady, however, a steady level of friction is observed. This is friction as a function of steady-state velocity and gives the Stribeck curve.
2. **Rising static friction.** The force required for breakaway (the transition from not-sliding to sliding) varies with the time spent at zero velocity (dwell time) and with force application rate. The physics underlying rising static friction are not well understood, but experimental data and empirical models are available from the tribology literature [2] [10]. Rising static friction interacts significantly with stick-slip and is represented as a function of dwell time in the model presented in Section 9.1.2, and as a function of force application rate in the model presented in Section 9.1.2.

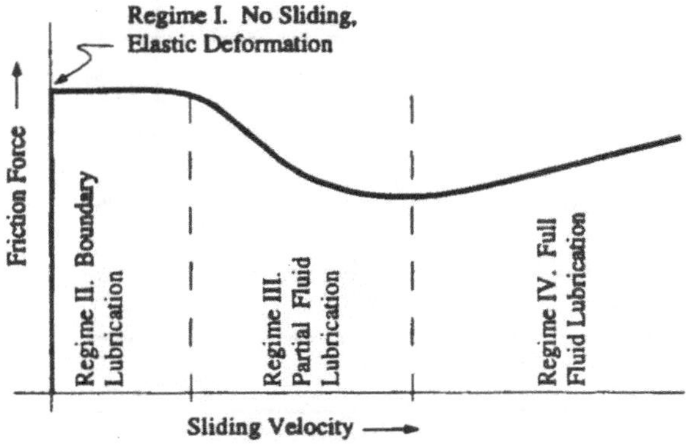

Figure 9.3 Friction as a function of steady-state velocity, the Stribeck curve. The four regimes of motion are described in section 5.1.a (from [2], courtesy of the publisher).

3. **Frictional memory** is the lag observed between changes in velocity or normal load and the corresponding change in friction force. It is most distinctly observed when there is partial fluid lubrication (partial fluid lubrication is a common condition in many machine elements, such as transmission and rolling bearings; see Section 9.1.5). As a result of frictional memory, instantaneous friction is a function of the history of sliding velocity and load as well as instantaneous velocity and load. An example of measured friction plotted against velocity is seen in Figure 9.4. The solid line indicates the friction as a function of *steady-state* sliding velocity and shows a single value for friction as a function of velocity. The solid line may be compared with Figure 9.3. The line marked *ooo* shows friction during intermittent sliding (stick-slip). The friction force is a multi-valued function of velocity: during acceleration, the friction force is higher, while during deceleration the friction force is lower. The hysteresis seen in Figure 9.4 indicates the presence of frictional memory.
4. **Presliding displacement** is the displacement of rolling or sliding contacts prior to true sliding. It arises due to elastic and/or plastic deformation of the contacting asperities and is seen in Figure 9.5. Because metal components make contact at small, isolated points called asperities, the tangential compliance of a friction contact (e.g., gear teeth in a transmission) may be substantially greater than the compliance of the bulk material. Presliding displacement in machine elements has been observed to be on the order of 1 to 5 micrometers. Because a small displacement in a bearing or between gear teeth may be amplified by a mechanism (for example, by a robot arm [1]), presliding displacement may give rise to significant output motions.

Friction models capturing these phenomena are described in Sections 9.1.2 and 9.1.2. In broad terms, these four dynamic

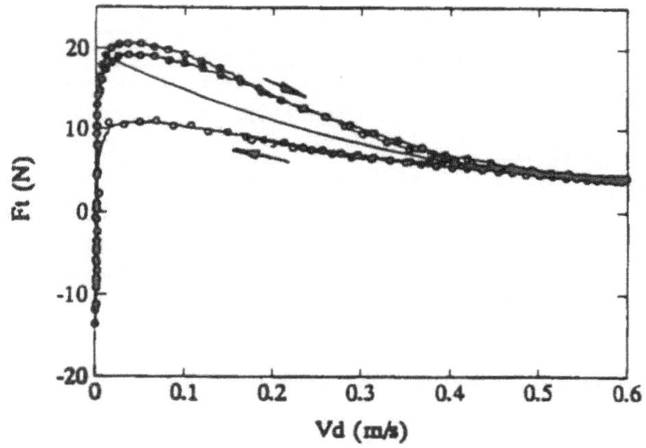

Figure 9.4 Measured friction versus sliding velocity: —, quasi-steady (equilibrium) sliding; *ooo*, intermittent sliding. (From Ibrahim, R.A. and Soom, A., Eds., *Friction-Induced Vibration, Chatter, Squeal, and Chaos, Proc. ASME Winter Annu. Meet.*, DE-Vol. 49, 139, 1992. With permission.)

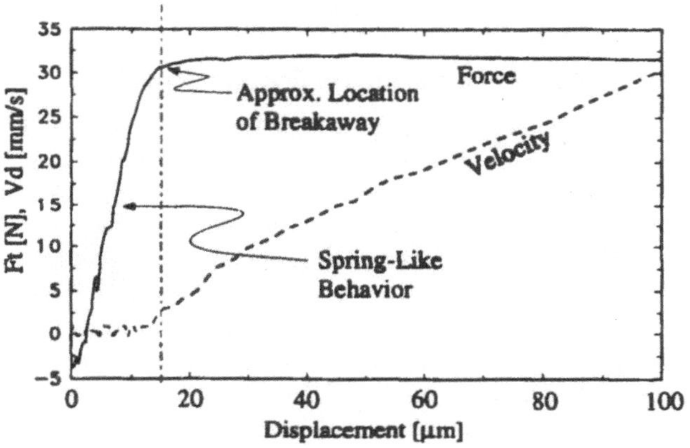

Figure 9.5 Force (—) and velocity (- -) versus displacement during the transition from static to sliding friction (breakaway). The spring-like behavior of the friction contact is seen as the linear force-displacement curve over the first 15μm of motion. (Adapted from [12].)

friction phenomena can be linked to behaviors observed in servo mechanisms [2]. These connections are presented in Table 9.1.

Additional Terms

Temporal friction phenomena: This term connotes both rising static friction and frictional memory.

Stick-slip: In the literature, stick-slip is used to refer to a broad range of frictional phenomena. Here, it is used to refer to a stable limit cycle arising during motion.

Hunting: With feedback control, a type of frictional limit cycle is possible that is not possible with passive systems: a limit cycle that arises while the net motion of the system is zero. This is called hunting and is most often associated with integral control.

TABLE 9.1 Observable Consequences of Dynamic Friction Phenomena

Dynamic Friction Phenomenon	Predicted/Observed Behavior
Stribeck friction	Needed to correctly predict initial conditions and system parameters leading to stick-slip
Rising static friction	Needed to correctly predict the interaction of velocity with the presence and amplitude of stick-slip
Frictional memory	Needed to correctly predict the interaction of stiffness (mechanical or feedback) with the presence of stick-slip
Presliding displacement	Needed to correctly predict motion during stick(e.g., during velocity reversal or before breakaway)

Standstill (or lost motion): Even where there is no stable limit cycle, the frictional disturbance to a servo system may be important. Standstill or lost motion refers to the frictional disturbance that arises when a system goes through zero velocity; it is seen in Figure 9.1. An ideal system would show continuous acceleration, but the system with friction may be arrested at zero velocity for a period of time. In machine tools, this is the frictional disturbance of greatest economic importance.

The Seven-Parameter Friction Model

The seven-parameter friction model [2] captures the four detailed friction phenomena, as well as coulomb, viscous, and static friction. The model has the advantage that the friction phenomena are explicitly represented with physically motivated model parameters. The model is presented in Equations 9.4 to 9.6, and the parameters are described in Table 9.2. The liability of the seven-parameter model is that it is not, in fact, a single integrated model, but a combination of three models: Equation 9.4 reflects the tangential force of constraint during stick; Equation 9.5 reflects frictional force during sliding; and Equation 9.6 represents a sampled process that reflects rising static friction.

Not-sliding (pre-sliding displacement):

$$F_f(x(t)) = -K_t x(t) \tag{9.4}$$

Sliding (coulomb + viscous + Stribeck curve friction with frictional memory):

$$F_f(v(t),t) = -\Bigg(F_C + F_v\,|v(t)| + F_s(\gamma,t_2)\frac{1}{1+\left(\frac{v(t-\tau_L)}{v_s}\right)^2}\Bigg)\operatorname{sgn}[v(t)] \tag{9.5}$$

Rising Static Friction (friction level at breakaway):

$$F_s(\gamma,t_2) = F_{s,a} + (F_{s,\infty} - F_{s,a})\frac{t_2}{t_2+\gamma} \tag{9.6}$$

where:

$F_f(.)$	=	instantaneous friction force
F_C	=	coulomb friction force*
F_v	=	viscous friction force*
F_s	=	magnitude of the Stribeck friction [frictional force at breakaway is given by $F_f(t = t_{\text{breakaway}}) = F_C + F_s$]
$F_{s,a}$	=	magnitude of the Stribeck friction at the end of the previous sliding period
$F_{s,\infty}$	=	magnitude of the Stribeck friction after a long time at rest (or with a slow application force)*
K_t	=	tangential stiffness of the static contact*
v_s	=	characteristic velocity of the Stribeck friction*
τ_L	=	time constant of frictional memory*
γ	=	temporal parameter of the rising static friction*
t_2	=	dwell time, time at zero velocity

(*) Marks friction model parameters; other variables are state variables.

Integrated Dynamic Friction Model

Canudas et al. [6] have proposed a friction model that is conceptually based on elasticity in the contact. They introduce the variable $z(t)$ to represent the average state of deformation in the contact. The friction is given by

$$F_f(t) = \sigma_0 z(t) + \sigma_1 \frac{dz(t)}{dt} + F_v v(t) \tag{9.7}$$

where $F_f(t)$ is the instantaneous friction and $v(t)$ is the contact sliding velocity. The state variable $z(t)$ is updated according to:

$$\frac{dz(t)}{dt} = v(t) - \frac{\sigma_0}{g(v(t))} z(t)|v(t)| \tag{9.8}$$

In steady sliding, $\dot{z}(t) = 0$, giving

$$z(t)|_{\text{steady-state}} = \frac{g(v(t))}{\sigma_0}\operatorname{sgn}[v(t)] \tag{9.9}$$

which then gives a steady-state friction:

$$F_f(t)\big|_{\dot{z}=0} = g(v(t)) + F_v v(t) \tag{9.10}$$

A parameterization that has been proposed to describe the nonlinear low-velocity friction is

$$g(v(t)) = F_C + F_s e^{-[v(t)/v_s]^2} \tag{9.11}$$

where F_c is the coulomb friction; F_s is magnitude of the Stribeck friction (the excess of static friction over coulomb friction); and v_s is the characteristic velocity of the Stribeck friction, approximately the velocity of the knee in Figure 9.3. With this description of $g(v(t))$, the model is characterized by six parameters: $\sigma_0, \sigma_1, F_v, F_C, F_s$, and v_s. For steady sliding, $[\dot{v}(t) = 0]$, the friction force (Figure 9.3) is given by:

$$\begin{aligned} F_{ss}(v(t)) &= g(v(t))\text{sgn}[v(t)] + F_v v \qquad (9.12) \\ &= F_c\text{sgn}[v(t)] + Fse^{-[v(t)/v_s]^2}\text{sgn}[v(t)] \\ &\quad + F_v v(t) \end{aligned}$$

When velocity is not constant, the dynamics of the model give rise to frictional memory, rising static friction, and presliding displacement.

The model has a number of desirable properties:

1. It captures presliding displacement, frictional memory, rising static friction, and the Stribeck curve in a single model without discontinuities.
2. The steady-state (friction–velocity) curve is captured by the function $g(v(t))$, which is chosen by the designer (e.g., Equation 9.11).
3. In simulation, the model is able to reproduce the data of a number of experimental investigations [6].

A difficulty associated with dynamic friction models lies with identifying the friction model parameters for a specific system. The static parameters involved in the function $g(v(t))$ and the viscous friction parameter F_v may be identified from experiments at constant velocity (see Section 9.1.4). The dynamic parameters σ_0 and σ_1 are more challenging to identify. These parameters are interrelated in their description of the physical phenomena, and their identification is made challenging by the fact that the state $z(t)$ is not physically measurable. A procedure for identifying these parameters is outlined in Section 9.1.4.

Magnitudes of Friction Parameters

The magnitudes of the friction model parameters naturally depend upon the mechanism and lubrication, but typical values may be offered, as seen in Table 9.2 (see [1], [2], [3], [12]). The friction force magnitudes, F_C, F_v, and $F_{s,\infty}$, are expressed as a function of normal force F_n; i.e., as coefficients of friction. Δ_x is the deflection before breakaway resulting from contact compliance.

In servo machines it is often impossible to know the magnitude of the normal force in sliding contacts. Examples of mechanisms with difficult-to-identify normal forces are motor brushes, gear teeth, and roller bearings, where the normal force is dependent on spring stiffness and wear, gear spacing, and bearing preload, respectively. For this reason, friction is often described for control design in terms of parameters with units of force, rather than as coefficients of friction.

In addition to the models described above, state variable friction models have been used to describe friction at very low velocities (μm/s). The state variable models are particularly suited to capture nonlinear low-velocity friction and frictional memory. Velocities of micrometers per second can be important in some control applications, such as wafer stepping or the machining of nonspherical optics (see [2][7]).

Friction that depends upon position has been observed in machines with gear-type transmissions. In general, mechanisms in which the normal force in sliding contacts varies during motion show position-dependent friction. Selected components of the Fourier transform and table lookup have been used to model the position-dependent friction. At least one study of friction compensation in an industrial robot has shown that it is important to model the position-dependent friction in order to accurately identify the detailed friction model parameters [1].

In practical machines, there are often many rubbing surfaces that contribute to the total friction: drive elements, seals, rotating electrical contacts, bearings, etc. In some mechanisms, a single interface may be the dominant contributor, as transmission elements often are. In other cases where there are several elements contributing at a comparable level, it may be impossible to identify their individual contributions without disassembling the machine. In these cases, it is often aggregate friction that is modeled.

The control designer faces a considerable challenge with respect to friction modeling. On the one hand, parameters of a dynamic friction model are at best difficult to identify while, on the other hand, recent theoretical investigations show that the dynamics of friction play an important role in determining frictional disturbances to motion and appropriate compensation [1], [5], [6], [7], [10], [12], [14]. Often, it is necessary to use a simplified friction model. The most common model used for control incorporates only coulomb friction, Equation 9.1. For machine tools, the Karnopp model, Equations 9.15 to 9.17, has been employed to represent static friction (e.g., [4]). And presliding displacement, Equation 9.4, has been modeled for precision control of pointing systems (e.g., [14]). Reference [1] provides an example showing how a detailed friction model can be used to achieve high-performance control, but also makes clear the technical challenges, including special sensing, associated with identifying the parameters of a detailed friction model. To date, there has been no systematic exploration of the trade-offs between model complexity and control performance.

9.1.3 Simulation

Simulation is the most widely used tool for predicting the behavior of friction compensation. The simulation of systems with friction is made challenging by the rapid change of friction as velocity goes through zero. In the simplest friction model, friction is discontinuous at zero velocity:

$$F_f(t) = -F_c\text{sgn}[v(t)] \qquad (9.13)$$

The discontinuity poses a difficulty for numerical integrators used in simulation. The difficulty can be addressed by "softening" the discontinuity, for example with

$$F_f(t) = -F_c\text{sgn}[v(t)]\left(1 - e^{-[|v(t)|/v_0]}\right) \qquad (9.14)$$

TABLE 9.2 Approximate Ranges of Detailed Friction Model Parameters for Metal-on-Metal Contacts Typical of Machines

	Parameter Range	Parameter Depends Principally Upon
F_C	$0.001 - 0.1x F_n$	Lubricant viscosity, contact geometry, and loading
F_v	0—very large	Lubricant viscosity, contact geometry, and loading
F_s, ∞	$0 - 0.1x F_n$	Boundary lubrication
k_t	$\frac{1}{\Delta_x} \times (F_s + F_C)$; $\Delta_x \simeq 1 - 50(\mu M)$	Material properties and surface finish
v_s	$0.00001 - 0.1$ (m/s)	Boundary lubrication, lubricant viscosity, material properties and surface finish, contact geometry, and loading
τ_L	$1 - 50$ (ms)	Lubricant viscosity, contact geometry, and loading
γ	$0 - 200$ (s)	Boundary lubrication

which produces a smooth curve through zero velocity. Models that are modified in this way, however, exhibit creep: small applied forces result in low but steady velocities. Some frictional contacts exhibit creep, but metal-on-metal contacts often exhibit a minimum applied force below which there is no motion.

The discontinuous friction model is a nonphysical simplification in the sense that a mechanical contact with distributed mass and compliance cannot exhibit an instantaneous change in force. Friction may be a discontinuous function of *steady-state* velocity (as are Figure 9.3, and Equation 9.11), but a system going through zero velocity is a transient event.

The integrated dynamic friction model (Section 9.1.2) uses an internal state (Equation 9.7) to represent the compliance of the contact and thereby avoids both the discontinuity in the applied forces and creep. The model has been used to reproduce, in simulation dynamic friction phenomena that have been experimentally observed. Another model that has been widely applied to simulation is the Karnopp friction model [11]. The Karnopp model solves the problems of discontinuity and creep by introducing a pseudo velocity and a finite neighborhood of zero velocity over which static friction is taken to apply. The pseudo velocity is integrated in the standard way

$$\dot{p}(t) = [F_a(t) - F_m(t)] \tag{9.15}$$

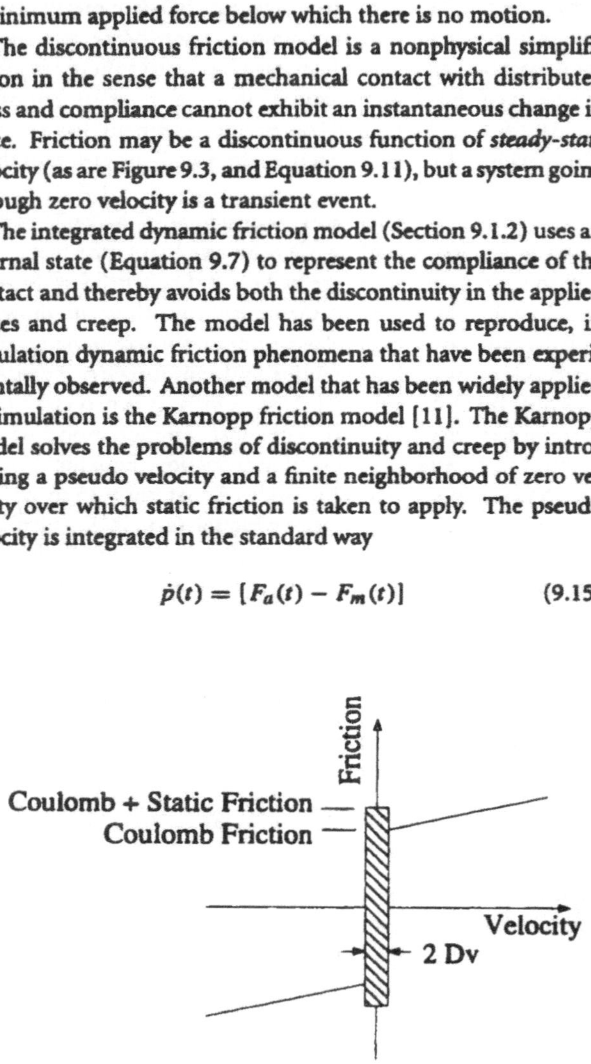

Figure 9.6 A friction-velocity representation of the Karnopp friction model (adapted from [11]).

where $p(t)$ is the pseudo velocity (identified with momentum in [11]), $F_a(t)$ is the applied force, and $F_m(t)$ is the modeled friction. The modeled friction is given by:

$$F_m(v(t), F_a(t)) = \begin{cases} -\text{sgn}[F_a(t)]\max[|F_a(t)|, (F_C + F_s)] & |v(t)| < D_V \\ -\text{sgn}[v(t)]F_C & |v(t)| \geq D_V \end{cases} \tag{9.16}$$

A small neighborhood of zero velocity is defined by D_V, as shown in Figure 9.6. Outside this neighborhood, friction is a function of velocity; coulomb friction is specified in Equation 9.16, but any friction-velocity curve could be used. Inside the neighborhood of zero velocity, friction is equal to and opposite the applied force up to the breakaway friction, and velocity is set to zero

$$v(t) = \begin{cases} 0, & |p(t)| < D_V \\ \frac{1}{M}p(t), & |p(t)| \geq D_V \end{cases} \tag{9.17}$$

where M is the mass.

The integration step must be short enough that at least one value of velocity falls in the range $|v(t)| < D_V$ At this point velocity is set to zero, according to Equation 9.17. If the applied force is less than the breakaway friction, $F_C + F_s$, then $\dot{p}(t) = 0$, and the system remains at rest. When $|F_a| > (F_C + F_s)$, $\dot{p}(t) \neq 0$ and, perhaps after some time, the condition of Equation 9.17 allows motion to begin. The Karnopp model represents static friction as applying over a region of low velocities rather than at the mathematical concept of zero velocity. The model thus eliminates the need to search for the exact point where velocity crosses zero.

While practical for simulation, and even feedback control (e.g., [4]), the Karnopp friction model is a simplification that neglects frictional memory, presliding displacement, and rising static friction. The simulation predictions are thus limited to gross motions; to accurately predict detailed motion, dynamic friction must be considered.

For any approach to simulating systems with friction, an integrator with variable time step size is important. The variable time step allows the integrator to take very short time steps near zero velocity, where friction is changing rapidly, and longer time steps elsewhere, where friction is more steady. Variable step size integration is standard in many simulation packages.

9.1.4 Off-Line Friction Parameter Identification

At this time, it is not possible to accurately predict the static and dynamic friction model parameters based on the specifications of a mechanism. Consequently, friction compensation methods that require a model require a method to identify the model parameters. The problem is one of nonlinear parameter identification, which has a large literature.

While roughly determining the coulomb and viscous friction parameters may entail only some straightforward constant force or constant velocity motions (see Section 9.1.4), determining the dynamic friction parameters often entails acceleration sensing or specialized approaches to parameter identification.

Position-Dependent Friction

Mechanisms that are spatially homogeneous, such as direct and belt-driven mechanisms, should not show a substantial position-dependent friction. But mechanisms with spatial inhomogeneities, such as gear drives, can show a large change in friction from one point to another, as seen in Figure 9.7 . The

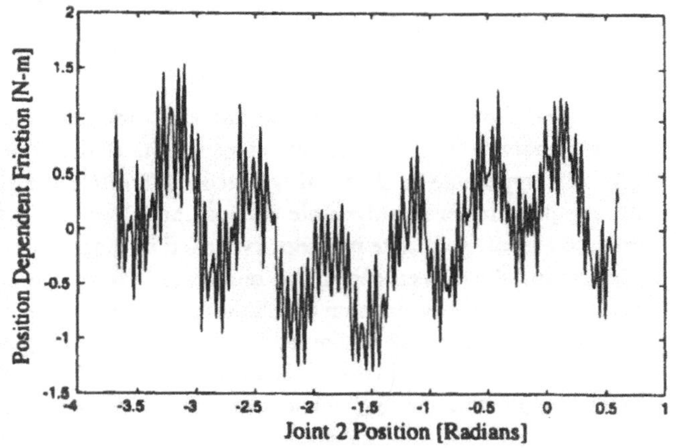

Figure 9.7 Position-dependent friction observed in joint 2 of a PUMA robot arm. The signal with a spatial period of approximately 0.7 radians corresponds to one rotation of the intermediate gear. One cycle of the predominant high-frequency signal corresponds to one rotation of the motor pinion gear. The 3 N-m peak-to-peak position dependent friction may be compared to the average Coulomb friction level of 12 N-m. (Adapted from [1].)

position-dependent friction can be measured by observing the breakaway friction, the level of force required to initiate motion, throughout the range of motion of the mechanism. To measure breakaway friction:

1. The mechanism is allowed to come to rest at the point where friction will be measured.
2. Force (torque) is applied according to a specified curve (generally a ramp).
3. The onset of motion is detected.

The applied force corresponding to the onset of motion is the breakaway friction. To achieve repeatable results, it is necessary to consider these factors:

1. In many machines, friction is observed to be higher after a period of inactivity and to decrease with the first few motions. Bringing lubricants into their steady-state condition accounts for this transient. To repeatably measure any friction parameter, the machine must be "warmed up" by typical motions prior to measurement.
2. The highest spatial frequencies present in the position-dependent friction may be quite high. Spatial sampling must be sufficient to avoid aliasing.
3. Force must be applied consistently. Because the breakaway force depends upon the dwell time and rate of force application (see Section 9.1.2) these variables must be controlled.
4. The method used to detect breakaway must be well selected. The onset of motion is not a simple matter, corresponding to the fact that motion occurs before true sliding begins (presliding displacement, see Figure 9.5). Detecting the first shaft encoder pulse after the beginning of force application, for example, is a very poor detection of sliding because the pulse can arise with presliding displacement at any point on the force application curve. Tests based on velocity or on total motion from the initial position have been used.

Gear drives can exhibit significant friction variation with motions corresponding to a tenth of the highest gear pitch. For example, in a mechanism with 15 : 1 reduction and 18 teeth on the motor pinion gear, $15x18x10 = 2700$ cycles of position-dependent friction per revolution of the mechanism. Capturing these variations during motion poses a significant sensing bandwidth challenge. Breakaway friction is a suitable tool for identifying position dependency in the sliding friction because it can be measured at many points, corresponding to a high spatial sampling.

Friction as a Function of Steady-State Velocity

The steady-state friction-velocity curve (Figure 9.3) can be observed with a series of constant velocity motions. The motions can be carried out with either open- or closed-loop control. If there is significant position-dependent friction, feedforward compensation of the position-dependent friction will improve the accuracy of the steady-sliding friction model. An example is shown in Figure 9.8, where the data show the nonlinear low-velocity friction. No measurements were taken in the region of the steep friction-velocity curve because stick-slip arises in the tested mechanism, even with acceleration feedback, and constant-velocity sliding at low velocities was not possible. The beginning of the negative velocity friction curve observed during constant-velocity sliding and the value of breakaway friction measured using the breakaway experiment (Section 9.1.4) would be sufficient to approximately identify the parameters of the Stribeck curve.

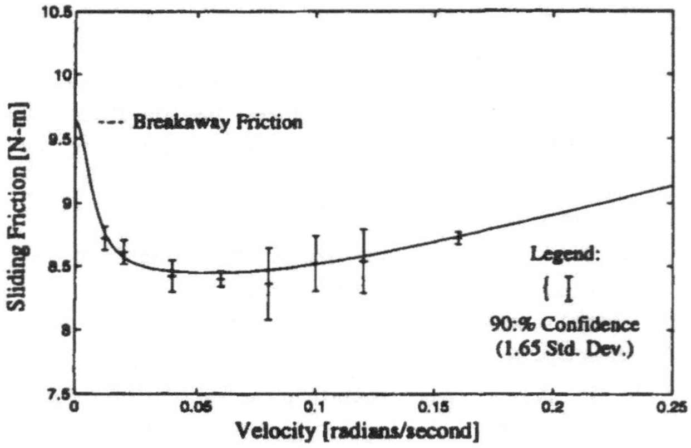

Figure 9.8 Friction as a function of velocity in joint 1 of a PUMA robot. The data points indicate the average level of friction observed in constant-velocity motions using a stiff velocity feedback. The solid curve is given by the seven parameter friction model, Equation 9.5. (Adapted from [1].)

Dynamic Friction Model Parameters

The dynamic friction phenomena–Stribeck friction, frictional lag, rising static friction, and presliding displacement–are difficult to observe directly. They operate over regions of very low velocity, in which mechanism motion may be unstable, or over short time intervals or small distances. Direct observation of the phenomena and measurement of the parameters is possible with acceleration sensing and careful elimination of sources of measurement error [1], [12].

In spite of the fact that sensitive sensing is required to directly observe the dynamic friction phenomena, their impact on motion may be substantial, motivating both their correct modeling during the design of compensation and providing a basis for identification from observable system motions. Figure 9.9 shows the simulated friction, as well as position and velocity curves, for a motor servo and two different sets of friction parameters. The integrated dynamic friction model (Section 9.1.2) was used for the simulation, with the parameters specified in Table 9.3.

TABLE 9.3 Parameters Used for the Simulations of Figure 9.9

$F_c = 0.292$(N·m)	$F_v = 0.0113$ (N·m-s/rad)
$F_s = 0.043$(N·m)	$v_s = 0.95$ (rad/s)

Case 1 :	$\sigma_0 = 2.0$	$\sigma_1 = 0.1$
Case 2 :	$\sigma_0 = 40.0$	$\sigma_1 = 2.0$

The two friction models are seen to give distinct motions, producing differences that can be used to identify the friction

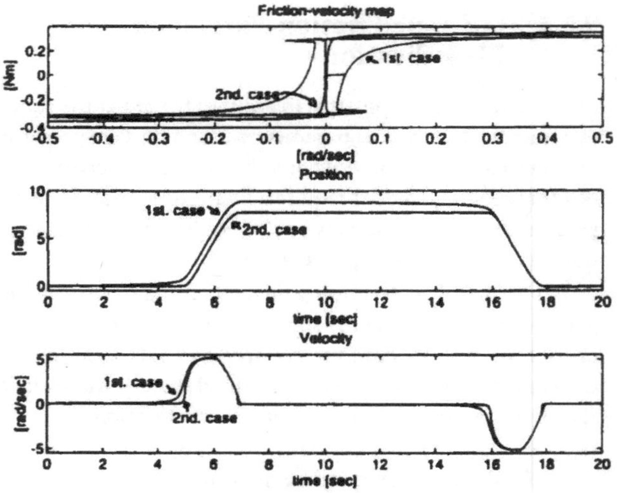

Figure 9.9 Friction and motion dependence on σ_0 and σ_1.

model parameters. The parameter identification is a nonlinear optimization problem; standard engineering software packages provide suitable routines.

The low-velocity portion of the friction-velocity curve, as well as frictional memory, presliding displacement, and rising static friction parameters, were identified in an industrial robot using open-loop motions and variable compliance. See [1] for the details of the experiment and identification.

In spite of progress in the area of modeling and demonstrations of parameter identification, there is not, at this time, a generally applicable or thoroughly tested method for identifying the parameters of a dynamic friction model. None of the methods reported to date has been evaluated in more than one application. Friction modeling thus remains, to a large extent, an application-specific exercise in which the designer identifies friction phenomena important for achieving specific design goals. Tribology and the models presented in Section 9.1.2 offer a guide to friction phenomena that will be present in a lubricated mechanism. Because of the challenges, sophisticated and successful friction compensation has been achieved in many cases without a dynamic friction model, but at the cost of entirely empirical development and tuning. One can say only that more research is needed in this area.

9.1.5 Friction Compensation

Friction compensation techniques are broken down into three categories: problem avoidance, nonmodel-based compensation techniques, and model-based compensation techniques. Problem avoidance refers to modifications to the system or its lubrication that reduce frictional disturbances. These changes often involve machine design or lubrication engineering and may not seem the domain of the control engineer. But system aspects that play a large role in the closed-loop performance, particularly the detailed chemistry of lubrication, may not have been adequately considered prior to the appearance of frictional disturbance to motion (imprecise control), and it may be up to the control engineer to suggest the use of friction modifiers in the lubricant. The division of control-based compensation techniques into model-based and nonmodel-based reflects the challenge associated with developing an accurate friction model.

Problem Avoidance

Lubricant Modification For reasons that relate to service life and performance, systematic lubrication is common in servo machines. The nature of sliding between lubricated metal contacts depends upon the sliding velocity and distance traveled. When a servo goes from standstill to rapid motion, the physics of friction transition from:

- Full solid-to-solid contact without sliding, motion by elastic deformation (presliding displacement) To solid-to-solid contact with sliding (boundary lubrication)
- A mix of fluid lubrication and solid-to-solid contact (partial fluid lubrication) To full fluid lubrication (oil or grease supports the entire load; elastohydrodynamic or hydrodynamic lubrication depending on contact geometry)

The physics of these processes are quite different from one another. The wide variety of physical processes and the difficulty of ascertaining which are active at any moment explain, in part, why a complete description of friction has been so elusive: the model must reflect all of these phenomena. Typical ranges for the friction coefficients in these different regimes are illustrated in Figure 9.10.

When sliding takes place at low velocities, actual shearing of solid material plays an important role in determining the friction. If the surfaces in contact are extraordinarily clean, the shearing takes place in the bulk material (e.g., steel) and the coefficient of friction is extremely high. More commonly, the sliding surfaces are coated with a thin boundary layer (typical thickness, 0.1 μm) of oxides or lubricants, and the shearing takes place in this layer.

Customarily, additives are present in machine lubricants, which bind to the surface and form the boundary layer, putting it under the control of the lubrication engineer. These additives constitute a small fraction of the total lubricant and are specific to the materials to which they bind. Friction modifiers are boundary lubricants that are specifically formulated to affect the coefficient of friction [9]. The control engineer should be aware of the possibilities, because lubrication is normally specified to maximize machine life, and friction modification is not always a priority of the lubrication engineer.

Hardware Modification The most common hardware modifications to reduce frictional disturbances relate to increasing stiffness or reducing mass. These modifications permit higher gains and help to increase the natural frequency of the mechanism in closed loop. The greater stiffness reduces the impact of friction directly, while a higher natural frequency interacts with frictional memory to reduce or eliminate stick-slip [7].

Other hardware modifications include special low-friction bearings and the use of "friction materials," such as Rulon^R, in machine tool slideways to eliminate stick-slip.

Nonmodel-Based Compensation Techniques

Modifications to Integral Control Integral control can reduce steady-state errors, including those introduced by friction in constant-velocity applications. When the system trajectory encounters velocity reversal, however, a simple integral control term can increase rather than reduce the frictional disturbance. A number of modifications to integral control are used reduce the impact of friction. Their application depends not only on system characteristics, but also upon the desired motions and that aspects of possible frictional disturbances that are most critical.

Position-error dead band. Perhaps the most common modification, a dead band in the input to the integrator eliminates hunting, but introduces a threshold in the precision with which a servo can be positioned. The dead band also introduces a nonlinearity, which complicates analysis. It can be modified in various ways, such as scaling the dead band by a velocity term to reduce its affect during tracking. The integrator with dead band does not reduce dynamic disturbances, such as quadrature glitch.

Lag compensation. Moving the compensator pole off the origin by the use of lag compensation with high but finite dc gain accomplishes something of the same end as a position-error dead band, without introducing a nonlinearity.

Integrator resetting. When static friction is higher than coulomb friction (a sign of ineffective boundary lubrication), the reduced friction following breakaway is overcompensated by the integral control action, and overshoot may result. In applications such as machine tools, which have little tolerance for overshoot, the error integrator can be reset when motion is detected.

Multiplying the integrator term by the sign of velocity. When there is a velocity reversal and the friction force changes direction, integral control may compound rather than compensate for coulomb friction. This behavior enlarges (but does not create) quadrature glitch. To compensate for this effect, integral control can be multiplied by the sign of velocity, a technique that depends upon coulomb friction dominating the integrated error signal. The sign of desired or reference model velocity is often used; if measured or estimated velocity is used, the modification introduces a high-gain nonlinearity into the servo loop.

High Servo Gains (Stiff Position and Velocity Control) Linearizing the Stribeck friction curve (Figure 9.3) about a point in the negatively sloped region gives a "negative viscous friction" effective during sliding at velocities in the partial fluid lubrication regime. Typically, the negative viscous friction operates over a small range of low velocities and gives a much greater destabilizing influence than can be directly compensated by velocity

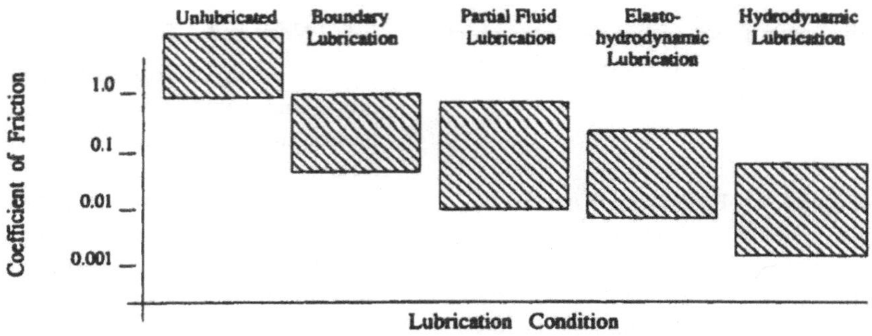

Figure 9.10 Typical ranges for friction in machines, corresponding to the friction process (adapted from [3]).

feedback. The negative viscous friction, not static friction, is often the greatest contributor to stick-slip.

As always, high servo gains reduce motion errors directly. In addition, stiff position control interacts with negative viscous friction and frictional memory to create a critical stiffness above which stick-slip is eliminated [7]. In a second-order system with proportional derivative (PD) control and frictional memory modeled as a time lag, as in Equation 9.5, a critical stiffness above which stick-slip is extinguished is given by

$$k_{cr} = M\frac{\pi^2}{\tau_L^2} \qquad (9.18)$$

where τ_L is the time lag of frictional memory in seconds, and M is the mechanism mass [2].

Qualitatively, the elimination of stick-slip at high stiffness can be understood as the converse of the destabilizing effect of transport lag in a normally stable feedback loop: the *destabilizing* effect of negative viscous friction operates through the delay of frictional memory, and the system is stabilized when the delay is comparable to the natural frequency of the system. Achieving the required feedback stiffness may require a very stiff mechanical system. When increasing stiffness extinguishes stick-slip, Equation 9.18 can be used to estimate the magnitude of the frictional memory.

In some cases, variable structure control with very high damping at low velocities counters the influence of nonlinear low-velocity friction and permits smooth motion where stick-slip might otherwise be observed. Machine and lubricant characteristics determine the shape of the low-velocity friction-velocity curve (see Table 9.1), which determines the required velocity feedback for smooth motion.

Learning Control Learning control, sometimes called repetitive control, generally takes the form of a table of corrections to be added to the control signal during the execution of a specific motion. The table of corrections is developed during repetitions of the specific motion to be compensated. This type of adaptive compensation is currently available on some machine tool controllers where the typical task is repetitive and precision is at a premium. The table of corrections includes inertial as well as friction forces. This type of adaptive control is grouped with nonmodel-based control because no explicit model of friction is present in the system. For further discussion and references, see [2].

Joint Torque Control Reductions of 30:1 in apparent friction have been reported using joint torque control [2], a sensor-based technique that encloses the actuator-transmission subsystem in a feedback loop to make it behave more nearly as an ideal torque source. Disturbances due to undesirable actuator characteristics (friction, ripple, etc.) or transmission behaviors (friction, flexibility, inhomogeneities, etc.) can be significantly reduced by sensing and high-gain feedback. The basic structure is shown in Figure 9.11; an inner torque loop functions to make the applied torque, T_a, follow the command torque, T_c.

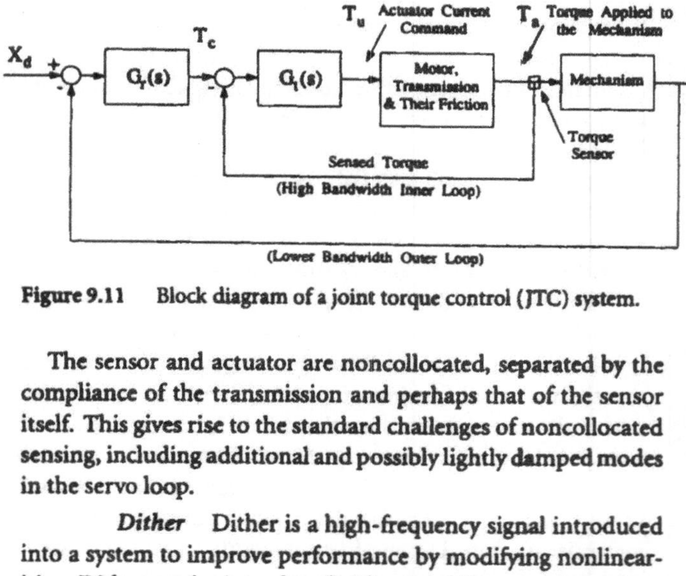

Figure 9.11 Block diagram of a joint torque control (JTC) system.

The sensor and actuator are noncollocated, separated by the compliance of the transmission and perhaps that of the sensor itself. This gives rise to the standard challenges of noncollocated sensing, including additional and possibly lightly damped modes in the servo loop.

Dither Dither is a high-frequency signal introduced into a system to improve performance by modifying nonlinearities. Dither can be introduced either into the control signal, as is often done with the valves of hydraulic servos, or by external mechanical vibrators, as is sometimes done on large pointing systems.

An important distinction arises in whether the dither force acts along a line parallel to the line of motion of the friction interface or normal to it, as shown in Figure 9.12. The effect of parallel dither is to modify the influence of friction (by averaging the nonlinearity); the effect of vibrations normal to the contact is to modify the friction itself (by reducing the friction coefficient). Dither on the control input ordinarily generates vibrations parallel to the friction interface, while dither applied by an external vibrator may be oriented in any direction.

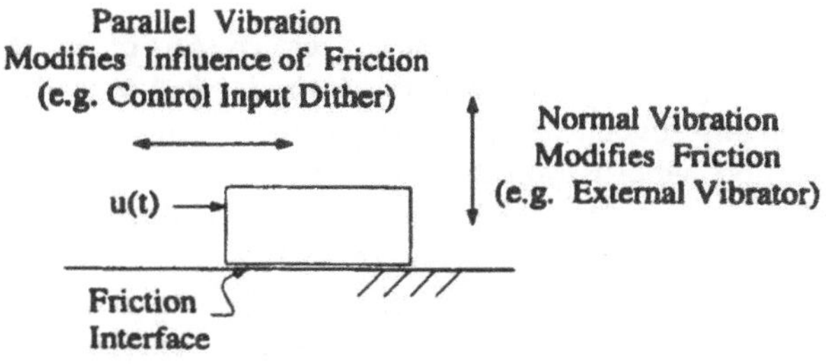

Figure 9.12 Direction and influence of dither.

Working with a simple coulomb plus viscous friction model, one would not expect friction to be reduced by normal vibrations, so long as contact is not broken; but because of contact compliance (the origin of presliding displacement), more sliding occurs during periods of reduced loading and less during periods of increased loading, which reduces the average friction. Reduction of the coefficient of friction to one third its undithered value has been reported.

Inverse Describing Function Techniques The describing function is an approximate, amplitude-dependent transfer function of a nonlinear element. The inverse describing function is a synthesis procedure that can be used when an amplitude-dependent transfer function is known and the corresponding time domain function is sought.

The inverse describing function has been used to synthesize nonlinear controllers that compensate for friction [13]. Nonlinear functions $f_P(e)$, $f_I(e)$, and $f_D(\dot{y})$ are introduced into the P, I, and D paths of proportional-integral-derivative (PID) control, as shown in Figure 9.13. The synthesis procedure is outlined in Figure 9.14. The method is included with nonmodel-based compensation because, while a friction model is required for the synthesis, it does not appear explicitly in the resulting controller.

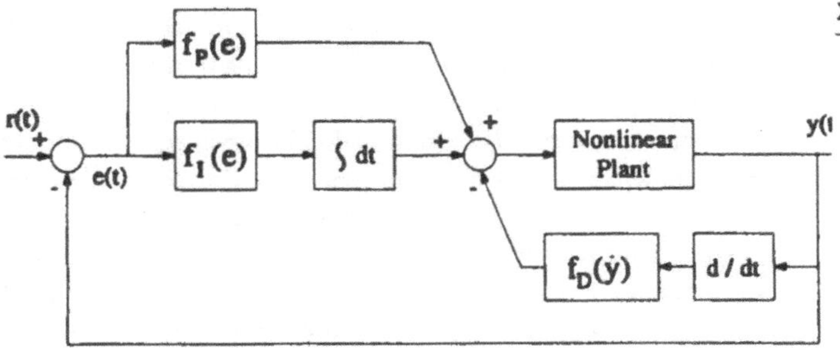

Figure 9.13 Nonlinear PI + tachometer structure.

The inverse describing function method offers the advantages that it has a systematic design procedure and it can be applied to systems with relatively complicated linear dynamics. Significant improvement in transient response is reported for a fourth-order system with flexible modes [13].

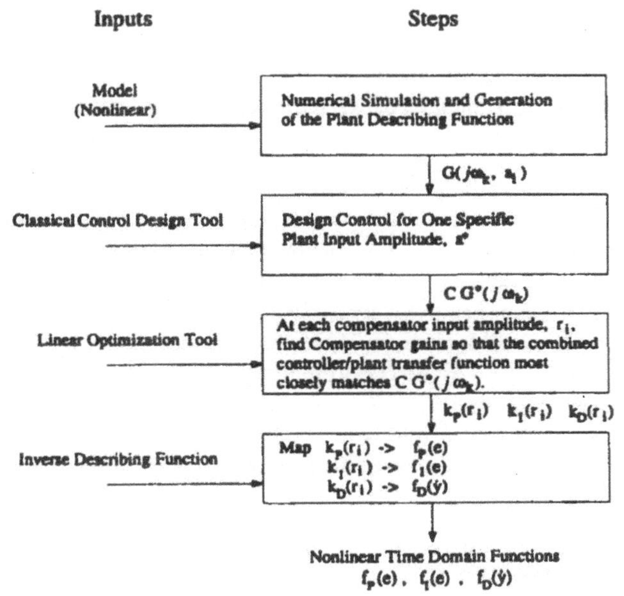

Figure 9.14 Outline of inverse describing function based synthesis of nonlinear friction compensation.

Model-Based Compensation Techniques

When friction can be predicted with sufficient accuracy, it can be compensated by feedforward application of an equal and opposite force. The basic block diagram for such a construction is shown in Figure 9.15. Examples of model-based compensation are provided in [1],[2] [4],[5],[6],[8],[14]. Compensation is addressed in the next section; questions of tuning the friction model parameters off-line and adaptively are discussed in Sections 9.1.4 and 9.1.5 respectively.

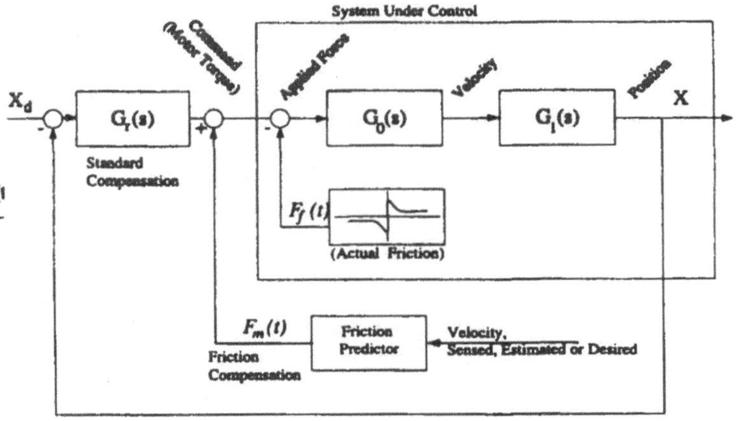

Figure 9.15 Model-based feedforward compensation for friction. (From [2], courtesy of the publisher.)

Compensation Three requirements for feedforward compensation are

1. An accurate friction model
2. Adequate control bandwidth
3. Stiff coupling between the actuator and friction element

The third item merits special consideration: significant friction often arises with the actuator itself, or perhaps with a transmission, in which case the friction may be stiffly coupled to the available control-based compensation. When the friction arises with components that are flexibly coupled to the actuator, it may not be possible to cancel friction forces with actuator commands. Some of the most successful applications of model-based friction compensation are in optical pointing and tracking systems, which can be made very stiff (e.g., [14]).

The important axes of distinction in model-based compensation are

1. The character and completeness of the friction model
2. Whether sensed, estimated, or desired velocity is used as input to the friction model

The most commonly encountered friction model used for compensation comprises only a coulomb friction term, with two friction levels, one for forward and the other for reverse motion. Industrially available machine tool controllers, for example, incorporate coulomb friction feedforward. To guard against instability introduced by overcompensation, the coulomb friction parameter is often tuned to a value less than the anticipated friction level.

The use of sensed or estimated velocity by the friction model closes a second feedback loop in Figure 9.15. If desired or reference model velocity is used, the friction compensation becomes a feedforward term. Because the coulomb friction model is discontinuous at zero velocity, and any friction model will show a rapid change in friction when passing through zero velocity, the use of sensed or estimated velocity can lead to stability problems. A survey of engineers in industry [2] indicates that the use of desired velocity is most common.

Additional friction model terms that have been used in compensation include:

1. Static friction
2. Frictional memory and presliding displacement
3. Position dependence

Static friction. Static friction may be represented in a model like the Karnopp model (see Section 9.1.3 and Figure 9.6). The most substantial demonstration of static friction compensation has been presented by Brandenburg and his co-workers [4] and is discussed in Section 9.1.5.

Frictional memory and presliding displacement. Examining Figure 9.5, it is seen that providing the full coulomb friction compensation for motions near the zero crossing of velocity would overcompensate friction. To prevent overcompensation, one industrial servo drive controller with coulomb friction compensation effectively implements the inverse of the Karnopp friction model

$$F_m(t) = \begin{cases} F_C \mathrm{sgn}[v(t)], & t_{zc} > t_{zc}^0 \\ 0, & t_{zc} \le t_{zc}^0 \end{cases} \tag{9.19}$$

where t_{zc} is the time since the last zero crossing of velocity, and t_{zc}^0 is a parameter that determines the interval over which the coulomb friction compensation is suppressed. The coulomb friction compensation could also be controlled by accumulated displacement since the last zero crossing of velocity (note that friction is plotted as a function of displacement in Figure 9.5) or as a function of time divided by acceleration (applied torque).

Friction compensation has also been demonstrated that incorporates a low-pass filter in the friction model and thus eliminates the discontinuity at zero velocity. One construction is given by

$$F_f(t) + \tau \frac{dF_f(t)}{dt} = F_C \mathrm{sgn}[v(t)] \tag{9.20}$$

where τ is a time constant. Walrath [14] reports an application of friction compensation based on Equation 9.20 that results in a 5:1 improvement in RMS pointing accuracy of a pointing telescope. Because of the accuracy requirement and the bandwidth of disturbances arising when the telescope is mounted in a vehicle, the tracking telescope pointing provides a very rigorous test of friction compensation. In the system described, τ is a function of the acceleration during the zero crossing, given by

$$\tau = \frac{1}{a_1 + a_2 \dot{v}(t)} \tag{9.21}$$

where a_1 and a_2 are empirically tuned parameters, and $\dot{v}(t)$ is estimated from the applied motor torque.

Position-dependent friction. In mechanisms with spatial inhomogeneities, such as gear teeth, friction may show substantial, systematic variation with position. Some machine tool controllers include an "anti-backlash" compensation that is implemented using table lookup. Friction is known to influence the table identification and, thus, the compensation. Explicit identification and feedforward compensation of position-dependent friction can also be done [1],[2].

Adaptive Control, Introduction Adaptive control has been proposed and demonstrated in many forms; when applied to friction compensation, adaptation offers the ability to track changes in friction. Adaptive friction compensation is indirect when an on-line identification mechanism explicitly estimates the parameters of the friction model. The advantage of indirect adaptive control is that the quality of the estimated parameters can be verified before they are used in the compensation loop. Typical examples of such verification are to test the sign of the parameters or their range of variation. By introducing an additional supervisory loop, the identified parameters can be restricted to a predetermined range.

Direct adaptive compensation rests on a different philosophy: no explicit on-line friction modeling is involved, but rather controller gains are directly adapted to minimize tracking error. Because there is no explicit friction prediction, it is more difficult to supervise the adaptive process to assure that the control reflects a physically feasible friction model or that it will give suitable transient behavior or noise rejection.

Direct Adaptive Control Model reference adaptive control (MRAC) is described elsewhere in this handbook. Perhaps the simplest implementation of model reference coulomb friction compensation is given by

$$\begin{aligned} u(t) &= (\text{PD control}) + \hat{F}_c(t)\text{sgn}[v(t)] \\ \hat{F}_c(t) &= C_1 \int_0^t \dot{e}(t)\text{sgn}[v(t)]dt \\ &\quad + C_2 \int_0^t e(t)\text{sgn}[v(t)]dt \\ \dot{e}(t) &= v_m(t) - v(t) \end{aligned} \tag{9.22}$$

where $v(t)$ is velocity; $v_m(t)$ is the reference model velocity; and $\hat{F}_c(t)$ is a coulomb friction compensation parameter. The parameters C_1 and C_2 are chosen by the designer and can be tuned to achieve the desired dynamics of the adaptive process. A Lyapunov function will show that the (idealized) process will converge for $C_1, C_2 > 0$.

Brandenburg and his co-workers [4] have carried out a thorough investigation of friction compensation in a two-mass system with backlash and flexibility (see also [2] for additional citations). They employ an MRAC structure to adapt the parameters of a coulomb friction-compensating disturbance observer. Without friction compensation, the system exhibits two stick-slip limit cycles. The MRAC design is based on a Lyapunov function; the result is a friction compensation made by applying a lag (PI) filter to $\dot{e}(t)$, the difference between model and true velocity. Combined with integrator dead band, their algorithm eliminates stick-slip and reduces lost motion during velocity reversal by a factor of 5.

Indirect Adaptive Control When the friction model is linear in the unknown parameters, it can be put in the form

$$F_f(t) = \theta^T \Phi(t) \tag{9.23}$$

where θ is the parameter vector and $\Phi(t)$ is the regressor vector. The regressor vector depends on the system state and can include nonlinear functions. For example, a model that includes the Stribeck curve is given by [5]:

$$\begin{aligned} \theta^T &= [F_c, F_s, F_v] \\ \Phi(t) &= \left[\text{sgn}(v(t)), \frac{1}{1 + (v(t)/v_s)^2}, v(t)\right] \end{aligned} \tag{9.24}$$

Assuming the simplest mechanism dynamics

$$M\dot{v}(t) = u(t) - F_f(t) \tag{9.25}$$

where M is the mass and $u(t)$ is the force or torque command, and sampling at time $t = kT$, the friction prediction error is given by:

$$e_k = u_k - M\dot{v}_k - \theta^T \Phi_k \tag{9.26}$$

To avoid the explicit measurement (or calculation) of the acceleration, the friction prediction error can be based on a filtered model. By applying a stable, low-pass filter, $F(s)$, to each side of Equation 9.25, the filtered mechanism dynamics are given, and the filtered prediction error is

$$e_k = \tilde{u}_k - M\tilde{\dot{v}}_k - \theta^T \tilde{\Phi}_k \tag{9.27}$$

where $\tilde{}$ indicates a low-pass filtered signal.

An update equation for the parameter estimate is constructed

$$\hat{\theta}_k = \hat{\theta}_{k-1} + \lambda_k P_k \tilde{\Phi}_k e_k \tag{9.28}$$

where λ_k is a rate gain and P_k is the inverse of the input correlation matrix for the recursive least squares algorithm (RLS) or the identity matrix for the least mean squared algorithm (LMS).

The control law is now implemented as

$$u(t) = (\text{Standard control}) + \hat{\theta}^T \Phi(t) \tag{9.29}$$

where $\Phi(t)$ is used for compensation, rather than the filtered $\tilde{\Phi}(t)$.

During the implementation of the estimation loop the following points are important:

1. The estimation should be frozen when operating conditions are not suitable for friction identification [e.g., $v(t) = 0$; this is the persistent excitation issue].
2. The sign of the estimated parameters should be always positive.
3. The compensation can be scaled down to avoid over-compensation.
4. The sign of the reference velocity can be used in place of the sign of the measured velocity in Equation 9.24, when $v(t)$ is small.

An alternative approach has been presented by Friedland and Park [8], who justify an update law that does not depend on acceleration measurement or estimation. The friction compensation is given by

$M\dot{v}(t) = u(t) - F_f(v(t), F_c^*$	[System Dynamics, single mass]
$F_f\left(v(t), \hat{F}_c\right) = \hat{F}_c\text{sgn}[v(t)]$	[Friction Model]
$u(t) = (\text{Standard Control}) + \hat{F}_c\text{sgn}[v(t)]$	[Control Law]
$\hat{F}_c = z(t) - k\lvert v(t)\rvert^{\mu}$	[Friction Estimator]
$\dot{z}(t) = k\mu\lvert v(t)\rvert^{\mu-1} \frac{1}{M}\left[u(t) - f\left(v(t), \hat{F}_c\right)\right]\text{sgn}[v(t)]$	[Friction Estimator Update Law]

where M is system mass; $u(t)$ is the control input; $z(t)$ is given by the friction estimator update law; $\hat{F}_c$ is the estimated coulomb friction; and μ and k are tunable gains. Defining the model misadjustment

$$e(t) = F_c^* - \hat{F}_c \tag{9.30}$$

one finds that

$$\dot{e}(t) = -k\mu\lvert v(t)\rvert^{\mu-1} e(t) \tag{9.31}$$

making $e = 0$ the stable fixed point of the process. The algorithm significantly improves dynamic response [4]; experimental results are presented in subsequent papers (see [2] for citations).

9.1.6 Conclusion

Achieving machine performance free from frictional disturbances is an interdisciplinary challenge; issues of machine design, lubricant selection, and feedback control all must be considered to cost effectively achieve smooth motion. Today, the frictional phenomena that should appear in a dynamic friction model are well understood and empirical models of dynamic friction are available, but tools for identifying friction in specific machines are lacking.

Many friction compensation techniques have been reported in both the research literature and industrial applications. Because a detailed friction model is often difficult to obtain, many of these compensation techniques have been empirically developed and tuned. Where available, reported values of performance improvement have been presented here. Reports of performance are not always available or easily compared because, to date, there has not been a controlled comparison of the effectiveness of the many demonstrated techniques for friction compensation. For many applications, it is up to the designer to seek out application-specific literature to learn what has been done before.

References

[1] Armstrong-Hélouvry, B., *Control of Machines with Friction*, Kluwer Academic Press, Boston, MA, 1991.

[2] Armstrong-Hélouvry, B., Dupont, P., and Canudas de Wit, C., A survey of models, analysis tools and compensation methods for the control of machines with friction, *Automatica*, 30(7), 1083–1138, 1994.

[3] Bowden, F.P. and Tabor, D. *Friction — An Introduction to Tribology*, Anchor Press/Doubleday, Reprinted 1982, Krieger Publishing Co., Malabar, 1973.

[4] Brandenburg, G. and Schäfer, U., Influence and partial compensation of simultaneously acting backlash and coulomb friction in a position- and speed-controlled elastic two-mass system, in *Proc. 2nd Int. Conf. Electrical Drives*, Poiana Brasov, September 1988.

[5] Canudas de Wit, C., Noel, P., Aubin, A., and Brogliato, B., Adaptive friction compensation in robot manipulators: low-velocities, *Int. J. Robotics Res.*, 10(3), 189–99, 1991.

[6] Canudas de Wit, C., Olsson, H., Aström, K.J., and Lischinsky, P., A new model for control of systems with friction, *IEEE Trans. Autom. Control*, in press, 1994.

[7] Dupont, P.E., Avoiding stick-slip through PD control, *IEEE Trans. Autom. Control*, 39(5), 1094–97, 1994.

[8] Friedland, B. and Park, Y.-J., On adaptive friction compensation, *IEEE Trans. Autom. Control*, 37(10), 1609–12, 1992.

[9] Fuller, D.D., *Theory and Practice of Lubrication for Engineers*, John Wiley & Sons, Inc., New York, 1984.

[10] Ibrahim, R.A. and Rivin, E., Eds., Special issue on friction induced vibration, *Appl. Mech. Rev.*, 47(7), 1994.

[11] Karnopp, D., Computer simulation of stick-slip friction in mechanical dynamic systems, *ASME J. Dynamic Syst., Meas., Control*, 107(1), 100–103, 1985.

[12] Polycarpou, A. and Soom, A., Transitions between sticking and slipping, in *Friction-Induced Vibration, Chatter, Squeal, and Chaos, Proc. ASME Winter Ann. Meet.*, Ibrahim, R.A. and Soom, A., Eds., DE-Vol. 49, 139–48, Anaheim: ASME; New York: ASME, 1992.

[13] Taylor, J.H. and Lu, J., Robust nonlinear control system synthesis method for electro-mechanical pointing systems with flexible modes, in *Proc. Am. Control Conf.*, AACC, San Francisco, 1993, 536–40.

[14] Walrath, C.D., Adaptive bearing friction compensation based on recent knowledge of dynamic friction, *Automatica*, 20(6), 717–27, 1984.

9.2 Motion Control Systems

Jacob Tal, *Galil Motion Control, Inc.*

9.2.1 Introduction

The motion control field has experienced significant developments in recent years. The most important one is the development of microprocessor-based digital motion controllers. Today, most motion controllers are digital in contrast to 20 years ago when most controllers were analog. The second significant development in this field is modularization, the division of motion control functions into components with well-defined functions and interfaces. Today, it is possible to use motion control components as building blocks and to integrate them into a system.

The following section describes the system components and operation including very simple mathematical models of the system. These models are adequate for the major emphasis of this chapter which discusses the features and capabilities of motion control systems. The discussion is illustrated with design examples.

9.2.2 System Elements and Operation

The elements of a typical motion control system are illustrated in the block diagram of Figure 9.16.

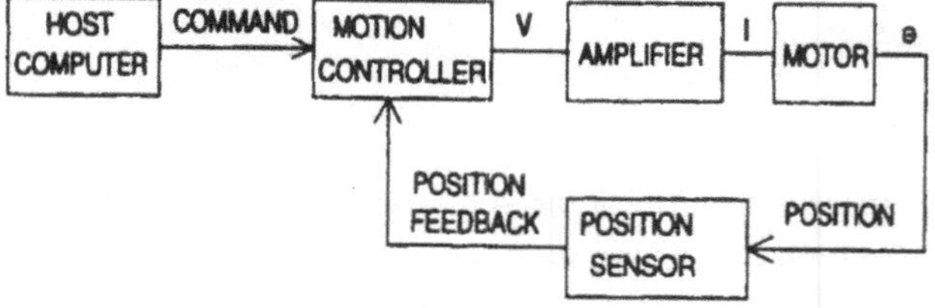

Figure 9.16 Elements of a motion control system.

The amplifier is the component that generates the current required to drive the motor. Amplifiers can be configured in different ways, thereby affecting their operation. The most common operating mode is the transconductance or current mode, where the output current, I, is proportional to the applied voltage, V. The ratio between the two signals, K_a, is known as the current gain.

$$I = K_a V \tag{9.32}$$

When current is applied to the motor, it generates a proportional torque, T_g.

$$T_g = K_t I \tag{9.33}$$

The constant K_t is called the torque constant. The effect of torque on motion is given by Newton's Second Law. Assuming negligible friction, the motor acceleration rate, α, is given by

$$\alpha = \frac{T_g}{J} \tag{9.34}$$

where J is the total moment of inertia of the motor and the load.

The motor position, Θ, is the second integral of the acceleration. This relationship is expressed by the transfer function

$$\frac{\Theta}{\alpha} = \frac{1}{s^2}. \tag{9.35}$$

The position of the motor is monitored by a position sensor. Most position sensors represent the position as a digital number with finite resolution. If the position sensor expresses the position as N units of resolution (counts) per revolution, the equivalent gain of the sensor is

$$K_f = \frac{N}{2\pi}\frac{\text{counts}}{\text{rad}}. \tag{9.36}$$

The component that closes the position loop is the motion controller. This is the "brain" of the system and performs all the computations required for closing the loop as well as providing stability and trajectory generation.

Denoting the desired motor position by R, the position feedback by C, and the position error by E, the basic equation of the closed-loop operation is

$$E = R - C. \tag{9.37}$$

The controller includes a digital filter which operates on the position error, E, and produces the output voltage, V. In order to simplify the overall modeling and analysis, it is more effective to approximate the operation of the digital filter by an equivalent continuous one. In most cases the transfer function of the filter is PID, leading to the equation,

$$H(s) = \frac{V}{E} = P + sD + \frac{I}{s}. \tag{9.38}$$

The mathematical model of the complete system can now be expressed by the block diagram of Figure 9.17. This model can be used as a basis for analysis and design of the motion control system. The analysis procedure is illustrated below.

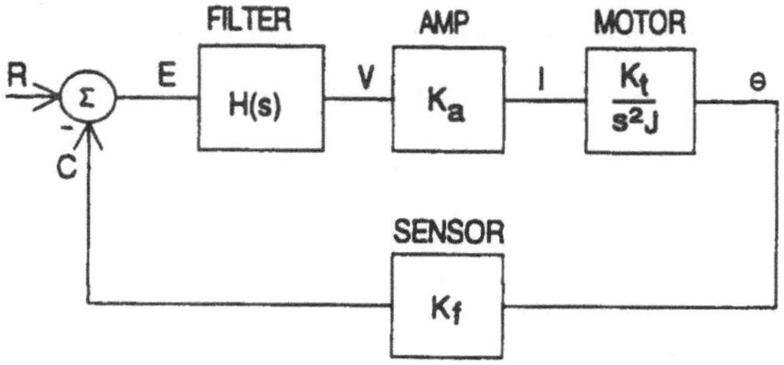

Figure 9.17 Mathematical model of a typical motion control system.

9.2.3 Stability Analysis Example

Consider a motion control system including a current source amplifier with a gain of $K_a = 2$ Amps per volt: the motor has a torque constant of $K_t = 0.1$ Nm/A and a total moment of inertia of $J = 2\cdot 10^{-4} K_g\cdot m^2$. The position sensor is an absolute encoder with 12 bits of binary output, implying that the sensor output varies between zero and $2^n - 1$ or 4095.

This means that the sensor gain is

$$K_f = \frac{4096}{2\pi} = 652 \quad \text{counts/rad}.$$

The digital controller has a sampling period of 1 msec. It has a single input, $E(K)$, and a single output $X(K)$. The filter equations are as follows:

$$F(K) = E(K)\cdot C + F(K-1), \quad \text{and}$$
$$X(K) = E(K)\cdot A + B[E(K) - E(K-1)] + F(K).$$

The signal $X(K)$ is the filter output signal which is applied to a digital-to-analog converter (DAC) with a 14-bit resolution, and an output signal range between -10V and 10V.

A mathematical model for the DAC is developed, noting that when $X(K)$ varies over a range ± 8192 counts, the output varies over ± 10V. This implies that the gain of the DAC is

$$G = \frac{10}{8192} = 1.22\cdot 10^{-3}\,\frac{\text{volts}}{\text{count}}$$

To model the equation of the filter, we note that this is a digital PID filter. This filter may be approximated by the continuous model of Equation 9.38 with the following equivalent terms:

$$P = AG$$
$$I = CG/T$$
$$D = BTG$$

where T is the sampling period.

For example, if filter parameters are

$$A = 20$$
$$B = 200$$
$$C = 0.15$$

and a sampling period is

$$T = 0.001\ s$$

the resulting equivalent continuous PID parameters are

$$\begin{aligned} P &= 0.0244 \\ I &= 0.183 \\ D &= 2.44 \cdot 10^{-4} \end{aligned}$$

Assuming the filter parameters shown above, we may proceed with the stability analysis.

The open loop transfer function of the control system is

$$L(s) = \frac{K_a K_t K_f}{Js^2}\left(P + sD + \frac{I}{s}\right)$$

For the given example, $L(s)$ equals

$$L(s) = \frac{159(s^2 + 100s + 750)}{s^3}$$

Start with the crossover frequency (the frequency at which the open-loop gain equals one) to determine the stability. This equals

$$\omega_c = 181 \quad \text{rad/s}$$

The phase shift at the frequency ω_c is

$$\Theta(\omega_c) = -120°$$

The system is stable with a phase margin of 60°.

The previous discussion focused on the hardware components and their operation. The system elements which form the closed-loop control system were described. In order to accomplish the required motion, it is necessary to add two more functions, motion profiling and coordination. Motion profiling generates the desired position function which becomes the input of the position loop. Coordination is the process of synchronizing various events to verify that their timing is correct. These functions are described in the following sections.

9.2.4 Motion Profiling

Consider the system in Figure 9.17 and suppose that the motor must turn 90°: The simplest way to generate this motion is by setting the reference position, R, to the final value of 90°. This results in a step command of 90°. The resulting motion is known as the step response of the system. Such a method is not practical as it provides little control on the motion parameters, such as velocity and acceleration.

An alternative approach is to generate a continuous time-dependent reference position trajectory, $R(t)$. The basic assumption here is that the control loop forces the motor to follow the reference position. Therefore, generating the reference trajectory results in the required motion.

The motion trajectory is generated outside the control loop and, therefore, has no effect on the dynamic response and the system stability. To illustrate the process of motion profiling, consider the case where the motor must turn 1 radian and come to a stop in 0.1 seconds. Simple calculation shows that if the velocity is a symmetric triangle, the required acceleration rate equals 400 rad/sec^2, and the position trajectory is

$$R(t) = \begin{cases} 200t^2 & 0 \le t \le 0.05 \\ 1 - 200(0.1 - t)^2 & 0.05 \le t \le 0.1 \end{cases}$$

The required move is accomplished by the motion controller computing the function $R(t)$ at the sample rate and applying the result as an input to the control loop.

9.2.5 Tools for Motion Coordination

Motion controllers are equipped with tools to facilitate the coordination between events. The tools may vary between specific controllers but their functions remain essentially the same. Coordination tools include

- stored programs
- control variables
- input/output interfaces
- trip points

The stored programs allow writing a set of instructions that can be executed upon command. These instructions perform a specific motion or a related function. Consider, as an example, the following stored program, written in the format of Galil controllers:

Instruction	Interpretation
#MOVE	Label
PR 5000	Relative distance of 5000 counts
SP 20000	Speed of 20,000 counts/sec
AC 100000	Acceleration rate of 100,000 counts/sec^2
BGX	Start the motion of the X-axis
EN	End the program

This program may be stored and executed in the motion controller, thus allowing independent controller operation without host intervention.

The capabilities of the motion controllers increase immensely with the use of symbolic variables. They allow the motion controller to perform mathematical functions and to use the results to modify the operation. Consider, for example, the following program which causes the X motor to follow the position of the Y motor: This is achieved by determining the difference in the motor positions, E, and by driving the X motor at a velocity, VEL, which is proportional to E. Both E and VEL are symbolic variables.

Instruction	Interpretation
#FOLLOW	Label
JG0	Set X in jog mode
AC 50000	Acceleration rate
BGX	Start X motion
#LOOP	Label
$E = \_TPY - \_TPX$	Position difference
$VEL = E * 20$	Follower velocity
JG VEL	Update velocity
JP #LOOP	Repeat the process
EN	End

The input/output interface allows motion controllers to receive additional information through the input lines. It also allows the controllers to perform additional functions through the output lines.

The input signals may be digital, representing the state of switches or digital commands from computers or programmable logic controllers (PLCs). The inputs may also be analog, representing continuous functions such as force, tension, temperature, etc.

The output signals are often digital and are aimed at performing additional functions, such as turning relays on and off. The digital output signals may also be incorporated into a method of communication between controllers or communication with PLCs and computers.

The following program illustrates the use of an input signal for tension control. An analog signal representing the tension is applied to the analog input port #1. The controller reads the signal and compares it to the desired level to form the tension error *TE*. To reduce the error, the motor is driven at a speed *VEL* that is proportional to the tension error.

Instruction	Interpretation
#TEN	Label
JG0	Zero initial velocity
AC 10000	Acceleration rate
BGX	Start motion
#LOOP	Label
TENSION = @AN [1]	Read analog input and define as tension
$TE = 6 -$ TENSION	Calculate tension error
$VEL = 3 * TE$	Calculate velocity
JG VEL	Adjust velocity
JP#LOOP	Repeat the process
EN	End

Synchronizing various activities is best achieved with trip points. The trip point mechanism specifies when a certain instruction must be executed. This can be specified in terms of motor position, input signal, or a time delay. The ability to specify the timing of events assures synchronization. Consider, for example, the following program where the motion of the X motor starts only after a start pulse is applied to input #1. After the X motor moves a relative distance of 5000 counts, the Y starts, and after the Y motor moves 3000 counts, an output signal is generated for 20 ms.

Instruction	Interpretation
#TRIP	Label
PR 10000, 10000	Motion distance for X and Y motors
SP 20000, 20000	Velocities for X and Y
AI1	Wait for start signal input 1
BGX	Start the motion of X
AD 5000	Wait until X moves 5000 counts
BGY	Start the Y motion
AD,3000	Wait until Y moves 3000 counts
SB1	Set output bit 1 high
WT20	Wait 20 ms
CB1	Clear output bit 1
EN	End

The following design example illustrates the use of the tools for programming motion.

9.2.6 Design Example – Glue Dispensing

Consider a two-axis system designed for glue dispensing, whereby an XY table moves the glue head along the trajectory of Figure 9.18. To achieve a uniform application rate of glue per unit length, it is necessary to move the table at a constant vector speed and to turn the glue valve on only when the table has reached uniform speed.

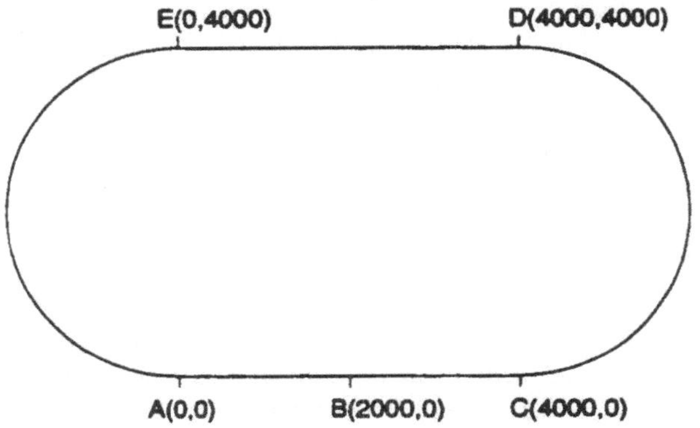

Figure 9.18 Motion path for the glue-dispensing system.

The motion starts at point A and ends at point C after a complete cycle to allow for acceleration and deceleration distances. The gluing starts and stops at point B.

The motion program is illustrated below. It is expressed in the format of Galil controllers. The instructions for straight lines and circular arcs are VP and CR, respectively. The glue valve is activated with the digital output signal #1. This signal is activated and deactivated when the X motor moves through point B. The timing of this event is specified by the trip point FM 2000 which waits until the X motor moves forward through the point $X = 2000$.

Instruction	Interpretation
#GLUE	Label
DP 0,0	Defined starting position as (0,0)
VP 4000,0	Move to point C
CR 2000,270,180	Follow the arc CD
VP 0,4000	Move to point E
CR 2000,90,180	Follow the arc EA
VP 4000,0	Move to point C
VE	End of motion
VS 20000	Vector speed
VA 100000	Vector acceleration
VD 100000	Vector deceleration
BGS	Start the motion
FM 2000	When $X = 2000$ (point B)
SB1	Set output 1–turn glue valve
WT 100	Wait 100 ms
FM 2000	When $X = 2000$ again
CB1	Turn the glue valve off
EN	End

References

[1] DC Motors, Speed Controls, Servo Systems, *Engineering Handbook*, Electrocraft Corp.

[2] Tal, J., *Step-by-Step Design of Motion Control Systems*, Galil Motion Control, 1994.

9.3 Ultra-High Precision Control

Thomas R. Kurfess, The George W. Woodruff School of Mechanical Engineering, The Georgia Institute of Technology, Atlanta, GA

Hodge Jenkins, The George W. Woodruff School of Mechanical Engineering, The Georgia Institute of Technology, Atlanta, GA

9.3.1 Introduction

Fierce international competition is placing an ever-increasing significance on precision and accuracy in ultra-high precision systems engineering. Control of dimensional accuracy, tolerance, surface finish and residual damage is necessary to achieve or inspect ultra-high precision components required for the latest technology. To achieve ultra-high precision, specialized control subsystems must be employed for modern machine tools and measurement machines. These systems are common in manufacturing today, and will become more widespread in response to continually increasing demands on manufacturing facilities. Today, ultra-high precision applications are found in the manufacture and inspection of items such as automobile bearings, specialized optics or mirrors, and platens for hard disk drives. The manufacture of x-ray optical systems for the next generation of microchip fabrication systems has also recently become internationally important.

An excellent example of an ultra-high precision machine is the large optic diamond turning machine at the Lawrence Livermore National Laboratory [8]. This machine operates with a dimensional error of less than 10^{-6} inches (1 μin) and has turned optical parts for many systems including the secondary mirror for the Keck telescope. Not only is the shape of the optics precise enough for use without further processing, but the surface finish is of high enough quality to eliminate polishing. In effect, diamond turning is capable of producing optical parts with extremely stringent optical and specular requirements.

In ultra-high precision machining, accuracies of 0.1μin and lower are targeted. To achieve these accuracies, many problems arise not commonly encountered when designing standard machine servocontrollers, such as low-velocity friction, axis-coupling, and phase lag.

Applications of Ultra-High Precision Systems

The small, but increasing, number of special applications requiring ultra-high precision control techniques include grinding, metrology, diamond turning machines, and the manufacture of semiconductors, optics and electronic mass media.

Diamond turning and grinding of optical quality surfaces on metal and glass require ultra-high precision movement. Accuracies for diamond turning of mirrors can be on the order of 10 nm. Optical grinding applications also require force control to promote ductile grinding, reducing subsurface damage of lenses [2],[13].

Another application for ultra-high precision control is metrology. Precision machined parts are either measured point by point using a vectored touch to include a sufficient amount of data to determine the part geometry, or they are measured in a continuous motion (scanning) along a part dimension using a constant force trajectory. In either case a force probe (or touch trigger) is used to make or maintain contact with the part surface. To have repeatable data, minimize the force-induced measurement errors, and extend the life of the scanning probe, the contact forces must be minimized and maintained as constant as possible.

The primary applications discussed in this chapter are diamond turning machines (DTM) and scanning coordinate measurement machines (CMMs). Both machines have many similarities including basic structure as well as velocity and position control designs. When addressing issues related to both types of machines, the term machine is used. When discussing a cutting machine (e.g., a DTM) the term machine tool is used; the term measurement machine (e.g., a CMM) is used for a machine used exclusively to measure.

History

Initial machine tool control began with simple numerical control (NC) in the 1950s and 1960s. Later (in the 1970s), computer numerical control (CNC) grew more prevalent as computer technology became less expensive and more powerful. These were either point-to-point controllers or tracking controllers, where the tool trajectory is broken up into a series of small straight-line movements. Algorithms for position feedback in most current machine tools are primarily proportional control with bandwidths of nominally 10 to 20 Hz. Increasing the accuracy and repeatability of the machine has been the driving force behind the development of NC and CNC technology.

However, not until the mid 1960s did the first diamond turning machines begin to take shape. These machines were developed at the Lawrence Livermore National Laboratory (LLNL) defense complex. DTM's differ widely from other machine tools. Two of the major differences critical for the control engineer are the mechanical design of the machine, and active analog compensators used in the machine control loop. The basic design philosophy employed by the engineers at LLNL is known as the deterministic theory [4]. The premise of this theory is that machines should be designed as well enough to eliminate the need for complicated controllers. Given such a system, the job of the control engineer is greatly simplified, however, to achieve the best possible performance, active analog compensators are still successfully employed on DTMs.

In the 1980s, other advances in machine tool and measurement system control were further developed such as parameter controllers used to servo force or feedrate. Such parameter controllers, which are fairly common today, operate at lower frequencies. For example, closed-loop force control system frequencies of less than 3 Hz are used in many applications [22].

Theoretical and experimental work has been and continues to be conducted using model reference adaptive control in machining (e.g., [24], [7], [11]. Both fixed gain and parameter adaptive controllers have been implemented for manipulating on-line feed rate to maintain a constant cutting force. Although such controllers have increased cutting efficiency, they typically reduce machine performance and may lead to instability. Although research in adaptive parameter controllers is fairly promising to date, these controllers have not been widely commercialized. Further details regarding machine tool control can be found in the literature [28].

Current Technology

The current technology for designing machines with the highest precision is based on a fundamental understanding of classical control, system identification, and precision components for the controller, drive and feedback subsystems. Control can compensate for poor equipment design only to a limited degree. A well-designed machine fitted with high precision hardware will ease the controller design task. This, of course, relates back to deterministic theory. Special hardware and actuators for ultra-high precision equipment such as laser interferometers, piezoelectric actuators, air and hydrostatic bearings are critical in achieving ultra-high precision. These subsystems and others are being used more frequently as their costs decrease and availability increases.

As stated above, the analog compensator design for precision systems is critical. For ultra-high precision surfaces, both position and velocity must be controlled at high bandwidths because air currents and floor vibrations can induce higher frequency noise. The specifications for the velocity compensator combine machine bandwidth and part requirements. Part geometry, in particular, surface gradients in conjunction with production rate requirements, dictate the velocity specifications for measurement machines and machine tools. For example, a part with high surface gradients (e.g., sharp corners) requires fast responses to velocity changes for cornering. Of course, higher production rates require higher bandwidth machines as feed rates are increased. Maximum servo velocity for a machine tool is also a function of initial surface roughness, and desired finish as well as system bandwidth. For measurement machines, only part geometry and machine bandwidth affect the system velocity specifications because no cutting occurs. One way to improve the performance of ultra-high precision machines is to use slow feed rates (on the order of 1 μin/sec), but such rates are usually unacceptable in high quantity production.

Until recently, force control was not used in many ultra-high precision machining applications, because the cutting forces are generally negligible due to the low feed rates. Furthermore, force control departs from standard precision machining approaches because force is controlled by servocontrol of feed velocity or position. Control is accomplished via position servocontrol using Hook's spring law, and damping relationships are used to control force if velocity servocontrol is used.

Unfortunately, force control in machining generates conflicting objectives between force trajectories and position trajectories (resulting in geometric errors) or velocity trajectories (resulting in undesired surface finish deviations). However, such conflicts can be avoided if the machine and its controllers are designed to decouple force from position and velocity. Such conflicts do not occur to as great an extent with measurement systems where contact forces must be kept constant between the machine measurement probe and the part being inspected.

Typical error sources encountered in precision machine control are machine nonlinearities, control law algorithms, structural response (frequency and damping), measurement errors, modeling errors, human errors, and thermal errors. Sources of machine nonlinearities include backlash, friction, actuator saturation, varying inertial loads, and machine/probe compliance. Methods for addressing these nonlinearities include high proportional gain, integral control, feedforward control, adaptive control, and learning algorithms.

Friction is a problem at the low feed rates needed to achieve high quality (optical) surface finishes. Low velocity friction (stiction) is a primary cause of nonlinearities affecting trajectory control of servo tables. At near zero velocities, stiction may cause axial motion to stop. In such a condition, the position tracking error increases and the control force builds. The axis then moves, leading to overshoot. Most industrial controllers use only proportional compensators that cannot effectively eliminate the nonlinearities of stiction [28]. Friction compensation in lead screws has been addressed by several authors concerned with coulomb friction and stiction [27]. Special techniques such as learning algorithms [25] and cross-coupling control [20] have been employed successfully in X-Y tables to reduce tracking errors caused by low velocity friction.

9.3.2 System Description

Because controller and system hardware directly affect machine performance, both must be thoroughly discussed to develop a foundation for understanding precision system control. This section describes the precision hardware components (e.g., servo drives and sensors) and controllers, in terms of their function and performance.

Basic Machine Layout

The control structure for a typical machine tool can be represented by a block diagram as shown in Figure 9.19. Typically, CNC controllers generate tool trajectories shown as the reference input in Figure 9.19. Given this reference trajectory and feedback from a position sensor, most CNC controllers (and other commercially available motion controllers) compute the position error for the system and provide an output signal (voltage) proportional to that error. A typical CNC output has a limited range of voltage (e.g., −10 to 10 V dc). The error signal is amplified to provide a power signal to the process actuators which, in turn, affect the process dynamics to change the output that is then measured by sensors and fed back to the CNC controller.

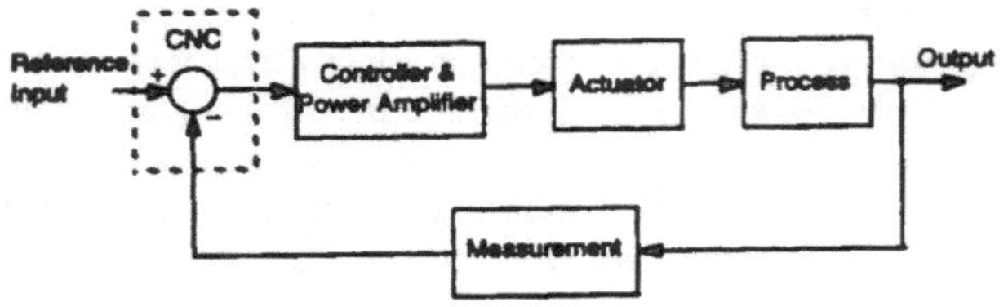

Figure 9.19 Control structure block diagram.

To improve system capability, various controllers may be placed between the position error signal and the power amplifier. This is discussed in detail in later sections. However, before beginning the controller design, it is valuable to examine the system components for adequacy. It may be desirable to replace existing motors, actuators, power amplifiers and feedback sensors, depending on performance specifications and cost constraints.

Transmissions and Drive Trains Typically, machines are based on a Cartesian or polar reference frame requiring converting the rotational displacements of actuators (i.e., servo motors) to linear displacements. Precision ball screws, friction drives or linear motors are used for this purpose. Nonlinearities, such as backlash, should be minimized or avoided in any drive system. In ball screws backlash can be greatly reduced, or eliminated in some cases, by preloading an axis. Preloading can be accomplished by suspending a weight from one side of the slide, ensuring that only one side of the screw is used to drive the slide. Backlash in gear reductions can be minimized with the use of harmonic drives. Greater precision requires friction (capstan) drives or linear motors, because position repeatability suffers from the recirculating ball screws. Capstan drives have direct contact so that no backlash occurs. However, they cannot transmit large forces, so that they are commonly used in CMMs and some DTMs where forces are low.

Rolling or antifriction elements are required to move heavy carriages along a guide way and to maintain linearity. Several configurations are used depending on the amount of precision required. Linear shafts and recirculating ball bearing pillow blocks are for less precise applications. For increased accuracy and isolation, noncontact elements are required at greater expense. Noncontact elements include air bearings, hydrostatic bearings and magnetic bearings. Air bearings are common on CMMs. Hydrostatic bearings are still a maturing technology but provide greater load carrying capability over air bearings [23].

Motors Several motor types are available for use in ultra-high precision control. Typical applications are powered by precision DC servomotors. Recent technology includes linear motors and microstepping motors.

Stepper motors provide an accurate and repeatable precise step motion, yielding a moderate holding torque. However, standard stepper motors have a fundamental step angle limitation (usually 1.8° per step). Stepper motors also have a low resonance frequency causing them to perform poorly at slow speeds, inducing vibrations on the order of 10 Hz. The step size and vibration limitations are somewhat reduced when using microstepping controllers. Finally, these motors can also dissipate large amounts of heat, warming their surroundings, and generating thermal errors due to thermal expansion of the machine.

Linear motors have the advantage of eliminating the lead screw elements of the actuator drive. They have no mechanical drive elements, but generally require rolling elements to suspend the "rotor" or slide. Heat is usually generated from a linear motor closer to the work piece and can be a significant source of thermal errors, depending on the application and machine configuration.

Power Amplifiers Power sources (amplifiers) can be either constant gain amplifiers or pulse width modulated amplifiers (for DC servo motors). Considerations for the amplifier selection are its linear range, maximum total harmonic distortion (THD) over the linear range, and maximum power output. All of these factors must be examined in designing ultra-high precision control systems. For high bandwidth systems, the amplifier dynamics may be critical. THD can be used to estimate nonlinear effects of the amplifier conservatively. The power limitations provide control limits to maintain controller linearity. (To protect the motors and amplifiers, circuit breakers or fuses should be placed appropriately. Transient currents (maximum current) and steady state current (stall current) must both be considered.)

Instrumentation: Sensors, Locations and Utilization

Instrumentation of many types is available for position, velocity and force feedback to the controller. The sensor choice depends on the particular application. For example, in diamond turning, cutting forces are generally negligible, so that only tool speed and position sensors (e.g., tachometers and encoders) are needed for control. However, in scanning metrology, the probe tip force must be controlled to a very high degree to limit probe deflections. Here force sensors must be chosen carefully to achieve the desired force response.

Because the tolerances typically held in ultra-high precision machining are on the order of 100 nm and lower, special sensors are needed for measurement and feedback. Displacement sensors used for ultra-high precision feedback include optical encoders, resolvers, precision potentiometers, linear variable differential transformers (LVDTs), eddy current sensors, capacitance gages, glass scales, magnetic encoders, and laser interferometers. Even if stepper motors are used, feedback transducers should be applied to eliminate a potential slip error. At this small measurement scale, thermal effects and floor/external vibrations not typically encountered in lower precision applications, appear more significant. Thus, the following points must be considered:

1. Allow the system to reach thermal equilibrium (attain a steady state temperature). Typically a 24-hour thermal drift test is required to assess the effect of temperature variations, [[1], Appendix C].
2. Provide environmental regulation. Once the machine has reached thermal equilibrium, it must be kept at a constant temperature to insure repeatability. The standard temperature as defined in [1] is 20°C (68°F). It is difficult, if not impossible, to keep the entire machine at a temperature of 20°C. Heat sources, such as motors and friction, will generate "hot spots" on the machine. Typically, the spatial thermal gradients are not detrimental to machine repeatability, but temporal thermal gradients are. Therefore, it is important to keep the machine temperature from varying with time.
   If laser interferometers are used, the temperature, relative humidity and pressure of the medium through which the laser beam passes (typically air) must be considered. If these quantities are not considered, the interferometer will experience a velocity of light (VOL) error, because the wavelength of light is modulated in a varying environment. It is not necessary to maintain these quantities at a fixed level (clearly, maintaining pressure is a difficult task), because interferometers can be equipped to compensate for VOL errors. In some critical applications, the laser beam is transmitted through a controlled atmosphere via separate beam ways. For the most precise machines developed, the beams are transmitted through a vacuum to eliminate all effects of temperature, humidity and pressure. This solution, however, is extremely difficult to realize.
3. The system must be isolated from floor vibrations. Granite bases, isolation damping pads, and air suspension are some of the techniques used to isolate a system from vibrations. Even the smallest disturbances, such as people conversing next to the machine, can adversely affect the surface finish.

For best results in precision control, is important to choose the appropriate feedback sensor(s) and mounting location. Several principles should be followed. Care must be taken to locate the sensor as close as possible to the line of action or contact, to avoid Abbe offset errors [3]. Of course it is always desirable to locate sensors and drives in a location that is insensitive to alignment errors and away from harm in case a crash occurs. With the appropriate design and consideration, a machine's accuracy can be improved to the level of the sensor's resolution.

It is critical to recognize that sensors have dynamics and, therefore, transfer functions. The sensor bandwidth must match or exceed the desired system frequency response. Some of these sensors will require their own compensation, such as the VOL error on laser interferometers caused by fluctuations in temperature, pressure, and humidity.

Discrete sensors, such as encoders (rotary and linear), are often used in precision machine systems. These sensors measure position by counting tick marks precisely ruled on the encoder. These discrete sensors produce two encoded phase signals for digital position indication (A and B channel quadrature). This increases their resolution by a factor of four and provides directional information. Rotary encoders and resolvers, used on the motor rotational axis, are not typically acceptable in ultra-high precision control because they are not collocated with the actual parameter being measured. Furthermore, their repeatability is affected by backlash and drive shaft (screw) compliance. However, linear versions of encoders (either magnetic or glass scales) are commonly used with good results. A limitation here is the minimum resolution size that is directly related to the spacing of the ticks ruled on the scale. Care must be taken to provide a clean environment, because dirt, smudges or other process by-products can reduce the resolution of these sensors, limiting their utility. The requirement for cleanliness may also require that these devices be kept isolated from potentially harsh environments.

Analog sensors are also widely used for feedback. Typical analog sensors used in precision control include LVDTs, capacitance gages and inductive (eddy current) gages. The least expensive analog displacement sensor is the LVDT, which uses an inductance effect on a moving ferrous core to produce a linear voltage signal proportional to the displacement. In most cases ultra-high precision air bearing LVDTs are necessary to overcome stiction of the core. The best resolution, for analog sensors, can be obtained with capacitance gages and eddy current gages which are widely used in ultra-high precision measurements. Capacitance gages can measure 10^{-12}m displacements [14]. However, at this resolution the range is extremely limited. These sensors require a measurement from a metallic surface, so that locations are typically in-line with the metal part or offset near the equipment ways. Because of the extremely high resolution and low depth of field of these sensors, it is preferable to keep them collinear to the actuators.

The interferometer is the most preferred position sensor in precision system design, because it utilizes the wavelength of light as its standard. Thus, it is independent of any physical artifact (such as gage blocks). In general, the laser interferometer is the best obtainable long-range measurement device, with measurements accurate to 0.25 nm, nominally. Because of size limitations, interferometers are typically located with an axial offset, making measurements susceptible to offset errors, which can be minimized by good metrological practices. Similar to encoders, interferometers use A and B channel quadrature as their output format. As previously stated, care must be taken to include temperature, humidity and pressure compensation to achieve the best possible resolution for the interferometer.

Typical sensors used in force measurement include LVDTs, strain gages, and piezoelectric transducers. LVDTs are used with highly calibrated springs or flexures, allowing forces to be estimated as a function of spring deflection. Bandwidth for this type of sensor is limited. Force measurements based on strain gages are also common. However, they too suffer from limited frequency response because they operate on the same principle as sensors employing flexures (Hook's Law).

Since most ultra-high precision applications require a higher bandwidth than available from flexures and strain gage systems, piezoelectric force transducers are typically used in these applications. Piezoelectric-based dynamometers use a quartz crystal to generate a charge proportional to a force. The dynamometer may be located either behind the tool or under the work piece to measure cutting or probe forces. Although these sensors have larger bandwidth, they exhibit some hysteresis and charge leakage which results in poor capability to measure lower frequency and DC forces due to drifting. However, they are generally stiffer than strain gage force sensors. Piezoelectric force sensors also tend to be large and reduce the effective tool stiffness.

Recent work in force sensors for cutting machines has led to the development of sensors that employ a piezoelectric film. The film is a 28μm of polyvinylidene fluoride placed on tool inserts close to the cutting edge [17]. Advantages of the film are that the force measurements are close to the cutting edge and there is essentially no loss of stiffness because of its small size. Newer magnetostrictive materials, such as metglass or Terfenol^R, are currently being examined for use as force sensors. These materials emit an electromagnetic field in proportion to the force applied.

Care must be given to cabling sensors and determining the expected signal-to-noise ratio. All cables must be shielded to prevent stray magnetic fields from inducing current noise in sensor cables. A means must be provided for independent calibration of sensors (e.g., a laser interferometer can be used to independently check the capacitance gage calibration). Some sensors are also calibrated to a specific cable length impedance.

Because many sensors are digitally based, anti-aliasing filters must be provided to ensure that signal components above the system Nyquist frequency do not cause aliasing problems. Details regarding the use of antialiasing filters can be found in Section 9.3.3. It is also prudent to remember that sensors themselves are dynamic systems, which vary in magnitude and phase with frequency. Specifications are typically provided by the manufacturer. However, it is strongly recommended that sensor system responses be verified. (System identification techniques used in system response verification are presented in Section 9.3.3.) This assists in designing and trouble shooting the controller.

Compensators: Position and Velocity Loops

Analog controllers can be added to existing CNC-type controllers to improve the speed and precision of existing machines if appropriate attention is paid to understanding the system dynamics. The controller and system block diagram are depicted in Figure 9.20, highlighting the additional compensator for improved performance. The inner loop is the velocity loop, and the outer loop is the position loop. Both position and velocity must be well-characterized and controlled in ultra-high precision processes. Because tool position relates directly to final part geometry the position compensator, $K_p(s)$, is used to control dimensional accuracy. Similarly, since the tool velocity is directly related to surface finish, the velocity compensator, $K_v(s)$, is used to control the surface finish. The plant (or process) is depicted as $G(s)$, with velocity feedback provided by a tachometer and position feedback supplied by a laser interferometer or linear encoders. Details of the control design are provided in Section 9.3.4.

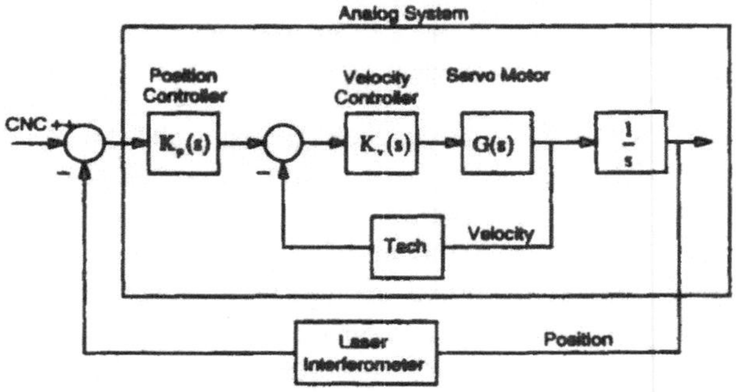

Figure 9.20 Machine with additional compensator.

Fast Tool Servos

Recent advances have been made in the use of piezoelectric actuators for achieving greater bandwidth on precision equipment. Piezoelectric actuators are high bandwidth actuators (on the order of 1 kHz) that can produce high forces and are relatively stiff. However, the stroke of these actuators is generally limited to about 100 μm or less. Piezoelectric actuators are used with conventional actuators (lead screws, capstan drives, etc.) to yield small, high frequency displacements and the conventional actuators provide the lower speed, longer range motions. The piezoelectric actuator is powered by a high voltage supply nominally generating 1000 volts. Actuators are run in a closed-loop mode, typically using a laser interferometer or a capacitance gage for displacement feedback. When used in a closed-loop, the piezoelectric actuator is referred to as a fast tool servo (FTS).

Several ultra-high precision applications using FTS have been developed [6]. The FTS has been used to compensate for cutter runout [18] and for negating the effects of low velocity friction in a series lead screw actuator. FTS's have also been successfully implemented on diamond turning machines [13], [9].

9.3.3 System Identification and Modeling

Successful ultra-high precision control design and implementation depend on the accuracy of system models. Thus, the dynamics of each component must be well-characterized. System identification techniques provide tools for this purpose [19]. A system or component transfer function, $G(s)$, is typically represented in the frequency domain, as shown in Figure 9.21. Here $X(s)$ is defined as the input to the system and $Y(s)$ is the system output.

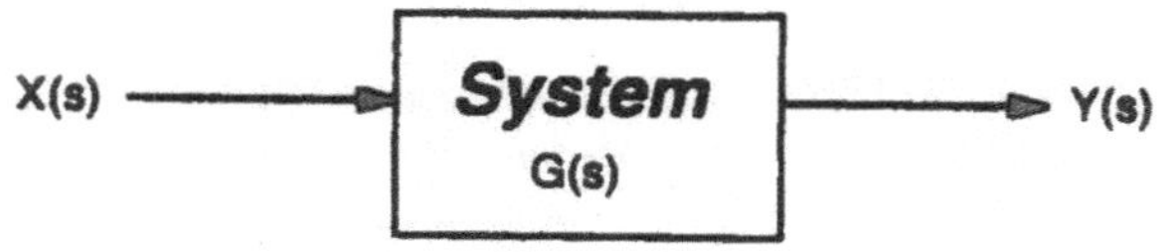

$$\frac{Y(s)}{X(s)} = G(s)$$

Figure 9.21 System input and output with resulting transfer function.

Model transfer functions, $G(s)$, are derived from time histories of the input signal, $x(t)$, and the output signal, $y(t)$, using techniques discussed in the following sections. The order of the model should be as small as possible, preferably related to physical characteristics of the machine (e.g., drive shaft compliance or carriage inertia). Several software packages provide routines that may be used to examine a range of model orders and determine the best fit for a given data set (e.g., MATLAB). However, this numerical approach should not be blindly used as a substitute for thoroughly understanding the process or physical system.

Multiple Axes – Cross-Coupling

Some machines have only a single axis or possess multiple axes that are not strongly coupled. For such machines single-axis system identification can be employed and is straightforward. The input to output transfer functions are easily found as discussed below. However, in many machining cases, such as grinding, several axes are involved in the machine control. If the dynamics of these axes are coupled, then cross-coupling effects must be considered. For example, one axis or more may be used to control force while another axis may be used to control velocity. Recent work addresses, in part, the decoupling of force and velocity loops in a grinding system by gain selection in each axis [12]. Several approaches for improved precision control with a cross-coupled system have been developed over the past 10 years. Another recent technique with superior performance is a cross-coupling controller, where the error in one axis is coupled to position control of the other axes to minimize contour errors [20].

Experimental Approaches

In the development of system models for the controllers, amplifiers and mechanical systems, several experimental approaches may be used to identify and verify component response. In this section the three most common approaches and the typical hardware configurations necessary to implement them are presented.

Stepped Sinusoidal Response Stepped sinusoidal input is the preferred method for determining transfer functions of all system components because it permits detailed system analysis at all frequencies. It is, however, the slowest of the three techniques discussed here. To use this method, sinusoids of various frequencies are fed to the system, one frequency at a time. The response of the system is measured at each frequency, recording the magnitude and phase variations between the input and output signals. The recorded data are then used to construct accurate Bode plots for the system that represent the plant transfer function, $G(s)$. Special purpose digital signal processing (DSP) equipment is available to sweep automatically through a range of frequencies and determine transfer functions in this manner. A coherence function of the data may be easily generated to assess the validity of the determined transfer function at various frequencies. The magnitude of the coherence function varies from zero to unity representing low and high confidence in results, respectively.

Step Input Response For first-order or second-order models, a step input response can be used to determine the system dynamics by applying system identification techniques on the input and output data. The sharpness of the actual step input will limit the frequency range of the identifiable response. Sharper steps have broader frequency spectrums and, therefore, may be used to generate models possessing broader frequency spectrums. Fairly sharp steps in voltage inputs can be easily generated, yielding excellent results for systems with electrical inputs, such as motors.

A fast Fourier transform (FFT) is used to determine the system model's magnitude and phase from the step input response. A frequency domain model (transfer function) or a state-space model can be derived using a least-squares parameter estimation technique. One such method is known as ARX, AutoRegressive eXternal input [19]. The ARX method minimizes the square of the error, $e(t)$, in the following set of equations

$$\begin{aligned} y(t) &= G(q)x(t) + H(q)e(t), \\ e(t) &= H^{-1}(q)[y(t) - G(q)x(t)] \equiv \text{error}, \end{aligned} \tag{9.39}$$

and

$$\left[\hat{G}_N, \hat{H}_N\right] = \min \sum_{t=1}^{N} e^2(t), \tag{9.40}$$

where $G(q)$ represents a discretized version of $G(s)$, $H(q)$ is a discretized weighting of the residual error of the model, N is the number of data points taken, and $\hat{G}_N$ and $\hat{H}_N$ are the best estimates (using the least-squares relationship given in Equation 9.40) of $G(q)$ and $H(q)$, respectively.

When using ARX procedures, several precautions must be taken. First, because ARX models tend to favor higher frequencies [26], the sampled system data should be filtered prior to identification, removing extraneous high frequency content. (The high frequency sources are typically from structural responses, disturbances, or noise, which cannot be controlled.) Also, simulation of the identified system should be compared with the actual data to verify model accuracy because the proper order model may not have been chosen. Some iteration on the model order may be required for an improved fit. As always, the lowest order model that acceptably incorporates the important system dynamics should be used.

White Noise Response Many digital signal processing boards use white noise input to determine the frequency response of a system. This method will generate Bode plots and will yield higher order systems than the step input approach. However, this technique may not be as accurate as the other techniques presented because the input signal may not be truly "white," that is, the generated input signal may not exhibit a uniform broad band spectrum. Frequency content may be lacking at certain frequencies in the input signal. Therefore, the model may be inaccurate at those frequencies.

White noise input and system output are also run through FFT algorithms to obtain Bode plots of the transfer function. Limitations on the valid frequency range of the transfer function are based on the sampling time. Windowing techniques (e.g., Hamming, Hanning, etc.) are used to decrease signal leakage in the Fourier analysis and may increase the accuracy of the attainable FFT results.

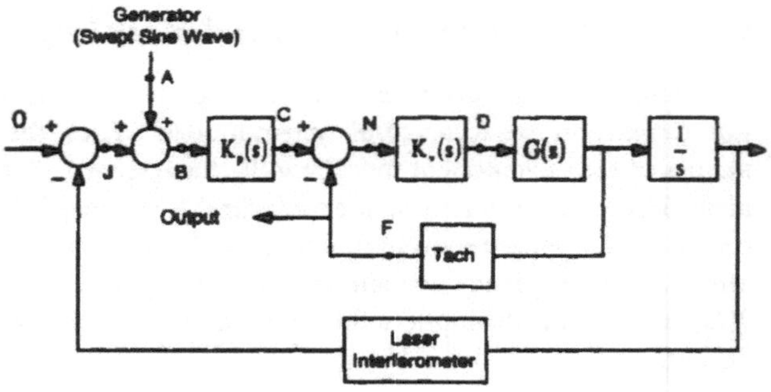

Figure 9.22 Location of input and output signals for system identification.

TABLE 9.4 Locations of Input and Output Signals for System Identification and Response

Plant/System to identify	Input location	Output location
$K_p(s)$	B	C
$K_v(s)$	N	D
$G(s)$	D	F
Closed-Loop velocity	C	F
Integrator (Laser interferometer)	F	J
Tach to position	F	J
Open-Loop position	B	J
Closed-Loop position	A	J

Experimental Set-Up

Only a few specialized hardware items are necessary to obtain transfer functions. The required equipment includes a signal generator (preferably computer based), a signal amplifier, sensors, and signal processing capability.

For the stepped-sine or white noise methods, a DSP instrument or DSP-PC board is useful for signal conditioning and transfer function analysis. The sensors required for these tests are the same as the feedback sensors listed in Section 9.3.2.

A system, shown in Figure 9.22, depicts the signal injection point from a DSP based instrument. The locations to collect input and output data for determining the component transfer functions are listed in Table 9.4. The system is connected to the CNC controller with a zero reference input (a regulator). The CNC controller is used to interpret the position from digital A and B quadrature (A quad B) signals out of the position encoder (seen in Figure 9.22 as the laser interferometer). Therefore, position measurements (point J in Figure 9.22) are obtained through the digital to analog (D/A) output of the CNC controller. This analog position value is the integral of the tachometer signal. It should be noted that the tachometer and interferometer dynamics are significantly faster than the other system components and may, therefore, be ignored.

Signal Conditioning

Because many sensor systems go through an analog to digital conversion (A/D), anti-aliasing filters are needed to prevent frequencies above the Nyquist frequency from being falsely interpreted as lower frequencies (aliasing). Although elliptic and Chebyshev filters have steeper drop-off rates, a low-pass Butterworth filter is recommended because of its flat response in the pass-band. To increase the attenuation roll-off, cascaded, higher order Butterworth filters are recommended. Because these filters have dynamics that can affect system performance, their dynamic responses must be experimentally characterized. For example, a Butterworth filter is maximally flat. However, when it is realized, it may slightly amplify or attenuate a signal and generate a slight error in a position or velocity command signal. Such slight errors may not be tolerable in ultra-high precision machines. These types of deviations, if not well understood, can reduce the performance of the machine.

9.3.4 Control Design

Good mechanical design and modeling are two important components of a precision system design. However, to achieve the highest level of performance, analog compensators must be employed to increase the bandwidth of the machine and to improve its robustness. Typically, controllers are used to improve closed-loop system performance of both the velocity and position loops, where the velocity control loop is nested inside the position loop shown in Figure 9.20. This section presents various analog controller types used in precision machines. Design methods for these controllers are presented and their implications for the actual precision servo system.

Discussion of Controllers

The controllers employed for ultra-high precision applications are typically embodied as standard analog configurations.

This approach eliminates all issues related to digital control such as sampling and quantization. The following discussion presents control elements used in both the velocity and position loops and outlines the basic the design procedures for these elements.

Velocity Loop Compensator Velocity control of a precision machine is critical in maintaining consistent and known surface finish during machining operations. Controlling velocity is also critical in maintaining consistent force trajectories while measuring or machining. The dynamic velocity loop compensator, $K_v(s)$, is a combination of a lead controller and a PI controller. The lead compensation provides additional phase margin for the system, because the integrator causes a 90° phase lag. Large amounts of phase and gain margin are usually specified for ultra-high precision systems to ensure their robustness to various outside disturbances. The integrator is used to eliminate steady-state velocity error. This is important because machines are usually servocontrolled at constant feed velocities.

Lead Compensator Design Procedure A lead compensator used to increase the phase margin for the velocity loop is depicted in Figure 9.23. Procedures for designing the controller to achieve the desired input to output voltage relationships are summarized in Equations 9.41 and 9.42 and Table 9.5. Such a compensator is designed in the following section for a diamond turning machine where actual specifications for phase and gain margin are provided.

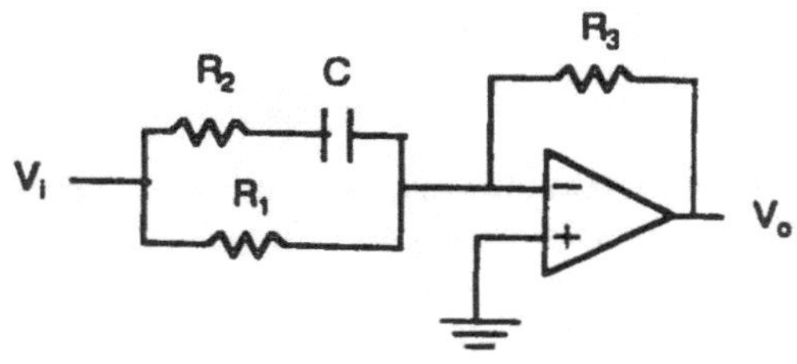

Figure 9.23 Lead compensator diagram.

$$\frac{V_0}{V_i} = -\frac{R_3}{R_1}\left(\frac{1+s(R_1+R_2)C}{1+sR_2C}\right) = -\frac{R_3}{R_1}\left(\frac{1+s\tau_1}{1+s\tau_2}\right) \quad (9.41)$$

$$\phi_{max} = \sin^{-1}\left(\frac{1-\alpha}{1+\alpha}\right); \quad \alpha = \frac{\tau_1}{\tau_2} = \frac{\omega_1}{\omega_2};$$
$$\omega_0 = \sqrt{\omega_1\omega_2} \quad (9.42)$$

TABLE 9.5 Lead Compensator Design Procedure

Procedure
Determine w_0 and ϕ_{max}.
Calculate α from ϕ_{max}.
Calculate w_1 and w_2 from α and w_0.
Calculate τ_1 and τ_2 from w_1 and w_2.
Choose C.
Calculate R_1, R_2 and R_3 from τ_1, τ_2 and C.

The procedure is started by selecting the maximum phase lead angle, ϕ_{max}, and the frequency where it is to occur, w_0 (the geometric mean of the two corner frequencies). Once these two quantities have been selected, the variables α and subsequently w_1, w_2, τ_1 and τ_2 are determined. Finally, the values for C, R_1, R_2 and R_3 are determined. Although the procedure is relatively straightforward, values for the three resistors and the capacitor must be chosen so that they are commercially available. For example, a resistance value of 10 kΩ, which is commercially available, is preferred to a value of 10.3456 kΩ, which is not commercially available.

PI Compensator Design Figure 9.24 represents a PI compensator design for the velocity loop. The design procedure is summarized in Table 9.6 and Equations 9.43 and 9.44. As previously stated, the PI compensator yields zero steady-state error to a velocity trajectory that is constant, in other words, a constant feed rate.

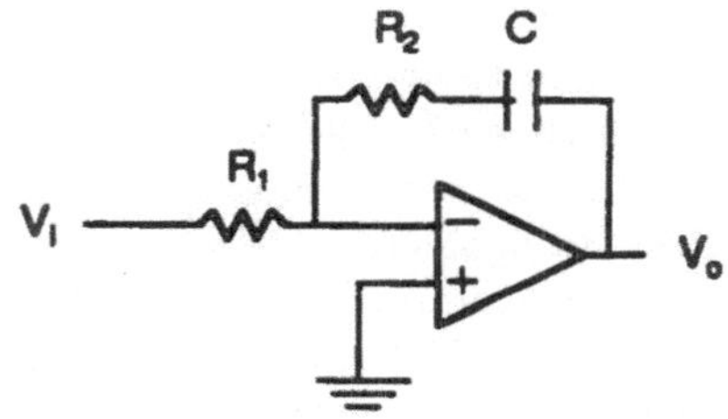

Figure 9.24 PI compensator diagram.

$$\frac{V_0}{V_i} = \frac{1+sR_2C}{sR_1C} = \frac{1+s\tau_2}{s\tau_1} \quad (9.43)$$

$$\omega_{break} = \frac{1}{\tau_2}; \left|\frac{V_0}{V_i}\right|_{\omega\to\infty} = \frac{\tau_2}{\tau_1} \quad (9.44)$$

TABLE 9.6 PI Compensator Design Procedure

Procedure
Determine break frequency, $\omega_{break} = 1/\tau_2$.
Choose high frequency gain.
Calculate τ_1 from τ_2.
Choose C.
Calculate R_1 and R_2.

Band Reject Filters for Structural Resonances Finally, it is worth mentioning that machine tools and measurement machines have a variety of structural modes or resonant frequencies. These are often detected during the system identification phase of the compensator design. To eliminate these modes, band reject filters are employed. These analog filters can be tuned to attenuate a specific range of frequencies and can be extremely beneficial in improving the system performance.

Input shaping is another available technique for suppressing resonance frequencies, often used in more flexible systems. The reader is referred to [21] for more detail.

Position Loop Compensator The position loop is a means of achieving the dimensional accuracy desired. For the position loop, the compensator, $K_p(s)$, is a series combination of a lead controller and a proportional control element. The lead compensator is designed in the same manner as the velocity loop. Because a natural integrator occurs in the position loop from the carriage velocity to the position sensor (see Figure 9.20), no additional integral control is necessary to achieve zero steady state to a constant command signal.

Advantages and Disadvantages

There are some advantages to this type of analog compensator scheme. Analog compensators can be used with most CNC type controllers to improve system bandwidth and precision, because most CNC controllers do not perform any dynamic compensation. Rather, they simply put out a voltage proportional to the position error of the machine tool. Another advantage of analog compensators is that their resolution is only limited to the tolerable signal-to-noise ratio, whereas digital systems are limited by quantization effects as well as noise.

There are, of course, some limitations to using analog compensators, the most significant of which is the inability to provide adaptive variable gains based on parameter estimation techniques. However, ultra-high precision plants tend to have less variability by design, eliminating the need for adaptive compensators. The other major limitation to analog compensators is that they are less flexible in design and implementation than digital systems .

Performance Criteria

System performance is typically measured in terms of bandwidth, steady-state error, rise time, overshoot, and damping. Bandwidth is important because a machine's bandwidth relates directly to its apparent stiffness. Steady-state error for both velocity and position loops must be eliminated to insure appropriate surface finishes and part geometries, respectively. Specifications for rise time assure that higher spatial frequency characteristics of the part geometry are not eliminated (i.e., the machine must be able to corner reasonably well). Damping and control of overshoot are particularly important in designing the position controller to avoid removing too much material from a part during machining.

Although the design approach presented here is nonadaptive, techniques can provide a robust compensator design. Robustness of design is necessary to assure stability and minimal performance variation of the closed-loop system in the presence of some uncertainty in the plants and environment. Robustness of the design in stability can be characterized in terms of gain and phase margins. Design techniques for obtaining compensators robust in performance include the use of root sensitivity and gain plots [16], and other parametric plots.

Root Sensitivity Gain plots are an alternate graphical representation of the Evans root locus plot [15] . They explicitly graph the eigenvalue magnitude vs. gain in a magnitude gain plot and the eigenvalue angle vs. gain in an angle gain plot. The magnitude gain plot employs a log-log scale whereas the angle gain plot uses a semilog scale (with logarithms base 10). Although gain is the variable of interest, any parameter may be used in the geometric analysis.

The root sensitivity of any system can be computed using the slopes of the gain plots [16]. In classical control theory, the root sensitivity, S_p, is defined as the relative change in a system root or eigenvalue, $\lambda_i (i = 1, ..., n)$, with respect to a system parameter, p. Most often, the parameter analyzed is the forward loop gain, k. The root sensitivity with respect to gain is given by

$$S_k = \frac{d\lambda(k)/\lambda(k)}{dk/k} = \frac{d\lambda(k)}{dk}\frac{k}{\lambda(k)}. \tag{9.45}$$

Equation 9.45 is often introduced in determining the break points of the Evans root locus plot for single-input, single-output systems. At the break points, S_k becomes infinite as at least two of the n system eigenvalues undergo a transition from the real domain to the complex domain or vice versa. This transition causes an abrupt change in the relationship between the eigenvalue angle $\angle\lambda$ and gain k, yielding an infinite eigenvalue derivative, $d\lambda/dk$.

The root sensitivity function, S_k, is a measure of the effect of parameter variations on the eigenvalues. An expression for the complex root sensitivity function is developed by employing a polar representation of the eigenvalues in the complex plane. Three assumptions are imposed: (i) the systems analyzed are lumped parameter, linear time-invariant (LTI) systems; (ii) there are no eigenvalues at the origin of the s-plane, i.e.,

$$\lambda_i \neq 0, \quad \forall i = 1, 2, \ldots, n \tag{9.46}$$

(although the eigenvalues may be arbitrarily close to the origin), and (iii) the forward scalar gain, k, is real and positive, i.e., $k \in \Re, k > 0$. Based on these assumptions, we draw the following observations: the real component of the sensitivity function is given by

$$Re\,|S_k| = \frac{d\ln|\lambda(k)|}{d\ln(k)}, \tag{9.47}$$

and the imaginary component of the sensitivity function is given by

$$Im\,|S_k| = \frac{d\angle\lambda(k)}{d\ln(k)}. \tag{9.48}$$

These observations may be proven as follows. Equation 9.45 may be rewritten in terms of the derivatives of natural logarithms as

$$S_k = \frac{d\ln[\lambda(k)]}{d\ln(k)}. \tag{9.49}$$

The natural logarithm of the complex value, λ, is equal to the sum of the logarithm of the magnitude of λ and the angle of λ multiplied by $j = \sqrt{-1}$. Thus, Equation 9.49 becomes

$$S_k = \frac{d[\ln|\lambda(k)| + j\angle\lambda(k)}{d\ln(k)}. \tag{9.50}$$

Because j is a constant, Equation 9.50 may be rewritten as

$$S_k = \frac{d\ln|\lambda(k)|}{d\ln(k)} + j\frac{d\angle\lambda(k)}{d\ln(k)}. \tag{9.51}$$

We next make the observation that the slope of the magnitude gain plot is the real component of S_k. The magnitude gain plot slope, M_m, is

$$M_m = \frac{d\log(|\lambda(k)|)}{d\log(k)}, \tag{9.52}$$

which may be rewritten as

$$M_m = \frac{d[\log(e)\ln(|\lambda(k)|)]}{d[\log(e)\ln(k)]} = \frac{d\ln(|\lambda(k)|)}{d\ln(k)} \tag{9.53}$$

corresponding to Equation 9.47.

Furthermore, the slope of the angle gain plot is the product of the imaginary component of S_k and the constant, $[log(e)]^{-1}$. The angle gain plot slope, M_a, is

$$M_a = \frac{d\angle\lambda(k)}{d\log(k)}, \tag{9.54}$$

which may be rewritten as

$$M_a = \frac{d\angle\lambda(k)}{d[\log(e)\ln(k)]} = \frac{1}{\log(e)}\frac{d\angle\lambda(k)}{d\ln(k)}, \tag{9.55}$$

and hence M_a is proportionally related to Equation 9.48 by $[log(e)]^{-1}$.

The complex root sensitivity function is now expressed with distinct real and imaginary components employing the polar form of the eigenvalues. It follows from assumption (ii) that $\ln(k)$ is real. (In general, most parameters studied are real and this proof is sufficient. If, however, the parameter analyzed is complex, as explored in [16], it is straightforward to extend the above analysis.)

The slopes of the gain plots provide a direct measure of the real and imaginary components of the root sensitivity and are available by inspection. The use of the gain plots with other traditional graphical techniques offers the control system designer important information for selecting appropriate system parameters.

Filtering and Optimal Estimation

In the presence of noisy sensor signals, special filtering techniques can be applied. Kalman filtering techniques can be used to generate a filter for optimal estimation of the machine state based on sensor data. This approach also provides an opportunity to use multiple sensor information to reduce the effects of sensor noise. However, Kalman filters are generally not necessary because much of the sensor noise can be eliminated by good precision system design and thorough control of the machine's environment.

Hardware Considerations

Selection of electronic components for the compensator is limited to commercially available resistors and capacitors. Although 1% tolerance is available on most components, the final design must be thoroughly tested to verify the response of the compensator. Off-the-shelf components may be used. However, their transfer functions must also be determined. For example if a Butterworth filter is used, it cannot be assumed that its transfer function is perfectly flat. The filter's transfer function must be mapped. Any deviation from a flat frequency response can generate positional or velocity errors that are unacceptable because of stringent requirements.

Software and Computing Considerations

Although this chapter is primarily concerned with analog control, there are some computing issues that are critical and must be addressed to assure peak performance of the closed-loop system. The tool (probe) trajectory in conjunction with the update rate must be carefully considered when designing a precision machine. If they are not an integral part of the control design, optimal system performance cannot be achieved. In many cases the CNC controller must have a fast update time corresponding to the trajectory and displacement profile desired. There are many CNC systems (including both stand-alone and PC-based controllers) on the market today, with servo update rates of 2000 Hz and higher, to perform these tasks. The controller must also be able to interface with the various sensors in the feedback design. Most of these controllers provide their own language or use g-codes, and can be controlled via PC-AT, PCMCIA, and VME bus structures. If care is not taken to interface the CNC controller to the analog portion of the machine, the resulting system will not perform to the specifications.

9.3.5 Example: Diamond turning machine

The most precise machine tools developed to date are the diamond turning machines (DTM) at the Lawrence Livermore National Laboratory (LLNL). These machine tools are single point turning machines, or lathes that are capable of turning optical finishes in a wide variety of nonferrous materials. This section presents a basic control design for such a machine. To avoid confusion, the machine vibration modes are not included in the analyses, and the experimental results have been smoothed out via filtering. The approach discussed here is the approach taken when designing analog compensators for diamond turning machines. Furthermore, the compensators developed here are in use in actual machine tools.

Figure 9.25 depicts a typical T-based diamond turning machine similar to the systems employed at LLNL. The machine consists of two axes of motion, X and Z, as well as a spindle on which the part is mounted. The machine is called a T-based machine because the cross axis (X) moves across a single column that supports it. Thus the carriage and the single column frame form a "T." In this particular T-based configuration, the entire spindle (with part) is servocontrolled in the Z direction which defines the longitudinal dimensions of the part; and the tool is servocontrolled in the X direction, defining the diametrical part dimensions.

The DTM is equipped with a CNC controller which plans the tool trajectory and determines any position error for the tool. The output of the CNC controller is a signal proportional to the tool position error, that is, the output of the summing junction for the outer (position) loop in Figure 9.20 (point J, Figure 9.22). The objective of this example is to demonstrate the design of the

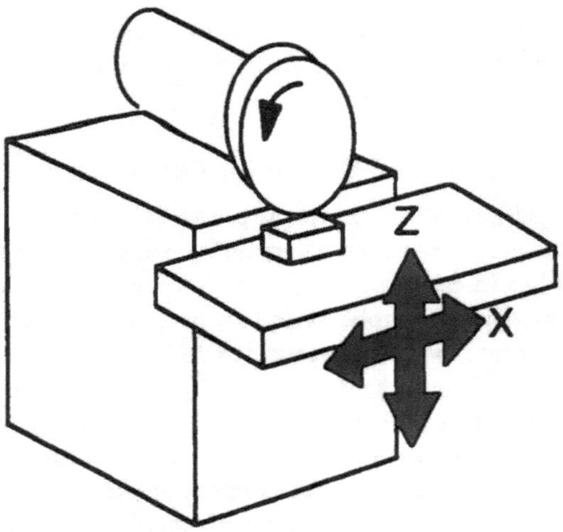

Figure 9.25 A T-Based diamond turning machine.

velocity and position controllers, $K_v(s)$ and $K_p(s)$, respectively, for the Z-axis. Usually, both $K_v(s)$ and $K_p(s)$ are simple gains. However, the performance required for this DTM requires active controllers in both the position and velocity loops.

Typically, the velocity compensator is designed first and, subsequently, the position compensator is designed. The velocity loop consists of the velocity compensator, a power amplifier (with unity gain that amplifies the power of the compensator signal), a unity gain tachometer mounted on the servomotor, and a summing junction. The position loop consists of the CNC controller, a position compensator, the dynamics from the closed-loop velocity system, a lead screw (with a 1 mm/rev pitch), and a (unity gain) laser interferometer providing position feedback. The lead screw acts as a natural integrator converting the rotational velocity of the servomotor into slide position. The laser interferometer provides a digital (A quad B) signal back to the CNC controller which, in turn, compares the actual location of the Z carriage to the desired location. This error is then converted to an analog signal by the CNC controller's digital to analog board and fed into the position compensator. For all practical purposes, the CNC controller may be considered a summing junction for the position loop. This assumption becomes invalid at frequencies that approach the controller's sample rate of 2 ms.

To determine the transfer function of the servo system, sinusoidal signals of various frequencies are injected into the servoamplifier, and the resulting tachometer outputs are recorded and compared to the inputs. Bode plots are then generated from the empirical data, and a transfer function is estimated from these plots. Figure 9.26 shows a typical Bode plot for the Z-axis of the DTM.

From Figure 9.26, the transfer function of the servo drive from the servo amplifier (not including the velocity compensator) input voltage, $V(s)$, to the Z-axis motor velocity, $\omega(s)$, is

$$G_p(s) = \frac{\omega(s)}{V(s)} = \frac{4500}{s+60}. \qquad (9.56)$$

The high gain of the system is due to the fact that the machine moves at relatively low velocities that must generate large feedback voltage signals.

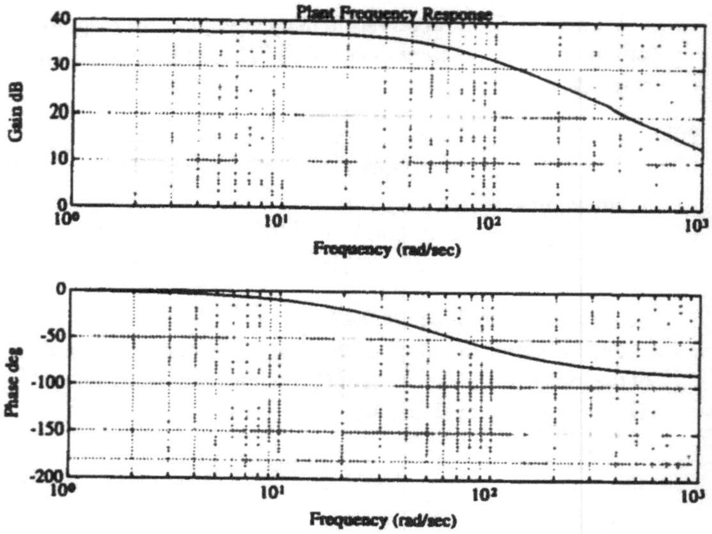

Figure 9.26 Bode plots for a diamond turning machine $G_p(s)$.

Specifications

This section presents typical specifications (and the reasons behind the specifications) for both the velocity and position loops for a DTM.

The specifications for the velocity loop are

1. Zero steady state error to a constant input in velocity, $V(s)$. This insures that the machine tool can hold constant feedrates without any steady-state error.
2. Maximum of 2 volts input into the servoamplifier. This is a limitation of the amplifier. Based on the current required to drive the motor, no more than 2 volts may be used.
3. The sensitivity to the forward loop gain should be less than 1.0. This reduces deviations in closed-loop system performance if the forward loop gain varies.
4. Damping ratio, $\zeta > 0.8$. A small amount of overshoot (2%) in velocity is acceptable. (Note that $\zeta > 0.707$ results in a closed-loop system that should not have any amplification. Thus the closed-loop Bode plot should be relatively flat below the bandwidth frequency.)
5. Maximize the closed loop bandwidth of the system. Target velocity loop bandwidth, 15 Hz $\approx$ 95 rad/s.
6. Minimum gain margin 40 dB.
7. Minimum phase margin 90°.

The specifications for the position loop are

1. Zero steady-state positioning error.
2. The sensitivity to the forward loop gain should be less than 1.5. This reduces deviations in closed-loop system performance if the forward loop gain varies.
3. Damping ratio, $\zeta > 0.9$. A small amount of overshoot (1%) in position is acceptable because the typical motion values are on the order of 0.1 μin.

4. Steady-state position error to a step should be zero.
5. Minimum gain margin 60 dB.
6. Minimum phase margin 75°.

Compensator Design: Velocity Loop

The design procedure of Section 9.3.4 is used to determine the velocity loop PI compensator. The final design of the velocity loop PI compensator is of the same form as the PI compensator of Figure 9.24, with a capacitance of 4μF and both resistance values at 1.33kΩ. These components are determined from the relationships of Equations 9.57 and 9.58. The measured frequency response of the PI compensator is shown in Figure 9.27. Note

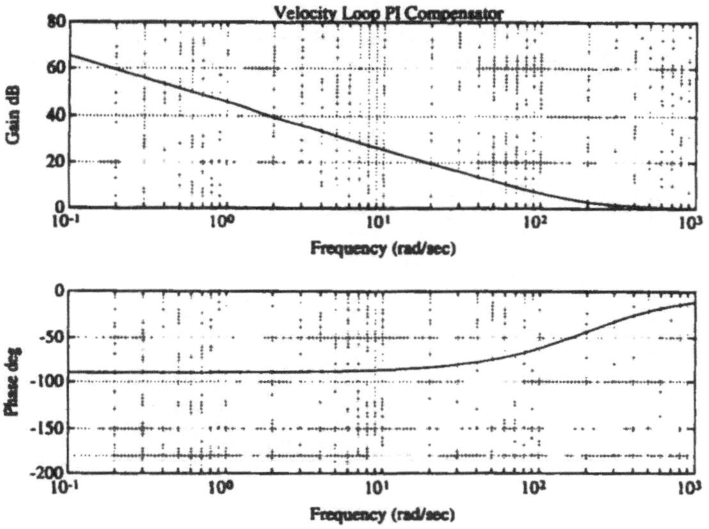

Figure 9.27 Velocity loop PI compensator response.

in Figure 9.27 that the integrator is visible with a low frequency slope of −20 dB/decade and a DC phase of −90°.

The lead compensator's final design is depicted in Figure 9.23, with $R_1 = 38.3k\Omega$, $R_2 = 19.6k\Omega$, $R_3 = 13.3k\Omega$, and $C = 0.0475\mu F$. The frequency response of the lead compensator is shown in Figure 9.28. Equations 9.59 and 9.60 are the governing relationships. A high impedance summing junction is placed ahead of the lead compensator. The lead compensator is designed to yield approximately 30° of lead at a frequency of 650 rad/s. As shown in Equation 9.60, this goal was accomplished within an acceptable error band. An inverting amplifier is used in the compensator design resulting in a phase shift of −180° in the Bode plot of Figure 9.28. This is corrected in the complete velocity compensator by the fact that the PI and lead compensators both invert their input signals, thus canceling the effects of the inversion when they are cascaded. Figure 9.29 represents the final complete velocity compensator.

All of the components shown for this controller are commercially available, a critical point in designing a compensator. If the components are not commercially available, special (and expensive) components would have to be fabricated. Thus, it is critical

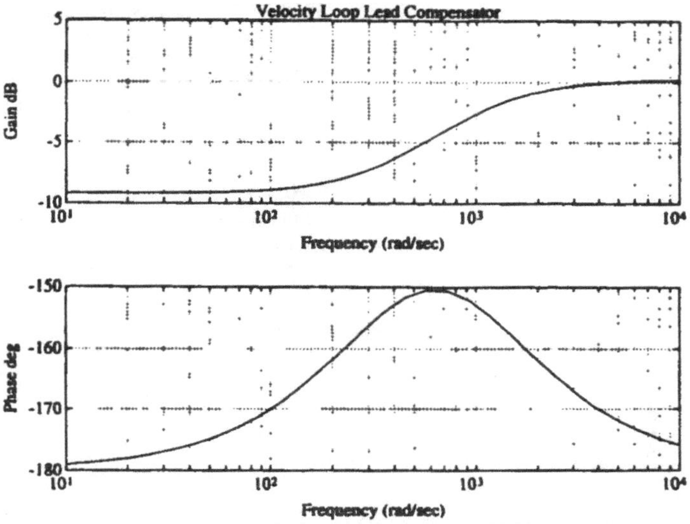

Figure 9.28 Velocity loop lead compensator frequency response.

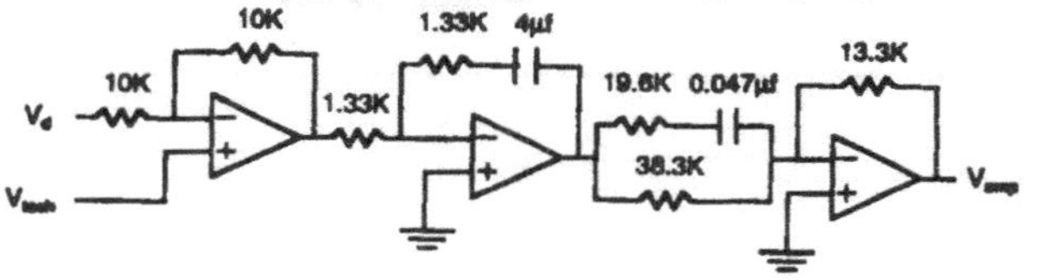

Figure 9.29 Velocity loop complete compensator.

to generate designs that can be realized with commercially available resistors and capacitors.

$$\frac{V_0}{V_i} = \frac{1+0.00532s}{0.00532s} \tag{9.57}$$

$$\omega_{\text{break}} = \frac{1}{0.00532} = 188(\text{rad/s}); \left|\frac{V_0}{V_i}\right|_{\omega\to\infty} = 1 \tag{9.58}$$

$$\frac{V_0}{V_i} = -0.347\frac{1+0.00272s}{1+0.000921s} \tag{9.59}$$

$$\phi_{\max} = 29.6°,\ \alpha = \frac{0.000921}{0.00272} = \frac{368}{1085} = 0.339,$$

$$\omega_0 = 632(\text{rad/s}) \tag{9.60}$$

The open-loop Bode plots for the velocity loop are given in Figure 9.30. From the open-loop Bode plots, it is clear that all of the gain and phase margin specifications have been met for the system. Also, the low frequency response of the system indicates a type I system, thus there will be no steady-state error to a constant velocity input.

Now the forward loop gain is determined using the root locus and parametric plot techniques. Figure 9.31 presents the root locus of the system as the loop gain is increased. The lead compensator's final design can be assessed by using the Bode plots in conjunction with the root locus plot. The pole and zero furthest in the left hand plane are from the lead compensator. The second zero and the pole at the origin are from the PI compensator. Clearly from the plot, there is a range of gains that results in an underdamped system and may yield unacceptable damping for the velocity loop. To investigate further the appropriate choice of the forward loop gain, the root locus is shown as a set

TABLE 9.7 Velocity Loop and Position Loop Controller Specification Summary

Specification	Velocity loop	Position loop
Min. closed-loop bandwidth	15 (Hz) (≈95 rad/s)	150 (Hz) (≈950 rad/s)
Steady-State error to a step	0	0
Min. damping ratio, ζ	0.8	0.9
Max. $\lvert S_k \rvert$	1.0	1.5
Min. phase margin, ϕ_m	90°	75°
Min. gain margin, g_m	40(dB)	60 (dB)

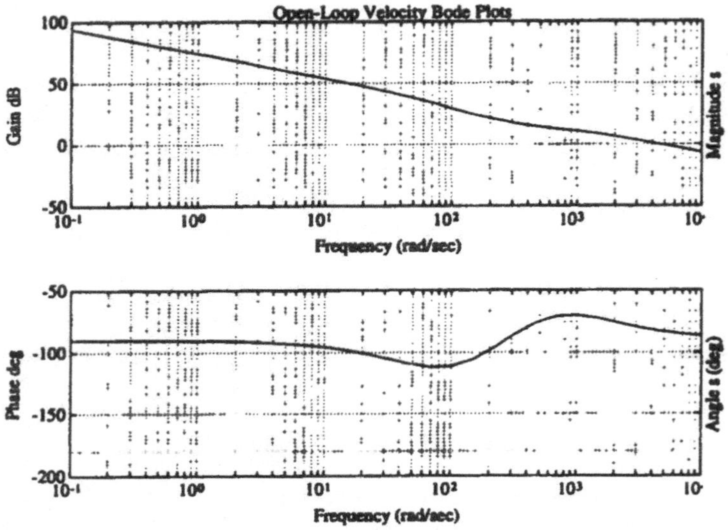

Figure 9.30 Velocity open-loop Bode plots.

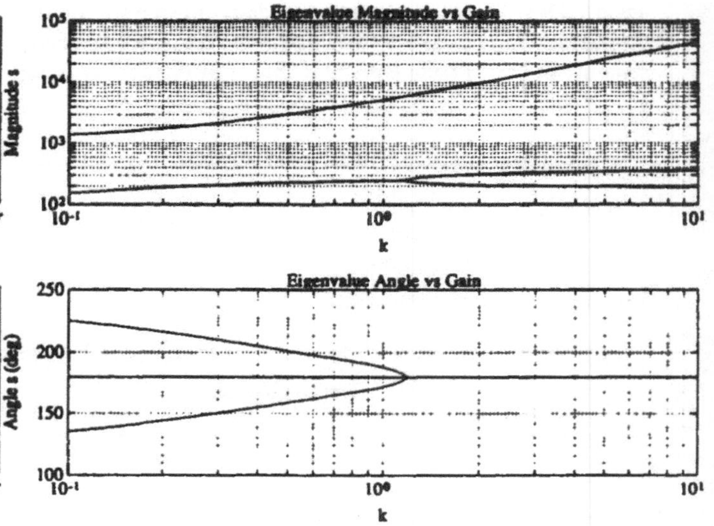

Figure 9.32 Velocity loop compensator gain plots.

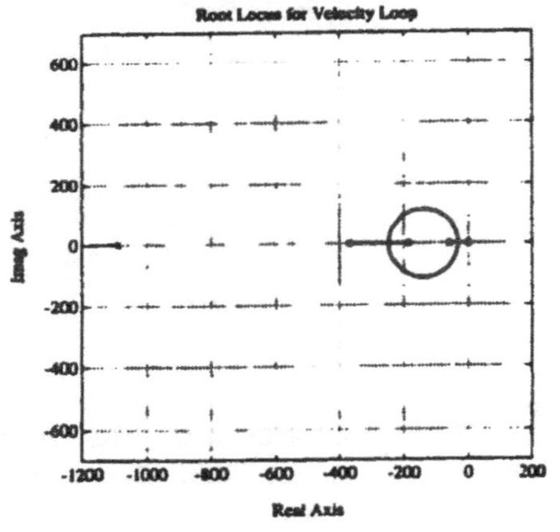

Figure 9.31 Velocity closed-loop root locus.

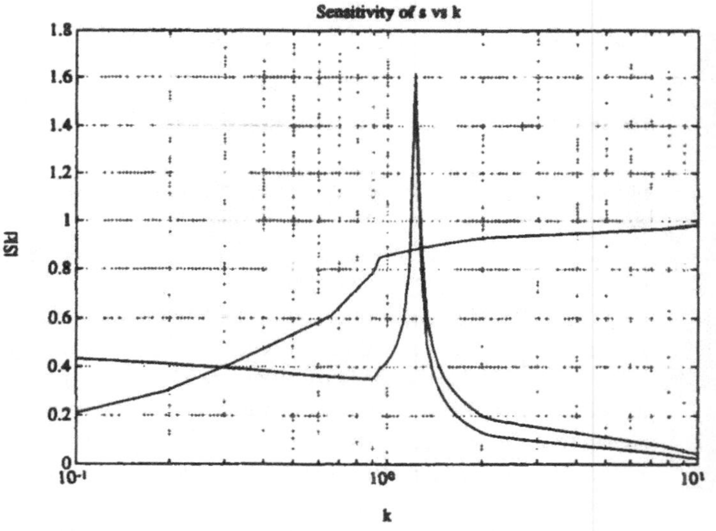

Figure 9.33 Velocity loop compensator root sensitivity.

of gain plots, plotting pole magnitude and angle as functions of loop gain. Figure 9.32 depicts the velocity loop gain plots, showing various important aspects of the loop gain effects on the closed-loop system dynamics. In particular, low and high gain asymptotic behaviors are evident as well as the gain at the break point of the root locus. Furthermore, because the natural frequency and damping can be related to the eigenvalue (or pole) magnitude and angle, the effects of loop gain variations on the system response speed and damping may be easily visualized. Clearly, from the gain plots, gains of slightly more than unity result in an overdamped closed-loop system. This places a lower limit on the loop gain. Figure 9.33 plots the magnitude of the root sensitivity as a function of gain. The only unacceptable loop gains are those in the vicinity of the break points. Thus, the design must not use a gain in the region of $1 < K_v < 2$.

Another parametric plot that can be generated is the phase margin as a function of loop gain (see Figure 9.34). Based on this plot, to achieve the appropriate phase margin the gain must be

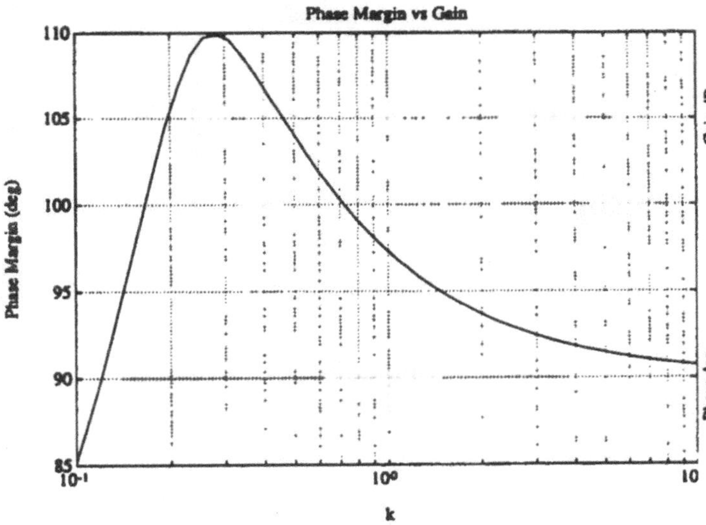

Figure 9.34 Velocity loop compensator phase margin vs. gain.

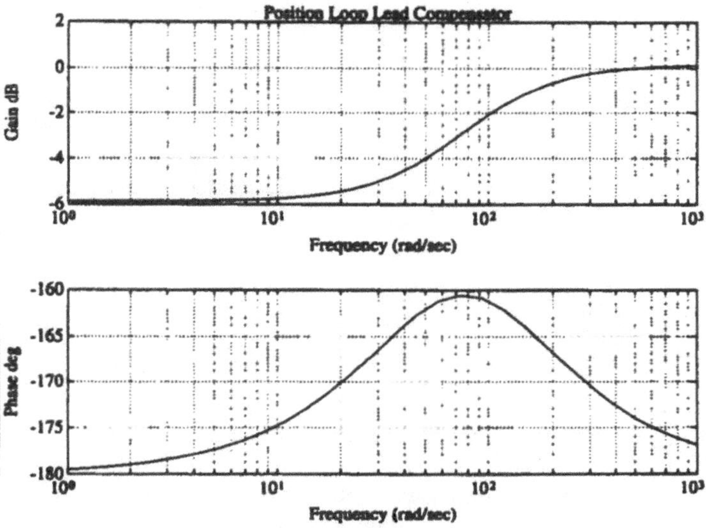

Figure 9.36 Lead compensator response.

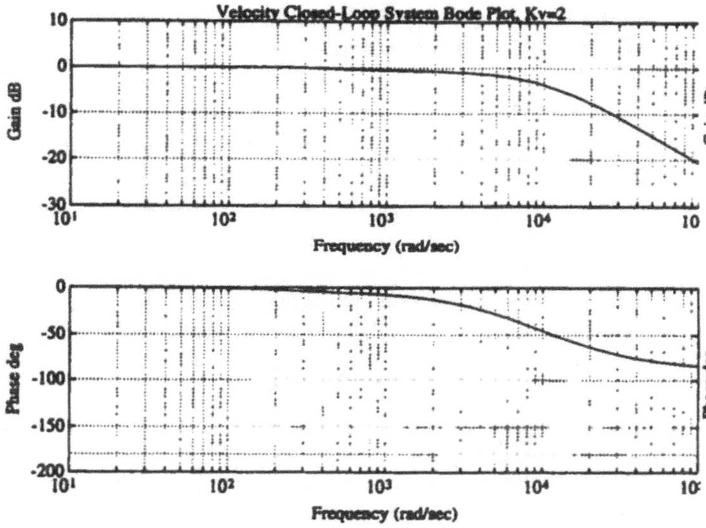

Figure 9.35 Velocity loop compensator closed-loop response, $K_V = 2$.

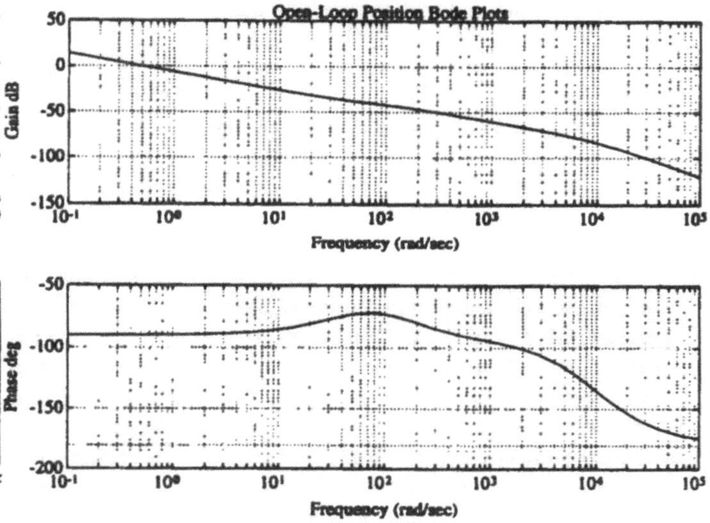

Figure 9.37 Position open-loop frequency response.

greater than 0.11. The maximum phase margin is at a gain of approximately 3. Also, at high gains, the phase margin approaches 90°, as expected, because the system has one more pole than zero. The gain margin of the system is infinite because the maximum phase shift is −90°. For this particular design, a value of 2 was chosen for the loop gain. It should be noted that the upper limit on the gain is power related, because higher gains require larger amounts of power. Thus, the target gain achieves all specifications with a minimal value. Figure 9.35 is the closed-loop Bode plot for the velocity loop.

Although gain plots and parametric plots are not necessary for the actual design of the compensators (in particular, the loop gain), they do offer insight into the various trade-offs available. For example, it is clear from the gain plots that gains much higher than 5 will not significantly affect the closed loop system response because both lower frequency poles are approaching the system transmission zeros. Higher gains only push the high frequency pole (which is dominated by the lower frequency dynamics) further out into the left hand plane. Thus higher gains, which are costly to implement, do not significantly enhance the closed-loop system dynamics beyond a certain limit and may excite higher frequency modes.

Compensator Design: Position Loop

The position loop is designed similarly to the velocity loop by determining values for the components of the lead compensator and the loop gain of the proportional controller. The final design of the lead compensator has the same form as Figure 9.23, with $R_1 = 42.2k\Omega$, $R_2 = 42.2k\Omega$, $R_3 = 21.5k\Omega$, and $C = 0.22\mu F$, and Equations 9.50 and 9.51 show its design values. The objective for the compensator is a maximum lead of 20° at a frequency of 75 rad/s. Figure 9.36 shows the lead compensator frequency response and Figure 9.37 plots the position open-loop

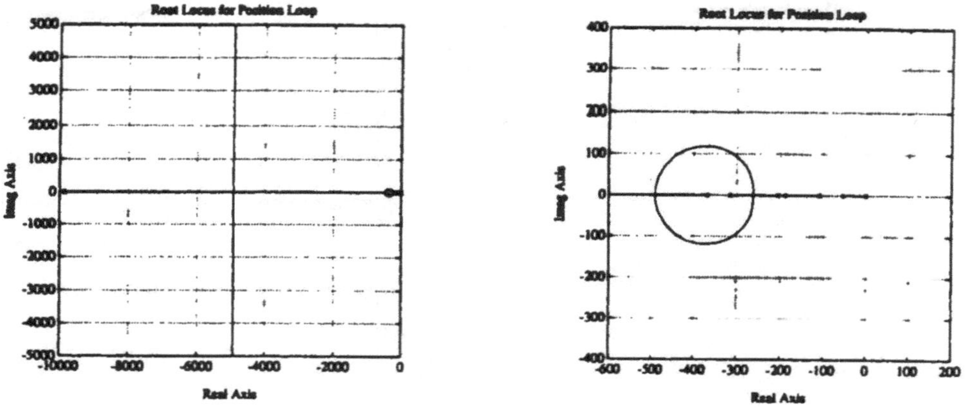

Figure 9.38 (a) and (b) - Position loop root locus.

response with unity gain on the proportional controller. Once again notice the low frequency slope of −20 dB/decade on the magnitude plot and −90° phase in the phase plot, indicate the natural free integrator in the position loop.

$$\frac{V_0}{V_i} = -0.510\frac{1+0.0186s}{1+0.00928s}. \quad (9.61)$$

$$\phi_{max} = 19.5^\circ,\ \omega_0 = 76.2\text{(rad/s)},$$
$$\alpha = \frac{0.00928}{0.0186} = \frac{53.8}{108} = 0.497. \quad (9.62)$$

To determine the appropriate position loop gain, K_p, for the system, a root locus plot in conjunction with gain plots and other parametric plots is used. Figure 9.38a is the root locus for the position loop. Figure 9.38b is a close-up view of the dominant poles and zeros located close to the origin. The pole at the origin is the natural integrator of the system, the other poles and zeros are from the lead compensator and the dynamics of the velocity loop. Clearly, from the root locus, there is a range of gains that yields an unacceptable damping ratio for this system. The gain plots of the position loop, shown in Figure 9.39, indicate that gains of greater than approximately 3000 are unacceptable. It is worthwhile noting that the gain plots show the two break-out points (at gain values of approximately 75 and 2600) and the break-in point at an approximate gain of 590. Due to power limitations, the gain of the position loop is limited to a maximum of 1000. Thus, the root locus shown in Figure 9.38b is more appropriate for the remainder of this design. Figure 9.40 shows the root sensitivity of the position loop. From the root sensitivity plot, it is clear that values of gain close to the break points on the root locus result in root sensitivities that do not meet specifications. Figure 9.40, also shows that the ranges $600 \leq K_p \leq 1000$ and $2000 \leq K_p \leq 3000$ are unacceptable.

Gain and phase margin can also be plotted as functions of the forward loop gain (Figures 9.41 and 9.42). The slope of the gain margin plot is −20 dB/decade as gain margin is directly related to the gain. Note that the maximum phase margin occurs at a gain of approximately 100. Based on these two plots, the gain must be lower than approximately 3000.

From the plots above, loop gains of either 300 or 1000 may be employed for the position compensator. The closed-loop frequency response is plotted in Figures 9.43 and 9.44 from the gain values of 300 and 1000, respectively. Either one of these designs results in an acceptable performance. Note that the higher value of gain yields a higher bandwidth system.

Summary Results

To summarize the results from the velocity and position compensator designs, the Tables 9.8 – 9.10 can be generated.

As can be seen from Tables 9.8 – 9.10, the performance objectives have been met for both the velocity and position loops. The only decision that remains is whether to use the lower or higher value for the position loop gain. For this particular case, it was decided to use the lower value which required a smaller and less expensive power supply.

9.3.6 Conclusions

Ultra-high precision systems are becoming more commonplace in the industrial environment as competition increases the demand for higher precision at greater speeds. The two most important factors involved in the successful implementation of an ultra-high precision system are a solid design of the open-loop system and a good model of the system and any compensator components. This chapter presented some basic concepts critical for the design and implementation of ultra-high precision control systems, including some instrumentation concerns, system identification issues and compensator design and implementation.

Clearly, this chapter is limited in its scope and there are many other details that must be considered before a machine can approach the accuracies of high performance coordinate measurement machines or diamond turning machines. Such machines have been developed over many years and continue to improve with time and technology. The interested reader is referred to the references at the end of this chapter for further details of designing and implementating ultra-high precision systems.

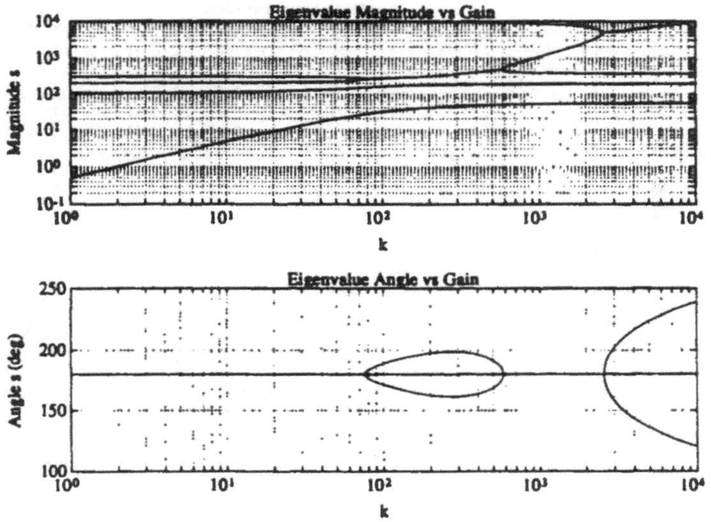

Figure 9.39 Position loop compensator gain plots.

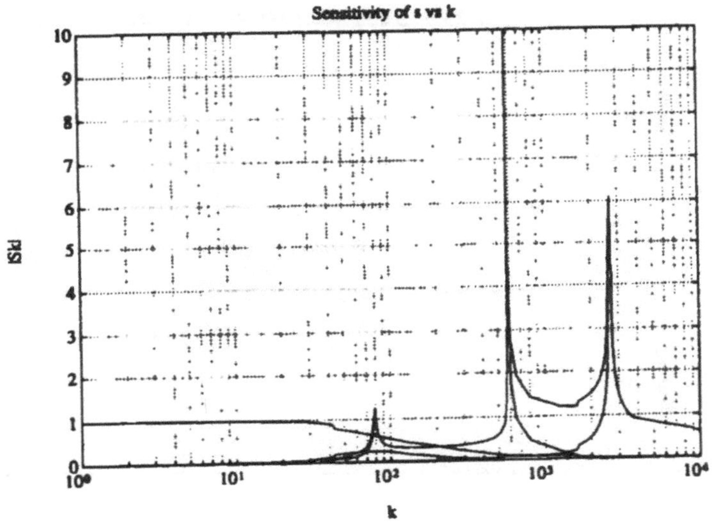

Figure 9.40 Position loop compensator root sensitivity.

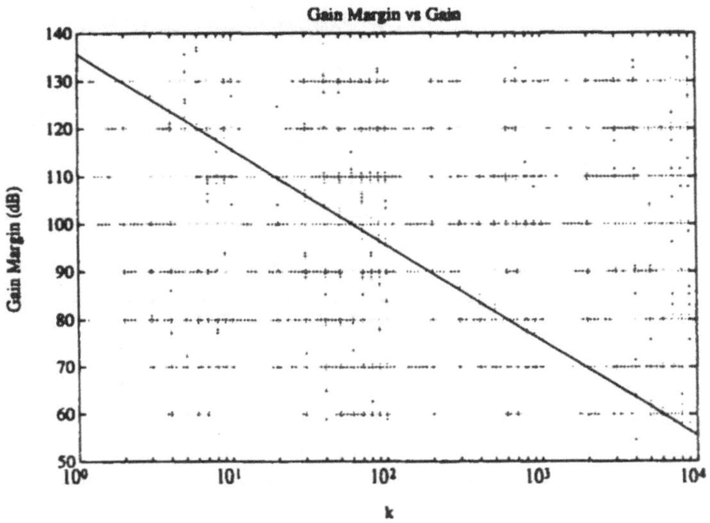

Figure 9.41 Position loop gain margin.

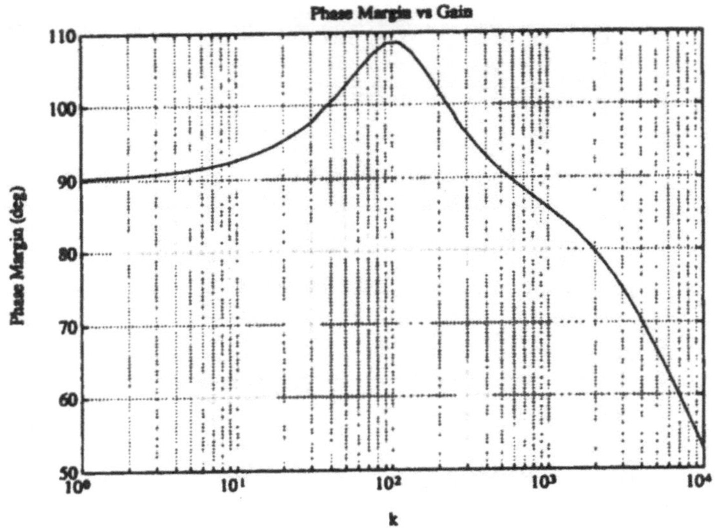

Figure 9.42 Position loop phase margin.

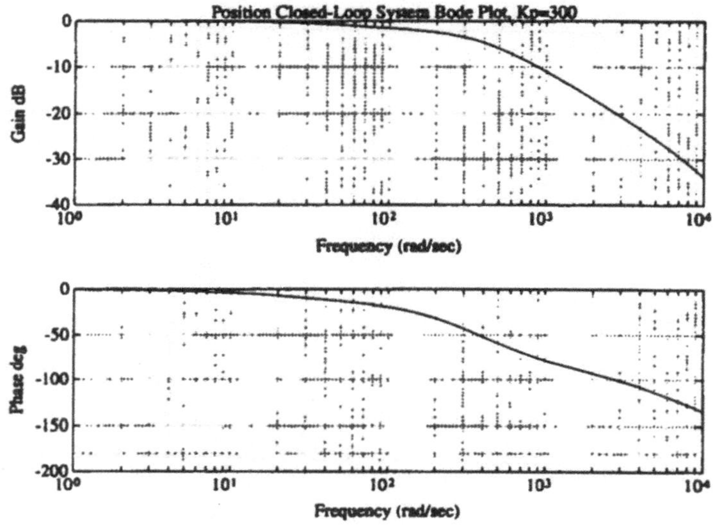

Figure 9.43 Position closed-loop frequency response, $K_p = 300$.

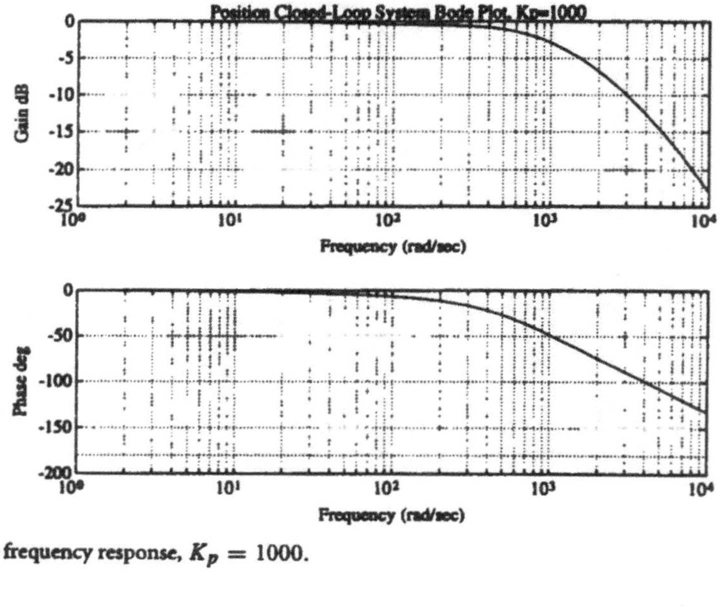

Figure 9.44 Position closed-loop frequency response, $K_p = 1000$.

TABLE 9.8 Results Velocity Loop

Specification	Design target	Value achieved
Min. closed-loop bandwidth	150 (Hz) (≈950 rad/s)	637 (Hz) (≈4000 rad/s)
Steady-State error to a step	0	0
Min. damping ratio, ζ	0.8	1.0
Max. $\|S_k\|$	1.0	0.92
Min. phase margin, ϕ_m	90°	93°
Min. gain margin, g_m	40 (dB)	∞ (dB)

TABLE 9.9 Position Loop $K_p = 300$

Specification	Design target	Value achieved
Min. closed-loop bandwidth	15 (Hz) (≈95 rad/s)	32 (Hz) (≈200 rad/s)
Steady-State error to a step	0	0
Min. damping ratio, ζ	0.9	0.93
Max. $\|S_k\|$	1.5	0.4
Min. phase margin, ϕ_m	75°	96°
Min. gain margin, g_m	60(dB)	87 (dB)

TABLE 9.10 Position Loop $K_p = 1000$

Specification	Design target	Value achieved
Min. closed-loop bandwidth	15 (Hz) (≈95 rad/s)	160 (Hz) (≈1000 rad/s)
Steady-State error to a step	0	0
Min. damping ratio, ζ	0.9	1.0
Max. $\|S_k\|$	1.5	1.2
Min. phase margin, ϕ_m	75°	85°
Min. gain margin, g_m	60(dB)	76 (dB)

9.3.7 Defining Terms

A quad B: Two phased (90 degrees) signals from an encoder.

A/D: Analog to digital conversion.

Abbe error: Measurement error which occurs when a gage is not collinear to the object measured.

aliasing: Identification of a higher frequency as a lower one, when the higher frequency is above the Nyquist frequency.

ARX: AutoRegressive eXternal input identification technique.

CMM: Coordinate measuring machine.

CNC: Computer numerical control.

cross-coupling: One axis affecting another.

D/A: Digital to analog conversion.

DTM: Diamond turning machine.

dynamometer: High precision and bandwidth multiaxis force sensor.

laser interferometer: Measurement instrument based on light wavelength interference.

LVDT: Linear variable differential transformer.

metglass: Magnetostrictive material (such as Terfenol^{R}).

metrology: The study of measurement.

piezoelectric effect: Strain-induced voltage, or voltage-induced strain.

stiction: Low velocity friction.

THD: Total harmonic distortion.

ultra-high precision: Dimensional accuracies of 0.1 μin (1 μin = 10^{-6} in).

white noise: Random signal with a uniform probability distribution.

References

[1] Methods for Performing Evaluation of Computer Numerically Controlled Machining Centers. *ANSI-ASME B5.54*, 1992.

[2] Blake, P., Bifano, T., Dow, T., and Scattergood, R., Precision Machining Of Ceramic Materials, *Am. Ceram. Soc. Bull.*, 67(6), 1038–1044, 1988.

[3] Bryan, J. B., The Abbe Principle Revisited-An Updated Interpretation, *Precision Eng.*, 1(3), 129–132, 1989.

[4] Bryan, J. B., The Power of Deterministic Thinking in Machine Tool Accuracy, UCRL-91531, 1984.

[5] Bryant, M.D. and Reeves, R.B., Precise Positioning Problems Using Piezoelectric Actuators with Force Transmission Through Mechanical Contact, *Precision Eng.*, 6, 129–134, 1984.

[6] Cetinkunt, S. and Donmez, A., CMAC Learning Controller for Servo Control of High Precision Machine Tools, *American Control Conference*, 2, 1976–80, 1993.

[7] Daneshmend, L.K. and Pak, H.A., Model Reference Adaptive Control of Feed Force in Turning, *J. Dyn. Syst. Meas. Control*, 108, 215–222, 1986.

[8] Dorf, R.C. and Kusiak, A., Eds., *Handbook of Design, Manufacturing and Automation*, John Wiley & Sons, New York, 1994.

[9] Falter, P. J. and Dow, T. A., Design And Performance Of A Small-Scale Diamond Turning Machine, *Precision Eng.*, 9(4), 185–190, 1987.

[10] Fornaro, R. J. and Dow, T. A., High-performance Machine Tool Controller, *Conf. Rec. - IEEE Ind. Appl. Soc. Annual Meeting*, 35(6), 1429–1439, 1988.

[11] Fussell, B.K. and Srinivasan, K., Model Reference Adaptive Control of Force In End Milling Operations, *Am. Control Conf.*, 2, 1189–94, 1988.

[12] Jenkins, H.E., Kurfess, T.R., and Dorf, R.C., Design of a Robust Controller for a Grinding System, *IEEE Conf. Control Appl.*, 3, 1579–84, 1994.

[13] Jeong, S. and Ro, P.I., Cutting Force-Based Feedback Control Scheme for Surface Finish Improvement in Diamond Turning, *Am. Control Conf.*, 2, 1981–1985, 1993.

[14] Jones, R. and Richardson, J., The Design and Application of Sensitive Capacitance Micrometers, *J. Phys. E: Scientific Instruments*, 6, 589, 1973.

[15] Kurfess, T. R. and Nagurka, M. L., Understanding the Root Locus Using Gain Plots, *IEEE Control Syst. Mag.*, 11(5), 37–40, 1991.

[16] Kurfess, T.R. and Nagurka, M.L., A Geometric Representation of Root Sensitivity, *J. Dyn. Syst. Meas. Control*, 116(2), 305–9, 1994.

[17] Li, C. J. and Li, S. Y., A New Sensor for Real-Time Milling Tool Condition Monitoring, *J. Dyn. Syst. Meas. Control*, 115, 285–290, 1993.

[18] Liang, S.Y. and Perry, S.A., In-Process Compensation For Milling Cutter Runout Via Chip Load Manipulation, *J. Eng. Ind. Trans. ASME*, 116(2), 153–160, 1994.

[19] Ljung, L., *System Identification: Theory for the User*, Prentice-Hall, Englewood Cliffs, NJ, 1987.

[20] Lo, C. and Koren, Y., Evaluation of Machine Tool Controllers, *Proc. Am. Control Conf.*, 1, 370–374, 1992.

[21] Meckl, P. H. and Kinceler, R., Robust Motion Control of Flexible Systems Using Feedforward Forcing Functions, *IEEE Trans. Control Syst. Technol.*, 2(3), 245–254, 1994.

[22] Pien, P.-Y. and Tomizuka, M., Adaptive Force Control of Two Dimensional Milling, *Proc. Am. Control Conf.*, 1, 399–403, 1992.

[23] Slocum, A., Scagnetti, P., and Kane, N., Ceramic Machine Tool with Self-Compensated, Water-Hydrostatic, Linear Bearings, *ASPE Proc.*, 57–60, 1994.

[24] Stute, G., Adaptive Control, *Technology of Machine Tools*, Machine Tool Controls, Lawrence Livermore Laboratory, Livermore, CA, 1980, vol. 4.

[25] Tsao, T.C. and Tomizuka, M., Adaptive and Repetitive

Digital Control Algorithms for Noncircular Machining, *Proc. Am. Control Conf.*, 1, 115–120, 1988.

[26] Tung, E. and Tomizuka, M., Feedforward Tracking Controller Design Based on the Identification of Low Frequency Dynamics, *J. Dyn. Syst. Meas. Control*, 115, 348–356, 1993.

[27] Tung, E., Anwar, G., Tomizuka, M., Low Velocity Friction Compensation and Feedforward Solution Based on Repetitive Control, *J. Dyn. Syst. Meas. Control*, 115, 279–284, 1993.

[28] Ulsoy, A.G. and Koren, Y., Control of Machine Processes, *J. Dyn. Syst. Meas. Control*, 115, 301–8, 1993.

9.4 Robust Control of a Compact Disc Mechanism

Maarten Steinbuch, Philips Research Laboratories, Eindhoven, The Netherlands

Gerrit Schootstra, Philips Research Laboratories, Eindhoven, The Netherlands

Okko H. Bosgra, Mechanical Engineering Systems and Control Group, Delft University of Technology, Delft, The Netherlands

9.4.1 Introduction

A compact disc (CD) player is an optical decoding device that reproduces high-quality audio from a digitally coded signal recorded as a spiral-shaped track on a reflective disc [2]. Apart from the audio application, other optical data systems (CD-ROM, optical data drive) and combined audio/video applications (CD-interactive, CD-video) have emerged. An important research area for these applications is the possibility of increasing the rotational frequency of the disc to obtain faster data readout and shorter access time. For higher rotational speeds, however, a higher servo bandwidth is required that approaches the resonance frequencies of bending and torsional modes of the CD mechanism. Moreover, the system behavior varies from player to player because of manufacturing tolerances of CD players in mass production, which explains the need for robustness of the controller.

Further, an increasing percentage of all CD-based applications is for portable use. Thus, additionally, power consumption and shock sensitivity play a decisive role in the performance assessment of controller design for CD systems.

In this chapter we concentrate on the possible improvements of both the track-following and focusing behavior of a CD player, using robust control design techniques.

9.4.2 Compact Disc Mechanism

A schematic view of a CD mechanism is shown in Figure 9.45. The mechanism is composed of a turntable dc motor for the rotation of the CD, and a balanced radial arm for track following. An optical element is mounted at the end of the radial arm. A diode located in this element generates a laser beam that passes through a series of optical lenses to give a spot on the information layer of the disc. An objective lens, suspended by two parallel leaf springs, can move in a vertical direction to give a focusing action.

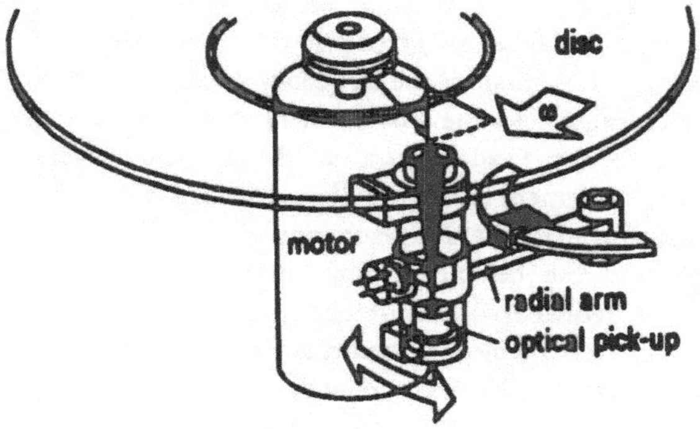

Figure 9.45 Schematic view of a rotating-arm compact disc mechanism.

Both the radial and the vertical (focus) position of the laser spot, relative to the track of the disc, have to be controlled actively. To accomplish this, the controller uses position-error information provided by four photodiodes. As input to the system, the controller generates control currents to the radial and focus actuator, which both are permanent-magnet/coil systems.

In Figure 9.46 a block diagram of the control loop is shown. The difference between the radial and vertical track position and the spot position is detected by the optical pickup; it generates a radial error signal (e_{rad}) and a focus error signal (e_{foc}) via the optical gain K_{opt}. A controller $K(s)$ feeds the system with the currents I_{rad} and I_{foc}. The transfer function from control currents to spot position is indicated by $H(s)$. Only the position-error signals after the optical gain are available for measurement. Neither the true spot position nor the track position is available as a signal.

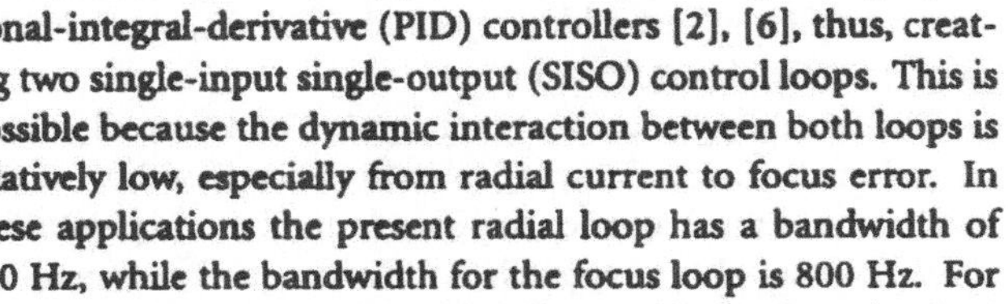

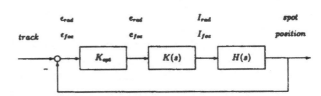

Figure 9.46 Configuration of the control loop.

In current systems, $K(s)$ is formed by two separate proportional-integral-derivative (PID) controllers [2], [6], thus, creating two single-input single-output (SISO) control loops. This is possible because the dynamic interaction between both loops is relatively low, especially from radial current to focus error. In these applications the present radial loop has a bandwidth of 500 Hz, while the bandwidth for the focus loop is 800 Hz. For more demanding applications (as discussed in the introduction) it is necessary to investigate whether improvements of the servo behavior are possible.

9.4.3 Modeling

A measured frequency response of the CD mechanism $G(s) = K_{opt}H(s)$ is given in Figure 9.47 (magnitude only). It has been determined by spectrum analysis techniques. At low frequencies, the rigid body mode of the radial arm and the lens-spring system (focus) can be easily recognized as double integrators in the 1,1 and 2,2 element of the frequency response, respectively. At higher frequencies the measurement shows parasitic dynamics, especially in the radial direction. Experimental modal analyses and finite element calculations have revealed that these phenomena are due to mechanical resonances of the radial arm, mounting plate, and disc (flexible bending and torsional modes).

With frequency-domain-based system identification, each element of the frequency response has been fitted separately using an output error model structure with a least-square criterion [10]. Frequency-dependent weighting functions have been used to improve the accuracy of the fit around the anticipated bandwidth of 1 kHz. The 2,1 element appeared to be difficult to fit because of the nonproper behavior in the frequency range of interest.

Combination of the fits of each element resulted in a 37th-order multivariable model. Using frequency-weighted balanced reduction [13], [3], this model was reduced to a 21st-order model, without significant loss in accuracy. The frequency response of the model is also shown in Figure 9.47.

Uncertainty Modeling

The most important system variations we want to account for are

1. Unstructured difference between model and measurement
2. Uncertain interaction
3. Uncertain actuator gain
4. Uncertainty in the frequencies of the parasitic resonances

The first uncertainty stems from the fact that our nominal model is only an approximation of the measured frequency response because of imperfect modeling. Further, a very-high-order nominal model is undesirable in robust control design since the design technique yields controllers with the state dimension of the nominal model plus weighting functions. For that reason, our nominal model describes only the rigid body dynamics, as well as the resonance modes that are most relevant in the controlled situation. Unmodeled high-frequency dynamics and the unstructured difference between model and measurement are modeled as a complex valued additive perturbation Δ_a, bounded by a high-pass weighting function.

The remaining uncertainty sources 2, 3 and 4 are all intended to reflect how manufacturing tolerances manifest themselves as variations in the frequency response from player to player. By so doing, we are able to appreciate the consequences of manufacturing tolerances on control design.

The uncertain interaction, item 2, is modeled using an antidiagonal output multiplicative parametric perturbation: $y = (I + \Delta_o)Cx$ where

$$\Delta_o = \begin{bmatrix} 0 & w_{o1}\delta_{o1} \\ w_{o2}\delta_{o2} & 0 \end{bmatrix}$$

The scalar weights w_{o1} and w_{o2} are chosen equal to 0.1, meaning 10% uncertainty.

Dual to the uncertain interaction, the uncertain actuator gains, item 3, are modeled as a diagonal input multiplicative parametric perturbation: $\dot{x} = Ax + B(I + \Delta_i)u$ where

$$\Delta_i = \begin{bmatrix} w_{i1}\delta_{i1} & 0 \\ 0 & w_{i2}\delta_{i2} \end{bmatrix} \tag{9.63}$$

The gain of each actuator is perturbed by 5%. With this value also non-linear gain variations in the radial loop due to the rotating-arm principle are accounted for along with further gain variations caused by variations in track shape, depth, and slope of the pits on the disc and varying quality of the transparent substrate protecting the disc [6].

Finally, we consider variations in the undamped natural frequency of parasitic resonance modes, item 4. From earlier robust control designs [11] it is known that the resonances at 0.8, 1.7, and 4.3 kHz are especially important. The modeling of the variations of the three resonance frequencies is carried out with the parametric uncertainty modeling toolbox [9]. The outcome of the toolbox is a linear fractional transformation description of the perturbed system with a normalized, block diagonal, parametric perturbation structure $\Delta_{par} = diag\{\delta_1 I_2, \delta_2 I_2, \delta_3 I_2\}$. Each frequency perturbation involves a real-repeated perturbation block of multiplicity two. The repeatedness stems from the fact that the frequencies ω_0 appear quadratically in the A-matrix. The lower and upper bounds are chosen 2.5% below and above the nominal value, respectively; see [8] for more details.

9.4.4 Performance Specification

A major disturbance source for the controlled system is track position irregularities. Based on standardization of CDs, in the radial direction the disc specifications allow a track deviation of 100 μm (eccentricity) and a track acceleration at scanning velocity of 0.4 m/s^2, while in the vertical direction these values are 1 mm and 10 m/s^2, respectively.

Apart from track position irregularities, a second important disturbance source is external mechanical shocks. Measurements show that during portable use disturbance signals occur in the frequency range from 5 up to 150 Hz, with accelerations (of the chassis) up to 50 m/s^2.

For the CD player to work properly, the maximum allowable position error is 0.1 μm in the radial direction and 1 μm in the focus direction. In the frequency domain, these performance specifications can be translated into requirements on the shape of the (output) sensitivity function $S = (I+GK)^{-1}$. Note that the track irregularities involve time-domain constraints on signals, which are hard to translate into frequency-domain specifications. To obtain the required track disturbance attenuation, the magnitude of the sensitivity at the rotational frequency should be less

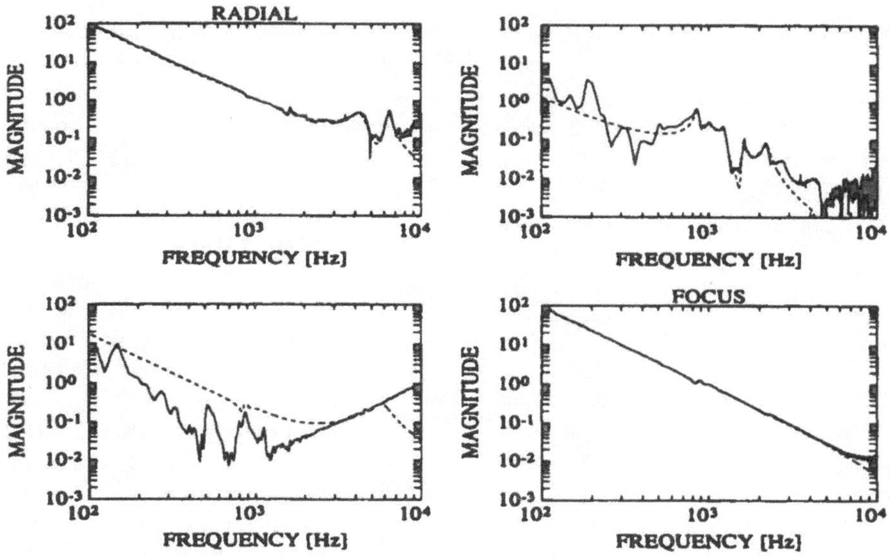

Figure 9.47 Measured frequency response of the CD mechanism (—) and of the identified 21st-order model (- -).

than 10^{-3} in both the radial and the focus direction. Further, for frequencies up to 150 Hz, the sensitivity should be as small as possible to suppress the impact of mechanical shocks and higher harmonics of the disc eccentricity.

To still satisfy these requirements for increasing rotational speed, the bandwidth has to be increased. As stated before, a higher bandwidth has implications for robustness of the design against manufacturing tolerances. But also, a higher bandwidth means a higher power consumption (very critical in portable use), generation of audible noise by the actuators, and poor playability of discs with surface scratches. Therefore, under the required disturbance rejection, we are striving towards the lowest possible bandwidth.

In addition to these conflicting bandwidth requirements, the peak magnitude of the sensitivity function should be less than 2 to create sufficient phase margin in both loops. This is important because the controller has to be discretized for implementation. Note that the Bode sensitivity integral plays an important role in the trade-off between the performance requirements.

9.4.5 μ-synthesis for the CD Player

To start with, the performance specifications on S can be combined with the transfer function KS associated with the complex valued additive uncertainty Δ_a. The performance trade-offs are then realized with a low-pass weighting function W_1 on S, while W_2 on KS reflects the size of the additive uncertainty and can also be used to force high roll-off at the input of the actuators.

In this context, the objective of achieving robust performance [1], [5] means that for all stable, normalized perturbations Δ_a the closed loop is stable and

$$\| W_1[I + (G + W_2\Delta_a)K]^{-1} \|_\infty < 1$$

This objective is exactly equal to the requirement that

$$\mu_\Delta[F_l(P, K)] < 1$$

with

$$F_l(P, K) = \begin{bmatrix} W_2KS & W_2KS \\ W_1S & W_1S \end{bmatrix} \quad (9.64)$$

and μ_Δ is computed with respect to the structured uncertainty block $\Delta = diag(\Delta_a, \Delta_p)$, in which Δ_p represents a fictitious 2×2 performance block. Note that this problem is related to the more common H_∞ problem

$$\left\| \begin{matrix} W_2KS \\ W_1S \end{matrix} \right\|_\infty < 1$$

However, with this H_∞ design formulation it is possible to design only for nominal performance and robust stability.

An alternative that we use is to specify performance on the transfer function SG instead of S, reducing the resulting controller order by 4 [11]. Then the following standard plant results

$$F_l(P, K) = \begin{bmatrix} W_2KS & W_2KSG \\ W_1S & W_1SG \end{bmatrix} \quad (9.65)$$

This design formulation has the additional advantage that it does not suffer from pole-zero cancellation as in the mixed sensitivity design formulations as in Equation 9.64.

Using the D-K iteration scheme [1], [5], μ-controllers are synthesized for several design problems. Starting with the robust performance problem of Equation 9.65, the standard plant is augmented step by step with the parametric perturbations 2, 3, and 4 listed earlier. In the final problem, with the complete uncertainty model, we thus arrive at a robust performance problem having nine blocks: $\Delta = diag\{\delta_1 I_2, \delta_2 I_2, \delta_3 I_2, \delta_{i1}, \delta_{i2}, \delta_{o1}, \delta_{o2}, \Delta_a, \Delta_p\}$.

The most important conclusions with respect to the use of μ-synthesis for this problem are

- Convergence of the D-K iteration scheme was fast (in most cases two steps). Although global convergence of the scheme cannot be guaranteed, in our case the resulting μ-controller did not depend on the starting controller.

- The final μ-controllers did have high order [39 for design (Equation 9.65) up to 83 for the full problem] due to the dynamic D-scales associated with the perturbations.
- Although most of the perturbations are real valued by nature, the assumption in design that all perturbations are complex valued did not introduce much conservativeness with respect to robust performance; see Figure 9.48.

  Note that the peak value of μ over frequency is 1.75, meaning that robust performance has not been achieved. This is due to the severe performance weighting W_1.
- The most difficult aspect of design appeared to be the shaping of weighting functions such that a proper trade-off is obtained between the conflicting performance and robustness specifications.

In Figure 9.49 the sensitivity transfer function is shown for the full problem μ-controller. For comparison, the result is also shown for two decentralized PID controllers achieving 800-Hz bandwidth in both loops. Clearly, the μ-controller achieves better disturbance rejection up to 150 Hz, has lower interaction, and a lower sensitivity peak value.

The controller transfer functions are given in Figure 9.50. Clearly, the μ-controller has more gain at low frequencies and actively acts upon the resonance frequencies.

9.4.6 Implementation Results

The digital implementation brings along a choice for the sampling frequency of the discretized controller. Based on experience with previous implementations of SISO radial controllers [11] and on the location of the fastest poles in the multiple-input multiple-output (MIMO) μ-controller (± 8 kHz), it is the intention to discretize the controller at 40 kHz. However, in the DSP environment used, this sampling frequency means that the order of the controller that can be implemented is, at most, 8. Higher order will lead to an overload of the DSP. This indicates a need for a dramatic controller order reduction since there is a large gap between the practically allowable controller order (8) and the controller order that have been found using the μ-synthesis methodology (83). It is, however, unlikely that the order of these controllers can be reduced to 8 without degrading the nominal and robust performance very much. To be able to implement more complex controllers there are two possibilities:

1. Designing for a sampling frequency below 40 kHz
2. Using more than one DSP system and keeping the sampling frequency at 40 kHz

In this research we chose the latter, for it is expected that the sampling frequency has to be lowered to such an extent that the performance will degrade too much (too much phase lag around the bandwidth leading to an unacceptable peak value in the sensitivity function). Although the second option introduces some additional problems, it probably gives a better indication of how much performance increase is possible with respect to PID-like control without letting this performance be influenced too much by the restrictions of implementation. To do this, model reduction of the controller is applied. The first step involves balancing and truncating of the controller states. This reduces the number of states from 83 to 53, without loss of accuracy. The next step is to split up the μ-controller into two 1-input 2-output parts: $K_{53} = \begin{bmatrix} K_{53}^1 & K_{53}^2 \end{bmatrix}$. Using a frequency-weighted closed-loop model reduction technique, the order of each of these single input multiple-output (SIMO) parts can be reduced while the loop around the other part is closed. Each of the reduced-order parts can be implemented at 40 kHz, in a separate DSP system, thus facilitating a maximum controller order of 16 (under the additional constraint that each part can be, at most, of 8th order). The resulting controller is denoted $K_8 = \begin{bmatrix} K_8^1 & K_8^2 \end{bmatrix}$.

The actual implementation of the controllers in the DSP systems has been carried out with the dSPACE Cit-Pro software [7]. Most problems occurring when implementing controllers are not essential to control theory, and most certainly not for H_∞ and μ theory, but involve problems such as scaling inputs and outputs to obtain the appropriate signal levels and to ensure that the resolution of the DSP is optimally used. These problems can be solved in a user-friendly manner using the dSPACE software.

The two DSP systems used are a Texas Instruments TMS320C30 16-bit processor with floating-point arithmetic and a Texas Instruments TMS320C25 16-bit processor with fixed-point arithmetic. The analog-to-digital converters (ADC) are also 16 bit and have a conversion time of 5 μs. The maximum input voltage is ± 10 V. The digital-to-analog converters (DAC) are 12 bit, have a conversion time of 3 μs, and also operate within the range of ± 10 V.

The first column $K_8^1(s) = [K_{11} \quad K_{21}]^T$ has been implemented in the TMS320C30 processor at a sampling frequency of 40 kHz. The dSPACE software provides for:

- A ramp-invariant discretization
- A modal transformation
- Generation of C code that can be downloaded to the DSP

The second column $K_8^2(s) = [K_{21} \quad K_{22}]^T$ is implemented in the TMS320C25 processor. Since this processor has fixed-point arithmetic, scaling is more involved; see also [12]. With the dSPACE software, the following steps have been carried out:

- A ramp-invariant discretization
- A modal transformation
- l_1 scaling for the nonintegrator states
- State scaling for the integrator states such that their contribution is, at most, 20%
- Generation of DSPL code that can be downloaded to the DSP

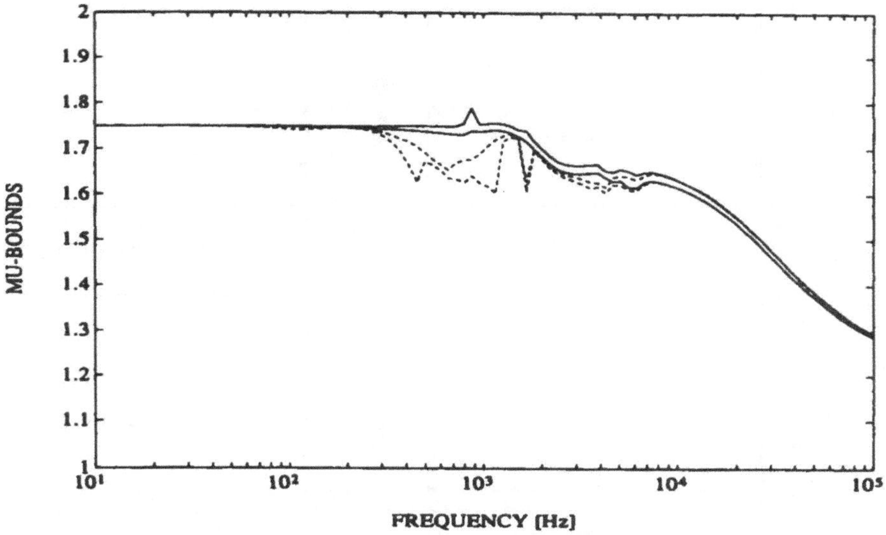

Figure 9.48 Complex (—) and real (- -) μ-bounds for the μ-controller on the standard plant, including all perturbations.

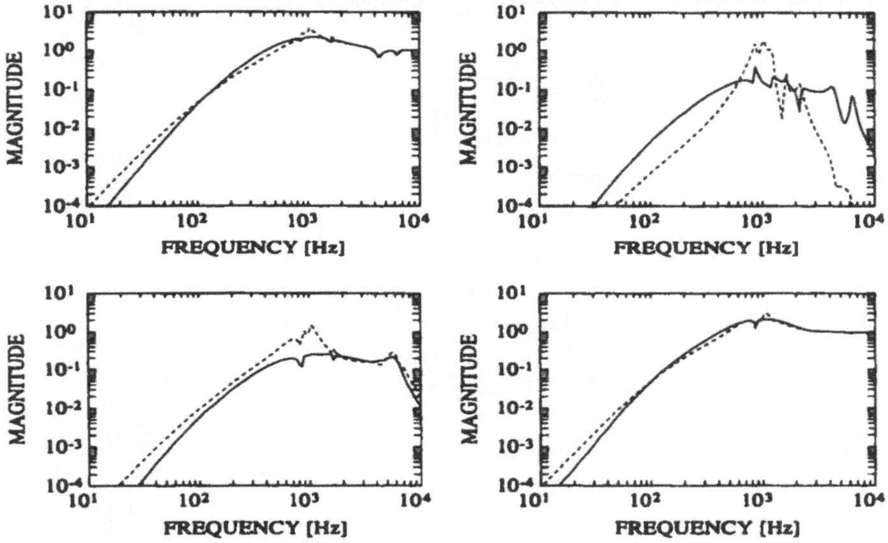

Figure 9.49 Nominal performance in terms of the sensitivity function with μ-controller (—) and two PID controllers (- -).

For the TMS320C25, a discretization at a sampling frequency of 40 kHz appeared to be too high since it resulted in a processor load of 115%. For that reason, the sampling frequency of this DSP has been lowered to 34 kHz, yielding a processor load of 98.6%.

The DSP systems have been connected to the experimental setup by means of two analog summing junctions that have been designed especially for this purpose; see Figure 9.51.

When the external 2 × 8 μ-controller of the complete design is connected to the experimental setup, we can measure the achieved performance in terms of the frequency response of the sensitivity function. The measurements have been carried out using a Hewlett Packard 3562 Dynamic Signal Analyzer. Because this analyzer can measure only SISO frequency responses, each of the four elements of the sensitivity function has been determined separately.

The measurements are started at the same position on the disc each time. This position is chosen approximately at the halfway point on the disc since the model is identified here and the radial gain is most constant in this region. In Figure 9.52 the measured and simulated frequency response of the sensitivity function is shown. The off-diagonal elements are not very reliable since the coherence during these measurements was very low because of small gains and nonlinearities, leading to bad signal-to-noise ratios.

The nominal performance has also been tested in terms of the possibility to increase the rotational frequency of the CD. It appeared possible to achieve an increase in speed of a factor 4 (with an open-loop bandwidth of 1 kHz).

Concluding this section, the measurements show that considerable improvements have been obtained in the suppression of track irregularities leading to the possibility of increasing the rotational frequency of the disc to a level that has not been achieved

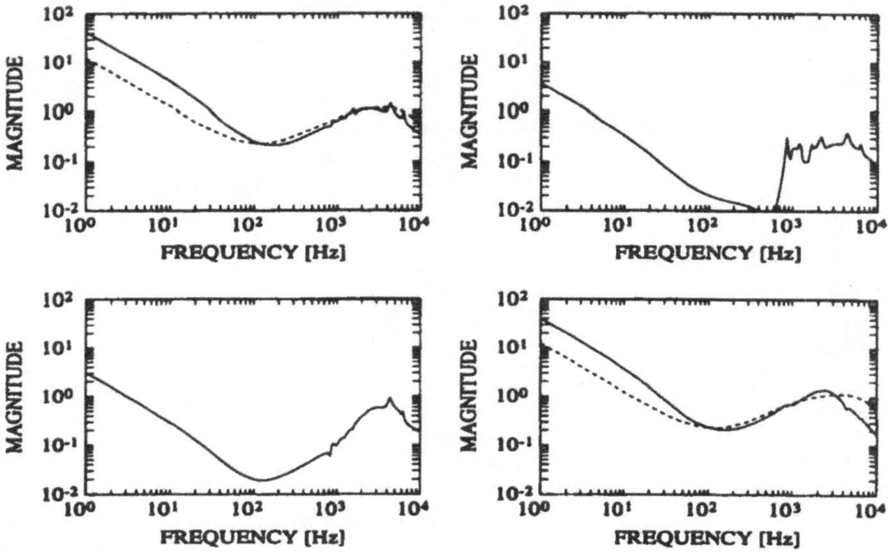

Figure 9.50 Frequency response of the μ-controller (—) and two PID controllers (- -).

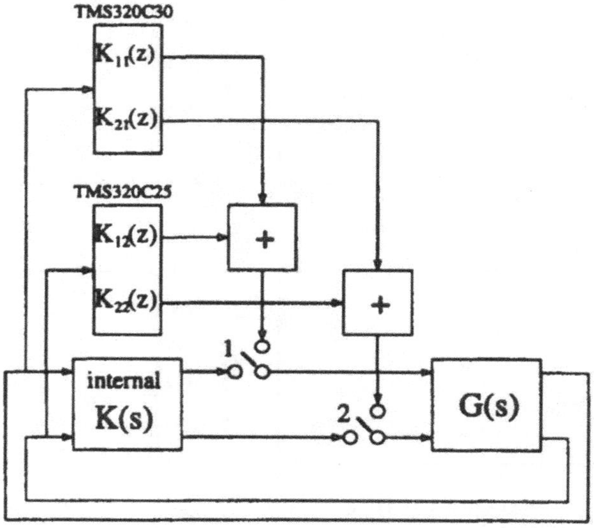

Figure 9.51 The connection of both DSP systems to the experimental setup.

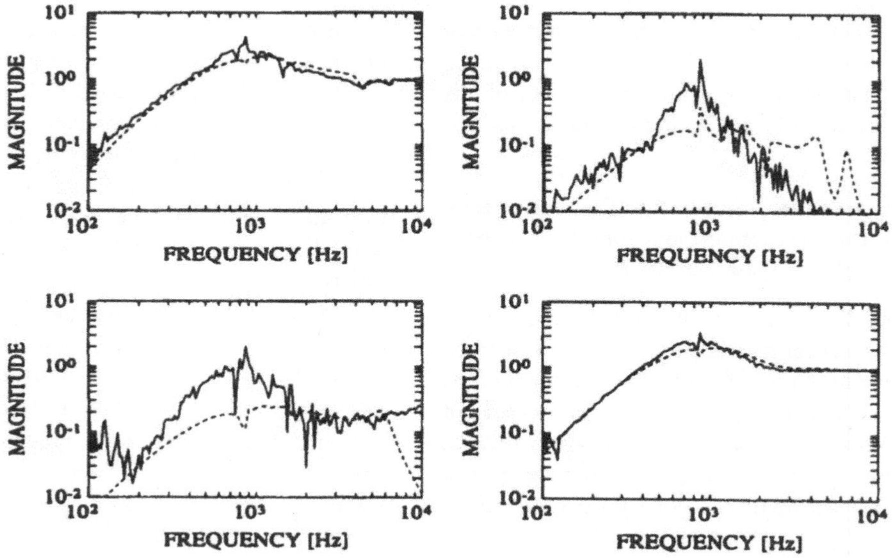

Figure 9.52 Measured frequency response of the input sensitivity function for the 2 × 8 reduced-order controller of the complete design (—) and the simulated output sensitivity function for the 83rd-order controller (- - -).

before. Nevertheless, the implemented performance differs on a few points from the simulated performance. It seems useful to exploit this knowledge to arrive at even better results in a next controller synthesis. Notice also the work in [4] where a controller for the radial loop has been designed using QFT, directly based on the measured frequency responses.

9.4.7 Conclusions

In this chapter μ-synthesis has been applied to a CD player. The design problem involves time-domain constraints on signals and robustness requirements for norm-bounded structured plant uncertainty. Several different uncertainty structures of increasing complexity are considered. A μ-controller has been implemented successfully in an experimental setup using two parallel DSPs connected to a CD player.

References

[1] Balas, G.J., Doyle, J.C., Glover, K., Packard, A.K., and Smith, R., *μ-Analysis and Synthesis Toolbox*, MUSYN Inc., Minneapolis, MN, 1991.

[2] Bouwhuis, G. et al., *Principles of Optical Disc Systems*, Adam Hilger Ltd., Bristol, UK, 1985.

[3] Ceton, C., Wortelboer, P., and Bosgra, O.H., Frequency weighted closed-loop balanced reduction, in *Proc. 2nd Eur. Control Conf.*, Groningen, June 26 - July 1 1993, 697–701.

[4] Chait, Y., Park, M.S., and Steinbuch, M., Design and implementation of a QFT controller for a compact disc player, *J. Syst. Eng.*, 4, 107–117, 1994.

[5] Doyle, J.C., Advances in Multivariable Control, lecture notes of the ONR/Honeywell Workshop, Honeywell, MN, 1984.

[6] Draijer, W., Steinbuch, M., and Bosgra, O.H., Adaptive control of the radial servo system of a compact disc player, *IFAC Automatica*, 28, 455–462, 1992.

[7] dSPACE GmbH, *DSPCitPro software package*, documentation of dSPACE GmbH, Paderborn, West Germany, 1989.

[8] Groos, P.J.M. van, Steinbuch, M., and Bosgra, O.H., Multivariable control of a compact disc player using μ synthesis, in Proc. 2nd Eur. Control Conf., Groningen, June 26 - July 1, 1993, 981–985.

[9] Lambrechts, P., Terlouw, J.C., Bennani, S., and Steinbuch, M., Parametric uncertainty modeling using LFTs, in Proc. 1993 Am. Control Conf., June 1993, 267–272

[10] Schrama, R., Approximate Identification and Control Design, Ph.D. thesis, Delft University of Technology, Delft, The Netherlands, 1992.

[11] Steinbuch, M., Schootstra, G., and Bosgra, O.H., Robust control of a compact disc player, IEEE 1992 Conf. Decision Control, Tucson, AZ, 2596–2600.

[12] Steinbuch, M., Schootstra, G., and Goh, H.T., Closed loop scaling in fixed-point digital control, *IEEE Trans. Control Syst. Technol.*, 2(4), 312–317, 1994.

[13] Wortelboer, P., Frequency Weighted Balanced Reduction of Closed-loop Mechanical Servo-systems, Ph.D. thesis, Delft University of Technology, Delft, The Netherlands, 1994.

III

Electrical and Electronic Control Systems

10

Power Electronic Controls

George C. Verghese
Massachusetts Institute of Technology

David G. Taylor
Georgia Institute of Technology, School of Electrical and Computer Engineering, Atlanta, GA

Thomas M. Jahns
GE Corporate R&D, Schenectady, NY

Rik W. De Doncker
Silicon Power Corporation, Malvern, PA

10.1 Dynamic Modeling and Control in Power Electronics 241
Introduction • Prototype Converters • Dynamic Modeling and Control • Extensions • Conclusion
References 251
Further Reading 252
10.2 Motion Control with Electric Motors by Input-Output Linearization 252
Introduction • Background • DC Motors • AC Induction Motors • AC Synchronous Motors • Concluding Remarks
References 265
10.3 Control of Electrical Generators 265
Introduction • DC Generators • Synchronous Generators • Induction Generators • Concluding Remarks
References 279

10.1 Dynamic Modeling and Control in Power Electronics

George C. Verghese, Massachusetts *Institute of Technology*

10.1.1 Introduction

This chapter[1] is written with the following purposes in mind:

- To describe the objectives, features, and constraints that characterize power electronic converters, emphasizing those aspects that are relevant to control.
- To outline the principles by which tractable dynamic models are obtained for power electronic converters; such models are required for application of the various control design approaches and techniques described throughout this handbook.
- To indicate how controls are typically designed and implemented in power electronics.
- To suggest to practitioners and researchers in both control and power electronics that power electronic systems constitute an interesting and important testbed for control.

Much of this chapter is distilled from the more detailed development in [3]. We begin, in this section, with an introduction to the issues that shape power electronics. The following section then describes some prototypical power electronic converters. The final two sections present more detailed discussions of dynamic modeling and control in power electronics, focusing first on a particular converter as a case study, and moving on to some extensions. The role of averaged models and sampled-data models is highlighted.

What is Power Electronics? Power electronics is concerned with high-efficiency conversion of electric power, from the form available at the input or power source, to the form required at the output or load. Most commonly, one talks of AC/DC, DC/DC, DC/AC, and AC/AC *converters*, where "AC" here typically refers to nominally *sinusoidal* voltage waveforms, while "DC" refers to nominally *constant* voltage waveforms. Small deviations from nominal are tolerable. An AC/DC converter (which has an AC power source and a DC load) is also called a *rectifier*, and a DC/AC converter is called an *inverter*. Applications of power electronics can be as diverse as high-voltage DC (HVDC) bulk-power transmission systems involving rectifiers and inverters rated in megawatts, or motor drives rated at a few kilowatts, or 50-W power supplies for electronic equipment.

The Dictates of High Efficiency High efficiency reduces energy costs, but as importantly it reduces the amount of dissipated heat that must be removed from the power converter. Efficiencies of higher than 99% can be obtained in large,

[1]The author is grateful to the following people for helpful comments: Steven Leeb, Bernard Lesieutre, Piero Maranesi, David Perreault, Seth Sanders, Aleksandar Stankovic, and Joseph Thottuvelil. For invaluable assistance with the figures, thanks are due to Deron Jackson and Steven Leeb.

0-8493-0054-3/00/$0.00+$.50

high-power systems, while small, low-power systems may have efficiencies closer to 80%. The goal of high efficiency dictates that the power processing components in the circuit be close to lossless. Switches, capacitors, inductors, and transformers are therefore the typical components in a power electronic converter.

The switches are operated cyclically, and serve to vary the circuit interconnections—or the "topological state" of the circuit—over the course of a cycle. The capacitors and inductors perform filtering actions, regulating power flows by temporarily storing or supplying energy. The transformers scale voltages and currents, and also provide electrical isolation between the source and load. *Ideal* switches, capacitors, inductors, and transformers do not dissipate power, and circuits comprising only such elements do not dissipate power either (provided that the switching operations do not result in impulsive currents or voltages, a constraint that is respected by power converters). In particular, an *ideal switch* has zero voltage across itself in its *on* (or closed, or conducting) state, zero current through itself in its *off* (or open, or blocking) state, and requires zero time to make a transition between these two states. Its power dissipation is therefore always zero. Of course, practical components depart from ideal behavior, resulting in some power dissipation.

Semiconductor Switches A switch in a power electronic converter is implemented via one or a combination of semiconductor devices. The most common such devices are

- diodes
- thyristors, which may be thought of as diodes with an additional gate terminal for control, and which are of various types (for example, silicon controlled rectifiers or SCRs, bidirectional thyristors called TRIACs that function as antiparallel SCRs, and gate turn-off thyristors or GTOs)
- bipolar junction transistors (BJTs, controlled via an appropriate drive at the base, and designed for operation in cutoff or saturation, rather than in their linear, active range), and insulated gate bipolar transistors (IGBTs, controlled via the gate)
- metal-oxide-semiconductor field effect transistors (MOSFETs, controlled via the gate)

The power loss associated with such switches comes from a nonzero voltage drop when they are closed, a nonzero leakage current when they are open, and a finite transition time from closed to open or vice versa, during which time both the voltage and current may be significant simultaneously.

A higher switching frequency generally implies a more compact converter, since smaller capacitors, inductors, and transformers can be used to meet the specified circuit characteristics. However, the higher frequency also means higher switching losses associated with the increased frequency of switch transitions, as well as other losses and limitations associated with high-frequency operation of the various components. Switching frequencies above the audible range are desirable for many applications.

Controlling the Switches Each type of semiconductor switch is amenable to a characteristic mode of control. Diodes are at one extreme, as they *cannot* be controlled; they conduct or block as a function solely of the current through them or the voltage across them, so neither their turn-on nor turn-off can be directly commanded by a control action. For SCRs and TRIACs, the turn-off happens as for a diode, but the turn-on is by command, under appropriate circuit conditions. For BJTs, IGBTs, MOSFETs and GTOs, both the turn-on and turn-off occur in response to control actions, provided circuit conditions are appropriate.

The choice of switch implementation depends on the requirements of each particular converter. For instance, the same circuit topology that is used for rectification can often be used for inversion, after appropriate modification of the switches and/or of the way they are controlled.

The only available control decisions are *when* to open and close the switches. It is by modulating the instants at which the switches are opened and closed that the dynamic behavior of a power electronic converter is regulated. Power electronic engineers have invented clever mechanisms for implementing the types of modulation schemes needed for feedback and feedforward control of power converters.

10.1.2 Prototype Converters

This section briefly describes the structure and operating principles of some basic power electronic converters. Many practical converters are directly derived from or closely related to one of the converters presented here. Also, many power electronic systems involve *combinations* of such basic converters. For instance, a high-quality power supply for electronic equipment might comprise a unity-power-factor, pulse-width-modulated (PWM) rectifier cascaded with a PWM DC/DC converter; a variable-frequency drive for an AC motor might involve a rectifier followed by a variable-frequency inverter.

High-Frequency PWM DC/DC Converters Given a DC voltage of value V (which can represent an input DC voltage, or an output DC voltage, or a DC difference between input and output voltages), we can easily use a controlled switch to "chop" the DC waveform into a *pulse waveform* that alternates between the values V and 0 at the switching frequency. This pulse waveform can then be lowpass-filtered with capacitors and/or inductors that are configured to respond to its average value, i.e., its DC component. By controlling the *duty ratio* of the switch, i.e., the fraction of time that the switch is closed in each cycle, we can control the fraction of time that the pulse waveform takes the value V, and thereby control the DC component of this waveform.

The preceding description applies directly to the simple *buck* (or voltage step-down) converter illustrated schematically in Figure 10.1. The load is taken to be just a resistor R, for simplicity. If the switch is operated periodically with a constant duty ratio D, and assuming the switch and diode to be ideal, it is easy to see that the converter settles into a periodic steady state in which the

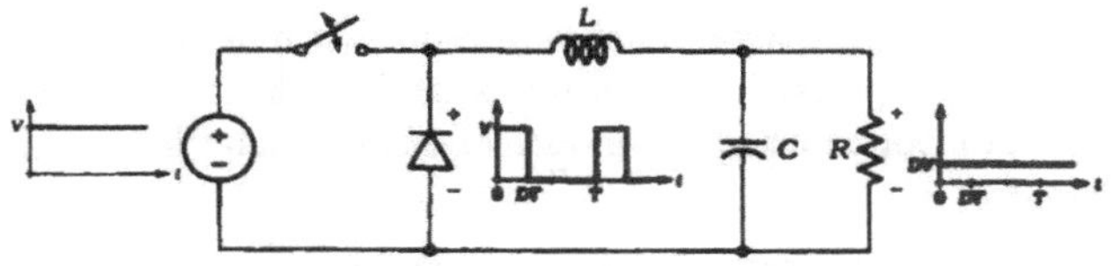

Figure 10.1 The basic principle of high-frequency PWM DC/DC conversion is illustrated here for the case of a buck converter with a resistive load. The switch operation converts the DC input voltage into a pulse waveform. The switching frequency is chosen much higher than the cutoff frequency of the lowpass LC filter, so the output voltage is essentially DC. If the switch is operated periodically with a constant duty ratio D, the converter settles into a periodic steady state in which the output voltage is DV plus a small switching-frequency ripple.

average output voltage is DV. If the switching frequency is high enough relative to the cutoff frequency of the lowpass filter, then the switching-frequency component at the output will be greatly attenuated. The output voltage will then be the DC value DV plus some small switching-frequency *ripple* superimposed on it.

The buck converter is the simplest representative of a class of DC/DC converters based on lowpass filtering of a high-frequency pulse waveform. These converters are referred to as *switching regulators* or *switched-mode* converters (to distinguish them from linear voltage regulators, which are based on transistors operating in their linear active range, and which are therefore generally less efficient.) *Pulse-width modulation* (PWM), in which the duty ratio is varied, forms the basis for the regulation of switched-mode converters, so they are also referred to as high-frequency PWM DC/DC converters. Switching frequencies in the range of 15 to 300 kHz are common.

The *boost* (or voltage step-up) converter in Figure 10.2 is a more complicated and interesting high-frequency PWM DC/DC converter, and we shall be examining it in considerably more detail. Here V_{in} denotes the voltage of a DC source, while the

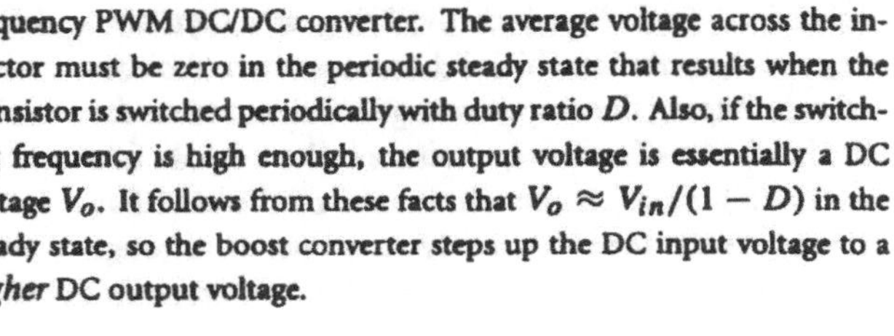

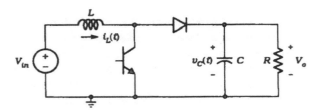

Figure 10.2 The boost converter shown here is a more complex high-frequency PWM DC/DC converter. The average voltage across the inductor must be zero in the periodic steady state that results when the transistor is switched periodically with duty ratio D. Also, if the switching frequency is high enough, the output voltage is essentially a DC voltage V_o. It follows from these facts that $V_o \approx V_{in}/(1-D)$ in the steady state, so the boost converter steps up the DC input voltage to a *higher* DC output voltage.

voltage across the load (again modeled for simplicity as being just a resistor) is essentially a DC voltage V_o , with some small switching-frequency ripple superimposed on it. The values of L and C are chosen such that the ripple in the output voltage is a suitably small percentage (typically < 5%) of the nominal load voltage. The left terminal of the inductor is held at a potential of V_{in} relative to ground, while its right terminal sees a pulse waveform that is switched between 0 (when the transistor is on, with the diode blocking) and the output voltage (when the transistor is off, with the diode conducting). In nominal operation, the transistor is switched periodically with duty ratio D, so the average potential of the inductor's right terminal is approximately $(1-D)V_o$. A periodic steady state is attained only when the inductor has zero average voltage across itself, i.e., when

$$V_o \approx \frac{V_{in}}{(1-D)} \tag{10.1}$$

Otherwise, the inductor current at the end of a switching cycle would not equal that at the beginning of the cycle, which contradicts the assumption of a periodic steady state. Since $0 < D < 1$, we see from Equation 10.1 that the output DC voltage is *higher* than the input DC voltage, which is why this converter is termed a "boost" converter.

Other High-Frequency PWM Converters Appropriate control of a high-frequency PWM DC/DC converter also enables conversion between waveforms that are not DC, but that are nevertheless slowly varying relative to the switching frequency. If, for example, the input is a slowly varying unidirectional voltage — such as the waveform obtained by rectifying a 60-Hz sinewave — while the converter is switched at a much higher rate, say 50 kHz, then we can still arrange for the output of the converter to be essentially DC.

The high-frequency *PWM rectifier* in Figure 10.3 is built around this idea, and comprises a diode bridge followed by a boost converter. The bridge circuit rectifies the AC supply, pro-

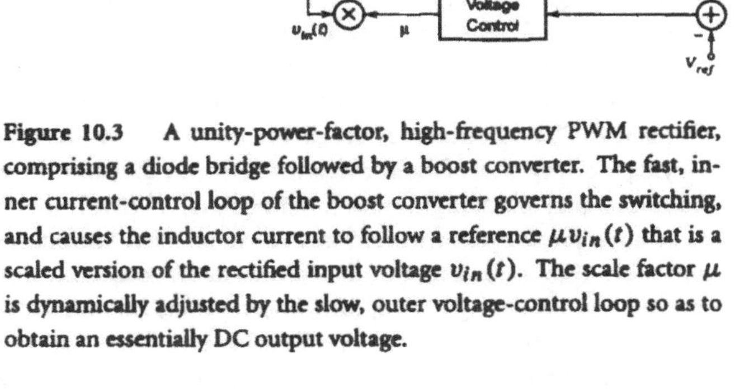

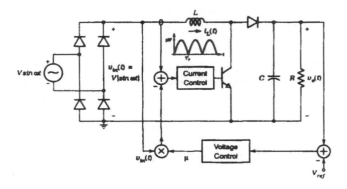

Figure 10.3 A unity-power-factor, high-frequency PWM rectifier, comprising a diode bridge followed by a boost converter. The fast, inner current-control loop of the boost converter governs the switching, and causes the inductor current to follow a reference $\mu v_{in}(t)$ that is a scaled version of the rectified input voltage $v_{in}(t)$. The scale factor μ is dynamically adjusted by the slow, outer voltage-control loop so as to obtain an essentially DC output voltage.

ducing (for the example of the 60-Hz supply mentioned above) the unidirectional voltage $v_{in}(t) = V|\sin 120\pi t|$ of period

$T_r = 1/120$ sec from the AC voltage $V \sin 120\pi t$. The resulting voltage is applied to the boost converter. The switching of the transistor (at 50 kHz, for example) is governed by a fast, inner current-control loop which causes the inductor current to follow a reference that is a scaled replica of $v_{in}(t)$, namely the rectified sinusoid $\mu v_{in}(t)$. The current drawn by the diode bridge from the AC supply is therefore sinusoidal and in phase with the supply voltage, leading to operation at essentially unity power factor. The scale factor μ that relates the inductor current reference to $v_{in}(t)$ is dynamically adjusted by the slow, outer voltage-control loop, so as to produce a DC output voltage (corrupted by a slight 120-Hz ripple and even less 50-kHz ripple). Unity-power-factor PWM rectifiers of this sort are becoming more common as front ends in power supplies for computers and other electronic equipment, because of the desire to extract as much power as possible from the wall socket, while meeting increasingly stringent power quality standards for such equipment.

In a high-frequency *PWM inverter*, the situation is reversed. The heart of it is still a DC/DC converter, and the input to it is DC. However, the switching is controlled in such a way that the filtered output is a slowly varying rectified sinusoid at the desired frequency. This rectified sinusoid can then be "unfolded" into the desired sinusoidal AC waveform, through the action of additional controllable switches arranged in a bridge configuration. In fact, both the chopping and unfolding functions can be carried out by the bridge switches, and the resulting high-frequency PWM bridge inverter is the most common implementation, available in single-phase and three-phase versions. These inverters are often found in drives for AC servo-motors, such as the permanent-magnet synchronous motors (also called "brushless DC" motors) that are popular in robotic applications. The inductive windings of the motor perform all or part of the electrical lowpass filtering in this case, while the motor inertia provides the additional mechanical filtering that practically removes the switching-frequency component from the mechanical motion.

Other Inverters Another common approach to constructing inverters again relies on a pulse waveform created by chopping a DC voltage of value V, but with the frequency of the pulse waveform now *equal* to that of the desired AC waveform, rather than much higher. Also, the pulse waveform is now generally caused (again through controllable switches configured in a bridge arrangement) to have a mean value of zero, taking values of V, 0, and $-V$, for instance. Lowpass filtering of this pulse waveform to keep only the fundamental and reject harmonics yields an essentially sinusoidal AC waveform at the switching frequency. The amplitude of the sinusoid can be controlled by varying the duty ratio of the switches that generate the pulse waveform; this may be thought of as low-frequency PWM. It is easy to arrange for the pulse waveform to have no even harmonics, and more elaborate design of the waveform can eliminate designated low-order (e.g., third, fifth, and seventh) harmonics, in order to improve the effectiveness of the lowpass filter. This sort of inverter might be found in variable-frequency drives for large AC motors, operating at power levels where the high-frequency PWM inverters described in the previous paragraph would not be practical (because of limitations on switching frequency that become dominant at higher power levels). The lowpass filtering again involves using the inductive windings and inertia of the motor.

Resonant Converters There is an alternative approach to controlling the output amplitude of a DC/AC converter, such as that presented in the previous paragraph. Rather than varying the duty ratio of the pulse waveform, a resonant inverter uses frequency variations. In such an inverter, a resonant bandpass filter (rather than a lowpass filter) is used to extract the sinewave from the pulse waveform; the pulse waveform no longer needs to have zero mean. The amplitude of the sinewave is strongly dependent on how far the switching frequency is from resonance, so control of the switching frequency can be used to control the amplitude of the output sinewave. One difficulty, however, is that variations in the load will lead to modifications of the resonance characteristics of the filter, in turn causing significant variations in the control characteristics.

If the sinusoidal waveform produced by a resonant inverter is rectified and lowpass filtered, what is obtained is a resonant DC/DC converter, as opposed to a PWM DC/DC converter. This form of DC/DC converter can have lower switching losses and generate less electromagnetic interference (EMI) than a typical high-frequency PWM DC/DC converter operating at the same switching frequency, but these advantages come at the cost of higher peak currents and voltages, and therefore higher component stresses.

Phase-Controlled Converters We have already mentioned a diode bridge in connection with Figure 10.3, and noted that it converts an AC waveform into a unidirectional or rectified waveform. Using *controllable* switches instead of the diodes allows us to *partially* rectify a sinusoidal AC waveform, with subsequent lowpass filtering to obtain an essentially DC waveform at a specified level. This is the basis for phase-controlled rectifiers, which are used as drives for DC motors or as battery charging circuits.

A typical configuration using thyristors is shown in Figure 10.4. The associated load-voltage waveform $v_o(t)$ is drawn for the case where the load current $i_o(t)$ stays strictly positive, as would happen with an inductive load. A thyristor can be turned on ("fired") via a gate signal whenever the voltage across the thyristor is positive; the thyristor turns off when the voltage across it reverses, or the forward current through it falls to zero. The control variable is the firing angle or *delay angle*, α, which is the (electrical) angle by which the firing of a thyristor is delayed, beyond the point at which it becomes forward biased. The role of α in shaping the output voltage is evident from the waveform in Figure 10.4. The average voltage at the output of the converter is $(2V/\pi)\cos\alpha$. If the admittance of the load has a lowpass characteristic, then the steady-state current through it will be essentially DC, at a level determined by the average output voltage of the converter.

The load current in the circuit of Figure 10.4 can never become negative, because of the orientation of the thyristors. Nevertheless, the converter can operate as an inverter too. For example, if

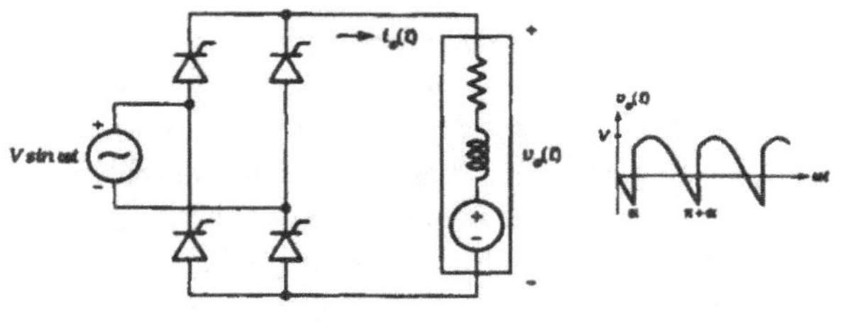

Figure 10.4 A phase-controlled converter with strictly positive load current, and the associated output voltage waveform, $v_o(t)$. The average value of $v_o(t)$ is $(2V/\pi)\cos\alpha$, where α is the firing angle or delay angle. Provided the load current does not vary significantly over a cycle, the converter functions as a rectifier when $\alpha < \pi/2$, and as an inverter when $\alpha > \pi/2$.

the load current is essentially constant over the course of a cycle (as it would be with a heavily inductive load) and if the average output voltage of the converter is made negative by setting $\alpha > \pi/2$, then the converter is actually functioning as an inverter. Such inversion might be used if the DC side of the converter is connected to a DC power source, such as a solar array or a DC generator, or a DC motor undergoing regenerative braking.

AC/AC Converters For AC/AC conversion between waveforms of the *same frequency*, we can use switches to window out sections of the source waveform, thereby reducing the fundamental component of the waveform in a controlled way; TRIACs are well suited to carrying out this operation. Subsequent filtering can be used to extract the fundamental of the windowed waveform. More intricate use of switches — in a *cycloconverter* — permits the construction of an approximately sinusoidal waveform at some specified frequency by "splicing" together appropriate segments of a set of three-phase (or multiphase) sinusoidal waveforms at a *higher* frequency; again, subsequent filtering improves the quality of the output sinusoid. While cycloconverters effect a direct AC/AC conversion, it is also common to construct an AC/AC converter as a cascade of a rectifier and an inverter (generally operating at *different* frequencies), forming a *DC-link converter*.

10.1.3 Dynamic Modeling and Control

Detailed Models Elementary circuit analysis of a power converter typically produces detailed, continuous-time, nonlinear, time-varying models in state-space form. These models have rather low order, provided one makes reasonable approximations from the viewpoint of control design: neglecting dynamics that occur at much higher frequencies than the switching frequency (for instance, dynamics due to parasitics, or to *snubber* elements that are introduced around the switches to temper the switch transitions), and focusing instead on components that are central to the power processing function of the converter. Much of this modeling phase — including the recognition of which elements are important, see [6] — is automatable. Various computer tools are also available for detailed dynamic simulation of a power converter model, specified in either circuit form or state-space form, see [9].

Simplified Models through Averaging or Sampling The continuous-time models mentioned in the preceding paragraph capture essentially all the effects that are likely to be significant for control design. However, the models are generally still too detailed and awkward to work with. The first challenge, therefore, is to extract from such a detailed model a simplified approximate model, preferably time-invariant, that is well matched to the particular control design task for the converter being considered. There are systematic ways to obtain such simplifications, notably through

- *averaging*, which blurs out the detailed switching artifacts
- *sampled-data modeling*, again to suppress the details internal to a switching cycle, focusing instead on cycle-to-cycle behavior

Both methods can produce time-invariant but still nonlinear models. Several approaches to nonlinear control design can be explored at this level. Linearization of averaged or sampled-data models around a constant operating point yields linear, time-invariant (LTI) models that are immediately amenable to a much larger range of standard control design methods.

In the remainder of this section, we illustrate the preceding comments through a more detailed examination of the boost converter that was introduced in the previous section.

A Case Study: Controlling The Boost Converter Consider the boost converter of Figure 10.2, redrawn in Figure 10.5 with some modifications. The figure includes a schematic illustration of a typical analog PWM control method that uses

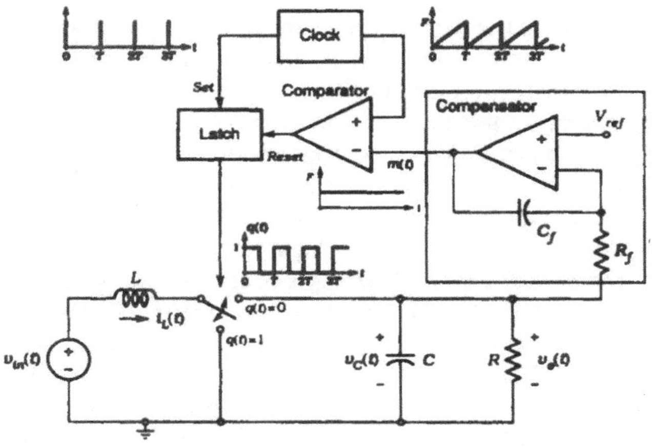

Figure 10.5 Controlling the boost converter. The operation of the switch is controlled by the latch. The switch is moved down every T seconds by a set pulse from the clock to the latch ($q(t) = 1$). The clock simultaneously initiates a ramp input of slope F/T to one terminal of the comparator. The modulating signal $m(t)$ is applied to the other terminal of the comparator. At the instant t_k in the kth cycle when the ramp crosses the level $m(t_k)$, the comparator output goes high, the latch resets ($q(t) = 0$), and the switch is moved up. The resulting duty ratio d_k in the kth cycle is $m(t_k)/F$.

output feedback. This control configuration is routinely and widely used, in a single-chip implementation; its operation will be explained shortly. We have allowed the input voltage $v_{in}(t)$ in the figure to be time varying, to allow for a source that is nominally DC at the value V_{in}, but that has some time-varying deviation or ripple around this value. Although a more realistic model of the converter for control design would also, for instance, include the equivalent series resistance — or ESR — of the output capacitor, such refinements can be ignored for our purposes here; they can easily be incorporated once the simpler case is understood. The rest of our development will therefore be for the model in Figure 10.5.

In typical operation of the boost converter under what may be called constant-frequency PWM control, the transistor in Figure 10.2 is turned on every T seconds, and turned off d_kT seconds later in the kth cycle, $0 < d_k < 1$, so d_k represents the duty ratio in the kth cycle. If we maintain a positive inductor current, $i_L(t) > 0$, then when the transistor is on, the diode is off, and vice versa. This is referred to as the *continuous conduction mode*. In the *discontinuous conduction mode*, on the other hand, the inductor current drops all the way to zero some time after the transistor is turned off, and then remains at zero, with the transistor and diode *both* off, until the transistor is turned on again. Limiting our attention here to the case of continuous conduction, the action of the transistor/diode pair in Figure 10.2 can be represented in idealized form via the double-throw switch in Figure 10.5.

We will mark the position of the switch in Figure 10.5 using a *switching function* $q(t)$. When $q(t) = 1$, the switch is down; when $q(t) = 0$, the switch is up. The switching function $q(t)$ may be thought of as (proportional to) the signal that has to be applied to the base drive of the transistor in Figure 10.2 to turn it on and off as desired. Under the constant-frequency PWM switching discipline described above, $q(t)$ jumps to 1 at the start of each cycle, every T seconds, and falls to 0 an interval d_kT later in its kth cycle. The average value of $q(t)$ over the kth cycle is therefore d_k; if the duty ratio is constant at the value $d_k = D$, then $q(t)$ is periodic, with average value D.

In Figure 10.5, $q(t)$ corresponds to the signal at the output of the latch. This signal is set to "1" every T seconds when the clock output goes high, and is reset to "0" later in the cycle when the comparator output goes high. The two input signals of the comparator are cleverly arranged so as to reset the latch at a time determined by the desired duty ratio. Specifically, the input to the "+" terminal of the comparator is a sawtooth waveform of period T that starts from 0 at the beginning of every cycle, and ramps up linearly to F by the end of the cycle. At some instant t_k in the kth cycle, this ramp crosses the level of the *modulating signal* $m(t)$ at the "−" terminal of the comparator, and the output of the comparator switches from low to high, thereby resetting the latch. The duty ratio thus ends up being $d_k = m(t_k)/F$ in the corresponding switching cycle. By varying $m(t)$ from cycle to cycle, the duty ratio can be varied.

Note that the *samples* $m(t_k)$ of $m(t)$ are what determine the duty ratios. We would therefore obtain the same sequence of duty ratios even if we added to $m(t)$ any signal that stayed negative in the first part of each cycle and crossed up through 0 in the kth cycle at the instant t_k. This fact corresponds to the familiar *aliasing* effect associated with sampling. Our standing assumption will be that $m(t)$ is not allowed to change significantly *within* a single cycle, i.e., that $m(t)$ is restricted to vary considerably more slowly than half the switching frequency. As a result, $m(t) \approx m(t_k)$ in the kth cycle, so $m(t)/F$ at any time yields the prevailing duty ratio (provided also that $0 \leq m(t) \leq F$, of course — outside this range, the duty ratio is 0 or 1).

The modulating signal $m(t)$ is generated by a feedback scheme. For the particular case of output feedback shown in Figure 10.5, the output voltage of the converter is compared with a reference voltage, and the difference is applied to a compensator, which produces $m(t)$. The goal of dynamic modeling, stated in the context of this example, is primarily to provide a basis for rational design of the compensator, by describing how the converter responds to variations in the modulating signal $m(t)$, or equivalently, to variations in the duty ratio $m(t)/F$. (Note that the ramp level F can also be varied in order to modulate the duty ratio, and this mechanism is often exploited to implement certain *feedforward* schemes that compensate for variations in the input voltage V_{in}.)

Switched State-Space Model for the Boost Converter

Choosing the inductor current and capacitor voltage as natural state variables, picking the resistor voltage as the output, and using the notation in Figure 10.5, it is easy to see that the following state-space model describes the idealized boost converter in that figure:

$$\begin{aligned} \frac{di_L(t)}{dt} &= \frac{1}{L}\Big[\Big(q(t)-1\Big)v_C(t) + v_{in}(t)\Big] \\ \frac{dv_C(t)}{dt} &= \frac{1}{C}\Big[\Big(1-q(t)\Big)i_L(t) - \frac{v_C(t)}{R}\Big] \\ v_o(t) &= v_C(t) \end{aligned} \quad (10.2)$$

Denoting the state vector by $\mathbf{x}(t) = [i_L(t) \;\; v_C(t)]'$ (where the prime indicates the transpose), we can rewrite the above equations as

$$\begin{aligned} \frac{d\mathbf{x}(t)}{dt} &= \Big[\Big(1-q(t)\Big)\mathbf{A}_0 + q(t)\mathbf{A}_1\Big]\mathbf{x}(t) + \mathbf{b}\, v_{in}(t) \\ v_o(t) &= \mathbf{c}\mathbf{x}(t) \end{aligned} \quad (10.3)$$

where the definitions of the various matrices and vectors are obvious from Equation 10.2. We refer to this model as the switched or instantaneous model, to distinguish it from the averaged and sampled-data models developed in later subsections.

If our compensator were to directly determine $q(t)$ itself, rather than determining the modulating signal $m(t)$ in Figure 10.5, then the above bilinear and time-invariant model would be the one of interest. It is indeed possible to develop control schemes directly in the setting of the switched model, Equation 10.3. In [1], for instance, a switching curve in the two-dimensional state space is used to determine when to switch $q(t)$ between its two possible values, so as to recover from a transient with a minimum number of switch transitions, eventually arriving at a periodic steady state. Drawbacks include the need for full state measurement and accurate knowledge of system parameters.

Various sliding mode schemes have also been proposed on the basis of switched models such as Equation 10.3, see for instance [13], [10], [7], and references in these papers. Sliding mode designs again specify a surface across which $q(t)$ switches, but now the (sliding) motion occurs on the surface itself, and is analyzed under the assumption of infinite-frequency switching. The requisite models are thus averaged models in effect, of the type developed in the next subsection. Any practical implementation of a sliding control must limit the switching frequency to an acceptable level, and this is often done via hysteretic control, where the switch is moved one way when the feedback signal exceeds a particular threshold, and is moved back when the signal drops below another (slightly lower) threshold. Constant-frequency implementations similar to the one in Figure 10.5 may also be used to get reasonable approximations to sliding mode behavior.

As far as the design of the compensator in Figure 10.5 is concerned, we require a model describing the converter's response to the modulating signal $m(t)$ or the duty ratio $m(t)/F$, rather than the response to the switching function $q(t)$. Augmenting the model in Equation 10.3 to represent the relation between $q(t)$ and $m(t)$ would introduce time-varying behavior and additional nonlinearity, leading to a model that is hard to work with. The models considered in the remaining sections are developed in response to this difficulty.

Nonlinear Averaged Model for the Boost Converter

To design the analog control scheme in Figure 10.5, we seek a tractable model that relates the modulating signal $m(t)$ or the duty ratio $m(t)/F$ to the output voltage. In fact, since the ripple in the instantaneous output voltage is made small by design, and since the details of this small output ripple are not of interest anyway, what we really seek is a continuous-time dynamic model that relates $m(t)$ or $m(t)/F$ to the *local average* of the output voltage (where this average is computed over the switching period). Also recall that $m(t)/F$, the duty ratio, is the local average value of $q(t)$ in the corresponding switching cycle. These facts suggest that we should look for a dynamic model that relates the local average of the switching function $q(t)$ to that of the output voltage $v_o(t)$.

Specifically, let us define the local average of $q(t)$ to be the lagged running average

$$d(t) = \frac{1}{T}\int_{t-T}^{t} q(\tau)\,d\tau \tag{10.4}$$

and call it the *continuous duty ratio* $d(t)$. Note that $d(kT) = d_k$, the actual duty ratio in the kth cycle (defined as extending from $kT - T$ to kT). If $q(t)$ is periodic with period T, then $d(t) = D$, the steady-state duty ratio. Our objective is to relate $d(t)$ in Equation 10.4 to the local average of the output voltage, defined similarly by

$$\bar{v}_o(t) = \frac{1}{T}\int_{t-T}^{t} v_o(\tau)\,d\tau \tag{10.5}$$

A natural approach to obtaining a model relating these averages is to take the local average of the state-space description in Equation 10.2. The local average of the derivative of a signal equals the derivative of its local average, because of the LTI nature of the local averaging operation we have defined. The result of averaging the model in Equation 10.2 is therefore the following set of equations:

$$\begin{aligned}\frac{d\bar{i}_L(t)}{dt} &= \frac{1}{L}\Big[\overline{qv_C}(t) - \bar{v}_C(t) + \bar{v}_{in}(t)\Big]\\ \frac{d\bar{v}_C(t)}{dt} &= \frac{1}{C}\Big[\bar{i}_L(t) - \overline{qi_L}(t) - \frac{\bar{v}_C(t)}{R}\Big]\\ \bar{v}_o(t) &= \bar{v}_C(t)\end{aligned} \tag{10.6}$$

where the overbars again denote local averages.

The terms that prevent the above description from being a state-space model are $\overline{qv_C}(t)$ and $\overline{qi_L}(t)$; the average of a product is generally *not* the product of the averages. Under reasonable assumptions, however, we can write

$$\begin{aligned}\overline{qv_C}(t) &\approx \bar{q}(t)\bar{v}_C(t) = d(t)\bar{v}_C(t)\\ \overline{qi_L}(t) &\approx \bar{q}(t)\bar{i}_L(t) = d(t)\bar{i}_L(t)\end{aligned} \tag{10.7}$$

One set of assumptions leading to the above simplification requires $v_C(\cdot)$ and $i_L(\cdot)$ over the averaging interval $[t - T, t]$ to not deviate significantly from $\bar{v}_C(t)$ and $\bar{i}_L(t)$, respectively. This condition is reasonable for a high-frequency switching converter operating with low ripple in the state variables. There are alternative assumptions that lead to the same approximations. For instance, if $i_L(\cdot)$ is essentially piecewise linear and has a slowly varying average, then the approximation in the second equation of Equation 10.7 is reasonable even if the ripple in $i_L(\cdot)$ is large; this situation is often encountered.

With the approximations in Equation 10.7, the description in Equation 10.6 becomes

$$\begin{aligned}\frac{d\bar{i}_L(t)}{dt} &= \frac{1}{L}\Big[\Big(d(t) - 1\Big)\bar{v}_C(t) + \bar{v}_{in}(t)\Big]\\ \frac{d\bar{v}_C(t)}{dt} &= \frac{1}{C}\Big[\Big(1 - d(t)\Big)\bar{i}_L(t) - \frac{\bar{v}_C(t)}{R}\Big]\\ \bar{v}_o(t) &= \bar{v}_C(t)\end{aligned} \tag{10.8}$$

What has happened, in effect, is that *all* the variables in the switched state-space model, Equation 10.2, have been replaced by their average values. In terms of the matrix notation in Equation 10.3, and with $\bar{\mathbf{x}}(t)$ defined as the local average of $\mathbf{x}(t)$, we have

$$\begin{aligned}\frac{d\bar{\mathbf{x}}(t)}{dt} &= \Big[\Big(1 - d(t)\Big)\mathbf{A}_0 + d(t)\mathbf{A}_1\Big]\bar{\mathbf{x}}(t) + \mathbf{b}\bar{v}_{in}(t)\\ \bar{v}_o(t) &= \mathbf{c}\bar{\mathbf{x}}(t)\end{aligned} \tag{10.9}$$

This is a nonlinear but *time-invariant* continuous-time state-space model, often referred to as the *state-space averaged* model [8]. The model is driven by the continuous-time control input $d(t)$ — with the constraint $0 \le d(t) \le 1$ — and by the exogenous input $\bar{v}_{in}(t)$. Note that, under our assumption of a slowly varying $m(t)$, we can take $d(t) \approx m(t)/F$; with this substitution, Equation 10.9 becomes an averaged model whose control input is the modulating signal $m(t)$, as desired.

The averaged model in Equation 10.9 leads to much more efficient simulations of converter behavior than those obtained using the switched model in Equation 10.3, provided only local averages of variables are of interest. This averaged model also forms a convenient starting point for various nonlinear control design approaches, see for instance [12], [14], [4], and references in these papers. The implementation of such nonlinear schemes would involve an arrangement similar to that in Figure 10.5, although the modulating signal $m(t)$ would be produced by some nonlinear controller rather than the simple integrator shown in the figure.

A natural circuit representation of the averaged model in Equation 10.8 is shown in Figure 10.6. This averaged circuit can actually be derived directly from the instantaneous circuit models in

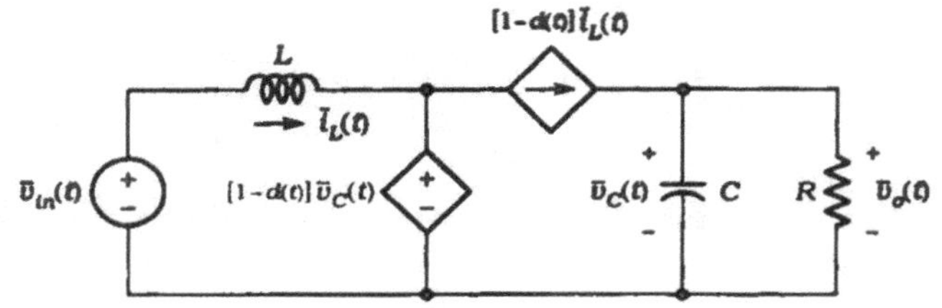

Figure 10.6 Averaged circuit model for the boost converter in Figure 10.2 and Figure 10.5, obtained by replacing instantaneous circuit variables by their local averages, and making the approximations $\overline{qv_C}(t) \approx \bar{q}(t)\bar{v}_C(t) = d(t)\bar{v}_C(t)$ and $\overline{qi_L}(t) \approx \bar{q}(t)\bar{i}_L(t) = d(t)\bar{i}_L(t)$. The transistor of the original circuit in Figure 10.2 is thereby replaced by a controlled voltage source of value $[1 - d(t)]\bar{v}_C(t)$, while the diode of the original circuit is replaced by a controlled current source of value $[1 - d(t)]\bar{i}_L(t)$.

Figure 10.2 or Figure 10.5, using "circuit averaging" arguments [3]. All the instantaneous circuit variables are replaced by their averaged values. The approximations in Equation 10.7 then cause the transistor of the original circuit in Figure 10.2 to be replaced by a controlled voltage source of value $[1-d(t)]\bar{v}_C(t)$, while the diode of the original circuit is replaced by a controlled current source of value $[1-d(t)]\bar{i}_L(t)$. This averaged circuit representation can be used as an input to circuit simulation programs, and again leads to more efficient simulations than those obtained using the original switched model.

Linearized Models for the Boost Converter The most common basis for control design in this class of converters is a linearization of the corresponding state-space averaged model. This linearized model approximately governs *small* perturbations of the averaged quantities from their values in some nominal (typically steady-state) operating condition. We illustrate the linearization process for the case of our boost converter operating in the vicinity of its periodic steady state.

Denote the *constant* nominal equilibrium values of the *averaged* state variables by I_L and V_C. These values can be computed from Equation 10.9 by setting the state derivative to zero, and replacing all other variables by their constant nominal values. The equilibrium state vector is thereby seen to be

$$\mathbf{X} = -[(1-D)\mathbf{A}_0 + D\mathbf{A}_1]^{-1}\mathbf{b}\,V_{in}$$

$$= \begin{pmatrix} I_L \\ V_C \end{pmatrix} = \begin{pmatrix} \frac{1}{R(1-D)^2} \\ \frac{1}{1-D} \end{pmatrix} V_{in} \qquad (10.10)$$

Denote the (small) deviations of the various averaged variables from their constant equilibrium values by

$$\begin{aligned} \tilde{i}_L(t) &= \bar{i}_L(t) - I_L \\ \tilde{v}_C(t) &= \bar{v}_C(t) - V_C \\ \tilde{d}(t) &= d(t) - D \\ \tilde{v}_{in}(t) &= \bar{v}_{in}(t) - V_{in} \\ \tilde{v}_o(t) &= \bar{v}_o(t) - V_o \end{aligned} \qquad (10.11)$$

Substituting in Equation 10.8 and neglecting terms that are of second order in the deviations, we obtain the following linearized averaged model:

$$\begin{aligned} \frac{d\tilde{i}_L(t)}{dt} &= \frac{1}{L}\Big[(D-1)\tilde{v}_C(t) + V_C\tilde{d}(t) + \tilde{v}_{in}(t)\Big] \\ \frac{d\tilde{v}_C(t)}{dt} &= \frac{1}{C}\Big[(1-D)\tilde{i}_L(t) - I_L\tilde{d}(t) - \frac{\tilde{v}_C(t)}{R}\Big] \\ \tilde{v}_o(t) &= \tilde{v}_C(t) \end{aligned} \qquad (10.12)$$

This is an LTI model, with control input $\tilde{d}(t)$ and disturbance input $\tilde{v}_{in}(t)$. Rewriting this model in terms of the matrix notation in Equations 10.9 and 10.10, with

$$\tilde{\mathbf{x}}(t) = \bar{\mathbf{x}}(t) - \mathbf{X} \qquad (10.13)$$

we find

$$\begin{aligned} \frac{d\tilde{\mathbf{x}}(t)}{dt} &= \Big[(1-D)\mathbf{A}_0 + D\mathbf{A}_1\Big]\tilde{\mathbf{x}}(t) \\ &\quad + [\mathbf{A}_1 - \mathbf{A}_0]\mathbf{X}\,\tilde{d}(t) + \mathbf{b}\,\tilde{v}_{in}(t) \\ \tilde{v}_o(t) &= \mathbf{c}\,\tilde{\mathbf{x}}(t) \end{aligned} \qquad (10.14)$$

Before analyzing the above LTI model further, it is worth remarking that we could have linearized the switched model in Equation 10.3 rather than the averaged model in Equation 10.9. The only subtlety in this case is that the perturbation $\tilde{q}(t) = q(t) - Q(t)$ (where $Q(t)$ denotes the nominal value of $q(t)$ in the periodic steady state) is still a 0/1 function, so one has to reconsider what is meant by a small perturbation in $q(t)$. If we consider a small perturbation $\tilde{q}(t)$ to be one whose *area* is small in any cycle, then we still arrive at a linearized model of the form of Equation 10.14, except that each averaged variable is replaced by its instantaneous version, i.e., D is replaced by $Q(t)$, $\tilde{d}(t)$ by $\tilde{q}(t)$, and so on. For linearization around a periodic steady state, the linearized switched model is (linear and) periodically varying.

Compensator Design for the Boost Converter The LTI model in Equation 10.12 that is obtained by linearizing the state-space averaged model around its constant steady state is the standard starting point for a host of control design methods. The transfer function from $\tilde{d}(t)$ to $\tilde{v}_o(t)$ is straightforward to compute, and turns out to be

$$H(s) = -\frac{I_L}{C}\,\frac{s - \frac{V_{in}}{LI_L}}{s^2 + \frac{1}{RC}s + \frac{(1-D)^2}{LC}} \qquad (10.15)$$

(where we have also used Equation 10.1 to simplify the expression in the numerator). The denominator of $H(s)$ is the characteristic polynomial of the system. For typical parameter values, the roots of this polynomial, i.e., the poles of $H(s)$, are oscillatory and lightly damped; note that they depend on D. Also observe that the system is non-minimum phase: the zero of $H(s)$ is in the right half-plane (RHP). The RHP zero correlates with the physical fact that the initial response to a step increase in duty ratio is a *decrease* in the average output voltage (rather than the increase predicted by the steady-state expression), because the diode conducts for a smaller fraction of the time. However, the buildup in average inductor current that is caused by the increased duty ratio eventually causes the average output voltage to increase over its prior level.

The lightly damped poles and the RHP zero of $H(s)$ signify difficulties and limitations with closed-loop control, if we use only measurements of the output voltage. Nevertheless, pure integral control, which is what is shown in Figure 10.5, can generally lead to acceptable closed-loop performance and adequate low-frequency loop gain (needed to counteract the effects of parameter uncertainty and low-frequency disturbances). For closed-loop stability, the loop crossover frequency must be made smaller than the "corner" frequencies (in the Bode plot) associated with the poles and RHP zero of $H(s)$. The situation for control can be significantly improved by incorporating measurements of the inductor current as well, as is done in current-mode control, which will be described shortly.

The open-loop transfer function from $\tilde{v}_{in}(t)$ to $\tilde{v}_o(t)$ has the same poles as $H(s)$, and no (finite) zeros; its low-frequency gain is $1/(1 - D)$, as expected from Equation 10.1. This transfer function is sometimes termed the open-loop *audio-susceptibility* of the converter. The control design has also to deal with rejection of the effects of input voltage disturbances on the output load voltage, i.e., with shaping the closed-loop audio-susceptibility.

In practice, a compensator designed on the basis of an LTI model would be augmented to take account of various large-signal contingencies. A *soft-start* scheme might be used to gradually increase the reference voltage V_{ref} of the controller, thereby reducing stresses in the circuit during start-up; *integrator windup* in the compensator would be prevented by placing back-to-back Zener diodes across the capacitor of the integrator, preventing large run-ups in integrator voltage during major transients; overcurrent protections would be introduced; and so on.

Current-Mode Control of the Boost Converter The attainable control performance may improve, of course, if additional measurements are taken. We have made some reference in the preceding sections to control approaches that use full state feedback. To design state feedback control for a boost converter, the natural place to start is again with the nonlinear time-invariant model (Equation 10.8 and 10.9) or the LTI model (Equations 10.12 and 10.14).

In this subsection, we examine a representative and popular state feedback scheme for high-frequency PWM converters such as the boost converter, namely current-mode control [2]. Its name comes from the fact that a fast inner loop regulates the inductor current to a reference value, while the slower outer loop adjusts the current reference to correct for deviations of the output voltage from its desired value. (Note that this is precisely what is done for the boost converter that forms the heart of the unity-power-factor PWM rectifier in Figure 10.3.) The current monitoring and limiting that are intrinsic to current-mode control are among its attractive features.

In constant-frequency peak-current-mode control, the transistor is turned on every T seconds, as before, but is turned off when the inductor current (or equivalently, the transistor current) reaches a specified reference or *peak* level, denoted by $i_P(t)$. The duty ratio, rather than being explicitly commanded via a modulating signal such as $m(t)$ in Figure 10.5, is now implicitly determined by the inductor current's relation to $i_P(t)$. Despite this modification, the averaged model in Equation 10.8 is still applicable in the case of the boost converter. (Instead of constant-frequency control, one could use hysteretic or other schemes to confine the inductor current to the vicinity of the reference current.)

A tractable and reasonably accurate continuous-time model for the dynamics of the outer loop is obtained by assuming that the average inductor current is approximately equal to the reference current:

$$\bar{i}_L(t) \approx i_P(t) \tag{10.16}$$

Making the substitution from Equation 10.16 in Equation 10.8 and using the two equations there to eliminate $d(t)$, we are left with the following first-order model:

$$\frac{d\,\bar{v}_C^2(t)}{dt} + \frac{2}{RC}\bar{v}_C^2(t) = -\frac{2 i_P(t)}{C}\left(L\frac{d\,i_P(t)}{dt} - \bar{v}_{in}\right) \tag{10.17}$$

This equation is simple enough that one can use it to explore various nonlinear control possibilities for adjusting $i_P(t)$ to control $\bar{v}_C(t)$ or $\bar{v}_C^2(t)$. The equation shows that, for constant $i_P(t)$ (or periodic $i_P(t)$, as in the nominal operation of the unity-power-factor PWM rectifier in Figure 10.3), $\bar{v}_C^2(t)$ approaches its constant (respectively, periodic) steady state exponentially, with time constant $RC/2$.

Linearizing the preceding equation around the equilibrium corresponding to a constant $i_P(t) = I_P$, we get

$$\begin{aligned}\frac{d\tilde{v}_C(t)}{dt} \quad + \quad \frac{2}{RC}\tilde{v}_C = & \\ - \frac{LI_P}{CV_C}\left(\frac{d\tilde{i}_P(t)}{dt} - \frac{V_{in}}{LI_P}\tilde{i}_P(t)\right) &+ \frac{I_P}{CV_C}\tilde{v}_{in}(t)\end{aligned} \tag{10.18}$$

It is a simple matter to obtain from Equation 10.18 the transfer functions needed for small-signal control design with disturbance rejection. We are still left with the same RHP zero as before, see Equation 10.15, in the transfer function from $\tilde{i}_P$ to $\tilde{v}_C$, but there is now only a single well-damped pole to deal with.

Current-mode control may in fact be seen as perhaps the oldest, simplest, and most common representative of a sliding mode control scheme in power electronics (we made reference earlier to more recent and more elaborate sliding mode controls). The inductor current is made to slide along the time-varying surface $i_L(t) = i_P(t)$. Equation 10.17 or 10.18 describes the system dynamics in the sliding mode, and also provides the basis for controlling the sliding surface in a way that regulates the output voltage as desired.

Sampled-Data Models for the Boost Converter Sampled-data models are naturally matched to power electronic converters, first because of the cyclic way in which power converters are operated and controlled, and second because such models are well suited to the design of digital controllers, which are used increasingly in power electronics (particularly for machine drives). Like averaged models, sampled-data models allow us to focus on cycle-to-cycle behavior, ignoring details of the intracycle behavior.

We illustrate how a sampled-data model may be obtained for our boost converter example. The state evolution of Equation 10.2 or 10.3 for each of the two possible values of $q(t)$ can be described very easily using the standard matrix exponential expressions for LTI systems. The trajectories in each segment can then be pieced together by invoking the continuity of the state variables. Under the switching discipline of constant-frequency PWM, where $q(t) = 1$ for the initial fraction d_k of the kth switching cycle, and $q(t) = 0$ thereafter, and assuming the input voltage is constant at V_{in}, we find

$$\mathbf{x}(kT+T) = e^{(1-d_k)\mathbf{A}_0 T}\left(e^{d_k \mathbf{A}_1 T}\mathbf{x}(kT) + \Gamma_1 V_{in}\right) + \Gamma_0 V_{in} \tag{10.19}$$

where

$$\begin{aligned} \Gamma_0 &= \int_0^{(1-d_k)T} e^{\mathbf{A}_0 t}\mathbf{b}\,dt \\ \Gamma_1 &= \int_0^{d_k T} e^{\mathbf{A}_1 t}\mathbf{b}\,dt \end{aligned} \tag{10.20}$$

The nonlinear, *time-invariant* sampled-data model in Equation 10.19 can directly be made the basis for control design. One interesting approach to this task is suggested in [5]. Alternatively, a linearization around the equilibrium point yields a discrete-time LTI model that can be used as the starting point for established methods of control design.

For a well-designed high-frequency PWM DC/DC converter in continuous conduction, the state trajectories in each switch configuration are close to linear, because the switching frequency is much higher than the filter cutoff frequency. What this implies is that the matrix exponentials in Equation 10.19 are well approximated by just the first two terms in their Taylor series expansions:

$$\begin{aligned} e^{(1-d_k)\mathbf{A}_0 T} &\approx \mathbf{I} + (1-d_k)\mathbf{A}_0 T \\ e^{d_k \mathbf{A}_1 T} &\approx \mathbf{I} + d_k \mathbf{A}_1 T \end{aligned} \tag{10.21}$$

If we use these approximations in Equation 10.19 and neglect terms in T^2, the result is the following approximate sampled-data model:

$$\mathbf{x}(kT+T) = \left(\mathbf{I} + (1-d_k)\mathbf{A}_0 T + d_k \mathbf{A}_1 T\right)\mathbf{x}(kT) + \mathbf{b}TV_{in} \tag{10.22}$$

This model is easily recognized as the forward-Euler approximation of the continuous-time model in Equation 10.9. Retaining the terms in T^2 leads to more refined, but still very simple, sampled-data models.

The sampled-data models in Equations 10.19 and 10.22 were derived from Equation 10.2 or 10.3, and therefore used samples of the natural state variables, $i_L(t)$ and $v_C(t)$, as state variables. However, other choices are certainly possible, and may be more appropriate for a particular implementation. For instance, we could replace $v_C(kT)$ by $\overline{v}_C(kT)$, i.e., the sampled local average of the capacitor voltage.

10.1.4 Extensions

The preceding development suggests how dynamic modeling and control in power electronics can be effected on the basis of either averaged or sampled-data models. Although a boost converter was used for illustration, the same general approaches apply to other converters, either directly or after appropriate extensions. To conclude this chapter, we outline some examples of such extensions.

Generalized Averaging It is often useful or necessary — for instance, in modeling the dynamic behavior of resonant converters — to study the *local fundamental* and *local harmonics* [11], in addition to local averages of the form shown in Equations 10.4 and 10.5. For a variable $x(t)$, the *local $\ell\omega_s$-component* may be defined as the following lagged running average:

$$<x>_\ell(t) = \frac{1}{T}\int_{t-T}^{t} x(\tau)e^{-j\ell\omega_s\tau}\,d\tau \tag{10.23}$$

In this equation, ω_s is usually chosen as the switching frequency, i.e., $2\pi/T$, and ℓ is an integer. The local averages in Equations 10.4 and 10.5 thus correspond to the choice $\ell = 0$; the choice $|\ell| = 1$ yields the local fundamental, while $|\ell| > 1$ yields the local ℓth harmonic. A key property of the local $\ell\omega_s$-component is that

$$\left\langle\frac{dx}{dt}\right\rangle_\ell = j\ell\omega_s <x>_\ell + \frac{d}{dt}<x>_\ell \tag{10.24}$$

where we have omitted the time argument t to keep the notation simple. For $\ell = 0$, we recover the result that was used in obtaining Equation 10.6 from Equation 10.2, namely that the local average of the derivative equals the derivative of the local average. More generally, we could evaluate the local $\ell\omega_s$-component of both sides of a switched state-space model such as Equation 10.3, for *several* values of ℓ. With suitable approximations, this leads to an augmented state-space model whose state vector comprises the local $\ell\omega_s$-components for *all* these values of ℓ. In the case of the boost converter, for instance, we could choose $\ell = +1$ and $\ell = -1$ in addition to the choice $\ell = 0$ that was used to get Equation 10.9 from Equation 10.3. The approximation that was used in Equation 10.7, which (with the time argument t still suppressed for notational convenience) we can rewrite as

$$\overline{q\mathbf{x}} \approx <q>_0<\mathbf{x}>_0 \tag{10.25}$$

can now be refined to

$$\overline{q\mathbf{x}} \approx <q>_0<\mathbf{x}>_0 + <q>_{-1}<\mathbf{x}>_1 + <q>_1<\mathbf{x}>_{-1} \quad (10.26)$$

The resulting state-space model will have a state vector comprising $<\mathbf{x}>_0$, $<\mathbf{x}>_1$, and $<\mathbf{x}>_{-1}$, and the solution $\mathbf{x}(t)$ will be approximated as

$$\mathbf{x}(t) \approx <\mathbf{x}>_0 + <\mathbf{x}>_1 e^{j\omega_s t} + <\mathbf{x}>_{-1} e^{-j\omega_s t} \quad (10.27)$$

Results in [11] show the significant improvements in accuracy that can be obtained this way, relative to the averaged model in Equation 10.9, while maintaining the basic simplicity and efficiency of averaged models relative to switched models. That paper also shows how generalized averaging may be applied to the analysis of resonant converters.

Generalized State-Space Models A sampled-data model for a power converter will almost invariably involve a state-space description of the form

$$\mathbf{x}(kT+T) = \mathbf{f}\Big(\mathbf{x}(kT), \mathbf{u}_k, \mathbf{T}_k, k\Big) \quad (10.28)$$

The vector $\mathbf{u}_k$ here comprises a set of parameters that govern the state evolution in the kth cycle (e.g., parameters that describe control choices and source variations during the kth cycle), and $\mathbf{T}_k$ is a vector of *switching times*, comprising the times at which switches in the converter open or close. The switching-time vector $\mathbf{T}_k$ will satisfy a set of constraints of the form

$$0 = \mathbf{c}\Big(\mathbf{x}(kT), \mathbf{u}_k, \mathbf{T}_k, k\Big) \quad (10.29)$$

If $\mathbf{T}_k$ can be solved for in Equation 10.29, then the result can be substituted in Equation 10.28 to obtain a standard sampled-data model in state-space form. However, there are many cases in which the constraint Equation 10.29 is *not* explicitly solvable for $\mathbf{T}_k$, so one is forced to take Equations 10.28 and 10.29 together as the sampled-data model. Such a pair, comprising a state evolution equation along with a side constraint, is what we refer to as a generalized state-space model. Note that the *linearized* version of Equation 10.29 will allow the switching-time *perturbations* to be calculated explicitly, provided the Jacobian matrix $\partial\mathbf{c}/\partial\mathbf{T}_k$ has full column rank. The result can be substituted in the linearized version of Equation 10.28 to obtain a linearized state-space model in standard — rather than generalized — form.

Hierarchical Modeling A hierarchical approach to modeling is mandated by the range of time scales encountered in a typical power electronic system. A single "all-purpose" model that captured all time scales of interest, from the very fast transients associated with the switch transitions and parasitics, to very slow transients spanning hundreds or thousands of switching cycles, would not be much good for any particular purpose. What is needed instead is a collection of models, focused at different time scales, suited to the distinct types of analysis that are desired at each of these scales, and capable of being linked to each other in some fashion.

Consider, for example, the high-power-factor PWM rectifier in Figure 10.3, which is examined further in [3] and [15]. The detailed evaluation of the switch stresses and the design of any snubbers requires circuit-level simulation and analysis over a short interval, comparable with the switching period T. The design of the inner current-control loop is conveniently done using a continuous-time averaged model, with averaging carried out over a window of length T; the model in Equation 10.8 is representative of this stage. The design of the outer voltage-control loop may be done using an averaged or sampled-data model computed over a window of length T_r, the period of the rectified sinusoidal input voltage; the model Equation 10.17 is representative of — or at least a precursor to — this stage. The lessons of each level have to be taken into account at all the other levels.

10.1.5 Conclusion

This has been a brief introduction to some of the major features and issues of dynamic modeling and control in power electronics. The intent was to be comprehensible rather than comprehensive; much that is interesting and relevant has inevitably been left out. Also, the emphasis here is undoubtedly skewed toward the material with which the author is most familiar. The interested reader who probes further will discover power electronics to be a vigorous, rewarding, and important domain for applications and extensions of the full range of control theory and methodology.

References

[1] Burns, W.W. and Wilson, T.G., Analytic derivation and evaluation of a state trajectory control law for DC-DC converters, in *IEEE Power Electronics Specialists Conf. Rec.*, 70–85, 1977.

[2] Hsu, S.P., Brown, A., Resnick, L., and Middlebrook, R.D., Modeling and analysis of switching DC-to-DC converters in constant-frequency current-programmed mode, in *IEEE Power Electronics Specialists Conf. Rec.*, 284–301, 1979.

[3] Kassakian, J.G., Schlecht, M.F., and Verghese, G.C., *Principles of Power Electronics*, Addison-Wesley, Reading, MA, 1991.

[4] Kawasaki, N., Nomura, H., and Masuhiro, M., The new control law of bilinear DC-DC converters developed by direct application of Lyapunov, *IEEE Trans. Power Electron.*, 10(3), 318–325, 1995.

[5] Khayatian, A. and Taylor, D.G., Multirate modeling and control design for switched-mode power converters, *IEEE Trans. Auto. Control*, 39(9), 1848–1852, 1994.

[6] Leeb, S.B., Verghese, G.C., and Kirtley, J.L., Recognition of dynamic patterns in DC-DC switching converters, *IEEE Trans. Power Electron.*, 6(2), 296–302, 1991.

[7] Malesani, L., Rossetto, L., Spiazzi, G., and Tenti, P., Performance optimization of Ćuk converters by sliding-mode control. *IEEE Trans. Power Electron.*, 10(3), 302–309, 1995.

[8] Middlebrook, R.D., and Ćuk, S., A general unified approach to modeling switching converter power stages, in *IEEE Power Electronics Specialists Conf. Rec.*, 18–34, 1976.

[9] Mohan, N., Robbins, W.P., Undeland, T.M., Nilssen, R., and Mo, O., Simulation of power electronic and motion control systems — an overview, *Proc. IEEE*, 82(8), 1287–1302, 1994.

[10] Sanders, S.R., Verghese, G.C., and Cameron, D.E., Nonlinear control of switching power converters, *Control — Theory Adv. Technol.*, 5(4), 601–627, 1989.

[11] Sanders, S.R., Noworolski, J.M., Liu, X.Z., and Verghese, G.C., Generalized averaging method for power conversion circuits. *IEEE Trans. Power Electron.*, 6(2), 251–259, 1991.

[12] Sanders, S.R. and Verghese, G.C., Lyapunov-based control for switched power converters, *IEEE Trans. Power Electron.*, 7(1), 17–24, 1992.

[13] Sira-Ramírez, H., Sliding motions in bilinear switched networks, *IEEE Trans. Circuits Syst.*, 34(8), 919–933, 1987.

[14] Sira-Ramírez, H. and Prada-Rizzo, M.T., Nonlinear feedback regulator design for the Ćuk converter, *IEEE Trans. Auto. Control*, 37(8), 1173–1180, 1992.

[15] Thottuvelil, V.J., Chin, D., and Verghese, G.C., Hierarchical approaches to modeling high-power-factor AC-DC converters, *IEEE Trans. Power Electron.*, 6(2), 179–187, 1991.

Further Reading

The list of references above suggests what are some of the journals and conferences relevant to the topic of this chapter. However, a large variety of other journals and conferences are devoted to, or occasionally contain, useful material. We limit ourselves here to providing just a few leads.

A useful perspective on the state-of-the-art in power electronics may be gleaned from the August 1994 special issue of the *Proceedings of the IEEE*, devoted to "Power Electronics and Motion Control."

Several papers on dynamic modeling and control for power electronics are presented every year at the IEEE Power Electronics Specialists Conference (PESC) and the IEEE Applied Power Electronics Conference (APEC). Many of these papers, plus others, appear in expanded form in the *IEEE Transactions on Power Electronics*. The April 1991 special issue of these *Transactions* was devoted to "Modeling in Power Electronics." In addition, the IEEE Power Electronics Society holds a biennial workshop on "Computers in Power Electronics," and a special issue of the *Transactions* with this same theme is slated to appear in May 1997.

The power electronics text of Kassakian et al. referenced above has four chapters devoted to dynamic modeling and control in power electronics. The first two of the following books consider the dynamics and control of switching regulators in some detail, while the third is a broader text that provides a view of a range of applications:

[1] Kislovski, A.S., Redl, R., and Sokal, N.O., *Dynamic Analysis of Switching-Mode DC/DC Converters*, Van Nostrand Reinhold, New York, 1991.

[2] Mitchell, D.M., *Switching Regulator Analysis*. McGraw-Hill, New York, 1988.

[3] Mohan, N., Undeland, T.M., and Robbins, W.P., *Power Electronics: Converters, Applications, and Design*, Wiley, New York, 1995.

10.2 Motion Control with Electric Motors by Input-Output Linearization[2]

David G. Taylor, Georgia Institute of Technology, School of Electrical and Computer Engineering, Atlanta, GA

10.2.1 Introduction

Due to the increasing availability of improved power electronics and digital processors at reduced costs, there has been a trend to seek higher performance from electric machine systems through the design of more sophisticated control systems software. There exist significant challenges in the search for improved control system designs, however, since the dynamics of most electric machine systems exhibit significant nonlinearities, not all state variables are necessarily measured, and the parameters of the system can vary significantly from their nominal values.

Electric machines are electromechanical energy converters, used for both motor drives and power generation. Nearly all electric power used throughout the world is generated by synchronous machines (operated as generators), and a large fraction of all this electric power is consumed by induction machines (operated as motors). The various types of electric machines in use differ with respect to construction materials and features, as well as the underlying principles of operation.

The first DC machine was constructed by Faraday around 1820, the first practical version was made by Henry in 1829, and the first commercially successful version was introduced in 1837. The three-phase induction machine was invented by Tesla around 1887. Although improved materials and manufacturing methods continue to refine electric machines, the fundamental issues relating to electromechanical energy conversion have been established for well over a century.

In such an apparently well-established field, it may come as a surprise that today there is more research and development

[2]This work was supported in part by the National Science Foundation under Grant ECS-9158037 and by the Air Force Office of Scientific Research under Grant F49620-93-1-0147.

activity than ever before. Included in a modern electric machine system is the electric machine itself, power electronic circuits, electrical and/or mechanical sensors, and digital processors equipped with various software algorithms. The recent developments in power semiconductors, digital electronics, and permanent-magnet materials have led to "enabling technology" for today's advanced electric machine systems. Perhaps more than any other factor, the increasing use of computers, for both the design of electric machines and for their real-time control, is enhancing the level of innovation in this field.

This chapter provides an overview of primarily one recent development in control systems design for electric machines operated as motor drives. The chapter takes a broad perspective in the sense that a wide variety of different machine types is considered, hopefully from a unifying point of view. On the other hand, in order to limit the scope substantially, an effort was made to focus on one more recent nonlinear control method, specifically input-output linearization, as opposed to the classical methods which have less potential for achieving high dynamic performance. An unavoidable limitation of the presentation is a lack of depth and detail beyond the specific topic of input-output linearization; however, the intention was to highlight nonlinear control technology for electric machines to a broad audience, and to guide the interested reader to a few appropriate sources for further study.

10.2.2 Background

Input-Output Linearization

The most common control designs for electric machines today, for applications requiring high dynamic performance, are based on forms of exact linearization [8]. The design concept is reflected by a two-loop structure of the controller: in the first design step, nonlinear compensation is sought which explicitly cancels the nonlinearities present in the motor (without regard to any specific control objective), and this nonlinear compensation is implemented as an inner feedback loop; in the second design step, linear compensation is derived on the basis of the resulting linear dynamics of the precompensated motor to achieve some particular control objective, and this linear compensation is implemented as an outer feedback loop. The advantage of linear closed-loop dynamics is clearly that selection of controller parameters is simplified, and the achievable transient responses are very predictable.

Not all nonlinear systems can be controlled in this fashion; the applicability of exact linearization is determined by the type and location of the model nonlinearities. Furthermore, exact linearization is not really a single methodology, but instead represents two distinct notions of linearizability, though in both cases the implementation requires full state feedback. In the first case, exact linearization of the input-output dynamics of a system is desired, with the output taken to be the controlled variables. This case, referred to as input-output linearization, is the more intuitive form of exact linearization, but can be applied in a straightforward way only to so-called minimum-phase systems (those systems with stable zero dynamics). A system can be input-output linearized if it has a well-defined relative degree (see [8] for clarification). In the second case, exact linearization of the entire state-space dynamics of a system is desired, and no output needs to be declared. This case, referred to as input-state linearization, has the advantage of eliminating any potential difficulties with internal dynamics but is less intuitively appealing and can be more difficult to apply in practice. Input-state linearization applies only to systems that are characterized by integrable, or nonsingular and involutive, distributions (see [8] for clarification).

Standard models of most electric machines are exactly linearizable, in the sense(s) described above. Prior literature has disclosed many examples of exact linearization applied to electric machines, including experimental implementation, for various machine types and various types of models. For any machine type, the most significant distinction in the application of exact linearization relates to the order of the model used. When full-order models are used, the stator voltages are considered to be the control inputs. When a reduced-order model is used, the assignment of the control inputs depends on how the order reduction has been performed: if the winding inductance is neglected, then voltage will still be the input but the mechanical subsystem model will be altered; if a high-gain current loop is employed, then current will be the input in an unaltered mechanical subsystem model. In either case, exact linearization provides a systematic method for designing the nonlinearity compensation within the inner nonlinear loop, so that the outer linear loop is concerned only with the motion control part of the total system.

Relation to Field Orientation

Prior to the development of a formal theory for exact linearization design, closely related nonlinear feedback control schemes had already been developed for the induction motor. The classical "field oriented control," introduced in [1] over 20 years ago, involves the transformation of electrical variables into a frame of reference which rotates with the rotor flux vector (the dq frame). This reference frame transformation, together with a nonlinear feedback, serves to reduce the complexity of the dynamic equations, provided that the rotor flux is not identically zero. Under this one restriction, the rotor flux amplitude dynamics are made linear and decoupled and, moreover, if the rotor flux amplitude is regulated to a constant value, the speed dynamics will also become linear and decoupled. Provided that the rotor flux amplitude may be kept constant, the field oriented control thus achieves an asymptotic linearization and decoupling, where the d-axis voltage controls rotor flux amplitude and the q-axis voltage controls speed.

Although the field oriented approach to induction motor control is widely used today and has achieved considerable success, the formal use of exact linearization design can provide alternative nonlinear control systems of comparable complexity, but achieving true (as opposed to asymptotic) linearization and decoupling of flux and torque or speed. For example, using a reduced-order electrical model (under the assumption that the rotor speed is constant), an input-state linearization static state-feedback design is reported in [6] that achieves complete decoupling of the rotor flux amplitude and torque responses. The full-order electromechanical model of an induction motor turns out not to be input-state linearizable [13]. However, as shown in [12] (see also [13]), input-output linearization methods do apply to the full-order electromechanical model.

With rotor flux amplitude and speed chosen as outputs to be controlled, simple calculations show that the system has well-defined relative degree {2, 2}, provided that the rotor flux is not identically zero. Hence, under the constraint of nonzero rotor flux, it is possible to derive a nonlinear static state-feedback that controls rotor flux amplitude and speed in a noninteracting fashion, with linear second-order transients for each controlled output (and with bounded first-order internal dynamics [13]). Although the full-order induction motor model is not input-state linearizable, the augmented system obtained by adding an integrator to one of the inputs does satisfy this property locally.

Performance Optimization

Although the difference between the classical control of [1] and the exact linearization control of [12, 13] may appear to be a minor one, the complete decoupling of speed and flux dynamics in the closed-loop system (during transients as well as in steady state) provides the opportunity to optimize performance. For example, as mentioned in [13], the flux reference will need to be reduced from nominal as the speed reference is increased above rated speed, in order to keep the required feed voltages within the inverter limits. Operation in this flux-weakening regime will excite the coupling between flux and speed in the classical field oriented control, causing undesired speed fluctuations (and perhaps instability).

There are other motivations for considering time-varying flux references as well. For instance, in [9, 10] the flux is adjusted as a function of speed in order to maximize power efficiency (i.e., only the minimum stator input power needed to operate at the desired speed is actually sourced). In [9], the flux reference is computed on the basis of predetermined relationships derived off-line and the control is implemented in a reference frame rotating with the stator excitation. In [10], the flux reference is computed on-line using a minimum power search method and the control is implemented in a fixed stator frame of reference. Yet another possibility, presented in [2], would be to vary the flux reference as a function of speed in order to achieve optimum torque (maximum for acceleration and minimum for deceleration) given limits on allowable voltage and current. In each of these references, high-gain current loops are used so that the exact linearization is performed with respect to current inputs rather than voltage inputs. Clearly, exact linearization permits optimization goals (which require variable flux references) and high dynamic performance in position, speed, or torque control, to be achieved simultaneously.

Wide Applicability

Exact linearization has been suggested for the control of many other types of electric machines as well. Both input-state linearization and input-output linearization are used to design controllers for wound-field brush-commutated DC motors, in [4, 5, 14]. For various permanent-magnet machines, there are many references illustrating the use of exact linearization. For instance, in [7], a three-phase wye-connected permanent-magnet synchronous motor with sinusoidally distributed windings is modeled with piecewise-constant parameters (which depend on current to account for reluctance variations and magnetic saturation), and an input-state linearization controller is derived from the rotor reference frame model. In [16], input-state linearization is applied to the hybrid permanent-magnet stepper motor with cogging torque accounted for, and it is further shown how constant load torques may be rejected using a nonlinear observer. This work was continued in [3], where the experimental implementation is described, including treatment of practical issues such as speed estimation from position sensors and operation at high speeds despite voltage limitations. Optimization objectives can be considered within exact linearization designs for these other types of machines too.

Review of Theory

In order to appreciate the concept of input-output linearization as it applies to electric motor drives, a brief review of the general theory is called for.

For the present purposes, it is sufficient to consider nonlinear multivariable systems of the form

$$\dot{x} = f(x) + \sum_{i=1}^{m} g_i(x)u_i \tag{10.30}$$

$$y_j = h_j(x), \quad j = 1, \ldots, m \tag{10.31}$$

where $x \in R^n$ is the state vector, $u \in R^m$ is the input vector (i.e., the control), and $y \in R^m$ is the output vector (i.e., to be controlled). Note that this is a square system, with the same number of inputs as outputs.

Given such a nonlinear system, it is said to possess relative degree $\{r_1, \ldots, r_m\}$ at x^0 if

$$L_{g_i} L_f^k h_j(x) \equiv 0 \quad \forall k < r_j - 1 \quad 1 \le i, j \le m \tag{10.32}$$

and

$$\operatorname{rank} \begin{bmatrix} L_{g_1} L_f^{r_1-1} h_1(x^0) & \cdots & L_{g_m} L_f^{r_1-1} h_1(x^0) \\ \vdots & \ddots & \vdots \\ L_{g_1} L_f^{r_m-1} h_m(x^0) & \cdots & L_{g_m} L_f^{r_m-1} h_m(x^0) \end{bmatrix} = m \tag{10.33}$$

The notation in Equations 10.32 and 10.33 stands for the Lie-derivative of a scalar function with respect to a vector function (see [8]). The first property implies that a chain structure exists, whereas the second property implies that input-output decoupling is possible.

For nonlinear systems possessing a well-defined relative degree, a simple diffeomorphic change of coordinates will bring the system into so-called normal form. In particular, under the change of variables

$$z_{1j} = h_j(x) \quad z_{2j} = L_f h_j(x) \quad \cdots \quad z_{r_j j} = L_f^{r_j-1} h_j(x) \tag{10.34}$$

it is easy to show that $y_j = z_{1j}$ where

$$\begin{aligned}
\dot{z}_{1j} &= L_f^1 h_j(x) + \sum_{i=1}^m \underbrace{L_{g_i} L_f^0 h_j(x)}_{\equiv 0} u_i = z_{2j} \\
\dot{z}_{2j} &= L_f^2 h_j(x) + \sum_{i=1}^m \underbrace{L_{g_i} L_f^1 h_j(x)}_{\equiv 0} u_i = z_{3j} \\
&\vdots \\
\dot{z}_{r_j j} &= L_f^{r_j} h_j(x) + \sum_{i=1}^m \underbrace{L_{g_i} L_f^{r_j - 1} h_j(x) u_i}_{\neq 0}
\end{aligned} \tag{10.35}$$

In other words, the output to be controlled, y_j, is the output of a chain of cascaded integrators, fed by a nonlinear but invertible forcing term. If $r_1 + \cdots + r_m < n$, then not all of the system's dynamics are accounted for in Equation 10.35. However, any remaining internal dynamics will not influence the input-output response, once the nonlinear feedback described below is applied.

More specifically, for nonlinear systems with well-defined relative degree, it is possible to solve for the control vector u from the system of algebraic equations

$$\begin{bmatrix} y_1^{(r_1)} \\ \vdots \\ y_m^{(r_m)} \end{bmatrix} = \underbrace{\begin{bmatrix} L_f^{r_1} h_1(x) \\ \vdots \\ L_f^{r_m} h_m(x) \end{bmatrix}}_{b(x)} \tag{10.36}$$

$$+ \underbrace{\begin{bmatrix} L_{g_1} L_f^{r_1 - 1} h_1(x) & \cdots & L_{g_m} L_f^{r_1 - 1} h_1(x) \\ \vdots & \ddots & \vdots \\ L_{g_1} L_f^{r_m - 1} h_m(x) & \cdots & L_{g_m} L_f^{r_m - 1} h_m(x) \end{bmatrix}}_{A(x)} \begin{bmatrix} u_1 \\ \vdots \\ u_m \end{bmatrix}$$

for all x near x^0, given some desired choice v for the term on the left-hand side where the superscript $^{(\cdot)}$ denotes time differentiation. Hence, the inner-loop nonlinearity compensation is performed by

$$u = \alpha(x) + \beta(x) v \tag{10.37}$$

where

$$\alpha(x) = -A^{-1}(x) b(x) \quad \beta(x) = A^{-1}(x) \tag{10.38}$$

which means that for each $j = 1, \ldots, m$

$$y_j^{(r_j)} = v_j \tag{10.39}$$

or, in state variable form,

$$\dot{z}_j = \bar{A}_j z_j + \bar{b}_j v_j \tag{10.40}$$

$$y_j = \bar{c}_j^T z_j \tag{10.41}$$

where

$$\bar{A}_j = \begin{bmatrix} 0 & 1 & \cdots & 0 \\ 0 & 0 & \ddots & 0 \\ \vdots & \vdots & \ddots & 1 \\ 0 & 0 & \cdots & 0 \end{bmatrix} \quad \bar{b}_j = \begin{bmatrix} 0 \\ \vdots \\ 0 \\ 1 \end{bmatrix}$$

$$\bar{c}_j^T = \begin{bmatrix} 1 & 0 & \cdots & 0 \end{bmatrix} \tag{10.42}$$

and

$$z_j = \begin{bmatrix} h_j(x) \\ \vdots \\ L_f^{r_j - 1} h_j(x) \end{bmatrix} \tag{10.43}$$

To implement this inner-loop design, it is necessary to measure the state vector x, and to have accurate knowledge of the system model nonlinearities $f, g_1, \ldots, g_m, h_1, \ldots, h_m$.

The outer-loop design, which is problem dependent, is very straightforward due to the linearity and complete decoupling of the input-output dynamics, as given in Equation 10.39. The outer-loop feedback design will be nonlinear with respect to the original state x, but linear with respect to the computed state of the normal form z. As an example of outer-loop design, to guarantee that $y_j \to y_{jd}$ as $t \to \infty$, given a desired output trajectory y_{jd} and its first r_j time derivatives, it suffices to apply

$$v_j = -k_j^T (z_j - z_{jd}) + y_{jd}^{(r_j)} \tag{10.44}$$

where z_j is computed from x according to Equation 10.43, the desired state trajectory vector z_{jd} is

$$z_{jd} = \begin{bmatrix} y_{jd} \\ \vdots \\ y_{jd}^{(r_j - 1)} \end{bmatrix} \tag{10.45}$$

and where the gain vector k_j is selected such that the matrix $\bar{A}_j - \bar{b}_j k_j^T$ has left-half plane eigenvalues. Implementation of this outer-loop design requires measurement of the state vector x, and accurate knowledge of the nonlinearities $f, h_1, \ldots, h_m$.

In the remainder of this chapter, the formalism outlined above on the theory of input-output linearization will be applied to a variety of DC and AC motor drive types. This review section has established the notion of relative degree as the critical feature of input-output linearization, and hence this chapter will now focus on this feature in particular and will bypass the explicit computation of the feedback, defined by Equations 10.37 and 10.44, for each motor. The objective is to provide, within a single self-contained document, a catalog of the main issues to be addressed when designing input-output linearizing controllers for electric motors. The consistent notation which will be used also adds to the clarity and usefulness of the results. The process begins with the fundamental modeling tools, such as those presented in [11], but ends with a formulation of the state variable models of various machines, and the corresponding relative degree checks. Implementation of input-output linearization for electric motors will thus be a straightforward extrapolation from the contents of this chapter.

10.2.3 DC Motors

The most logical point of departure for this catalog of results on electric motor input-output linearizability would be DC motors with separate field and armature windings, since these motors are simpler to model than AC motors yet possess a significant nonlinearity. Field coils are used to establish an air gap flux between stationary iron poles and the rotating armature. The armature has axially directed conductors which are connected to a brush commutator (a mechanical switch), and these conductors are continuously switched such that those located under a pole carry similarly directed currents. Interaction of axially directed armature currents and radially directed field flux produces a shaft torque.

Dynamic Model

The mechanical dynamic equation which models the rotor velocity ω is the same for all types of electric motors. Under the assumption that the mechanical load consists only of a constant inertia J, viscous friction with friction coefficient B, and a constant load torque T_l, the mechanical dynamics are given by

$$J\dot{\omega} = T_e - B\omega - T_l \tag{10.46}$$

Each type of electric motor has its own unique expression for electrical torque T_e, which for DC motors is

$$T_e = Mi_f i_a \tag{10.47}$$

where M designates the mutual inductance between the field and armature windings which carry currents i_f and i_a, respectively. The electrical dynamic equations describing the flow of currents in the field and armature windings are

$$v_f = R_f i_f + L_f \frac{di_f}{dt} \tag{10.48}$$

$$v_a = R_a i_a + Mi_f\omega + L_a \frac{di_a}{dt} \tag{10.49}$$

where v_f and v_a are the voltages applied to the field and armature, R_f and R_a are the resistances of the field and armature, and L_f and L_a are the self inductances of the field and armature, respectively. The above model is complete for the case where separate voltage sources are used to excite the field and armature windings. For further modeling details, see [11].

When it is desired to operate the DC motor from a single voltage source v, the two windings must be connected together in parallel (shunt connection) or in series (series connection). In either of these configurations, operation of the motor at high velocities without exceeding the source limits is made possible by including an external variable resistance to limit the field flux. The electrical dynamic equations then become

$$v = (R_f + R_x)i_f + L_f \frac{di_f}{dt} \tag{10.50}$$

$$v = R_a i_a + Mi_f\omega + L_a \frac{di_a}{dt} \tag{10.51}$$

for the shunt-wound motor, where external resistance R_x is in series with the field winding, and

$$0 = (R_f + R_x)i_f - R_x i_a + L_f \frac{di_f}{dt} \tag{10.52}$$

$$v = (R_a + R_x)i_a - R_x i_f + Mi_f\omega + L_a \frac{di_a}{dt} \tag{10.53}$$

for the series-wound motor, where external resistance R_x is in parallel with the field winding. Operation without field weakening requires $R_x = 0$ for the shunt-wound motor, and $R_x = \infty$ for the series-wound motor. For the latter case, as $R_x \to \infty$ an order reduction occurs since the currents flowing in the field and armature windings are identical in the limit.

In order to determine the extent to which the various operating modes of DC motors are input-output linearizable, it is necessary to determine the state-variable models and to assess the relative degree of these models. For sake of brevity, only position control will be taken as a primary control objective; speed control and torque control follow in an obvious manner. Hence, all models will include a state equation for rotor position θ (although this is typically unnecessary for speed and torque control).

Separately Excited DC Motor

Consider first the separately excited DC motor, in which the field and armature windings are fed from separate voltage sources. In order to match the common notation used for the review of input-output linearization principles, the variables of the separately excited DC motor are assigned according to

$$x_1 = \theta \quad x_2 = \omega \quad x_3 = i_f \quad x_4 = i_a \quad u_1 = v_f \quad u_2 = v_a \quad y_1 = \theta \tag{10.54}$$

Note that only one output, associated with the position control objective, is specified at this point. Using Equations 10.46 to 10.49, these assignments lead to the state variable model defined by

$$f(x) = \begin{bmatrix} x_2 \\ \frac{1}{J}(Mx_3x_4 - Bx_2 - T_l) \\ -\frac{1}{L_f}(R_f x_3) \\ -\frac{1}{L_a}(R_a x_4 + Mx_2x_3) \end{bmatrix} \qquad g_1(x) = \begin{bmatrix} 0 \\ 0 \\ \frac{1}{L_f} \\ 0 \end{bmatrix}$$

$$g_2(x) = \begin{bmatrix} 0 \\ 0 \\ 0 \\ \frac{1}{L_a} \end{bmatrix} \quad h_1(x) = x_1 \tag{10.55}$$

In order to check the relative degree according to the definition given in Equations 10.30 to 10.33, a second output to be controlled needs to be declared, so that the system will be square. Nevertheless, the various Lie-derivative calculations associated with the first output (rotor position) are easily verified to be

$$\begin{aligned} L_{g_1}h_1(x) &\equiv 0 & L_{g_2}h_1(x) &\equiv 0 \\ L_{g_1}L_f h_1(x) &\equiv 0 & L_{g_2}L_f h_1(x) &\equiv 0 \\ L_{g_1}L_f^2 h_1(x) &= \frac{M}{JL_f}x_4 & L_{g_2}L_f^2 h_1(x) &= \frac{M}{JL_a}x_3 \end{aligned} \tag{10.56}$$

If the second output is chosen to be the field current i_f, i.e., if $h_2(x) = x_3$, then the remaining Lie-derivative calculations are given by

$$L_{g_1}h_2(x) = \frac{1}{L_f} \quad L_{g_2}h_2(x) = 0 \tag{10.57}$$

In this case, the calculation

$$\det\begin{bmatrix} \frac{M}{JL_f}x_4 & \frac{M}{JL_a}x_3 \\ \frac{1}{L_f} & 0 \end{bmatrix} = -\frac{M}{JL_aL_f}x_3 \qquad (10.58)$$

indicates that the decoupling matrix is nonsingular almost globally, and that the relative degree is well defined almost globally, i.e.,

$$\{r_1, r_2\} = \{3, 1\} \; (\text{if } i_f \neq 0) \qquad (10.59)$$

Note that the singularity at $i_f = 0$ corresponds to a particular value of one of the controlled outputs, namely $y_2 = 0$. Consequently, this singularity is easily avoidable during operation and can be handled at start-up without difficulty.

If instead the second output is chosen to be the armature current i_a, i.e., if $h_2(x) = x_4$, then the remaining Lie-derivative calculations

$$L_{g_1}h_2(x) = 0 \quad L_{g_2}h_2(x) = \frac{1}{L_a} \qquad (10.60)$$

lead to a decoupling matrix

$$\det\begin{bmatrix} \frac{M}{JL_f}x_4 & \frac{M}{JL_a}x_3 \\ 0 & \frac{1}{L_a} \end{bmatrix} = \frac{M}{JL_aL_f}x_4 \qquad (10.61)$$

which again is nonsingular almost globally. This implies that the relative degree is again well defined almost globally, i.e.,

$$\{r_1, r_2\} = \{3, 1\} \; (\text{if } i_a \neq 0) \qquad (10.62)$$

with a singularity when $i_a = 0$. Again the singularity corresponds to a particular value of one of the controlled outputs, namely $y_2 = 0$. Consequently, this singularity is easily avoidable during operation and can be handled at start-up without difficulty.

Provided that some alternative start-up procedure is used to establish a nonzero current in the appropriate winding and that the commanded second output is chosen to be away from zero, the nonlinearity compensation defined by $\alpha(x)$ and $\beta(x)$ in Equations 10.37 and 10.38 is well defined and easily implemented for the separately excited DC motor. Unfortunately, the value of input-output linearization for the single-source excitation strategies is significantly more limited.

Shunt Wound DC Motor

Consider now the shunt wound DC motor, in which a single source excites both field and armature windings due to the parallel connection between the two windings (with external resistance in series with the field winding to limit the field flux if desired). The variable assignments

$$x_1 = \theta \quad x_2 = \omega \quad x_3 = i_f \quad x_4 = i_a \quad u = v \quad y = \theta \qquad (10.63)$$

are essentially the same as before, with the exception that just one source voltage v is present. With this variable assignment, the state variable model of the shunt wound DC motor from Equations 10.50 and 10.51 becomes

$$f(x) = \begin{bmatrix} x_2 \\ \frac{1}{J}(Mx_3x_4 - Bx_2 - T_l) \\ -\frac{1}{L_f}(R_f + R_x)x_3 \\ -\frac{1}{L_a}(R_ax_4 + Mx_2x_3) \end{bmatrix} \qquad (10.64)$$

$$g(x) = \begin{bmatrix} 0 \\ 0 \\ \frac{1}{L_f} \\ \frac{1}{L_a} \end{bmatrix} \qquad h(x) = x_1$$

The evaluation of relative degree is simpler than before, since the shunt wound DC motor is a single-input system, and there is no need to consider defining a second output. The Lie-derivative calculations

$$L_gh(x) \equiv 0 \quad L_gL_fh(x) \equiv 0 \quad L_gL_f^2h(x) = \frac{M}{J}\left(\frac{x_4}{L_f} + \frac{x_3}{L_a}\right) \qquad (10.65)$$

indicate that this system has a well-defined relative degree, except when the field and armature currents satisfy a particular algebraic constraint. In particular, it is clear that

$$r = 3 \quad \left(\text{if } i_f \neq -\frac{L_a}{L_f}i_a\right) \qquad (10.66)$$

Note that the singularity occurs in a region of the state-space which is not directly defined by the value of the output variable (rotor position). Hence, singularity avoidance is no longer a simple matter. Most important, though, is the fact that the interconnected windings impose the constraint of unipolar torque (i.e., either positive torque or negative torque); hence, it is customary to use just a unipolar voltage source. Consequently, the shunt wound DC motor is not nearly as versatile in operation as the separately excited DC motor, despite the fact that it possesses well-defined relative degree for a large subset of the state-space.

Series Wound DC Motor

Consider now the series wound DC motor, in which a single source excites both field and armature windings due to the series connection between the two windings (with external resistance in parallel with the field winding to limit the field flux if desired). The variable assignments

$$x_1 = \theta \quad x_2 = \omega \quad x_3 = i_f \quad x_4 = i_a \quad u = v \quad y = \theta \qquad (10.67)$$

are the same as for the shunt wound DC motor, with the exception that when field weakening is unnecessary ($R_x = \infty$) the order of the model effectively drops (x_3 and x_4 are not independent). With this variable assignment, the state variable model of the series wound DC motor from Equations 10.52 and 10.53 becomes

$$f(x) = \begin{bmatrix} x_2 \\ \frac{1}{J}(Mx_3x_4 - Bx_2 - T_l) \\ -\frac{1}{L_f}((R_f + R_x)x_3 - R_xx_4) \\ -\frac{1}{L_a}((R_a + R_x)x_4 - R_xx_3 + Mx_2x_3) \end{bmatrix}$$

$$g(x) = \begin{bmatrix} 0 \\ 0 \\ 0 \\ \frac{1}{L_a} \end{bmatrix} \qquad h(x) = x_1 \tag{10.68}$$

The evaluation of relative degree for this system is completed by computing the Lie-derivatives

$$L_g h(x) \equiv 0 \quad L_g L_f h(x) \equiv 0 \quad L_g L_f^2 h(x) = \frac{M}{J L_a} x_3 \tag{10.69}$$

and the result is that relative degree is well defined provided that the field current is nonzero, i.e.,

$$r = 3 \quad (\text{if } i_f \neq 0) \tag{10.70}$$

Again the singularity occurs in a region of the state-space which is not directly defined by the value of the output variable (rotor position). Hence, singularity avoidance is again not so simple. Moreover, as before, the interconnected windings impose the constraint of unipolar torque (i.e., either positive torque or negative torque), so a unipolar voltage source would be used. Consequently, the series wound DC motor is also not nearly as versatile in operation as the separately excited DC motor, despite the fact that it possesses well-defined relative degree for a large subset of the state-space.

10.2.4 AC Induction Motors

The most appropriate AC motor to consider first would be the induction motor, due to its symmetry. This motor is without doubt the most commonly used motor for a wide variety of industrial applications.

Dynamic Model

The three-phase, wye-connected induction motor is constructed from a magnetically smooth stator and rotor. The stator is wound with identical sinusoidally distributed windings displaced 120°, with resistance R_s, and these windings are wye-connected. The rotor may be considered to be wound with three identical short-circuited and wye-connected sinusoidally distributed windings displaced 120°, with resistance R_r.

The voltage equations in machine variables may be expressed by

$$\begin{bmatrix} v_s \\ 0 \end{bmatrix} = \begin{bmatrix} \mathcal{R}_s & 0 \\ 0 & \mathcal{R}_r \end{bmatrix} \begin{bmatrix} i_s \\ i_r \end{bmatrix} + \begin{bmatrix} \dot{\lambda}_s \\ \dot{\lambda}_r \end{bmatrix} \tag{10.71}$$

where v_s is the vector of stator voltages, i_s is the vector of stator currents, λ_s is the vector of stator flux linkages, i_r is the vector of induced rotor currents, and λ_r is the vector of induced rotor flux linkages. The resistance matrices for the stator and rotor windings are

$$\mathcal{R}_s = \begin{bmatrix} R_s & 0 & 0 \\ 0 & R_s & 0 \\ 0 & 0 & R_s \end{bmatrix} \quad \mathcal{R}_r = \begin{bmatrix} R_r & 0 & 0 \\ 0 & R_r & 0 \\ 0 & 0 & R_r \end{bmatrix} \tag{10.72}$$

Denoting any of the above stator or rotor vectors (voltage, current, flux) by generic notation f_s or f_r, respectively, the vector structure will be

$$f_s = \begin{bmatrix} f_{as} & f_{bs} & f_{cs} \end{bmatrix}^T \tag{10.73}$$

$$f_r = \begin{bmatrix} f_{ar} & f_{br} & f_{cr} \end{bmatrix}^T \tag{10.74}$$

where components are associated with phases a, b, c in machine variables. Assuming magnetic linearity, the flux linkages may be expressed by

$$\begin{bmatrix} \lambda_s \\ \lambda_r \end{bmatrix} = \begin{bmatrix} \mathcal{L}_s & \mathcal{L}_m(\theta) \\ \mathcal{L}_m^T(\theta) & \mathcal{L}_r \end{bmatrix} \begin{bmatrix} i_s \\ i_r \end{bmatrix} \tag{10.75}$$

with inductance matrices

$$\mathcal{L}_s = \begin{bmatrix} L_s & M_s & M_s \\ M_s & L_s & M_s \\ M_s & M_s & L_s \end{bmatrix} \tag{10.76}$$

$$\mathcal{L}_m(\theta) = \tag{10.77}$$

$$M \begin{bmatrix} \cos(N\theta) & \cos(N\theta + \frac{2\pi}{3}) & \cos(N\theta - \frac{2\pi}{3}) \\ \cos(N\theta - \frac{2\pi}{3}) & \cos(N\theta) & \cos(N\theta + \frac{2\pi}{3}) \\ \cos(N\theta + \frac{2\pi}{3}) & \cos(N\theta - \frac{2\pi}{3}) & \cos(N\theta) \end{bmatrix}$$

$$\mathcal{L}_r = \begin{bmatrix} L_r & M_r & M_r \\ M_r & L_r & M_r \\ M_r & M_r & L_r \end{bmatrix} \tag{10.78}$$

where L_s is the stator self-inductance, M_s is the stator-to-stator mutual inductance, L_r is the rotor self-inductance, M_r is the rotor-to-rotor mutual inductance, M is the magnitude of the stator-to-rotor mutual inductances which depend on rotor angle θ, and N is the number of pole pairs.

For the mechanical dynamics, the differential equation for rotor velocity ω is

$$J\dot{\omega} = T_e - B\omega - T_l \tag{10.79}$$

where the torque of electrical origin may be determined from the inductance matrices using the general expression

$$T_e = \frac{1}{2} i^T \frac{d\mathcal{L}(\theta)}{d\theta} i = -\frac{1}{2} \lambda^T \frac{d\mathcal{L}^{-1}(\theta)}{d\theta} \lambda \tag{10.80}$$

where $\mathcal{L}(\theta)$, i, λ denote the complete inductance matrix, current vector, and flux linkage vector appearing in Equation 10.75. For further modeling details, see [11].

Reference Frame Transformation

Because of the explicit dependence of the voltage equations on rotor angle θ, direct analysis of induction motor operation using machine variables is quite difficult. Even the determination of steady-state operating points is not straightforward. Hence, it is customary to perform a nonsingular change of variables, called a reference frame transformation, in order to effectively replace the variables associated with the physical stator and/or rotor windings with variables associated with fictitious windings oriented within the specified frame of reference.

For the symmetrical induction motor, the interaction between stator and rotor can be easily understood by considering a reference frame fixed on the stator, fixed on the rotor, rotating in synchronism with the applied stator excitation, or even rotating with an arbitrary velocity. The generality with which reference frame transformations may be applied with success to the induction motor is due to the assumed symmetry of both the stator and rotor windings.

For present purposes, it suffices to consider a stationary frame of reference located on the stator. In the new coordinates, there will be a q-axis aligned with phase a on the stator, an orthogonal d-axis located between phase a and phase c on the stator, and a 0-axis which carries only trivial information. The transformation matrix used to express rotor variables in this reference frame is

$$K_{rs}(\theta) = \tag{10.81}$$

$$\sqrt{\frac{2}{3}}\begin{bmatrix} \cos(N\theta) & \cos(N\theta + \frac{2\pi}{3}) & \cos(N\theta - \frac{2\pi}{3}) \\ -\sin(N\theta) & -\sin(N\theta + \frac{2\pi}{3}) & -\sin(N\theta - \frac{2\pi}{3}) \\ \frac{1}{\sqrt{2}} & \frac{1}{\sqrt{2}} & \frac{1}{\sqrt{2}} \end{bmatrix}$$

which is naturally θ-dependent, and the transformation matrix used to transform the stator variables is the constant matrix

$$K_{ss} = \sqrt{\frac{2}{3}}\begin{bmatrix} 1 & -\frac{1}{2} & -\frac{1}{2} \\ 0 & -\frac{\sqrt{3}}{2} & \frac{\sqrt{3}}{2} \\ \frac{1}{\sqrt{2}} & \frac{1}{\sqrt{2}} & \frac{1}{\sqrt{2}} \end{bmatrix} \tag{10.82}$$

Both matrices are orthonormal, meaning that they are constructed using orthogonal unit vectors, and hence their inverses are equal to their transposes. Formally stated, the change of variables considered is defined by

$$\tilde{f}_r = \begin{bmatrix} f_{qr} & f_{dr} & f_{0r} \end{bmatrix}^T \quad \tilde{f}_r = K_{rs}(\theta) f_r \tag{10.83}$$

$$\tilde{f}_s = \begin{bmatrix} f_{qs} & f_{ds} & f_{0s} \end{bmatrix}^T \quad \tilde{f}_s = K_{ss} f_s \tag{10.84}$$

where the tilde represents the appropriate stator or rotor variable in the new coordinates.

Since the windings on both the stator and rotor are wye-connected, the sum of the stator currents, as well as the sum of the rotor currents, must always be equal to zero. In other words, $i_{as} + i_{bs} + i_{cs} = 0$ and $i_{ar} + i_{br} + i_{cr} = 0$. Note that the reference frame transformation, primarily intended to eliminate θ from the voltage equations, will also satisfy the algebraic current constraint by construction.

After some tedious but straightforward algebra, it can be shown that the transformed voltage equations become

$$v_{qs} = R_s i_{qs} + \dot{\lambda}_{qs} \tag{10.85}$$

$$v_{ds} = R_s i_{ds} + \dot{\lambda}_{ds} \tag{10.86}$$

$$v_{0s} = R_s i_{0s} + \dot{\lambda}_{0s} \tag{10.87}$$

$$0 = R_r i_{qr} - N\omega\lambda_{dr} + \dot{\lambda}_{qr} \tag{10.88}$$

$$0 = R_r i_{dr} + N\omega\lambda_{qr} + \dot{\lambda}_{dr} \tag{10.89}$$

$$0 = R_r i_{0r} + \dot{\lambda}_{0r} \tag{10.90}$$

and the transformed flux linkage equations become

$$\begin{bmatrix} \lambda_{qs} \\ \lambda_{ds} \\ \lambda_{0s} \\ \lambda_{qr} \\ \lambda_{dr} \\ \lambda_{0r} \end{bmatrix} = \begin{bmatrix} L_s - M_s & 0 & 0 & \frac{3}{2}M & 0 & 0 \\ 0 & L_s - M_s & 0 & 0 & \frac{3}{2}M & 0 \\ 0 & 0 & L_s + 2M_s & 0 & 0 & 0 \\ \frac{3}{2}M & 0 & 0 & L_r - M_r & 0 & 0 \\ 0 & \frac{3}{2}M & 0 & 0 & L_r - M_r & 0 \\ 0 & 0 & 0 & 0 & 0 & L_r + 2M_r \end{bmatrix} \begin{bmatrix} i_{qs} \\ i_{ds} \\ i_{0s} \\ i_{qr} \\ i_{dr} \\ i_{0r} \end{bmatrix} \tag{10.91}$$

Note that all dependence on rotor angle θ has been eliminated and, hence, all analysis is substantially simplified. On the other hand, this change of variables does not eliminate nonlinearity entirely. Note also that $i_{0s} = i_{0r} = 0$ so that the 0-axis equations from above can be completely ignored.

The transformed electrical model is presently expressed in terms of the two orthogonal components of current and flux linkage, on both the stator and the rotor. Any state variable description will require that half of the transformed electrical variables be eliminated. Six possible permutations are available to select from, namely (i_s, i_r), (λ_s, λ_r), (i_s, λ_r), (λ_s, i_r), (i_s, λ_s), and (i_r, λ_r). Substituting the transformed inductances of Equation 10.91 into the torque expression (Equation 10.80), these permutations result in six possible torque expressions, namely

$$T_e = \begin{cases} \frac{3}{2}MN(i_{qs}i_{dr} - i_{ds}i_{qr}) \\ \frac{3}{2}MN(\lambda_{qs}\lambda_{dr} - \lambda_{ds}\lambda_{qr}) \\ \frac{\frac{3}{2}MN}{L_r - M_r}(i_{qs}\lambda_{dr} - i_{ds}\lambda_{qr}) \\ \frac{\frac{3}{2}MN}{L_s - M_s}(\lambda_{qs}i_{dr} - \lambda_{ds}i_{qr}) \\ N(i_{qs}\lambda_{ds} - i_{ds}\lambda_{qs}) \\ N(\lambda_{qr}i_{dr} - \lambda_{dr}i_{qr}) \end{cases} \tag{10.92}$$

Any of these expressions is valid, provided that the voltage equations are rewritten in terms of the same set of electrical state variables.

Input-Output Linearizability

With the above modeling background, it is now possible to proceed with the main objective of determining the extent to which the induction motor is input-output linearizable. The only remaining modeling step is to select which permutation of electrical state variables to use, and then to construct the state variable model. Taking stator current and rotor flux as state variables

$$x_1 = \theta \quad x_2 = \omega \quad x_3 = \lambda_{qr} \quad x_4 = \lambda_{dr} \quad x_5 = i_{qs} \quad x_6 = i_{ds} \tag{10.93}$$

and selecting the orthogonal components of stator voltage as the two inputs, rotor angle θ as the primary output and rotor flux magnitude squared $\lambda_{qr}^2 + \lambda_{dr}^2$ as the secondary output, the remaining standard notation will be

$$u_1 = v_{qs} \quad u_2 = v_{ds} \quad y_1 = \theta \quad y_2 = \lambda_{qr}^2 + \lambda_{dr}^2 \tag{10.94}$$

Using the notations assigned above, the resulting state variable model is defined by

$$f(x) = \begin{bmatrix} x_2 \\ \frac{1}{J}(k(x_4x_5 - x_3x_6) - Bx_2 - T_l) \\ -\alpha_r x_3 + Nx_2x_4 + \beta_r x_5 \\ -\alpha_r x_4 - Nx_2x_3 + \beta_r x_6 \\ -\alpha_s x_5 - \gamma x_2 x_4 + \beta_s x_3 \\ -\alpha_s x_6 + \gamma x_2 x_3 + \beta_s x_4 \end{bmatrix}$$

$$g_1(x) = \begin{bmatrix} 0 \\ 0 \\ 0 \\ 0 \\ \delta \\ 0 \end{bmatrix} \qquad g_2(x) = \begin{bmatrix} 0 \\ 0 \\ 0 \\ 0 \\ 0 \\ \delta \end{bmatrix} \tag{10.95}$$

$$h_1(x) = x_1 \qquad h_2(x) = x_3^2 + x_4^2$$

with constant coefficients given by

$$\begin{aligned} k &= \frac{\frac{3}{2}MN}{L_r - M_r} \\ \alpha_r &= \frac{R_r}{L_r - M_r} \\ \beta_r &= \frac{R_r(\frac{3}{2}M)}{L_r - M_r} \\ \alpha_s &= \frac{R_s(L_r-M_r)^2 + R_r(\frac{3}{2}M)^2}{((L_s-M_s)(L_r-M_r) - (\frac{3}{2}M)^2)(L_r-M_r)} \\ \beta_s &= \frac{R_r(\frac{3}{2}M)}{((L_s-M_s)(L_r-M_r) - (\frac{3}{2}M)^2)(L_r-M_r)} \\ \gamma &= \frac{\frac{3}{2}MN}{(L_s-M_s)(L_r-M_r) - (\frac{3}{2}M)^2} \\ \delta &= \frac{L_r-M_r}{(L_s-M_s)(L_r-M_r) - (\frac{3}{2}M)^2} \end{aligned} \tag{10.96}$$

The nonlinearities of the induction motor are clearly apparent in Equation 10.95.

In order to check relative degree, the Lie-derivative calculations

$$\begin{aligned} L_{g_1}h_1(x) &\equiv 0 & L_{g_2}h_1(x) &\equiv 0 \\ L_{g_1}L_f h_1(x) &\equiv 0 & L_{g_2}L_f h_1(x) &\equiv 0 \\ L_{g_1}L_f^2 h_1(x) &= \frac{k\delta}{J}x_4 & L_{g_2}L_f^2 h_1(x) &= -\frac{k\delta}{J}x_3 \end{aligned} \tag{10.97}$$

for the first output and

$$\begin{aligned} L_{g_1}h_2(x) &\equiv 0 & L_{g_2}h_2(x) &\equiv 0 \\ L_{g_1}L_f h_2(x) &= 2\beta_r\delta x_3 & L_{g_2}L_f h_2(x) &= 2\beta_r\delta x_4 \end{aligned} \tag{10.98}$$

for the second output lead to a decoupling matrix with singularity condition

$$\det\begin{bmatrix} \frac{k\delta}{J}x_4 & -\frac{k\delta}{J}x_3 \\ 2\beta_r\delta x_3 & 2\beta_r\delta x_4 \end{bmatrix} = \frac{2k\beta_r\delta^2}{J}\left(x_3^2 + x_4^2\right) \tag{10.99}$$

and to the conclusion that

$$\{r_1, r_2\} = \{3, 2\} \ (\text{if } y_2 \neq 0) \tag{10.100}$$

Hence, input-output linearization may be applied to the transformed model of the induction motor (and, hence, to the machine variable model of the induction motor via inverse reference frame transformations) provided that the rotor flux is nonzero. Since the singularity at zero rotor flux corresponds to a particular value of the secondary output variable, i.e., to $y_2 = 0$, it is easy to avoid this singularity in operation by commanding rotor fluxes away from zero and by using a start-up procedure to premagnetize the rotor prior to executing the input-output linearization calculations on-line.

10.2.5 AC Synchronous Motors

An important class of AC machines frequently used as actuators in control applications is the class of synchronous machines. Though these machines essentially share the same stator structure with the induction motor, the construction of the rotor is quite different and accounts for the asymmetry present in the modeling.

Dynamic Model

The class of synchronous machines contains several specific machines that are worth covering in this chapter, and these specific cases differ with respect to their rotor structures. These specific cases can all be considered to be special cases of a general synchronous machine. Hence, this section will begin with a presentation of the basic modeling equations for the general synchronous machine, and then will specialize these equations to the specific cases of interest prior to evaluating input-output linearizability for each specific case.

The general synchronous machine, which is commonly used as a generator of electric power, consists of a magnetically smooth stator with identical three-phase wye-connected sinusoidally distributed windings, displaced 120°. The rotor may or may not possess magnetic saliency in the form of physical poles. It may or may not possess a rotor cage (auxiliary rotor windings) for the purpose of providing line-start capability and/or to damp rotor oscillations. Finally, it may or may not possess the capability of establishing a rotor field flux, via either a rotor field winding or permanent magnets mounted on the rotor; however, if no provision for rotor field flux exists, then necessarily the rotor must have a salient pole construction.

In machine variables, the general expression for the voltage equations will involve the symmetric stator phases (designated by subscripts *as*, *bs*, and *cs* for phases *a*, *b*, and *c*) and the asymmetric rotor windings (designated by subscripts *kq* and *kd* for the *q*-axis and *d*-axis auxiliary windings and by subscript *fd* for the field winding, which is assumed to be oriented along the d-axis). Hence, the voltage equations are written

$$\begin{bmatrix} v_s \\ v_r \end{bmatrix} = \begin{bmatrix} \mathcal{R}_s & 0 \\ 0 & \mathcal{R}_r \end{bmatrix} \begin{bmatrix} i_s \\ i_r \end{bmatrix} + \begin{bmatrix} \dot{\lambda}_s \\ \dot{\lambda}_r \end{bmatrix} \quad (10.101)$$

where v_s is the vector of stator voltages, i_s is the vector of stator currents, λ_s is the vector of stator flux linkages, v_r is the vector of rotor voltages (zero for the auxiliary windings), i_r is the vector of rotor currents, and λ_r is the vector of rotor flux linkages. The stator and rotor resistance matrices are given by

$$\mathcal{R}_s = \begin{bmatrix} R_s & 0 & 0 \\ 0 & R_s & 0 \\ 0 & 0 & R_s \end{bmatrix} \quad \mathcal{R}_r = \begin{bmatrix} R_{kq} & 0 & 0 \\ 0 & R_{fd} & 0 \\ 0 & 0 & R_{kd} \end{bmatrix} \quad (10.102)$$

When denoting any of the above stator or rotor vectors (voltage, current, flux) by generic notation f_s or f_r, respectively, the vector structure will be

$$f_s = \begin{bmatrix} f_{as} & f_{bs} & f_{cs} \end{bmatrix}^T \quad (10.103)$$

$$f_r = \begin{bmatrix} f_{kq} & f_{fd} & f_{kd} \end{bmatrix}^T \quad (10.104)$$

where stator components are associated with phase windings in machine variables, and rotor components are associated with the two auxiliary windings and the field winding in machine variables. Assuming magnetic linearity, the flux linkages may be expressed by

$$\begin{bmatrix} \lambda_s \\ \lambda_r \end{bmatrix} = \begin{bmatrix} \mathcal{L}_s(\theta) & \mathcal{L}_m(\theta) \\ \mathcal{L}_m^T(\theta) & \mathcal{L}_r \end{bmatrix} \begin{bmatrix} i_s \\ i_r \end{bmatrix} \quad (10.105)$$

with inductance matrices

$$\mathcal{L}_s(\theta) = \begin{bmatrix} L_s & M_s & M_s \\ M_s & L_s & M_s \\ M_s & M_s & L_s \end{bmatrix} \quad (10.106)$$

$$- L_m \begin{bmatrix} \cos(2N\theta) & \cos(2N\theta - \frac{2\pi}{3}) & \cos(2N\theta + \frac{2\pi}{3}) \\ \cos(2N\theta - \frac{2\pi}{3}) & \cos(2N\theta + \frac{2\pi}{3}) & \cos(2N\theta) \\ \cos(2N\theta + \frac{2\pi}{3}) & \cos(2N\theta) & \cos(2N\theta - \frac{2\pi}{3}) \end{bmatrix}$$

$$\mathcal{L}_m(\theta) = \quad (10.107)$$

$$\begin{bmatrix} M_q \cos(N\theta) & M_f \sin(N\theta) & M_d \sin(N\theta) \\ M_q \cos(N\theta - \frac{2\pi}{3}) & M_f \sin(N\theta - \frac{2\pi}{3}) & M_d \sin(N\theta - \frac{2\pi}{3}) \\ M_q \cos(N\theta + \frac{2\pi}{3}) & M_f \sin(N\theta + \frac{2\pi}{3}) & M_d \sin(N\theta + \frac{2\pi}{3}) \end{bmatrix}$$

$$\mathcal{L}_r = \begin{bmatrix} L_{qr} & 0 & 0 \\ 0 & L_{fr} & M_r \\ 0 & M_r & L_{dr} \end{bmatrix} \quad (10.108)$$

where L_s is (average) the stator self-inductance, M_s is the (average) stator-to-stator mutual inductance, L_m is the stator inductance coefficient that accounts for rotor saliency, L_{qr} is the self-inductance of the q-axis auxiliary winding, L_{dr} is the self-inductance of the d-axis auxiliary winding, L_{fr} is the self-inductance of the field winding, M_r is the mutual inductance between the two d-axis rotor windings, M_q, M_d, and M_f are the magnitudes of the angle-dependent mutual inductance between the stator windings and the various rotor windings, and N is the number of pole pairs. Note that this model does not account for the possibility of stator saliency (which gives rise to magnetic cogging).

For the mechanical dynamics, the differential equation for rotor velocity ω is

$$J\dot{\omega} = T_e - B\omega - T_l \quad (10.109)$$

where the torque of electrical origin may be determined from the inductance matrices using the general expression

$$T_e = \frac{1}{2} i^T \frac{d\mathcal{L}(\theta)}{d\theta} i = -\frac{1}{2} \lambda^T \frac{d\mathcal{L}^{-1}(\theta)}{d\theta} \lambda \quad (10.110)$$

where $\mathcal{L}(\theta)$, i, λ denote the complete inductance matrix, current vector, and flux linkage vector appearing in Equation 10.105. For further modeling details, see [11].

Reference Frame Transformation

The model derived above is not only nonlinear, it is also not in a form convenient for determining the steady-state conditions needed for achieving constant velocity operation, due to the model's periodic dependence on position. This dependence on position can be eliminated by a nonsingular change of variables, which effectively projects the stator variables onto a reference frame fixed to the rotor. Although it is possible to construct transformations to other frames of reference, these would not eliminate the position dependence due to the asymmetry present in the rotor. Since the asymmetric rotor windings are presumed to be aligned with the rotor frame of reference just one transformation matrix is necessary, which transforms circuit variables from the stator windings to fictitious windings which rotate with the rotor, and it is given by

$$K_{sr}(\theta) = \sqrt{\frac{2}{3}} \begin{bmatrix} \cos(N\theta) & \cos(N\theta - \frac{2\pi}{3}) & \cos(N\theta + \frac{2\pi}{3}) \\ \sin(N\theta) & \sin(N\theta - \frac{2\pi}{3}) & \sin(N\theta + \frac{2\pi}{3}) \\ \frac{1}{\sqrt{2}} & \frac{1}{\sqrt{2}} & \frac{1}{\sqrt{2}} \end{bmatrix} \quad (10.111)$$

This matrix is orthonormal and, hence, its inverse is equal to its transpose. Formally stated, the change of variables considered is defined by

$$\tilde{f}_s = \begin{bmatrix} f_{qs} & f_{ds} & f_{0s} \end{bmatrix}^T \quad \tilde{f}_s = K_{sr}(\theta) f_s \quad (10.112)$$

where the tilde represents the stator variables in the new coordinates.

Since the windings on the stator are wye-connected, the sum of the stator currents must always be equal to zero. In other words, $i_{as} + i_{bs} + i_{cs} = 0$. Note that the reference frame transformation, primarily intended to eliminate θ from the voltage equations, will also satisfy the algebraic current constraint by construction.

After some tedious but straightforward algebra, it can be shown that the transformed voltage equations become

$$v_{qs} = R_s i_{qs} + N\omega\lambda_{ds} + \dot{\lambda}_{qs} \quad (10.113)$$

$$v_{ds} = R_s i_{ds} - N\omega\lambda_{qs} + \dot{\lambda}_{ds} \quad (10.114)$$

$$v_{0s} = R_s i_{0s} + \dot{\lambda}_{0s} \quad (10.115)$$

$$0 = R_{kq}i_{kq} + \dot{\lambda}_{kq} \quad (10.116)$$
$$v_{fd} = R_{fd}i_{fd} + \dot{\lambda}_{fd} \quad (10.117)$$
$$0 = R_{kd}i_{kd} + \dot{\lambda}_{kd} \quad (10.118)$$

and the transformed flux linkage equations become

$$\begin{bmatrix} \lambda_{qs} \\ \lambda_{ds} \\ \lambda_{0s} \\ \lambda_{kq} \\ \lambda_{fd} \\ \lambda_{kd} \end{bmatrix} = \begin{bmatrix} L_s - M_s - \frac{3}{2}L_m & 0 & 0 & \kappa M_q & 0 & 0 \\ 0 & L_s - M_s + \frac{3}{2}L_m & 0 & 0 & \kappa M_f & \kappa M_d \\ 0 & 0 & L_s + 2M_s & 0 & 0 & 0 \\ \kappa M_q & 0 & 0 & L_{qr} & 0 & 0 \\ 0 & \kappa M_f & 0 & 0 & L_{fr} & M_r \\ 0 & \kappa M_d & 0 & 0 & M_r & L_{dr} \end{bmatrix} \begin{bmatrix} i_{qs} \\ i_{ds} \\ i_{0s} \\ i_{kq} \\ i_{fd} \\ i_{kd} \end{bmatrix} \quad (10.119)$$

where $\kappa = \sqrt{\frac{3}{2}}$. Note that all dependence on rotor angle θ has been eliminated and, hence, all analysis is substantially simplified. However, nonlinearity has not been entirely eliminated. Note also that $i_{0s} = 0$ so that the 0-axis equation from above can be completely ignored.

Due to the rotor asymmetry, the most convenient set of electrical variables for expressing the electrical torque consists of stator current and stator flux. Using these variables, the torque expression becomes

$$T_e = N(i_{qs}\lambda_{ds} - i_{ds}\lambda_{qs}) \quad (10.120)$$

This and other expressions will now be specialized to cover three common types of synchronous motors: the rotor-surface permanent magnet motor, the rotor-interior permanent magnet motor, and the reluctance motor.

SPM Synchronous Motor

The surface-magnet PM synchronous motor is obtained when the rotor field winding is replaced by permanent magnets attached to the surface of a smooth rotor, with auxiliary rotor windings removed. In other words, this is the special case where i_{fd} = constant, $i_{kq} = i_{kd} = 0$, and $L_m = 0$. Of course, incorporating the effects of auxiliary windings is straightforward, but this is not pursued here.

In order to simplify notation, the new coefficients

$$\lambda_m = \kappa M_f i_{fd} \quad L = L_s - M_s \quad (10.121)$$

concerning magnet flux and inductance are defined. Note that these new coefficients allow the torque to be expressed by

$$T_e = N\lambda_m i_{qs} \quad (10.122)$$

Variables are assigned standard notation for states, inputs, and outputs according to

$$x_1 = \theta \quad x_2 = \omega \quad x_3 = i_{qs} \quad x_4 = i_{ds}$$
$$u_1 = v_{qs} \quad u_2 = v_{ds} \quad y_1 = \theta \quad y_2 = i_{ds} \quad (10.123)$$

Since for this motor the torque depends only on i_{qs} but not on i_{ds}, the d-axis current is the appropriate choice of the second output.

With the above variable assignments, the state variable model is

$$f(x) = \begin{bmatrix} x_2 \\ \frac{1}{J}(N\lambda_m x_3 - Bx_2 - T_l) \\ -\frac{1}{L}(R_s x_3 + N(\lambda_m + Lx_4)x_2) \\ -\frac{1}{L}(R_s x_4 - NLx_3x_2) \end{bmatrix} \quad (10.124)$$

$$g_1(x) = \begin{bmatrix} 0 \\ 0 \\ \frac{1}{L} \\ 0 \end{bmatrix} \quad g_2(x) = \begin{bmatrix} 0 \\ 0 \\ 0 \\ \frac{1}{L} \end{bmatrix}$$
$$h_1(x) = x_1 \quad h_2(x) = x_4$$

The presence of nonlinearity in the electrical subdynamics is clear. In checking for relative degree, the Lie-derivative calculations

$$\begin{aligned} L_{g_1}h_1(x) &\equiv 0 & L_{g_2}h_1(x) &\equiv 0 \\ L_{g_1}L_f h_1(x) &\equiv 0 & L_{g_2}L_f h_1(x) &\equiv 0 \\ L_{g_1}L_f^2 h_1(x) &= \frac{N\lambda_m}{JL} & L_{g_2}L_f^2 h_1(x) &= 0 \end{aligned} \quad (10.125)$$

for the first output and

$$L_{g_1}h_2(x) = 0 \quad L_{g_2}h_2(x) = \frac{1}{L} \quad (10.126)$$

for the second output lead to a globally nonsingular decoupling matrix, as confirmed by

$$\det\begin{bmatrix} \frac{N\lambda_m}{JL} & 0 \\ 0 & \frac{1}{L} \end{bmatrix} = \frac{N\lambda_m}{JL^2} \quad (10.127)$$

Hence, the conclusion is that the relative degree is globally well defined and

$$\{r_1, r_2\} = \{3, 1\} \text{ (globally)} \quad (10.128)$$

IPM Synchronous Motor

The interior-magnet PM synchronous motor is obtained when the rotor field winding is replaced by permanent magnets mounted inside the rotor, thus introducing rotor saliency, and with auxiliary rotor windings removed. In other words, this is the special case where i_{fd} = constant, $i_{kq} = i_{kd} = 0$, and $L_m \neq 0$. Of course, incorporating the effects of auxiliary windings is straightforward, but this is not pursued here.

In order to simplify notation, the new coefficients

$$\lambda_m = \kappa M_f i_{fd} \quad L_q = L_s - M_s - \frac{3}{2}L_m \quad L_d = L_s - M_s + \frac{3}{2}L_m \quad (10.129)$$

concerning magnet flux and inductance are defined. Note that these new coefficients allow the torque to be expressed by

$$T_e = N\lambda_m i_{qs} + N(L_d - L_q)i_{qs}i_{ds} \quad (10.130)$$

Variables are assigned standard notation for states, inputs, and outputs according to

$$x_1 = \theta \quad x_2 = \omega \quad x_3 = i_{qs} \quad x_4 = i_{ds}$$
$$u_1 = v_{qs} \quad u_2 = v_{ds} \quad y_1 = \theta \tag{10.131}$$

Since torque depends on both the q-axis and d-axis currents, the appropriate choice of the second output is nonunique and should account for the relative magnitude of the two torque production mechanisms.

With the above variable assignments, the state variable model is

$$f(x) = \begin{bmatrix} x_2 \\ \frac{1}{J}\left(N\lambda_m x_3 + N(L_d - L_q)x_3x_4 - Bx_2 - T_l\right) \\ -\frac{1}{L_q}\left(R_s x_3 + N(\lambda_m + L_d x_4)x_2\right) \\ -\frac{1}{L_d}\left(R_s x_4 - NL_q x_3 x_2\right) \end{bmatrix}$$

$$g_1(x) = \begin{bmatrix} 0 \\ 0 \\ \frac{1}{L_q} \\ 0 \end{bmatrix} \quad g_2(x) = \begin{bmatrix} 0 \\ 0 \\ 0 \\ \frac{1}{L_d} \end{bmatrix} \quad h_1(x) = x_1 \tag{10.132}$$

For this synchronous motor, nonlinearity also exists in the torque expression.

In checking for relative degree, each of the currents will be individually selected as the second output. The Lie-derivative calculations which are common to both cases are

$$\begin{aligned} L_{g_1}h_1(x) &\equiv 0 \qquad L_{g_2}h_1(x) \equiv 0 \\ L_{g_1}L_f h_1(x) &\equiv 0 \qquad L_{g_2}L_f h_1(x) \equiv 0 \\ L_{g_1}L_f^2 h_1(x) &= \frac{N}{JL_q}\left(\lambda_m + (L_d - L_q)x_4\right) \\ L_{g_2}L_f^2 h_1(x) &= \frac{N}{JL_d}(L_d - L_q)x_3 \end{aligned} \tag{10.133}$$

For the case when $h_2(x) = x_4$, the remaining Lie-derivative calculations

$$L_{g_1}h_2(x) = 0 \quad L_{g_2}h_2(x) = \frac{1}{L_d} \tag{10.134}$$

provide the decoupling matrix

$$\det\begin{bmatrix} \frac{N}{JL_q}\left(\lambda_m + (L_d - L_q)x_4\right) & \frac{N}{JL_d}(L_d - L_q)x_3 \\ 0 & \frac{1}{L_d} \end{bmatrix} = \frac{N}{JL_dL_q}\left(\lambda_m + (L_d - L_q)x_4\right) \tag{10.135}$$

which yields

$$\{r_1, r_2\} = \{3, 1\} \ \left(\text{if } y_2 \neq -\frac{\lambda_m}{L_d - L_q}\right) \tag{10.136}$$

For the case when $h_2(x) = x_3$, the remaining Lie-derivative calculations

$$L_{g_1}h_2(x) = \frac{1}{L_q} \quad L_{g_2}h_2(x) = 0 \tag{10.137}$$

provide a decoupling matrix

$$\det\begin{bmatrix} \frac{N}{JL_q}\left(\lambda_m + (L_d - L_q)x_4\right) & \frac{N}{JL_d}(L_d - L_q)x_3 \\ \frac{1}{L_q} & 0 \end{bmatrix} = -\frac{N}{JL_dL_q}(L_d - L_q)x_3 \tag{10.138}$$

which yields

$$\{r_1, r_2\} = \{3, 1\} \ (\text{if } y_2 \neq 0) \tag{10.139}$$

Hence, for this motor, it is clear that input-output linearization is a viable control strategy provided that isolated singularities are avoided, and this is simply achieved due to the fact that the singularities correspond to particular values of the second controlled output variable.

Synchronous Reluctance Motor

The synchronous reluctance motor is obtained when the rotor field winding and auxiliary windings are removed, but rotor saliency is introduced. In other words, this is the special case where $i_{fd} = i_{kq} = i_{kd} = 0$ and $L_m \neq 0$. Of course, incorporating the effects of auxiliary windings is straightforward, but this is not pursued here.

In order to simplify notation, the new coefficients

$$L_q = L_s - M_s - \frac{3}{2}L_m \quad L_d = L_s - M_s + \frac{3}{2}L_m \tag{10.140}$$

concerning inductance are defined. Note that these new coefficients allow the torque to be expressed by

$$T_e = N(L_d - L_q)i_{qs}i_{ds} \tag{10.141}$$

Variables are assigned standard notation for states, inputs, and outputs according to

$$x_1 = \theta \quad x_2 = \omega \quad x_3 = i_{qs} \quad x_4 = i_{ds}$$
$$u_1 = v_{qs} \quad u_2 = v_{ds} \quad y_1 = \theta \tag{10.142}$$

Since torque depends on both the q-axis and d-axis currents in a symmetric way, the appropriate choice of the second output is nonunique and either of these currents would serve the purpose equally well.

With the above variable assignments, the state variable model is

$$f(x) = \begin{bmatrix} x_2 \\ \frac{1}{J}\left(N(L_d - L_q)x_3x_4 - Bx_2 - T_l\right) \\ -\frac{1}{L_q}\left(R_s x_3 + NL_d x_4 x_2\right) \\ -\frac{1}{L_d}\left(R_s x_4 - NL_q x_3 x_2\right) \end{bmatrix}$$

$$g_1(x) = \begin{bmatrix} 0 \\ 0 \\ \frac{1}{L_q} \\ 0 \end{bmatrix} \quad g_2(x) = \begin{bmatrix} 0 \\ 0 \\ 0 \\ \frac{1}{L_d} \end{bmatrix} \quad h_1(x) = x_1 \tag{10.143}$$

Again, nonlinearity is present in both the electrical and mechanical subdynamics. In checking for relative degree, two cases are considered, corresponding to the two obvious choices for the second output. The common Lie-derivative calculations are given by

$$\begin{aligned} L_{g_1}h_1(x) &\equiv 0 \quad L_{g_2}h_1(x) \equiv 0 \\ L_{g_1}L_f h_1(x) &\equiv 0 \quad L_{g_2}L_f h_1(x) \equiv 0 \\ L_{g_1}L_f^2 h_1(x) &= \frac{N(L_d - L_q)}{JL_q}x_4 \\ L_{g_2}L_f^2 h_1(x) &= \frac{N(L_d - L_q)}{JL_d}x_3 \end{aligned} \quad (10.144)$$

For the case when $h_2(x) = x_4$, the remaining calculations

$$L_{g_1}h_2(x) = 0 \quad L_{g_2}h_2(x) = \frac{1}{L_d} \quad (10.145)$$

provide the decoupling matrix

$$\det\begin{bmatrix} \frac{N(L_d-L_q)}{JL_q}x_4 & \frac{N(L_d-L_q)}{JL_d}x_3 \\ 0 & \frac{1}{L_d} \end{bmatrix} = \frac{N(L_d - L_q)}{JL_dL_q}x_4 \quad (10.146)$$

which suggests that

$$\{r_1, r_2\} = \{3, 1\} \ (\text{if } y_2 \neq 0) \quad (10.147)$$

For the case when $h_2(x) = x_3$, the remaining calculations

$$L_{g_1}h_2(x) = \frac{1}{L_q} \quad L_{g_2}h_2(x) = 0 \quad (10.148)$$

provide the decoupling matrix

$$\det\begin{bmatrix} \frac{N(L_d-L_q)}{JL_q}x_4 & \frac{N(L_d-L_q)}{JL_d}x_3 \\ \frac{1}{L_q} & 0 \end{bmatrix} = -\frac{N(L_d - L_q)}{JL_dL_q}x_3 \quad (10.149)$$

which suggests that

$$\{r_1, r_2\} = \{3, 1\} \ (\text{if } y_2 \neq 0) \quad (10.150)$$

Hence, for this motor, it is clear that input-output linearization is a viable control strategy provided that isolated singularities are avoided, and this is simply achieved due to the fact that the singularities correspond to particular values of the second controlled output variable.

10.2.6 Concluding Remarks

This chapter has described how input-output linearization may be used to achieve motion control for a fairly wide class of motors. Although the most crucial issues of state-variable modeling and relative degree have been adequately covered, a few remaining points need to be made. In several cases, namely the shunt-connected and series-connected DC motors and the induction motor, the choice of outputs has led to first-order internal dynamics. It is not difficult to show, however, that these internal dynamics present no stability problems for input-output linearization. Also, it should be emphasized that the control possibilities for shunt-connected and series-connected DC motors are rather limited because of their unipolar operation, but that the isolated singularities found for all the other motors should not pose any real problems in practice.

The material presented in this chapter can be extended in several directions, beyond the obvious step of completely specifying the explicit feedback controls. Input-output linearization may be applied also to motors with nonsinusoidal winding distribution, to motors operating in magnetic saturation, or to motors with salient pole stators and concentrated windings (e.g., the switched reluctance motor). Moreover, there are still many supplementary issues that are important to mention, such as the augmentation of the basic controllers with algorithms to estimate unmeasured states and/or to identify unknown parameters.

Since model-based nonlinear controllers depend on parameters that may be unknown or slowly varying, an on-line parameter identification scheme can be included to achieve indirect adaptive control. Especially parameters describing the motor load are subject to significant uncertainty. There are several motives for agreeing to the additional complexity required for implementation of indirect adaptive control, including the potential for augmented performance, and improved reliability due to the use of diagnostic parameter checks. Typical examples of parameter identification schemes and adaptive control may be found in [13] for induction motors, and in [15] for permanent-magnet synchronous motors.

One of the challenges in practical nonlinear control design for electric machines is to overcome the need for full state measurement. The simplest example of this is commutation sensor elimination (e.g., elimination of Hall-effect devices from inside the motor frame). For electronically commutated motors, such as certain permanent-magnet synchronous motors, some applications do not require control of instantaneous torque, but can get by with simple commutation excitation with a variable firing angle. Even at this level of control, the need for a commutation sensor is crucial in order to maintain synchronism. The literature on schemes used to drive commutation controllers without commutation sensors is quite extensive.

A more difficult problem in the same direction is high-accuracy position estimation suitable for eliminating the high-resolution position sensor required by most of the nonlinear controls discussed earlier. In some applications, despite the need for high dynamic performance, the use of a traditional position sensor is considered undesirable due to cost, the volume and/or weight of the sensor, or the potential unreliability of the sensor in harsh environments. In this situation, the only alternative is to extract position (and/or speed) information from the available electrical terminal measurements. Of course, this is not an easy thing to do, precisely because of the nonlinearities involved. Some promising results have been reported in the literature on this topic, for permanent-magnet synchronous motors, switched reluctance motors, and for induction motors.

For the induction motor specifically, there is also a need to estimate either the rotor fluxes or rotor currents, in order to implement the input-output linearizing controller discussed earlier. Thorough treatments of this estimation problem are also available in the literature.

References

[1] Blaschke, F., The principle of field orientation applied to the new transvector closed-loop control system for rotating field machines, *Siemens Rev.*, 39, 217–220, 1972.

[2] Bodson, M., Chiasson, J., and Novotnak, R., High-performance induction motor control via input-output linearization, *IEEE Control Syst. Mag.*, 14(4), 25–33, 1994.

[3] Bodson, M., Chiasson, J.N., Novotnak, R.T., and Rekowski, R.B., High-performance nonlinear feedback control of a permanent magnet stepper motor, *IEEE Trans. Control Syst. Technol.*, 1(1), 5–14, 1993.

[4] Chiasson, J. and Bodson, M., Nonlinear control of a shunt DC motor, *IEEE Trans. Autom. Control*, 38(11), 1662–1666, 1993.

[5] Chiasson, J., Nonlinear differential-geometric techniques for control of a series DC motor, *IEEE Trans. Control Syst. Technol.*, 2(1), 35–42, 1994.

[6] De Luca, A. and Ulivi, G., Design of an exact nonlinear controller for induction motors, *IEEE Trans. Autom. Control*, 34(12), 1304–1307, 1989.

[7] Hemati, N., Thorp, J.S., and Leu, M.C., Robust nonlinear control of brushless dc motors for direct-drive robotic applications, *IEEE Trans. Ind. Electron.*, 37(6), 460–468, 1990.

[8] Isidori, A., *Nonlinear Control Systems*, 2nd ed., Springer-Verlag, New York, 1989.

[9] Kim, D.I., Ha, I.J., and Ko, M.S., Control of induction motors via feedback linearization with input-output decoupling, *Int. J. Control*, 51(4), 863–883, 1990.

[10] Kim, G.S., Ha, I.J., and Ko, M.S., Control of induction motors for both high dynamic performance and high power efficiency, *IEEE Trans. Ind. Electron.*, 39(4), 323–333, 1992.

[11] Krause, P.C., *Analysis of Electric Machinery*, McGraw-Hill, New York, 1986.

[12] Krzeminski, Z., Nonlinear control of induction motor, *Proc. 10th IFAC World Congress*, Munich, Germany, 1987, pp. 349–354.

[13] Marino, R., Peresada, S., and Valigi, P., Adaptive input-output linearizing control of induction motors, *IEEE Trans. Autom. Control*, 38(2), 208–221, 1993.

[14] Oliver, P.D., Feedback linearization of dc motors, *IEEE Trans. Ind. Electron.*, 38(6), 498–501, 1991.

[15] Sepe, R.B. and Lang, J.H., Real-time adaptive control of the permanent-magnet synchronous motor, *IEEE Trans. Ind. Appl.*, 27(4), 706–714, 1991.

[16] Zribi, M. and Chiasson, J., Position control of a PM stepper motor by exact linearization, *IEEE Trans. Autom. Control*, 36(5), 620–625, 1991.

10.3 Control of Electrical Generators

Thomas M. Jahns, GE Corporate R&D, Schenectady, NY

Rik W. De Doncker, Silicon Power Corporation, Malvern, PA

10.3.1 Introduction

Electric machines are inherently bidirectional energy converters that can be used to convert electrical energy into mechanical energy during motoring operation, or mechanical into electrical energy during generating operation. Although the underlying principles are the same, there are some significant differences between the machine control algorithms developed for motion control applications and those for electric power generation. While shaft torque, speed, and position are the controlled variables in motion control systems, machine terminal voltage and current are the standard regulated quantities in generator applications.

In this section, control principles will be reviewed for three of the most common types of electric machines used as electric power generators — DC machines, synchronous machines, and induction machines. Although abbreviated, this discussion is intended to introduce many of the fundamental control principles that apply to the use of a wide variety of alternative specialty electric machines in generating applications.

10.3.2 DC Generators

Introduction

DC generators were among the first electrical machines to be deployed reliably in military and commercial applications. Indeed, DC generators designed by the Belgian Professor F. Nollet and built by his assistant J. Van Malderen were used as early as 1859 as DC sources for arc lights illuminating the French coast [8]. During the French war in 1870, arc lights powered by the same magneto-electro machines were used to illuminate the fields around Paris to protect the city against night attacks [25]. Another Belgian entrepreneur, Zenobe Gramme, who had worked with Van Malderen, developed the renowned Gramme winding and refined the commutator design to its present state in 1869. In that year he demonstrated that the machine was capable of driving a water pump at the World Exposition held in Vienna.

Indeed, DC generators are constructed identically to DC motors, although they are controlled differently. Whereas DC motors are controlled by the armature current to regulate the torque production of the machine [23], DC generators are controlled by the field current to maintain a regulated DC terminal voltage. Typical applications for DC generators have included power supplies for electrolytic processes and variable-voltage supplies for DC servo motors, such as in rolling mill drives. Rotating machine uninteruptible power supplies (UPS), which use batteries for energy storage, require AC and DC generators, as well as AC and DC motors.

Other important historical applications include DC substations for DC railway supplies. However, since the discovery of DC rectifiers, especially silicon rectifiers (1958), rotating DC generators have been steadily replaced in new DC generator installations by diode or thyristor solid-state rectifiers. In those situations where mechanical-to-electrical energy conversion takes place, DC generators have been replaced by synchronous machines feeding DC bridge rectifiers, particularly in installations at high power, i.e., installations requiring several tens of kilowatts. Synchronous generators with DC rectifiers are also referred to as brushless DC generators.

There are multiple reasons for the steady decline in the number of DC generators. Synchronous machines with solid-state rectifiers do not require commutators or brushes that require maintenance. The commutators of DC machines also produce arcs that are not acceptable in mining operations and chemical plants. In a given mechanical frame, synchronous machines with rectifiers can be built with higher overload ratings. Furthermore, as the technology of solid-state rectifiers matured, the rectifier systems and the brushless DC generator systems became more economical to build than their DC machine counterparts.

Despite this declining trend in usage, it is still important to understand the DC generator's basic operating principles because they apply directly to other rotating machine generators and motors (especially DC motors). The following subsections describe the construction and equivalent model of the DC generator and the associated voltage control algorithms.

DC Machine Generator Fundamentals

Machine Construction and Vector Diagram Most rotating DC generators are part of a rotating machine group where AC induction machines or synchronous machines drive the DC generator at a constant speed. Figure 10.7 illustrates a typical rotating machine lineup that converts AC power to DC power. Note that this rotating machine group has bidirectional power flow capability because each machine can operate in its opposite mode. For example, the DC generator can operate as a DC motor. Meanwhile, the AC motor will feed power back into the AC grid and operate as an AC generator.

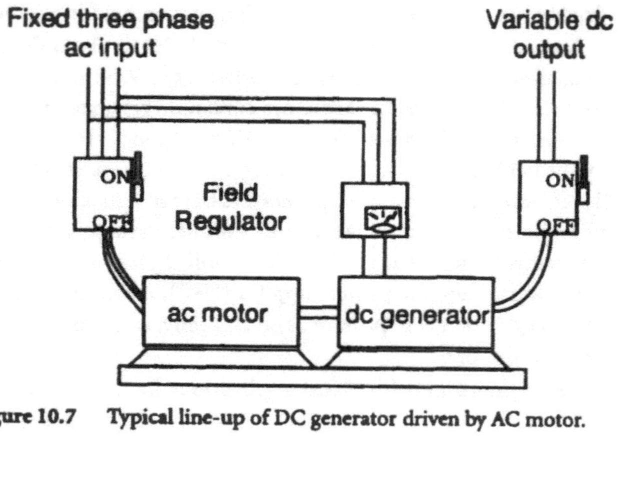

Figure 10.7 Typical line-up of DC generator driven by AC motor.

Figure 10.8(a) shows the DC generator construction, which is similar to a DC shunt motor. The space flux vector diagram of Figure 10.8(b) shows the amplitude and relative spatial position of the stator and armature fluxes that are associated with the winding currents. The moving member of the machine is called the armature, while the stationary member is called the stator. The armature current i_a is taken from the armature via a commutator and brushes. A stationary field is provided by an electromagnet or a permanent magnet and is perpendicular to the field produced by the armature. In the case of an electromagnet, the field current can be controlled by a DC power supply.

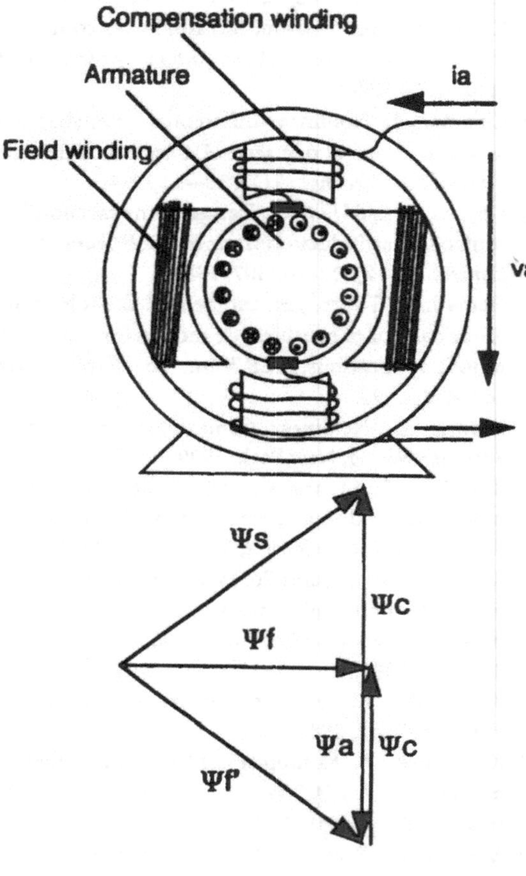

Figure 10.8 (a) Construction of DC generator, showing the stator windings and armature winding. (b) Vector diagram representing flux linked to field winding (Ψ_f), armature winding (Ψ_a), and series compensation winding (Ψ_c).

Most DC machines are constructed with interpole windings to compensate for the armature reaction. Without this compensation winding the flux inside the machine would increase to Ψ_f' when the armature current (and flux Ψ_a) increases. Due to the magnetic saturation of the pole iron, this increased flux leads to a loss in internal back-emf voltage. Hence, the output voltage would drop faster than predicted by the internal armature resistance whenever load increases. Conducting the armature current through the series compensation winding creates a compensation flux Ψ_c which adds to the flux Ψ_f' such that the total flux of the machine remains at the original level Ψ_f determined by the field current.

The vector diagram of Figure 10.8(b) shows another way to illustrate the flux linkages inside the DC generator. One can construct first the so-called stator flux Ψ_s of the DC machine. The stator flux can be defined as the flux produced solely by the stator windings, i.e., field winding and compensation winding. Adding the armature flux Ψ_a to the stator flux Ψ_s yields the field winding flux Ψ_f which represents the total flux experienced by the magnetic circuit.

Another advantage of the compensation winding is a dramatic reduction of the armature leakage inductance because the flux inside the DC machine does not change with varying armature current. In other words, the magnetic stored energy inside the machine is greatly decoupled from changes of load current, making the machine a stiffer voltage source during transient conditions.

DC Generator Equivalent Circuit and Equations The DC generator equivalent circuit is depicted in Figure 10.9. The rotating armature windings produce an induced voltage that is proportional to the speed and the field flux of the machine (from the *Bvl* rule [6]):

$$e_a = k\omega_m \Psi_f \tag{10.151}$$

Parasitic armature elements are the armature resistance R_a (re-

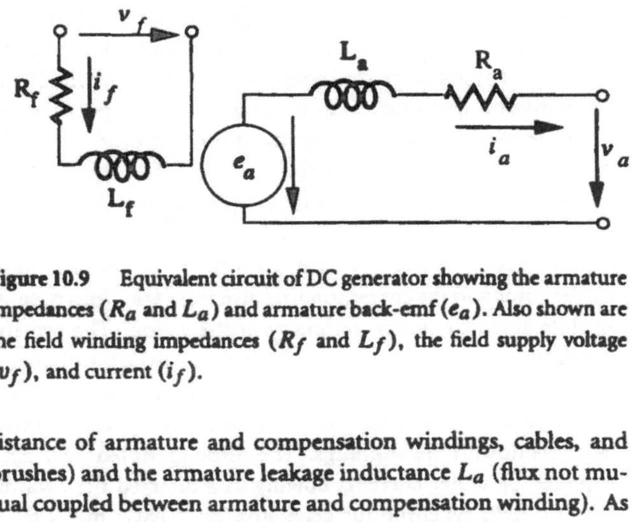

Figure 10.9 Equivalent circuit of DC generator showing the armature impedances (R_a and L_a) and armature back-emf (e_a). Also shown are the field winding impedances (R_f and L_f), the field supply voltage (v_f), and current (i_f).

sistance of armature and compensation windings, cables, and brushes) and the armature leakage inductance L_a (flux not mutual coupled between armature and compensation winding). As a result, the armature voltage loop equation is given by:

$$v_a = e_a - L_a \frac{di_a}{dt} - R_a i_a \tag{10.152}$$

The field winding is perpendicular to the armature winding and the compensation winding and experiences no induced voltages. Hence, its voltage-loop equation simplifies to:

$$v_f = L_f \frac{di_f}{dt} + R_f i_f = \frac{d\Psi_f}{dt} + \frac{R_f}{L_f}\Psi_f \tag{10.153}$$

The electromagnetic torque T_a produced by the DC generator is proportional to the field flux and the armature currents (from the *Bil* rule [6]) and can be expressed as:

$$T_a = k\Psi_f i_a \tag{10.154}$$

Note that this torque T_a acts as a load on the AC motor that drives the DC generator, as shown in Figure 10.7. Hence, to complete the set of system equations that describe the dynamic behavior of the DC generator, a mechanical motion equation describing the interaction between the DC generator and the AC drive motor is necessary. We assume that the mechanical motion equation of a DC generator fed by an electric motor can be characterized by the dynamic behavior of the combined inertia of the generator armature and the rotor of the motor and their speed-proportional damping. Speed-proportional damping is typically caused by a combination of friction, windage losses, and electrical induced losses in the armature and rotor:

$$T_m - T_a = J\frac{d\omega_m}{dt} + D\omega_m \tag{10.155}$$

where J is the total combined inertia of the DC generator and drive motor, ω_m is the angular velocity of the motor-generator shaft, and D is the speed-proportional damping coefficient.

The motor torque T_m is determined by the type of prime mover used and its control functions. For example, synchronous machines do not vary their average speed whenever the load torque (produced by the DC generator) varies. On the other hand, the speed of AC induction machines slips a small amount at increased load torque. The torque response of speed-controlled DC machines or AC machines depends greatly on the speed or position feedback loop used to control the machine. Assuming a proportional-integral speed control feedback loop, the torque of the driving motor can be described by the following function:

$$T_m = K_p(\omega_m^* - \omega_m) + K_i \int (\omega_m^* - \omega_m)dt \tag{10.156}$$

where ω_m is the measured speed of the drive motor, ω_m^* is the desired speed of the drive motor, K_p is the proportional gain of the drive motor feedback controller, and K_i is the integral gain of the drive motor feedback controller.

Assuming no field regulation and assuming a constant armature speed, the DC generator output voltage varies in steady state as a function of the load current according to:

$$v_a = e_a - R_a i_a \tag{10.157}$$

Figure 10.10 illustrates the DC generator steady-state load characteristic. In some applications the drop of the output voltage with increasing load current is not acceptable and additional control is required. Also, during dynamic conditions, e.g., a step change in the load current, the armature voltage may respond too slowly or with insufficient damping. The following section analyzes the dynamic response and describes a typical control algorithm to improve the DC generator performance.

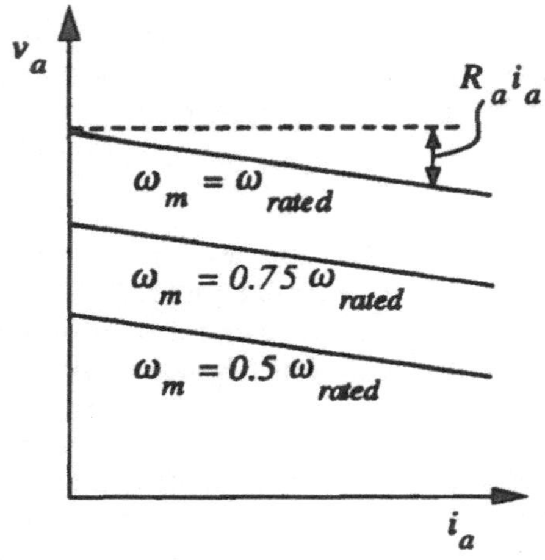

Figure 10.10 DC generator steady-state load characteristic.

DC Generator Control Principles

Basic Generator System Control Characteristics A control block diagram can be derived from the DC generator equations and is illustrated in Figure 10.11. As indicated by this block diagram, the DC generator system exhibits some complicated coupled dynamics.

For example, whenever the generator load increases quickly (e.g., due to switching of resistive-inductive loads), a rapid increase of the armature current will occur because the armature time constant is typically less than 100 msec. This leads to an equally fast increase in the generator load torque T_a. The speed regulator of the drive motor will respond to this fast torque increase with some time delay (determined by the inertia of the drive and the speed PI controller). Some speed variation can be expected as a result of this delay. Such speed variations directly influence the armature-induced voltage e_a. This causes the armature voltage and the armature current to change, thereby altering the generator load torque.

In conclusion, the DC generator drive behaves as a system of higher order. Figure 10.12 shows that a step change in load current has an impact on the armature voltage, the torque of the DC generator, and the speed of the rotating group. In this example, the output voltage settles to its new steady-state value after an oscillatory transient. Under certain circumstances (e.g., low inertia) the drive may ultimately become unstable.

Generator Voltage Regulation A field controller can be added to enhance the dynamic performance of the generator system. However, this controller acts on the proportionality factor between the back-emf voltage and the speed of the drive. Hence, this control loop makes the system nonlinear, and simple feedback control can be difficult to apply. Feedforward back-emf decoupling (from speed) combined with feedback control is therefore often used in machine controllers because of its simplicity and fast dynamic response.

Figure 10.13 shows the block diagram of a field controller which utilizes feedback and feedforward control. The feedforward loop acts to decouple the speed variations from the generator back-emf e_a. This controller is obtained by inverting the model equations of the back-emf and the field winding.

Figure 10.14 shows the response of the generator system when only feedback control is applied. The armature voltage does not have a steady-state error but still shows some oscillatory behavior at the start of the transient. To improve further the dynamic response of the generator it is essential to decouple the influence of the speed variation on the armature back-emf e_a. This can be achieved using feedforward control by measuring the drive speed and computing the desired field voltage to maintain a constant back-emf e_a as shown in Figure 10.13. A lead network compensates for the field winding inductive lag (within power limits of the field winding power supply). In practice, the field power supply has a voltage limit which can be represented by a saturation function in the field controller block diagram.

Note that the feedforward control loop relies on good estimates of the machine parameters. The feedback control loop does not require these parameters and guarantees correct steady-state operation. Figure 10.15 illustrates the response of the DC generator assuming a 10% error in the control parameters in the feedforward controller. Clearly, a faster response is obtained with less transient error.

10.3.3 Synchronous Generators

Synchronous Machine Fundamentals and Models

Introduction Synchronous machines provide the basis for the worldwide AC electric power generating and distribution system, earning them recognition as one of the most important classes of electrical generators. In fact, the installed power-generating capacity of AC synchronous generators exceeds that of all other types of generators combined by many times. The scalability of synchronous generators is truly impressive, with available machine ratings covering at least 12 orders of magnitude from milliwatts to gigawatts.

A synchronous generator is an AC machine which, in comparison to the DC generator discussed in Section 10.3.2, has its key electrical components reversed, with field excitation mounted on the rotor and the corresponding armature windings mounted in slots along the inner periphery of the stator. The armature windings are typically grouped into three phases and are specially distributed to create a smoothly rotating magnetic flux wave in the generator's airgap when balanced three-phase sinusoidal currents flow in the windings. The rotor-mounted field can be supplied by either currents flowing in a directed field winding (wound-field synchronous generator, or WFSG) or, in special cases, by permanent magnets (PMSG). Figure 10.16 includes a simplified cross-sectional view of a wound-field machine showing the physical relationship of its key elements. A wealth of technical literature is available which addresses the basic operating principles and construction of AC synchronous machines in every level of desired detail [7], [22], [24].

Synchronous Generator Model There are many alternative approaches which have been proposed for modeling synchronous machines, but one of the most powerful from the standpoints of both analytical usefulness and physical insight is

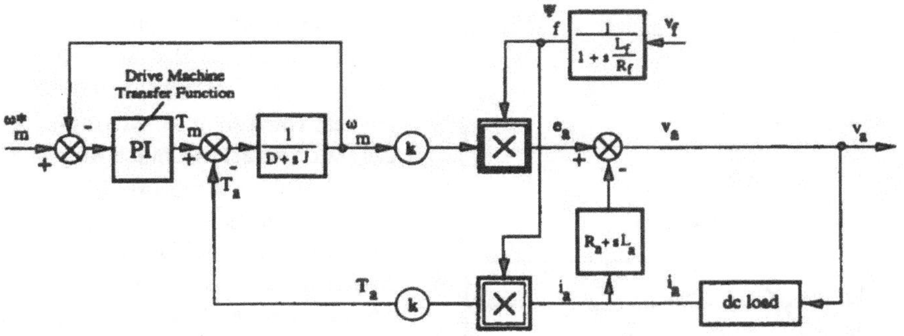

Figure 10.11 Block diagram of basic DC generator system.

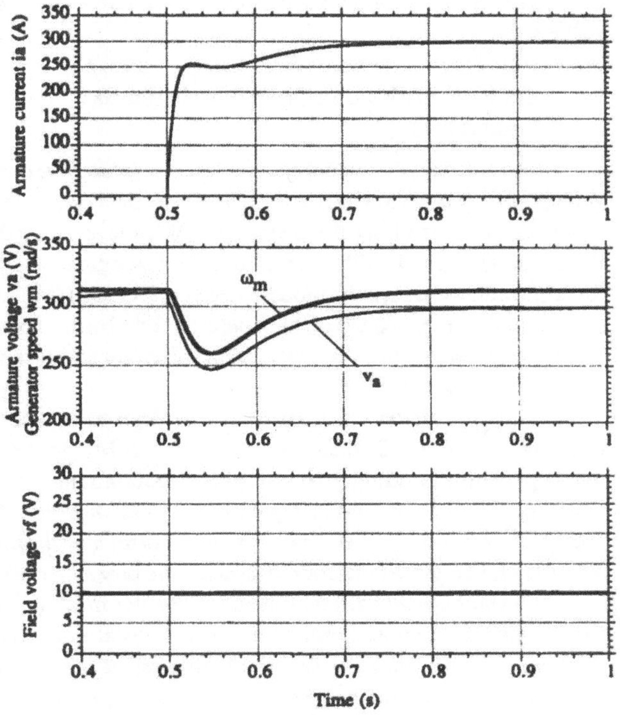

Figure 10.12 Response of DC generator with constant field v_f.

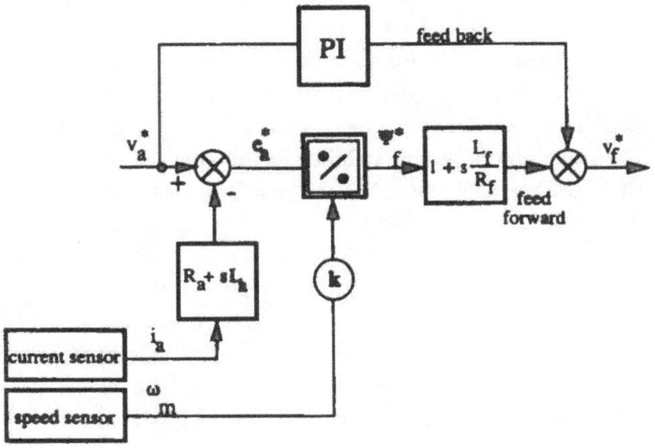

Figure 10.13 Feedforward and feedback control loops for DC generator field control.

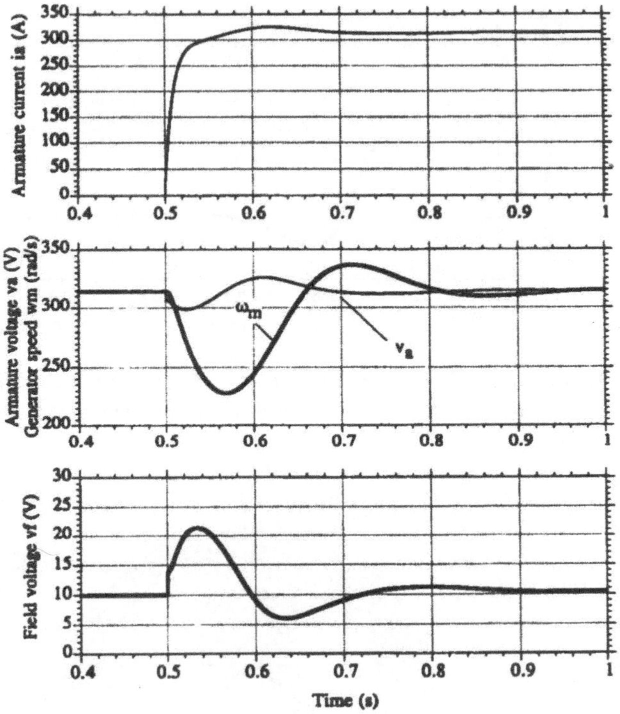

Figure 10.14 Response of DC generator with feedback control on field excitation v_f.

the dq equivalent circuit model shown in Figure 10.17. These coupled equivalent circuits result from application of the Park-Blondel dq transformation [15] which transforms the basic electrical equations of the synchronous machine from the conventional stationary reference frame locked to the stator windings into a rotating reference frame revolving in synchronism with the rotor. The direct (d) axis is purposely aligned with the magnetic flux developed by the rotor field excitation, as identified in Figure 10.16, while the quadrature (q) axis is orthogonally oriented at 90 electrical degrees of separation from the d-axis. As a result of this transformation, AC quantities in the stator-referenced equations become DC quantities during steady-state operation in the synchronously rotating reference frame.

Each of the d- and q-axis circuits in Figure 10.17 takes the form of a classic coupled-transformer equivalent circuit built around the mutual inductances L_{md} and L_{mq} representing the coupled

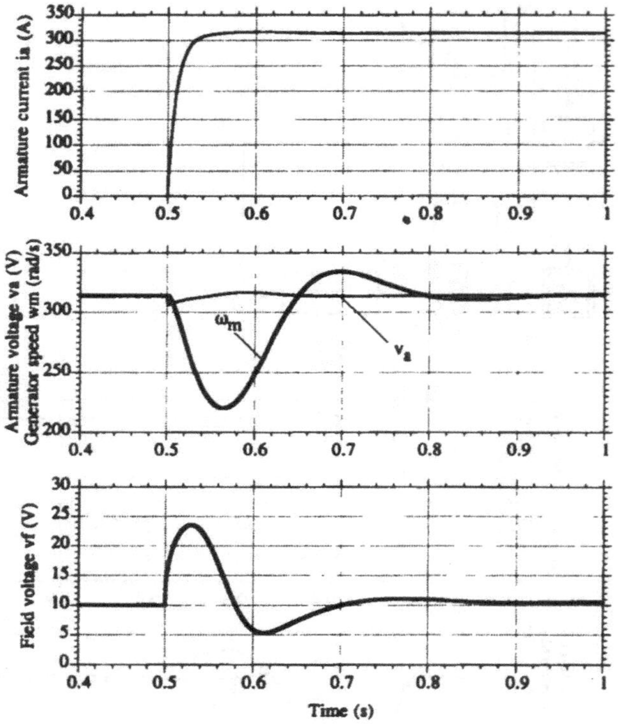

Figure 10.15 Response of DC generator with feedforward and feedback control on field excitation v_f with a 10% error in the feedforward control parameters.

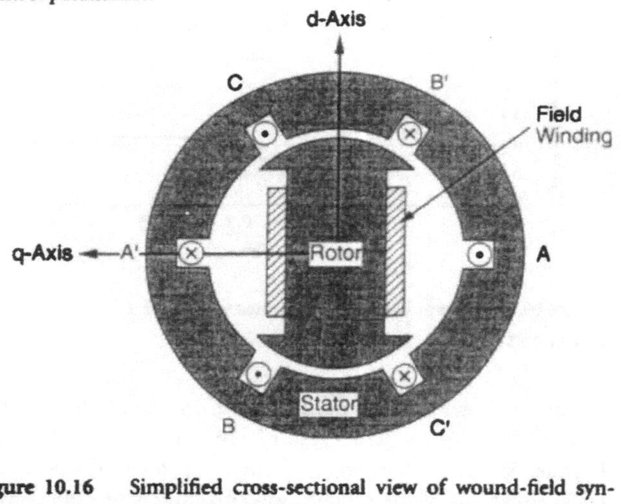

Figure 10.16 Simplified cross-sectional view of wound-field synchronous generator, including identification of d-q axes in rotor-oriented reference frame.

stator-rotor magnetic flux in each of the two axes. The principal difference between the d- and q-axis circuits is the presence of an extra set of external excitation input terminals in the d-axis rotor circuit modeling the field excitation. The externally controlled voltage source v_f which appears in this rotor excitation circuit represents the principal input port for controlling the output voltage, current, and power of a wound-field synchronous generator. The other rotor circuit legs which appear in both the d- and q-axis circuits are series resistor-inductor combinations modeling the passive damping effects of special rotor-mounted damper windings or eddy-current circuits in solid-iron rotor cores.

When analyzing the interaction of a synchronous generator (or several generators) with the associated power system, it is often very convenient to reduce the higher-order dq synchronous generator model to a much simpler Thevenin equivalent circuit model, as shown in Figure 10.18, consisting of a voltage source behind an inductive impedance [10]. The corresponding values of equivalent inductance and source voltage in this model change quite significantly depending on whether the analysis is addressing either steady-state operation or transient response. In particular, the generator's transient synchronous inductance L_s' is significantly smaller than the steady-state synchronous inductance L_s in most cases, while the transient model source voltage E_s' is, conversely, noticeably larger than the steady-state source voltage E_s.

Exciters for Wound-Field Synchronous Generators

The exciter is the functional block that regulates the output voltage and current characteristics of a wound-field synchronous generator by controlling the instantaneous voltage (and, thus, the current) applied to the generator's field winding. Exciters for large utility-class synchronous generators ($\geq$ 1000 MW) must handle on the order of 0.5% of the generators' rated output power, thereby requiring high-power excitation equipment with ratings of 5 MW or higher [27].

Typically, the exciter's primary responsibility is to regulate the generator's output AC voltage amplitude for utility power system applications, leading to the simplified representation of the resulting closed-loop voltage regulation system shown in Figure 10.19. Almost all of the basic exciter control algorithms in use today for large synchronous generators are based on classical control techniques which are well suited for the machine's dynamic characteristics and limited number of system inputs and outputs. The dynamics of such large generators are typically dominated by their long field winding time constants, which are on the order of several seconds, making it possible to adequately model the generator as a very sluggish low-pass filter for small-signal excitation analyses.

While such sluggish generator dynamics may simplify the steady-state voltage regulator design, it complicates the task of fulfilling a second major responsibility of the exciter which is to improve power system stability in the event of large electrical disturbances (e.g., faults) in the generator's vicinity. This large-disturbance stabilization requires that the exciter be designed to increase its output field voltage to its maximum (ceiling) value as rapidly as possible in order to help prevent the generator from losing synchronism with the power grid (i.e., "pull-out") [21]. Newer "high initial response (HIR)" exciters are capable of increasing their output voltages (i.e., the applied field voltage V_f') from rated to ceiling values in less than 0.1 sec. [12].

Exciter Configurations A wide multitude of exciter designs have been successfully developed for utility synchronous generators and can be found in operation today. These designs vary in such key regards as the method of control signal amplifi-

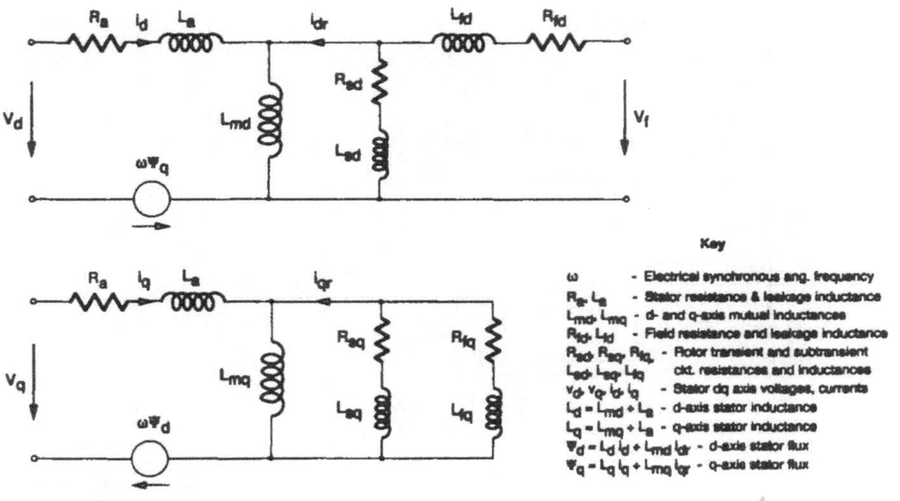

Figure 10.17 Synchronous generator d- and q-axis equivalent circuits.

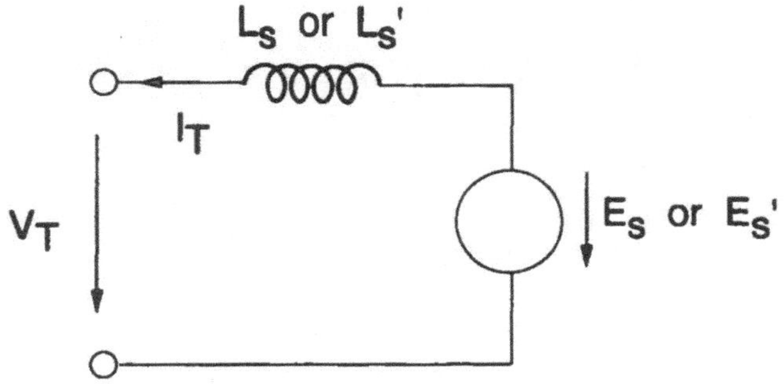

Figure 10.18 Synchronous generator Thevenin-equivalent circuit model (unprimed variables identify steady-state circuit model variables; primed variables identify transient model).

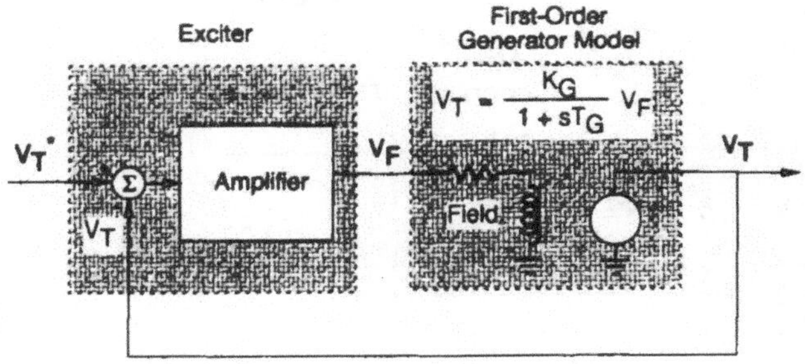

Figure 10.19 Simplified block diagram of synchronous generator closed-loop voltage regulation system.

cation to achieve the required field winding power rating, and the selected source of the excitation power (i.e., self-excitation from the generator itself vs. external excitation using an independent power source). Despite such diversity, the majority of modern exciters in use today can be classified into one of two categories based on the means of power amplification control: "rotating exciters" which use rotating machines as the control elements, and "static exciters" which use power semiconductors to perform the amplification.

Rotating Exciters Rotating exciters have been in use for many years and come in many varieties using both DC commutator machines and AC alternators as the main exciter machines [1], [4]. A diagram of one typical rotating exciter using an AC alternator and an uncontrolled rectifier to supply the generator field is shown in Figure 10.20(a). The output of the exciter alternator is controlled by regulating its field excitation, reflecting the cascade nature of the field excitation to achieve the necessary power amplification. One interesting variation of this scheme is the so-called "brushless exciter" which uses an inverted AC alternator as the main exciter with polyphase armature windings on the rotor and rotating rectifiers to eliminate the need for slip rings to supply the main generator's field winding [26].

The standard control representation developed by IEEE for the class of rotating exciters typified by the design in Figure 10.20(a) is shown in the accompanying Figure 10.20(b) [11]. The dynamics of the exciter alternator are modeled as a low-pass filter $[1/(K_E + T_E)]$ dominated by its field constant, just as in the case of the main generator discussed earlier (see Figure 10.19). The time constant (T_A) of the associated exciter amplifier is considerably shorter than the exciter's field time constant and plays a relatively minor role in determining the exciter's principal control characteristics.

The presence of the rate feedback block $[K_F, T_F]$ in the exciter control diagram is crucial to the stabilization of the overall voltage regulator. Without it, the dynamics of the voltage regulating loop in Figure 10.19 are dominated by the cascade combination of two low-frequency poles which yield an oscillatory response as the amplifier gain is increased to improve steady-state regulation. The rate feedback (generally referred to as "excitation control system stabilization") provides adjustable lag-lead compensation for the main regulating loop, making it possible to increase the regulator loop gain crossover frequency to 30 rad/sec or higher.

Static Exciters In contrast, a typical static exciter configuration such as the one shown in Figure 10.21(a) uses large silicon controller rectifiers (SCRs) rather than a rotating machine to control the amount of generator output power fed back to the field winding [17]. The voltage regulating function is performed

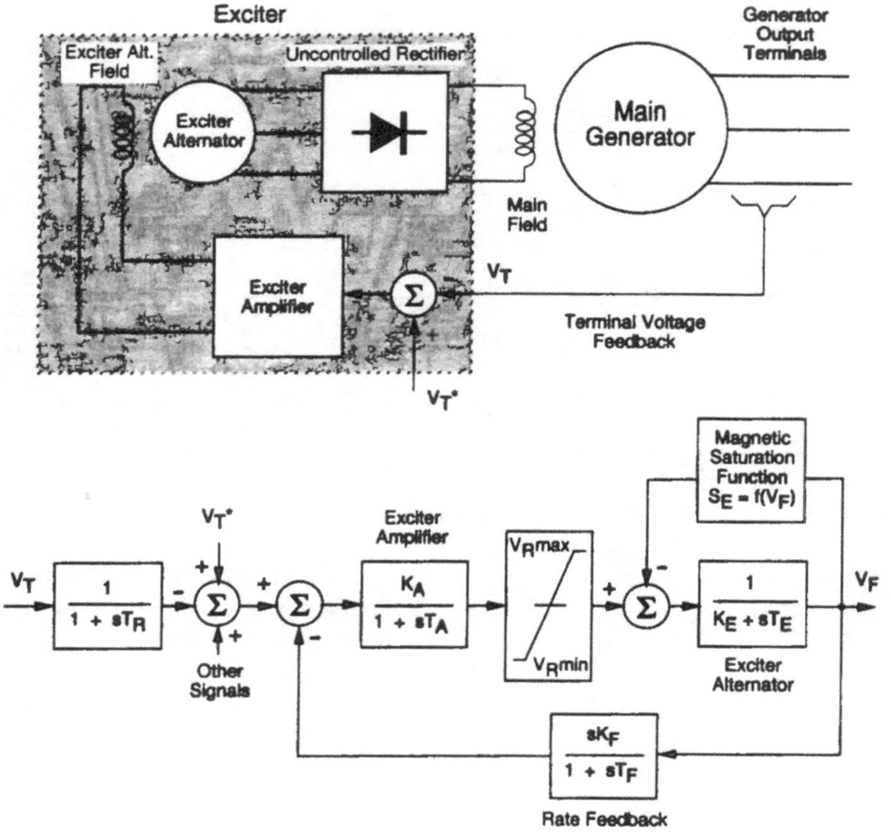

Figure 10.20 Block diagrams of a rotating exciter scheme using an uncontrolled rectifier showing (a) physical configuration, and (b) corresponding exciter standard control representation [12].

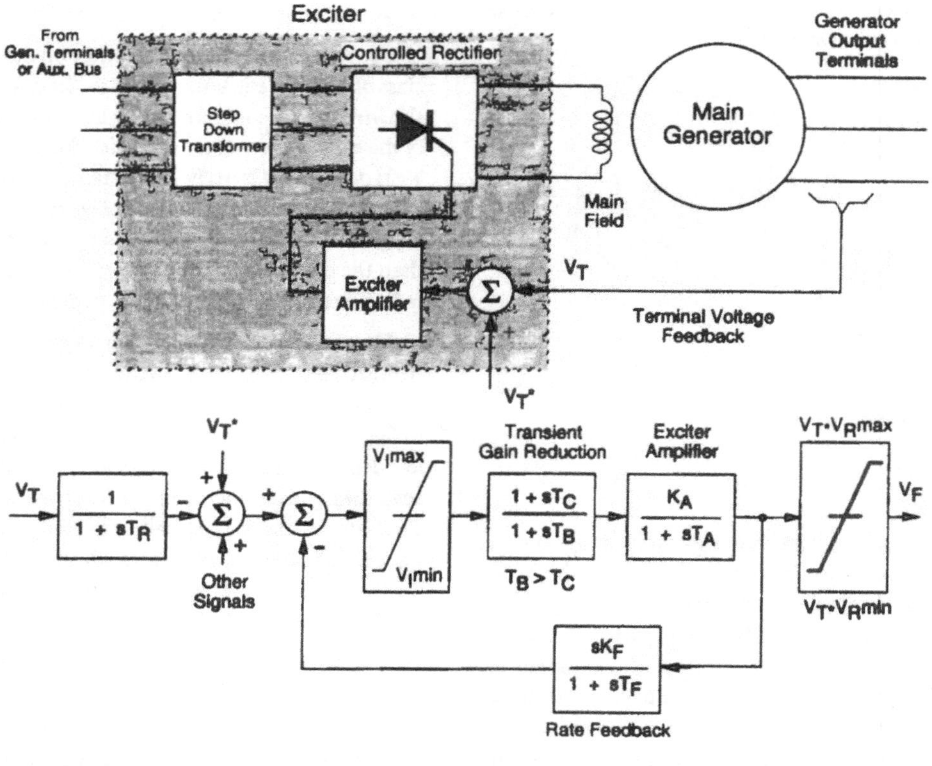

Figure 10.21 Block diagrams of a static exciter scheme using phase-controlled SCRs showing (a) physical configuration, and (b) corresponding exciter standard control representation [18]. (Note that either the "Transient Gain Reduction" block or the "Rate Feedback" block is necessary for system stabilization, but not both.)

using classic phase control principles [14] which determine when the SCRs are gated on during each 60-Hz cycle to supply the desired amount of field excitation.

Static exciters generally enjoy the advantage of much faster response times than rotating exciters, making them excellent candidates for "high initial response" exciters as discussed above. This is reflected in the standard static exciter control representation shown in Figure 10.21(b), which lacks the sluggish field time constant T_E that dominates the dynamics of the rotating exciter [13]. As a result, the task of stabilizing the main voltage regulating loop is simplified in the absence of this low-frequency pole.

Nevertheless, lag-lead compensation is often added to the exciter control in order to prevent the generator's high-frequency regulator gain from reducing the stability margin of the interconnected power system [13]. This compensation (referred to as "transient gain reduction") is provided in the Figure 10.21(b) control diagram using either the $T_B - T_C$ block in the forward gain path or the same type of $K_F - T_F$ rate feedback block introduced previously in Figure 10.20(b) (only one or the other is necessary in a given installation).

Additional Exciter Responsibilities In addition to its basic responsibilities for regulating the generator's output voltage described above, the exciter is also responsible for additional important tasks including power system stabilization, load or reactive power compensation, and exciter/generator protection. A block diagram of the complete generator-plus-exciter system identifying these supplementary functions is provided in Figure 10.22. Each of these exciter functions will be addressed briefly in this section.

Power System Stabilization Since a utility generator is typically one component in a large power grid involving a multitude of generators and loads distributed over a wide geographic area, the interactions of all these mechanical and electrical systems give rise to complex system dynamics. In some cases, the result is low-frequency dynamic instabilities in the range of 0.1 to 2 Hz involving one or more generators "swinging" against the rest of the grid across intervening transmission lines [5]. One rather common source of "local" mode instabilities is a remote generator located at the mouth of a mine that is connected to the rest of the power grid over a long transmission line with relatively high per-unit impedance. Other examples of dynamic instabilities involve the interactions of several generators, giving rise to more complicated "inter-area" modes which can be more difficult to analyze and resolve.

Unfortunately, the steps that are taken to increase exciter response for improved generator transient stability characteristics (i.e., "high initial response" exciters) tend to aggravate the power system stability problems. Weak transmission systems which result in large power angles between the generator internal voltage and the infinite bus voltage exceeding 70 degrees tend to demonstrate a particular susceptibility to this type of power system instability while under voltage regulator control.

The typical approach to resolving such dynamic instability problems is to augment the basic exciter with an additional feedback control loop [13] known as the power system stabilizer (PSS) as shown in simplified form in Figure 10.23. As indicated in this figure, alternative input signals for the PSS that are presently being successfully used in the field include changes in the generator shaft speed ($\Delta\omega_r$), generator electrical frequency ($\Delta\omega_e$), and the electrical power (ΔP_e). The primary control function performed by the PSS is to provide a phase shift using one or more adjustable lead-lag stages, which compensates for the destabilizing phase delays accumulated in the generator and exciter electrical circuits. Additional PSS signal processing in Figure 10.23 is typically added to filter out undesired torsional oscillations and to prevent the PSS control loop from interfering with the basic exciter control actions during major transients caused by sudden load changes or power system faults.

Proper settings for the primary PSS lead-lag gain parameters vary from site to site depending on the characteristics of the generator, its exciter, and the connected power system. Since the resulting dynamic characteristics can get quite complicated, a combination of system studies and field tests are typically required in order to determine the proper PSS gain settings for the best overall system performance. Effective empirical techniques for setting these PSS control gains in the field have gradually been developed on the basis of a significant experience base with successful PSS installations [16].

Load or Reactive Power Compensation A second auxiliary function provided in many excitation systems is the tailored regulation of the generator's terminal voltage to compensate for load impedance effects or to control the reactive power delivered by the generator [13]. One particularly straightforward version of this type of compensation is shown in Figure 10.24. As indicated in this figure, the compensator acts to supplement the measured generator's terminal voltage that is being fed back to the exciter's summing junction with extra terms proportional to the generator output current. The current-dependent compensation terms are added both in phase and 90° out of phase with the terminal voltage, with the associated compensation gains, R_c and X_c, having dimensions of resistive and reactive impedance.

Depending on the values and polarities of these gains, this type of compensation can be used for different purposes. For example, if R_c and X_c are negative in polarity, this block can be used to compensate for voltage drops in power system components such as step-up transformers that are downstream from the generator's terminals where the voltage is measured. Alternatively, positive values of R_c and X_c can be selected when two or more generators are bussed together with no intervening impedance in order to force the units to share the delivered reactive power more equally.

Although not discussed here, more sophisticated compensation schemes have also been developed which modify the measured terminal voltage based on calculated values of the real and reactive power rather than the corresponding measured current components. Such techniques provide means of achieving more precise control of the generator's output power characteristics and find their origins in exciter development work that was completed several decades ago [20].

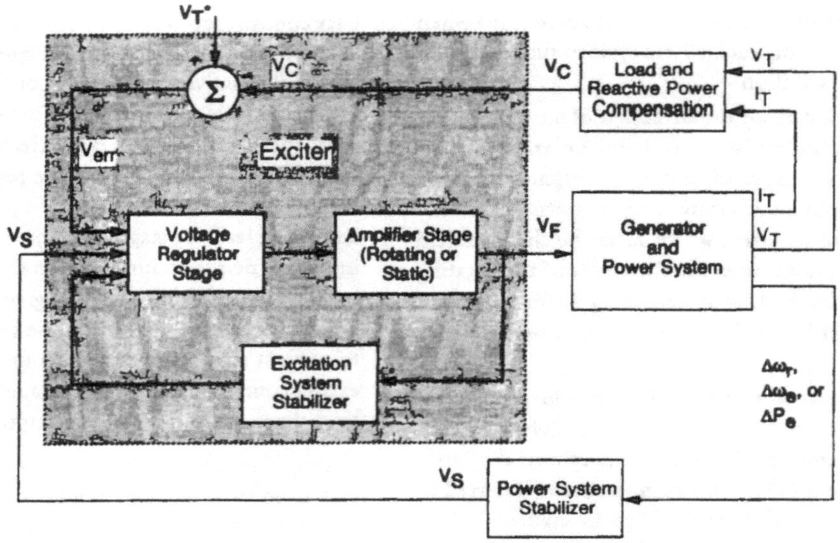

Figure 10.22 Block diagram of complete generator-plus-exciter control system, identifying supplementary control responsibilities.

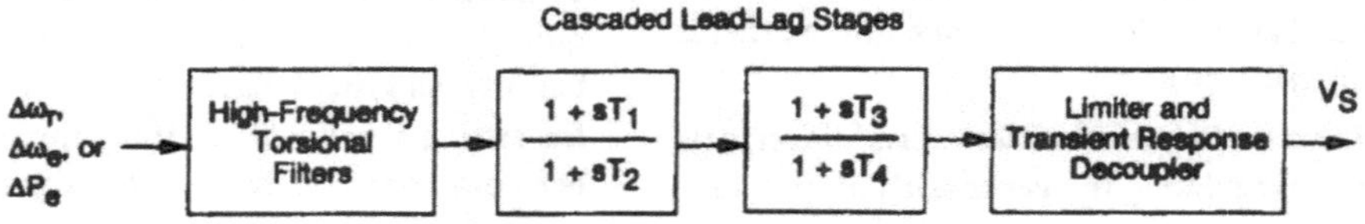

Figure 10.23 Basic control block diagram of power system stabilizer (PSS).

Exciter/Generator Protection Although a general review of the important issue of generator protection is well beyond the scope of this chapter [2], the specific role of the exciter in the generator's protection system deserves a brief discussion. Some of these key responsibilities include the following:

- Underexcited reactive ampere limit (URAL) — the minimum excitation level is limited as a function of the output reactive current since excessive underexcitation of the generator can cause dangerous overheating in the stator end turns
- Generator maximum excitation limit — at the other extreme, the maximum excitation current is limited to prevent damage to the exciter equipment and to the generator field winding due to overheating
- Volts-per-Hertz limiter — excessive magnetic flux levels in the generator iron which can cause internal overheating are prevented by using the exciter to limit the generator's output voltage as a function of output frequency (i.e., shaft speed).

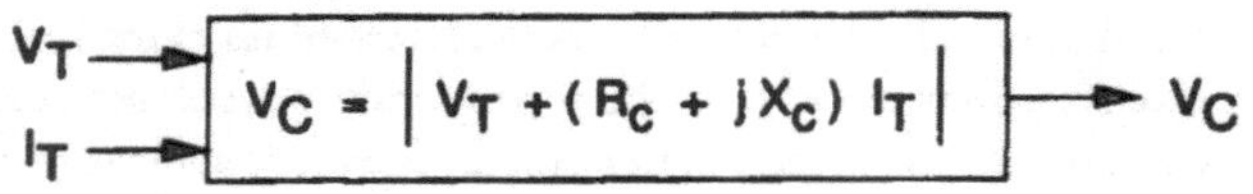

Figure 10.24 Example of generator load/reactive power compensation.

10.3.4 Induction Generators

Induction Generator Fundamentals and Models

Introduction Induction generators are induction machines that operate above synchronous speed and thereby convert mechanical power into electrical power. Induction generators have been extensively used in applications such as wind turbines and hydroelectric storage pumping stations. Frequent direct line starting is required in both of these applications. The AC induction machine offers the advantage that it can be designed for direct line-start operation, thereby avoiding additional synchronization machines and control. Furthermore, induction machines tolerate the thermal load associated with direct AC line-starting transients better than synchronous machines.

At high power levels above 1 MVA, efficiency considerations favor synchronous generators, while at very low power levels below 1 kVA, permanent magnet synchronous generators (e.g., automobile alternators) are more cost effective. One can conclude that induction generators are preferred in generator applications that require frequent starting and that are in a power range of 10 to 750 kVA.

Induction generators can operate in two distinctively different modes. In the first mode, the induction generator is connected to a fixed-frequency AC voltage source (e.g., utility line voltage) or a variable-frequency voltage-controlled AC source, such as pulse-width-modulated (PWM) inverters. The AC source provides the excitation (i.e., the magnetization current) for the induction machine. In this mode, the magnetizing flux is determined or controlled by the AC source voltage.

The second mode of operation is the so-called self-excited

mode. During self-excitation the magnetizing current for the induction generator is provided by external reactive elements (usually capacitors) or voltage-source inverters operating in six-step waveform mode. Neither of these schemes make it convenient to regulate the terminal voltage of the machine. The output voltage of the generator depends on many variables and parameters such as generator speed, load current, magnetization characteristics of the machine, and capacitor values.

The induction generator itself must deliver the necessary power to offset losses induced by the circulating reactive currents during self-excited operation, and the associated stator copper losses are typically high. As a result, self-excited induction generators are rarely used for continuous operation because they do not achieve high efficiency. Moreover, it is difficult to start the excitation process under loaded conditions. Nevertheless, self-excited operation with capacitors is sometimes used to brake induction motors to standstill in applications demanding rapid system shutdowns. In addition, six-step voltage-source inverters are often used in traction motor drive applications to apply regenerative braking.

The control principles of induction generators feeding power into a controlled AC voltage source will be discussed in the following sections. Self-excited operation of induction generators has been analyzed extensively, and interested readers are referred to the available technical literature for more details [18].

Induction Generator Model Induction generators are constructed identically to induction motors. A typical induction machine has a squirrel-cage rotor and a three-phase stator winding. Figure 10.25 illustrates the construction of a two-pole (or one pole pair), two-phase induction machine. The stator winding consists of two windings (d and q) that are magnetically perpendicular. The squirrel-cage rotor winding consists of rotor bars that are shorted at each end by rotor end rings.

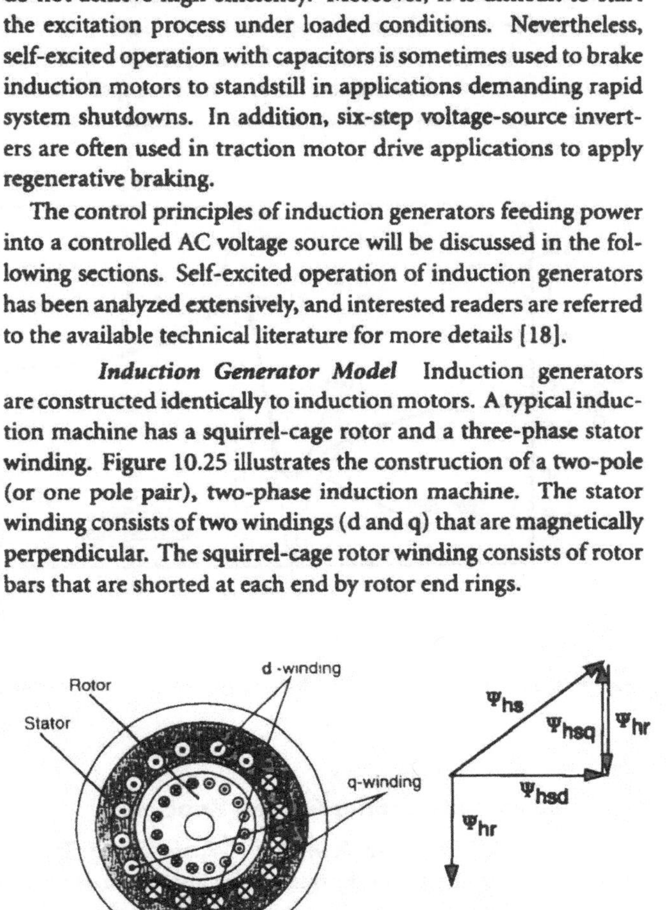

Figure 10.25 Two-phase induction machine, showing stator and rotor windings and flux diagram associated with stator and rotor currents.

The flux linkages associated with currents in each set of windings (i.e., the stator magnetizing flux $\underline{\Psi}_{hs}$, the rotor magnetizing flux $\underline{\Psi}_{hr}$, and the d- and q-axis stator magnetizing flux components $\underline{\Psi}_{hsd}$ and $\underline{\Psi}_{hsq}$,) are shown in the flux diagram. (Underlined variables designate vector quantities.) Note that each magnetizing flux vector represents a flux component produced by the current in a particular winding. These flux components are not to be confused with the total stator or rotor flux linkages $\underline{\Psi}_s$ and $\underline{\Psi}_r$ which represent the superimposed flux coupling from all windings, as discussed later in this section.

The basic equivalent circuit of an induction machine for steady-state operation is illustrated in Figure 10.26. This equivalent circuit shows that each phase winding has parasitic resistances and leakage inductances and that the stator and the rotor are magnetically coupled. However, other equivalent circuits can be derived for an induction machine, as illustrated in Figure 10.27. These equivalent circuits are obtained by transforming stator or rotor current and voltage quantities with a turns ratio "a." Figure 10.27 also specifies the different turns ratios and the corresponding flux vector diagrams which identify the flux reference vector used for each of the three equivalent circuits.

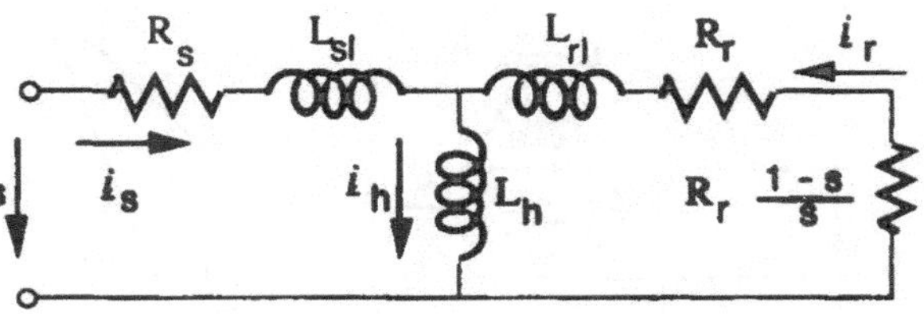

Figure 10.26 Single-phase equivalent (steady-state) circuit of induction machine. L_h, main (magnetizing) inductance; L_{sl}, stator leakage inductance; L_{rl}, rotor leakage inductance; $L_s = L_h + L_{sl}$, stator inductance; $L_r = L_h + L_{rl}$, rotor inductance; R_s, stator resistance; R_r, rotor resistance; $\underline{v}_s = v_{sd} + jv_{sq}$, stator (line-to-neutral) voltage, dq component; $\underline{i}_s = i_{sd} + ji_{sq}$, stator current, dq component; $\underline{i}_r = i_{rd} + ji_{rq}$, rotor current, dq component; $\underline{i}_h = \underline{i}_s + \underline{i}_r$, magnetizing current; s, slip of the induction machine.

Some equivalent circuits are simpler for analysis because one leakage inductance can be eliminated [3]. For example, a turns ratio $a = L_h/L_r$ transforms the equivalent circuit of Figure 10.26 into the topmost circuit of Figure 10.27, which has no leakage inductance in the rotor circuit. Hence, the rotor flux is selected here as the main flux reference vector. Also, the d-axis d_a of the dq synchronous reference frame that corresponds with this turns ratio "a" is linked to the rotor flux so that $d_a = d_r$.

Torque-Slip Characteristics The power the induction generator delivers depends on the slip frequency or, equivalently, the slip of the machine. Slip of an induction machine is defined as the relative speed difference of the rotor with respect to the synchronous speed set by the excitation frequency:

$$s = \frac{f_e - f_m}{f_e} = \frac{\omega_e - \omega_m}{\omega_e} = \frac{n_e - n_m}{n_e} \qquad (10.158)$$

with s being the slip of the induction machine, f_e the stator electrical excitation frequency (Hz), f_m the rotor mechanical rotation frequency (Hz), ω_e the stator excitation angular frequency (rad/s), ω_m the rotor mechanical angular frequency (rad/s), n_e the rotational speed of excitation flux in airgap (r/min), n_m the rotor mechanical shaft speed (r/min).

In the case of machines with higher pole-pair numbers, the

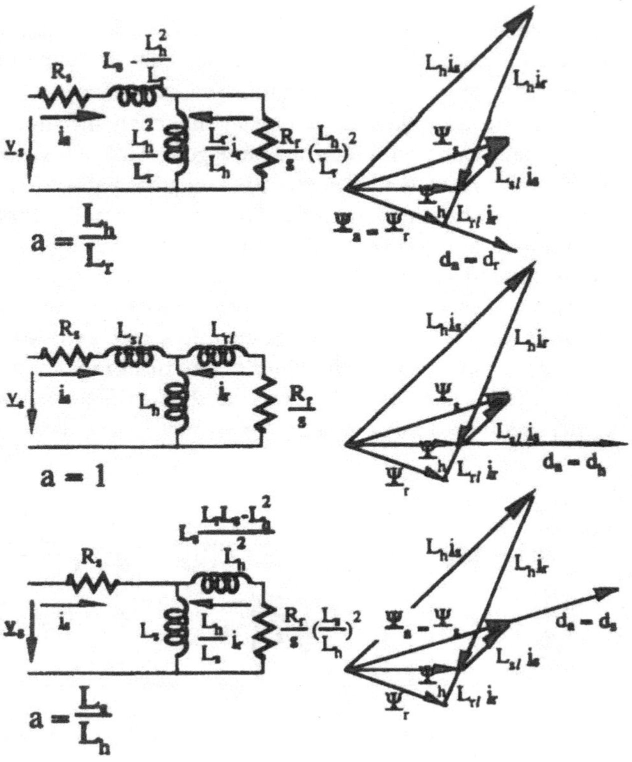

Figure 10.27 Modified equivalent circuits of induction machine, showing equivalent circuits and corresponding flux vector diagrams identifying the flux reference vector.

$$\underline{\Psi}_s = L_s\underline{i}_s + L_h\underline{i}_r$$

$$\underline{\Psi}_r = L_r\underline{i}_r + L_h\underline{i}_s$$

$$\underline{\Psi}_h = L_h\underline{i}_h = L_h\underline{i}_s + L_h\underline{i}_r = \underline{\Psi}_{hs} + \underline{\Psi}_{hr}$$

mechanical speed is usually defined in electrical degrees according to:

$$n_m = pn_{rotor} \tag{10.159}$$

where p is the pole pair number and n_{rotor} is the rotor speed as measured by observer in mechanical degrees.

The stator resistance R_s of medium and large induction machines can usually be neglected because the designer strives to optimize the efficiency of the induction machine by packing as much copper in the stator windings as possible. Using this approximation together with the equivalent circuit of Figure 10.27 that eliminates the stator leakage inductance ($a = L_s/L_h$), the steady-state torque per phase of the AC induction machine can easily be calculated as a function of the supply voltage and the slip frequency, yielding:

$$\frac{T_{em}}{T_k} = \frac{2}{\frac{s}{s_k} + \frac{s_k}{s}} \tag{10.160}$$

where

$$T_k = \frac{pV_s^2}{2\omega_e^2 L_l} \tag{10.161}$$

$$s_k = \frac{R_2}{\omega_e L_l} \tag{10.162}$$

$$L_l = L_s\frac{L_sL_r - L_h^2}{L_h^2} \tag{10.163}$$

where T_{em} is the electromagnetic torque per phase (Nm), T_k is the per-phase pull-out (maximum) torque (Nm), s_k is the pull-out slip associated with T_k, V_s is the rms stator line-to-neutral supply voltage (V), and L_l is the leakage inductance (H).

Figure 10.28 illustrates a typical torque-slip characteristic of an induction machine. According to Equation 10.160, the torque of the induction machine at high slip values varies approximately inversely with the slip frequency. Operating an induction machine in this speed range beyond the pull-out slip is inefficient and unstable in the absence of active control, and, hence, this operating regime is of little interest. Stable operation is achieved in a narrow speed range around the synchronous speed ($s = 0$) between $-s_k$ and $+s_k$ which are identified in Figure 10.28.

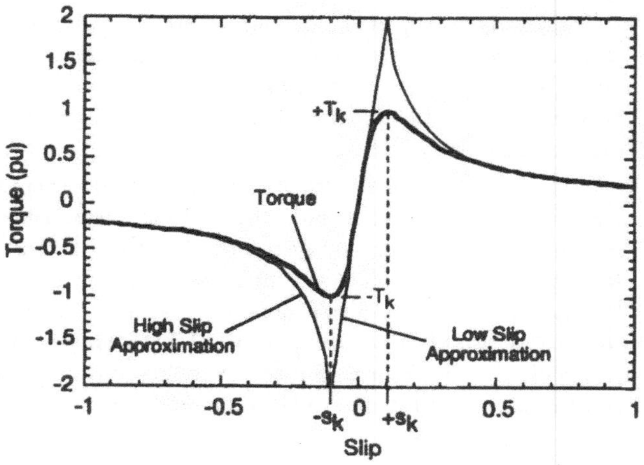

Figure 10.28 Typical slip-torque characteristic of induction machine.

Induction Generator Control Principles

Basic Slip Control Principle Whenever the slip is below the pull-out slip s_k, the torque varies approximately linearly with slip. Hence, the induction machine behaves similarly to a DC generator with constant armature voltage and constant field excitation. As soon as the speed of the generator exceeds the no-load (synchronous) speed (i.e., negative slip), mechanical power is transformed into electrical power (i.e., negative torque). Conversely, the machine operates as a motor with positive torque when the speed is below the no-load speed (i.e., positive slip).

Clearly, control of the slip frequency provides a direct means for controlling the AC induction generator's output torque and power. In applications where the electrical supply frequency is practically constant (e.g., utility systems), slip control can be realized by sensing the rotor speed and driving the generator shaft with the prime mover at the desired slip value with respect to the measured stator excitation frequency.

New induction generator systems use inverters connected to the machine's stator terminals to control the generator power, as illustrated in Figure 10.29. Inverters are power electronic devices that transform DC power to polyphase AC power or vice versa. Both the amplitude and frequency of the output waveforms delivered by the inverter are independently adjustable. In operation, the inverter provides the magnetization energy for the induction generator while the generator shaft is driven by a motor or some other type of prime mover. The AC-to-DC inverter converts the generator AC power to DC power, and this DC power can be transformed to AC power at fixed frequency (e.g., 50 or 60 Hz) using a second DC-to-AC inverter as shown in Figure 10.29. Both inverters are constructed as identical bidirectional units but are controlled in opposite power-flow modes.

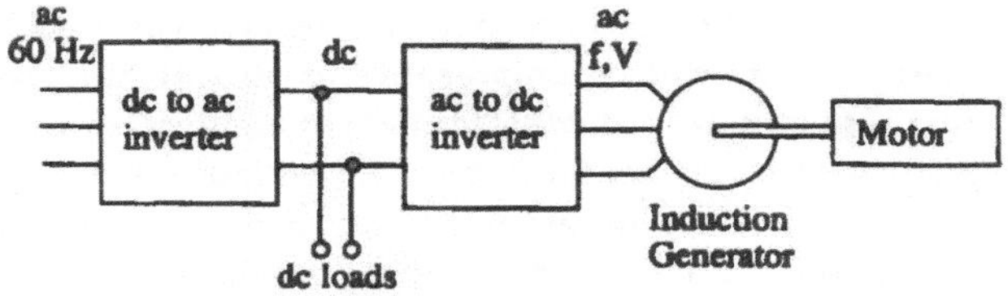

Figure 10.29 Inverter-fed induction generator system.

Induction generator applications that have a wide speed range or operate under fast varying dynamic load conditions require variable-frequency control to allow stable operation. Indeed, typical rated slip of induction machines is below 2%. Hence, a 2% speed variation around the synchronous speed changes torque from zero to 100% of its rated value. This poses significant design challenges in applications such as wind turbines operating in the presence of strong wind gusts. The high stiffness of the induction generator's torque-speed characteristic makes it very difficult to implement a speed governor to control the pitch of the turbine blades that is sufficiently fast-acting and precise to adequately regulate the machine's slip within this narrow range. On the other hand, an inverter can rapidly adjust the generator's electrical excitation frequency and slip to provide stable system operation with constant output power under all operating conditions.

Field-Oriented Control Principles The fundamental quantity that needs to be controlled in an induction generator is torque. Torque control of the inverter-fed induction machine is usually accomplished by means of field-oriented control principles to ensure stability. With field-oriented control, the torque and the flux of the induction generator are independently controlled in a similar manner to a DC generator with a separately excited field winding, as discussed earlier in this chapter.

The principles of field orientation can best be explained by recognizing from the flux vector diagrams shown in Figure 10.27 that the rotor current vector $\underline{i}_r$ is perpendicular to the rotor flux $\underline{\Psi}_r$. Furthermore, these vector diagrams illustrate that the stator current space vector $\underline{i}_s$ is composed of the rotor current $\underline{i}_r$ and the magnetizing current $\underline{i}_h$. Note that the space vectors in the vector diagrams are drawn in a reference frame that is rotating at the synchronous excitation frequency ω_e. By aligning a dq coordinate system with the rotating rotor flux vector $\underline{\Psi}_r$, one can prove that under all conditions (including transient conditions) the torque of the induction machine per phase is given by [22]:

$$T_{em} = -\frac{p}{2} i_{rq} \Psi_r = \frac{p}{2} \frac{L_h}{L_r} i_{sq} \Psi_r \tag{10.164}$$

Positive torque signifies motoring operation, while a negative torque indicates generator operation. Hence, in a synchronous dq reference frame linked to the rotor flux (top diagram in Figure 10.27), the q-axis component of the stator current i_{sq} corresponds to the torque-producing stator current component, being equal in amplitude to the rotor current component i_{rq}, with the opposite sign. The d-axis component of the stator current i_{sd} equals the magnetizing current, corresponding to the rotor flux-producing current component.

A control scheme that allows independent control of rotor flux and torque in the synchronously rotating rotor flux reference frame can now be derived, and the resulting control block diagram is shown in Figure 10.30. A Cartesian-to-polar coordinate transformation calculates the amplitude and the angle γ_{rs} of the stator current commands (in the synchronous reference frame) corresponding to the desired flux- and torque-producing components. (Controller command signals are marked with superscript $*$ in Figure 10.30, with negative torque commands corresponding to generator operation.) It is very important to note that the torque and the flux commands are decoupled (i.e., independently controlled) using this field-oriented control scheme. The controller can be seen as an open-loop disturbance feedforward controller, the disturbance signal being the variation of the flux position γ_r.

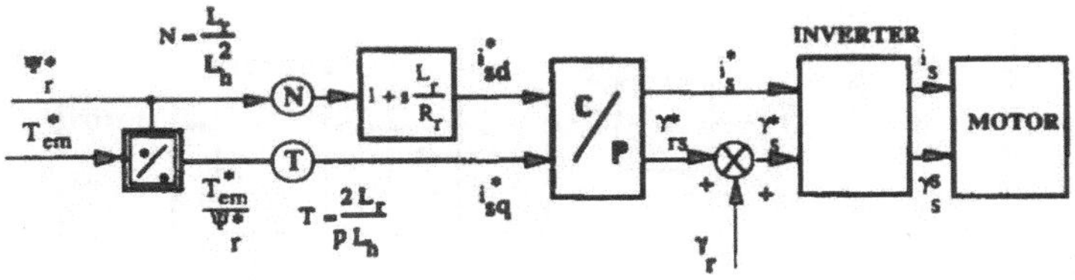

Figure 10.30 Field oriented controller allowing independent flux and torque control. γ_r, angular position of rotor flux with respect to stationary reference; γ_s, angular position of stator current with respect to stationary reference; γ_{rs}, angle between stator current and rotor flux.

The C/P block in Figure 10.30 indicates a Cartesian-to-polar coordinate transformation according to the following generalized equations:

$$x = \sqrt{x_d^2 + x_q^2} \tag{10.165}$$

$$\tan\alpha = \frac{x_q}{x_d} \tag{10.166}$$

with x_d the d-component or real component of space vector $\underline{x} = x_d + jx_q$, x_q the q-component or imaginary component of $\underline{x}$, x the amplitude of the space vector $\underline{x}$, and α the angular position of space vector $\underline{x}$.

Direct vs. Indirect Field Orientation To complete the controller calculation loops, one needs the rotor flux position γ_r to calculate the stator current vector position in the stationary reference frame that is linked to the stator of the machine. In other words, one needs to determine the orientation of the rotating field flux vector. It is for this reason that the control method illustrated in Figure 10.30 was called "field orientation."

Two field orientation strategies have been derived to detect the rotor flux position. Direct field orientation methods use sensors to directly track the flux position. Hall sensors are seldom used because of the high temperature inside the induction machine. Typical flux sensors are flux coils (sensing induced voltage) followed by integrators. The latter gives satisfactory results at frequencies above 5 to 10 Hz.

The direct field orientation control block diagram can be completed as shown in Figure 10.31. However, control problems can arise because most flux sensors are positioned on the stator and not on the rotor. As a result, these sensors monitor the stator flux Ψ_s^s in a stationary reference frame (marked with superscript "s") and not the rotor flux which is used in the decoupling network. This makes it necessary to use the flux linkage equations to derive the rotor flux from the flux sensor measurements. The required calculations introduce estimated machine parameters (leakage inductances) into the disturbance feedforward path [3] leading to detuning errors. Another approach is to decouple the machine equations in the stator flux reference frame in which the flux sensors are actually operating. This method requires a decoupling network of greater complexity but achieves high accuracy and potentially zero detuning error under steady-state conditions [9], [19].

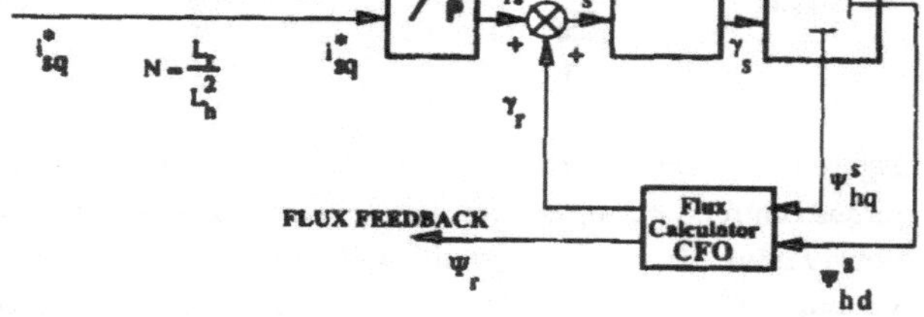

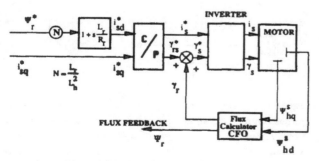

Figure 10.31 Direct field orientation method.

The second category of field orientation methods are called indirect field orientation because they derive the flux position using a calculated or estimated value of the angle γ_{mr} between the flux and the rotor position. This angle is nothing other than the rotor flux "slip" angle which varies at the slip frequency ω_{mr}. The rotor position γ_m is measured using a shaft position sensor, while the flux slip angle is derived from the slip frequency by integration. The dynamic field-oriented system equations of the induction machine are used to derive the slip frequency ω_{mr} and the slip angle γ_{mr}, as follows:

$$\omega_{mr} = \frac{L_h}{L_r}\frac{R_r}{p}\frac{i_{sq}^*}{\Psi_r^*} \tag{10.167}$$

$$\gamma_{mr} = \int \omega_{mr} dt \tag{10.168}$$

Figure 10.32 illustrates how a controller can be constructed to calculate the slip frequency command ω_{mr}^*, the slip angle command γ_{mr}^*, and the rotor flux position command γ_r^*. As Figure 10.32 shows, most indirect field-oriented controllers are constructed as open-loop disturbance feedforward controllers using the commanded current components instead of measured current quantities. This approach is justified because state-of-the-art, high-frequency, current-regulated inverters produce relatively precise current waveforms with respect to the current commands.

Note that indirect field orientation depends on the rotor time constant L_r/R_r which is composed of estimated machine parameters. As stated above, direct field orientation also needs machine parameters to calculate the flux vector position and most direct flux sensors do not operate at low frequencies. To circumvent these problems, both methods can be combined using special field-oriented algorithms to ensure greater accuracy.

Control Method Comparison Field orientation is a more advanced control technique than the slip controller discussed above. Field orientation offers greater stability during fast-changing transients because it controls torque while maintaining constant flux. Hence, the trapped magnetic energy in the machine does not change when speed or torque variations occur. This decoupled control strategy allows field orientation to control an AC induction machine exactly the same as a separately excited DC machine with a series compensation armature winding (see the section on DC generators).

The reader is invited to compare the vector diagram of Figure 10.8 in the DC generator section (illustrating the independent flux and torque control of a DC generator) and the space vector diagram of an induction generator illustrated in Figure 10.25 or Figure 10.27. The rotor flux vector $\underline{\Psi}_{hr}$ of the induction machine corresponds to the armature flux Ψ_a in the DC machine, while the stator magnetizing flux d-component $\underline{\Psi}_{hsd}$ corresponds to the field winding flux Ψ_f. The compensation winding flux Ψ_c of the DC machine relates to the q-component of the stator magnetizing flux vector $\underline{\Psi}_{hsq}$ of the AC machine. While the spatial orientation of the flux vectors in the DC machine is fixed, the space vectors of the induction machine rotate at synchronous speed. As a result, independent control of the individual space vectors in the AC machine can only be achieved by controlling the amplitude and the phase angle of the stator flux (i.e., stator voltage and current vectors).

Another approach to understanding the difference between a field-oriented controller and a slip controller is to consider the fact that field-oriented controllers control the flux slip *angle* and the stator current vector *angle* while slip controllers only regulate the *frequency* of these vectors. Controlling the angle of space vectors in electrical systems is analogous to position control

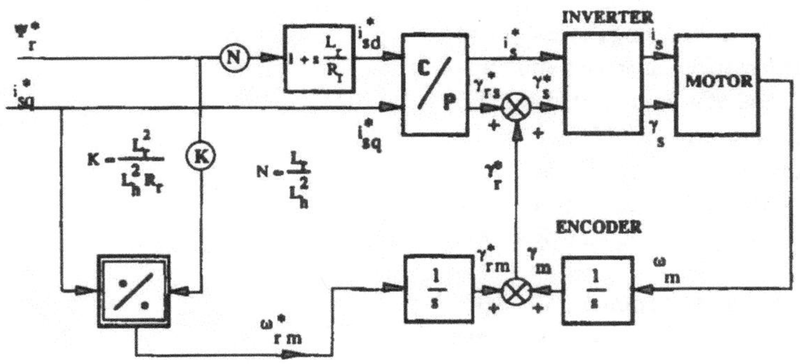

Figure 10.32 Indirect field orientation method.

in mechanical systems, while frequency control corresponds to speed control. It is immediately recognized that position control always offers greater stiffness than speed control. Hence, field orientation can offer the same stiffness improvements compared to slip-frequency controllers during transient conditions.

One disadvantage of field-oriented control is that it requires considerably more computation power than the simpler slip control algorithms. Many field-oriented controllers are implemented using digital signal processors (DSPs). Consequently, increased use of field orientation for induction generators will depend on future trends in the cost of digital controllers and sensors as well as the development of new control algorithms that decrease the controller's field installation time (e.g., machine parameter autotuning) while optimizing generator efficiency.

10.3.5 Concluding Remarks

This chapter has attempted to provide a concise overview of the major classes of electrical generators and their associated control principles. Despite the notable differences between the three types of electrical machines reviewed—DC, synchronous, and induction—there are some important underlying control aspects which they share in common. These include the single-input/single-output nature of the basic regulating control problem in each case, with dynamic response typically dominated by a long magnetic flux (i.e., field) time constant. This long time constant is responsible for the slow dynamic response which characterizes the majority of generator regulating systems now in the field.

As pointed out in the chapter, the introduction of power electronics provides access to generator control variables which can circumvent the limitations imposed by the flux time constant, leading to significant improvements in the regulator's dynamic response and other performance characteristics. Such advances have already had a significant impact in many applications, and work is continuing in many locations to extend these techniques to achieve further improvements in generator control performance and economics.

References

[1] Barnes, H. C., Oliver, J.A., Rubenstein, A.S., and Temoshok, M., Alternator-rectifier exciter for cardinal plant, *IEEE Trans. Power Appar. Syst.*, 87, 1189–1198, 1968.

[2] Berdy, J., Crenshaw, M.L., and Temoshok, M., Protection of large steam turbine generators during abnormal operating conditions, *Proc. CIGRE Int. Conf. on Large High Tension Electric Systems*, Paper No. 11-05, 1972.

[3] Blaschke, F. and Bayer, K. H., Die Stabilität der Feldorientierten Regelung von Asynchron-Maschinen, *Siemens Forsch. Entwick. Ber.*, 7(2), 77–81, 1978.

[4] Bobo, P.O., Carleton, J. T., and Horton, W.F., A new regulator and excitation system, *AIEE Trans. Power Appar. Syst.*, 72, 175–183, 1953.

[5] Bollinger, K.E. (Coordinator), *Power System Stabilization via Excitation on Control*, IEEE Tutorial Course Notes, Pub. No. 81 EHO 175-0 PWR, IEEE Press, New York, 1981.

[6] Brown, D. and Hamilton, E.P., *Electromechanical Energy Conversion*, Macmillan, New York, 1984.

[7] Concordia, C., *Synchronous Machines*, John Wiley & Sons, New York, 1951.

[8] Daumas, M., Ed., *Histoire Générale des Techniques*, Vol. 3, 1978, p. 330–335.

[9] De Doncker, R.W. and Novotny, D.W., The universal field oriented controller, *IEEE Trans. Ind. Appl.*, 30(1), 92–100, 1994.

[10] Fitzgerald, A.E., Kingsley, C., and Umans, S.D., *Electric Machinery*, McGraw-Hill, New York, 1983.

[11] IEEE Committee Report, Computer representation of excitation systems, *IEEE Trans. Power Appar. Syst.*, 87, 1460–1464, 1968.

[12] IEEE Standard 421-1972, Criteria and Definitions for Excitation Systems for Synchronous Machines, Institute of Electrical and Electronics Engineers, New York.

[13] IEEE Committee Report, Excitation system models for power system stability studies, *IEEE Trans. Power Appar. Syst.*, 100, 494–507, 1981.

[14] Kassakian, J.G., Schlecht, M.F., and Verghese, G.C., *Principles of Power Electronics*, Addison-Wesley, Reading, MA, 1991.

[15] Krause, P.C., Wasynczuk, O., and Sudhoff, S.D., *Analysis of Electric Machines*, IEEE Press, New York, NY, 1995.

[16] Larsen, E.V. and Swann, D.A., Applying Power System Stabilizers, Part II. Performance Objectives and Tuning Concepts, Paper 80 SM 559-5, presented at IEEE PES Summer Meeting, Minneapolis, 1980.

[17] McClymont, K.R., Manchur, G., Ross, R.J., and Wilson, R.J., Experience with high-speed rectifier excitation systems, *IEEE Trans. Power Appar. Syst.*, 87, 1464–1470, 1968.

[18] Novotny, D., Gritter, D., and Studtman, G., Self-excitation in inverter driven induction machines, *IEEE Trans. Power Appar. Syst.*, 96(4), 1117–1183, 1977.

[19] Profumo, F., Griva, G., Pastorelli, M., Moreira, J., and De Doncker, R., Universal field oriented controller based on air gap sensing via third harmonic stator voltage, *IEEE Trans. Ind. Appl.* 30(2), 448–455, 1994.

[20] Rubenstein, A.S. and Walkey, W.W., Control of reactive kVA with modern amplidyne voltage regulators, *AIEE Trans. Power Appar. Syst.*, 76, 961–970, 1957.

[21] Sarma, M., *Synchronous Machines (Their Theory, Stability, and Excitation Systems)*, Gordon and Breach, New York, NY, 1979.

[22] Say, M.G., *Alternating Current Machines*, 5th ed., Pitman, Bath, U.K., 1983.

[23] Sen, P., *Thyristor DC Drives*, John Wiley & Sons, New York, 1981.

[24] Slemon, G.R. and Straughen, A., *Electric Machines*, Addison-Wesley, Reading, MA, 1980.

[25] Tissandier, G., *Causeries Sur La Science*, Librairie Hachette et Cie, Paris, 1890.

[26] Whitney, E.C., Hoover, D.B., and Bobo, P.O., An electric utility brushless excitation system, *AIEE Trans. Power Appar. Syst.*, 78, 1821–1824, 1959.

[27] Wildi, T., *Electrical Machines, Drives, and Power Systems*, Prentice Hall, Englewood Cliffs, NJ, 1991.

11

Control of Electrical Power

Harry G. Kwatny
Drexel University

Claudio Maffezzoni
Politecnico Di Milano

John J. Paserba, Juan J. Sanchez-Gasca, and Einar V. Larsen
GE Power Systems Engineering, Schenectady, NY

11.1 Control of Electric Power Generating Plants 281
Introduction • Overview of a Power Plant and its Control Systems • Power Plant Modeling and Dynamical Behavior • Control Systems: Basic Architectures • Design of a Drum Level Controller
References 309
Further Reading 310
11.2 Control of Power Transmission 311
Introduction • Impact of Generator Excitation Control on the Transmission System • Power System Stabilizer (PSS) Modulation on Generator Excitation Systems • Practical Issues for Supplemental Damping Controls Applied to Power Transmission Equipment • Examples • Recent Developments in Control Design • Summary
References 322

11.1 Control of Electric Power Generating Plants

Harry G. Kwatny, Drexel University
Claudio Maffezzoni, Politecnico Di Milano

11.1.1 Introduction

This chapter provides an overview of the dynamics and control of electric power generating plants. The main goals are to characterize the essential plant physics and dynamical behavior, summarize the principle objectives of power plant control, and describe the major control structures in current use. Because of space limitations the discussion will be limited to fossil-fueled, drum-type steam generating plants. Much of it, however, is also relevant to once-through and nuclear powered plants.

The presentation is organized into four major sections. Section 11.1.2 provides a description of a typical plant configuration, explains in some detail the specific objectives of plant control, and describes the overall control system architecture. The control system is organized in a hierarchy, based on time scale separation, in which the highest level establishes set points for lower level regulators so as to meet the overall unit operating objectives.

Section 11.1.3 develops somewhat coarse linear models which qualitatively portray the small signal process behavior, characterize the essential interactions among process variables, and can be used to explain and justify the traditional regulator architectures. They are also useful for obtaining initial estimates of control system parameters which can then be fine-tuned using more detailed, nonlinear simulations of the plant.

The configurations commonly used in modern power plants for the main process variables are described in Section 11.1.4. These include controllers for pressure and generation, evaporator (drum level) temperature, and combustion control. The discussion in Section 11.1.4 is mainly qualitative, based on the understanding of plant behavior developed in Section 11.1.3.

Once a control configuration is chosen, the various compensator design parameters are established by applying analytical control design methods combined with extensive simulation studies. Because space is limited, it is not possible to provide such an analysis for each of the plant subsystems. However, in Section 11.1.5 we do so for the drum level regulator. Drum level control is chosen for illustration because it is particularly important to plant operation and because it highlights the difficulties associated with low load plant dynamics and control. There are many important and outstanding issues regarding automation at low load steam generation levels. In practice, most plants require considerable manual intervention when maneuvering at low load. The most important concerns relate to the evaporation process (the circulation loop) and to the combustion process (furnace). Section 11.1.5 revisits the circulation loop, examines the behavioral changes that take place as generation level is reduced, and explains the consequences for control.

0-8493-0054-3/00/$0.00+$.50

11.1.2 Overview of a Power Plant and its Control Systems

Overall Plant Structure

A typical power plant using fossil fuel as its energy source is organized into three main subsystems, corresponding to the three basic energy conversions taking place in the process: the steam generator (SG) (or boiler), the turbine (TU) integrated with the feed-water heater train, and the electric generator (EG) (or alternator). The SG converts the chemical energy available in the fuel (either oil, or natural gas or coal) into internal energy of the working fluid (the steam). The TU transforms the internal energy of steam flowing from the SG into mechanical power and makes it available at the shaft for the final conversion into electrical power in the EG.

The interactions among the principal subsystems are sketched in Figure 11.1, where only the mass and energy flows at the subsystem's boundaries are displayed.

The overall process can be described as follows: the feed-water coming from the feed-water heater train enters the SG where, due to the heat released by fuel combustion, ShS is generated and admitted into the HPT through a system of control valves (TV-hp). Here, the steam expands down to the reheat pressure, transferring power to the HPT shaft, and is discharged into a steam reheater (part of SG) which again superheats the steam (RhS). RhS is admitted into the RhT through the control valve TV-rh, normally working fully open; the steam expands successively in RhT and LPT down to the condenser pressure, releasing the rest of the available power to the turbine shaft. Condensed water is extracted from the condenser and fed to low-pressure feed-water heaters, where the feed-water is preheated using the steam extractions from RhT and LPT. Then the pressure is increased to its highest value by FwP and the feed-water gets its final preheating in the high-pressure feed-water heaters using steam extractions from HPT and RhT. The mechanical power released by the entire compound turbine is transferred to the EG, which converts that power into electrical power delivered to the grid via a three-phase line.

The control objectives in such a complex process can be synthesized as follows: transferring to the grid the demanded electrical power P_e with the maximum efficiency, with the minimum risk of plant trip, and with the minimum consumption of equipment life. As is usual in process control, such a global objective is transformed into a set of simpler control tasks, based on two principal criteria: (1) the outstanding role of certain process variables in characterizing the process efficiency and the operating constraints; (2) the weakness of a number of process interactions, which permits the decomposition of the overall process into subprocesses.

Referring to Figure 11.1, we observe that the EG, under normal operating conditions, is connected to the grid and is consequently forced to run at synchronous speed. Under those conditions it acts as a mechanical-electrical power converter with almost negligible dynamics. So, neglecting high frequency, we may assume that the EG merely implies $P_e = P_m$ (where P_m is the mechanical power delivered from the turbine). Of course, the EG is equipped with its own control, namely voltage control, which has totally negligible interactions with the control of the rest of the system (it works at much higher bandwidth).

Moreover, the turbines have very little storage capacity, so that, neglecting high frequency effects, turbines may be described by their steady-state equations:

$$P_m = P_{HP} + P_{LP} \tag{11.1}$$

$$P_{HP} = \alpha_T w_T (h_T - h_{tR}) \tag{11.2}$$

$$P_{LP} = \alpha_R w_R (h_R - h_0) \tag{11.3}$$

where P_{HP} and P_{LP} are the mechanical power released by the HPT and the RhT and LPT, respectively. w_T is the ShS mass flow-rate, h_T the corresponding enthalpy, h_{tR} the steam enthalpy at the HPT discharge, w_R is the RhS mass flow-rate, h_R the corresponding enthalpy, h_0 the fluid enthalpy at the LPT discharge, and α_T, α_R are suitable constants (≤ 1) accounting for the steam extractions from the HPT and the RhT and LPT, respectively.

With the aim of capturing the fundamental process dynamics, one may observe that the enthalpy drops $(h_T - h_{tR})$ and $(h_R - h_0)$ remain approximately unchanged as the plant load varies, because turbines are designed to work with constant pressure ratios across their stages, while the steam flow varies. This means that the output power P_m consists of two contributions, P_{HP} and P_{LP}, which are approximately proportional to the ShS flow and to the RhS flow, respectively. In turn, the flows w_T and w_R are determined by the state of the SG (i.e., pressures and temperatures) and by the hydraulic resistances that the turbines (together with their control valves) present at the SG boundaries.

Steam extractions (see Figure 11.1) mutually influence subsystems SG and TU: any variations in the principal steam flow w_T create variation in SE flow and, consequently, a change in the feed-water temperature at the inlet of the SG. Feed-water mass flow-rate, on the contrary, is essentially imposed by the FwP, which is generally equipped with a flow control system which makes the FwP act as a "flow-generator." Fortunately, the overall gain of the process loop due to the steam extractions is rather small, so that the feed-water temperature variations may be considered a small disturbance for the SG, which is, ultimately, the subprocess where the fundamental dynamics take place.

In conclusion, power plant control may be studied as a function of steam generator dynamics with the turbine flow characteristics acting as boundary conditions at the steam side, the feed-water mass flow-rate and the feed-water temperature acting as exogenous variables, and Equations 11.1, 11.2, and 11.3 determining the power output. To understand the process dynamics, it is necessary to analyze the internal structure of the SG. In the following, we will make reference to a typical drum boiler [1] ; once-through boilers are not considered for brevity.

Typical Structure of a Steam Generator

A typical scheme of a fossil-fueled steam generator, in Figure 11.2 depicts the principal components.

In Figure 11.2, the air-gas subsystem is clearly recognizable; the combustion air is sucked in by the fan (1) and conveyed

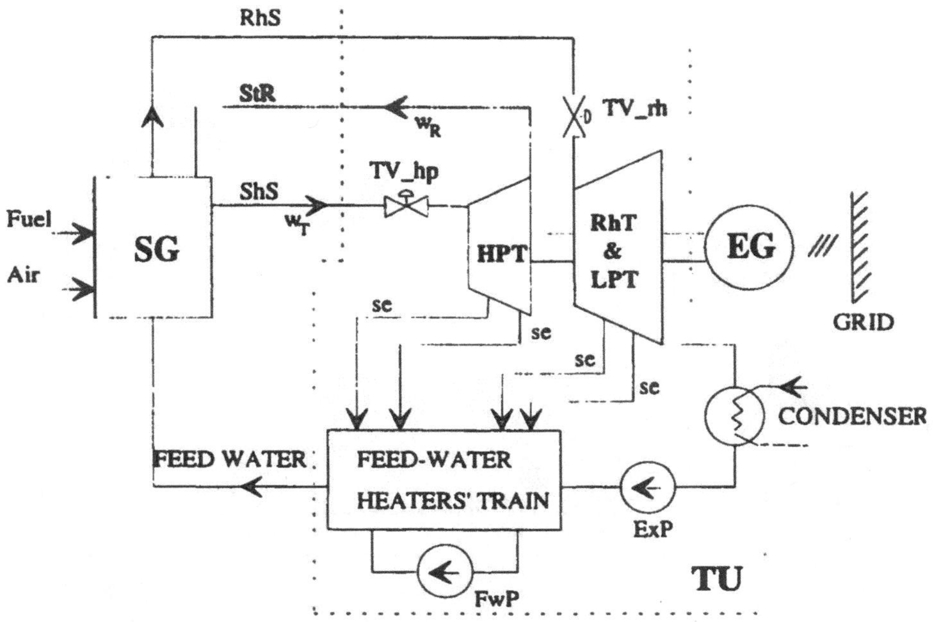

Figure 11.1 Subsystems interaction. RhS = Reheated steam; StR = Steam to reheat, ShS = Superheated steam; HPT = High pressure turbine; RhT = Reheat turbine; LPT = Low pressure turbine; se = Steam extraction; ExP = Extraction pump; FwP = Feed-water pump; TV-hp = Turbine valve, high pressure; TV-rh = Turbine valve, reheated steam.

through the air heaters (2) (using auxiliary steam) and (3) (exchanging heat counter flow with the flue gas leaving the furnace backpass) to the furnace wind box (4), where air is distributed to the burners, normally arranged in rows. Fuel and air, mixed at the burner nozzles, produce hot combustion gas in the furnace (5), where heat is released, principally by radiation, from the gas (and the luminous flame) to the furnace walls, usually made of evaporating tubes. The hot gas releases almost 50% of its available heat within the furnace and leaves it at high temperature; the rest of the internal energy of the hot gas is transferred to the steam through a cascade of heat exchangers in the back-pass of the furnace ((6) and (9) superheat the steam to high pressure, while (7) and (8) reheat steam and, at the end of the backpass, to the feed-water in the economizer (10). The gas is finally used in a special air heater (3) (called Ljungstroem) to capture the residual available energy. The flue gas is conveyed to the stack (12), possibly through induced draft fans (11), which are employed with coal-fired furnaces to keep the furnace pressure slightly below the atmospheric pressure.

The heat exchangers making up the furnace walls and the various banks arranged along the flue-gas path are connected on the steam side to generate superheated steam; this can be split into four subprocesses; water preheating, boiling, superheating, and reheating. The flow diagram of the water-steam subsystems is shown in Figure 11.3, where common components are labeled with the same numbers as in Figure 11.2.

In the scheme of Figure 11.3, the evaporator is the natural circulation type (also called drum-boiler). It consists of risers, the tubes forming the furnace walls where boiling takes place, and the steam separator (drum), where the steam-water mixture from the risers is separated into dry steam (flowing to superheating) and saturated water which, after mixing with feed-water, feeds the downcomers. There are two special devices (called spray desuperheaters), one in the high-pressure section, the other in the reheat section, which regulate superheated and reheated steam temperatures.

Process dynamics in a steam boiler is determined by the energy stored in the different sections of the steam-water system, especially in the working fluid and in the tube walls containing the fluid. Storage of energy in the combustion gas is practically negligible, because hot gas has a very low density. For those reasons, it is natural to focus on the steam-water subsystem, except for those special control issues where (fast) combustion dynamics are directly involved.

Control Objectives

A generation unit of the type described in Figures 11.1, 11.2, and 11.3 is asked to supply a certain power output P_e, that is (see Equations 11.1–11.3) certain steam flows to the turbines, while insuring that the process variables determining process efficiency and plant integrity are optimal. Because efficiency increases as the pressure and the temperature at the turbine inlet (i.e., at the throttle) increase, whereas stress on machinery goes in the opposite direction, the best trade-off between steam cycle efficiency and plant life results in prescribing certain values to throttle pressure p_T and temperature T_T and to reheat temperature T_R (reheat pressure is not specified because there is no throttling along the reheating section under normal conditions).

Moreover, proper operation of the evaporation section requires correct steam separator conditions, meaning a specified

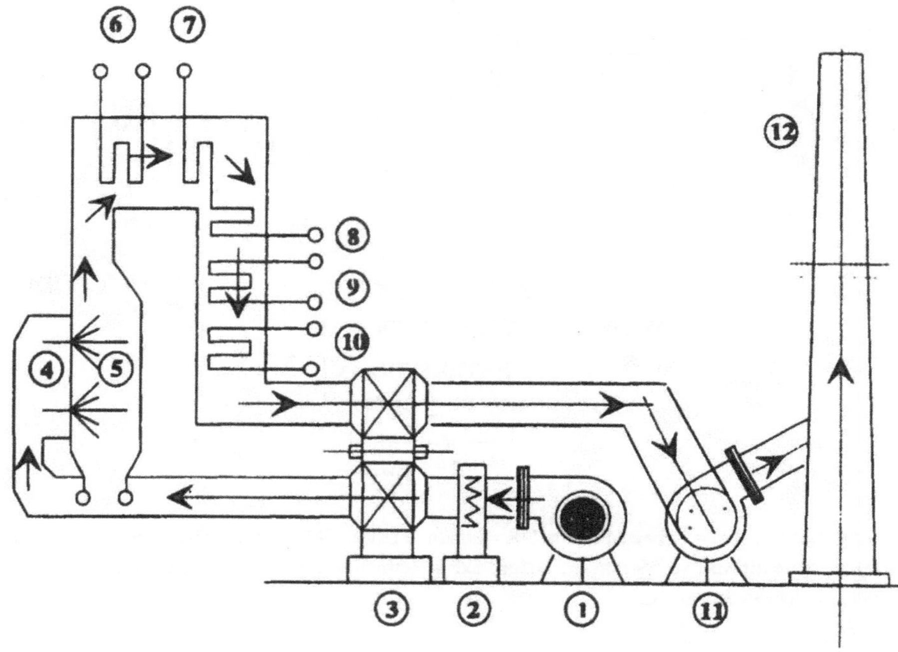

Figure 11.2 Typical scheme of the steam generator. (1) Air fan; (2) Auxiliary air heater; (3) Principal air heater; (4) Wind box; (5) Furnace (with burners); (6) High-temperature superheater; (7) High-temperature part of the reheater; (8) Low-temperature part of the reheater; (9) Low-temperature superheater; (10) Economizer; (11) Flue-gas fan (present only with balanced draft furnace); (12) Stack.

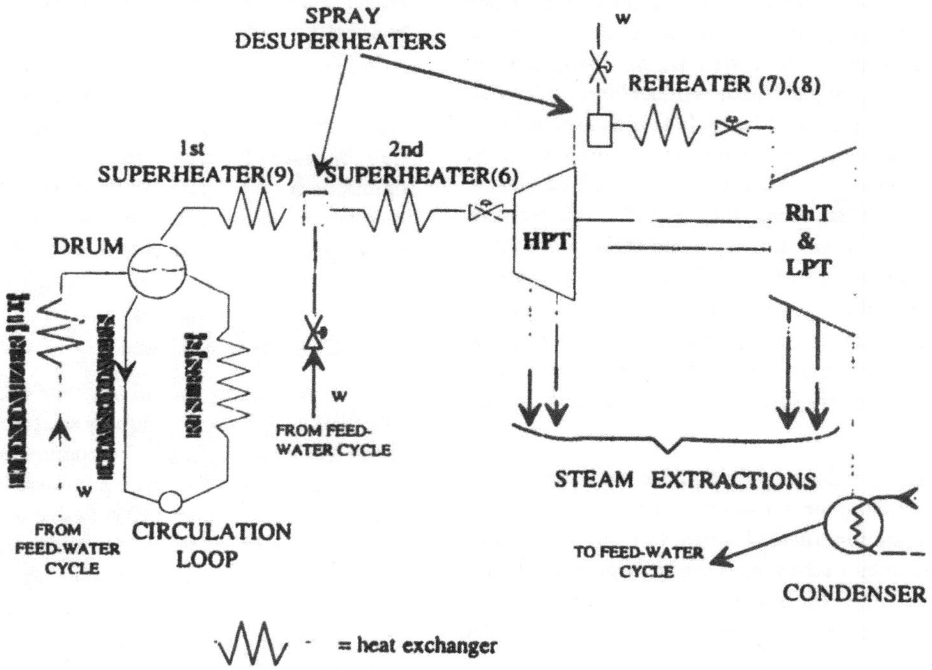

Figure 11.3 Steam-water subsystem.

water level y_D in the drum.

Overall efficiency is substantially affected by combustion quality and by the waste of energy in the flue gas. Because operating conditions also have environmental impact, they are controlled by properly selecting the air-to-fuel ratio, which depends on the condition of the firing equipment (burners etc.). In coal fired units, the furnace needs to be operated slightly below atmospheric pressure to minimize soot dispersion to the environment. Furnace pressure requires careful control, integrated with combustion control.

Early control systems gave a static interpretation of the above requirements, because the process variables and control values were set at the design stage based on the behavior expected. More recent control systems allow some adaptation to off-design conditions experienced in operation. This produces a hierarchically structured control system, whose general organization is shown in Figure 11.4.

In the scheme of Figure 11.4 are three main control levels:

- The *unit control level*, where the overall unit objective in meeting the power system demand is transformed into more specific control tasks, accounting for the actual plant status (partial unavailability of components, equipment stress, operating criteria); the decomposition into control subtasks is generally achieved by computation of set points for the main process variables.
- The *principal regulation level*, where the main process variables are controlled by a proper combination of feedforward (model based) and feedback actions. Decoupling of the overall control into independent controllers is based on the special nature of the process.
- The *dependent loop level*, where the physical devices allowing the modulation of basic process variables are controlled in a substantially independent manner with a control bandwidth much wider than the upper level regulation. These loops are the means by which the principal regulations may be conceived and designed to control process variables (like feed-water flow) rather than acting as positioning devices affected by sensitive nonlinearities (e.g., the Voigt speed control of the feed-water pump).

The tendency to decentralize control actions is quite common in process control and should be adopted generally to allow system operability. To avoid conflict with overall unit optimization, most recent control systems have extended the role of the unit coordinator, which does not interfere with the individual functionality of lower loops, but acts as a set point computer finding the optimal solution within the operation allowed by plant constraints.

When assessing control objectives, one of the crucial problems is to define system performance. For power plants, one needs to clarify the following:

- the kinds of services the unit is required to perform, usually defined in terms of maximal rate for large ramp load variations, the participation band for the power-frequency control of the power system, and the maximum amplitude and response time for the primary speed regulation in case of contingencies;
- the maximal amplitude of temperature fluctuations during load variations, to limit equipment stress due to creep or fatigue;
- maximal transient deviation of throttle pressure and drum level, to avoid potentially dangerous conditions, evaluated for the largest disturbances (e.g., in case of load rejection).

There are a few control issues still debated. The first is whether it is more convenient to operate the unit at fixed pressure (nominal constant pressure at the throttle), to let the pressure slide with a fully open turbine control valve, or to operate the unit with controlled sliding pressure. The second is the question of how much pressure variation during plant transients should be considered deleterious to some aspect of plant performance. The two questions are connected, because adopting a pure sliding pressure control strategy contradicts the concept of pressure control.

Consider the first issue. When the turbine load (i.e., the steam flow) is reduced, steam pressures at the different turbine stages are approximately proportional to the flow. Therefore, it is natural to operate the turbine at sliding pressure. On the other hand, the pressure in the steam generator is the index of the energy stored in the evaporator. Because drum boilers have a very large energy capacitance in the evaporator, boiler pressure is very difficult to change. Therefore, sliding pressure in the boiler affects load variation slowly (not the case of once-through boilers). The best condition would be to keep the pressure fixed in the boiler while sliding pressure in the turbine: this strategy would require significant throttling on the control valves with dramatic loss of efficiency in the overall steam cycle. The most popular strategy for drum boiler power plants is, therefore, controlled sliding pressure, where the boiler is operated at constant pressure above a certain load (this may be either the technical minimum or 50–60% of the MCR)[1] and pressure is reduced at lower loads. To insure high efficiency at any load, the turbine is equipped with a control stage allowing partial arc admission (i.e., with the control valves always opening in sequence to limit the amount of throttling).

The second issue is often the source of misleading design. There is no evidence that loss of performance is directly related to transient pressure deviations within safety limits, which may be very large. On the other hand, it has been demonstrated [2] that, because throttle pressure control can cause furnace over-firing, too strict pressure control can substantially disturb steam temperature.

In the following, we will consider the most common operating condition for a drum boiler, that is, with throttle pressure con-

[1] MCR = Maximum Continuous Rate

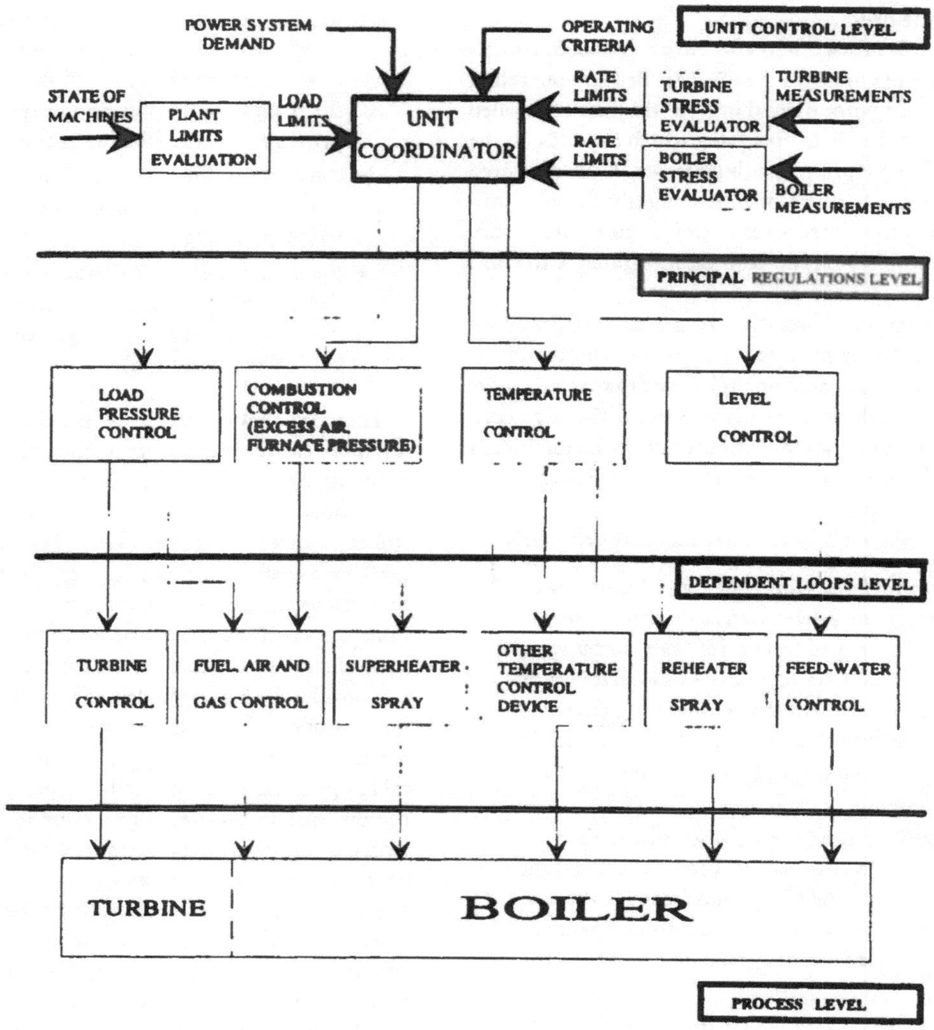

Figure 11.4 Hierarchical organization of power plant control.

trolled at a constant value during load variation, with the main objective of returning pressure to the nominal value within a reasonable time after the disturbance (e.g., load variation), while strictly insuring that it remains within safety limits (which may also depend on the amplitude of the disturbance).

11.1.3 Power Plant Modeling and Dynamical Behavior

Models for Structural Analysis

Investigating the dynamics of power plants [3] requires detailed models with precise representation of plant components. These models are generally used to build plant simulators, from which control strategies may be assessed. Large scale models are generally based on first principle equations (mass, momentum, and energy balances), combined with empirical correlations (like heat transfer correlations), and may be considered as *knowledge models*, i.e., models through which process dynamics can be thoroughly ascertained and understood. *Knowledge models* are the only reliable way, beside extensive experimentation, to learn about power plant dynamics, in particular, the many interactions among process variables and their relevance.

Today power plant simulators are broadly accepted: overall control system testing and tuning has been carried out successfully with a real-time simulator of new and novel generating unit design [4]. These detailed models are built by considering many "small" fluid or metal volumes containing matter in homogeneous conditions and writing balance equations for each volume. The resulting system can include from hundreds up to some thousands of equations.

A different kind of model has proved very helpful in establishing and justifying the basic structure of power plant control systems. Only first-cut dynamics are captured revealing the essential input-output interactions. These models, called *interpretation models*, are based on extremely coarse lumping of mass and energy balances, whose selection is guided by previous knowledge of the fundamental process dynamics. *Interpretation models* are credible because they have been derived from and compared with *knowledge models* and should be considered as useful tutorial tools to explain fundamental dynamics. Because the scope of this presentation is modeling to support control analysis, only simple *interpretation models* will be developed. However, these simple models are not useful for dynamic performance evaluation of control systems, because the dynamics they account for are only first order approximations. Nevertheless, they account for gross process interactions, into which they give good qualitative insight.

We may start developing the process model by referring to the considerations on the overall plant features presented in Section 11.1.2 and summarized in Figure 11.5. According to Section 11.1.2, the effect of the feed-water cycle on the main steam-water subsystem (SWS) variables is accounted for by including the feed-water temperature (or enthalpy) as a disturbance among the inputs of the SWS. There are some drastic simplifications in the scheme of Figure 11.5. Thermal energy released from the hot gas to the SWS walls is not totally independent of the SWS state (i.e., of wall temperatures); there is almost full independence of Q_{ev} (because heat is transferred by radiation from the hot combustion gas). Q_{SH} and Q_{RH} are more sensitive to the wall temperature because the temperature of the combustion gas is progressively decreasing; even more sensitive is Q_{ECO}, where the gas temperature is quite low. However, because the economizer definitely plays a secondary role in boiler dynamics, the scheme of Figure 11.5 is substantially correct. The dominant inputs affecting the thermal energy transferred to the SWS are the fuel and air flows and other inputs to the combustion system.

In Section 11.1.2, it was also noted that the SWS dynamics (due to massive storages of mass and energy) are far slower than combustion and air-gas (C&AG) dynamics; for that reason C&AG dynamics are negligible when the response of the main SWS variables is considered. C&AG dynamics are only relevant for specific control and dynamic problems regarding either combustion stability (relevant at low load in coal fired furnaces) or the control of the furnace pressure p_g (of great importance to protect the furnace from implosion in case of fuel trip). Then, in most control problems, we may consider the C&AG system together with its feeding system as a nondynamic process segment whose crucial role is determining energy release (and energy release partition) to different sections of the SWS. In this regard, it is important to identify how C&AG inputs may be used to influence steam generation.

Increasing the total fuel flow into the furnace will simultaneously increase all heat inputs to the different boiler sections; air flow is varied in strict relation to fuel flow so as to insure the "optimal" air-to-fuel ratio for combustion.

Because of the nonlinearity of heat transfer phenomena, Q_{EC}, Q_{EV}, Q_{SH}, and Q_{RH} do not vary in proportion to the fuel input, i.e., while varying the total heat input to the boiler, the partition of heat release is also changed. For instance, when the fuel input is increased, the heat input to the evaporator Q_{EV} (that is released in the furnace) increases less than the other heat inputs (Q_{EC}, Q_{SH}, and Q_{RH}) which are released in the upper section and backpass of the furnace. Thus, while raising the steam generation (roughly proportional to Q_{EV}), steam superheating and reheating would generally increase if proper corrective measures were not applied to rebalance heat distribution. Those measures are usually viewed as temperature control devices, because they allow superheating and reheating control in off-design conditions. One type of control measure acts on the C&AG system: the most popular approaches are 1) the recirculation of combustion gas from the backpass outlet to the furnace bottom, 2) tilting burners for tangentially fired furnaces, and 3) partitioning of the backpass by a suitable screen, equipped with gas dampers to control the gas flow partition between the two branches. The first two approaches influence the ratio between Q_{EV} and the rest of the heat release. The last varies the ratio between Q_{SH} and Q_{RH}.

The second type of temperature control measure acts on the SWS. This is the spray desuperheaters (see Figure 11.3), which balances heat input variations by injecting spray water into the superheating and the reheating path. Although superheater spray does not affect the global efficiency, reheater spray worsens efficiency so that it is only used for emergency control (when it is not possible to keep reheater temperature below a limit value by other control measures). Typical drum boiler power plants provide desuperheating sprays in the high pressure and reheat parts of the SWS and, in addition, gas recirculation as a " normal" means of controlling reheat temperature.

From this discussion, it should also be clear that modulation of heat input to the furnace and variation of recirculation gas flow simultaneously affect all process variables, because they influence heat release to all sections of the SWS. This is the principal source of interactions in the process. Because the air-to-fuel ratio is varied within a very narrow range to optimize combustion, we may assume that the boiler heat rate is proportional to w_f. Thus, to analyze the SWS dynamics, we may consider w_f and the recirculation gas mass flow-rate w_{rg} as the equivalent inputs from the gas side, because they determine the heat transfer rates Q_{EC}, Q_{EV}, Q_{SH}, and Q_{RH}.

Pressure Dynamics

A very simple *interpretation model* of evaporator dynamics can be derived with the following "drastic" assumptions:

1. The fluid in the whole circulation loop (drum, risers, and downcomers) is saturated.
2. The metal walls in the entire circulation loop are at saturation temperature.
3. The steam quality in the risers is, at any time, linearly dependent on the tube abscissa.
4. The fluid pressure variations along the circulation loop can be neglected for evaluating mass and energy storage.

The first three assumptions can be roughly considered low-frequency approximations because, excluding rapid pressure variations, water subcooling at the downcomers' inlet is very small. Moreover, because of the very high value of the heat transfer coefficient in the risers (in the order of 100 kW/m^{2}K), the metal wall very quickly follows any temperature variation in the fluid. Finally, steam quality is nearly linear at steady state because the heat flux to the furnace wall is evenly distributed. The last

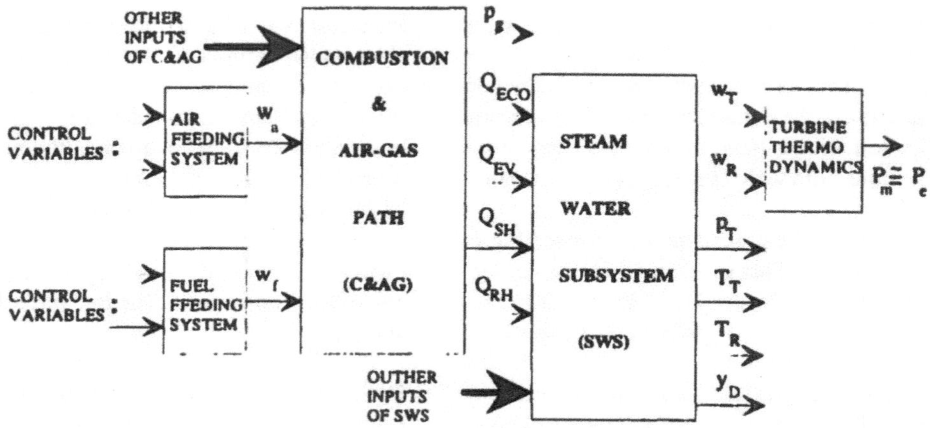

Figure 11.5 Input-output structure of the process. w_a = air mass flow-rate; w_f = fuel mass flow-rate; p_g = gas pressure in the furnace; Q_{ECO} = thermal energy to ECOnomizer; Q_{EV} = thermal energy to EVaporator; Q_{SH} = thermal energy to SuperHeaters; Q_{RH} = thermal energy to ReHeater.

assumption is based on the fact that pressure differences along the loop (which are essential for circulation) are of the order of 1% of the absolute fluid pressure in the drum so that we may identify the pressure in the evaporator with the pressure p_D in the drum. Then the global energy balance and the global mass balance in the evaporator are

$$\frac{dE_{EV}(p_D, \alpha)}{dt} = w_w h_E - w_V h_{VS}(p_D) + Q_{EV}, \quad (11.4)$$
$$\frac{dM_{EV}(p_D, \alpha)}{dt} = w_w - w_V, \quad (11.5)$$

where w_w is the feed-water mass flow-rate (mfr), w_V is the steam mfr at the drum outlet, h_E is the water enthalpy at the economizer outlet, $h_{VS}(p)$ is the steam saturation enthalpy at the pressure p, Q_{EV} is the heat-rate to the evaporator, E_{EV} is the total energy stored in the evaporator (fluid and metal of drum, downcomers and risers), and M_{EV} is the total mass stored in the evaporator. E_{EV} and M_{EV}, beside obvious geometrical parameters like volumes, depend on two process variables: the pressure p_D and the void fraction α in the evaporator, defined as the ratio of the volume occupied by steam and the total volume of the evaporator.

A better index for the mean energy level in the evaporator is obtained by subtracting the total mass multiplied by the inlet enthalpy h_E from Equation 11.4.

$$\frac{dE_{EV}(p_D,\alpha)}{dt} - h_E \frac{dM_{EV}(p_D,\alpha)}{dt} =$$
$$- w_V \left[h_{VS}(p_D) - h_E\right] + Q_{EV}. \quad (11.6)$$

Noting that h_E is subject to limited and very slow variations (the economizer is a huge heat exchanger exploiting a limited temperature difference between flue gas and water), so that dh_E/dt is usually small, Equation 11.6 can be interpreted by introducing the net energy storage in the evaporator, $E^*_{EV} := E_{EV} - h_E M_{EV}$: the difference between the input heat transfer rate Q_{EV} and the power spent for steam generation, $P_{sg} := w_V \left[h_{VS}(p_D) - h_E\right]$, in transient conditions, is balanced by the storage of the net energy E^*_{EV} in the evaporator. Moreover, whereas the mass M_{EV} depends mainly on α, the net energy E^*_{EV} depends mainly on p_D, because about 50% of the energy is stored in the metal walls, which are insensitive to α.

Equation 11.6 can be rewritten approximately as

$$C_{EV} \frac{dp_D}{dt} = Q_{EV} - w_V(h_{VS}(p_D) - h_E), \quad (11.7)$$

where

$$C_{EV} = \frac{\partial E^*_{EV}(p_D, \bar{\alpha})}{\partial p_D},$$

$\bar{\alpha}$ being the nominal void fraction. Equation 11.7 yields the fundamental dynamics of drum pressure and justifies the popular claim that drum pressure is associated with the stored evaporator energy. Equation 11.7 may be usefully rewritten in normalized per unit (p.u.) form, i.e., referring energy and pressure to the nominal conditions, $Q^\circ_{EV} = w^\circ_V(h_{VS}(p^\circ_D) - h^\circ_E)$, where the superscript ° denotes nominal value:

$$\tau_{EV}\, \dot{p}_{Dn} = Q_{EVn} - w_{Vn} \frac{h_{VS}(p_D) - h_E}{h_{VS}(p^0_D) - h^0_E} \quad (11.8)$$

with the subscript n denoting the variable expressed in p.u.. Typical values for the normalized "capacitance" τ_{EV} are 200–300 sec. It may also be observed that τ_{EV} is a function of the drum pressure p_D and is roughly inversely proportional to the pressure; thus, pressure dynamics will slow down while reducing the operating pressure.

Although for the evaporator h_E can be considered a slowly varying exogenous variable, w_V depends on the drum pressure p_D and on the total hydraulic resistance opposed by the cascade of superheaters and turbine.

Let's first characterize the turbine. For simplicity, assume that the turbine control valves are governed in full arc mode (i.e., with parallel modulation). Then the control stage of the turbine (generally of impulse type) can be viewed as the cascade of a throttle valve and a nozzle. This implies

$$w_T = C_V(x)\sqrt{\rho_T p_T}\,\chi_V(p_N/p_T), \quad (11.9)$$
$$w_T = K_N\sqrt{\rho_N p_N}\,\chi_N(p'/p_N), \quad (11.10)$$

where $C_V(x)$ is the flow coefficient of the control valve set (dependent on the valve's position χ), ρ_T is the steam density at throttle, p_N is the valve outlet pressure, $\chi_V(\beta)$ is a suitable function of the valve pressure ratio, K_N is a nozzle flow constant, ρ_N the density at the nozzle inlet, p' the pressure at the nozzle outlet, and χ_N a function similar to χ_V.

Because the HPT consists of many cascaded stages, the pressure ratio (p'/p_N) across the control stage will remain nearly constant with varying flow w_T. Then $\chi_N(p'/p_N = constant$.

Bearing in mind that superheated steam behaves like an ideal gas and that valve throttling is an isenthalpic expansion, so that $p_N/\rho_T \cong p_N/p_T$, eliminate the pressure ratio p_N/p_T by dividing Equation 11.9 by Equation 11.10. The ratio p_N/p_T is a monotonic function of $C_V(x)$. Substituting this function in Equation 11.9 results in

$$w_T = f_T\,(C_V(x))\,\sqrt{\rho_T p_T} = f_T^*(x)\sqrt{\rho_T p_T}, \qquad (11.11)$$

where $f_T(\cdot)$ is a monotonic function of its argument. Equation 11.11 says that the cascade of the turbine and its control valves behave like a choked-flow valve with a "special" opening law $f_T^*(x)$.

Even when the turbine control valves are commanded in partial-arc mode (i.e., with sequential opening), one arrives at a flow equation of the same type as Equation 11.11, but differently dependent on the valve opening command signal x. To summarize, the HPT with its control valves determines a boundary condition for the steam flow given by Equation 11.11; with typical opening strategies of full-arc and partial-arc admission, the global flow characteristic $f_T^*(x)$ looks like that in Figure 11.6. A more elaborate characterization of sequentially opened valves may be found in [7].

To obtain flow conditions for w_V instead of w_T (i.e., at the evaporator outlet), flow through superheaters must be described (see Figure 11.3). First-cut modeling of superheaters' hydrodynamics is based on the following remarks:

1. Mass storage in the superheaters is very limited (as compared with energy storage in the evaporator) because steam has low density and desuperheating spray flow w_{ds} is small compared with w_V. Thus, head losses along superheaters may be computed with the approximation $w_V \approx w_T$.
2. Head losses in the superheaters develop in turbulent flow, so that

$$p_D - p_T = k_{SH}\frac{w_T^2}{\rho_{SH}} \qquad (11.12)$$

where ρ_{SH} is a mean density of the superheated steam and a k_{SH} constant.

Equations 11.11 and 11.12 can be combined with Equation 11.7 or Equation 11.8 to build a simple model of the fundamental pressure dynamics. To this end, we derive a linearized model for small variations about a given steady state condition, identified as follows:

- assuming that the unit is operated (at least in the considered load range) at constant throttle pressure, p_T at any steady state equals the nominal pressure p_T°;
- the unit is usually operated at constant throttle temperature ($T_T + T_T^\circ$, so that the temperature profile along the superheaters does not change significantly; we may therefore assume that the mean superheating temperature T_{SH}, at any steady state, equals its value in nominal conditions T_{SH}°;
- based on Equations 11.1–11.3 and the related remarks, the load L (in p.u.) of the plant at any steady state equals the ratio between the steam flow w_T and its nominal value w_T°.

Moreover, the following assumptions are made:
(a) superheated steam behaves like an ideal gas:

$$\rho_{SH} = \frac{p_{SH}}{RT_{SH}} \approx \frac{p_T}{RT_{SH}}, \rho_T = \frac{p_T}{RT_T}$$

where R is the gas constant;
(b) in nominal conditions, desuperheating spray mass flow-rate is zero:

$$w_{ds}^\circ = 0,$$

so that $w_{d_s}^\circ = w_T^\circ$.
The model will be expressed in p.u. variables by defining

$$\delta_p = \Delta_p/p_T^\circ,$$

for any pressure p,

$$\delta_w = \Delta_w/w_T^\circ$$

for any mass flow-rate w,

$$\delta_h = \Delta_h/\left(h_{VS}(p_D^\circ) - h_E^\circ\right)$$

for any enthalpy h,

$$\delta_T = \Delta_T/T^\circ$$

for any temperature T, and

$$\delta Q_{ev} = \Delta Q_{ev}/Q_{ev}^\circ$$

(° denotes, as usual, nominal conditions). Then Equations 11.8, 11.11, and 11.12 yield the following linearized system:

$$\begin{aligned}\tau_{EV}\,\dot{\delta p_D} &= \delta Q_{ev} - \alpha\delta w_V + L\beta_1\delta p_D \\ &\quad + L\delta h_E \qquad (11.13)\\ \delta w_T &= \delta Y + L\delta p_T, \quad \text{and} \qquad (11.14)\\ \delta p_T &= \frac{1}{1-\gamma^\circ L^2}\delta p_D \\ &\quad - \frac{2\gamma^\circ L}{(1-\gamma^\circ L^2)}\delta w_T \\ &\quad - \gamma^\circ L^2\delta T_{SH}, \qquad (11.15)\end{aligned}$$

where

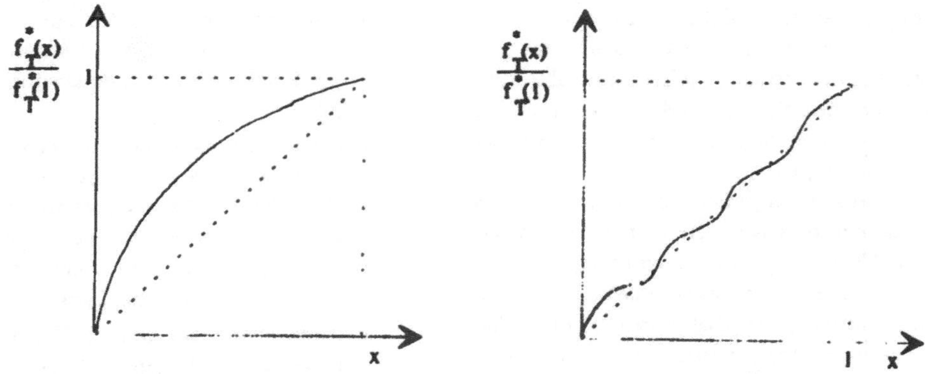

(a) Full arc command mode

(b) Partial-arc command mode
(with 4 control valves)

Figure 11.6 Flow characteristic of HPT.

$$\begin{aligned}
\tau_{EV} &= \tau_{EV}(1+\gamma^\circ),\\
\gamma^\circ &= \frac{p_D^\circ}{p_T^\circ} - 1,\\
\alpha &= \frac{h_{VS}(p_D^\circ) - h_E}{h_{VS}(p_D^\circ) - h_E^\circ},\\
\beta_1 &= -\left.\frac{dh_{VS}}{dp_D}\right|_\circ \frac{p_T^\circ}{(h_{VS}(p_D^\circ) - h_E^\circ)},\\
Y &:= f_T^*(x)/\sqrt{RT_T}
\end{aligned}$$

is the turbine "admittance," and $\bar{h}_E$ is the value of h_E at the linearization steady state.

Note that γ° is usually about 0.05 or less, $\bar{h}_E$ undergoes limited variations (α is very close to 1), β_1 is positive for $p_D^\circ > 30$ bar and is generally small because $h_{VS}(p)$ is a flat thermodynamic function. Moreover temperature variations δT_{SH} are generally slow and limited amplitude, so that $\gamma^\circ L^2 \delta T_{SH}$ is totally negligible.

Then, pressure dynamics may be approximately represented by the very simple block diagram of Figure 11.7, where

$$\mu_p = \frac{1}{1+2\gamma^\circ L^2} \quad \text{and}$$

$$\mu_Y = \frac{2\gamma^\circ L}{1+2\gamma^\circ L^2}.$$

There is a seeming inconsistency in Figure 11.7, because the variable Y is considered as an input variable, but its definition, $Y := w_T/p_T$, implies that it depends on the control variable x and also on the throttle temperature T_T.

However, it is a common practice to equip the turbine control valve with a wide band feedback loop of the type shown in Figure 11.8. Because valve servomotors today are very fast and no other lags are in the loop, at any frequency of interest for the model of Figure 11.7, $Y \cong \bar{Y}$. So the turbine admittance actually becomes a control variable and the loop of Figure 11.8 serves two complementary purposes: first, it linearizes the nonlinear characteristics of Figure 11.6 and, second, it rejects the effect of temperature fluctuations on the steam flow to the turbine, thereby decoupling pressure dynamics and temperature dynamics.

Let's analyze the scheme of Figure 11.7, bearing in mind that $0 \le \alpha < 1$, $0 < \beta_1 \le 0.1$ (for $p_D^\circ > 30$ bar, $0.9 < \mu_P < 1$, $0 < \mu_Y < 0.1$ and, of course, $L^* \le L \le 1$, where L is the minimal technical load (typically $L^* \approx 0.3$). We may observe that the pressure dynamics are characterized by a time constant τ_P, given by

$$\tau_P = \frac{\tau_{ev}}{L(\mu_P\alpha - \beta_1)} \approx \frac{\tau_{ev}}{L} \tag{11.16}$$

that is, the ratio between the evaporator energy capacity and the load. Thus, the open-loop response of the pressure to exogenous variables slows down as the load decreases.

Neglecting the effects of the small disturbances δh_E and δw_{ds}, a natural way to follow plant load demand in the fixed-pressure operating strategy is to let the turbine admittance δY vary according to the load demand ΔL_d and let the heat transfer rate δQ_{ev} vary so as to balance the power spent for steam generation, i.e., $\alpha\delta V_V$. This means that

$$\delta Y = \Delta L_d \tag{11.17}$$

$$\delta Q_{ev} = \alpha \delta w_v. \tag{11.18}$$

As a consequence,

$$\delta p_d = 0$$

$$\delta p_T = -\mu_Y \delta Y,$$

$$\text{and} \quad \delta w_T = \delta Y(1 - L\mu_Y).$$

Because of the head losses along the superheaters ($\mu_Y \neq 0$), the strategy expressed by Equation 11.18, keeping the energy storage (i.e., p_D) constant in the evaporator, actually determines a drop $-\mu_Y\delta Y$ of the throttle pressure p_T and, consequently, a reduced power output ($\delta w_T = \delta Y(1 - L\mu_Y) < \Delta L_D$). In other words, if one wants to keep the throttle pressure p_T constant when the load is increased, the energy storage in the evaporator also needs to be slightly increased because of the head losses:

$$\begin{aligned}
\delta p_D \mu_p &= \delta Y \mu_Y, \\
\delta p_d &= 2\gamma^\circ L \delta Y = 2\gamma^\circ L \Delta L_d.
\end{aligned} \tag{11.19}$$

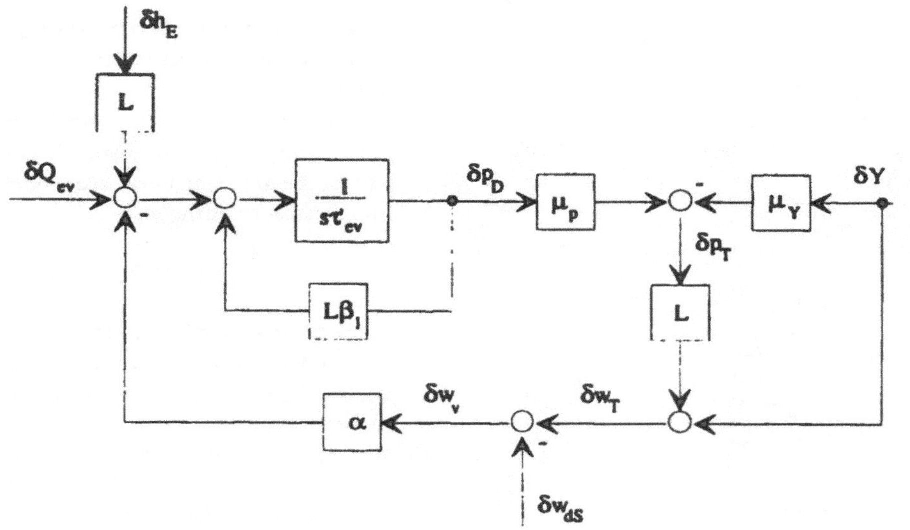

Figure 11.7 Block diagram of the linearized pressure dynamics.

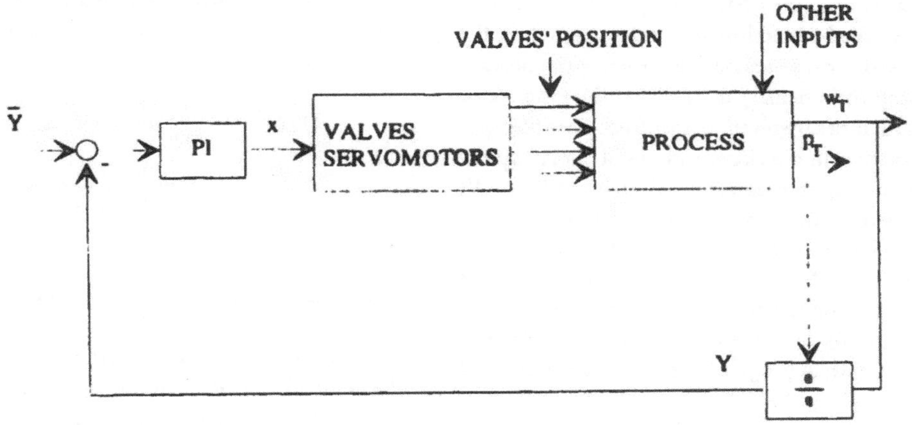

Figure 11.8 Turbine admittance feedback loop.

In Figure 11.7, observe that only Equation 11.19 implies a boiler overfiring, i.e., a transient extra fuel supply during load increase to adjust the energy stored in the evaporator.

If the feedback compensation Equation 11.18 is applied to the boiler, the pressure dynamics become slightly unstable because of the "intrinsic" positive feedback due to β_1. The same result happens if the feedback loop of Figure 11.8 is realized, as is sometime the case, not as an admittance loop but as a simple mass flow-rate loop (i.e., omitting dividing by p_T). Thus, when applying either mass flow-rate feedback, boiler stabilization must be provided.

Drum Level Dynamics

Computing drum pressure by the scheme of Figure 11.7, we return to Equation 11.5 that establishes the global mass balance in the evaporator. Equation 11.5 may be linearized as

$$\sigma_p \, \delta \dot{p}_D - \sigma_\alpha \Delta \dot{\alpha} = \delta w_w - \delta w_V, \qquad (11.20)$$

where

$$\sigma_p = \frac{V p_T^\circ}{w_T^\circ}\left[(1-\bar{\alpha})\left(\frac{d\bar{\rho}_{LS}}{dp}\right) + \bar{\alpha}\left(\frac{d\bar{\rho}_{VS}}{dp}\right)\right]$$

and $\sigma_\alpha = \dfrac{V(\bar{\rho}_{LS} - \bar{\rho}_{VS})}{w_T^\circ}$.

V is total fluid volume in the evaporator, ρ_{LS} and ρ_{VS} are the liquid and vapor densities as functions of the pressure, and the upper script $^{-}$ denotes the steady state of linearization. If the unit is operated at constant pressure, σ_α is independent of the load, and σ_p only slightly dependent.

However, Equation 11.20 determines only the global void fraction α, while the relevant variable for the control is the level in the drum.

We may write

$$\alpha = \frac{V_r}{V}\alpha_r + \frac{V_D}{V}\alpha_D \qquad (11.21)$$

where V_r and V_D are the volumes of the risers and of the drum, respectively, and α_r and α_d are the separate void fractions relative to V_r and V_D. If we assume that, at the considered steady state, the level y_D is equal to the drum radius R_D, then,

$$\Delta\alpha_D = -\frac{2}{\pi}\delta y_D, \qquad (11.22)$$

with

$$\delta y_D \quad := \quad \Delta y_D / R_D.$$

Combining Equations 11.20, 11.21, and 11.22, the following equation is obtained:

$$\begin{aligned} \delta \dot{y}_D &= \frac{1}{\tau_L}(\delta w_w - \delta w_v) + k_p \delta \dot{p}_D \\ &\quad + k_r \Delta \dot{\alpha}_r \qquad (11.23) \\ \tau_L &= \frac{V_D(\bar{\rho}_{LS} - \bar{\rho}_{VS})}{w_T^{\circ}} \frac{2}{\pi}, \\ k_p &= -\frac{\sigma_p}{\sigma_\alpha} \frac{\pi}{2} \frac{V}{V_D}, \\ \text{and} \quad k_r &= \frac{\pi}{2} \frac{V_r}{V_D}. \end{aligned}$$

Equation 11.23 shows that the drum level is subject to three different kinds of variations: the first, of integral type, is due to the imbalance between feed-water flow and outlet steam flow; the second, of proportional type, arises because, even with constant stored mass, the mean fluid density depends on the evaporator pressure p_D; the third, more involved, comes from possible variations of the void fraction in the risers and might occur rapidly because any variation, $\Delta\alpha_r$ immediately reflects onto δy_D. We need to understand where $\Delta\alpha_r$ comes from. Recall the assumptions at the beginning of Section 11.1.3, write equations similar to Equations 11.4 and 11.5 but limited to the circulation tubes (i.e., to the downcomers and the risers), and derive the "net energy" stored corresponding to Equation 11.6, where, instead of enthalpy h_E, the inlet enthalpy h_{LS} of the downcomer tubes is used:

$$\begin{aligned} &\frac{dE_{ct}(p_D, \alpha_r)}{dt} - h_{LS}(p_D)\frac{dM_{ct}(p_D, \alpha_r)}{dt} \\ &\quad = Q_{EV} - \chi_r w_r \left[h_{VS}(p_D) - h_{LS}(p_D)\right], \qquad (11.24) \end{aligned}$$

where E_{ct} is the total energy (fluid + metal) stored in the circulation tubes, M_{ct} the corresponding fluid mass, h_{LS} and h_{VS} the liquid and vapor saturation enthalpies, and χ_r and w_r the steam quality and the mass flow-rate at the risers' outlet.

Then, based on assumption (3) stated at the beginning of Section 11.1.3, the following relationship is obtained:

$$\begin{aligned} \alpha_r &= (1+\beta)\left[1 - \frac{\beta}{\chi_r}\ln\left(1 + \frac{\chi_r}{\beta}\right)\right], \qquad (11.25) \\ \beta &= \rho_{VS}/(\rho_{LS} - \rho_{VS}). \end{aligned}$$

To solve the model, we need to derive the circulation mass-flow rate w_r, which is obtained from the momentum equation applied to the circulation tubes:

$$\alpha_r(\rho_{LS} - \rho_{VS}) = w_r^2\left(\frac{C_{dc}}{\rho_{LS}} + \frac{C_r}{\rho_r}\right), \qquad (11.26)$$

where

$$\rho_r = \rho_{LS} - \alpha_r(\rho_{LS} - \rho_{VS})$$

is the mean density in the risers and C_{dc}, C_r are suitable constants yielding the head losses in the downcomers and risers tubes, respectively.

Equations 11.25 and 11.26 may be used to eliminate w_r and x_r from Equation 11.24; through trivial but cumbersome computations, the following linearized model can be obtained for $\Delta\alpha_r$ (in $\mathcal{L}$-transform form):

$$\Delta\alpha_r = \frac{1}{1 + sT_2}\left[\lambda_2(\delta Q_{EV} - \tau_{rt} s \delta p_D) + \lambda_1 \delta p_D\right], \qquad (11.27)$$

where τ_{rt} is a normalized capacitance similar to τ_{EV} in Equation 11.8 but related only to the circulation tubes (typically $\tau_{rt} \approx 0.7\tau_{EV}$), T_2 is a small time-constant (a few seconds) associated with the dynamics of the void fraction within the risers, and λ_2 and λ_1 are suitable constants. The difference $\delta Q_{EV} - \tau_{rt} s \delta p_D$ is the heat transfer rate available for steam generation in the risers, given by the (algebraic) sum of the input thermal energy δQ_{EV} and the energy $-\tau_{rt}\delta \dot{p}_D$ released in the case of pressure decrease and corresponding to a reduction of the stored energy. The L-transformation of Equation 11.23 and substitution of Equation 11.27 give

$$\begin{aligned} \delta y_D &= \frac{1}{s\tau_L}(\delta w_v - \delta w_w) + \frac{k_2}{1 + sT_2}\delta Q_{EV} \\ &\quad + k_1 \frac{1 - sT_1}{1 + sT_2}\delta p_D, \qquad (11.28) \end{aligned}$$

where

$$k_2 = k_r\lambda_2, \quad k_1 = k_p + k_r\lambda_1, \quad T_1 = \frac{k_2\tau_{rt} - k_p T_2}{k_1}.$$

The parameters of model Equation 11.28 are dimensionless, with the following typical values: $\tau_L \approx 130$ sec., $k_2 \approx 0.25$, $k_1 \approx 0.5$, $T_2 \approx 4$ sec., $\tau_{rt} \approx 0.7$, $\tau_{EV} \approx 150$ sec. (at nominal pressure), $k_p \approx 0.8$.

Since $T_2 \ll \tau_{rt}$, the time constant $T_1 \approx 70$ sec. is always positive and is essentially determined by the "capacitance effect" (τ_{rt}). The model Equation 11.28 clearly accounts for the well-known shrink and swell effect due to the nonminimum phase zero $(1 - sT_1$. Because T_2 is very small, any perturbation producing sudden pressure derivatives causes a sudden variation of the drum level in the direction opposite to the long-term trend. To this aim, referring to Figure 11.7, consider a step perturbation of δY, with, e.g., $\delta Y = \Delta/s$. Then

$$\delta p_D = -\frac{\mu'\Delta}{s}\frac{1}{1 + sT_3}, \quad \delta w_v = -\mu''\frac{\Delta}{s}\frac{1 - sT_4}{1 + sT_3}, \qquad (11.29)$$

with μ and T_3 suitable constants. At nominal load ($L = 1$) and with typical values, ($\mu_Y = 0.1, \alpha = 1, \mu_p = 0.9, \beta_1 = 0.05, \tau_{EV} = 200$ sec.) $\mu' \cong 1.06$, $T_3 \cong 248$ sec., $\mu'' \cong 0.054$, $T_4 \cong 4600$ sec. The effect of the step variation with $\Delta = 0.1$ is depicted in Figure 11.9. Equation 11.29 is a very useful model to conceive level control structure.

Reheat and Superheat Steamside Dynamics

Superheaters and reheaters are large heat exchangers with steam flowing into the tubes and gas crossing the tube banks

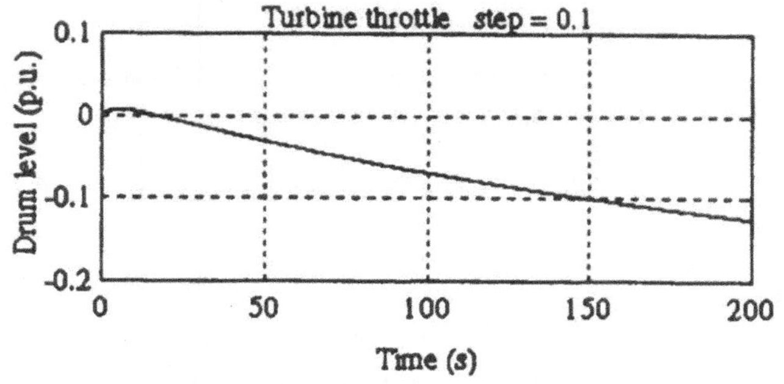

Figure 11.9 Drum level dynamic.

in cross-flow. There are some general properties that are worth recalling:

1. The heat transfer coefficient on the gas side is much smaller that the one on the steam side, so that steady state behavior is nearly independent of steamside coefficients;
2. The dynamics of these heat exchangers are essentially due to the considerable energy stored in the metal wall, because flue gas has negligible density and steam has much lower capacitance than the corresponding metal wall;
3. The mass stored creates much faster dynamics than energy stored, because only steam is involved.

Property 3 can easily be checked bearing in mind that the fundamental time constant of mass storage is,

$$\tau_{MS} = M_V / w_V,$$

where M_V is the mass of steam within the heat exchanger and w_V the mass flow-rate flowing through it, whereas the fundamental time constant of energy storage is ,

$$\tau_{ES} = \frac{M_v C_v + M_m C_m}{c_p w_v},$$

where C_v and C_p are the specific heats of steam at constant volume and pressure, M_m and C_m the mass and the specific heat of metal wall. Because $M_m C_m \gg M_v C_v$ and $C_p \approx 1.3 C_v$, $\tau_{ES} \gg \tau_{MS}$. For a typical superheater τ_{MS} is a few seconds, but τ_{ES} is more than 20 times τ_{MS}. Then, when considering temperature dynamics, which, because of property (1) are due to the metal wall capacitance, mass storage may be neglected (i.e., one can consider the steam flow independent of the tube abscissa).

Superheater or reheater outlet temperature T_{ox} is influenced by three different variables, the heat transfer rate Q_x to the external wall, the steam flow w_x, and the inlet temperature T_{ix}. Pressure fluctuations within the heat exchanger have a limited influence on T_{ox} and may be neglected. For small variations the situation is described in Figure 11.10.

It is relevant to control design to characterize the transfer functions $G_T(s)$, $G_w(s)$, and $G_Q(s)$. It is known [3] that adequate modeling of superheaters and reheaters requires a distributed parameters approach. However, reasonable lumped parameter approximation may be used for G_T, G_w, and G_Q.

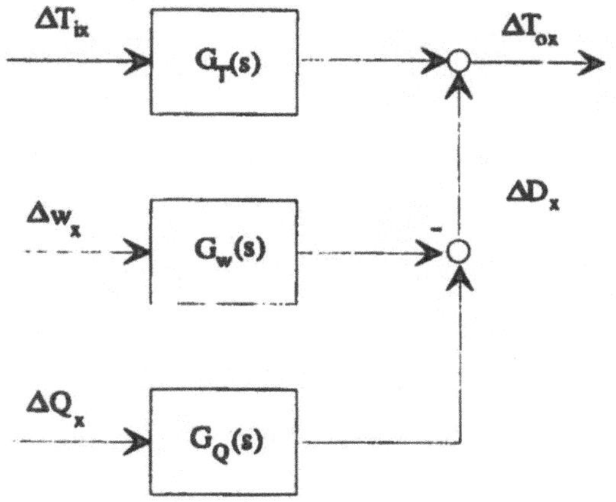

Figure 11.10 Conceptual scheme of temperature dynamics.

G_w and G_Q behave like first-order transfer functions, similar to each other, dominated by a time constant not far from τ_{ES} and with gain essentially dependent on the gas-to-wall heat transfer coefficient. The function $G_T(s)$ is, on the contrary, approximated well by:

$$G_T(s) \approx \mu_T \frac{1}{(1 + s\tau_{ES}/N)^N}, \tag{11.30}$$

where N is the integer nearest to $S_i \gamma_i / 2 w_x c_p$ (with S_i and γ_i the steam-to-wall exchange surface and heat transfer coefficient) and μ_T is less than 1. Typically, secondary superheaters are short heat exchangers with $N \approx 2$. Larger reheaters may have $N \approx 3-4$.

Moreover, it appears that the process is nonlinear, because τ_{ES} is nearly proportional to the inverse of the load. N is only slightly dependent on w_x because $\gamma_i \equiv w_x^{0.8}$.

The temperature dynamics are affected by multiple lags, varying with the load. In addition, transducers for steam temperature are generally affected by a small (a few second) and a larger (some tens of seconds) time lag due to the thermal inertia of the cylinder where the sensor is placed.

Of course, multiple lags are in the loop when ΔT_{ix} is the control variable. This is the case when desuperheating spray is used to achieve mixing between the superheated steam at the outlet of the preceding component (e.g., the primary superheater) and the water spray is modulated by a suitable valve. Because the attemperator has a very small volume, storage in it is negligible, and its equations are given by steady-state mass and energy balances (see Figure 11.11; referring to the superheating section):

$$w_v + w_{ds} = w_T \quad \text{and} \tag{11.31}$$

$$w_v h_v + w_{ds} h_{ds} = w_T h_i. \tag{11.32}$$

In normal plant operation, the steam flow w_T in the secondary superheater is imposed (over a wide band) by the load controller, h_v is determined by the upstream superheater, and h_{ds} is nearly constant. The second superheater inlet temperature T_i is given by the following variation equation:

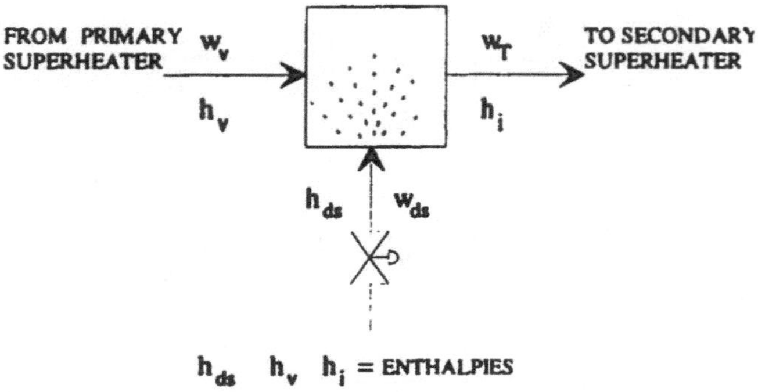

Figure 11.11 Desuperheating spray.

$$\Delta T_i = \frac{1}{C_p}\Delta h_i = \frac{(\bar{h}_v - \bar{h}_i)}{C_p \bar{w}_T}\Delta w_T + \frac{\bar{w}_v}{\bar{w}_T}\Delta T_v - \frac{(\bar{h}_v - \bar{h}_{ds})}{C_p \bar{w}_T}\Delta w_{ds}, \quad (11.33)$$

where

$$\Delta T_i = \Delta h_v / \dot{c}_p,$$

and the upper script $^-$ denotes the steady-state of linearization. In Equation 11.33 Δw_{ds} is the control variable which directly modulates temperature T_i, and Δw_T and ΔT_v represent disturbances to the temperature. Because the influence of Δw_{ds} on pressure is very small, temperature control via desuperheating spray does not significantly influence boiler pressure.

In Figure 11.10, the heat transfer rate ΔQ_x may also be used to control temperature. This would be very effective because $G_Q(s)$ incorporates less phase lags than $G_T(s)$. Unfortunately (see Section 11.1.3) it is impossible to modulate the heat transfer rate to a single heat exchanger in the boiler (e.g., by varying fuel flow or recirculation gas flow) without simultaneously influencing the heat released to all of the other heat exchangers in the boiler. For instance, when recirculation gas dampers are modulated to control reheat temperature, the heat transfer rates to the evaporator and to the superheaters are also simultaneously varied. This fact generates interaction among the different process variables, to the extent that it is often necessary to introduce feedforward decoupling actions to achieve acceptable control performance. When spray is used to control superheater temperature, then Δw_x and ΔQ_x constitute disturbances for the temperature control induced, for instance, by varying fuel flow required by load-pressure control.

Fortunately (see Figure 11.7), when the heat transfer rate to the evaporator is increased, the steam generation is also increased so that Δw_x and ΔQ_x grow nearly as much. Because $G_w(s)$ and $G_Q(s)$ are similar, the global disturbance ΔD_x is much smaller than the two individual disturbances. However, as recalled in Section 11.1.3, when the heat transfer rate Q_{EV} to the evaporator is varied by the fuel flow, the heat transfer rates to the superheaters and reheater do not vary in the same percentage, so that Δw_x (nearly proportional to ΔQ_{EV}) does not exactly balance ΔQ_x. This means that $\Delta D_x \neq 0$.

Power Generation

According to Equations 11.1–11.3 and the subsequent remarks, it can be approximately assumed that

$$\delta P_e = \delta P_m \cong k_{HP}\delta w_T + k_{RH}\delta w_r, \quad (11.34)$$

where

$$k_{HP} := \alpha_T(h_T^\circ - h_{TR}^\circ)w_T^\circ / P_e^\circ,$$
$$k_{RP} := \alpha_R(h_R^\circ - h_0^\circ)w_R^\circ / P_e^\circ,$$

with the upper script $^\circ$ denoting nominal values, and $\delta_z = \Delta z / z^\circ$, for any variable z. We know from Section 11.1.3 that w_T is given by the scheme of Figure 11.7 and is sensitive only to the high pressure part of the steam-water subsystem. To understand the factors influencing w_R, let's refer to Figure 11.12.

Because HPT has negligible storage

$$w_T + w_B = w_{se}. \quad (11.35)$$

The reheater is a large steam heat exchanger; feed-water heaters (FWH) fed by steam extractions are large tube and shell heat exchangers where steam extractions are condensed to heat feedwater. Both of these components have significant mass storage. The FWH represented in Figure 11.12 accounts (in an equivalent way) for the overall capacitance of FWHs. Neglecting the desuperheating spray (normally zero), the relevant mass balances are

$$\frac{dM_R}{dt} = w_B - w_R, \quad (11.36)$$
$$\frac{dM_{FH}}{dt} = w_{se} - w_e, \quad (11.37)$$

where M_R is the steam mass in the reheater, M_{FH} is the steam mass in FWHs, and w_e is the condensation mass flow-rate in FWHs.

Pressure losses in the reheater are small and can be neglected. Because V_E is normally fully open, we may assume that the pressure in the entire reheater is the same as at the RHT inlet. Moreover, reheater temperature has much slower dynamics than mass storage (see Section 11.1.3), so that it may be considered constant while evaluating dM_R/dt.

Applying an equation similar to Equation 11.10 to RHT and considering superheated steam as an ideal gas, the RHT flow equation is

$$w_R = k'_{RHT} p_B / \sqrt{RT_{RH}}, \quad (11.38)$$

where T_{RH} is the reheater outlet temperature, R is the gas constant, and k'_{RHT} is a suitable constant. Then, taking variations of Equations 11.36, 11.37, and 11.38,

$$\delta w_R = \frac{1}{1 + s\tau_R}\left(\frac{w_T^\circ}{w_R^\circ}\delta w_T - \eta \delta w_w\right), \quad (11.39)$$

where τ_R is a time constant resulting from the sum of the storage capacitance of the reheater and of FWHs, multiplied by the flow resistance of RHT, and the last term results from considering that

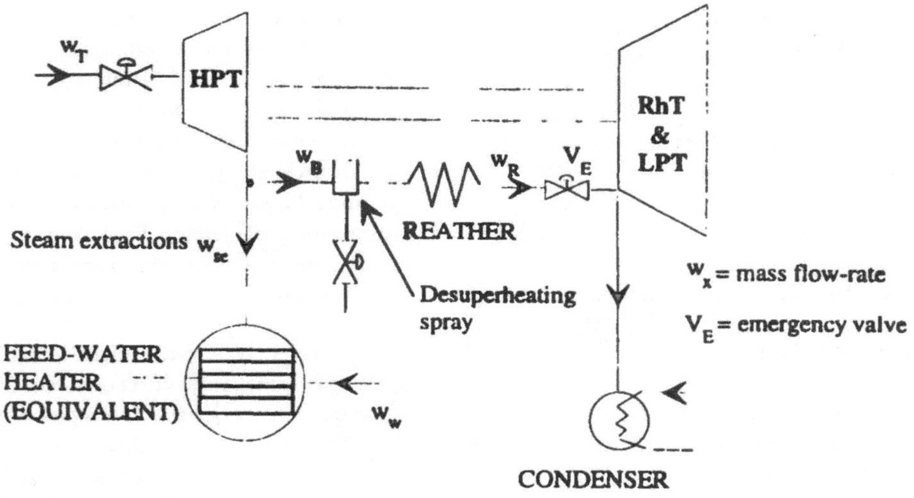

Figure 11.12 Steam reheating.

condensation flow-rate variation, Δw_c is essentially due to feed-water flow variation Δw_w. In controlled conditions δw_w strictly follows δw_T, so that Equation 11.39 becomes

$$\delta w_R = \frac{1}{1+s\tau_R}\delta w_T. \tag{11.40}$$

Equation 11.39 deserves a couple of remarks [5]: the time constant τ_R has values of about 10–12 sec., nearly 50% due to FWH's capacitance; the possibility of varying the principal steam flow δw_R by changing the feed-water flow ($\eta\delta w_w$) has suggested one of the most recent expedients to realize quick power variations even without HPT throttle reserve [6].

When special control of feed-water is not applied, Equation 11.40 can be substituted in Equation 11.34 with the following conclusion:

$$\delta P_E \approx \left(k_{HP} + \frac{k_{RH}}{1+s\tau_R}\right)\delta w_T, \tag{11.41}$$

realizing that about 1/3 ($k_{HP} \approx 0.3 - 0.4$) of electric power output is strictly proportional to the steam flow and about 2/3 ($k_{HP} + k_{RH} = 1$) is affected by a time lag τ_R, which cannot be negligible when load varies rapidly (as in the case of load rejection incidents or turbine speed control problems). Finally, observe that the parameters of Equation 11.41 are slightly dependent on plant load.

Combustion and Air-Gas Dynamics

Referring to Figure 11.2, we see that the air-gas subsystem forms a complex circuit, where the largest storage (of mass and, thus, of energy) is the furnace. Apart from air preheaters (2) and (3), the dynamics of the air-gas subsystem are very fast, so that it is substantially decoupled from the dynamics of the steam-water subsystem. Air-gas dynamics are relevant only to two special problems:

1. control of furnace pressure in a balanced draft furnace (recognizable from the induced draft fan (11) of Figure 11.2), and
2. flame stability at low loads in coal fired plants.

The second problem requires a very complex analysis and is relevant only in very particular situations.

To analyze (1), two dynamical phenomena must be studied:

- combustion kinetics, i.e., the chemical process governing fuel oxidation and its heat release, and
- mass and energy storage in the furnace, possibly augmenting the furnace storage to account for the rest of the air-gas circuit.

Distributed parameter modeling would be required to describe these phenomena accurately. Again, a simple lumped model may be used to explain basic concepts. A simple way to account for combustion kinetics is to introduce a combustion time constant τ_e relating fuel flow-rate w_f to heat transfer rate Q_f in the furnace,

$$\Delta Q_f = \frac{1}{1+s\tau_e}H_f\Delta w_f, \tag{11.42}$$

where H_f is the heat value of the fuel. The time constant τ_e is of the order of a second, smaller for oil or gas, and larger for pulverized coal.

Mass and energy balances for the furnace can be derived from the following assumptions:

- the gas pressure p_g is uniform in the furnace,
- heat transfer from gas to wall in the furnace is computed from the mean gas temperature, T_g, and
- the combustion gas behaves as an ideal gas.

Then, without considering gas recirculation:

$$\frac{dE_g}{dt} = w_a h_a + Q_f - w_{go}h_{go} - Q_r, \tag{11.43}$$

and

$$\frac{dM_g}{dt} = w_a + w_f - w_{go}, \tag{11.44}$$

where w_a is the air-flow rate with enthalpy h_a, Q_r is the heat transfer rate radiated to the wall, w_{go} is the outlet gas flow-rate with enthalpy h_{go}, $M_g = V_f \rho_g$ is the total mass of gas in the furnace (ρ_g is the mean density), and $E_g = c_{vg} T_g M_g$ is the total energy of the gas (c_{vg} is the specific heat at constant volume). Ideal gas law and radiation law equations are

$$\frac{p_g}{\rho_g} = R_g T_g, \tag{11.45}$$

$$Q_r = k_{rr}\left(T_g^4 - T_w^4\right) \approx k_{rr} T_g^4, \tag{11.46}$$

where k_{rr} is a constant and T_w is the wall temperature ($T_w^4 \ll T_g^4$).

Boundary conditions are determined by head losses along the air-gas circuit and by a forced and induced draft fan. Considering the outlet boundary conditions,

$$p_g - p_o \cong -p_v + (k_a + k_f(z))\, w_{go}, \tag{11.47}$$

where p_o is the atmospheric pressure, p_v is the head at $w_{go} = 0$ of the induced draft fan, k_a is the constant yielding the head losses along the gas circuit, and $k_f(z)$ is the constant yielding the head losses of the fan depending on the control inlet vane position z.

Assuming that the forced draft fans are controlled with air flow-rate w_a and that, in the mass equation, $w_f \ll w_a$ may be approximated by Q_f/h_f, linearization of the model Equations 11.43–11.47 yields

$$\delta p_g = \frac{r_g^\circ L}{(1+s\tau_1)(1+s\tau_2)}\left\{\left(\frac{w_f}{w_{go}}\right)^* (1+s\tau_3)\delta Q_f + \left(\frac{w_a}{w_{go}}\right)^* (1+s\tau_4)\delta w_a - \mu_v L(1+s\tau_5)\delta z\right\}, \tag{11.48}$$

where all the variables δ_y are expressed in p.u., the superscript * denotes the value at the linearization steady state, L is the plant load, and $r_g^\circ, \tau_1, \tau_2, \tau_3, \tau_4, \tau_5, \mu_v$ are suitable constants computed from design or operating data.

Omitting cumbersome computations for brevity, Equation 11.48 deserves some remarks. The time constants $\tau_j, j = 2\ldots,5$, are proportional to the furnace crossing time $t_f = (M_g/w_g)^*$. M_g^* is nearly insensitive to the load, above the technical minimum, w_g^* is nearly proportional to the load, and the time constants $\tau_j, j = 2\ldots,5$ become larger as the load decreases. The smallest time constant, on the contrary, decreases as the load decreases. And because it is only a few tenths of second at the maximum load, it may be neglected. Typical values are $t_f \approx 2-4$ sec., $\tau_1 < 0.1t_t$, $\tau_2 \approx 0.25t_f$, $\tau_3 \approx 5 - 6t_f$, $\tau_4 \approx 0.1t_f$, r_g° is about 0.1.

Due to the large value of τ_3, fuel trip from maximum load ($\delta Q_f = 1$) causes large pressure drops, with extreme risk of implosion. Inlet vane control is introduced to attenuate the effect of such a disturbance.

Concluding Remarks on Dynamics

To summarize the analysis of the preceding Sections, it is useful to identify the interactions in the system. Consider the input control variables, namely, the fuel flow-rate w_f, the air flow-rate w_a, the turbine admittance Y, the feed-water flow-rate w_w, the recirculation gas flow-rate w_{rg}, desuperheating spray flows w_{ds} and w_{dr}, induced fan vane position z, and the output variables to be controlled, namely, electric power P_e, throttle pressure p_T, drum level y_D, throttle temperature T_T, reheater outlet temperature T_{RH}, furnace pressure p_g, and air-to-fuel ratio λ_{af} (or combustion efficiency). From Figure 11.7 and Equation 11.41, P_e and p_T are strictly related and simultaneously affected by the heat transfer rate (i.e., w_f) and turbine admittance Y.

Flow rates w_{ds} and w_{dr} slightly affect P_e and p_T because they have only a small mass effect on steam flow. Similarly, feed-water flow w_w only marginally affects power (see Equation 11.39). Air flow is nearly proportionally to fuel flow, so that trimming actions to optimize λ_{af} have little influence on the heat transfer rate Q_{EV} to the evaporator. Only w_{rg} changes the heat release partition in the boiler; the control band width of this variable (used to control reheat temperature) is, however, rather narrow for that reason. Therefore, pressure and power form a (2x2) subsystem, tightly coupled but with limited disturbance from outside.

Drum level y_D is the only output variable markedly influenced by feed-water flow w_w; y_D is also "disturbed" (see Equation 11.28) by steam flow w_v, by drum pressure p_o, and by the evaporator heat transfer rate Q_{EV}, so that w_w does not influence the load-pressure subsystem, but y_D is considerably influenced by the control variables of that subsystem.

Superheated temperature T_T, according to Figure 11.10 and Equation 11.33, is influenced by w_{ds} and is "disturbed" by steam flow and heat transfer rate Q_{SH}, which, in turn, follows fuel flow variations. So, disturbance of the power-pressure subsystem to temperature is relevant, even though there is a natural partial compensation due to the boiler behavior (see ΔD_x in Figure 11.10).

Reheater spray follows a rule similar to superheater spray.

Furnace pressure is dynamically decoupled from all of the variables related to the steam-water side subsystems and, according to Equations 11.42 and 11.48 is affected by air flow, fuel flow, and induced draft fan inlet vane position (normally used as its control variable). A similar criterion applies to the fuel-air ratio.

Input-output relations are summarized in Figure 11.13, where only the major interactions are considered.

Finally, in coal fired units fuel is supplied by pulverizers, not influenced by the rest of the plant. The pulverizers have sluggish dynamics due to the dead time of the grinding process and to the uncertain behavior of such machines caused by coal quality variation and machine wear. For those plants, where fuel flow cannot be considered a directly manipulated variable, the slow response of coal pulverizers must be cascaded with the evaporator dynamics of Figure 11.7 often creating severe problems for system stability and control.

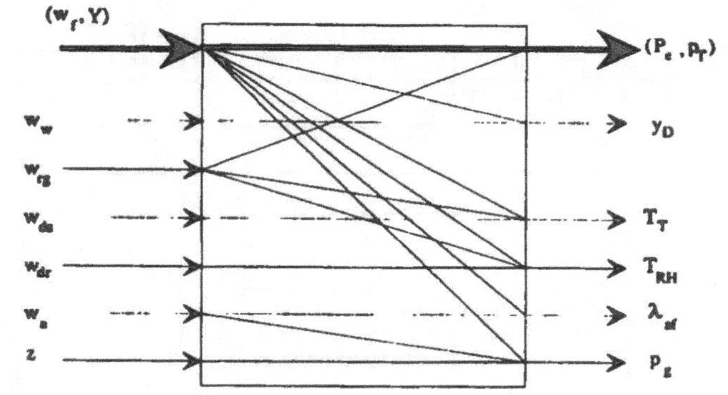

Figure 11.13 Process interactions.

11.1.4 Control Systems: Basic Architectures

Introductory Remarks on Control

This section discusses plant control at the "principal regulations level" (see Figure 11.4) which is organized around four control subsystems:

- load and pressure control: the regulation of power generation and steam pressure at the throttle,
- drum level control: the regulation of water level in the drum (steam/water separator),
- temperature control: the regulation of steam temperature at the superheater and reheater outlets, and
- combustion control: the regulation of heating rate (fuel flow), excess oxygen (air flow), and furnace pressure.

Power plant control systems have evolved over many decades. Today there are thousands of electric generating plants operating throughout the world. It would be difficult to find more than a few with identical control systems. Yet certain basic configurations are almost universally employed with relatively minor variations. These are described in the following paragraphs.

Control of Pressure and Generation

As discussed in Section 11.1.3, steam pressure and power generation are tightly coupled process variables. Both are strongly affected by energy (fuel) input and throttle valve position. This two-input, two-output system must be considered as such. Even though single-input, single-output compensation arrangements are successful, they must be designed (tuned) as a unit. Three basic architectures are commonly employed:

- turbine following: generation is paired with fuel rate and pressure with throttle valve position,
- boiler following: generation is paired with throttle valve position and pressure with fuel rate, and
- coordinated control: a true two-input, two-output configuration of which there are variations.

The turbine-following arrangement, shown in Figure 11.14 has distinctive attributes. First, the control of the energy input to the boiler is relatively slow compared with the positioning of the throttle valve. As a result, turbine-following control allows rapid regulation of throttles pressure and slow, but stable, regulation of generation. Consequently, turbine-following control is preferred for plants not used for load following.

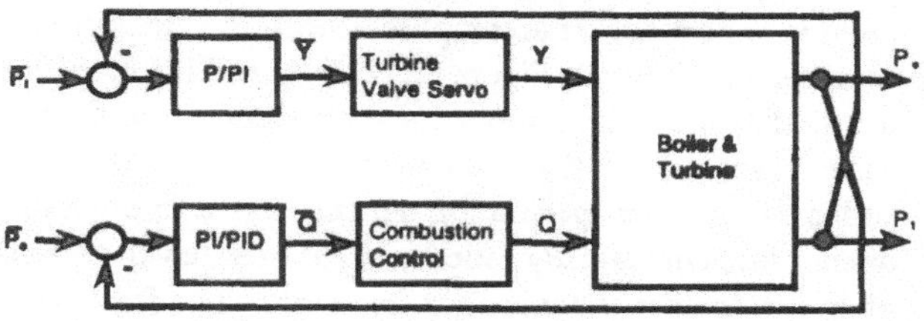

Figure 11.14 The turbine following configuration for pressure and generation control.

The boiler-following architecture is illustrated in Figure 11.15. It produces substantially more rapid responses to generation commands but they can be quite oscillatory. Moreover, pressure response is typically oscillatory.

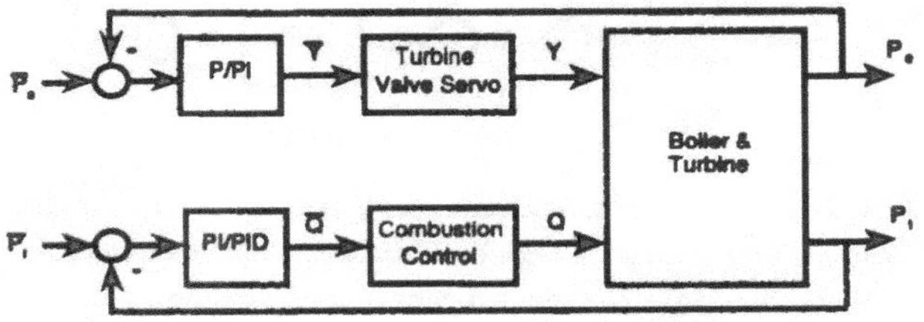

Figure 11.15 The boiler following configuration for pressure and generation control.

Modern requirements for load following have led to the widespread use of two-input, two-output pressure and generation control. There are a number of approaches to coordinated control. One configuration (commonly referred to as "coordinated control" or "integrated control") is shown in Figure 11.16. Properly designed coordinated-control systems can provide excellent response to load demand changes.

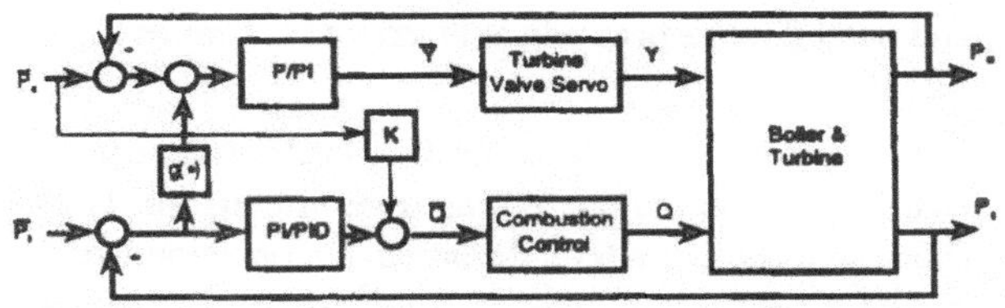

Figure 11.16 A coordinated-control configuration for pressure and generation control.

Generator speeds naturally synchronize because of their interconnection via the electrical network. Ultimately, the (steady-state) network synchronous speed is regulated by a system level controller through the assignment of generation commands to individual units. Nevertheless, speed governing on a substantial fraction of the network's generating units is essential to damping the power system's electromechanical oscillations. As a result, in many plants, the goal of turbine flow control includes speed governing as well as regulating power output. This dual requirement is almost always accomplished with the "frequency bias" arrangement shown in Figure 11.17. Here turbine speed error is fed directly through a proportional compensator to the turbine valve servo and simultaneously a frequency error correction is added to the power generation demand signal through the frequency bias constant B_f. Ideally, B_f is precisely the sensitivity of system load to synchronous frequency. The frequency bias arrangement can be incorporated in either the boiler-following or coordinated-control configurations.

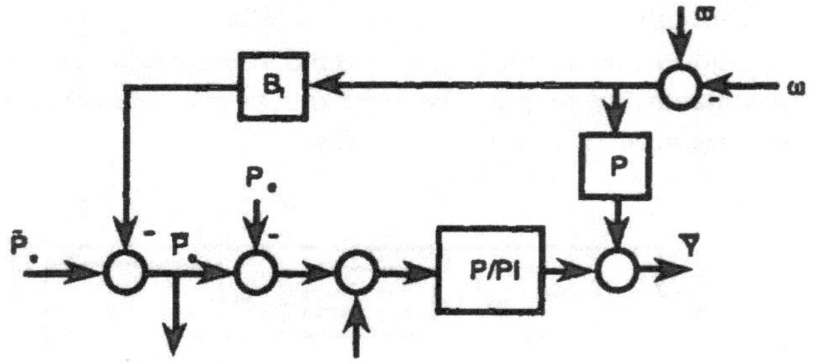

Figure 11.17 The turbine valve may be used to stabilize turbine speed and regulate generation with the "frequency bias" modification of the boiler-following or coordinated-control configurations.

Drum Level Control

The goal of the drum level controller is to manipulate the flow of feedwater into the drum so that the drum water level remains sufficiently close to a desired value. Feedwater flow is typically regulated by a flow control valve or by adjusting the speed of the feedwater pump. Drum level controllers are classified as single-, two-, or three-element. A single-element level controller utilizes feedback of a drum level measurement as illustrated in Figure 11.18(a). Two- and three-element controllers include "feedforward" measurements of steam flow and both steam and water flow, respectively, as illustrated in Figures 11.18(b) and 11.18(c).

The importance of the steam flow feedforward can be appreciated by examining the leading term in Equation 11.28 which shows that the drum level deviation is proportional to the integral of the difference between steam and water flow. Any sustained difference between steam flow and water flow can quickly empty or fill the drum. In current practice, three-element controllers are typically used during normal operation but are not suitable at very low loads, where it is common to switch to single-element configurations. Drum level dynamics are also nonminimum phase (the so-called "shrink and swell" effect) as can be observed in the last term of Equation 11.28. Drum level control

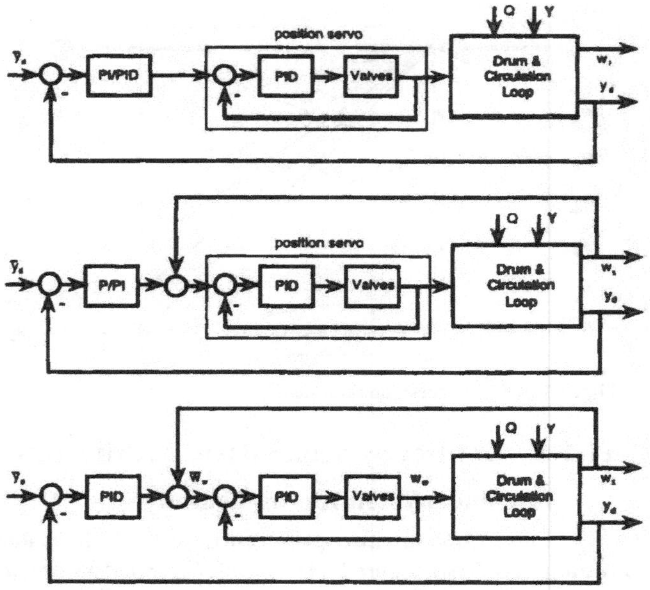

Figure 11.18 (a). A typical single-element drum level configuration. (b). A typical two-element drum level configuration. (c). A typical three-element drum level configuration.

will be examined in more detail in Section 11.1.5.

Temperature Control

An important goal of plant control is to regulate steam temperature at the turbine entry points, i.e., at the superheater and reheater outlets. There are a number of control means for accomplishing this. The most direct are attemperators which inject water at the heat exchanger inlet. By moderating the fluid temperature entering the heat exchanger, it is possible to control the outlet temperature. Other possibilities are associated with adjusting the heat transferred to the fluid as it passes through the exchanger. This can be accomplished by changing the mass flow rate of gas past the heat transfer surfaces with recirculated gas or excess air flow, or the gas temperature at the exchanger surfaces by altering the burner positions or "tilt" of the burners. Sometimes a combination of these methods is employed. In the following discussion it is assumed that control is affected by attemperators.

The dynamics of superheaters and reheaters have been discussed in Section 11.1.3. Recall that the response of the outlet temperature to a change in inlet temperature is characterized by series of first-order lags with time constants that vary inversely with the steam flow rate through the heat exchanger. Because of the significant time delay of the outlet temperature response, a cascade control arrangement, as illustrated in Figure 11.19, is typically required for temperature regulation. The attemperator outlet temperature is a convenient intermediate feedback variable although, depending on the heat exchanger construction, other intermediate steam temperatures may also be available for measurement. Because of the strong dependence of the time lag on steam flow, parameterization of the regulator parameters on steam flow is necessary for good performance over a wide load range. Some control systems incorporate disturbance feedforward.

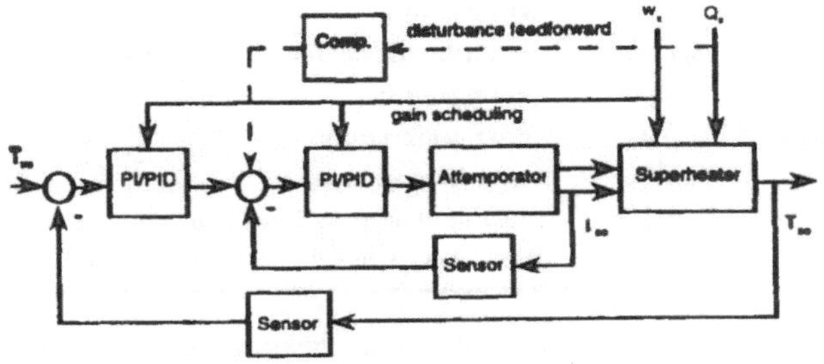

Figure 11.19 A fairly sophisticated control temperature arrangement employs a cascade arrangement, gain scheduling and disturbance feed forward.

Combustion Control

The main purpose of the combustion control system is to regulate the fuel and air inputs into the furnace to maintain the desired heat input into the steam generation process while assuring appropriate combustion conditions (excess oxygen). In most instances, regulating the furnace gas pressure is a secondary, but important, function of the combustion controller. A typical combustion control configuration is illustrated in Figure 11.20. Recall that the heating rate command signal $\bar{Q}$ is generated by the pressure-generation controller (Figure 11.14).

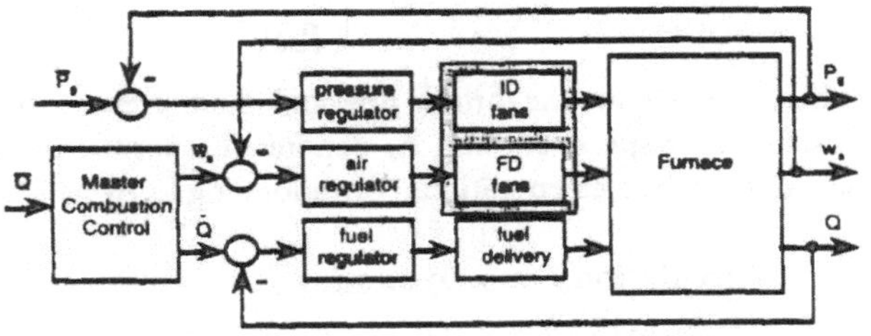

Figure 11.20 Combustion control includes regulation of fuel and air flow and furnace pressure.

The details of the combustion control system depend significantly on the type of fuel. Oil and gas are typically regulated with flow control valves. These controls are usually fast. Also, oil and gas flow and the caloric content of the fuel can be reliably measured. Pulverized coal presents a different situation. Fuel flow is regulated by adjusting the "feeder speed" (which directly changes the flow rate of coal into the pulverizer) and the primary air flow (the air flow through the pulverizer that carries the pulverized coal into the furnace). The pulverizing process is quite slow and adds a delay of 100–300 sec. in the fuel delivery process. Moreover, the flow rate of coal is difficult to estimate accurately and coal's caloric content varies.

The master combustion control proportions the fuel and air requirements and establishes set points for the lower level controllers. Calorimetry information (notably excess oxygen) is provided on a sampled or continuous basis. Minimum fuel and air (primary and secondary) flow constraints are also accommodated.

11.1.5 Design of a Drum Level Controller

Issues Involving Design Over a Wide Load Range

The control architectures described in Section 11.1.4 represent a starting point for detailed design of the plant control system. Once the configuration is chosen, it is necessary to determine compensator parameters that produce acceptable performance over a wide range of operating conditions. Doing this involves applying various analytical control design tools in conjunction with detailed simulation studies and generally terminating with field tuning during plant commissioning. This section provides an example of the analytical phase in designing a drum level controller.

Certain necessary plant operations, e.g., startup, rapid load following and controlled runback, evoke behavior that can only be attributed to the nonlinearity of the steam generation process. Deleterious nonlinear effects are particularly evident at low generation rates and when relatively large changes in generation level are made. Regulating drum level and combustion stability are particularly problematic at low loads. Nonlinear behavior can be readily observed in two ways: by comparing large and small excursions from a given equilibrium point or by comparing small signal behavior at different equilibria corresponding to distinct load levels. In the following paragraphs, we will examine the steam generation process from the latter point of view. This is particularly useful when designing control systems based on small signal (linear) behavior.

Power plant controllers are designed to provide good performance at or near rated conditions. At off-design conditions, e.g., as load is decreased, performance deteriorates because the plant behaves differently than at the design point. Typically, satisfactory performance can be achieved down to about 30% of rated generation. At some point performance becomes unsatisfactory, requiring manual intervention or a switch to a retuned or even restructured control system. At low loads the nonlinearities associated with the evaporation and combustion processes [8], in particular, are quite severe. Compounding this complexity, plant operation over a wide generating range may require configuration changes. For example, as generation is reduced, the number of burners and fuel pumps or pulverizers will be reduced (see, for example, [9]), and steam by-pass systems may be employed, particularly, in large once-through fossil plants and nuclear plants [10].

The assumptions behind the model of evaporator dynamics developed in Section 11.1.3 are not suitable for capturing dynamics at low steam generation rates. In Section 11.1.5, following the assumptions Equations 11.3 and 11.4 are relaxed to allow more general variations of steam quality and fluid pressure through the evaporator. This model is then used to investigate the small signal dynamical behavior at various load levels. Control system design is considered in Section 11.1.5. Much of the discussion herein is based on [11] which emphasizes the nonlinear behavior of the circulation loop. Earlier discussions on level control system design included [12] and [13].

Circulation Loop Dynamics Revisited

The system (see Figure 11.3) is comprised of a natural circulation loop whose main components are the drum which separates the steam and water, the downcomer piping which carries water from the drum to the bottom of the furnace, and the riser tubes which are exposed to the burning furnace gas and in which boiling takes place. A feedwater valve regulates the flow of water into the system and the throttle valve regulates the steam flow out of it. The heat absorbed by the fluid in the risers is a third input. The drum dynamical equations are considered first and then the remainder of the circulation loop.

Drum Dynamics

The dynamical equations for the drum can be formulated in a number of different ways depending on the variables chosen to characterize the thermodynamic state of the drum. Two thermodynamic variables must be selected from the four possible pairs: (T, v), (s, P), (s, v), (T, P), where T, P, v, and s denote temperature, pressure, specific volume, and specific entropy, respectively.[2] Because the fluid in the drum is in the saturated state, the pair (T, P) is not suitable but any of the other choices is valid. As an example, we give the equations for the (s, P) formulation. The following assumptions are made:

1. Drum liquid and gas are in the saturated state.
2. Pressure is uniform throughout the drum.
3. All liquid resides at the bottom of the drum and all gas at the top.

Then, the drum dynamics, derivable from mass and energy balance equations, (see [9]) are

$$\frac{dP_d}{dt} = -\frac{[w_e+(1-x_r)w_r-w_{dc}]\frac{v_{df}^2}{v_{dg}}+[x_r w_r - w_s]\frac{1}{v_{df}}}{(V-V_w)\frac{v_{df}}{v_{dg}^2}\frac{\partial v_g}{\partial P_d}+V_w\frac{1}{v_{dg}}\frac{\partial v_f}{\partial P_d}}, \tag{11.49}$$

and

$$\frac{dV_w}{dt} = -\frac{[w_e+(1-x_r)w_r-w_{dc}](V-V_w)}{(V-V_w)\frac{v_{df}}{v_{dg}^2}\frac{\partial v_g}{\partial P_d}+V_w\frac{1}{v_{dg}}\frac{\partial v_f}{\partial P_d}}$$
$$\frac{\left(\frac{v_{df}}{v_{dg}}\right)^2\frac{\partial v_g}{\partial P_d}-V_w[x_r w_r - w_s]\frac{\partial v_f}{\partial P_d}}{(V-V_w)\frac{v_{df}}{v_{dg}^2}\frac{\partial v_g}{\partial P_d}+V_w\frac{1}{v_{dg}}\frac{\partial v_f}{\partial P_d}}, \tag{11.50}$$

where the following nomenclature has been adopted:

w_r, w_{dc}, w_s	mass flow rates, riser, downcomer and turbine, respectively,
v_{df}, v_{dg}	drum specific volume, liquid and gas, respectively,
P_d	drum pressure,
T_d	drum temperature,
V	total drum volume,
V_w	volume of water in drum, and
x_d	net drum quality, $x_d = V_w/V$.

In addition to these differential equations we require the constitutive relations (coexistence curve)

$$v_f = v_f(P) \quad \text{and} \quad v_g = v_g(P) \tag{11.51}$$

and the drum level equation

$$y_D = f(V_w). \tag{11.52}$$

Under the stated assumptions, the drum thermodynamic state, entropy and pressure (s, P), is equivalent to (x_d, P) or (V_w, P) because

$$x_d = v_f(P) + x_d[v_g(P) - v_f(P)]. \tag{11.53}$$

Hence, we refer to the above equations as the (s, P) formulation even though s does not explicitly appear.

The steam flow out of the drum to the turbine is governed by the relationship

$$w_s = w_{s0}A_1\left(\frac{P_d}{P_{d0}}\right) \tag{11.54}$$

where w_{s0}, P_{d0} are the throttle flow and drum pressure at rated conditions, respectively, and A_t denotes the normalized valve position, with rated conditions corresponding to $A_t = 1$.

Circulation Loop Dynamics

The main deficiency of the model in Section 11.1.3 was the simplified treatment of the circulating fluid flow and of the complex two-phase flow dynamics of the riser loop. Consequently, it does not adequately represent the nonminimum phase characteristics associated with shrink-swell phenomenon at lower load levels. Another approach to riser modeling is based on discretizing the time-dependent, nonlinear partial differential equations of one-dimensional, two-phase flow using the method of collocation by splines, a form of finite element analysis.

The circulation loop, composed of the downcomer and riser, is assumed to be characterized by homogeneous single or two-phase flow. The conservation of mass, energy, and momentum lead to three first-order partial differential equations which are then discretized using collocation by linear splines. We use a single element for the downcomer containing fluid in a single phase (liquid) and N elements of equal length for the riser which consists of two-phase flow. In the latter case the fluid properties change significantly along the spatial coordinate. In the entropy-pressure (s, P) formulation, the downcomer equations are

[2] In the standard convention, specific entropy is denoted by the symbol s which is also used to denote the Laplace transform variable. This will not lead to confusion in the subsequent discussion.

$$\frac{dw_1}{dt} = -A\left(\frac{P_1 - P_0}{L_{do}}\right) - \frac{2w_1v_1}{A_{do}}\left(\frac{w_1 - w_0}{L_{do}}\right) - \frac{w_1^2}{A_{do}}\left(\frac{v_1 - v_0}{L_{do}}\right) - A_{do}\left(-\frac{g}{v_1} + f_{do}w_1^2\right), \quad (11.55)$$

$$\frac{ds_1}{dt} = v_1\left\{\frac{1}{A_{do}T_1}\frac{\partial q}{\partial z} - \frac{w_1}{A_{do}}\left(\frac{s_1 - s_0}{L_{do}}\right) + \frac{w_1v_1}{A_{do}T_1}f_{do}w_1^2\right\}, \quad (11.56)$$

and

$$\frac{dP_0}{dt} = \frac{1}{\gamma_A}\left\{\frac{v_0^2}{A_{do}}\left(\frac{w_1 - w_0}{L_{do}}\right) - \gamma_B v_{1-1}\left(-\frac{w_0}{A_{do}}\left(\frac{s_1 - s_0}{L_{do}}\right) + \frac{w_0v_0}{A_{do}T_0}f_{do}w_0^2\right)\right\}. \quad (11.57)$$

The riser equations are

$$\frac{dw_i}{dt} = -A\left(\frac{P_i - P_{i-1}}{L}\right) - \frac{2w_iv_i}{A}\left(\frac{w_i - w_{i-1}}{L}\right) - \frac{w_i^2}{A}\left(\frac{v_i - v_{i-1}}{L}\right) - A\left(\frac{g}{v_i} + f_rw_i^2\right), \quad (11.58a)$$

$$\frac{ds_i}{dt} = v_i\left\{\frac{1}{AT_i}\frac{\partial q}{\partial z} - \frac{w_i}{A}\left(\frac{s_i - s_{i-1}}{L}\right) + \frac{w_iv_i}{AT_i}f_rw_i^2\right\}, \quad (11.58b)$$

and

$$\frac{dP_{i-1}}{dt} = \frac{1}{\gamma_A}\left\{\frac{v_{i-1}{}^2}{A}\left(\frac{w_i - w_{i-1}}{L}\right) - \gamma_B v_{i-1}\left(\frac{1}{AT_{i-1}}\frac{\partial q}{\partial z} - \frac{w_{i-1}}{A}\left(\frac{s_i - s_{i-1}}{L}\right) + \frac{w_{i-1}v_{i-1}}{AT_{i-1}}f_rw_{i-1}^2\right)\right\} \quad (11.58c)$$

for $i = 2, \ldots, N + 1$. The following nomenclature is employed:

N	number of riser sections,
L_{do}, L	downcomer length and riser section length (total riser length/N),
A_{do}, A	downcomer, riser cross section areas,
w_i	mass flow rate at ith node,
P_i	pressure at ith node,
T_i	temperature at ith node,
s_i	aggregate entropy at ith node, and
v_i	specific volume at ith node.

Once again, we need constitutive relations to complete the model. These are required in the form

$$v = v(s, P), T = T(s, P) \quad (11.59a)$$

$$\gamma_A = \left(\frac{\partial v}{\partial P}\right)_s, \quad \text{and} \quad \gamma_B = \left(\frac{\partial v}{\partial S}\right)_P. \quad (11.59b)$$

Reduction of Circulation Loop Equations

The circulation loop model described above contains fast dynamics irrelevant to the control problem. In general terms, fast dynamics are associated with certain pressure-flow dynamics (hydraulic/acoustic oscillations) and slow dynamics are associated with thermal (entropy) transients. Formally, we can approach the problem of identifying and approximating fast dynamics using asymptotic analysis, because fast dynamics are associated with the fact that the parameter γ_A is small. Our analysis is based on two assumptions:

- only the slowest mode of fast pressure-flow dynamics is significant to the control problem; we shall refer to this as hydraulic dynamics.
- spatial variations in flow and pressure along the circulation loop are negligible as far as the hydraulic dynamics are concerned.

The flow and pressure equations can be written

$$a_i\frac{dw_i}{dt} = -(P_i - P_{i-1}) + F_i(w - i, w_{i-1}, s_i, s_{i-1}, P_i, P_{i-1}), \quad a_i := L_i/A_i \quad \text{and} \quad (11.60a)$$

$$b_i\frac{dP_{i-1}}{dt} = (w_i - w_{i-1}) + g_i(w_{i-1}, s_i, s_{i-1}, P_{i-1}), \quad b_i := \gamma_A A_i L_i / v_{i-1}^2. \quad (11.60b)$$

Now, we define the average circulation loop flow and pressure:

$$w_{av} := \sum_{i=1}^{N+1} \alpha_i w_i, \alpha_i = \frac{a_i}{\sum_{i=1}^{N+1} a_i} \quad (11.61)$$

$$\text{and} \quad P_{av} := \sum_{i=1}^{N+1} \beta_i P_i, \beta_i = \frac{b_i}{\sum_{i=1}^{N+1} b_i}. \quad (11.62)$$

Let us also define the functions

$$f_i(w_{av}, s_i, s_{i-1}, P_{av}) := F_i(w_{av}, w_{av}, s_i, s_{i-1}, P_{av}, P_{av}). \quad (11.63)$$

We can state the key assumption:

Assumption 1: For w_i, s_i and $P_i, i = 0, \ldots, N + 1$, the following approximations are valid in the slow time scale:

$$f_i(w_{av}, s_i, s_{i-1}, P_{av}) \approx F_i(w_i, w_{i-1}, s_i, s_{i-1}, P_i, P_{i-1}) \quad (11.64)$$

and

$$g_i(w_{av}, s_i, s_{i-1}, P_{av}) \approx g_i(w_{i-1}, s_i, s_{i-1}, P_{i-1}). \quad (11.65)$$

It is easy to validate this assumption in the equilibrium state. Invoking these approximations and simply adding Equations 11.60 results in

$$\frac{dw_{av}}{dt} = \left\{-(P_d - P_0) + \sum_{i=1}^{N+1} f_i(w_{av}, s_i, s_{i-1}, P_{av})\right\} \tag{11.66a}$$

and

$$\frac{dP_{av}}{dt} = \left\{-(w_r - w_0) + \sum_{i=1}^{N+1} g_i(w_{av}, s_i, s_{i-1}, P_{av})\right\}. \tag{11.66b}$$

Assumption 2: The slow time scale flow and pressure distribution through the circulation loop is adequately approximated by equilibrium conditions of Equations 11.60 and the approximations of Assumption 1, i.e.,

$$-(P_i - P_{i-1}) + f_i(w_{av}, s_i, s_{i-1}, P_{av}) = 0, \quad i = 1, \ldots, N+1, \tag{11.67a}$$

and

$$(w_i - w_{i-1}) + g_i(w_{av}, s_i, s_{i-1}, P_{av}) = 0, \quad i = 1, \ldots, N+1. \tag{11.67b}$$

These equations are important because they allow the computation of P_0 and $\omega_r = \omega_{N+1}$ which are necessary to establish the interface of the circulation loop with the drum. From Equations 11.20 and the definitions of ω_{av}, P_{av}, we can derive the following relationships:

$$P_0 = P_{av} - \sum_{i=1}^{N}\left(\sum_{j=1}^{N+1-i} \alpha_j\right) f_i(w_{av}, s_i, s_{i-1}, P_{av}) \tag{11.68a}$$

and

$$w_r = w_{av} - \sum_{i=2}^{N+1}\left(\sum_{j=1}^{i-1} \beta_j\right) g_i(w_{av}, s_i, s_{i-1}, P_{av}). \tag{11.68b}$$

The remaining interface equation is

$$w_0 = \sqrt{|P_d - P_0| sign(P_d - P_0)/f_{de}}. \tag{11.69}$$

Equilibria and Perturbation Dynamics

First, we consider the open-loop behavior. The procedure followed is

1. Trim the system at load levels ranging from near 5% to 100%.
2. Compute the linear perturbation equations.
3. Analyze the pole-zero patterns as a function of load level.

Equilibrium values are computed by specifying the desired load, drum pressure, and drum level and then computing the required control inputs and the remaining state variables. A Taylor linearization at each equilibrium point yields a linear model of the perturbation dynamics. Thus, it is possible to determine the system poles and zeros and to examine how they change as a function of load. Figure 11.21 gives a sample of the results obtained from solving the equilibrium equations.

Figure 11.22 is an eigenvalue plot showing how the plant dynamics vary with load. Table 11.2 summarizes a complete plant modal analysis at 100% load.

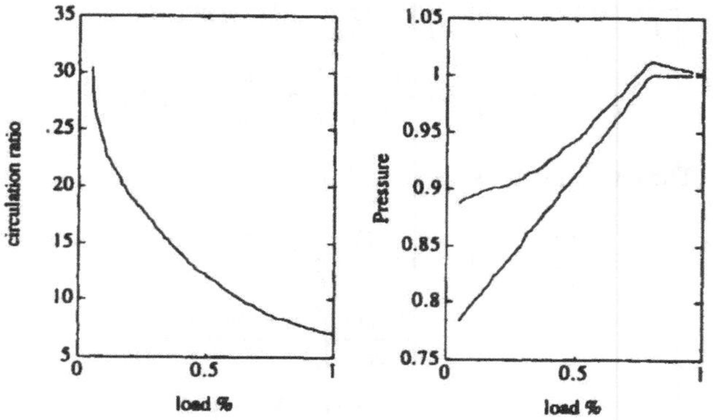

Figure 11.21 Typical equilibrium curves show the circulation ratio, average loop pressure (upper), and drum pressure (lower) as a function of load.

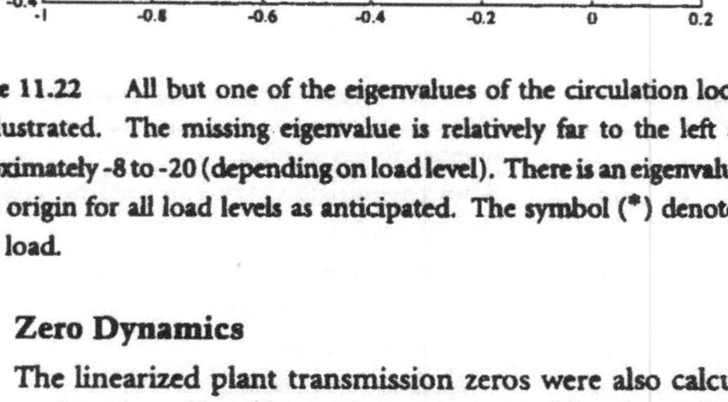

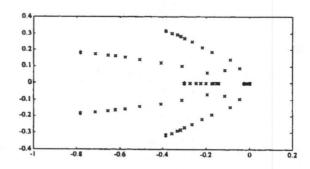

Figure 11.22 All but one of the eigenvalues of the circulation loop are illustrated. The missing eigenvalue is relatively far to the left at approximately -8 to -20 (depending on load level). There is an eigenvalue at the origin for all load levels as anticipated. The symbol (*) denotes 100% load.

Zero Dynamics

The linearized plant transmission zeros were also calculated as a function of load level using various combinations of inputs and outputs. These results are summarized in Figures 11.23 and 11.24. These figures, which characterize the relation between

TABLE 11.1 Summary of Dynamical Equations

$$u_1 = q, u_2 = w_e, u_3 = A_1$$
$$\frac{dw_{av}}{dt} = f_1(w_{av}, s_1, s_2, s_3, P_{av}, P_d)$$
$$\frac{ds_1}{dt} = f_2(w_{av}, s_1, P_{av}) + g_{21}(P_{av}, s_1)u_1 + g_{22}(w_{av}, P_d)u_2$$
$$\frac{ds_2}{dt} = f_3(w_{av}, s_1, s_2, P_{av}) + g_{31}(P_{av}, s_2)u_1$$
$$\frac{ds_3}{dt} = f_4(w_{av}, s_2, s_3, P_{av}) + g_{41}(P_{av}, s_3)u_1$$
$$\frac{ds_4}{dt} = f_5(w_{av}, s_3, s_4, P_{av}) + g_{51}(P_{av}, s_4)u_1$$
$$\frac{dP_{av}}{dt} = f_6(w_{av}, s_1, s_2, s_3, s_4, P_{av}, P_d) + g_{61}(w_{av}, s_1, s_2, s_3, s_4, P_{av})u_1$$
$$\frac{dP_d}{dt} = f_7(w_{av}, s_1, s_2, s_3, s_4, P_{av}, P_d, V_w) + g_{71}(w_{av}, s_1, s_2, s_3, s_4, P_{av}, P_d, V_w)u_1 + g_{72}(P_d, V_w)u_2 + g_{73}(P_d, V_w)u_3$$
$$\frac{dV_w}{dt} = f_8(\omega_{av}, s_1, s_2, s_3, s_4, P_{av}, P_d, V_w) + g_{81}(\omega_{av}, s_1, s_2, s_3, s_4, P_{av}, P_d, V_w)u_1 + g_{82}(P_d, V_w)u_2 + g_{83}(P_d, V_w)u_3$$
$$y_1 = P_d, y_2 = y_D = f(V_w), y_3 = w_s = h_3(P_d)$$

TABLE 11.2 Eigenvalues and Eigenvectors at 100% Load

Mode	1	2&3	4&5	6	7	8
Mode description	Drum pressure-circulation flow rebalance	Drum-Riser mass balance oscilation	Riser flow-density oscilation	Circulation flow-drum level rebalance	Energy-coupled drum-level	Drum level mass balance
Eigenvalue	−7.7758	−.7817±.1835*i*	−.3854±.3145*i*	−0.3023	−0.0274	0.0000
ω_{av}	0.5288	.6866±.5787*i*	−.9500±.1684*i*	0.9692	0.4173	0.0000
s_1	−0.0066	−.0052±.0063*i*	.0034±.0046*i*	0.0045	−0.0193	0.0000
s_2	0.0005	.0025±.0148*i*	.0097±.0058*i*	0.0058	−0.0212	0.0000
s_3	0.0001	.0186±.0222*i*	.0029±.0207*i*	0.0071	−0.0230	0.0000
s_4	0.0001	−.0933±.0099*i*	−.0178±.0280*i*	0.0087	−0.0247	0.0000
P_{av}	−0.0323	.0451±.0178*i*	−.0107±.0270*i*	−0.0267	−0.1285	0.0000
P_d	0.7644	.0561±.0283*i*	−.0321±.0170*i*	0.0187	−0.1117	0.0000
V_d	0.3675	−.3925±.1526*i*	.0879±.2397*i*	0.2436	0.8916	1.0000

feedwater flow and drum level, and heat rate and drum level, respectively, show the nonminimum phase behavior typical of such systems, with zeros in the right half-plane.

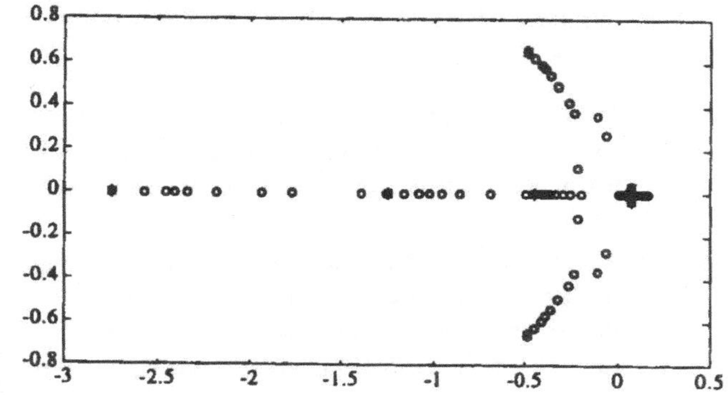

Figure 11.23 The zeros of the transfer function $\omega_e \rightarrow$ lev show the expected nonminimum phase characteristic of drum level dynamics.

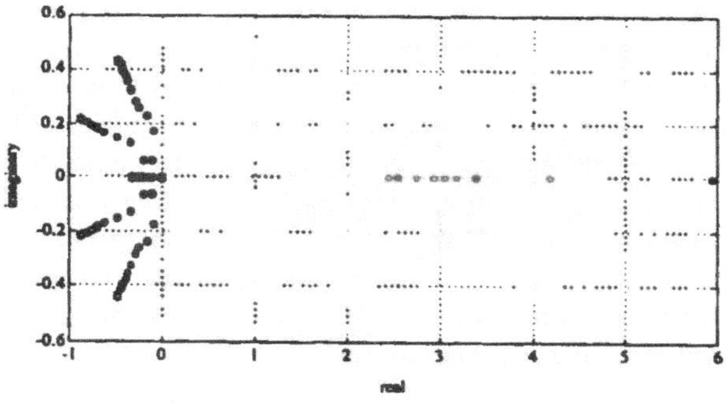

Figure 11.24 Zeros of the transfer function $Q \rightarrow$ lev. The nonminimum phase characteristic which produces the shrink/swell phenomenon is clearly evident.

Drum Level Control at High and Low Loads

The control configurations of interest in this section are shown in Figure 11.18, which illustrates typical one-, two-, and three-element feedwater regulators with proportional-integral-derivative (PID) compensation applied to the water level error signal and "feedforward" of the steam flow. This is the control structure most commonly used for feedwater regulation in drum type power plants. The unfortunate terminology "feedforward" arises from the view that steam flow is a disturbance vis-a-vis the drum level control loop, in which case such terminology would be appropriate. From the point of view here, however, because throttle valve position is the disturbance and steam flow is an output, we will refer to steam flow "feedback." This seemingly innocuous distinction is critical to understanding low load performance. If steam flow feedback is omitted, the result is a single-element controller.

With the configuration specified, the control system design problem reduces to selecting PID parameters for the compensator. It would be desirable to identify one set of parameters to provide acceptable performance at all load levels but it is well-known that this is not possible. Open-loop analysis shows that the dynamics vary dramatically over the load range and, in particular, the position of the open-loop zeros, which limit achievable performance at all loads, is particularly poorly located at low loads. Thus, it is common practice to use two different sets of parameter values, one for high loads and the other for low loads. Even so, it will be seen that the achievable low load performance is not very good and that feedback of steam flow degrades performance at the very low end.

Compensator Design

Performance Limitations There is a basic performance trade-off associated with the PID compensator for feedwater regulation. The regulator introduces a (second) pole at the origin and allows for placement of two zeros. The three design parameters may be viewed as these two zero locations and the loop gain. There are four dominant modes: the three modes 6, 7, and 8 described in Table 11.2 and the mode introduced by the compensator and corresponding to the new pole at the origin. The latter will be referred as the "drum level trim" mode. This terminology is consistent with the intent of compensator integral action, to eliminate steady-state errors from the drum level. One classical design approach is to fix the compensator zero locations and to examine the root locus with respect to loop gain. Figure 11.25 through Figure 11.27 illustrate a root locus plot at 100% load. One of the compensator zeros is placed on the real axis close to the origin because the undesirable, destabilizing phase lag introduced by the integral action is compensated for by the neighboring zero. Of course this traps a pole near the origin. This pole corresponds to the trim mode and the result is a very slow trimming of the steady-state drum level error.

The placement of the second zero is much more critical. Here it is placed on the real axis between the poles corresponding to modes 7 and 8. The logic is to pull mode 8, the critical drum mass balance mode, from the origin to the left, the farther the better, which translates into high gain. As seen in the figures, this draws modes 6 and 7 together, coupled into an oscillatory mode which approaches instability as the loop gain is increased. These poles are attracted by the nonminimum phase zeros. This represents one basic trade-off. A gain must be chosen which provides acceptable speed of response for mode 8, while retaining reasonable damping for modes 6 and 7.

Performance Under Load Variation It is important to examine the performance of this controller at lower load levels. In terms of pole location, performance degrades as load is reduced:

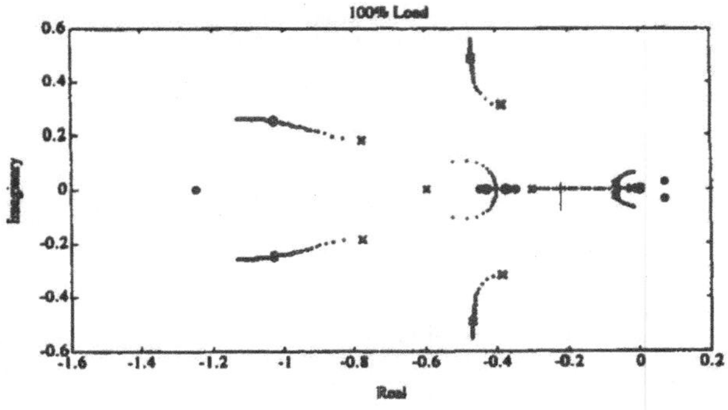

Figure 11.25 This figure provides an overview of the root locus at 100% load. The designations are: open loop poles (x), open loop zeros (o), design poles (*). All but one (the leftmost) pole is shown. There are also two additional zeros, both real and to the left. Notice the two right half-plane zeros. Two of the four dominant poles (near the origin) are not clearly visible on this scale.

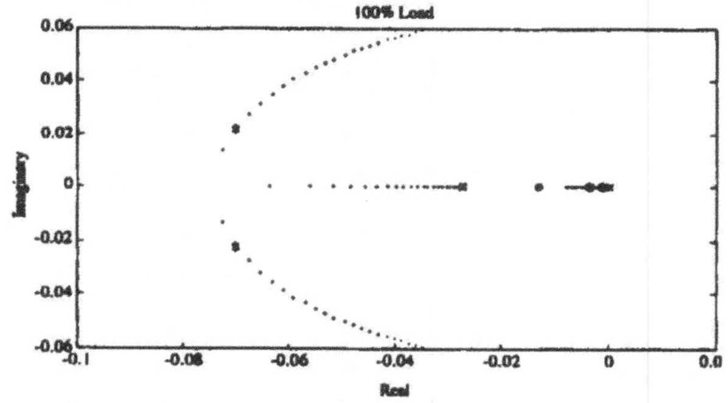

Figure 11.26 A closer view of the dominant poles is provided by this figure. Notice that the selected value of the gain provides closed-loop poles for the disturbance-coupled drum level mode, which are slightly underdamped. Further increases in gain reduce the damping, and eventually these poles move into the right half-plane attracted by the right half-plane zeros. Although some additional gain would be acceptable (in fact, desirable) at this load level, it has adverse affects at lower loads, as will be seen.

- At 60% load the disturbance-coupled drum dynamics have significantly degraded. Higher gain is called for, but once again, lower load dynamics would suffer.
- At 50% load the disturbance-coupled dynamics have further degraded, although the drum level mode response time is somewhat improved. This is a critical load which marks a transition in the qualitative closed-loop behavior. As load is reduced, the open-loop poles associated with modes 6 and 7 move to the right along the real axis. At 50% load the mode 7 pole is precisely located at one of the compensator zeros. Because of this, mode 7 is effectively unregulated and remains fixed as gain is varied. In modern terminology, this is an input decoupling zero. It is possible, of course, to place the compensator zero more to the right in order to lower the load level (50%) at which the transition takes place. This strat-

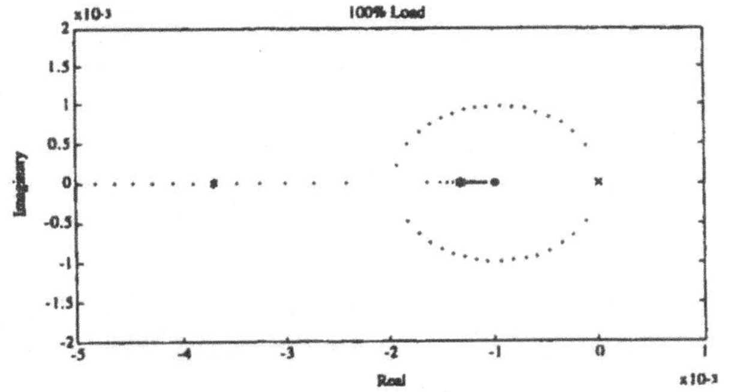

Figure 11.27 This figure provides an even closer look at the poles associated with the drum level mode and the level trim mode. These are by far the slowest modes. At the design condition, the drum level mode has a time constant of about 200 sec. This response time could be improved by increasing the gain, but at the expense of lower load stability. The trim mode is considerably slower.

egy reduces the achievable response time of mode 8. Consequently, this is a second basic trade-off which must be addressed.

- As load is further reduced to 40%, mode 7 couples with mode 8, rather than mode 6, to form a slow, but acceptably damped, oscillatory mode. The damping, however, diminishes as load is further reduced because the (open-loop) mode 7 pole migrates to the right and also because one of the right half-plane zeros approaches the origin from the right. There is a distinct change in root locus behavior in the transition from 60%–50%–40% load levels. Figure 11.28 illustrates the dominant poles at 40% load.
- At 30% load there is a fairly dramatic degradation of performance. Although the system remains stable, the dominant dynamic is a slow oscillation (a period of over 600 sec.) which is very lightly damped. A reduction in gain would improve performance, although it would be marginal at best.
- At 20% load the closed-loop system is unstable. A reduction in gain is required to stabilize it. Performance can be optimized by adjusting all three controller parameters, but even the best achievable performance is not very good. Moreover, even stable performance is attainable only with substantial degradation of high load performance.

Low Load Regulator Eventually, it is necessary to face the dilemma that stable low load operation is not achievable with the conventional control configuration tuned to provide reasonably good performance at higher loads. Thus, a new set of PID parameters is needed for low load operation. This is accomplished by drastically moving both compensator zeros to the right and choosing a new loop gain by examining the root locus. Figure 11.29 shows the root locus for a low load design at 20% load. It is important to note that the critical mode 8 has a closed-loop time constant 4 to 5 times slower than obtainable at

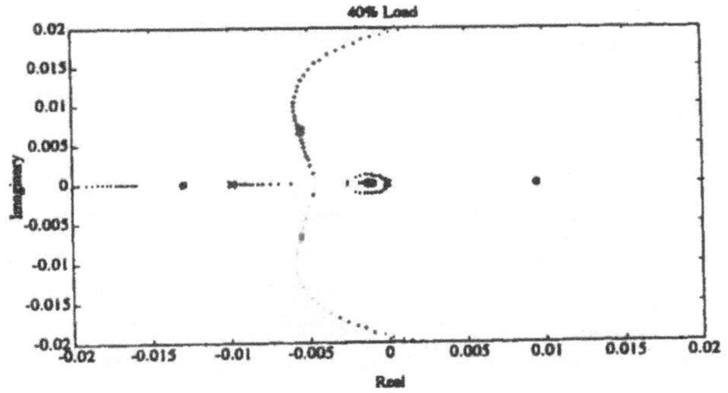

Figure 11.28 This figure provides a closer look at the dominant poles at 40% load. The damping ratio is already somewhat less than ideal so that increased gain would not be desirable.

high load. Here again a transition for this regulator, of exactly the same type as described earlier at 50% load, takes place between 10% and 5% load.

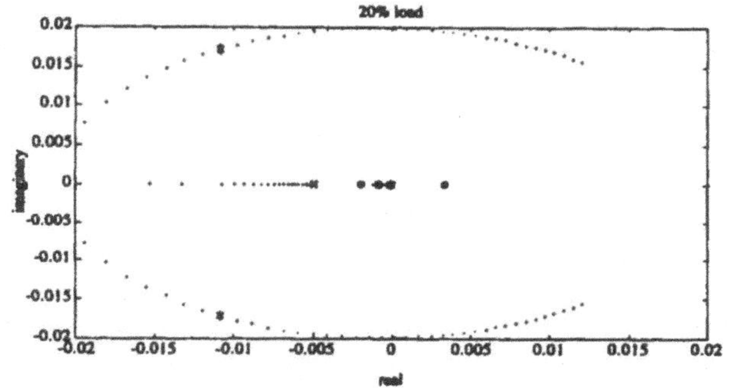

Figure 11.29 The low load regulator root locus at 20% load shows that the system is stable but has much slower response of the dominant modes.

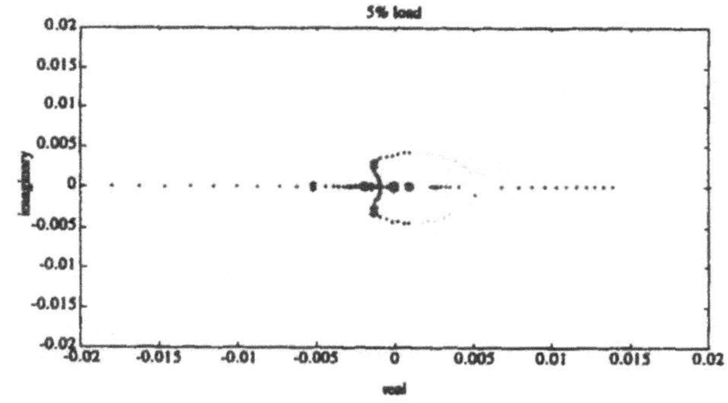

Figure 11.30 At 5% load a transition has taken place and, even though the system is still stable, the damping is substantially reduced.

Effect of Steam Flow Feedback So far the effect of steam flow feedback has not been addressed. It is possible to examine the effect of the steam flow feedback loop on the otherwise open-loop plant. Computation shows that the effect on most of the eigenvalues is negligible. Only mode 7 is significantly influenced. At all load levels the steam flow feedback moves the eigenvalues to the right. Although the largest changes are associated with the high load levels, the consequences are important only at low load levels. It is necessary to redesign the compensator with the steam flow feedback in place.

Figures 11.31 and 11.32 show the closed-loop pole location for the redesigned high load controller with steam flow feedback for load levels of 100% and 40%. Once again 50% load is a critical transition point. At 30% load stability, margins are poor and probably not acceptable. The system is unstable at 20% load. Figures 11.33 and 11.34 show the eigenvalue locations when the low load controller defined above is employed with steam flow feedforward. By comparison with Figures 11.29 and 11.30, steam flow feedback marginally degrades stability at 20% (and also at 10%, not shown) load, but the system is unstable at 5%. The choice not to use steam flow feedback with the low load controller is made to retain stability at very low loads.

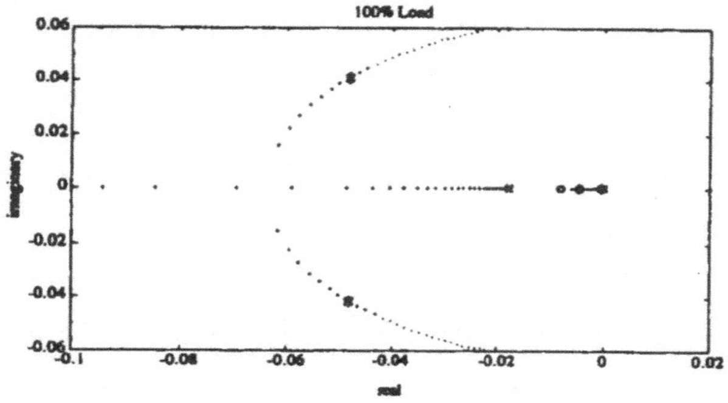

Figure 11.31 Root locus for redesigned regulator using steam flow feedback (three-element control) at 100% load.

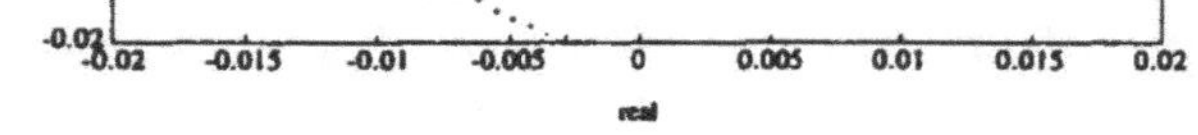

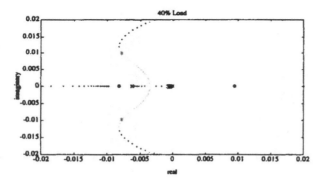

Figure 11.32 Root locus using three-element control at 40% load. Stability is somewhat degraded.

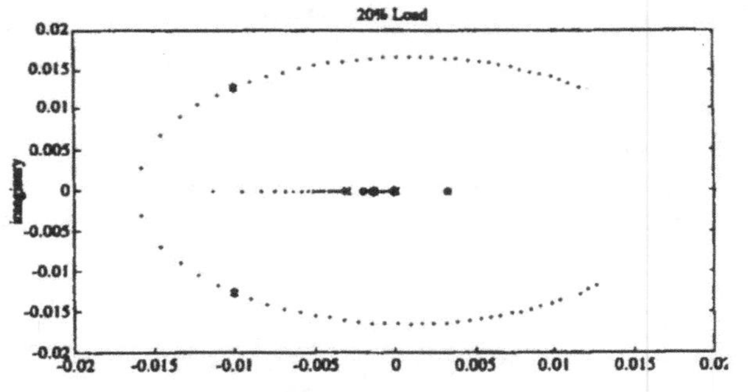

Figure 11.33 A three element-controller retuned for low load real operation provides reasonable stability at 20% load which is only marginally inferior to a single control shown above in Figure 11.29.

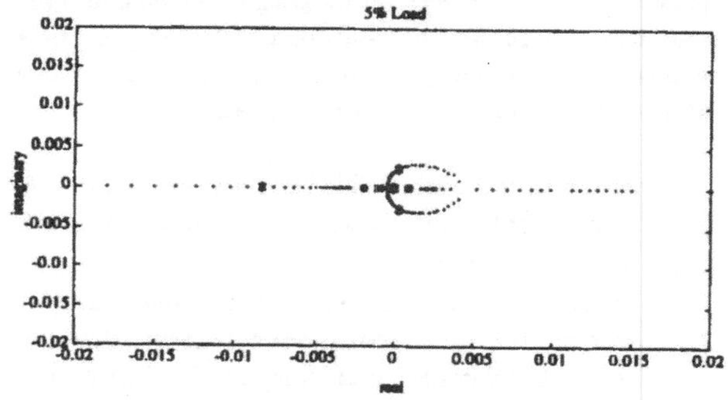

Figure 11.34 At 5% load the three-element controller results in an unstable system. The essential point is that at some load level the three-element controller will lead to an unstable response unless the controller is again retuned. Even retuning for stability, however, is not likely to result in acceptable performance.

Disturbance Response It is useful to assess the relative importance of each of the closed-loop dynamical modes in terms of their appearance in the drum level response to disturbances, either heat rate disturbance or throttle valve disturbance, or to drum level set point change. To do this, we look at the residues associated with each closed-loop pole when the system is subjected to a disturbance input or commanded set point change. Tables 11.3 and 11.4 summarize the closed loop eigenvalues at high and low loads, respectively. Table 11.3 corresponds to the high load controller and Table 11.4 to the low load controller. Table 11.3 contains data corresponding to the three-element high load controller whereas Table 11.4 is generated using the low load single-element controller. Thus, the 30% column in the two tables can be compared for the effect of the change in controllers on the closed-loop eigenvalues.

Notice that there are 11 closed-loop eigenvalues or poles. The open-loop modes 6, 7, and 8 correspond to closed-loop poles 8, 9, and 10. Closed loop pole 11 is the slow drum level trim mode introduced by the controller integral action. In the ramp disturbance case, there is a 12th pole at the origin which contributes a constant offset in the steady-state drum level because the controller is only of type 1. The drum level response is a weighted sum of the time responses corresponding to the simple poles, and the coefficients are the transfer function residues at the respective poles. Consequently, the relative significance of each mode in the drum level response can be identified by comparing the residue magnitudes. These computations (see Figures 11.36 and

TABLE 11.3 High Load Closed-Loop Eigenvalues

100%	85%	70%	60%	50%	40%	30%
−6.2157	−6.2375	−6.2468	−6.2386	−6.2401	−6.2794	−6.4753
−1.0655	−1.0013	−0.9477	−0.9165	−0.8764	−0.8126	−0.7230
−0.2552*i*	−0.2387*i*	−0.2216*i*	−0.2093*i*	−0.1909*i*	−0.1539*i*	−0.0577*i*
−1.0655	−1.0013	−0.9477	−0.9165	−0.8764	−0.8126	−0.7230
+0.2552*i*	+0.2387*i*	+0.2216*i*	+0.2093*i*	+0.1909*i*	+0.1539*i*	+0.0577*i*
−0.4714	−0.4362	−0.4050	−0.3851	−0.3601	−0.3221	−0.2700
−0.5156*i*	−0.4874*i*	−0.4609*i*	−0.4436*i*	−0.4190*i*	−0.3764*i*	−0.3134*i*
−0.4714	−0.4362	−0.4050	−0.3851	−0.3601	−0.3221	−0.2700
+0.5156*i*	+0.4874*i*	+0.4609*i*	+0.4436*i*	+0.4190*i*	+0.3764*i*	+0.3134*i*
−0.4357	−0.3937	+0.3844	−0.3773	−0.3680	−0.3536	−0.3309
	−0.0159*i*	−0.0370*i*	−0.0445*i*	−0.0503*i*	−0.0533*i*	−0.0483*i*
−0.3689	−0.3937	+0.3844	−0.3773	−0.3680	−0.3536	−0.3309
	+0.0159*i*	+0.0370*i*	+0.0445*i*	+0.0503*i*	+0.0533*i*	+0.0483*i*
−0.0482	+0.0441	−0.0372	−0.0304	−0.0225	−0.0236	−0.0265
−0.0413*i*	−0.0290*i*	−0.0207*i*	−0.0182*i*	−0.0140*i*		
−0.0482	−0.0441	−0.0372	−0.0304	−0.0225	−0.0078	−0.0006
+0.0413*i*	+0.0290*i*	+0.0207*i*	+0.0182*i*	+0.0140*i*	−0.0099*i*	−0.0091*i*
−0.0046	−0.0050	−0.0056	−0.0063	−0.0083	−0.0078	−0.0006
					+0.0099*i*	+0.0091*i*
−0.0005	−0.0005	−0.0005	−0.0005	−0.0005		−0.0005

TABLE 11.4 Low Load Closed-Loop Eigenvalues

30%	20%	10%	5%
−6.4923	−7.2444	−8.8627	−12.3819
−0.7237 − 0.0570*i*	−0.7765	−0.7189	−0.6922
−0.7237 + 0.0570*i*	−0.5246	−0.5187	−0.5185
−0.2710 − 0.3133*i*	−0.2269 − 0.2594*i*	−0.1247 − 0.2111*i*	−0.1734 − 0.0660*i*
−0.2710 + 0.3133*i*	−0.2269 + 0.2594*i*	−0.1247 + 0.2111*i*	−0.1734 + 0.0660*i*
−0.3300 − 0.0458*i*	−0.2970 − 0.0372*i*	−0.2612	−0.0625 − 0.1397*i*
−0.3300 + 0.0458*i*	−0.2970 + 0.0372*i*	−0.2058	−0.0625 + 0.1397*i*
−0.0130 − 0.0224*i*	−0.0088 − 0.0182*i*	−0.0045 − 0.0104*i*	−0.0009 − 0.0036*i*
−0.0130 + 0.0224*i*	−0.0088 + 0.0182*i*	−0.0045 + 0.0104*i*	−0.0009 + 0.0036*i*
−0.0008	−0.0009	−0.0012	−0.0039
−0.0001	−0.0001	−0.0001	−0.0001

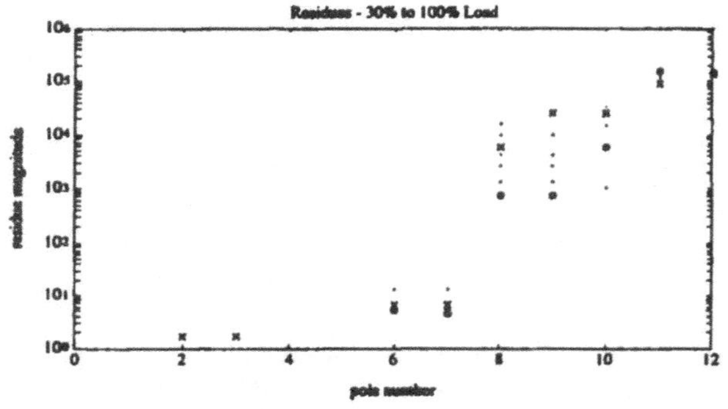

Figure 11.35 Drum level residues for the closed-loop system responding to a ramp firing rate disturbance at load levels from 30% (x) to 100% (o). Recall that the 12th pole is at the origin and is contributed by the ramp disturbance. The nonzero residue merely confirms that the type 1 controller leads to a constant drum level offset when subjected to a ramp input. Otherwise, the dominant modes in the response are the three open-loop modes 6, 7, and 8 and the trim mode. The trim mode clearly dominates at 100% load, but, at lower loads, poles 9 and 10 are quite important also. These poles correspond to open-loop modes 7 and 8 which, through the feedback loop, have been coupled into a lightly damped oscillation.

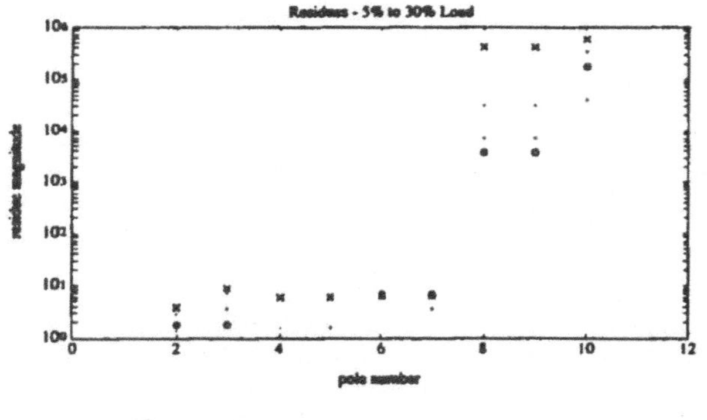

Figure 11.36 The low load controller is subjected to the same ramp firing rate disturbance. The residues are shown for the load range 5% (x) to 30% (o). Although there is a 12th pole at the origin, its residue is very small because the controller zeros are much closer to the origin and, as load is dropped, an open-loop plant zero moves toward the origin from the right. The trim mode residue is also extremely small for the same reason. The remaining dominant modes are the same as the high load case (open-loop modes 6, 7, and 8). Notice that the residue value values are considerably larger, however.

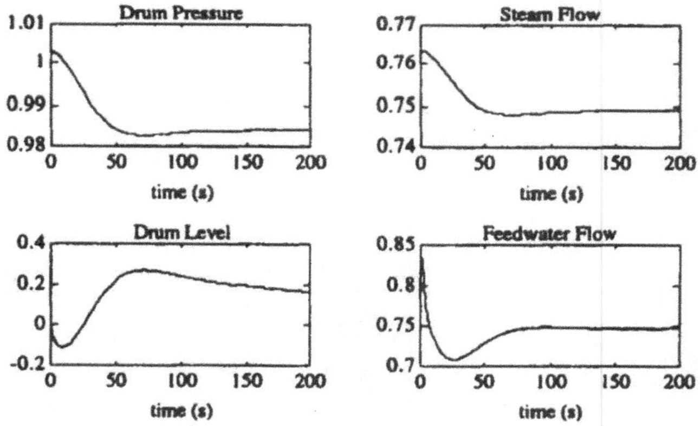

Figure 11.37 Three-element controller response to a 5% step command changing load from 80% to 75%.

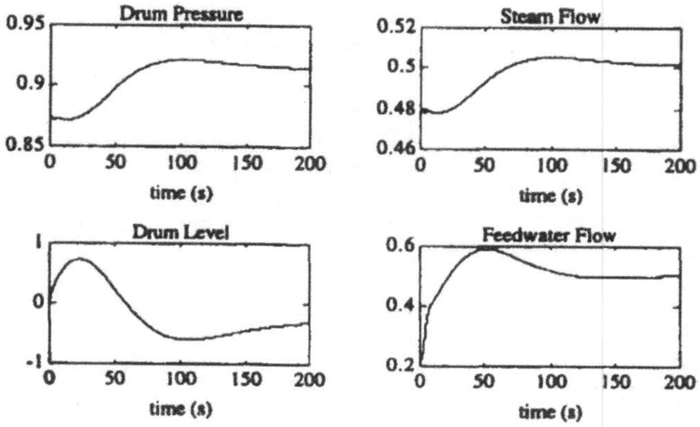

Figure 11.38 Three-element controller response to a 10% step command changing load from 40% to 50%.

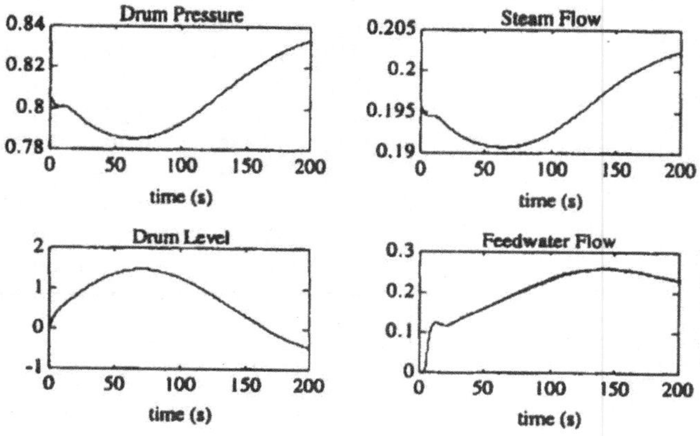

Figure 11.39 Single-element controller response to a 5% step command changing load from 15% to 20%. Below 30% load, a single-element controller is used. For this particular plant, the 15%–20% range is particularly troublesome.

11.37) confirm that the dominant dynamics are the four slowest modes identified above in Table 11.3: circulation flow-drum level rebalance (mode 6), energy coupled drum level (mode 7), drum level mass rebalance (mode 8), and the drum level trim mode. Control system design must focus on the regulation of these modes.

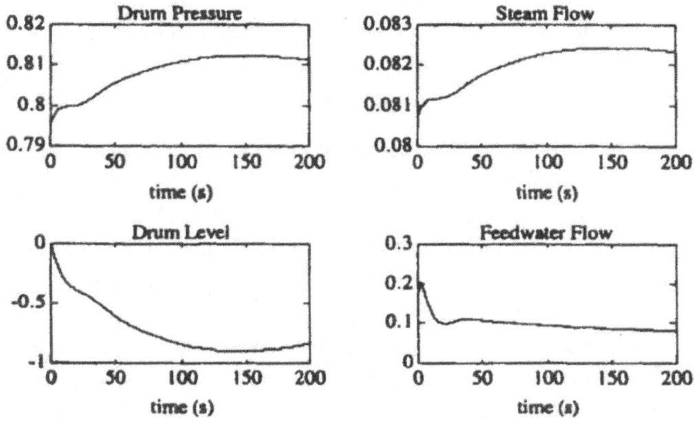

Figure 11.40 Single-element controller response to a 2% step command changing load from 10% to 8%. At this very low load, larger step changes are not possible.

The effect of steam flow feedforward can be seen in the 30% load results in Figures 11.35 and 11.36, which correspond to the three-element and single-element controllers, respectively. Consider pole number 10 which corresponds to the mass balance mode (open-loop mode 8). The 30% residue of this pole is the '*x*' in Figure 11.35 and the '*o*' in Figure 11.36. There is an order of magnitude improvement with the three-element controller in this important mode.

Simulation Responses

Another perspective is obtained from the closed-loop time responses. The following figures correspond to a feedback regulator structure that employs a coventional three element feedwater controller above 30% load and a single element controller below 30% load. These controllers cannot tolerate load level step changes of more than about 5%–10%, and somewhat smaller step changes at very low loads. The following figures present a sequence of simulation results of step load change commands spanning the range from high load to very low load. The three-element controller used at higher loads responds much better to load disturbances than the single element controller because the residues associated with the dominant closed-loop modes are much smaller. However, the destabilizing effect of steam flow feedback (disturbance feedforward) precludes its use at low loads.

References

[1] *Steam, its Generation and Use*, 39th Ed., *Babcock & Wilcox*, New York.

[2] Bolis, V., Maffezzoni, C., and Ferrarini, L., *Synthesis of the overall boiler-turbine control system by single loop auto-tuning technique*, Proc. 12th World IFAC Congress, 3, 409–414, 1993 (also to appear in *Contr. Eng. Practice*, 1995).

[3] Maffezzoni, C., *Issues in modeling and simulation of power plants*. Proc. IFAC Symposium on Control of Power Plants and Power Systems, 1, 19–27, 1992.

[4] Groppelli, P., Maini, M., Pedrini, G., and Radice, A., *On plant testing of control systems by a real-time simulator*, 2nd Annual IDSA/EPRI Joint Control and Instrumentation Conf., Kansas City, 1992.

[5] Colombo, F., De Marco, A., Ferrari, E., and Magnani, G., *Considerations upon the representation of turbine and boiler in the dynamic response of fossil-fired electrical units*. Proc. Cigrè-IFAC Symposium on Control Applications for Power System Security, Florence (Italy), 1983.

[6] Fuetterer, B., Lausterer, G.K., and Leibbrandt, S.R., *Improved unit dynamic response using condensate stoppage*, Proc. IFAC Symposium on Control of Power Plants and Power Systems, 1, 129–146, 1992.

[7] Kalnitsky, K.C. and Kwatny, H.G., First Principle Model for Steam Turbine Control Analysis. *J. Dyn. Syst., Meas. Contr.*, 103, 61–68, 1981.

[8] Kwatny, H.G. and Bauerle, J., *Simulation Analysis of the Stability of Coal Fired Furnaces at Low Load*, Proc. 2nd IFAC Workshop on Power Plant Dynamics and Control, Philadelphia, 1986.

[9] Kwatny, H.G., McDonald, J.P., and Spare, J.H., *Nonlinear Model for Reheat Boiler Turbine-Generator Systems, Part 1-General Description and Evaluation, Part 2-Development*. 12th Joint Automat. Contr. Conf., 1971.

[10] Kwatny, H.G. and Fink, L.H., Acoustics, Stability and Compensation in Boiling Water Reactor Pressure Control Systems. *IEEE Trans. Automat. Contr.* AC-20, 727–739, 1975.

[11] Kwatny, H.G. and Berg, J., *Drum Level Regulation at All Loads: A Study of System Dynamics and Conventional Control Structures*. Proc. 12th IFAC World Congress, Sydney, 1993.

[12] Schulz, R., The Drum Water Level in The Multivariable Control of a Steam Generator, *IEEE Trans. Ind. Electron. Contr. Instrum. IECI-20*, 164–169, 1973.

[13] Nahavandi, A.N. and Batenburg, A., Steam Generator Water Level Control, *Trans. ASME J. Basic Eng.*, 343–354, 1966.

[14] Chien, K.L., Ergin, E.I., Ling, C., and Lee, A., Dynamic Analysis of a Boiler. *Trans. ASME*, 80, 1809–1819, 1958.

[15] Profos, P., *Die Regulung von Dampenflagen*, (The Control of Thermal Power Plants), Springer, Berlin, 1962, (in German).

[16] Dolezal, R. and Varcop, L., *Process Dynamics*, Elsevier, Amsterdam, 1970.

[17] Klefenz, G., *La regulation dans les central thermiques* (The control of thermal power plants). Edition Eyrolles, Paris, 1974 (in French).

[18] Friedly, J.C., *Dynamic Behavior of Processes*, Prentice Hall, Englewood Cliffs, NJ, 1972.

[19] Kecman, V., *State-Space Models of Lumped and Distributed Systems*, Springer, New York, 1988.

[20] Dukelow, S.G., *The Control of Boilers*, ISA Press, 1986.

[21] Maffezzoni, C., *Dinamica dei generatori di vapore* (The dynamics of steam generators), Masson Ed., 1989 (also printed by Hartmann and Braun, 1988), (in Italian).

[22] Maffezzoni, C., *Controllo dei generatori di vapore* (The control of steam generators), Masson Ed., 1990 (also printed by Hartmann and Braun, 1990), (in Italian).

[23] Quazza, G. and Ferrari, E ., Role of Power Station Control in Overall System Operation. In *Real-Time Control of Electric Power Systems*, Handschin, E., Ed., Elsevier, Amsterdam, 1972

[24] Nicholson, H., *Dynamic Optimization of a Boiler*, Proc. IEEE, 111, 1478–1499, 1964.

[25] Anderson, J.H., *Dynamic Control of a Power Boiler*, Proc. IEEE, 116, 1257–1268, 1969.

[26] McDonald, J.P. and Kwatny, H.G., Design and Analysis of Boiler-Turbine-Generator Controls Using Optimal Linear Regulator Theory. *IEEE Trans. Automat. Contr.* AC-18, 202–209, 1973.

[27] Maffezzoni, C., *Concepts, practice and trends in fossil fired power plant control.* Proc. IFAC Symposium on Control of Power Plants and Power Systems, Pergamon Press, pp. 1-9, 1986.

[28] Calvaer, A., Ed., *Power systems. Modeling and control applications.* Proc. IFAC Symposium, Pergamon, 1988.

[29] Welfonder, E., Lausterer, G.K., and Weber H., Eds., *Control of power plants and power systems*, Proc. IFAC Symposium, Pergamon, 1992.

[30] *10th World Cong. Automat. Contr.*, IFAC Proceedings, Pergamon, 1987.

[31] *11th World Cong. Automat. Contr.*, IFAC Proceedings, Pergamon, 1990.

[32] *12th World Cong. Automat. Contr.*, IFAC Proceedings, Pergamon, 1993.

[33] Uchida, M. and Nakamura, H., *Optimal control of thermal power plants in Kyushu electric power company*, Proc. 1st IFAC Workshop on Modeling and Control of Electric Power Plants, Pergamon, 1984.

[34] Mann, J., *Temperature control using state feedback in a fossil-fired power plant*, Proc. IFAC Symposium on Control of Power Plants and Power Systems, Pergamon, 1992.

Further Reading

Power plant automatic control systems have evolved over many decades as have the generating plants themselves and the environment within which they operate. In the early years the main issues revolved around the control devices themselves, along with actuators and sensors. Indeed the physical design and construction of the early pneumatic, mechanical, and analog electrical control components were quite remarkable and intricate. In contrast, the overall control architectures themselves were very simple and the setting of controller parameter values was always done in the field with carefully controlled tests carried out by engineers possessing keen knowledge of process analysis and feedback concepts combined with extensive operating experience.

In this environment, most of the work done to improve the control systems of electric power plants was performed by the major suppliers of control equipment in close cooperation with the utility companies that operated them. Much of the expertise was recorded in numerous obscure corporate documents that generally did not find their way into the public domain. So, assembling a complete technical and historical record is very difficult and the present work does not attempt that task. By the late 1950s, the maturation of the field of automatic control, both theory and devices, and the demands for improved generating plant performance through automation coalesced. A serious effort at model-based control system analysis and design for power generating plants began and with it has emerged a sizable technical literature. Our goal is to provide a connection to it. While there are undoubtedly antecedents, much of the literature on boiler modeling, related to the special needs of control system design, traces back to the 1958 paper of Chien et al. [14] which deserves special mention.

In the open literature are some books, with different points of view, that deal with boiler dynamics and control providing good balance between theory and practice. Again, for its historical significance, we note the book by Profos [15]. Another important one (1970) is Dolezal and Varcop [16], where attention concentrates on boiler dynamics. The analysis is strongly based on mathematical modeling, with much analytical detail. It is very useful in explaining the origin of basic dynamical behavior and identifying the process parameters on which they depend. The book is a milestone in boiler dynamics, though, perhaps, not so "friendly" to read. A book (1974) by Klefenz [17], which is the French translation of the original German version, has a limited circulation. Most of the concepts illustrated there are still valid and applied in practice. An excellent general text for methods and examples of process dynamics, with many references, is the book by Friedly [18]. Another is by Kecman [19].

The book by Dukelow [20], is an excellent condensation of engineering skill and field experience. Without equations, boiler control concepts and schematics are plainly discussed and explained. It represents a good first reference for anyone approaching the subject without practical experience. Several books (1989-1990), published in Italian by one of the authors, propose a systematic approach to (simplified) boiler dynamics [21] based on first principles as a prerequisite to understanding the fundamentals of boiler control [22], and the most popular control concepts and their underlying motivations.

The books above cited all have a "classical" approach to power plant control, following the approach typical of process control, i.e., structuring the control system in a number of hierarchically arranged loops and trying to minimize interactions among them. Nonclassical control strategies have been studied and proposed in the last quarter century with the application of "modern" control techniques, such as optimal control, adaptive control, robust control, and variable structure control. Much of the motivation for improved dynamic performance can be attributed to the plant's importance to overall power system operation [23]. Early investigations that applied modern control methods to power plants include [24]–[26]. The interested reader is referred to the survey paper [27] summarizing that research effort up to 1986 and to the Proceedings of subsequent dedicated events (primarily [28], [29] and the target area sessions within the world IFAC Congresses [30], [31], and [32]). There are many outstanding contributions among the many interesting papers in the literature on nonclassical control concepts applied to power plants; we want to mention a couple of cases where the new concepts have been introduced in continuously operating in commercial units: the steam temperature control of once-through boilers based on optimal control theory (see [33] and related references) and an observer-based state feedback scheme applied to superheated steam temperature control [34].

Concluding this short survey of contributed work, testing and installing improved control systems in commercial power plants is very expensive, so that it is often impossible for researchers to test their findings; yet, it is very important to validate new concepts on a realistic test-bench (possibly the plant). Complex, large simulators are now available that allow credible testing [4].

11.2 Control of Power Transmission

John J. Paserba, Juan J. Sanchez-Gasca, and Einar V. Larsen, GE Power Systems Engineering, Schenectady, NY

11.2.1 Introduction

The power transmission network serves to deliver electrical energy from power plant generators to consumer loads. While the behavior of the power system is generally predictable by simulation, its nonlinear character and vast size lead to challenging demands on planning and operating engineers. The experience and intuition of these engineers is generally more important to successful operation of the system than elegant control designs.

In most developed countries, reliability of electrical supply is essential to the national economy, therefore, reliability plays a large role in all aspects of designing and operating the transmission grid. Two key aspects of reliability are adequacy and security [1]. Adequacy means the ability of the power system to meet the power transfer needs within component ratings and voltage limits. Security means being able to cope with sudden major system changes (contingencies) without uncontrolled loss of load. In this section, the focus will be mostly on the security aspect, emphasizing power system stability.

There are two main categories of power system performance problems. One category involving steady-state issues, such as thermal and voltage operating limits, is not discussed in this section. The second broad category involves power system dynamics, dealing with transient, oscillatory, and voltage stability. The following three subsections provide an overview of these stability issues.

Transient Stability

Transient stability has traditionally been the performance issue most limiting on power systems, and thus, the phenomenon is well understood by power system engineers. Transient stability refers to the ability of all machines in the system to maintain synchronism following a disturbance such as a transmission system fault, a generator or transmission line trip, or sudden loss of load. Unlike the other stability issues described here, transient stability is a relatively fast phenomenon, with loss of synchronism typically occurring within two seconds of a major disturbance.

Traditional stability analysis software programs are designed to study and identify transient stability problems. These software programs provide the ability to model electrical and mechanical dynamics of generators, excitation systems, turbine/governors, large motors, and other equipment such as static Var compensators (SVC) and high voltage direct current (HVDC) systems. Scenarios studied typically involve major events such as faults applied to a transmission line with line trips. A transient instability is identified by examining the angles (or speeds) of all machines in the system. If the angle (or speed) of any machine does not return toward an acceptable steady-state value within one to three seconds following a power system disturbance, the machine is said to have lost synchronism with the system. Although not always modeled for transient stability analysis, this loss of synchronism will typically result in the generator being tripped off-line by a protective relay.

Another way of looking at the transient stability issue is to consider voltage swings in the transmission network. For example, many utilities require that the voltage swing during a transient event remains above a pre-determined value (e.g., 80% of nominal) because relays may trip generators or loads if the voltage is too low for too long. If the voltage swing does not meet this criterion, the event causing the disturbance, while not unstable in the classical definition of transient stability, is still considered transiently unacceptable.

In many systems, discrete, open-loop action is applied to insure stable operation in the power system following a major disturbance. These so-called "Special Protection Systems" or "Remedial-Action Schemes" can be as simple as undervoltage load shedding on selected loads or as complex as wide area generator tripping and intentional network separation [2]. Such schemes have proven highly useful, and will likely always be needed. Designing such schemes and keeping them updated as systems evolve is an ongoing challenge. To date automation is not applied to developing such schemes; experience and large-scale simulation tools are relied upon.

Voltage Stability

With the evolution of modern power systems, voltage stability has emerged as the limiting consideration in many systems. Voltage stability refers to the power systems' ability to maintain acceptable voltages under normal operating conditions and after a major system disturbance. A system is said to be voltage unstable (or in voltage collapse) when an increase in load demand, a major disturbance, or a change in the system causes a progressive and uncontrollable decrease in voltage. Means of protecting and reinforcing power systems against voltage collapse depend mostly on the supply and control of reactive power in the system. The phenomenon of voltage instability (collapse) is dynamic, yet frequently evolves very slowly, from the perspective of a transient stability. A voltage collapse initiated by loss of infeed into a load center, for example, can progress over a one to five minute period. This type of phenomenon presents two challenges to the power system engineer. The first challenge is in identifying and simulating the voltage collapse. The second challenge is selecting the system reinforcements to correct the problem in the most economic manner. A recently emerging class of computer simulation software provides utility engineers with powerful new tools for analyzing long-term dynamic phenomena. The ability to perform long-term dynamic simulations either with detailed dynamic modeling or simplified quasi-steady-state modeling permits assessment of critical power system problems more accurately than with conventional power flow and stability programs. The voltage stability issue is discussed in great detail in [3].

Oscillatory Stability

Power systems contain electromechanical modes of oscillation due to synchronous machine rotor masses swinging relative to one another. A power system having several such machines will act like a set of masses (rotating inertia of machines) interconnected by a network of springs (ac transmission), and will exhibit multiple modes of oscillation. These "power-swing modes" usually occur in the frequency range of 0.1 to 2 Hz. Particularly troublesome are the so-called inter-area oscillations which usually occur in the frequency range of 0.1 to 1 Hz. The inter-area modes are usually associated with groups of machines swinging relative to other groups across a relatively weak transmission path. The higher frequency modes (1 to 2 Hz) usually involve one or two machines swinging against the rest of the power system or electrically close machines swinging against each other [4].

Because there is great incentive to minimize transmission losses in the power system, power swing modes have very little inherent damping. Damping which is present is usually due to steam or water flow against the turbine blades of generating units and to special conductors placed in the rotor surface of the generators, known as amortisseur (damper) windings. High power flows in the transmission system create conditions where swing modes can experience destabilization. The effect of these oscillations on the power system may be quite disruptive if they become too large or are underdamped. The actual power (watts) swings may themselves not be overly troublesome, but they can result in voltage oscillations in the power system adversely affecting the system's performance. Limitations may be imposed on the power transfer between areas to reduce the possibility of sustained or growing oscillations, or special controls may be added to damp these oscillations.

Once lightly damped, sustained, or growing oscillations have been observed or predicted by time simulation for a specific system condition, the most common and effective way to investigate them in detail is with linear or small-signal, frequency-domain analysis. With the linear approach to studying power systems, the engineer can view the stability problem from a different perspective than time-domain simulations. This approach enhances the overall understanding of the system's performance. The basic elements and tools of linear analysis commonly applied to power systems include eigenvalues (poles, modes, or characteristic roots), root locus, transfer functions, Bode plots, Nyquist plots, eigenvectors (mode shapes), participation factors, Prony processing, and others. While power systems are nonlinear in nature, analysis based on small-signal linearization around a specific operating point can be important in solving damping problems. When combined with practical engineering experience, time-domain simulations, and field measurements, linear analysis can assure reasonable stability margins under many operating conditions.

Considering that the majority of closed-loop supplemental controls in the power system are for damping oscillations, the remainder of this section on "Control of Power Transmission" will focus on the closed-loop controls which affect power system oscillatory stability.

11.2.2 Impact of Generator Excitation Control on the Transmission System

The main control function of the excitation system is to regulate the generator terminal voltage. This is accomplished by adjusting the field voltage in response to terminal voltage variations. A typical excitation system includes a voltage regulator, an exciter, protective circuits, limiters, and measurement transducers. The voltage regulator processes the voltage deviations from a desired set point and adjusts the required input signals to the exciter, which provides the dc voltage and current to the field windings, to take corrective action. Figure 11.41 represents a single generator with its excitation system, supplying power through a transmission line to an "infinite bus" (a power grid having significantly more capacity than the generator under study).

Early excitation systems were manually controlled, i.e., an operator manually adjusted the excitation system current with a rheostat to obtain a desired voltage. Research and development in the 1930s and 1940s showed that applying a continuously acting proportional control in the voltage regulator significantly increased the generator steady-state stability limits. Beginning in the late 1950s and early 1960s most of the new generating units were equipped with continuously acting voltage regulators. As these units became a larger percentage of the generating capacity, it became apparent that the voltage regulator action could have a detrimental impact on the overall stability of the power system. Low frequency oscillations often persisted for long periods of time and in some cases presented limitations to the system's power transfer capability. In the 1960s, power oscillations were observed following the interconnection of the Northwest and Southwest U.S. power grids. Power oscillations were also detected between Saskatchewan, Manitoba, and Ontario systems in the Canadian power system. It was found that reducing the voltage regulator gain of excitation systems improved the system stability.

Great effort has been directed toward understanding the effect of the excitation system on the dynamic performance of power systems. The single generator infinite bus system shown in Figure 11.41 provides a good vehicle to develop concepts related to the stability of a synchronous generator equipped with an excitation system. Figure 11.42 is the block diagram associated with the linear representation of the system shown in Figure 11.41. This diagram has become a standard tool to gain insight into the stability characteristics of the system [5]. The stability of the linearized system can be studied by considering the torques that comprise the accelerating torque T_a: the synchronizing torque T_s, which is in phase with the rotor angle deviations, the damping torque T_d which is in phase with the rotor speed deviations, and the torque T_{ex} due to the excitation system acting through the field winding of the machine. The mechanical torque, T_m, is assumed constant. In Figure 11.42 the phase relations are indicated for conditions of positive synchronizing torque which tends to restore the rotor to steady-state conditions by accelerating or decelerating the rotor inertia. The damping torque is also positive and tends to damp rotor oscillations. T_{ex} is shown as contributing positive synchronizing torque and negative damping torque. The primary synchronizing effect is exhibited in the term K_1 of Figure 11.42, which represents the change in torque mostly in phase with angle. The damping parameter D represents turbine-generator friction, windage, and the impact of steam or water flow through the turbine. Typically, the torque-angle loop is stable due to the inherent damping and restoring forces in the power system. However, unstable oscillations can result from the introduction of negative damping torques added through the excitation system loop [5], [6].

Phase lags are introduced by the voltage regulator and the generator field dynamics so that the resulting torque is out of phase with both rotor angle and speed deviations. The phase characteristics of this path depend on the generator and voltage regulator characteristics, and the parameters K_4, K_5, and K_6. K_5 plays a dominant role in the phase relationships with respect to the damping torque. For weak systems and heavy loads, K_5 is typically negative. Coupled with the phase lags due to the voltage regulator and generator field winding, the path via K_5 can produce a torque T_{ex} which contributes positive synchronizing torque and a negative damping torque. This negative damping component can cancel the small inherent positive damping due to the damping torque and lead to an unstable system. A detailed description of these parameters is given in [5].

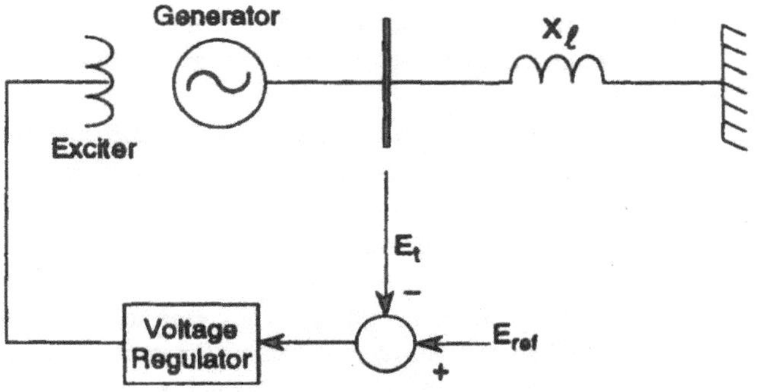

Figure 11.41 Single generator and excitation system.

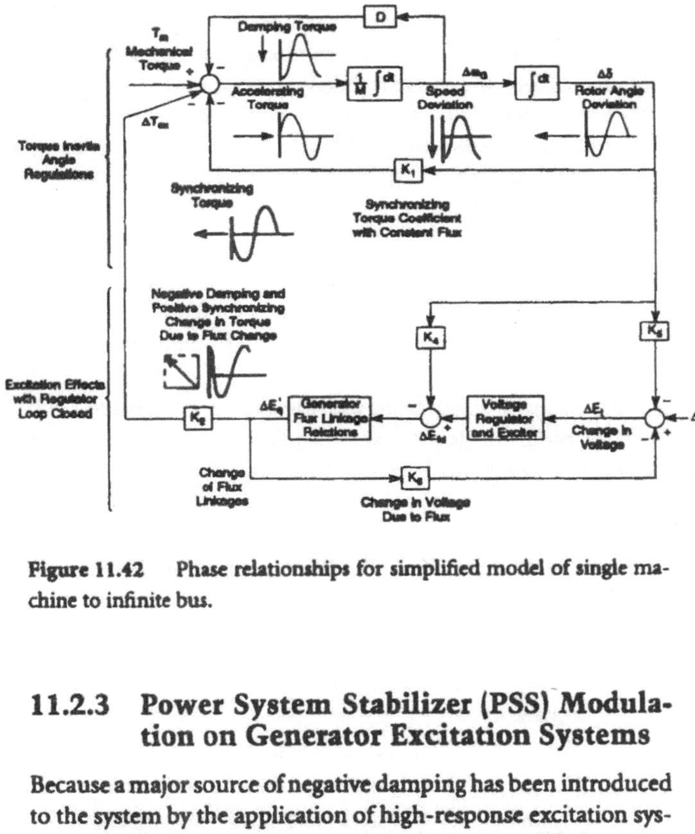

Figure 11.42 Phase relationships for simplified model of single machine to infinite bus.

11.2.3 Power System Stabilizer (PSS) Modulation on Generator Excitation Systems

Because a major source of negative damping has been introduced to the system by the application of high-response excitation systems, an effective way to increase damping is to modify the action of these excitation systems. A Power System Stabilizer (PSS) is often used to provide a supplementary signal to the excitation system. The basic function of the PSS is to extend stability limits by modulating generator excitation to provide positive damping torque to power swing modes.

To provide damping, a PSS must produce a component of electrical torque on the rotor in phase with speed deviations. The implementation details differ, depending upon the stabilizer input signal employed. PSS input signals which have been used include generator speed, frequency, and power [6]. However, for any input signal, the transfer function of the PSS must compensate for the gain and phase characteristics of the excitation system, the generator, and the power system. These collectively determine the transfer function from the stabilizer output to the component of electrical torque which can be modulated via excitation control.

Figure 11.43 is an extension of Figure 11.42 to include a speed-input PSS (PSS(s)). The transfer function $G(s)$ represents the characteristics of the generator, the excitation system, and the power system. A PSS utilizing shaft speed as an input must compensate for the lags in $G(s)$ to produce a component of torque in phase with speed changes so as to increase the damping of the rotor oscillations. An ideal PSS characteristic would therefore be inversely proportional to $G(s)$. Such a stabilizer would be impractical because perfect compensation for the lags of $G(s)$ requires pure differentiation with its associated high gain at high frequencies. A practical speed PSS must utilize lead/lag stages to compensate for phase lags in $G(s)$ over the frequency range of interest. The gain must be attenuated at high frequencies to limit the impact of noise and minimize interaction with torsional modes of turbine-generator shaft vibration. Low-pass and possibly band reject filters are required. A washout stage is included to prevent steady-state voltage offsets as system frequency changes. A typical transfer function of a practical PSS which meets the above criteria is given by

$$PSS(s) = K_s \frac{T_W s(1+sT_1)(1+sT_3)}{(1+T_W s)(1+sT_2)(1+sT_4)} F(s)$$

where $F(s)$ represents a filter designed to eliminate torsional frequencies.

A PSS must be tuned to provide the desired system performance under the power system condition which requires stabilization, typically weak systems with heavy power transfer, while at the same time being robust in that undesirable interactions are avoided for all system conditions. To develop such a design, it is important to understand the path $G(s)$ through which the PSS operates:

1. The phase characteristics of $G(s)$ are nearly identical to the phase characteristics of the closed-loop voltage regulator.
2. The gain of $G(s)$ increases with generator load.
3. The gain of $G(s)$ increases as the ac system strength increases. This effect is amplified with high-gain voltage regulators.
4. The phase lag of $G(s)$ increases as the ac system becomes stronger. This has the greatest influence with high-gain exciters, because the voltage regulator crossover frequency approaches that of the frequency of the swing modes.

$G(s)$ has the highest gain and greatest phase lag under conditions of full load on the generator and the strongest transmission system (i.e., higher short circuit strength). These conditions therefore represent the limiting case for achievable gain with a speed-input PSS and constitute the base condition for stabilizer design. Because the gain of the plant decreases as the system becomes weaker, the damping contribution for the strong system should be maximized to insure best performance with a weakened system.

In general, the highest compensation center frequency which provides adequate local mode damping will yield the greatest contribution to intertie modes of oscillation. The following are guidelines for setting the lead/lag stages to achieve adequate local mode damping with maximum contribution to intertie modes of oscillation. Two basic criteria in terms of phase compensation are

1. It is most important to maximize the bandwidth within which the phase lag remains less than 90°. This is true even though less than perfect phase compensation results at the local model frequency.
2. The phase lag at the local mode frequency should be less than about 45°. This can be improved somewhat by decreasing the washout time constant, but too low a washout time constant will add phase lead and an associated desynchronizing effect to the intertie oscillations. In general, it is best to keep the washout time constant greater than one second.

The gain and frequency at which an instability occurs also provide an indication of appropriate lead/lag settings. The relationship of these parameters to performance is useful in root locus analysis and in field testing. The following observations hold:

3. The frequency at which an instability occurs is highest for the best lead/lag settings. This is related to maximizing the bandwidth within which the phase lag remains less than 90°.
4. The optimum gain for a particular lead/lag setting is consistently about one-third of the instability gain.

11.2.4 Practical Issues for Supplemental Damping Controls Applied to Power Transmission Equipment

In many power systems, equipment is installed in the transmission network to improve various performance issues such as transient, oscillatory, or voltage stability. Often this equipment is based on power electronics which generally means that the device can be rapidly and continuously controlled. Examples of such equipment include high voltage dc systems (HVDC), static Var compensators (SVC), and thyristor-controlled series compensation (TCSC). To improve damping in a power system, a supplemental damping controller can be applied to the primary regulator of a device. The supplemental control action should

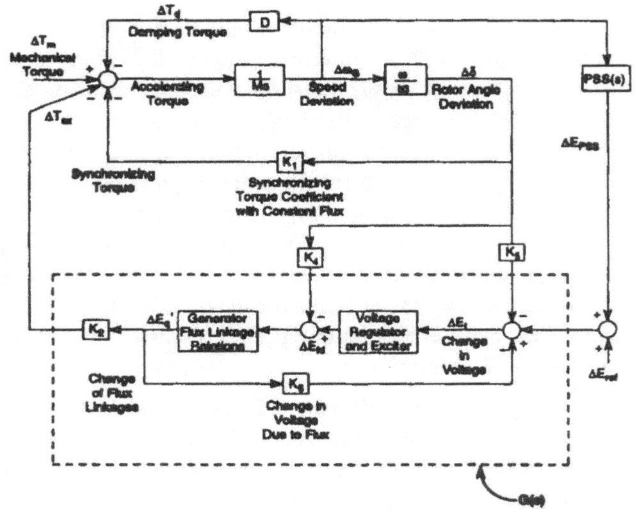

Figure 11.43 Simplified model of single machine to infinite bus.

modulate the output of a device, in turn affecting power transfer and adding damping to the power system swing modes. This section provides an overview of some practical issues affecting the ability of damping controls to improve power system dynamic performance [7]. Section 11.2.5 on "Examples" illustrates the application of these concepts [8].

Siting

Siting plays an important role in the ability of a device to stabilize a swing mode. Many controllable power system devices are sited based on issues unrelated to stabilizing the network (e.g., HVDC transmission and generators). In other situations (e.g., SVC or TCSC), the equipment is installed primarily to help support the transmission system, and siting will be heavily influenced by its stabilizing potential. Device cost represents an important driving force in selecting a location. In general, there will be one location which makes optimum use of the controllability of a device. If the device is located at a different location, a larger sized device would be needed to achieve a desired stabilization objective. In some cases, overall costs may be minimized with non-optimum locations of individual devices, because other considerations must also be taken into account, such as land price and availability, environmental regulations, etc.

The inherent ability of a device to achieve a desired stabilization objective in a robust manner, while minimizing the risk of adverse interactions, is another consideration which can influence the siting decision. Most often, these other issues can be overcome by appropriate selection of input signals, signal filtering, and control design discussed later in this section. This is not always possible, however, so these issues should be included in the decision-making process for choosing a site. For many applications, it will be desirable to apply the devices in a distributed manner. This approach helps to maintain a more uniform voltage profile across the network, during both steady-state operation and after transient events. Greater security may also be possible with distributed devices because the overall system can more likely tolerate the loss of one of the devices.

Objectives for Control Design and Operation

Several aspects of control design and operation must be satisfied during both transient and steady-state operation of the power system before and after a major disturbance. These aspects suggest that controls applied to the power system should meet these requirements:

1. Survive the first few swings after a major system disturbance with some degree of safety. The safety factor is usually built into a Reliability Council's criteria (e.g., keeping voltages above some threshold during the swings).
2. Provide some minimum level of damping in the steady-state condition after a major disturbance (post-contingent operation) because, in addition to providing security for contingencies, some applications will require "ambient" damping to prevent spontaneous growth of oscillations in steady-state operation.
3. Minimize the potential for adverse side effects, which can be classified as follows:
   (a) Interactions with high-frequency phenomena on the power system, such as turbine-generator torsional vibrations and resonances in the ac transmission network.
   (b) Local instabilities within the bandwidth of the desired control action.
4. Be robust. This means that the control will meet its objectives for a wide range of operating conditions encountered in power system applications. The control should have minimum sensitivity to system operating conditions and component parameters because power systems operate over a wide range of operating conditions and there is often uncertainty in the simulation models used for evaluating performance. The control should also have minimum communication requirements.
5. Be highly dependable. This means that the control has a high probability of operating as expected when needed to help the power system. This suggests that the control should be testable in the field to ascertain that the device will act as expected in a contingency. The control response should be predictable. The security of system operations depends on knowing, with a reasonable certainty, what the various control elements will do in a contingency.

Closed-Loop Control Design

Closed-loop control is utilized in many power system components. Voltage regulators are commonplace in generator excitation systems, capacitor and reactor banks, tap-changing transformers, and SVCs. Modulation controls to enhance power system stability have been applied extensively to generator exciters and to HVDC and SVC systems. A notable advantage of closed-loop control is that stabilization objectives can often be met with less equipment and impact on the steady-state power flows than with open-loop controls.

Typically, a closed-loop controller is always active. One benefit of such a continuing response to low-level motion in the system is that it is easy to test continuously for proper operation. Another benefit is that, once the system is designed for the worst-case contingency, the chance of a less severe contingency causing a system breakup is lower than if only open-loop controls are applied. Disadvantages of closed-loop control involve mainly the potential for adverse interactions. Another possible drawback is the need for small step sizes, or vernier control in the equipment, which will have some impact on cost. If communication is needed, this could also be a problem. However, experience suggests that adequate performance should be attainable using only locally measurable signals.

One of the most critical steps in control design is to select an appropriate input signal. The other issues are determining the input filtering and control algorithm to assure attainment of the stabilization objectives in a robust manner with minimal risk of adverse side effects. The following paragraphs discuss design approaches for closed-loop stability controls, so that the potential benefits of such control can be realized in the power system.

Input-Signal Selection The choice of local signals as inputs to a stabilizing control function is based on several considerations:

1. The input signal must be sensitive to the swings on the machines and lines. In other words, the swing modes must be "observable" in the input signal selected. This is mandatory for the controller to provide a stabilizing influence.
2. The input signal should have as little sensitivity as possible to other swing modes in the power system. For example, in a transmission line device, the control action will benefit only those modes which involve power swings on that line. Should the input signal respond to local swings within an area at one end of the line, then valuable control range would be wasted in responding to an oscillation over which the damping device has little or no control.
3. The input signal should have little or no sensitivity to its own output in the absence of power swings. Similarly, there should be as little sensitivity as possible to the action of other stabilizing controller outputs. This decoupling minimizes the potential for local instabilities within the controller bandwidth.

These considerations have been applied to a number of modulation control designs, which have proven themselves in many actual applications. The application of PSS controls on generator excitation systems was the first such study. The study concluded that speed or power were best and that frequency of the generator substation voltage was an acceptable choice as well [6]. When applied to SVCs, the magnitude of line current flowing past the SVC was the best choice [9]. For torsional damping controllers on HVDC systems, it was found that the frequency of a synthesized voltage close to the internal voltage of the nearby generator (calculated with locally measured voltages and currents) was best [10]. In the case of a series device in a transmission line, these considerations lead to the conclusion that frequency of a synthesized remote voltage to "find" the center of an area involved in a swing mode is a good choice. Synthesizing voltages at either end of the line allows determining the frequency difference across the line at the device location, which can then be used to make the line behave like a damper across the inherent "spring" nature of the ac line. Synthesizing input signals is discussed further in the section on "Examples."

Input-Signal Filtering To prevent interactions with phenomena outside the desired control bandwidth, low-pass and high-pass filtering is used for the input signal. In certain applications, notch filtering is needed to prevent interactions with specific lightly damped resonances. This has been the case with SVCs interacting with ac network resonances and modulation controls interacting with generator torsional vibrations. On the low-frequency end, the high-pass filter must have enough attenuation to prevent excessive response during slow ramps of power or during the long-term settling following a loss of generation or load. This filtering must be considered while designing the overall control, as it will strongly affect performance and the potential for local instabilities within the control bandwidth. However, finalizing such filtering usually must wait until the design for performance is completed, after which the attenuation needed at specific frequencies can be determined. During the control design work, a reasonable approximation of these filters needs to be included. Experience suggests a high-pass break near 0.05 Hz (three-second washout time constant), and a double low-pass break near 4 Hz (40-msec. time constant) as shown in Figure 11.44, is suitable as a starting point. A control design which adequately stabilizes the power system with these settings for the input filtering will probably be adequate after the input filtering parameters are finalized.

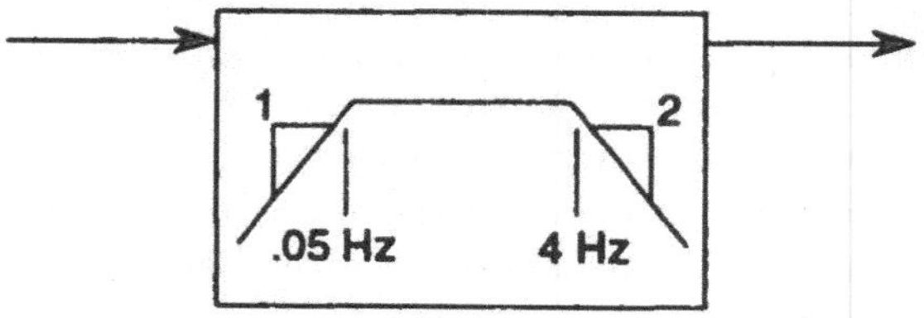

Figure 11.44 Initial input-signal filtering.

Control Algorithm Typically, the control algorithm for damping controllers leads to a transfer function which relates an input signal(s) and a device output. When the input is a speed or frequency type and the output affects the real power, the control algorithm should approach a proportional gain to provide a pure damping influence. This is the starting point for understanding how deviations in the control algorithm affect system performance.

In general, the transfer function of the control and input-signal filtering is most readily discussed in terms of its gain and phase relationship versus frequency. A phase shift of 0° in the transfer function means that the output is simply proportional to the input, and, for discussion, is assumed to represent a pure damping effect on a lightly damped power swing mode. Phase lag in the transfer function (up to 90°), translates to a positive synchronizing effect, tending to increase the frequency of the swing mode when the control loop is closed. The damping effect will decrease with the sine of the phase lag. Beyond 90°, the damping effect will become negative. Conversely, phase lead is a desynchronizing influence and will decrease the frequency of the swing mode when the control loop is closed. Generally, the desynchronizing effect should be avoided, so the preferred transfer function is one which lags between 0° and 45° in the frequency range of the swing modes the control is designed to damp.

After the shape of the transfer function meeting the desired control phase characteristics is designed, the gain of the transfer function is selected to obtain the desired level of damping. The gain should be high enough to insure full utilization of the device for the critical disturbances, but no greater, so that risks of adverse effects are minimized. Typically, the gain selection is done with root locus or Nyquist analysis. This handbook presents many other control design methods that can be utilized to design supplemental controls for the power transmission system.

Performance Evaluation Good simulation tools are essential for applying damping controls to power transmission equipment for system stabilization. The controls must be designed and tested for robustness with these simulation tools. For many operating conditions, the only feasible means of testing the system is by simulation, so that confidence in the power system model is crucial. A typical large-scale power system model may contain up to 15,000 state variables or more. For design purposes, a reduced-order model of the power system is often desirable. If the size of the study system is excessive, the large number of system variations and parametric studies required becomes tedious and prohibitively resource limited for many linear analysis techniques in general use. Good understanding of the system performance can be obtained with a model of only the relevant dynamics for the problem under study. The key conditions which establish that controller performance is adequate and robust can be identified from the reduced-order model, and then tested with the full-scale model. Recent improvements have been made in deriving meaningful reduced-order equivalent power system models [11].

Field testing is also essential for applying supplemental controls to power systems. Testing needs to be performed with the controller open loop, comparing the measured response at its own input and the inputs of other planned controllers against the simulation models. Once these comparisons are acceptable, the system can be tested with the control loop closed. Again, the test results should correlate reasonably with the simulation program. Methods have been developed for performing testing of the overall power system to provide benchmarks for validating the full-system model. Testing can also be done on the simulation program to obtain the reduced-order models for the advanced control design methods [12]. Methods have also been developed to improve the modeling of individual components.

Adverse Side Effects Historically in the power industry, each major advance in improving system performance has had adverse side effects. For example, adding high-speed excitation systems more than 40 years ago caused the destabilization known as the "hunting" mode of the generators. The solution was power system stabilizers, but it took more than 10 years to learn to tune them properly, and there were some unpleasant surprises involving interactions with torsional vibrations on the turbine-generator shaft [6].

HVDC systems also interacted adversely with torsional vibrations (the so-called subsynchronous torsional interaction (SSTI) problem), especially when augmented with supplemental modulation controls to damp power swings. Similar SSTI phenomena exist with SVCs, although to a lesser degree than with HVDC. Detailed study methods have since been established which permit designing systems with confidence that these effects will not disturb normal operation. Protective relaying exists to cover unexpected contingencies [10], [13].

Another potentially adverse side effect is with SVC systems which can interact unfavorably with network resonances. This has caused problems in the initial application of SVCs to transmission systems. Design methods now exist to deal with this phenomenon, and protective functions exist within SVC controls to prevent continuing exacerbation of an unstable condition [9].

As technologies continue to evolve, such as the current industry focus on Flexible AC Transmission Systems (FACTS), new opportunities arise for improving power system performance. FACTS devices introduce capabilities that may be an order of magnitude greater than existing equipment for stability improvement. Therefore there may be much more serious consequences if new devices fail to operate properly. Robust, non-interacting controls for FACTS devices are critically important for stability of the power system [14].

11.2.5 Examples

Large power-electronic devices have been applied to power transmission grids for many years. One of the most common and well-established uses of large-scale power electronics is high voltage dc transmission systems (HVDC). As noted above, current industry activity is focusing on new types of equipment which can affect ac transmission systems, using the acronym "FACTS." This acronym arises from the concept of "Flexible AC Transmission Systems," where power-electronic devices can improve flexibility beyond what is possible with conventional switched compensating equipment. Because these devices are thyristor-controlled, their operating point can be adjusted within a few milliseconds, and there is no mechanical wear associated with rapid cycling of the operating point. These characteristics are the primary features which made power-electronic devices attractive for stabilization functions [8].

In this section, three types of power-electronic systems are described and two examples are presented, emphasizing their ability to add damping to power system modes of oscillation.

High-Voltage DC (HVDC)

High-voltage dc (HVDC) transmission is used to interconnect asynchronous ac grids for power transfer over very long distances. The system includes line-commutated converters at both the sending and receiving ac terminals with a dc link between. In long-distance transmission schemes, this dc link is a transmission line or cable. In back-to-back schemes, both converters are in the same building and the dc link is simply a short length of busbar.

HVDC provides a means to control real power transfer directly between ac networks, independent of ac phase angle. This feature is sometimes utilized to provide a modulating function to the controls for damping power-swing oscillations. Because the converters are line commutated, they draw a reactive power load from the ac networks. This reactive power is related to the real power and both ac and dc voltages, which tends to be a detriment to adding modulation controls. Considerable study has been directed at supplemental controls for HVDC systems. Some schemes have benefited from modulation [15].

Static Var Compensator (SVC)

The SVC provides rapid control of an effective shunt susceptance on the transmission grid and is typically used to regulate voltage at a bus. This action is similar to conventional switched reactors or shunt capacitors which have been used for many years, except that with the SVC, the control can be accomplished quickly and continuously. The voltage-regulating function of an SVC is sometimes augmented by a supplemental control to modulate the voltage set point to add damping to power-swing modes.

Generally, the best location for shunt compensation is the point where the voltage swings are greatest for the dominant swing mode. In a simple remote generation system, this is at the electrical midpoint between the internal voltage of the generator and that of the receiving system. In an interconnected system, the siting decision is complicated by the fact that there are multiple modes of oscillation. Knowing the general characteristics of the dominant swing modes, (based on which generators swing relative to the others), siting decisions may be made using the same basic concept as for the simple remote generation system just described. In addition, properly selecting input signals can allow the damping benefits for a specific mode, while not risking instability of others. These important issues are described in detail in [9].

When an SVC is located near a load area, modulation may be ineffective or even detrimental to power-swing damping. The difficulty lies in the fact that the change in power of the generators in the load area is affected in two ways by a variation in voltage. One is due to the change in power flow across the transmission lines from the sending system, and the other is due to the change in local load. These effects counteract each other, and the leverage for controlling the swing mode is very small. Worse than being small, however, is that the sign of the effect can change with operating point, so that a control designed for a beneficial effect in some operating conditions may have a detrimental effect in others. This inherent characteristic implies that a robust damping control is difficult to achieve on an SVC sited in a load area for swing modes which involve that area. Note that there is usually value in simply stiffening the voltage in a load area, however, so applying an SVC with only voltage control may help stability.

As an example of the effect of SVC on damping power system oscillations, consider the three-area system in Figure 11.45. This system has two modes of oscillation, one near 0.4 Hz and the other near 0.9 Hz. The lower frequency mode involves Area 1 swinging against Areas 2 and 3, while the higher frequency mode involves Area 2 swinging against Area 3. With the SVC located midway along the major intertie (Bus 1 to Bus 5 is the intertie, Bus 4 is midway as shown in Figure 11.45), the leverage of the SVC on damping (known as controllability) the two modes varies with intertie power transfer. The controllability is greatest at higher power transfer and is different for each of the modes, based on the participation of generators in the specific swing mode. The best input signal for the SVC power-swing damping control, based on work presented in [9], is line current magnitude. Figure 11.46 shows the input-signal filtering and controller configuration, and Figure 11.47 shows the corresponding transfer function.

Figure 11.48 shows the power system response following a disturbance in Area 1, while Area 1 is exporting power. The thin curves are for the system with no damping controller on the SVC, and the thicker curves are the case with a damping control. The swings of Area 1 angle reflect primarily the 0.4-Hz mode, while the Area 2 angle contains some of both modes. Figure 11.49 shows the simulation results for a disturbance in Area 2 with Area 1 importing power. Here, Area 2 shows primarily the 0.9-Hz mode, and Area 1 shows primarily the 0.4-Hz mode. In both cases, the simulation results show that significant damping is added to the system with a control added to modulate the SVC voltage set point, even for greatly varying power system operating conditions. These results suggest that the control design is robust.

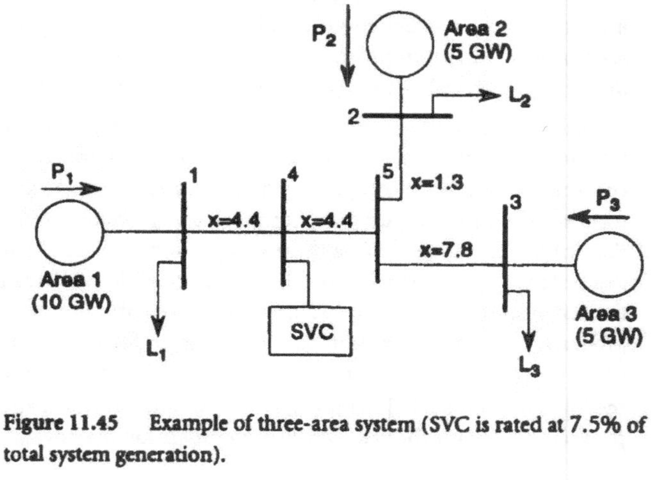

Figure 11.45 Example of three-area system (SVC is rated at 7.5% of total system generation).

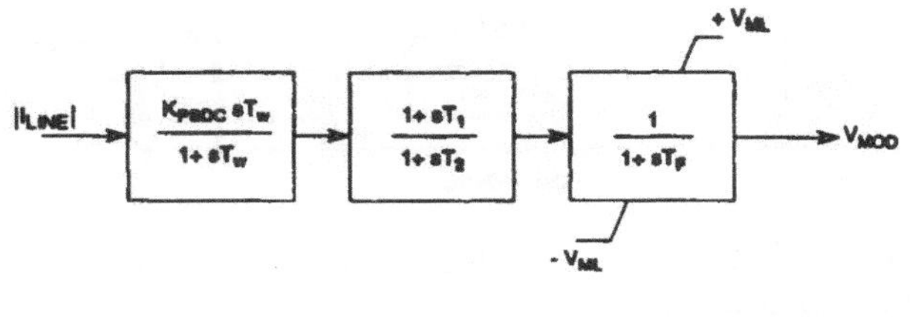

Figure 11.46 Simple power-swing damping control structure for an SVC.

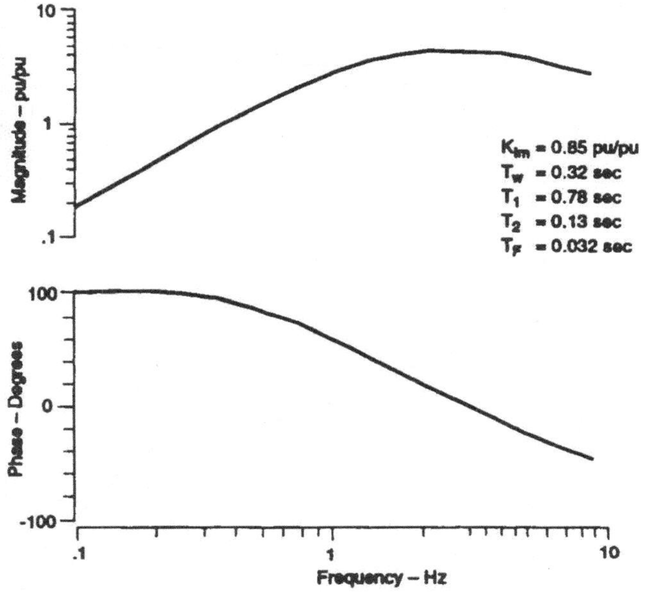

Figure 11.47 Sample transfer function of a simple SVC damping control.

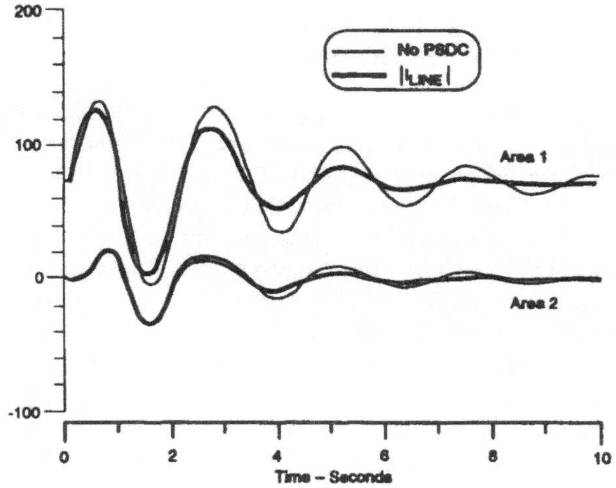

Figure 11.48 Angle swings for a disturbance in Area 1 with Area 1 exporting power (Area 3 is reference).

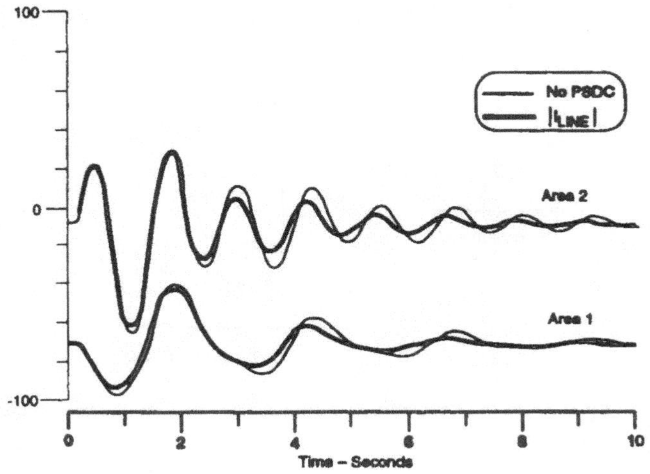

Figure 11.49 Angle swings for a disturbance in Area 2 with Area 1 importing power (Area 3 is reference).

Thyristor-Controlled Series Compensation (TCSC)

Series capacitors have long been applied to transmission systems for reducing the effective impedance between load and source to increase power transfer capability. Adding thyristor control provides significant leverage in directing steady-state power flow along desired paths. In addition, this controllability can be highly effective for damping power swings. The thyristor control also mitigates the adverse side effects from the resonance between the series capacitor and the inductance of the transmission lines, transformers, and generators (known as subsynchronous resonance or SSR) [16].

A TCSC module consists of a series capacitor and a parallel path with a thyristor switch and a surge inductor. Also in parallel, as is typical with series capacitor applications, is a metal-oxide varistor (MOV) for overvoltage protection. A complete compensation system may be made up of several of these modules in series and may also include a conventional (fixed) series capacitor bank as part of the overall scheme, as shown in Figure 11.50. The TCSC can be rapidly and continuously controlled in a wide band of operating points that can make it appear as a net capacitive or net inductive impedance in a transmission line [17].

The following is an example of a damping control for an actual TCSC installation. The objective of the control was to continuously adjust the series reactance (X_C) of the TCSC to add damping to critical electromechanical modes of oscillation. An appropriate input signal with the proper characteristics to meet the stabilization objective of this design was selected. The signal that proved most appropriate for this application is the frequency difference across the system obtained by synthesizing voltages behind reactances in both directions away from the TCSC. This concept is illustrated in Figure 11.51 and described in [14] and [18]. This signal is synthesized from *local* voltage and current measurements and eliminating the need for long-distance communication of signals. To prevent interaction with phenomena outside the desired control bandwidth, low-pass, high-pass, and torsional filtering is included in the modulation control path, as well as some phase compensation. The control structure for one specific installation [16] is shown in Figure 11.52. The transfer function of this control and input signal filtering are shown in Figure 11.53. Note that the 6-Hz and 13-Hz notch filters are in the input-signal filtering path to minimize the potential for site-specific torsional frequency interaction.

The system examined here is a radial machine swinging against the rest of a power system at a mode near 1 Hz. This configuration is significant because actual TCSC controls were factory tested and field tested on a system with a topology such as this [16]. The TCSC for this system has an *rms* line-to-line voltage rating of 500 *kV*, *rms* line current rating of 2900 *Amps*, and a nominal reactance rating of 8Ω. Figure 11.54 shows analog simulator results for a severe fault applied on one of the system buses. At approximately one-half second, a fault is placed on the system and the frequency shows significant oscillations without the supplemental damping control. At six seconds into this test, the TCSC power swing damping control is engaged and the oscillations are eliminated within three cycles of the 1-Hz oscillation. The remainder of the simulator test results show the ambient damping of the system. The control deadband prohibits further reduction at this point. These results clearly demonstrate the potential leverage of a TCSC for damping power system oscillations.

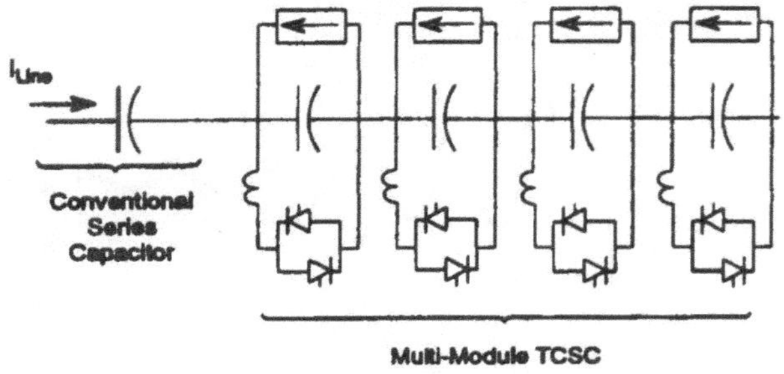

Figure 11.50 TCSC compensation scheme including a multimodule TCSC.

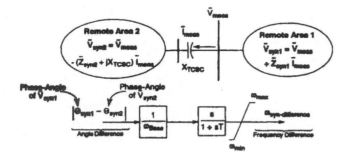

Figure 11.51 Synthesis of remote voltage phasers, angle difference, and frequency difference.

11.2.6 Recent Developments in Control Design

Traditional approaches to aid the damping of power swings include the use of Power System Stabilizers (PSS) to modulate the generator excitation control, for which much experience and insight exist in the industry. Unlike PSS control at a generator location, the speed deviations of the machines of interest are not readily available to a controller sited in the transmission path. Further, since the intent is to damp complex swings involving large numbers of generators, speed signals themselves are not necessarily the best choice for an input signal. It is desired to extract an input signal from the locally measurable quantities at the controller location. Finding an appropriate combination of measurements is the most important aspect of control design.

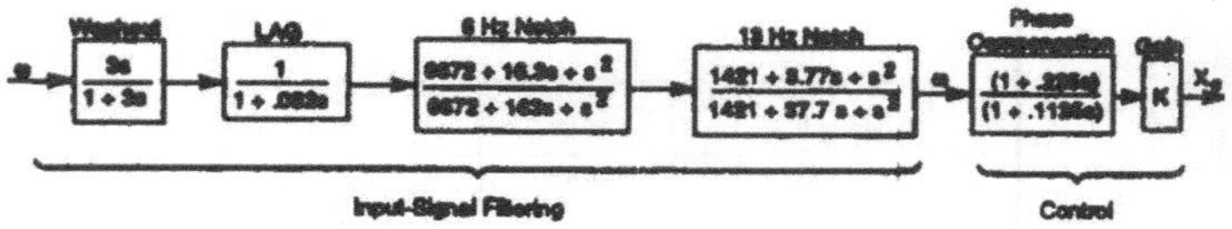

Figure 11.52 Input-Signal filtering and control structure for a specific TCSC installation.

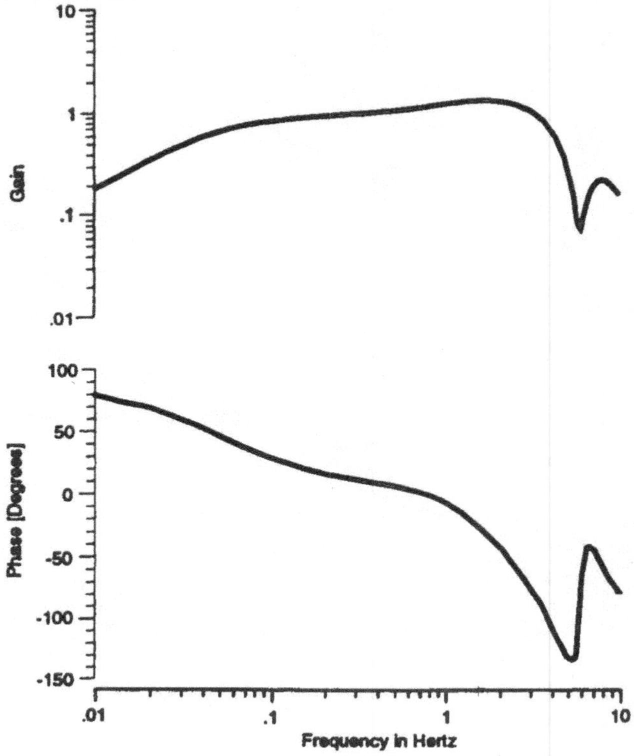

Figure 11.53 Transfer function of a TCSC damping control.

Approximate Multimodal Decomposition

The approach outlined in this section and described in [18] makes a few approximations to develop engineering insight in the control design process. One key approximation is that the modes of interest exhibit light damping. Another is that the impact of any individual control on the frequency and mode shape of the power swing is small. These assumptions permit breaking the system apart based upon the approximate mode shape information, and determining the incremental effect of controllers on each mode separately. The effect of the controllers upon themselves also becomes apparent, and the design can proceed using

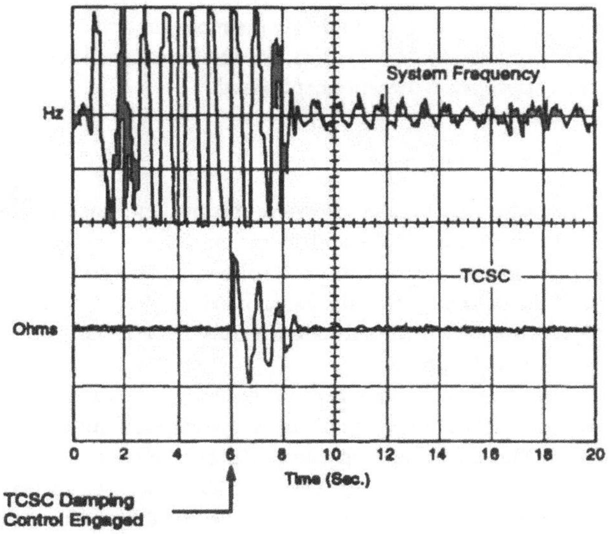

Figure 11.54 Damping benefits of a TCSC as shown on an analog simulator.

this information to assure minimum self-interaction effects on the final response.

Formulation

In the single-machine model described earlier, the mechanical swing mode is represented in terms of a synchronizing and a damping torque with control loops built around it. The same type of representation for each of the swing modes can be obtained by transforming the linearized representation of the power system into the equivalent form [18]:

$$\begin{bmatrix} \Delta\dot{\delta}_{mi} \\ \Delta\dot{\omega}_{mi} \\ \dot{Z}_{mi} \end{bmatrix} = \begin{bmatrix} 0 & \omega_b & 0 \\ -k_{mi} & -d_{mi} & -A_{d23} \\ A_{d31} & A_{d32} & A_{d33} \end{bmatrix} \begin{bmatrix} \Delta\delta_{mi} \\ \Delta\omega_{mi} \\ Z_{mi} \end{bmatrix} + \begin{bmatrix} 0 \\ -B_{d2} \\ B_{d3} \end{bmatrix} u,$$

$$y = \begin{bmatrix} C_{d1} & C_{d2} & C_{d3} \end{bmatrix} \begin{bmatrix} \Delta\delta_{mi} \\ \Delta\omega_{mi} \\ Z_{mi} \end{bmatrix} + Du,$$

where $\Delta\delta_{mi}$ and $\Delta\omega_{mi}$ represent the modal angle and modal speed, respectively, associated with a swing mode λ_i, k_{mi} and d_{mi} represent modal synchronizing and damping coefficients, and Z_{mi} consists of all the other state variables. Based on this representation, a block diagram, similar to the one developed in [5], can be constructed for mode λ_i. Such a block diagram is shown in Figure 11.55. In this figure, $K_{mi}(s)$, $K_{ci}(s)$, $K_{oi}(s)$, and $K_{ILi}(s)$ denote *modal, controllability, observability,* and *inner loop* transfer functions, respectively. These transfer functions evaluated at $s = j\omega$ are complex, providing both gain and phase information which can be used to select effective transfer functions for feedback control.

From Figure 11.55, the effective control action can be described

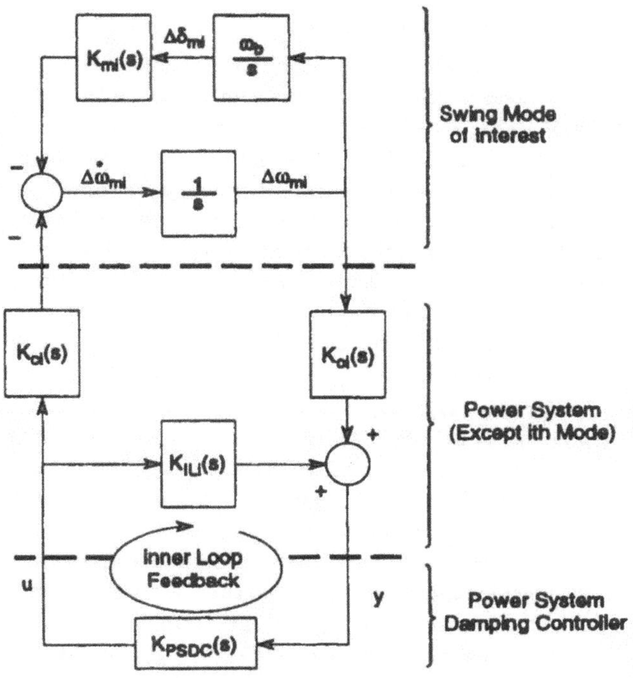

Figure 11.55 Multimodal decomposition block diagram.

by the transfer function:

$$K_{ei}(s) = K_{ci}(s)\frac{K_{PSDC}(s)}{1 - K_{ILi}(s)K_{PSDC}(s)}K_{oi}(s)$$

This relationship describes the impact of a given damping controller, $K_{PSDC}(s)$, on the modal system and is useful in estimating the eigenvalue sensitivity of the ith swing mode. This relationship shows that the effective control action $K_{ei}(s)$ directly relates to the controllability $K_{ci}(s)$ and observability $K_{oi}(s)$ functions. Assuming the perturbation of the complex pair of eigenvalues corresponding to the ith swing mode as $\Delta\lambda_i = -\Delta\sigma \pm j\Delta\omega_i$, the following relationships can be shown [18]:

$$\begin{aligned} \Delta\sigma_i &\approx \text{Real}\{\{K_{ei}(j\omega_i)/2\}, \\ \Delta\omega_i &\approx -\text{Imag}\{\{K_{ei}(j\omega_i)/2\} \end{aligned}$$

Thus, for an ideal damping controller design, $K_{ei}(s)$ is desired to be real for the frequency range of interest. For a practical controller design, some phase lag at the swing mode frequency is acceptable, since it tends to increase synchronizing torque. Also, a good bandwidth is desired, such that high-frequency interactions can be limited by the controller design, as described earlier in this section.

This direct relationship of modal sensitivity to the controller and power system characteristics provides the key to achieving the desired insight into control design. Subsequent discussion will focus on each of the terms in the above equations, to show how they relate to control performance and how they can be used to select effective measurements and controls.

Interpretations

Controllability The effect of a controller on a given swing mode is defined as the controllability function $K_{ci}(s)$, and $K_{ei}(s)$ is directly proportional to $K_{ci}(s)$. In PSS design, $K_{ci}(s)$ depends mostly on the excitation system, generator flux dynamics and network impedances. For network control devices such as TCSC, $K_{ci}(s)$ depends mostly on the network structure and loads. When evaluated at $s = j\omega_i$, $K_{ci}(j\omega_i)$ provides a measure of how controllable the ith mode is by the control signal u. If $K_{ci}(j\omega_i)$ is zero, then mode i is not affected by u. When more than one damping controller is used, $K_{ci}(s)$ is a vector transfer function.

In general, $K_{ci}(j\omega_i)$ is different for each swing mode. For example, a TCSC sited on a tie line would have significant controllability on the associated inter-area mode, but much smaller controllability over local modes. In cases with multiple inter-area modes, the controllability may be nearly 180° out of phase from one mode to the next. Such a condition would mean that, if the machine speeds were averaged and transmitted to the controller, then the action of improving the damping on one mode would simultaneously decrease the damping of another mode. This situation must be compensated for by using an appropriate set of measurements so that this inherent adverse impact will be minimized or eliminated.

The quantity $K_{ci}(j\omega_i)$ is a good indicator for evaluating effective locations to apply damping controllers, because the larger this is, the greater the leverage on the swing mode. For example, an SVC on a bus needing voltage support will be more effective for damping control than one close to a generator terminal bus.

Observability The effective control action $K_{ei}(s)$ is also directly proportional to the observability function $K_{oi}(s)$. The function $K_{oi}(s)$ relates the measured signal y to the ith modal speed $\Delta\omega_{mi}$. For a PSS using the machine speed as the input signal, $K_{oi}(s) = 1$ for the local mode of that machine. When evaluated at $s = j\omega_i$, $K_{oi}(j\omega_i)$ gives an indication of the modal content of the ith swing mode in the measured signal y. Its magnitude can be used to assess the effectiveness of measurements y for damping control applications. In a multimodal system, because $K_{oi}(s)$ is defined with respect to the modal speed, measurements directly related to machine speeds will have observability gains that are predominantly real. Signals more closely related to angular separation, such as power flow, will have an integral characteristic, i.e., nearly 90° of lag with respect to the speed. If $K_{oi}(j\omega_i)$ is small, then the ith mode is weakly observable from the measurement y. Thus having large $K_{oi}(j\omega_i)$ for the dominant modes of interest is one of the criteria in selecting an input signal for a damping controller.

Inner Loop The control design must also consider the effect of the controller output on its input (i.e., the component of the measured signal y due to the control u), other than via the swing mode of interest. This effect may be considered a "feed-forward" term, but here we have called it the "inner-loop" effect, symbolized by the transfer function $K_{ILi}(s)$.

The inner loop transfer function $K_{ILi}(s)$ is extremely important in damping controller design using input signals other than generator speeds. In [9] and [18], simple analysis based on the constraint imposed by $K_{ILi}(s)$, are developed to aid the selection of an appropriate measurement signal.

11.2.7 Summary

This section provides an overview of some key issues for control design, implementation, and operation. Basic stability issues for power systems, such as transient, voltage, and oscillatory stability were introduced. Given the focus of this book, the remainder of this section was on closed-loop controls that affect power system oscillatory stability.

Basic concepts and a historic overview of generator excitation controls as they affect power system stability were introduced and described. Supplemental controls typically applied to generator excitation systems were also discussed (power system stabilizers (PSS)). Furthermore, practical issues for applying supplemental controls to power system transmission equipment (such as HVDC, SVC, and TCSC) were presented. The issues addressed included siting, objectives for control design and operation, closed-loop control design, input-signal selection, input-signal filtering, control algorithms, performance evaluation, and a discussion on potentially adverse side effects due to the application of supplemental controls. Two detailed examples of power electronic devices (SVC and TCSC) were provided illustrating these basic concepts. Finally, a presentation on recent developments in control design was included. Several references were provided for further reading on the issues presented in this section.

This information shows that supplemental control can be beneficial in increasing the stability and utilization of electric power systems.

References

[1] Bertoldi, O. and CIGRE WG 37.08, *Adequacy and Security of Power Systems at the Planning Stage: Main Concepts and Issues,* Paper 1A-05, Symposium on Electric Power System Reliability, Montreal, Canada, 1991.

[2] Anderson, P.M. and LeReverend, B.K., Industry Experience with Special Protection Schemes, *IEEE Power Eng. Soc.,* (PES) Paper 94-WM184-2 PWRS, New York, 1994.

[3] Voltage Stability of Power Systems: Concepts, Analytical Tools and Industry Experiences, *IEEE PES* Special Publication 90TH0358-2-PWR, 1990.

[4] Kundur, P., *Power System Stability and Control,* McGraw-Hill, New York, 1994.

[5] deMello, F.P. and Concordia, C., Concepts of Synchronous Machine Stability as Affected by Excitation Control, *IEEE Trans. Power Apparatus Syst.,* PAS-88, 316–329, 1969.

[6] Larsen, E.V. and Swann, D.A., Applying Power System Stabilizers, Parts I, II, and III, *IEEE Trans. Power Apparatus Syst.,* PAS-100, 3017–3046, 1981.

[7] Paserba, J.J., Larsen, E.V., Grund, C.E., and Murdoch, A., Mitigation of Inter-Area Oscillations by Control, *IEEE PES* Special Publication 95-TP-101 on Inter-Area Oscillations in Power Systems, 1995.

[8] Paserba, J.J., et al., *Opportunities for Damping Oscillations by Applying Power Electronics in Electric Power Systems*, CIGRE Symp. Power Electron. Electr. Power Systems, Tokyo, Japan, 1995.

[9] Larsen, E.V. and Chow, J.H., SVC Control Design Concepts for System Dynamic Performance, in *Application of Static Var Systems for System Dynamic Performance*, IEEE PES Special Publication No. 87TH1087-5-PWR, 1987, 36–53.

[10] Piwko, R.J. and Larsen, E.V., HVDC System Control for Damping Subsynchronous Oscillations, *IEEE Trans. Power Apparatus Syst.*, PAS-101(7), 2203–2211, 1982.

[11] Chow, J.H., Date, R.A., Othman, H.A., and Price, W.W., *Slow Coherency Aggregation of Large Power Systems*, IEEE Symp. Eigenanalysis and Frequency Domain Methods for System Dynamic Performance, IEEE PES Special Publication 90TH0292-3-PWR, 1990, 50–60.

[12] Hauer, J.F., Application of Prony Analysis to the Determination of Model Content and Equivalent Models for Measured Power Systems Response, *IEEE Trans. Power Syst.*, 1062–1068, 1991.

[13] Bahrman, M.P., Larsen, E.V., Piwko, R.J., and Patel, H.S., Experience with HVDC Turbine-Generator Torsional Interaction at Square Butte, *IEEE Trans. Power Apparatus Syst.*, PAS-99, 966–975, 1980.

[14] Clark, K., Fardanesh, B., and Adapa, R., *Thyristor-Controlled Series Compensation Application Study—Control Interaction Considerations*, IEEE PES Paper 94-SM-478-8-PWRD, San Francisco, CA, 1994.

[15] Cresap, R.L., Mittelstadt, W.A., Scott, D.N., and Taylor, C.W., Operating Experience with Modulation of the Pacific HVDC Intertie, *IEEE Trans. Power Apparatus Syst.*, PAS-9, 1053–1059, 1978.

[16] Piwko, R.J., Wegner, C.A., Damsky, B.L., Furumasu, B.C., and Eden, J.D., *The Slatt Thyristor-Controlled Series Capacitor Project — Design, Installation, Commissioning, and System Testing*, CIGRE Paper 14-104, Paris, France, 1994.

[17] Larsen, E.V., Clark, K., Miske, S.A., and Urbanek, J., Characteristics and Rating Considerations of Thyristor Controlled Series Compensation, *IEEE Trans. Power Delivery*, 992–1000, 1994.

[18] Larsen, E.V., Sanchez-Gasca, J.J., and Chow, J.H., *Concepts for Design of FACTS Controllers to Damp Power Swings*, IEEE PES Paper 94-SM-532-1-PWRS, San Francisco, CA, 1994.

IV

Control Systems Including Humans

12

Human-in-the-Loop Control

12.1 Introduction 327
12.2 Frequency-Domain Modeling of the Human Operator 328
The Crossover Model of the Human Operator • A Structural Model of the Human Operator
12.3 Time-Domain Modeling of the Human Operator 329
An Optimal Control Model of the Human Operator
12.4 Alternate Modeling Approaches 330
Fuzzy Control Models of the Human Operator
12.5 Modeling Higher Levels of Skill Development 331
12.6 Applications 331
An Input Tracking Problem
12.7 Closure 334
12.8 Defining Terms 334
References 334
Further Reading 335

R. A. Hess
University of California, Davis

12.1 Introduction

Interest in modeling the behavior of a human as an active feedback control device began during World War II, when engineers and psychologists attempted to improve the performance of pilots, gunners, and bombardiers. To design satisfactory manually controlled systems these researchers began analyzing the neuromuscular characteristics of the human operator. Their approach, e.g., [19] was to consider the human as an inanimate servomechanism with a well-defined input and output. Figure 12.1(a) is a schematic representation of a tracking task in which the human is attempting to keep a moving target within the reticle of a gun sight. Figure 12.1(b) is a feedback block diagram of the gunnery task of Figure 12.1(a) in which the human has been represented as an error-activated compensation element.

The input to the human in Figure 12.1 is a visual "signal" indicating the error between desired and actual system output, the gun azimuth. The output of the human is a command to the device which drives the gun in azimuth. This device might be a simple gearing mechanism linked to a wheel which the human turns or an electric motor which transforms a joystick input into a proportional rate of change of the gun barrel's azimuth angle. What mathematical representation should be used for the block labeled "human controller"? The early researchers in the discipline known as "manual control" hypothesized this answer to the question: the same types of mathematical equations used to describe linear servomechanisms, namely, sets of linear, constant-coefficient differential equations. The hypothesis

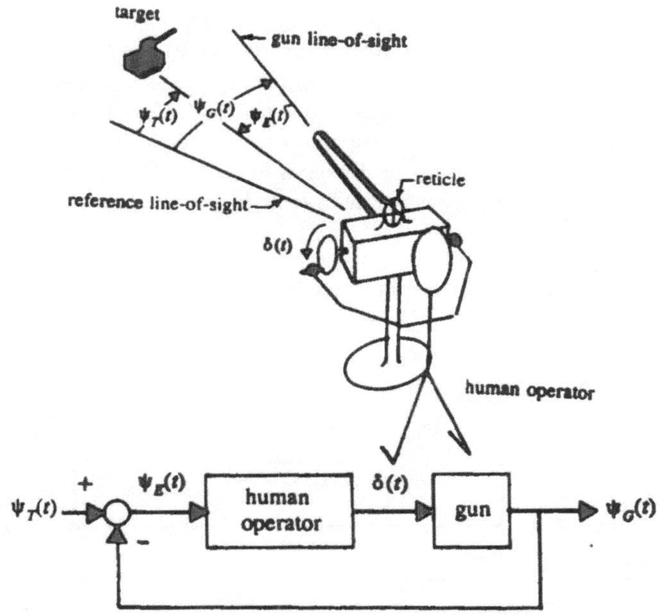

Figure 12.1 (a) A human-in-the-loop control problem. (b) A block diagram representation of the task of Figure (a).

of these early researchers turned out to be true, and the *control-theory paradigm* for quantifying human control behavior was born. This paradigm has become fundamental for manual control engineers [13].

The evolution of the control-theory paradigm for the human

0-8493-0054-3/00/$0.00+$.50

controller or operator paralleled the development of new synthesis techniques in feedback control. Thus, "optimal control models" (OCMs) of the human operator [8] appeared as linear quadratic Gaussian (LQR) control system design techniques were being developed. "Fuzzy controller" models [12] and "H-infinity" models [1] of the human operator closely followed the appearance of these design techniques.

The models discussed here will primarily be those which have successfully solved human-in-the-loop control problems. This approach will neglect some of the more recent human operator models, which are incompletely tested.

12.2 Frequency-Domain Modeling of the Human Operator

Modeling the human operator as a set of linear, constant-coefficient differential equations suggests representing the human as a transfer function. This approach, generalized to describing function descriptions, captured the attention of some of the earliest and most influential manual control engineers [14]. Figure 12.2 shows a describing function representing the human operator or controller in a single-input, single-output (SISO) tracking task, such as an aircraft pitch-attitude tracking task. Here the transfer function representation of the human has been generalized as a quasi-linear describing function [4] by the addition of an additive "remnant" signal, $n_e(t)$. This signal represents the portion of the system error signal $e(t)$ unexplainable by linear operator behavior, and not linearly correlated with the system input $c(t)$. The spectral measurements of remnant coalesced best when the remnant was assumed to be injected into the displayed error $e(t)$ rather than the operator's output $\delta(t)$. For this reason, the remnant portion of the quasi-linear describing function is almost universally shown with error-injected remnant.

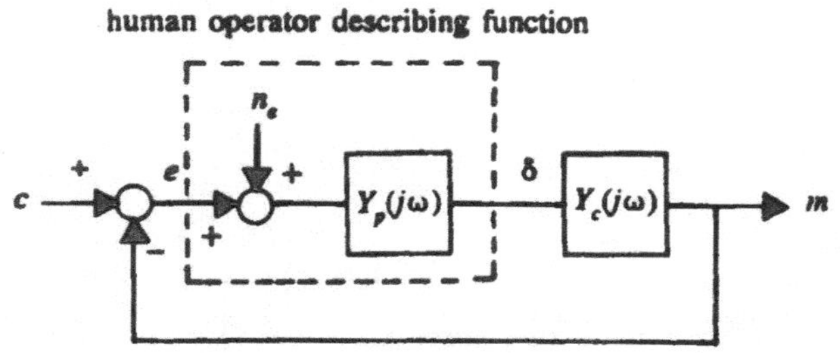

Figure 12.2 A quasi-linear describing function representation of the human operator.

Frequency-domain identification of human operator describing functions in simple laboratory tracking tasks has been actively studied over the past three decades [15]. In these experiments, the describing functions identified were $Y_p(j\omega)$ (or $Y_p(j\omega)Y_c(j\omega)$) and $\Phi_{n_e n_e}(\omega)$, where the latter quantity is defined as the power spectral density of the remnant signal $n_e(t)$. The controlled element or plant was a member of a set of "stereotypical" controlled-element dynamics, i.e., $Y_c(s) = \frac{K_c}{s^k}$, $k = 0, 1, 2$, and the inputs or disturbances were random-appearing signals, often generated as sums of sinusoids. The results led to one of the first true engineering models of the human operator, referred to as the crossover model.

12.2.1 The Crossover Model of the Human Operator

The crossover model is based on the following experimentally verifiable fact: In a Bode diagram representing the loop transmission $Y_p(j\omega)\cdot Y_c(j\omega)$ of the system, such as shown in Figure 12.2, the human adopts dynamic characteristics $Y_p(j\omega)$ so that

$$Y_p(j\omega)\cdot Y_c(j\omega) \approx \frac{\omega_c e^{-\tau_e j\omega}}{j\omega}, \quad \text{for} \quad \omega \quad \text{near} \quad \omega_c. \tag{12.1}$$

The crossover frequency, ω_c, is defined as the frequency where $|Y_p Y_c(j\omega)| = 1.0$. Equation 12.1 is valid in a broad frequency range (1 to 1.5 decades) around the crossover frequency ω_c. The factor τ_e, referred to as an effective time delay, represents the cumulative effect of actual time delays in the human information processing system (e.g., visual detection times, neural conduction times, etc.), the low-frequency effects of higher frequency human operator dynamics (e.g., muscle actuation dynamics), and higher frequency dynamics in the controlled element, itself. Here, "higher frequency" refers to frequencies well above ω_c.

Associated with Equation 12.1 is a model of $\Phi_{n_e n_e}(\omega)$, the power spectral density of the error-injected remnant. Again, extensive experimental evidence suggests the following form:

$$\Phi_{n_e n_e}(\omega) \approx \frac{R\bar{e}^2}{\omega^2+\omega_R^2} \tag{12.2}$$

where $\bar{e}^2$ represents the mean-square value of the error signal $e(t)$ in Figure 12.2. Table 12.1 shows approximate parameter values for the crossover and remnant models with related equations. In applying the crossover model, plant dynamics as simple as those in Table 12.1 will be rare. In these cases, one should interpret the stereotypical dynamics in the table to reflect the actual plant characteristics in the crossover region.

A detailed summary of empirically derived rules for selecting crossover model parameters, given the controlled-element dynamics and the bandwidth of the input signal, are given by [5]. In addition, simplified techniques are given for estimating human-machine performance (e.g., root-mean-square tracking error $\sqrt{\bar{e}^2}$) in continuous tasks with random-appearing inputs.

The reason for beginning this discussion with the crossover model is that it is basic for manual control modeling. Any valid model of the human operator in continuous tasks with random-appearing inputs *must* exhibit the characteristics of Equation 12.1.

The loop transmission prescribed by Equation 12.1 is similar to that which an experienced control system designer would select in a frequency-domain synthesis of a control system with an inanimate compensation element and performance requirements similar to the manually controlled system [16].

TABLE 12.1 Parameters and Relations for Crossover Model

Y_c (around crossover)	τ_0 (s)	ω_{c_0} (rad/s)	R	ω_R (rad/s)
K	0.30	5.0	0.1 to 0.5	3.0
$\frac{K}{s}$	0.35	4.5	0.1 to 0.5	3.0
$\frac{K}{s^2}$	0.50	3.0	0.1 to 0.5	1.0

$$\begin{aligned} \omega_C &\approx \omega_{c_0} + 0.18\omega_{BW_c} \\ \tau_e &\approx \tau_0 - 0.08\omega_{BW_c} \\ \tau_0 &= \frac{\pi}{2\omega_{c_0}} \end{aligned}$$

12.2.2 A Structural Model of the Human Operator

A model of the human operator which follows the crossover model, but provides a more detailed representation of human operator dynamics is offered in [7] and is referred to as a **structural model of the human operator**. The model is shown in Figure 12.3. Table 12.2 lists the model parameter values which depend on the order of the controlled-element dynamics in the crossover region. The parameter "k" in Table 12.2 refers to the controlled-element order around crossover, i.e., $k = 0$ for zeroth order, $k = 1$ for first order, etc.

TABLE 12.2 Nominal Parameters for Structural Model.

k	K_e	K_1	K_2	T_1 (s)	T_2 (s)	τ (s)	ζ_n	ω_n (rad/s)
0	1.0[a]	1.0	2.0	5.0	[b]	0.15	0.707	10.0
1	1.0	1.0	2.0	5.0	[c]	0.15	0.707	10.0
2	1.0	1.0	10.0	2.5	[b]	0.15	0.707	10.0

[a] K_e chosen to provide desired crossover frequency
[b] selected to achieve K/s–like crossover characteristics
[c] Parameter not applicable

The structural model of the human operator is called an "isomorphic model," because the model's internal feedback structure reflects hypothesized proprioceptive feedback activity in the human.

If the control engineer can reasonably estimate the crossover frequency for the manual control task at hand, Figure 12.3 and Table 12.2 can provide a model of the linear human operator dynamics ($Y_p(j\omega)$) adequate for many engineering applications. The empirically derived rule for determining crossover frequency for the crossover model can be used to define the crossover frequency for the structural model. Remnant parameter estimates can also be obtained in a similar fashion.

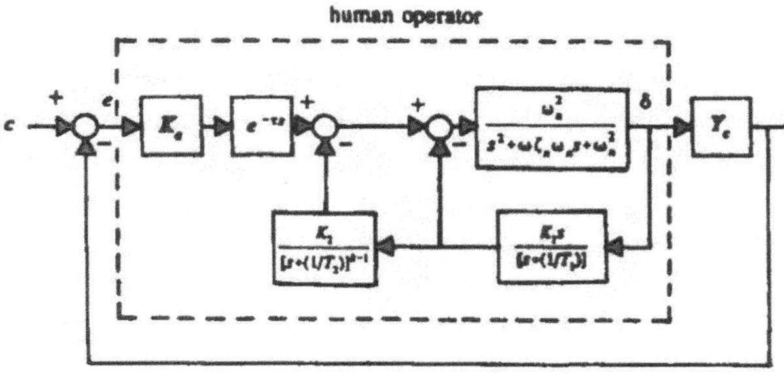

Figure 12.3 The structural model of the human operator.

12.3 Time-Domain Modeling of the Human Operator

12.3.1 An Optimal Control Model of the Human Operator

The advent of a time-domain control synthesis technique in the mid-1960s, referred to as linear quadratic Gaussian (LQG) design led to a powerful model of the human operator called the **Optimal Control Model** (OCM). This model differs from those defined in the preceding because it is algorithmic and is based on a time-domain optimization procedure. The model is algorithmic because the quantitative specification of certain human operator information processing limitations, such as signal-to-noise ratios on observed and control variables and sensory-motor time delays, together with an objective function which the human is assumed to be minimizing in the task at hand, can lead to direct computation of the linear human operator dynamics and remnant. In addition, this algorithmic capability is not limited to SISO systems but also can be extended to human control of multi-input, multioutput (MIMO) systems.

Figure 12.4 shows the basic OCM structure. Focusing for the moment on a SISO system, the elements in Figure 12.4 from time delay to neuromuscular dynamics form the human operator dynamics. Strictly, speaking, the OCM is never a SISO model, because a fundamental hypothesis in the model formulation is that, if a variable $d(t)$ is displayed to the operator, its time derivative $\dot{d}(t)$ is also sensed, both signals corrupted with white observation noise. Thus, in its simplest form, the $y(t)$ in Figure 12.4 is a column vector whose two elements are the displayed signal and

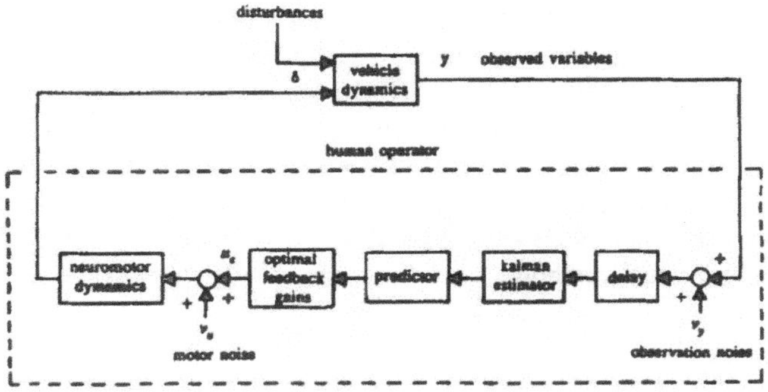

Figure 12.4 The optimal control model of the human operator.

its time derivative. Likewise, the $v_y(t)$ in Figure 12.4 is also a column vector whose elements are the observation noise signals. The covariances of these observation noise signals are assumed to scale with the covariances of the displayed/observed signals $y(t)$ and $\dot{y}(t)$. In addition to observation noise, the signal $u_c(t)$ is also assumed to be corrupted with "motor" noise. This noise provides performance predictions which are more realistic than those produced when motor noise is absent.

Table 12.3 shows "nominal" values of the noise-signal ratios and time delay for typical applications of the OCM to tracking or disturbance regulation tasks with random-appearing inputs or disturbances. The remaining user-defined element in application of the OCM is the quadratic index of performance which the operator is assumed to be minimizing. In many SISO applications, the following index of performance has been employed:

$$J = \int_0^\infty \left[y^2(t) + \rho\dot{\delta}^2(t)\right] dt. \tag{12.3}$$

TABLE 12.3 Nominal Parameter Values for Optimal Control Model (OCM).

$\frac{V_{y_i}}{\bar{\sigma}^2_{y_i}}$ [a]	$\frac{V_u}{\bar{u}^2_c}$ [b]	τ (s)
0.01π	0.001π	0.15

[a] noise-to-signal ratio for observation noise $v_{y_i}(t)$.
[b] noise-to-signal ratio for motor noise $v_u(t)$.

Inclusion of control rate in the index of performance will produce first-order lag dynamics in the OCM [8]. These constitute the neuromotor dynamics shown in Figure 12.4. As in any LQG synthesis application, selection of the index of performance weighting coefficients is nontrivial. One procedure for selecting ρ in Equation 12.3 is a trial and error technique in which ρ is varied until the time constant of the first-order lag (neuromotor dynamics) of approximately 0.1 s. Another procedure for choosing ρ which does not require trial and error iteration has been used successfully [6] and is referred to as an "effective time constant" method.

The power of the OCM lies in its ability to provide accurate representations of human control behavior given "standard" values for quantifying the information processing limitations mentioned in the preceding.

Finally, it should be noted that other human operator models have been developed which are progeny of the OCM. The biomorphic model of the human pilot [3], is one example, as is the OCM extension and refinement described by [21], and the more recent H-infinity model of [1].

12.4 Alternate Modeling Approaches

12.4.1 Fuzzy Control Models of the Human Operator

Fuzzy set theory leads to a description of cause and effect relationships which differ considerably from the control theoretic approaches discussed in the preceding [20]. In classical set theory, there is a distinct difference between elements which belong to a set and those which do not. Fuzzy set theory allows elements to belong to more than one set and assigns each element a membership value, M, between 0 and 1 for each set of which it is a member. Consider, for example, the use of fuzzy sets to model the manner in which a pilot might control the pitch attitude of an aircraft, creating what can be termed a **fuzzy control model** of the human. Assume that the aircraft's pitch attitude, $\theta(t)$, can vary within $-20° \le \theta(t) \le +20°$. This range is often referred to as the "universe of discourse" in fuzzy set theory. On can define a number of membership functions over this range of discourse as in Figure 12.5(a) where five functions are shown. The shape defining each function as well as the number of functions spanning the universe of discourse is entirely up to the analyst. The five triangular functions of Figure 12.5(a) have been chosen for simplicity. The functions have also been given linguistic definitions, i.e., "$\theta(t)$ very negative, $\theta(t)$ negative," etc. Now, as in Figure 12.5(a), an aircraft pitch attitude of $-12.5°$ is a member of five sets with membership values of: $M(-12.5) = 0.25$ (very negative), 0.75 (negative), 0 (around zero), 0 (positive), and 0 (very positive).

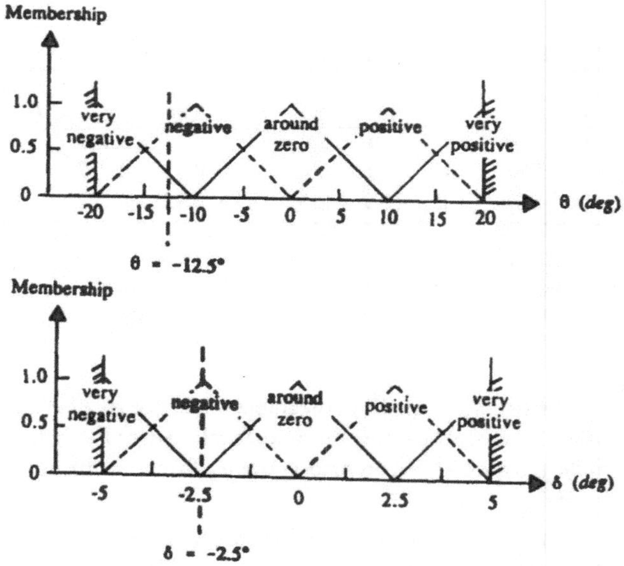

Figure 12.5 (a) Defining input membership functions for a fuzzy control model of the human operator. (b) Defining output membership functions for a fuzzy control model of the human operator.

Modeling the human operator with fuzzy sets involves using fuzzy relations between input and output values. Let us assume the output set here consists of elevator deflections which the pilot commands in response to an observed pitch attitude deviation from zero. Output fuzzy sets can be defined over the universe of discourse for the elevator deflections, here defined as $-5° \leq \delta(t) \leq +5°$, using membership functions shown in Figure 12.5(b). Again, linguistic definitions have been employed to describe these membership functions. Perhaps the pitch attitude of $-12.5°$ produced a pilot commanded elevator angle of $-2.5°$, as shown in Figure 12.5(b). Now, as will be discussed in a later section, by observing the pitch-attitude change and resultant control actions of an experienced pilot, or group of experienced pilots in a flight simulator, the control engineer can create a "relational matrix" R between $\theta(t)$ and $\delta(t)$ as $R = M(\theta)XM(\delta)$, where "X" denotes a Cartesian product, or minimum for each element pairing of $M(\theta)$ and $M(\delta)$. Assume that, on the basis of simulator observations, the relational matrix below is obtained:

$$R = \begin{bmatrix} 0.6 & 0.5 & 0 & 0 & 0 \\ 0.2 & 0.75 & 0.5 & 0 & 0 \\ 0 & 0.4 & 1 & 0.4 & 0 \\ 0 & 0 & 0.5 & 0.75 & 0.0 \\ 0 & 0 & 0 & 0.4 & 0.6 \end{bmatrix}. \quad (12.4)$$

The R of Equation 12.4 forms the basis of a "fuzzy control" model of the pilot. In exercising the model, one can assume that the pilot makes observations every ΔT sec. For example, at some observation instant assume $\theta(n\Delta T) = -15.0°$. As shown in Figure 12.5(a), the memberships of the input are

> $M(\theta) = 0.5$ (very negative pitch), 0.5 (negative pitch), 0 (around zero pitch), 0 (positive pitch), and 0 (very positive pitch)

Using R and $M(\theta)$, the memberships of the fuzzy output can be obtained [17] as

> $M(\delta)$ = max of {min[0.6, 0.5], min[0.2, 0.5], min[0, 0], min[0, 0], min[0, 0]} for very negative elev.
>
> = max of {min[0.5, 0.5], min[0.75, 0.5], min[0.4, 0], min[0, 0], min[0, 0]} for negative elev.
>
> = max of{min[0, 0.5], min[0.5, 0.5], min[1, 0], min[0.5, 0], min[0, 0]} for around zero elev.
>
> = max of{min[0, 0.5], min[0, 0.5], min[0.4, 0], min[0.75, 0], min[0.4, 0]} for positive elev.
>
> = max of{min[0, 0.5], min[0, 0.5], min[0., 0], min[0, 0], min[0.6, 0]} for very positive elev.

The right hand sides of each of the five equations above constitute the membership values for $\delta(t)$ at this instant. These values have been obtained by pairing the elements of R with elements of $M(\theta)$ as would be done in the matrix multiplication $M\Theta R$. The minimum value of each pairing was found, and finally, the maximum of the j elements selected. This yields

> $M(\delta) = 0.5$ (very negative elev.), 0.5 (negative elev.), 0.5 (around zero elev.), 0 (positive elev.), and 0 (very positive elev.)

Notice that the output of this fuzzy relation or fuzzy mapping is not a definite value of $\delta(n\Delta t)$, but a collection of membership values. In order to implement this fuzzy pilot model, say, in a simulation of an aircraft, one needs to "defuzzify" model outputs such as the $M(\delta)$ above into nonfuzzy $\delta(n\Delta T)$, which can then serve as useful inputs to the simulation.

It is worth emphasizing that determining fuzzy models of human operator behavior is rarely as simple as described. For example, human operator actions are often predicated upon system error and upon error rate. This obviously complicates the definition of the human's input membership functions. Despite these complications, fuzzy models of the human operator have been successful in modeling human-in-the-loop control in a variety of tasks. Examples include modeling motorcycle drivers [12], ships helmsmen [18], and automobile drivers [9].

12.5 Modeling Higher Levels of Skill Development

In the models described briefly in the preceding paragraphs, it has been tacitly assumed that the human was operating upon displayed system error (and error rate in case of the OCM). Such tasks are referred to as **compensatory tracking tasks**. Often displays for human operators contain error information and output information as well. Such displays are referred to as "pursuit" displays, and the resulting tasks are referred to as **pursuit tracking tasks**. A discussion of such tasks and their associated human operator models can be found in [5]. However, another, more subtle modeling issue can arise in which the human can actually develop pursuit tracking behavior with only a compensatory display. Such human behavior has been the subject of the "Successive Organizations of Perception" (SOP) theory, [10]. The highest level of skill development is referred to as "precognitive" behavior and implies human execution of preprogrammed responses to certain stimuli without the necessity of continuous feedback information. Modeling any of these higher levels of skill development is beyond the scope of this chapter.

12.6 Applications

The preceding sections have outlined approaches for modeling the human operator in SISO control systems in which the human is sensing system error and where the system input is random or random appearing. The various modeling approaches have met with success in engineering analyses. It is useful at this juncture to consider applying a subset of these models to a very simple but illustrative human-in-the-loop control example.

12.6.1 An Input Tracking Problem

Consider again the block diagram of Figure 12.2 representing a human-in-the-loop SISO control problem. Here, the human's task is to null a displayed error signal (difference between command input and system output) by manipulating a control device producing an output $\delta(t)$. The plant dynamics are very simple, consisting of an integrator, i.e., the plant transfer function is $Y_C(s) = \frac{1}{s}$. Such dynamics are often referred to as rate-command, because a constant control input $\delta(t)$ produces a constant output rate, $\dot{m}(t)$. The command input here is filtered white noise, wherein the filter transfer function is $F(s) = \frac{1}{(s+1)^2}$. Three models of the human will be generated here: 1) a crossover model, 2) a structural model, and 3) an Optimal Control Model. In addition, a brief discussion of how a fuzzy control model might be developed is also included.

Crossover Model

Development of the crossover model of Equation 12.1 begins by estimating the system crossover frequency, ω_C, and the effective time delay, τ_e. These parameters have been found empirically to depend upon the plant dynamics and input bandwidth [15], via the equations shown in Table 12.1, i.e.,

$$\begin{aligned} \omega_C &\approx \omega_{C_0} + 0.18\omega_{BW_c}, \\ \tau_e &\approx \tau_0 - 0.08\omega_{BW_c}, \end{aligned}$$

and

$$\tau_0 = \frac{\pi}{2\omega_{C_0}}, \tag{12.5}$$

where ω_{BW_c} refers to the input bandwidth, here approximated by the break frequency of the input filter $F(s)$ as 1 rad/s. Thus, Equations 12.5 yield

$$\begin{aligned} \omega_C &= 4.5 + 0.18(1.0) \quad \text{rad/s} \\ &= 4.68 \quad \text{rad/s} \\ \tau_e &= 0.35 - 0.08(1.0) \quad \text{s} \\ &= 0.27 \quad \text{s} \end{aligned} \tag{12.6}$$

Equation 12.1 becomes

$$Y_P Y_C(j\omega) \approx \frac{4.68e^{-0.27s}}{(j\omega)} \tag{12.7}$$

Now the power spectral density of the error-injected remnant is given by Equation 12.2. Using Table 12.1, the remnant power spectral density becomes

$$\Phi_{n_e n_e}(\omega) = \frac{0.4\bar{e}^2}{\omega^2+(0.3)^2}. \tag{12.8}$$

In Equation 12.8 an intermediate value of R has been used. Simple block diagram algebra, fundamental spectral analysis techniques and the "$\frac{1}{3}$ power law" can be used to derive the following equation for estimating human-in-the-loop tracking performance [5]:

$$\bar{e}^2 \approx \frac{\frac{1}{3}\bar{c}^2\left(\frac{\omega_{BW_c}}{\omega_C}\right)^2}{1-\frac{R}{\omega_R^2\tau_e}I_1} \tag{12.9}$$

where $\bar{c}^2$ refers to the mean-square value of the system input, here assumed to be unity. The I_1 in Equation 12.9 can be obtained from Figure 12.6 [15]. In this case, $I_1 \approx 3.6$ and Equation 12.9 gives an estimate of tracking performance as $\bar{e}^2 = 0.037$ with root-mean-square (RMS) tracking error $\sqrt{\bar{e}^2} = 0.19$. The one caveat that should accompany the use of Equation 12.9 is that it assumes a rectangular input spectra. In the task at hand, the input was formed by passing white noise through a second-order filter, and thus was not rectangular. However, to obtain a preliminary estimate of tracking performance, one can use the cutoff frequency of the continuous-input spectrum as the cutoff of the rectangular spectrum. Tracking performance can be more accurately predicted by a computer simulation of the crossover model, with injected remnant. Some iteration will be required, because the magnitude of the injected remnant has to scale with $\bar{e}^2$, the quantity one is trying to obtain from the simulation. Nonetheless, the author has found that the iterative process is fairly brief.

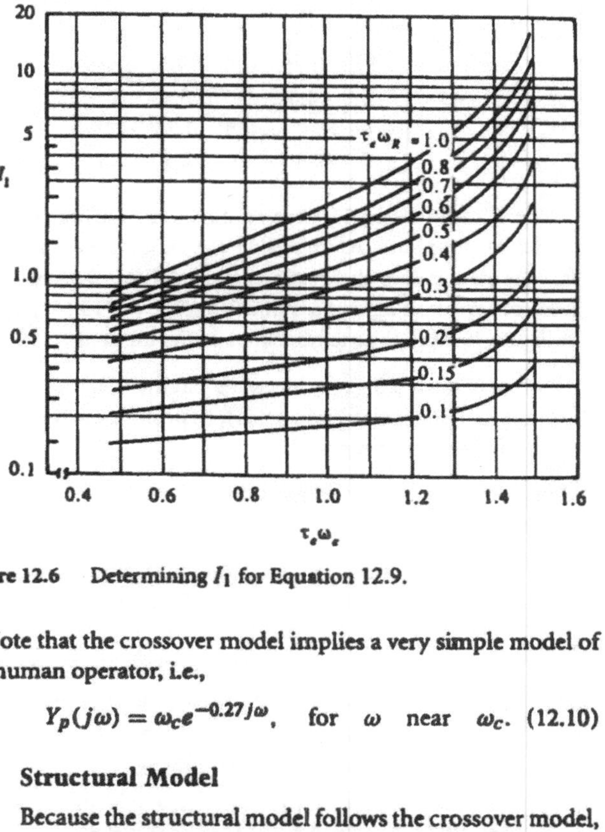

Figure 12.6 Determining I_1 for Equation 12.9.

Note that the crossover model implies a very simple model of the human operator, i.e.,

$$Y_P(j\omega) = \omega_C e^{-0.27j\omega}, \quad \text{for} \quad \omega \quad \text{near} \quad \omega_C. \tag{12.10}$$

Structural Model

Because the structural model follows the crossover model, one can use the first of Equations 12.5 to determine the crossover frequency as $\omega_C = 4.68$ rad/s. Referring to Table 12.2, and recalling that the plant exhibits first-order dynamics at all frequencies (i.e., the slope of the magnitude portion of its Bode diagram is −20 dB/dec at all frequencies), one uses the structural model parameters corresponding to $k = 1$. Thus, all of the structural model parameters can be chosen except K_e which can be selected by requiring $|Y_P(j\omega_C)Y_C(j\omega_C)| = 1.0$. This yields $K_e = 18.1$.

As opposed to Equation 12.10, the structural model offers a more detailed representation of human operator dynamics. In this application

$$Y_P(j\omega) = \frac{1810(j\omega+.2)e^{-0.15j\omega}}{(j\omega+.0497)[(j\omega)^2+14.29(j\omega)+402.1]}. \tag{12.11}$$

One can also utilize the same remnant model, as discussed, with the crossover model and approximate tracking performance with Equation 12.9 and Figure 12.6.

Optimal Control Model

Given the nominal OCM parameters from Table 12.3, only the selection of the weighting coefficient ρ in Equation 12.3 remains before the LQG synthesis technique produces a pilot model. One approach to selecting ρ has been mentioned in the preceding as an "effective time constant" method [6]. However, a simpler approach will be adopted which again calls on the empirical relation in the first of Equations 12.5 to define the crossover frequency ω_C. It can be shown that a relatively simple relationship exists between the plant dynamics and the coefficient ρ in an optimal regulator problem [11], namely,

$$\omega_{BW} \approx \left[\left(\frac{1}{\rho}\right)^{\frac{1}{2}}\right]^{\frac{1}{n-m+1}} \qquad (12.12)$$

where ω_{BW} here is the bandwidth of the closed-loop, human-vehicle system and the parameters m and n are obtained from the plant dynamics when expressed as

$$Y_c(s) = \frac{K(s^m+a_{m-1}s^{m-1}+\cdots+a_1s+a_0)}{s^n+b_{n-1}s^{n-1}+\cdots+b_1+b_0}. \qquad (12.13)$$

In more precise terms, ω_{BW} is the magnitude of that closed-loop pole closest to the frequency where the magnitude of the closed-loop system transfer function is 6 dB below its zero-frequency value. Now one can approximate the crossover frequency, ω_C, given ω_{BW} from Equation 12.12 by

$$\omega_C \approx 0.56\omega_{BW}. \qquad (12.14)$$

Using the nominal parameter values given in Table 12.3, the OCM can now be applied to the tracking task. With $Y_c = \frac{1}{s}$, $K = 1, m = 0$, and $n = 1$ in Equation 12.13. Equations 12.12 and 12.14 and $\omega_C = 4.68$ rad/s. used in the previous models, yield $\rho = \left(\frac{0.56}{\omega_C}\right)^4 = \left(\frac{0.56}{4.68}\right)^4 = 2.05 \cdot 10^{-4}$.

The OCM computer program utilized in this study is described in [2]. As opposed to the crossover and structural models, error-injected remnant and RMS tracking error are obtained as part of the OCM model solution. The RMS tracking error predicted by the OCM was $\sqrt{\bar{e}^2} = 0.21$, which compares favorably with the 0.19 value obtained with the $\frac{1}{3}$ power law in the crossover and structural models. Finally, Figure 12.7 compares the Bode diagrams for the loop transmission $Y_pY_c(j\omega)$ obtained with each of the three modeling approaches. Again, the comparison is favorable.

Fuzzy Control Model Formulation

Developing a fuzzy control model for the human operator in the input tracking task defined in the preceding would require actual simulator tracking data or a less attractive alternative involving detailed discussions with trained human operators who might describe how they select their control actions given certain displayed errors. This procedure assumes that the control engineer is unwilling or unable to employ any of the models just discussed. This might be the case, for example, if the plant dynamics were unknown, were suspected of being highly nonlinear, or if the task were sufficiently removed from that of tracking random-appearing input signals, e.g., tracking transient signals.

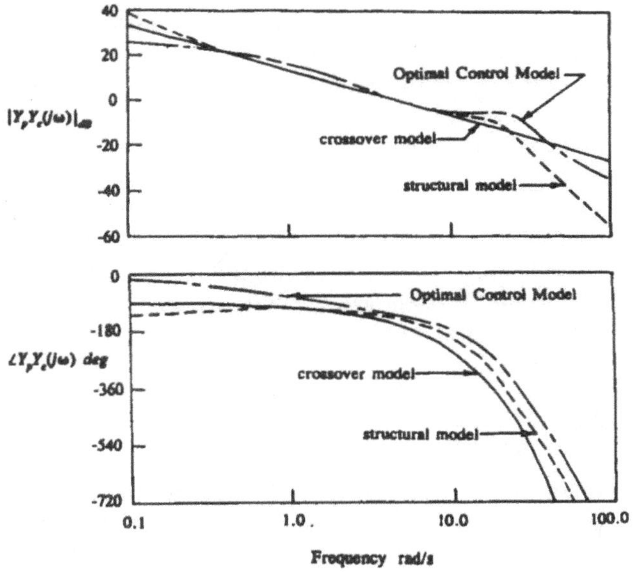

Figure 12.7 Bode diagrams of loop transmissions ($Y_pY_c(j\omega)$) for application example.

Space does not permit a detailed discussion of the fuzzy control approach to this problem, other than a very general, rudimentary outline.

Assume that a set of human operator input and output time histories ($e(t)$ and $\delta(t)$ respectively in Figure 12.2) are available from a human-in-the-loop control simulation of the task at hand. Let us further assume that, on the basis of discussions with the operator, the manual control engineer has been convinced that only error and not some combination of error and error rate or error and integral error are used by the operator in generating appropriate control inputs, $\delta(t)$.

Next, the engineer would create a set of membership functions for the error and output signals, similar to those shown in Figure 12.5(a). Choosing the number and shapes of the functions are part of the art in applying fuzzy control. Next, human operator time delay, τ, would be estimated. This delay would be considerably smaller than the effective delay, τ_e, discussed in the crossover model as it is intended only to approximate the delay between a visual stimulus and the initiation of a control response by the operator. Values on the order of 0.1 s. would be appropriate. The engineer would then tabulate pairs of error and delayed output values $e(n\Delta T)$ and $\delta(n\Delta T + \tau)$.

Using the paired input and output values, a relational matrix such as Equation 12.4 might be created as follows: For each input/output pair, form a relational matrix R_i as $R_i = M(e_i)XM(\delta_i)$, i.e., the Cartesian product of $M(e_i)$ and $M(\delta_i)$, or minimum for each element pairing of $M(e_i)$ and $M(\delta_i)$. After this has been done for all of the input-output pairs, the final relational matrix R is obtained by using, as each element $R(i, j)$, the largest corresponding element found in any R_i. This last operation is sometimes referred to as finding the "union" of the relational matrices.

With the relational matrix R thus obtained, one has the basis of a fuzzy control model which can produce membership numbers for a $\delta(n\Delta T + \tau)$ for any error $e(n\Delta T)$. This result is still fuzzy (recall the $M(\delta)$ obtained after Equation 12.4) and needs to be "defuzzified" for appropriate use in any simulation employing the fuzzy control model.

12.7 Closure

This treatment of human-in-the-loop control has been brief, and many important topics have gone untouched. Two examples of such unexplored topics are the modeling of the human operator in tasks where motion cues are important, and the modeling of human control of MIMO systems. It is hoped that the foregoing material has provided the reader with sufficient introductory background to explore these omitted topics.

12.8 Defining Terms

Compensatory tracking task: A tracking task in which the human is presented with a display of system error only.

Crossover model: A model of the human operator/plant combination (loop transmission) for compensatory tasks stating that the loop transmission in manually controlled SISO systems can be approximated by an integrator and time delay around the crossover frequency.

Fuzzy control model: A model of the human operator derived from fuzzy set theory. In its simplest form, this model is defined by input and output membership functions and a relational matrix describing mappings between these functions.

Optimal control model: An algorithmic, time-domain based model of the human operator based upon linear quadratic Gaussian control system design.

Pursuit tracking task: A tracking task in which the human is presented with a display of system error and system output.

Structural model: A model of the human operator which follows the crossover model, but provides a more detailed representation of human operator dynamics.

References

[1] Anderson, M., Standard optimal pilot models, *AIAA*, 94–3627, 1994.

[2] Curry, R. E., Hoffman, W. C., and Young, L. R., Pilot modeling for manned simulation, Air Force Flight Dynamics Lab., AFFDL-TR-76-124, 1976, Vol. 1.

[3] Gerlach, O. H., The biomorphic model of the human pilot, Delft University of Technology, Dept. of Aerospace Engineering, Report LR-310, 1980.

[4] Graham, D. and McRuer, D. T., *Analysis of nonlinear control systems*, Dover, New York, 1971, Chapter 10.

[5] Hess, R. A., Feedback control models, in *Handbook of Human Factors*, Salvendy, G., Ed., John Wiley & Sons, New York, 1212–1242, 1987.

[6] Hess, R. A. and Kalteis, R., Technique for Predicting Longitudinal Pilot-Induced Oscillations, *J. Guidance, Control, Dyn.*, 14(1), 198–204, 1991.

[7] Hess, R. A., Methodology for the Analytical Assessment of Aircraft Handling Qualities, in *Control and Dynamic Systems, Vol. 33, Advances in Aerospace Systems Dynamics and Control Systems*, Part 3, Leondes C.T., Ed., Academic, San Diego, 1990, 129–150.

[8] Kleinman, D. L., Levison, W. H., and Baron, S., An optimal control model of human response, part I: theory and validation, *Automatica*, 6(3), 357–369, 1970.

[9] Kramer, U., On the application of fuzzy sets to the analysis of the system-driver-vehicle environment, *Automatica*, 21(1), 101–107, 1985.

[10] Krendel, E. S. and McRuer, D. T., A servomechanism approach to skill development, *J. Franklin Inst.*, 269(1), 24–42, 1960.

[11] Kwakernaak, J. and Sivan, R., *Linear optimal control systems*, John Wiley & Sons, New York, 1972.

[12] Liu, T. S. and Wu, J. C., A model for a rider-motorcycle system using fuzzy control, *IEEE Trans. Syst., Man, and Cybernetics*, 23(1), 267–276, 1993.

[13] McRuer, D. T., Human dynamics in man-machine systems, *Automatica*, 16(3), 237–253, 1980.

[14] McRuer, D. T. and Krendel, E. S., Dynamic Response of Human Operators, Wright Air Development Center, WADC TR 56-524, 1957.

[15] McRuer, D. T., Graham, D., Krendel, E. S., and Reisener, W., Jr., Human pilot dynamics in compensatory systems, Air Force Flight Dynamics Lab., AFFDL-65-15, 1965.

[16] Nise, N., *Control Systems Engineering*, 1st ed., Benjamin/Cummings, Redwood City, CA, 1992, Chapter 11.

[17] Sheridan, T. B., *Telerobotics, Automation, and Human Supervisory Control*, MIT, Cambridge, MA, 1992, Chapter 1.

[18] Sutton, R. and Towill, D. R., *Modelling the Helmsman in a Ship Steering System Using Fuzzy Sets*, Proc. of the IFAC Conference on Man-Machine systems: Analysis, Design and Evaluation, Oulu, Finland, 1988.

[19] Tustin, A., The nature of the operator's response in manual control and its implication for controller design, *J. IEE*, 94, IIa(2), 1947.

[20] Zadeh, L., Outline of a new approach to the analysis of

complex systems and decision processes, *IEEE Trans. Syst., Man, Cybernetics*, SMC-3(1), 28–44, 1973.

[21] Wewerinke, P. H., Models of the human observer and controller of a dynamic system, Ph.D. Thesis, University of Twente, the Netherlands, 1989.

Further Reading

An excellent summary of application of human operator models to modeling human pilot behavior, circa 1974, is available in the AGARD report referenced herein, *Mathematical Models of Human Pilot Behavior*, by McRuer and Krendel.

The excellent text by Thomas Sheridan and William Ferrell, *Man-Machine Systems*, published by MIT Press in 1974, provides a broad view of the topics involved with human-machine interaction.

The 1987 Wiley publication *Handbook of Human Factors*, [5] contains a chapter by the author entitled Feedback Control Models. This chapter covers some MIMO applications of the classical and modern (OCM) human operator models. A new edition of this Handbook is currently in press.

The book, *Modelling Human Operators in Control System Design*, by Robert Sutton and published by Wiley in 1990 offers a very readable treatment of human-in-the-loop control systems. This book includes chapters devoted to modeling the human operator using fuzzy sets.

Index

μ-synthesis, 233

A/F maldistribution, 92
ac induction motors, 258
ac synchronous motors, 260
ac/ac converters, 245
acceleration resolved control, 181
acid
  diprotic, 32
acids, 31
actuator dynamics, 169
actuator inertia, 168
actuator nonlinearities, 147
actuators, 16, 117
  control, 130
adaptive control, 64, 172, 183, 206
adaptive controller, 103
adaptive feedback linearization, 173
admittance, 184
admittance control, 184
agent
  monoprotic, 31
air-fuel ratio control, 89
air-gas dynamics, 295
algebra
  control Lie, 189
  Lie, 189
algorithm, 113
  control, 95, 316
algorithms, 63
  feedback control, 21
  path planning, 189
aliasing effect, 246
alternator, 282
amplifiers
  power, 214
analog filter, 22
analysis
  dynamic, 120
  performance, 71
  stability, 209
  structural, 286
  trim, 116
angle
  delay, 244
  flux slip, 278
  stator current vector, 278
angular velocity, 138
approximate multimodal decomposition, 320
architectures, 297
attitude control, 139
attitude-hold configuration, 151
audio-susceptibility, 249
autotuning, 39
averaging, 245
  generalized, 250
axes
  multiple, 217
axis
  inertial, 137, 139

backmixing, 36
band reject filters, 219
base frame, 166
base space, 189
bases, 31
basic slip control, 276
batch ph control, 40
behavior
  dynamical, 286
bleaching, 45
blend chest dynamics, 56
blocking state, 242
boiler, 282
boost converter, 243, 245, 247, 248, 250
boundary layer system, 172
buck converter, 242

cartesian force, 180
cartesian manipulator velocities, 179
cartesian space, 179
cascade, 63
causticizing area, 45
cellulose, 44
chain form, 190
chemical pulping, 45
Chow's theorem, 189
Christoffel symbols, 167
circuit
  dc generator equivalent, 267
circular orbit, 139
circulation loop dynamics, 300
circulation loop equations, 301
closed-loop control design, 315
closed-loop time constant, 59
closure
  involutive, 188
combustion, 295
combustion control, 299
command tracking, 155
compact disc mechanism, 231
compensation, 195, 205
  load power, 273
  model-based, 205

reactive power, 273
compensator
lead, 219
pi, 219
position loop, 220
velocity loop, 219
compensator design, 223, 225, 248, 304
compensators, 216
compensatory tracking tasks, 331
compliance frame, 179
composite control, 171, 172
computing, 221
condition
pseudogradient, 111
second-order nonholonomic, 186
conditions
controllability, 143
conducting state, 242
configuration space, 166, 178
constant
closed-loop time, 59
constant nominal equilibrium values, 248
constant voltage waveforms, 241
continuous conduction mode, 246
continuous disconduction mode, 246
continuous duty ratio, 247
control
acceleration resolved, 181
adaptive, 64, 172, 183, 206
admittance, 184
air-fuel ratio, 89
attitude, 139
basic slip, 276
batch ph, 40
combustion, 299
composite, 171, 172
current-mode, 249
damping, 184
direct adaptive, 207
drum level, 298, 303
energy-efficient, 24
engine, 87
feedback, 92
feedforward, 40, 99
feedforward ph, 40
field-oriented, 277
flight, 113, 119, 121, 127
flow, 20
generator excitation, 312
generator system, 268
hybrid force/position, 180
idle speed, 89, 96
ignition, 88
impedance, 178, 183, 184
indirect adaptive, 207
integral, 203
inverse damping, 184
joint torque, 204
learning, 176, 204
linear, 56
local-loop, 21
logical, 23
mimo, 64
multivariable, 128
operational space, 171
passivity based robust, 174
passivity-based adaptive, 174
pd, 169
position, 178
pressurization, 20
repetitive, 176
resolved acceleration, 171
robust, 172, 231
siso, 63
spacecraft attitude, 129
stiffness, 184
temperature, 298
time-optimal, 175
toilet, 5
tracking, 122, 123
ultra-high precision, 212
velocity, 203
velocity-resolved, 182
control actuators, 130
control algorithm, 95, 316
control design, 64, 76, 218, 315
control equipment, 53, 56
control law, 137
control Lie algebra, 189
control system, 153, 171
control system design, 37, 171
control system structure, 153
control systems, 282, 297
control technology, 4
control-theory paradigm, 327
controllability, 143, 189, 321, 322
controllability conditions, 143
controller
drum level, 299
dynamic dissipative, 148
lqg, 146
self-tuning, 40
strictly proper dissipative, 149
controller design, 144, 158
controller tuning procedures, 38
controllers, 218
dissipative, 151
dynamic dissipative, 148
linear, 38
model-based, 145
nonlinear, 38
passivity-based, 147

static dissipative, 147
controls, 20
supplemental damping, 314
converter
boost, 243, 245, 247, 248, 250
buck, 242
dc-link, 245
voltage step-down, 242
voltage step-up, 243
converters, 241
ac/ac, 245
high-frequency pwm dc/dc, 242
phase-controlled, 244
resonant, 244
switched-mode, 243
coordination
motion, 210
cross-coupling, 217
crossover model, 328, 332
current-mode control, 249
curve
Stribeck, 196
strong acid-strong base, 31
titration, 30
curves
process titration, 30
cycle
kraft liquor, 45
cyclic input, 189
cycloconverter, 245
cylinder air charge, 90

dampers, 17
damping control, 184
dc generator, 268
dc generator equivalent circuit, 267
dc generators, 265
dc machine generator, 266
dc motor
separately excited, 256
series wound, 257
shunt wound, 257
dc motors, 255
dc-link converter, 245
dead band, 31
deadtime, 59, 60
deadtime compensation, 63
deadzone nonlinearity, 22
decomposition
approximate multimodal, 320
delay angle, 244
dependent loop level, 285
derivative
partial, 103
design
closed-loop control, 315
compensator, 223, 225, 248, 304
control, 64, 218, 315
control system, 37, 171
controller, 144, 158
feedback, 99
feedforward control, 99
filter, 162
Kalman filter, 158
linear quadratic regulator, 158
lq regulator, 161
steady-state process, 46
vessel, 35
design model, 158, 160
destabilizing, 204
devices
discrete-state, 23
modulating, 23
diagram
vector, 266
diamond turning machine, 221
digester house, 45
digital filters, 22
diprotic acid, 32
direct adaptive control, 207
direct field orientation, 278
discrete-state devices, 23
disease, 9
displacement
presliding, 197, 206
dissipative controllers, 151
distribution
involutive, 187
disturbance response, 306
disturbances
self-induced torque, 156
dither, 204
double integrator, 170
driftless systems, 186
drive trains, 214
drum dynamics, 300
drum level control, 298, 303
drum level controller, 299
drum level dynamics, 291
duct static pressure, 27
duty ratio, 242
dynamic analysis, 120
dynamic dissipative controller, 148
dynamic dissipative controllers, 148
dynamic friction, 202
dynamic friction model, 202
dynamic model, 256
dynamic modeling, 241
dynamical behavior, 286
dynamics, 167
actuator, 169
air-gas, 295

blend chest, 56
circulation loop, 300
drum, 300
drum level, 291
equilibria & perturbation, 302
inverse, 170
joint space inverse, 170
lagrangian, 167
nonlinear, 116
pressure, 287
pulp and paper process, 58
reheat steamside, 292
spacecraft rotational, 132
superheat steamside, 292
task space inverse, 171
zero, 302

economizer cooling, 24
effect
aliasing, 246
efficiency
high, 241
elasticity
joint, 169
electric generator, 282
electric power generating plants, 281
electrical demand limiting, 25
electrical generators, 265
electrodes, 30
electronics
power, 241
elements
snubber, 245
end-effector frame, 166
end-effector space, 166
energy
kinetic, 167
potential, 167
energy-efficient control, 24
engine control, 87
engine models, 97
environment
rigid, 178, 181
soft, 179, 181
environment forces, 169
environments
exogenous input, 156
equations
circulation loop, 301
Euler, 115
force, 115
kinematic, 115
Lagrange, 171
moment, 115
nonlinear spacecraft, 138
system, 178
trim, 117
equations of motion, 149
equilibria & perturbation dynamics, 302
equipment
control, 53, 56
power transmission, 314
error
tuned, 110
estimation
optimal, 221
Euler equations, 115
evaporators
multiple effect, 45
exciter, 270
exciter/generator protection, 273
exciters, 270
rotating, 271
static, 271
exhaust gas recirculation (EGR), 88
exogenous input environments, 156
explorer i, 133

fans, 19
fast tool servos, 216
feasible control signals, 123
feedback
steam flow, 305
feedback control, 92
feedback control algorithms, 21
feedback design, 99
feedback linearization, 170
feedforward, 63
feedforward control, 40, 99
feedforward control design, 99
feedforward ph control, 40
feedforward schemes, 246
fiber, 44
field
vector, 187
field orientation, 253
direct, 278
indirect, 278
field-oriented control, 277
filter
analog, 22
digital, 22
noise spike, 22
pseudosensitivity, 103
sensitivity, 104
filter design, 162
filtering, 221
input-signal, 316
filters
band reject, 219
flexible space structures, 142
flight control, 113, 119, 121, 127

flight mechanics, 114
flow control, 20
flushing, 7
flux slip angle, 278
force
  cartesian, 180
  hybrid, 178
force equations, 115
force of constraint, 180
force of motion, 180
force-controlled subspace, 179
forces
  environment, 169
form
  chain, 190
frame
  base, 166
  compliance, 179
  end-effector, 166
  task, 166, 179
  world, 166
frequency-domain modeling, 328
friction, 168, 201
  dynamic, 202
  off-line, 201
  position-dependent, 201, 206
  rising static, 196
  static, 206
  Stribeck, 196
friction compensation, 202
friction materials, 196
friction modeling, 195, 196
friction parameters, 199
frictional memory, 197, 206
function
  inverse describing, 205
  sensitivity, 103
  switching, 246
functions
  inner loop transfer, 321
fundamental
  local, 250
fuzzy control model, 333
fuzzy control models, 330

gains
  high servo, 203
generalized averaging, 250
generalized state-space models, 251
generator
  dc, 268
  dc machine, 266
  electric, 282
  induction, 276
  steam, 282
generator excitation control, 312
generator excitation systems, 313
generator system control, 268
generator voltage, 268
generators
  dc, 265
  electrical, 265
  induction, 274
  synchronous, 268
  wound-field synchronous, 270
generic loop, 192
graphical programming, 27
groundwood mills, 45
group
  special euclidean, 166
  special orthogonal, 166

h_∞- and μ synthesis methods, 146
hardware, 221
hardware modification, 203
harmonics
  local, 250
health, 10
heat exchangers, 16, 292
hemicellulose, 44
hierarchical modeling, 251
high efficiency, 241
high servo gains, 203
high-frequency pwm dc/dc converters, 242
high-frequency pwm inverter, 244
high-frequency pwm rectifier, 243
high-voltage dc (hvdc), 318
high-voltage dc (hvdc) bulk-power transmission systems, 241
historical approaches to systems, 11
human operator, 328
hybrid force, 178
hybrid force/position control, 180
hygiene, 10
hyperstatic situation, 178

ICAFC, 95
ideal switch, 242
idle speed control, 89, 96
ignition control, 88
impedance, 184
impedance control, 178, 183, 184
indirect adaptive control, 207
indirect field orientation, 278
indoor toilet, 10
induction generator, 276
induction generator model, 275
induction generators, 274
inertia matrix, 167
inertial axis, 137, 139
infeasible control signals, 123
inner loop, 322
inner loop transfer functions, 321

inner loop/outer loop architecture, 170
input
cyclic, 189
input signal, 316
input-output linearizability, 259
input-output linearization, 253
input-signal filtering, 316
integrable, 187
integral control, 203
integral windup, 37
integrated dynamic friction model, 198
integrator, 60
double, 170
integrator windup, 249
interpretation models, 286
inverse damping control, 184
inverse describing function, 205
inverse dynamics, 170
inverse kinematics, 166
inverse stiffness, 184
inverter, 241
high-frequency pwm, 244
inverters, 244
involutive closure, 188
involutive distribution, 187
ipm synchronous motor, 262
isa terminology, 47
ISC, 97, 98
calibration, 99
isopotential point, 30

joint elasticity, 169
joint space, 166
joint space inverse dynamics, 170
joint torque control, 204
joints, 166
prismatic, 166

Kalman filter design, 158
kinematic equations, 115
kinematic reciprocity relationship, 179
kinematics, 166
inverse, 166
velocity, 166
kinetic energy, 167
knowledge models, 286
kraft liquor cycle, 45
kraft process, 45

Lagrange equations, 171
lagrangian, 167
lagrangian dynamics, 167
lambda tuning, 56, 58
law
control, 137
nonlinear feedback control, 150
lead compensator, 219
learning control, 176, 204
level
dependent loop, 285
principal regulation, 285
unit control, 285
Lie algebra, 189
Lie bracket, 187
lignin, 44
lime kiln, 45
line-of-sight pointing, 152
linear control, 56
linear controllers, 38
linear quadratic regulator design, 158
linear truth model, 157
linearizability
input-output, 259
linearization
adaptive feedback, 173
feedback, 170
input-output, 253
robust feedback, 172
linearized mathematical model, 142
linearized model, 134
linearized models, 248
links, 166
liquor
white cooking, 45
load power compensation, 273
local fundamental, 250
local harmonics, 250
local-loop control, 21
logical control, 23
loop
generic, 192
inner, 322
position, 222
velocity, 222
loop tuning, 55
low load regulator, 305
lq regulator design, 161
lqg controller, 146
lubricant modification, 203

machine
synchronous, 268
maneuver tracking, 155
maneuvering
vehicle, 156
manipulator jacobian, 166
manufacturing
uniform, 64
manufacturing processes, 45
mathematical model, 149
mathematical models, 21
matrix

inertia, 167
orientation, 166
unity, 180
weighting, 118
mechanical pulping, 45
memory
frictional, 197, 206
methods
h_∞- and μ synthesis, 146
microsynthesis, 233
mills
groundwood, 45
mimo control, 64
mode
continuous conduction, 246
discontinuous conduction, 246
model
crossover, 328, 332
design, 158, 160
dynamic, 256
dynamic friction, 202
fuzzy control, 333
induction generator, 275
integrated dynamic friction, 198
linear truth, 157
linearized, 134
linearized mathematical, 142
mathematical, 149
nonlinear averaged, 247
optimal control, 329, 333
seven-parameter friction, 198
state space, 94
state-space averaged, 247
structural, 329, 332
switched state-space, 246
synchronous generator, 268
time-invariant continuous-time state-space, 247
time-invariant sampled-data, 250
model discretization, 91
model-based compensation, 205
model-based controllers, 145
modeling, 70, 129, 155, 157, 217, 232
dynamic, 241
frequency-domain, 328
friction, 195, 196
hierarchical, 251
power plant, 286
sampled-data, 245
time-domain, 329
uncertainty, 232
models
engine, 97
fuzzy control, 330
generalized state-space, 251
interpretation, 286
knowledge, 286
linearized, 248
mathematical, 21
sampled-data, 250
singular perturbation, 169
models for control, 72
modulating devices, 23
modulating signal, 246
modulation
power system stabilizer (pss), 313
pulse-width (pwm), 243
moment equations, 115
monoprotic agent, 31
motion
pilot head, 156
motion control systems, 208
motion coordination, 210
motion profiling, 210
motor
ipm synchronous, 262
spm synchronous, 262
synchronous reluctance, 263
motors, 214
ac induction, 258
ac synchronous, 260
dc, 255
multibody flexible space systems, 149
multiple axes, 217
multiple effect evaporators, 45
multivariable control, 128

natural period, 35
Neal–Smith criterion, 127
noise
white, 218
noise spike filter, 22
nonholonomic path planning problem, 188
nonholonomic systems, 185
nonholonomy
test, 187
nonlinear averaged model, 247
nonlinear control law, 138
nonlinear controllers, 38
nonlinear dynamics, 116
nonlinear feedback control law, 150
nonlinear inner loop, 125
nonlinear spacecraft equations, 138
nonlinearities
actuator, 147
sensor, 147
nonmodel-based compensation, 203
normal solution, 29

observability, 143, 321, 322
off-line friction, 201
open kinematic chain, 166
operational space control, 171

operator
human, 328
optimal control model, 329, 333
optimal estimation, 221
optimization
performance, 254
orientation
field, 253
orientation matrix, 166
oscillatory stability, 312
output stabilization, 193

paper making, 46
papermaking, 49
paradigm
regulator, 122
parameters
friction, 199
partial derivative, 103
passivity based robust control, 174
passivity property, 168
passivity-based adaptive control, 174
passivity-based controllers, 147
path planning algorithms, 189
path space, 191
pd control, 169
PD plus feedforward acceleration, 170
pd plus feedforward acceleration, 170
performance
system, 153
performance analysis, 71
performance optimization, 254
performance requirements, 153
period
natural, 35
persistency of excitation, 174
perturbation
singular, 169
perturbations
switching-time, 251
ph measuring system, 29
ph scale, 29
phase condition, 104
phase-controlled converters, 244
physical system, 155
pi, 63
pi compensator, 219
pid, 63
pid.f, 63
pilot head motion, 156
pilot-induced oscillations, 127
plant
power, 282
structure, 282
plants
electric power generating, 281
point
isopotential, 30
pointing
line-of-sight, 152
poses, 166
position control, 178
position loop, 222, 225
position loop compensator, 220
position loops, 216
position-dependent friction, 201, 206
potential energy, 167
power amplifiers, 214
power electronics, 241
power generation, 294, 297
power plant, 282
power plant modeling, 286
power system stabilization, 273
power system stabilizer (pss) modulation, 313
power transmission, 311
power transmission equipment, 314
presliding displacement, 197, 206
pressure dynamics, 287
pressurization control, 20
principal regulation level, 285
principle of virtual work, 169
prismatic joints, 166
procedures
controller tuning, 38
process
kraft, 45
sulfite, 45
process titration curves, 30
processes
manufacturing, 45
product variability, 46
profiling
motion, 210
programming
graphical, 27
textual, 27
proof-of-concept testing, 77
property
passivity, 168
protection
exciter/generator, 273
pseudogradient condition, 111
pseudosensitivity filter, 103
pulp and paper mill, 46
pulp and paper mill culture, 55
pulp and paper process dynamics, 58
pulping
chemical, 45
mechanical, 45
pulse waveform, 242
pulse-width modulation (pwm), 243
pumps, 19

pursuit tracking tasks, 331

quadrature glitch, 195
quasi-steady-state system, 172
quaternion, 150

ratio
    continuous duty, 247
reactive power compensation, 273
reagent delivery systems, 37
recovery boiler, 45
rectified waveform, 244
rectifier, 241
    high-frequency pwm, 243
reference frame transformation, 258, 261
reference signal extrapolation, 124
refining, 46
regressor, 168
regulator
    low load, 305
regulator paradigm, 122
regulators
    switching, 243
reheat steamside dynamics, 292
reheaters, 292
relationship
    kinematic reciprocity, 179
repetitive control, 176
residence time, 34
resolved acceleration control, 171
resonance, 59
resonances
    structural, 219
resonant converters, 244
response
    disturbance, 306
    step input, 217
    stepped sinusoidal, 217
    white noise, 218
revolute, 166
rigid environment, 178, 181
rising static friction, 196
robot, 165
robust control, 172, 231
robust feedback linearization, 172
robustness, 59
root sensitivity, 220
rotating exciters, 271

sampled-data modeling, 245
sampled-data models, 250
sampling, 245
sanitation system, 8
scale
    ph, 29
scheme
    soft-start, 249
second-order nonholonomic condition, 186
self-induced torque disturbances, 156
self-tuning controller, 40
semiconductor switches, 242
sensitivity
    root, 220
sensitivity filters, 104
sensitivity function, 103
sensor nonlinearities, 147
sensors, 19, 214
separately excited dc motor, 256
series wound dc motor, 257
setpoint resetting, 26
seven-parameter friction model, the, 198
shape space, 189
short-circuiting, 30
shunt wound dc motor, 257
signal
    input, 316
    modulating, 246
signal conditioning, 218
signals
    feasible control, 123
    infeasible control, 123
simulation, 70, 199
single quadratic cost functional, 102
single-body flexible spacecraft, 142
singular perturbation, 169
singular perturbation models, 169
sinusoidal voltage waveforms, 241
siso control, 63
siting, 315
situation
    hyperstatic, 178
slow manifold, 107
smoothness
    tracking, 155
snubber elements, 245
soft environment, 179, 181
soft-start scheme, 249
software, 221
solution
    normal, 29
space
    base, 189
    cartesian, 179
    configuration, 166, 178
    end-effector, 166
    joint, 166
    path, 191
    shape, 189
    task, 166
spacecraft
    single-body flexible, 142
spacecraft attitude control, 129

spacecraft attitude sensors, 130
spacecraft rotational dynamics, 132
spacecraft rotational kinematics, 130
special euclidean group, 166
special orthogonal group, 166
spm synchronous motor, 262
stability, 147, 148, 151
    oscillatory, 312
    transient, 311
    voltage, 312
stability analysis, 209
stabilization, 155, 163, 193
    output, 193
    power system, 273
    state, 193
stabilization control system, 152
stabilization control system structure, 157
state
    blocking, 242
    conducting, 242
state space model, 94
state stabilization, 193
state-space averaged model, 247
static dissipative controllers, 147
static exciters, 271
static friction, 206
static mixer, 34
static var compensator (svc), 318
stator current vector angle, 278
steady-state process design, 46
steady-state velocity, 201
steam flow feedback, 305
steam generator, 282
steam pressure, 297
step input response, 217
stepped sinusoidal response, 217
stiffness
    inverse, 184
stiffness control, 184
Stribeck curve, 196
Stribeck friction, 196
strictly proper dissipative controller, 149
strong acid-strong base curve, 31
structural analysis, 286
structural model, 329, 332
structural resonances, 219
structure
    control system, 153
    stabilization control system, 157
structures
    flexible space, 142
subspace
    force-controlled, 179
    velocity-controlled, 179
sulfite process, 45
superheat steamside dynamics, 292
superheaters, 292
supplemental damping controls, 314
switch
    ideal, 242
switched state-space model, 246
switched-mode converters, 243
switches
    semiconductor, 242
switching function, 246
switching regulators, 243
switching times, 251
switching-time perturbations, 251
symbols
    Christoffel, 167
synchronous generator model, 268
synchronous generators, 268
synchronous machine, 268
synchronous reluctance motor, 263
system
    boundary layer, 172
    control, 153, 171
    ph measuring, 29
    physical, 155
    quasi-steady-state, 172
    sanitation, 8
    stabilization control, 152
    transmission, 312
system equations, 178
system performance, 153
systems, 151
    control, 282, 297
    driftless, 186
    generator excitation, 313
    high-voltage dc (hvdc) bulk-power transmission, 241
    historical approaches to, 11
    motion control, 208
    multibody flexible space, 149
    nonholonomic, 185
    reagent delivery, 37
    ultra-high precision, 212

task frame, 166, 179
task space, 166
task space inverse dynamics, 171
tasks
    compensatory tracking, 331
    pursuit tracking, 331
technology
    control, 4
temperature control, 298
textual programming, 27
theorem
    Chow's, 189
thermal storage, 26
thermal-mechanical pulping (tmp), 45
thyristor-controlled series compensation (tcsc), 319

time
  residence, 34
time-domain approach, 122
time-domain modeling, 329
time-domain specifications, 20
time-invariant continuous-time state-space model, 247
time-invariant sampled-data model, 250
time-optimal control, 175
time-scale separation, 121
times
  switching, 251
titration curve, 30
toilet, 4
  indoor, 10
toilet control, 5
torque-slip, 275
tracking
  command, 155
  maneuver, 155
tracking control, 122, 123
tracking smoothness, 155
trajectories, 167
transformation
  reference frame, 258, 261
transient stability, 311
transmission
  power, 311
transmission system, 312
transmissions, 214
transmitters, 19
trim analysis, 116
trim equations, 117
tuned error, 110
tuning, 23
  lambda, 56, 58
  loop, 55
turbine, 282
two-dimensional contour following, 183

ultra-high precision control, 212
ultra-high precision systems, 212
uncertainty modeling, 232
unidirectional waveform, 244
uniform manufacturing, 64
unit control level, 285
unity matrix, 180

validation, 91
values
  constant nominal equilibrium, 248
valve positioners, 31
valves, 17
vector diagram, 266
vector field, 187
vehicle maneuvering, 156
vehicle vibration, 156
velocities
  cartesian manipulator, 179
velocity
  angular, 138
velocity control, 203
velocity kinematics, 166
velocity loop, 222, 223
velocity loop compensator, 219
velocity loops, 216
velocity of constraint, 180
velocity of freedom, 180
velocity-controlled subspace, 179
velocity-resolved control, 182
vessel design, 35
vibration
  vehicle, 156
voltage
  generator, 268
voltage stability, 312
voltage step-down converter, 242
voltage step-up converter, 243

waveform
  pulse, 242
  rectified, 244
  unidirectional, 244
waveforms
  constant voltage, 241
  sinusoidal voltage, 241
weighting matrix, 118
white cooking liquor, 45
white noise response, 218
windup, 40
  integrator, 249
world frame, 166
wound-field synchronous generators, 270

zero dynamics, 302
zone temperature, 20

9780367399061
T - #0254 - 071024 - C0 - 279/216/19 - PB - 9780367399061 - Gloss Lamination